Methods in Molecular Biology

Series Editor
John M. Walker
School of Life and Medical Sciences
University of Hertfordshire
Hatfield, Hertfordshire AL10 9AB, UK

For further volumes:
http://www.springer.com/series/7651

Autophagy

Methods and Protocols

Edited by

Nicholas Ktistakis and Oliver Florey

Signalling Programme, Babraham Institute, Cambridge, UK

Humana Press

Editors
Nicholas Ktistakis
Signalling Programme
Babraham Institute
Cambridge, UK

Oliver Florey
Signalling Programme
Babraham Institute
Cambridge, UK

ISSN 1064-3745 ISSN 1940-6029 (electronic)
Methods in Molecular Biology
ISBN 978-1-4939-8872-3 ISBN 978-1-4939-8873-0 (eBook)
https://doi.org/10.1007/978-1-4939-8873-0

Library of Congress Control Number: 2018965163

This Humana Press imprint is published by the registered company Springer Science+Business Media, LLC, part of Springer Nature.
The registered company address is: 233 Spring Street, New York, NY 10013, U.S.A.

Preface

The process of self-eating as a way to survive periods of starvation has been known for over 150 years, and this response was from the beginning designated as "autophagy." Following early progress in establishing the morphological and some biochemical parameters of the autophagic response in the 1950s and 1960s, the field of autophagy was transformed when Yoshinori Ohsumi and others applied a genetic approach to isolate a near complete set of autophagy genes in yeast, which allowed the characterization of equivalent genes and processes in other organisms including mammals.

We now know that autophagy is a highly conserved catabolic and quality control pathway across all eukaryotes. Its dysregulation can underpin disease and developmental problems, frequently in ways not immediately predictable. Despite tremendous progress in researching the autophagic response in a variety of experimental settings, much remains to be discovered, from the basic mechanisms to the physiological functions.

This volume hopes to be a comprehensive and extensive set of protocols for the study of autophagy in vitro and in vivo, primarily in mammals but also in various model organisms including yeast, Drosophila, and *C. elegans*. Protocols are also presented for the study of autophagy-related processes outside of the canonical autophagy pathways.

We first present 11 chapters describing protocols for studying autophagy in vitro, i.e., using pure or semi-pure components; these studies have contributed a great deal toward understanding autophagy in molecular terms, and it is likely that they will continue to do so in the future. The next 8 chapters describe protocols for the study of autophagy (canonical and noncanonical) by imaging methods, either using light microscopy or EM. Being able to follow the formation of autophagosomes and to study their unique architecture has traditionally been a productive area of research, and it is likely that novel insights from such studies will continue to be derived in the future. In the last few years, much effort has been devoted to the establishment of assays and screening platforms for large-scale discovery efforts to identify genes and chemicals affecting autophagy; here we have included 10 chapters describing a set of up-to-date protocols for such studies. The next 9 chapters deal with the study of autophagy not in tissue culture but in whole organisms. We believe that work along those lines is likely to come to dominate the autophagy field in the future, and it is therefore very useful to have a set of protocols mapping out these useful approaches. The final 10 chapters describe assays and techniques for studying selective autophagy, i.e., the induction of autophagy for the elimination of specific cargoes. Although autophagy is sometimes solely considered a starvation response, it can be argued that its function as a specific eliminator of unwanted substances is a major physiological function and therefore likely to continue to be studied intensely.

We wish to thank all authors who took time off from their own work in order to provide the protocols included in this book.

Cambridge, UK

Nicholas Ktistakis
Oliver Florey

Contents

Contributors

HAGAI ABELIOVICH • *Biochemistry, Food Science and Nutrition, Hebrew University of Jerusalem, Rehovot, Israel*

CHRISTINE ABERT • *Department of Biochemistry and Cell Biology, Max F. Perutz Laboratories (MFPL), Vienna Biocenter (VBC), University of Vienna, Vienna, Austria*

JAIME AGUDO-CANALEJO • *Rudolf Peierls Centre for Theoretical Physics, University of Oxford, Oxford, UK; Department of Chemistry, The Pennsylvania State University, University Park, PA, USA*

SONJA AITS • *Cell Death and Lysosomes Group, Experimental Neuroinflammation Laboratory, Department of Experimental Medical Science, Lund University, Lund, Sweden; Peter MacCallum Cancer Centre, Melbourne, Australia*

ALICIA ALONSO • *Instituto Biofisika (CSIC, UPV/EHU), Bilbao, Spain; Departamento de Bioquímica y Biología Molecular, Universidad del País Vasco, Bilbao, Spain*

ZURIÑE ANTÓN • *Instituto Biofisika (CSIC, UPV/EHU), Bilbao, Spain; Departamento de Bioquímica y Biología Molecular, Universidad del País Vasco, Bilbao, Spain*

RITSUKO ARAI • *Department of Anatomy and Histology, Fukushima Medical University, School of Medicine, Fukushima, Japan*

MEENAKSHI BANERJEE • *Department of Molecular and Cellular Biochemistry, University of Kentucky, Lexington, KY, USA*

DARIO BARBONE • *Zuckerberg San Francisco General Hospital and Trauma Center, University of California San Francisco, San Francisco, CA, USA*

NASTASSIA BAYLAC • *Molecular and Integrative Biosciences Research Program, University of Helsinki, Helsinki, Finland*

PETRA BELI • *Institute of Molecular Biology (IMB), Mainz, Germany*

ÅSA BIRNA BIRGISDOTTIR • *Molecular Cancer Research Group, Department of Medical Biology, University of Tromsø – The Arctic University of Norway, Tromsø, Norway*

KEITH B. BOYLE • *Division of Protein and Nucleic Acid Chemistry, MRC Laboratory of Molecular Biology, Cambridge, UK*

RAPHAEL BUENO • *Division of Thoracic Surgery, Brigham and Women's Hospital, Harvard Medical School, Boston, MA, USA*

MAEVE A. CALDWELL • *Trinity College Institute for Neuroscience, Trinity College, Dublin, Ireland*

YANG CHEN • *Center for Precision Medicine Multi-Omics Research, Peking University, Health Science Center, Beijing, China*

XIGUI CHEN • *Department of Neuropathology, Medical Research Institute and Center for Brain Integration Research, Tokyo Medical and Dental University, Tokyo, Japan*

MIRUNA CHIPARA • *Institute of Cancer and Genomic Sciences, Institute of Biomedical Research, College of Medical and Dental Sciences, University of Birmingham, Birmingham, UK*

PATRICE CODOGNO • *Institut Necker-Enfants Malades (INEM), INSERM U1151-CNRS UMR 8253, Université Paris Descartes-Sorbonne Paris Cité, Paris, France*

V. COURTNEY BROADDUS • *Zuckerberg San Francisco General Hospital and Trauma Center, University of California San Francisco, San Francisco, CA, USA*

LUCY CROMPTON • *Cell Biology Laboratories, School of Biochemistry, University of Bristol, Bristol, UK*

A. M. CUERVO • *Department of Developmental and Molecular Biology, Bronx, NY, USA; Institute for Aging Studies, Albert Einstein College of Medicine, Bronx, NY, USA*

JÖRN DENGJEL • *Department of Biology, University of Fribourg, Fribourg, Switzerland; Department of Dermatology, Medical Center—University of Freiburg, Freiburg, Germany*

BRITTA DIEDRICH • *Agilent Technologies, Waldbronn, Germany*

IVAN DIKIC • *Institute of Biochemistry II, Goethe University School of Medicine, Frankfurt, Germany; Buchmann Institute for Molecular Life Sciences, Goethe University, Frankfurt, Germany*

AUDREY DONG • *Department of Neuroscience, Perelman School of Medicine at the University of Pennsylvania, Philadelphia, PA, USA*

NICOLAS DUPONT • *Institut Necker-Enfants Malades (INEM), INSERM U1151-CNRS UMR 8253, Université Paris Descartes-Sorbonne Paris Cité, Paris, France*

NIKOLAI ENGEDAL • *Centre for Molecular Medicine Norway (NCMM), Nordic EMBL Partnership, University of Oslo, Oslo, Norway*

EEVA-LIISA ESKELINEN • *Molecular and Integrative Biosciences Research Program, University of Helsinki, Helsinki, Finland*

KATHERINE FLETCHER • *Signalling Programme, Babraham Institute, Cambridge, UK*

OLIVER FLOREY • *Signalling Programme, Babraham Institute, Cambridge, UK*

CARLO FOLLO • *Zuckerberg San Francisco General Hospital and Trauma Center, University of California San Francisco, San Francisco, CA, USA*

IAN G. GANLEY • *MRC Protein Phosphorylation and Ubiquitylation Unit, School of Life Sciences, University of Dundee, Dundee, UK*

LIANG GE • *State Key Laboratory of Membrane Biology, Beijing, China; Tsinghua-Peking Center for Life Sciences, Beijing, China; School of Life Sciences, Tsinghua University, Beijing, China*

A. GIATROMANOLAKI • *Department of Pathology, Democritus University of Thrace, and University General Hospital of Alexandroupolis, Alexandroupolis, Greece*

DANIEL GRASSO • *Pathophysiology Department, Institute of Biochemistry and Molecular Medicine (CONICET), School of Pharmacy and Biochemistry, University of Buenos Aires, Buenos Aires, Argentina*

SIGURDUR GUDMUNDSSON • *Molecular and Integrative Biosciences Research Program, University of Helsinki, Helsinki, Finland*

AUDREY GUESDON • *Structural and Computational Biology Unit, European Molecular Biology Laboratory (EMBL), Heidelberg, Germany*

JENS C. HAMANN • *Meyer Cancer Center, Weill Cornell Medicine, New York, NY, USA; Cell Biology Program, Sloan Kettering Institute for Cancer Research, New York, NY, USA; Louis V. Gerstner, Jr. Graduate School of Biomedical Sciences, Memorial Sloan Kettering Cancer Center, New York, NY, USA*

MAHO HAMASAKI • *Andor Technology Ltd., Newark, DE, USA*

SIVAN HENIS-KORENBLIT • *The Mina and Everard Goodman Faculty of Life Sciences, Bar-Ilan University, Ramat-Gan, Israel*

JAVIER H. HERVÁS • *Instituto Biofisika (CSIC, UPV/EHU), Bilbao, Spain; Departamento de Bioquímica y Biología Molecular, Universidad del País Vasco, Bilbao, Spain*

Erika L. F. Holzbaur • *Department of Physiology, University of Pennsylvania Perelman School of Medicine, Philadelphia, PA, USA*
Michael Howell • *Tumour Cell Death Laboratory, Cancer Research UK Beatson Institute, Glasgow, UK; High Throughput Screening, The Francis Crick Institute, London, UK*
Cheng-Wei Hsieh • *Institute of Biological Chemistry, Academia Sinica, Taipei, Taiwan*
Yunjie Huang • *Division of Pulmonary Medicine, Cincinnati Children's Hospital Medical Center, Cincinnati, OH, USA*
Nazmul Huda • *Department of Pathology and Laboratory Medicine, Indiana University School of Medicine, Indianapolis, IN, USA*
Anne-Claire Jacomin • *School of Life Sciences, University of Warwick, Coventry, UK*
Elise Jacquin • *Signalling Programme, Babraham Institute, Cambridge, UK; INSERM, U1231, Université de Bourgogne Franche Comté, Dijon, France*
Ming Jiang • *Tumour Cell Death Laboratory, Cancer Research UK Beatson Institute, Glasgow, UK*
Natalia Jimenez-Moreno • *Cell Biology Laboratories, School of Biochemistry, University of Bristol, Bristol, UK*
Terje Johansen • *Molecular Cancer Research Group, Department of Medical Biology, University of Tromsø – The Arctic University of Norway, Tromsø, Norway*
Smita Joshi • *Department of Molecular and Cellular Biochemistry, University of Kentucky, Lexington, KY, USA*
Gábor Juhász • *Department of Anatomy, Cell and Developmental Biology, Eötvös Loránd University, Budapest, Hungary; Institute of Genetics, Biological Research Centre, Szeged, Hungary*
Thomas Juretschke • *Institute of Molecular Biology (IMB), Mainz, Germany*
Y. R. Juste • *Department of Developmental and Molecular Biology, Bronx, NY, USA; Institute for Aging Studies, Albert Einstein College of Medicine, Bronx, NY, USA*
Stéphanie Kaeser-Pebernard • *Department of Biology, University of Fribourg, Fribourg, Switzerland*
Jenny Kahlhofer • *Division of Cell Biology, Biocenter, Medical University of Innsbruck, Innsbruck, Austria*
D. Kalamida • *Department of Radiotherapy/Oncology, Democritus University of Thrace, and University General Hospital of Alexandroupolis, Alexandroupolis, Greece*
Katri Kallio • *Molecular and Integrative Biosciences Research Program, University of Helsinki, Helsinki, Finland*
Eleftherios Karanasios • *Department of Basic and Clinical Neuroscience, Institute of Psychiatry, Psychology and Neuroscience, King's College London, Maurice Wohl Clinical Neuroscience Institute, London, UK*
Bilon Khambu • *Department of Pathology and Laboratory Medicine, Indiana University School of Medicine, Indianapolis, IN, USA*
Sung Eun Kim • *Helen Diller Family Comprehensive Cancer Center, University of California, San Francisco, CA, USA; Cell Biology Program, Sloan Kettering Institute for Cancer Research, New York, NY, USA*
Roland L. Knorr • *Department of Theory and Bio-Systems, Max Planck Institute of Colloids and Interfaces, Potsdam, Germany; Department of Biochemistry and Molecular Biology, Graduate School and Faculty of Medicine, The University of Tokyo, Tokyo, Japan*
Panagiota Kolitsida • *Biochemistry, Food Science and Nutrition, Hebrew University of Jerusalem, Rehovot, Israel*

MASAAKI KOMATSU • *Department of Physiology, Juntendo University Graduate School of Medicine, Tokyo, Japan*
KANOH KONDO • *Department of Neuropathology, Medical Research Institute and Center for Brain Integration Research, Tokyo Medical and Dental University, Tokyo, Japan*
M. I. KOUKOURAKIS • *Department of Radiotherapy/Oncology, Democritus University of Thrace, and University General Hospital of Alexandroupolis, Alexandroupolis, Greece*
TANJA KUHM • *Structural and Computational Biology Unit, European Molecular Biology Laboratory (EMBL), Heidelberg, Germany*
VINEET VINAY KULKARNI • *Department of Neuroscience, Perelman School of Medicine at the University of Pennsylvania, Philadelphia, PA, USA*
OUIN KUNITAKI • *Department of Genetics, Osaka University Graduate School of Medicine, Osaka, Japan*
JON D. LANE • *Cell Biology Laboratories, School of Biochemistry, University of Bristol, Bristol, UK*
CÉLINE LARGEAU • *Institute for Integrative Biology of the Cell (I2BC), CEA, CNRS, Université Paris-Sud, Université Paris-Saclay, Gif-sur-Yvette cedex, France*
MARIA M. LAZARI • *Institut Necker-Enfants Malades (INEM), INSERM U1151-CNRS UMR 8253, Université Paris Descartes-Sorbonne Paris Cité, Paris, France*
RENAUD LEGOUIS • *Institute for Integrative Biology of the Cell (I2BC), CEA, CNRS, Université Paris-Sud, Université Paris-Saclay, Gif-sur-Yvette cedex, France*
LAURE-ANNE LIGEON • *Viral Immunobiology, Institute of Experimental Immunology, University of Zürich, Zürich, Switzerland*
EIRINI LIONAKI • *Institute of Molecular Biology and Biotechnology, Foundation for Research and Technology Hellas, Crete, Greece*
MONICA LOI • *Viral Immunobiology, Institute of Experimental Immunology, University of Zürich, Zürich, Switzerland*
JACLYN S. LONG • *High Throughput Screening, The Francis Crick Institute, London, UK*
MORTEN LUHR • *Centre for Molecular Medicine Norway (NCMM), Nordic EMBL Partnership, University of Oslo, Oslo, Norway*
CHRISTIAN MÜNZ • *Viral Immunobiology, Institute of Experimental Immunology, University of Zürich, Zürich, Switzerland*
SANDRA MADAY • *Department of Neuroscience, Perelman School of Medicine at the University of Pennsylvania, Philadelphia, PA, USA*
SASCHA MARTENS • *Department of Biochemistry and Cell Biology, Max F. Perutz Laboratories (MFPL), Vienna Biocenter (VBC), University of Vienna, Vienna, Austria*
PETER MAYRHOFER • *Unit Membrane Biochemistry and Transport, Institut Pasteur, Paris, France*
THOMAS G. MCWILLIAMS • *MRC Protein Phosphorylation and Ubiquitylation Unit, School of Life Sciences, University of Dundee, Dundee, UK; Translational Stem Cell Biology Research Program, Research Programs Unit, Faculty of Medicine, University of Helsinki, Helsinki, Finland*
ALICIA MELÉNDEZ • *Department of Biology, Queens College, The City University of New York, Flushing, NY, USA; Biology and Biochemistry PhD Programs, The Graduate Center of the City University of New York, New York, NY, USA*
ZOLTAN METLAGEL • *Thermo Fisher Scientific, Hillsboro, OR, USA*
ANDREW S. MOORE • *Department of Physiology, University of Pennsylvania Perelman School of Medicine, Philadelphia, PA, USA*

AMELIE J. MUELLER • *Interfaculty Institute of Cell Biology, Eberhard Karls University Tuebingen, Tübingen, Germany; International Max Planck Research School "From Molecules to Organisms", Tübingen, Germany*
MARIA NEW • *Molecular Cell Biology of Autophagy Laboratory, The Francis Crick Institute, London, UK*
IOANNIS P. NEZIS • *School of Life Sciences, University of Warwick, Coventry, UK*
PAUL NISTOR • *School of Clinical Medicine, University of Bristol, Bristol, UK*
NOBUO N. NODA • *Laboratory of Structural Biology, Institute of Microbial Chemistry, Tokyo, Japan*
KAZUTO OHASHI • *Institute for Molecular and Cellular Regulation, Gunma University, Gunma, Japan*
HITOSHI OKAZAWA • *Department of Neuropathology, Medical Research Institute and Center for Brain Integration Research, Tokyo Medical and Dental University, Tokyo, Japan*
IDIL ORHON • *Department of Cell Biology, University of Groningen, University Medical Center Groningen, Groningen, The Netherlands*
CHINATSU OTOMO • *Department of Integrative Structural and Computational Biology, The Scripps Research Institute, La Jolla, CA, USA*
TAKANORI OTOMO • *Department of Integrative Structural and Computational Biology, The Scripps Research Institute, La Jolla, CA, USA*
MADHU M. OUSEPH • *Department of Pathology and Laboratory Medicine, Brown University, Providence, RI, USA*
MICHAEL OVERHOLTZER • *Cell Biology Program, Sloan Kettering Institute for Cancer Research, New York, NY, USA*
KONSTANTINOS PALIKARAS • *Institute of Molecular Biology and Biotechnology, Foundation for Research and Technology Hellas, Crete, Greece*
JANA PETRI • *Department of Animal Behavior, Freie Universität Berlin, Berlin, Germany*
DUŠAN PETROVIĆ • *Department of Chemistry, BMC, Uppsala University, Uppsala, Sweden; Institute of Complex Systems ICS-6 (Structural Biochemistry), Forschungszentrum Jülich, Jülich, Germany*
IRINA POKROVSKAYA • *Department of Physiology and Biophysics, University of Arkansas for Medical Sciences, Little Rock, AR, USA*
TASSULA PROIKAS-CEZANNE • *Interfaculty Institute of Cell Biology, Eberhard Karls University Tuebingen, Tübingen, Germany; International Max Planck Research School "From Molecules to Organisms", Tübingen, Germany*
FELIX RANDOW • *Division of Protein and Nucleic Acid Chemistry, MRC Laboratory of Molecular Biology, Cambridge, UK; Department of Medicine, Addenbrooke's Hospital, University of Cambridge, Cambridge, UK*
MADS SKYTTE RASMUSSEN • *Molecular Cancer Research Group, Department of Medical Biology, University of Tromsø – The Arctic University of Norway, Tromsø, Norway*
WILLIAM G. RICHARDS • *Division of Thoracic Surgery, Brigham and Women's Hospital, Harvard Medical School, Boston, MA, USA*
ALEJANDRO ROPOLO • *Pathophysiology Department, Institute of Biochemistry and Molecular Medicine (CONICET), School of Pharmacy and Biochemistry, University of Buenos Aires, Buenos Aires, Argentina*
TATIANA R. ROSENSTOCK • *Department of Physiological Science, Santa Casa de São Paulo School of Medical Science, São Paulo, Brazil*
KEVIN M. RYAN • *Tumour Cell Death Laboratory, Cancer Research UK Beatson Institute, Glasgow, UK; High Throughput Screening, The Francis Crick Institute, London, UK*

CARSTEN SACHSE • *Structural and Computational Biology Unit, European Molecular Biology Laboratory (EMBL), Heidelberg, Germany; Ernst-Ruska Centre for Microscopy and Spectroscopy with Electrons (ER-C-3/Structural Biology), Forschungszentrum Jülich, Jülich, Germany*

JUN-ICHI SAKAMAKI • *High Throughput Screening, The Francis Crick Institute, London, UK*

SOVAN SARKAR • *Institute of Cancer and Genomic Sciences, Institute of Biomedical Research, College of Medical and Dental Sciences, University of Birmingham, Birmingham, UK*

REBECCA SAUNDERS • *Tumour Cell Death Laboratory, Cancer Research UK Beatson Institute, Glasgow, UK*

MELANIE SCHWARTEN • *Institute of Complex Systems ICS-6 (Structural Biochemistry), Forschungszentrum Jülich, Jülich, Germany*

PER O. SEGLEN • *Centre for Molecular Medicine Norway (NCMM), Nordic EMBL Partnership, University of Oslo, Oslo, Norway*

ELENA SERANOVA • *Institute of Cancer and Genomic Sciences, Institute of Biomedical Research, College of Medical and Dental Sciences, University of Birmingham, Birmingham, UK*

PETROS STATHAKOS • *Cell Biology Laboratories, School of Biochemistry, University of Bristol, Bristol, UK*

BRIAN STORRIE • *Department of Physiology and Biophysics, University of Arkansas for Medical Sciences, Little Rock, AR, USA*

BIRGIT STRODEL • *Institute of Complex Systems ICS-6 (Structural Biochemistry), Forschungszentrum Jülich, Jülich, Germany; Institute of Theoretical and Computational Chemistry, Heinrich Heine University Düsseldorf, Düsseldorf, Germany*

QIAN PETER SU • *Institute for Biomedical Materials & Devices (IBMD), Faculty of Science, University of Technology Sydney, New South Wales, Australia*

PAULA SZALAI • *Centre for Molecular Medicine Norway (NCMM), Nordic EMBL Partnership, University of Oslo, Oslo, Norway*

GYŐZŐ SZENCI • *Department of Anatomy, Cell and Developmental Biology, Eötvös Loránd University, Budapest, Hungary*

SAROLTA TÓTH • *Department of Anatomy, Cell and Developmental Biology, Eötvös Loránd University, Budapest, Hungary*

SZABOLCS TAKÁTS • *Hungarian Academy of Sciences, Premium Postdoctorate Research Program, Budapest, Hungary; Department of Anatomy, Cell and Developmental Biology, Eötvös Loránd University, Budapest, Hungary*

ABUL K. TARAFDER • *Structural and Computational Biology Unit, European Molecular Biology Laboratory (EMBL), Heidelberg, Germany*

NEKTARIOS TAVERNARAKIS • *Institute of Molecular Biology and Biotechnology, Foundation for Research and Technology Hellas, Crete, Greece; Faculty of Medicine, Department of Basic Sciences, University of Crete, Crete, Greece*

SHARON A. TOOZE • *Molecular Cell Biology of Autophagy Laboratory, The Francis Crick Institute, London, UK*

PANAGIOTIS TSAPRAS • *School of Life Sciences, University of Warwick, Coventry, UK*

HIROYUKI UEDA • *Department of Genetics, Osaka University Graduate School of Medicine, Osaka, Japan; Department of Intracellular Membrane Dynamics, Graduate School of Frontier Biosciences, Osaka University, Osaka, Japan*

TAKASHI UENO • *Laboratory of Proteomics and Biomolecular Science, Research Support Center, Juntendo University Graduate School of Medicine, Tokyo, Japan*

MARIA I. VACCARO • *Pathophysiology Department, Institute of Biochemistry and Molecular Medicine (CONICET), School of Pharmacy and Biochemistry, University of Buenos Aires, Buenos Aires, Argentina*
TIM VAN ACKER • *Molecular Cell Biology of Autophagy Laboratory, The Francis Crick Institute, London, UK*
SATOSHI WAGURI • *Department of Anatomy and Histology, Fukushima Medical University, School of Medicine, Fukushima, Japan*
QING JUN WANG • *Department of Ophthalmology and Visual Sciences, University of Kentucky, Lexington, KY, USA*
CARL WARD • *Institute of Cancer and Genomic Sciences, Institute of Biomedical Research, College of Medical and Dental Sciences, University of Birmingham, Birmingham, UK*
YASUNORI WATANABE • *Department of Bioscience, Graduate School of Agriculture, Ehime University, Matsuyama, Japan; Laboratory of Structural Biology, Institute of Microbial Chemistry, Tokyo, Japan*
OLIVER H. WEIERGRÄBER • *Institute of Complex Systems ICS-6 (Structural Biochemistry), Forschungszentrum Jülich, Jülich, Germany*
SIDNEY W. WHITEHEART • *Department of Molecular and Cellular Biochemistry, University of Kentucky, Lexington, KY, USA*
DIETER WILLBOLD • *Institute of Complex Systems ICS-6 (Structural Biochemistry), Forschungszentrum Jülich, Jülich, Germany; Institut für Physikalische Biologie, Heinrich Heine University Düsseldorf, Düsseldorf, Germany*
MARTINA WIRTH • *Molecular Cell Biology of Autophagy Laboratory, The Francis Crick Institute, London, UK*
THOMAS WOLLERT • *Unit Membrane Biochemistry and Transport, Institut Pasteur, Paris, France*
AKINORI YAMASAKI • *Laboratory of Structural Biology, Institute of Microbial Chemistry, Tokyo, Japan*
SHENGMIN YAN • *Department of Pathology and Laboratory Medicine, Indiana University School of Medicine, Indianapolis, IN, USA*
WEI YUAN YANG • *Institute of Biological Chemistry, Academia Sinica, Taipei, Taiwan; Institute of Biochemical Sciences, College of Life Sciences, National Taiwan University, Taipei, Taiwan*
XIAO-MING YIN • *Department of Pathology and Laboratory Medicine, Indiana University School of Medicine, Indianapolis, IN, USA*
LI YU • *State Key Laboratory of Membrane Biology, Biomedical Pioneering Innovation Center (BIOPIC), School of Life Sciences, Peking University, Beijing, China*
MIN ZHANG • *State Key Laboratory of Membrane Biology, Beijing, China; Tsinghua-Peking Center for Life Sciences, Beijing, China; School of Life Sciences, Tsinghua University, Beijing, China*
JINCHAO ZHANG • *Department of Molecular and Cellular Biochemistry, University of Kentucky, Lexington, KY, USA*
HUI ZOU • *Department of Pathology and Laboratory Medicine, Indiana University School of Medicine, Indianapolis, IN, USA; College of Veterinary Medicine, Yangzhou University, Yangzhou, People's Republic of China*

Part I

Autophagry In Vitro

Chapter 1

Recombinant Expression, Purification, and Assembly of p62 Filaments

Abul K. Tarafder, Audrey Guesdon, Tanja Kuhm, and Carsten Sachse

Abstract

This chapter describes the recombinant overexpression of the canonical selective autophagy receptor p62/SQSTM1 in *E. coli* and affinity purification. Also described is the method to induce p62 filament assembly and their visualization by negative stain electron microscopy (EM). In cells, p62 forms large structures termed p62 bodies and has been shown to be aggregation prone. This tendency to aggregate poses problems for expression and purification in vitro, which is a prerequisite for structural analysis. Here, we describe the method to express and purify soluble p62, using the solubility tag, MBP, in conjunction with autoinduction. Furthermore, we describe the protocol to assemble p62 into filaments by controlling the ionic strength of its buffer, as well as the preparation of negative stain EM grids to visualize the filaments. In vitro formed p62 filaments can be used to study receptor cargo interactions in minimal reconstituted autophagy model systems.

Key words p62, Autophagy, Filaments, Affinity chromatography, Negative stain electron microscopy

1 Introduction

p62/Sequestosome 1 (SQSTM1), hereafter termed p62, is a 440 amino acid multifunctional scaffold protein that is conserved in higher eukaryotes. It takes its name from its apparent molecular weight on SDS-PAGE gels despite its predicted size of 47 kDa. Sequence analysis of p62 reveals it is comprised of three domains and a number of binding motifs. The N-terminal PB1 (Phox1 and Bem1p) domain mediates homo and hetero-dimerization and is followed by a ZZ-type zinc finger, a TRAF6-binding domain, LC3-interacting region (LIR), Keap1-interacting region (KIR), and a C-terminal ubiquitin-associated (UBA) domain which is known to dimerize in isolation [1–4].

Two overarching roles can be assigned to p62. First, it functions as a signaling adaptor involved in stress and growth response pathways such as NFκ-B, Nrf2, and mTOR via interaction with protein kinases such as atypical kinase C and MEK5, among others

Nicholas Ktistakis and Oliver Florey (eds.), *Autophagy: Methods and Protocols*, Methods in Molecular Biology, vol. 1880, https://doi.org/10.1007/978-1-4939-8873-0_1,

[5, 6]. Second, as described further in this chapter, p62 is an autophagic receptor that confers specificity to cargo engulfment in selective autophagy by linking the autophagic machinery to cargo destined for degradation [7]. Selective autophagy is a cellular degradation pathway that encapsulates large cargo fated for degradation, such as protein aggregates or pathogens, in double-membrane vesicles termed autophagosomes and delivers them to the lysosome. The multidomain structure of p62 allows it to function as a selective autophagy receptor with the PB1 domain-mediated oligomerization, while the LIR domain interacts with LC3 on the autophagosomal membrane and the UBA domain with ubiquitinylated cargo simultaneously [8]. There is mounting evidence that the signaling and autophagy functions of p62 combine to regulate a diverse set of processes within the cell such as inflammation, cell death, and tumorigenesis [9].

As well as being a selective autophagy receptor, p62 is also a substrate for autophagy. p62 accumulates in ubiquitin-positive inclusions in neurodegenerative diseases and proteinopathies [10, 11], and p62-transfected cells show protein dense arrays when observed by electron microscopy (EM) [7]. In vitro, MBP or GST-p62 fusion proteins were shown to form large aggregates [12]. All these findings are indicative of p62 forming oligomeric structures. Indeed, in 2015, our laboratory demonstrated that p62 is able to form organized filaments when purified and assembled in vitro. Using cryo-EM and successively longer p62 constructs, it was shown that the p62 PB1 domain alone is able to form long (>1 μm) flexible filamentous structures. Furthermore, when full-length (FL) p62 is purified and assembled, it forms significantly shorter filaments than the PB1 domain in isolation (0.1–0.3 μm), presumably due to regions of predicted disorder in the C-terminus of the protein. Importantly, these p62 filaments were capable of binding to LC3, and poly-ubiquitin indicating these p62 filaments can function as an autophagic receptor [13].

Here, the method to overexpress and purify p62 FL as well as a number of p62 truncations is described, along with the protocol to assemble p62 filaments and preparation of negative stain EM grids to visualize the filaments by transmission EM. In vitro prepared p62 filaments serve as model systems to reconstitute aspects of the selective autophagy and signaling machinery.

2 Materials

Chemicals were obtained from Sigma-Aldrich (St. Louis, Missouri, USA) or VWR international (Radnor, Pennsylvania, USA) unless otherwise stated. All solutions were prepared using ultrapure water. Prepare and store all reagents at room temperature unless otherwise indicated.

Table 1
Construct details used in this manuscript

Construct	Protein encoded	Domains/motifs	Cleavage site
pOPTM- p62 PB1 1-102	MBP-p62 PB1 a.a 1-102	PB1	TEV
pOPTM- p62 PB1 1-122	MBP-p62 PB1 a.a 1-122	PB1	TEV
pOPTM- p62 PB1 ZZ 1-167	MBP-p62 PB1 ZZ a.a 1-167	PB1, ZZ	TEV
pETM43-p62ΔUBA	MBP-p62 a.a 1-389-his$_6$	PB1, ZZ, TRAF6, LIR, KIR	3C
pETM43- p62 FL	MBP-p62 a.a 1-440-his$_6$	PB1, ZZ, TRAF6, LIR, KIR, UBA	3C

2.1 Overexpression of p62 Proteins

1. A glycerol stock of *E. coli* BL21(DE3) CodonPlus-RIL cells (Agilent, Santa Clara, CA, USA) contains the plasmid encoding the p62 construct of interest. The plasmids used are listed in Table 1. All encode *H. sapiens* p62 and have an N-terminal MBP, followed by either a TEV or 3C cleavage site. For longer p62 constructs, a C-terminal his$_6$ tag was also included to allow separation of full-length proteins from C-terminal degradation products.
2. Luria Broth (1 L): dissolve 10 g bacto tryptone, 5 g bacto yeast extract, and 5 g NaCl to 1 L of H_2O, and sterilize by autoclaving. 100 mg/mL ampicillin (1000×, pOPTM selection), dissolve 1 g ampicillin, sodium salt in 10 mL H_2O, filter sterilize, and store at −20 °C. 34 mg/mL kanamycin (1000×, pETM43 selection), dissolve 340 mg kanamycin in 10 mL H_2O, filter sterilize, and store at −20 °C. 34 mg/mL chloramphenicol (1000×, pRIL selection), dissolve 340 mg chloramphenicol in absolute ethanol, and store at −20 °C.
3. ZY autoinduction media: this media is made up of four components: ZY (1 L), dissolve 10 g tryptone and 5 g yeast extract in H_2O (NB. ZY is Luria Broth (LB) without NaCl). 20× NPS (1 L), dissolve 66 g $(NH4)_2SO_4$, 136 g KH_2PO_4, and 142 g Na_2HPO_4 in H_2O, adjust pH to 6.75, and make up to 1 L. 50× 5052 (1 L), dissolve 250 g glycerol, 25 g glucose, and 100 g α-lactose in H_2O. 1 M $MgSO_4$ (100 mL), dissolve 12.04 g $MgSO_4$ in 100 mL H_2O. Autoclave all solutions separately.

   For 1 L ZY autoinduction media, mix components in the order stated: 928 mL ZY, 1 mL 1 M $MgSO_4$, 20 mL 50× 5052, and 50 mL 20× NPS. Add 1 mL of the appropriate antibiotic (*see* above) prior to use [14].

4. Temperature-controlled shaking incubators with the capability to house 500 mL and 2 L flasks.
5. 0.2 μm Millex GP polyethersulfone syringe filters.
6. Phosphate buffered saline.
7. Avanti J-20 XP centrifuge and JLA8.1000 rotor (Beckman Coulter, Brea, CA, USA) or equivalent.
8. Heraeus Megafuge 16R Benchtop centrifuge (ThermoScientific, Waltham, MA, USA) or equivalent.

2.2 Affinity Purification of p62 Proteins

1. 1 M HEPES pH 8.0: dissolve 238.3 g HEPES (free acid) in 800 mL H_2O, adjust the pH to 8.0 with NaOH and bring the volume to 1 L, and store at 4 °C.
2. 5 M NaCl: dissolve 292.2 g of NaCl in 1 L H_2O.
3. 1 M imidazole: dissolve 68.08 g imidazole in 1 L H_2O.
4. 0.5 M 1,4-dithio-DL-threitol (TCEP): dissolve 2.87 g TCEP in 15 mL H_2O, pH the solution to 7.0 with 10 M NaOH, and bring the solution to 20 mL with H_2O, filter sterilize, and store at −20 °C.
5. 100 mM maltose: dissolve 0.34 g maltose monohydrate in 10 mL H_2O.
6. Complete mini EDTA-free protease inhibitor tablets.
7. MBP lysis buffer: 50 mM HEPES pH 8.0, 150 mM NaCl, 0.5 mM TCEP, 1 complete protease inhibitor tablet per 50 mL lysis buffer.
8. MBP wash buffer: 50 mM HEPES pH 8.0, 150 mM NaCl, 0.5 mM TCEP.
9. MBP elution buffer: 50 mM HEPES pH 8.0, 150 mM NaCl, 0.5 mM TCEP, 10 mM maltose.
10. His_6 lysis buffer: 50 mM HEPES pH 8.0, 1 M NaCl, 0.5 mM TCEP, 1 complete protease inhibitor tablet per 50 mL lysis buffer.
11. His_6 affinity buffer: 50 mM HEPES pH 8.0, 1 M NaCl, 0.5 mM TCEP.
12. His_6 wash buffer: 50 mM HEPES pH 8.0, 1 M NaCl, 0.5 mM TCEP, 50 mM imidazole.
13. His_6 elution buffer: 50 mM HEPES pH 8.0, 1 M NaCl, 0.5 mM TCEP, 250 mM imidazole.
14. Filament formation buffer: 50 mM HEPES pH 8.0, 50 mM NaCl, 0.5 mM TCEP.
15. A mechanical device to disrupt *E. coli* cells (e.g., a sonicator, French press, or cell homogenizer).
16. Avanti J-20 XP centrifuge and JA25.50 rotor (Beckman Coulter, Brea, CA, USA) or equivalent.

17. Econo-Pac 20 mL chromatography columns.
18. Amylose resin.
19. Ni-NTA resin.
20. Rotating wheel.
21. 96 well plate and protein assay dye reagent.
22. Novex SureLock SDS-PAGE apparatus and Novex Bis-Tris precast gels (Life Technologies, Carlsbad, CA, USA) or equivalent.
23. Snakeskin dialysis tubing (3.5 kDa molecular weight cutoff [MWCO]) and dialysis clips.
24. Magnetic stirrer.
25. A Dewar flask filled with liquid nitrogen.

2.3 MBP-Tag Cleavage and Filament Formation

1. His_6-TEV (Tobacco etch virus) and his_6-3C proteases (*see* **Note 1**).
2. Biotech Cellulose Ester (CE) 100 kDa MWCO dialysis tubing.
3. PD10 desalting column.

2.4 Sucrose Gradient

1. Sucrose.
2. TL 100 ultracentrifuge and TLS55 rotor (Beckman Coulter, Brea, CA, USA).
3. Polyallomer tubes, 2.2 mL.

2.5 Negative Stain Electron Microscopy

1. 2% Uranyl acetate: dissolve 0.2 g uranyl acetate powder in 10 mL H_2O (Can take hours). Filter solution with a 0.2 μm syringe filter. Wrap tube in aluminum foil and store in the dark at room temperature (RT) (*see* **Note 2**).
2. Pelco easy Glow discharger or equivalent.
3. Copper EM Grids coated with a continuous carbon film (Electron Microscopy Sciences, Hatfield, PA, USA) or equivalent.
4. Glass slide.
5. Inverted tweezers, type N5, or equivalent.
6. Parafilm.
7. Whatman grade 1 qualitative blotting paper.
8. Transmission electron microscope.

3 Methods

3.1 Overexpression of p62 Proteins

To prevent p62 aggregation and partitioning into the insoluble fraction upon overexpression, all p62 constructs were expressed as a fusion protein with the solubility factor maltose binding protein

(MBP), and expression from the T7 promoter was induced by autoinduction at a low temperature rather than the canonical addition of IPTG.

1. Inoculate 50–150 mL LB broth containing 100 μg/mL ampicillin (pOPTM vectors) or 34 μg/mL kanamycin (pETM43 vectors) and 34 μg/mL chloramphenicol in a 500 mL baffle-bottomed flask with a glycerol stock of *E. coli* BL21(DE3) CodonPlus-RIL cells transformed with the p62 construct of interest. Place in an incubator at 37 °C overnight with agitation at 200 rpm.
2. Add 25 mL of the saturated overnight culture to each 1 L of ZY autoinduction media containing 100 μg/mL ampicillin (pOPTM vector) or 34 μg/mL kanamycin (pETM43 vector) and 34 μg/mL chloramphenicol in a 2 L baffle bottom flask. Ordinarily, 2–6 L of cells are grown at a time to ensure sufficient yield of purified protein.
3. Cells were incubated at 37 °C for 6 h with agitation at 200 rpm. Cells were removed from the incubator and allowed to rest at room temperature. The temperature of the shaking incubator was reduced to 20 °C, and when this temperature was reached, the cells were returned to the incubator and incubated overnight at 20 °C with agitation at 200 rpm.
4. Cells were harvested by centrifugation in a Beckman Avanti J-20 XP centrifuge with rotor JA8.1000 at 4000 × *g* for 10 min at 4 °C.
5. The supernatant was decanted, and the cell pellets were resuspended in ice-cold PBS and transferred to 50 mL falcon tubes.
6. The cells were centrifuged at 4000 × *g* for 10 min at 4 °C in a Heraeus Megafuge 16R Benchtop centrifuge, the supernatant decanted, and the cell pellets either processed immediately or stored at −80 °C. A 1 L culture typically yields 5–7 g of cell pellet.

3.2 Affinity Purification of MBP-p62 PB1 a.a 1-102, MBP-p62 PB1 a.a 1-122, and MBP-p62 PB1 ZZ (a.a 1-167)

p62 truncations can be purified by affinity chromatography using amylose resin via the N-terminal MBP tag (Fig. 1a). Although it is not necessary to work at 4 °C, the sample, all reagents, and tubes should be kept ice-cold, and heating of the sample should be kept to a minimum. As MBP-p62 ZZ (a.a 1-167) contains a zinc-binding domain, the addition of $ZnCl_2$ to all purification buffers stabilizes this protein.

1. For MBP-p62 PB1 ZZ (1-167), supplement all buffers with 20 mM $ZnCl_2$.
2. Thaw cell pellets slowly by incubating tubes in water at RT.
3. Resuspend cell pellet thoroughly with ice-cold MBP lysis buffer (10 mL/g of cell pellet).

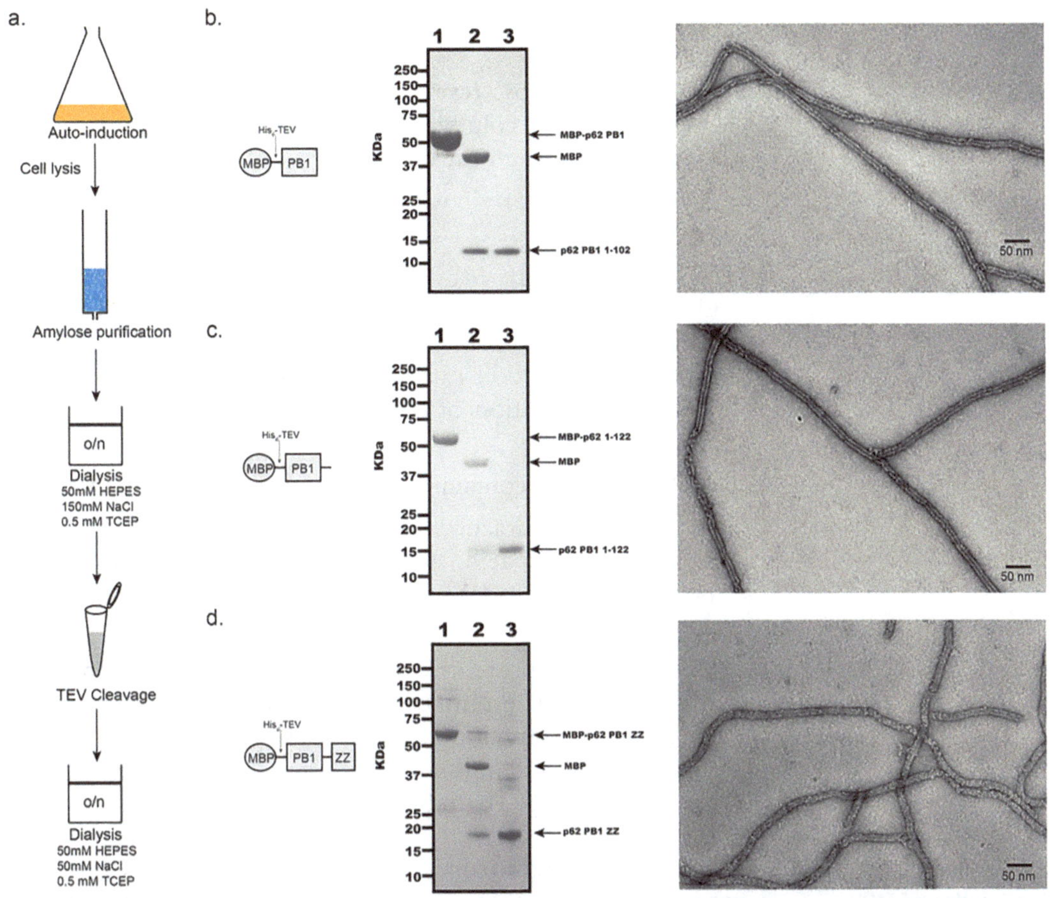

Fig. 1 Purification and filament assembly of p62 PB1 a.a 1-102, p62 PB1 a.a 1-122, and p62 PB1 ZZ a.a 1-167. (**a**) Schematic representation of purification protocol. (**b**) Left, schematic representation of MBP-p62 PB1 a.a 1-102; middle, SDS-PAGE of MBP-p62 PB1 a.a 1-102 purification and cleavage; lane 1, purified MBP-p62 PB1 a.a 1-102; lane 2, after TEV cleavage; lane 3, after 100 kDa MWCO dialysis; right, negative stain electron micrograph of purified p62 PB1 a.a 1-102 filaments. Scale bar = 50 nm. (**c**) Left, schematic representation of MBP-p62 PB1 a.a 1-122; middle, SDS-PAGE of MBP-p62 PB1 a.a 1-122 purification and cleavage; lane 1, purified MBP-p62 PB1 a.a 1-122; lane 2, after TEV cleavage; lane 3, after 100 kDa MWCO dialysis; right, negative stain electron micrograph of purified p62 PB1 a.a 1-122 filaments. Scale bar = 50 nm. (**d**) Left, schematic representation of MBP-p62 PB1 ZZ a.a 1-167; middle, SDS-PAGE of MBP-p62 PB1 a.a 1-167 purification and cleavage; lane 1, purified MBP-p62 PB1 a.a 1-167; lane 2, after TEV cleavage; lane 3, after 100 kDa MWCO dialysis; right, negative stain electron micrograph of purified p62 PB1 ZZ a.a 1-167 filaments. Scale bar = 50 nm

4. Lyse the cell suspension using a French press or by sonication (*see* **Note 3**).
5. Clear the lysate by centrifugation at 48,000 × *g* for 45 min at 4 °C using a Beckman Avanti J-20 XP centrifuge and JA25.50 rotor.
6. Equilibrate amylose resin with MBP wash buffer. Use 2 mL dry resin per 50 mL of cell lysate.

7. Incubate amylose resin with cell lysate for 1 h at 4 °C with end-over-end rotation.
8. The lysate/amylose resin mix is transferred to an Econo-Pac chromatography column. The resin will be retained as the lysate flows through the column.
9. Wash the amylose resin with 40 column volumes (CV) of MBP wash buffer.
10. Elute protein with MBP elution buffer in fractions of 1CV. Elution can be monitored by setting up a 96 well plate with 60 μL of Bio-Rad protein assay dye reagent per well. 5 μL of each elution fraction can be added to a well, mixed well by pipetting. Formation of blue color indicates the presence of protein.
11. Analyze fractions containing protein by SDS-PAGE (*see* **Note 4**).
12. Pool appropriate fractions and dialyze against 2 L of MBP wash buffer overnight at 4 °C with gentle stirring using dialysis membranes with a MWCO below 30 kDa. Recover the sample after dialysis.
13. Ascertain the protein concentration using a NanoDrop or equivalent. The purified protein can be used immediately for filament formation (*see* Subheading 3.4) or can be aliquoted, snap-frozen in liquid nitrogen, and stored at −80 °C (*see* **Note 5**) (Fig. 1b–d, SDS-PAGE, Lane 1).

3.3 Affinity Purification of MBP-p62Δ UBA-his$_6$ and MBP-p62 FL-his$_6$

Full-length and longer truncations of p62 suffer from C-terminal degradation during the purification procedure. To combat this, a C-terminal his$_6$ tag was introduced and utilized for affinity purification using Ni-NTA resin in order to reduce the amount of degraded protein purified (Fig. 2a). Full-length p62 also shows a greater propensity to aggregate and precipitate than the shorter p62 PB1 truncations. To combat this, in conjunction with the N-terminal MBP solubility tag, the purification buffers contained high salt.

1. Follow **steps 2–5** from Subheading 3.2 using his$_6$ lysis buffer.
2. Equilibrate Ni-NTA resin with his$_6$ affinity buffer. Use 2 mL dry resin per 50 mL of cell lysate.
3. Incubate equilibrated Ni-NTA resin with cell lysate for 1 h at 4 °C with end over end rotation.
4. The lysate/Ni-NTA resin mix is transferred to an Econo-Pac chromatography column. The resin will be retained as the lysate flows through the column.
5. Wash the Ni-NTA resin with 10 CV of his$_6$ affinity buffer.
6. Wash the Ni-NTA resin with 10 CV of his$_6$ wash buffer.

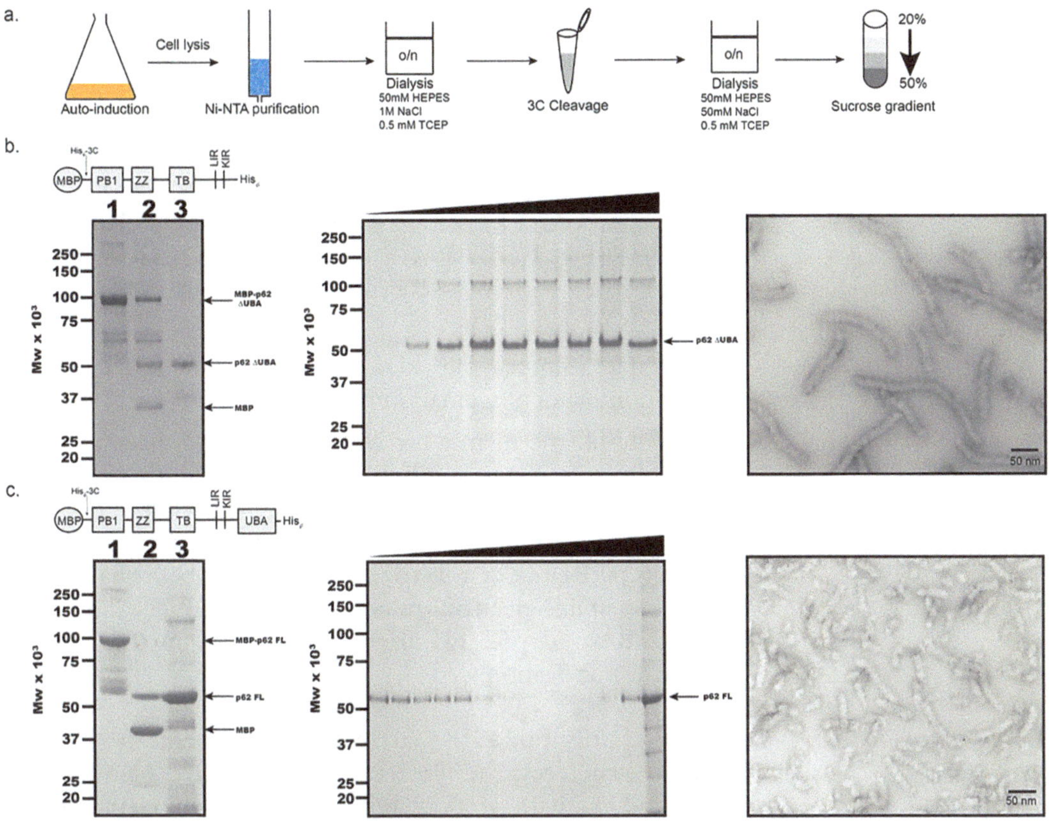

Fig. 2 Purification and filament assembly of p62ΔUBA and p62 FL. (**a**) Schematic representation of purification protocol. (**b**) Left, SDS-PAGE of MBP-p62ΔUBA purification and cleavage; lane 1, purified MBP-p62ΔUBA; lane 2, after TEV cleavage; lane 3, after 100 kDa MWCO dialysis; middle, SDS-PAGE of sucrose gradient fractions; right, negative stain electron micrograph of purified p62ΔUBA filaments. Scale bar = 50 nm. (**c**) Left, SDS-PAGE of MBP-p62FL purification and cleavage; lane 1, purified MBP-p62FL, lane 2, after TEV cleavage; lane 3, after 100 kDa MWCO dialysis; middle, SDS-PAGE of sucrose gradient fractions; right, negative stain electron micrograph of purified p62 FL filaments. Scale bar = 50 nm

7. Follow **steps 10–13** from Subheading 3.2 using His_6 elution buffer for elution and his_6 affinity buffer for dialysis (Fig. 2b, c, SDS-PAGE, Lane 1).

3.4 Cleavage of N-Terminal MBP Tag and Filament Formation

In order to form filaments, the N-terminal MBP tag must be cleaved from the p62 proteins and removed and the NaCl concentration of the buffer lowered. This is performed by cleavage with either TEV or 3C proteases followed by dialysis using a 100 kDa MWCO membrane. The high MWCO of the membrane allows the removal of the cleaved MBP, protease, and monomeric p62, while retaining p62 filaments which have a far greater MW than 100 kDa.

1. If proteins were stored at −80 °C, thaw in water at RT. In the case of p62 PB1 ZZ (a.a 1-167), the $ZnCl_2$ must be removed

before cleavage as Zn^{2+} ions will inhibit the TEV protease. To achieve this, the protein is passed through a PD10 desalting column, equilibrated with MBP wash buffer (*see* Subheading 3.2, **step 1**).

2. Incubate p62 protein with the appropriate protease for 4 h at RT at a 1/100 molar ratio. p62 PB1 a.a 1-102, p62 PB1 a.a 1-120, and p62 PB1 ZZ a.a 1-167 should be incubated with his_6-TEV (Fig. 1, SDS-PAGE, Lane 2) and p62ΔUBA and p62 FL incubated with his_6-3C protease (Fig. 1b–d and Fig. 2b, c, SDS-PAGE, Lane 2).
3. Equilibrate Biotech 100 kDa dialysis membrane for at least 30 min in filament formation buffer.
4. Carefully fill the membrane with sample, and dialyze against 2 L of filament formation buffer at 4 °C overnight with gentle stirring. Recover the sample after dialysis (Fig. 1b–d and Fig. 2b, c, Lane 3). p62 PB1 a.a 1-102, p62 PB1 a.a 1-120, and p62 PB1 ZZ a.a 1-167 filaments can be visualized by negative stain electron microscopy (*see* Subheading 3.6). p62ΔUBA and p62 FL filaments require a further sucrose gradient purification step.

3.5 Sucrose Gradient Purification of p62Δ UBA and p62 FL Filaments

To further purify p62ΔUBA and p62 FL filaments from other oligomeric structures, the filaments were subjected to a 20–50% sucrose gradient with a 5% increment. The larger filaments will travel further through the gradient than the other smaller oligomeric species leading to purification of the longest filaments.

1. Prepare 50% sucrose solution (50 mL): dissolve 25 g sucrose in filament formation buffer (*see* Subheading 3.4, **step 1**).
2. Prepare the following sucrose solutions by dilution of the 50% sucrose solution with filament formation buffer: 45, 40, 35, 30, 25, and 20%.
3. In a 2.2 mL polyallomer tube, carefully layer 150 μL of each sucrose solution commencing with the 50% sucrose solution and ending with 20%.
4. Carefully apply the sample at the top of the gradient.
5. Centrifuge at 160,000 × *g* for 4 h 30 min at 4 °C using a Beckman TL 100 ultracentrifuge and TLS55 rotor.
6. Gradient fractions are collected from the top of the gradient in 100 μL fractions. The pellet is resuspended in 50 μL filament formation buffer. Fractions are analyzed by SDS-PAGE (Fig. 2b, c, SDS-PAGE, middle panel) and negative stain electron microscopy. The longest p62 filaments are obtained in the highest sucrose concentration fractions and the pellet.

3.6 Negative Stain Electron Microscopy

p62 filament formation can be assessed by negative stain EM. This process involves the application of the sample to carbon-coated copper EM grids and staining with a heavy metal stain such as uranyl acetate. The heavy metal stain surrounds the sample but is excluded from the volume occupied by the sample, hence the term "negative stain." As the heavy metal stain has a high atomic number, it will scatter electrons stronger than the protein filament, which excludes the stain, leading to the contrast in the image.

1. Pick up carbon-coated copper EM grids by the rim with tweezers and place onto a glass slide.
2. Place glass slides into Pelco easy Glow discharger and glow discharge the grids at 15 mA for 45 s at a vacuum of 0.26 mbar (*see* **Note 6**). Glow discharging is necessary to make the hydrophobic carbon film on the grid more hydrophilic to allow spreading of the sample on the grid.
3. On a piece of parafilm, put two drops of filament formation buffer and two drops of 2% uranyl acetate (approximately 20 μL) per grid to be processed.
4. Pick up a glow-discharged grid by the rim with tweezers. Add 2.5 μL of sample to the grid. Allow to absorb to the grid for 45 s.
5. Blot off excess sample by touching the grid at a right angle to Whatman No. 1 filter paper.
6. Touch the face of the grid where the sample was applied to a drop of filament formation buffer. Blot off the excess. Repeat with second drop.
7. Touch the face of the grid where the sample was applied to a drop of 2% uranyl acetate. Blot off the excess.
8. Touch the face of the grid where the sample was applied to a drop of 2% uranyl acetate. Incubate for 45 s before blotting off excess stain thoroughly.
9. Leave grid to dry at RT for at least 10 min.
10. Visualize grid in a transmission electron microscope (Figs. 1b–d and 2b, c micrographs).

4 Notes

1. We produce his_6-TEV and his_6-3C proteases in house, but both proteases are available from commercial sources.
2. Uranyl acetate is radioactive. Follow local health and safety guidelines for its use and disposal.
3. We lyse cells either by three to four passes through a pre-cooled M110-L microfluidizer at 10–15,000 psi or by sonication at

4 °C with a Branson W250D Sonifier (10% amplitude, 50% duty cycle, 45 s sonication followed by 45 s rest, repeated four times).

4. We use the XCell SureLock Mini-cell electrophoresis system together with precast gradient gels (e.g., Novex NuPAGE 10-well, 4–12% bis-tris gels) run with MES buffer (Life Technologies, Carlsbad, CA, USA).
5. The protein concentration needs to be above 1 mg/mL for the subsequent filament formation steps. If the concentration is less than this, concentrate the protein using a protein concentrator. For p62ΔUBA and p62 FL, do not concentrate to over 3 mg/mL as the protein begins to precipitate.
6. Glow discharge settings will vary dependent on the unit available and will need to be optimized with the setup available.

Acknowledgments

We acknowledge the assistance by the EMBL's Protein Expression and Purification Core Facility for reagents and Electron Microscopy Core Facility for the instrument support.

References

1. Lamark T, Perander M, Outzen H, Kristiansen K, Øvervatn A, Michaelsen E, Bjørkøy G, Johansen T (2003) Interaction codes within the family of mammalian Phox and Bem1p domain-containing proteins. J Biol Chem 278:34568–34581
2. Jain A, Lamark T, Sjøttem E, Larsen KB, Awuh JA, Øvervatn A, McMahon M, Hayes JD, Johansen T (2010) p62/SQSTM1 is a target gene for transcription factor NRF2 and creates a positive feedback loop by inducing antioxidant response element-driven gene transcription. J Biol Chem 285:22576–22591
3. Pankiv S, Clausen TH, Lamark T, Brech A, Bruun JA, Outzen H, Øvervatn A, Bjørkøy G, Johansen T (2007) p62/SQSTM1 binds directly to Atg8/LC3 to facilitate degradation of ubiquitinated protein aggregates by autophagy. J Biol Chem 282:24131–24145
4. Long J, Garner TP, Pandya MJ, Craven CJ, Chen P, Shaw B, Williamson MP, Layfield R, Searle MS (2010) Dimerisation of the UBA domain of p62 inhibits ubiquitin binding and regulates NF-kappaB signalling. J Mol Biol 396:178–194
5. Moscat J, Diaz-Meco MT (2009) p62 at the crossroads of autophagy, apoptosis, and cancer. Cell 137:1001–1004
6. Wilson MI, Gill DJ, Perisic O, Quinn MT, Williams RL (2003) PB1 domain-mediated heterodimerization in NADPH oxidase and signaling complexes of atypical protein kinase C with Par6 and p62. Mol Cell 12:39–50
7. Johansen T, Lamark T (2011) Selective autophagy mediated by autophagic adapter proteins. Autophagy 7(3):279–296
8. Mizushima N, Yoshimori T, Ohsumi Y (2011) The role of Atg proteins in autophagosome formation. Annu Rev Cell Dev Biol 27:107–132
9. Johansen T, Sachse C (2015) The higher-order molecular organization of p62/SQSTM1. Oncotarget 6:16796–16797
10. Kuusisto E, Salminen A, Alafuzoff I (2001) Ubiquitin-binding protein p62 is present in neuronal and glial inclusions in human tauopathies and synucleinopathies. Neuroreport 12:2085–2090

11. Zatloukal K, Stumptner C, Fuchsbichler A, Heid H, Schnoelzer M, Kenner L et al (2002) p62 is a common component of cytoplasmic inclusions in protein aggregation diseases. Am J Pathol 160:255–263
12. Paine MG, Babu JR, Seibenhener ML, Wooten MW (2005) Evidence for p62 aggregate formation: role in cell survival. FEBS Lett 579:5029–5034
13. Ciuffa R, Lamark T, Tarafder AK, Guesdon A, Rybina S, Hagen WJH, Johansen T, Sachse C (2015) The selective autophagy receptor p62 forms a flexible filamentous helical scaffold. Cell Rep 11:748–758
14. Studier FW (2005) Protein production by auto-induction in high-density shaking cultures. Protein Expr Purif 41:207–234

Chapter 2

Structural Studies of Autophagy-Related Proteins

Melanie Schwarten, Oliver H. Weiergräber, Dušan Petrović, Birgit Strodel, and Dieter Willbold

Abstract

Information about the structure and dynamics of proteins is crucial for understanding their physiological functions as well as for the development of strategies to modulate these activities. In this chapter we will describe the work packages required to determine the three-dimensional structures of proteins involved in autophagy by using X-ray crystallography or nuclear magnetic resonance (NMR) spectroscopy. Further we will provide instructions how to perform a molecular dynamics (MD) simulation using GABARAP as example protein.

Key words X-ray crystallography, NMR spectroscopy, MD simulation

1 Introduction

Like any complex cellular pathway, the autophagy machinery comprises proteins of various structural classes, which contain very different types of domains, such as kinase domains, protein interaction modules, and intrinsically disordered segments. Owing to this diversity, it is conceptually impossible to devise protocols which describe procedures in full detail while, at the same time, being applicable to this entire group of polypeptides without modification. In other words, investigating the structure of an individual autophagy-related protein may turn out to be just as straightforward or just as demanding as for any other molecule of interest. We will therefore approach the subject by outlining the processes of structural characterization by experimental as well as in silico methods at an appropriate level of generalization and provide examples as suitable, thus illustrating both the underlying strategies and the variety of conditions that researchers may encounter in their own studies.

Nicholas Ktistakis and Oliver Florey (eds.), *Autophagy: Methods and Protocols*, Methods in Molecular Biology, vol. 1880, https://doi.org/10.1007/978-1-4939-8873-0_2,

2 Methods

2.1 X-Ray Crystallography

2.1.1 Crystallization

1. Prepare a solution containing the protein of interest in a suitable buffer system. This sample should be of the highest quality that is reasonably achievable (*see* **Note 1**).
2. Select conditions to be used for initial screening. A multitude of screening sets are available from commercial suppliers, implementing empirical sparse-matrix approaches or systematic variation of select parameters (*see* **Note 2**).
3. Prepare protocols on a robotic liquid handling system (in case a robot is not available, plates for initial screening can of course be set up manually). Most instruments support the setup of vapor diffusion experiments in sitting-drop geometry. Define a suitable range of volumes in accordance with the properties of the target plate as well as the capabilities of the instrument.
4. Place sample, screening solutions, and crystallization plate into the system.
5. Set up crystallization experiments. For initial screening, prepare plates with as many conditions as feasible (determined by amount of protein available, capacity for storage, etc.). These should include at least several hundred screening solutions as well as different protein concentrations. As an alternative to the use of separate protein samples, the mixing ratio of protein and screening solution (default 1:1) may be varied, which will change both initial and equilibrium conditions. Whenever possible, prepare replicates to be stored at different temperatures.
6. Seal plates tightly with a clear foil, which should also be UV-transparent.
7. Store plates at the desired temperature(s). Reasonable constancy of temperature is important in order to avoid defects in crystal growth.
8. Observe the evolution of the experiment at regular intervals. Since equilibration of drop and reservoir will roughly follow an exponential, the interval between observations may be increased over time.
9. Evaluate promising results by optical properties. Crystalline material will usually display birefringence (unless the crystal system is cubic or a crystal is viewed along an isotropic axis), which can be observed under crossed polarizers. Protein crystals can often be differentiated from salt by their intrinsic tryptophan fluorescence or other spectroscopic properties.
10. Optimize initial hits. Set up fine screens by varying the initial reservoir composition (concentration of precipitant, pH, etc.) and protein concentration, possibly also temperature. More

extensive changes in crystal morphology may be achieved by screening additives together with a single reservoir solution which has been optimized beforehand.

11. Treat with phasing compounds, as required. These include heavy atom salts or small organic compounds containing suitable marker atoms.

    Examples: For crystallizing the complex of human GABARAP and an artificial dodecapeptide, optimization of an initial hit yielded a reservoir solution containing 30% (w/v) PEG 3350, 800 mM NaCl, 50 mM MES, pH 5.9. The experiment was set up at a temperature of 12 °C with a protein solution containing the complex at 15 mg/mL [1]. The same protein bound to a calreticulin-derived undecapeptide was found to crystallize under significantly different conditions: 27% (v/v) PEG MME 550, 10 mM $ZnSO_4$, 100 mM MES, pH 6.5, at 20 °C and with a protein concentration of 11 mg/mL. Notably, Zn^{2+} ions played a critical role by establishing a lattice contact [2]. The related Atg8 family protein GATE-16 was initially observed to crystallize in a concentrated solution at 4 °C in the presence of 100 mM phosphate buffer, pH 7. In this case, the supersaturation driving the phase transition was largely afforded by the protein itself, which greatly reduced the precipitant concentration required. Diffraction-quality samples were obtained in 100 mM phosphate buffer, 50 mM KCl, 10 mM DTT at 20 °C, applying the protein at 5 mg/mL [3]. Finally, native human ATG101 was crystallized in 8% (w/v) PEG 3350, 50 mM NaCl, 10 mM β-mercaptoethanol, 50 mM MES, pH 5.6, with a 2.2 mg/mL protein solution at 20 °C. Incorporation of selenomethionine in place of methionine was found to reduce the solubility of ATG101, and the PEG 3350 concentration was decreased to 5% (w/v) for best results [4].

2.1.2 Crystal Harvesting and Cryoprotection

1. Prepare tools required for crystal manipulation. These include a scalpel for cutting the sealing foil, cryoloops of appropriate size for handling crystals, and other microtools, if available (*see* **Note 2**).
2. Prepare solutions for cryoprotection during flash cooling. These are typically modifications of the respective reservoir solution including cryoprotectants such as polyols or certain salts (ideally, other solutes should not be diluted). If such compounds are already present, their concentrations may be increased as appropriate.
3. Open the well containing the crystals, typically by cutting the seal of a multiwell plate.

4. Using a cryoloop or similar tool, transfer a crystal into the cryosolution and incubate for a few minutes. Alternatively, the cryosolution may be added directly to the drop containing the crystal. This in situ treatment is often more gentle since it avoids mechanical stress to the crystal at this stage. If the concentration of cryoprotectant needs to be increased significantly, doing so in steps is advisable. On the other hand, sensitive crystals can sometimes be handled successfully by only shortly submerging them in the cryosolution, or by using a non-penetrating oil, so as to only replace the solvent outside the crystal.
5. Catch the crystal of interest with a cryoloop or similar tool. Be careful to harvest a single crystal and not to include other material (precipitate, etc.), if possible. When using a cryoloop, it is desirable to have the crystal suspended in a thin film of solvent, without direct contact to the nylon fiber.
6. For flash cooling, mount the crystal in a cryostream of gaseous nitrogen at about 100 K, or plunge into liquid nitrogen. The rate of temperature decrease throughout the sample is critical.
7. Store the crystal in an appropriate vial at liquid nitrogen temperature until data collection.

   Examples: Crystals of GABARAP bound to an artificial peptide, as well as crystals of GATE-16, were flash-cooled after incubation with reservoir solution containing 20% and 35%, respectively, of glycerol [1, 3]. For ATG101 (native and selenomethionyl protein), 25% glycerol or PEG 400 were found to be suitable [4]. Finally, in the case of GABARAP-calreticulin peptide co-crystals the concentration of PEG MME 550 was increased to 29% and 5% glycerol added [2].

2.1.3 Data Collection

1. Mount the crystal on a goniostat, either manually or via a robotic sample changer. During the measurement, the sample is typically kept under cryogenic conditions, i.e., in gaseous nitrogen at approximately 100 K.
2. Center the crystal on the intersection of the spindle axis with the beam, using motorized or manual controls.
3. Collect test diffraction images, typically two 1° X-ray exposures separated by 90° of rotation. This step serves two main purposes: (1) judging the diffraction quality of the crystal, including an estimate of the achievable data resolution, (2) indexing of the diffraction pattern, which allows for evaluation of possible crystal systems, implying a minimum expected point group symmetry in the diffraction data, as well as determination of crystal orientation.
4. Determine a strategy for collection of a complete dataset. The required rotation range and the optimal starting point are

determined by the presumed symmetry and the orientation of the crystal, whereas the maximum angular increment per exposure is additionally limited by its mosaicity. These values are typically provided by the software used for evaluating the test images.

5. Collect datasets. Depending on the phasing procedure of choice, criteria of data quality are prioritized differently. In short, molecular replacement (MR) benefits from data with good completeness, especially at low resolution, whereas in experimental phasing, which relies on small differences in intensities, the accuracy of measurements is more critical and high redundancy for both constituents of each Friedel pair is advisable.

2.1.4 Structure Determination by Molecular Replacement

1. Evaluate and integrate the dataset, using your data processing software of choice. Since anomalous differences are not used in routine MR, Friedel mates may often be merged. On the other hand, anomalous difference maps can sometimes be helpful in identifying bound nonprotein atoms (ions) at a later stage (*see* **Note 3**).
2. Select search model(s) for MR. This decision needs to consider both the coverage of the target sequence by the model and the similarity of the two sequences.
3. Modify the model(s) as appropriate. This usually involves removal of parts which are not present in the structure of interest, adjustment of the amino acid sequence, and optionally a B-factor modification to account for increased uncertainty in surface residues.
4. Perform the actual MR search. Available software offers different levels of automation and support for intelligent decision-making.
5. Select the best solution(s), based on correlation scores between observed and calculated structure factor amplitudes or similar measures. The standard crystallographic residual R is often unsuitable for this purpose, due to its poor discrimination among unrefined solutions.
6. Perform an initial refinement with the candidate structure(s). In case of a reasonable solution, refinement statistics should reveal a decrease of crystallographic residuals. Importantly, the resulting electron density ($2F_o - F_c$ and $F_o - F_c$ maps) should be examined for sensible protein features and indications of gross errors. The behavior during refinement offers the most reliable criterion for telling correct from wrong MR solutions.

   Examples: For GABARAP complexed with the artificial peptide [1] and with the calreticulin-derived peptide [2], initial phasing was achieved by MR using search models derived from

GABARAP crystal structures (PDB IDs 1GNU and 1KJT, respectively). In the case of GATE-16 [3], a previously determined GATE-16 structure (PDB ID 1EO6) was adopted.

2.1.5 Structure Determination by Experimental Phasing

1. Evaluate and integrate the dataset, using your data processing software of choice. Since anomalous differences are useful in the vast majority of scenarios, Friedel mates should not be merged (*see* **Note 4**).
2. In case differences between datasets are to be evaluated (most notable in isomorphous replacement techniques), ensure that these datasets are on the same scale. Certain software packages for phasing may rescale internally, making this step unnecessary.
3. Determine the marker atom substructure. This can be accomplished either by examination of (isomorphous or anomalous) difference Patterson maps or by direct methods approaches involving dual-space recycling. Available software offers different levels of automation and intelligent decision-making.
4. Calculate initial phases. This is often performed by the software used for marker atom localization.
5. Improve initial phases by density modification techniques. Several algorithms are available; their efficiency depends on the (expected) content of the asymmetric unit, such as non-crystallographic symmetry (NCS), solvent fraction, etc.
6. Develop a first protein model. Most software packages used for phasing today include some flavor of automated tracing algorithm which can provide a polyalanine or even a complete polypeptide model. The success of such methods critically depends on the quality of the data and available phases. Evaluate the initial model together with the experimentally phased (and modified) electron density; build missing segments if possible, and adjust residue numbering.
7. Improve the model iteratively. If the initial model is reasonably complete, one may directly proceed to a standard refinement workflow (*see* below). Otherwise, model-derived phases may be combined with the previous experimental phases, followed by density modification and another instance of automated or manual rebuilding. This iterative procedure, which often leads to a progressive improvement of model quality and completeness, is implemented in current phasing packages.

   Examples: The structure of human ATG101 was determined by single-wavelength anomalous diffraction phasing with a selenomethionyl analog [4]. Five out of six expected selenium sites could be identified based on a dataset recorded using 0.979 Å synchrotron radiation, which corresponds to a Se X-ray absorption wavelength (K edge). After a reasonably complete model had been obtained, refinement was continued using native data extending to higher resolution (Fig. 1).

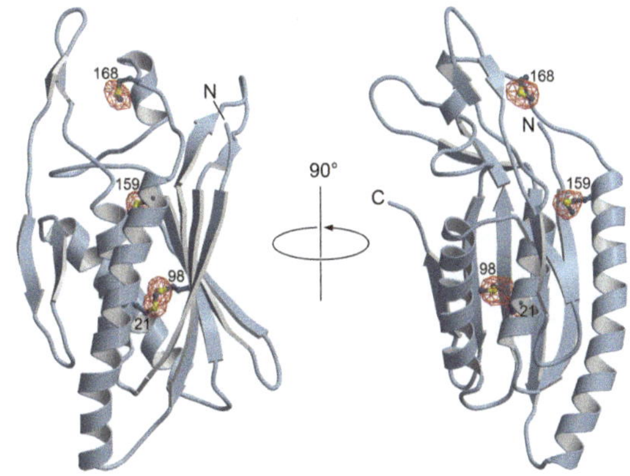

Fig. 1 Crystal structure of human ATG101, a component of the ULK1 complex involved in autophagy initiation. The structure has been determined via Se-SAD phasing of a selenomethionyl derivative. An anomalous difference map (calculated using phases of the native ATG101 structure and contoured at 3.5 σ) features prominent peaks at sites of selenium incorporation; the respective methionine side chains of the native structure are shown in ball-and-stick mode and their residue numbers indicated. ATG101 displays an $\alpha + \beta$ fold resembling the open conformation of a canonical HORMA domain

2.1.6 Refinement and Validation

1. Refine the structure, using your software package of choice, against the highest quality dataset available. A conservative parameterization of the model, ensuring a high observations-to-parameters ratio, is recommended for initial refinement (*see* **Note 5**).
2. Validate the structure w.r.t. the electron density as well as stereochemical criteria. Make manual adjustments as required, and add missing segments.
3. In order to achieve convergence, refinement and rebuilding/validation are usually iterated several times. In the course of this process, the number and weight of restraints may be reduced and/or refinable model parameters added, e.g., by applying anisotropic displacement parameters. Nonprotein compounds such as water molecules, ions, or organics—if and to the extent to which they are discernible in the electron density—need to be included as well.
4. Refinement may be terminated as soon as all of these criteria are fulfilled: (1) no further improvement of refinement statistics is observed, (2) electron density features are accounted for as completely and plausibly as possible, and (3) the structure meets stereochemical standards.
5. Deposit the structure along with experimental data in a publicly accessible database, usually in the Protein Data Bank (PDB).

   Examples: In all structures discussed here, atomic coordinates were refined along with individual B-factors. For GABARAP in complex with the artificial peptide, a number of groups for translation-libration-screw (TLS) refinement were defined, whereas a single group was found to be sufficient in the case of ATG101. The GATE-16 structure (two copies per asymmetric unit) was refined in the presence of NCS restraints.

2.1.7 Interpretation

1. Evaluate the structure in terms of the biological system under study (autophagy in this case), considering complementary data if available. These may originate from NMR or MD studies (also covered in this chapter), as well as biochemical and cell biology experiments. Bear in mind that each observation is strictly valid only under the conditions of the respective experiment, requiring caution when matching and merging results (*see* **Note 6**).

2.2 NMR Spectroscopy

2.2.1 Sample Preparation

1. Prepare a [U-^{13}C, U-^{15}N]-enriched protein sample. Typically a concentration around 1 mM is used. If an organic buffer or additives are needed, it can be helpful to replace them with deuterated ones, especially for recording the side chain experiments and the NOESYs. Add 5–10% of D_2O to the sample for locking (*see* **Note 7**).

   Examples: For the structure determination of GABARAP, the NMR samples contained 0.8 mM GABARAP in 25 mM sodium phosphate, 100 mM NaCl, 100 mM KCl, 100 μM PMSF, 0.02% (w/v) sodium azide, and 50 μM EDTA, pH 6.9 [5]. The NMR sample of the yeast homologue Atg8 contained 500 μM Atg8 in 20 mM sodium phosphate, 150 mM sodium chloride, 5 mM dithiothreitol, 1 mM EDTA, and 0.05% (w/v) sodium azide, pH 6.4 [6].

2.2.2 Recording of NMR Spectra

1. Place the NMR sample in the magnet. Wait until the desired temperature is stable. Lock on the deuterium signal of the solvent to ensure that the magnetic field is stable during the experiments. Tune and match the probe. As there is often cross talk between the capacitors, it is recommended to start with the lowest gyromagnetic ratio RF channel first (^{15}N) and continue with ^{13}C and ^{1}H. Check the channels in an interleaved way. Generally, the ^{1}H tuning and matching strongly depends on the ionic strength of the sample. Shim the magnetic field to ensure high magnetic field homogeneity. This can be done by adjusting the shims manually or nowadays by using a gradient shimming routine. Depending on the NMR system used the order of locking, tuning and shimming may vary.
2. Calibrate the pulse lengths and powers for the individual nuclei.
3. Record the spectra. Often 2D (^{1}H-^{15}N)-HSQC, (^{1}H-^{13}C)-HSQCs for the aliphatic and the aromatic region, 3D

HNCO, HNCA, HNCACB, HNHA, HCCH-COSY, HCCH-TOCSY, as well as (^{1}H-^{1}H-^{15}N)-NOESY-HSQC and (^{1}H-^{13}C-^{1}H)-HSQC-NOESY spectra are used. Additional experiments can be used to facilitate the assignment. Further different variants of the experiments can give a higher resolution (e.g., TROSY spectra for large proteins at high magnetic fields [7]) or increase the time efficiency (BEST experiments [8, 9]).

Example: For the structure determination of GABARAP, spectra were recorded at 298 K, and the following experiments were used: (^{1}H-^{15}N)-HSQC, HNCACB, CBCA(CO)NH, HNCO, HC(CO)NH-TOCSY, HCCH-TOCSY, HCCH-COSY, and ^{15}N- and ^{13}C-edited NOESY spectra [10].

2.2.3 Sequence-Specific Resonance Assignments

To obtain backbone resonance assignments, a [U-^{13}C, U-^{15}N]-enriched protein sample is necessary. Use a combination of different triple resonance experiments for linking the individual amino acid residues to a chain. Using a HNCACB spectrum is a way often used for proteins up to a size of about 20 kDa like proteins of the Atg8 family. Bigger proteins often do not give spectra with sufficient quality. In the HNCACB, the amide group resonances of one amino acid residue are correlated with the Cα and the Cβ resonances of that residue as well as with the Cα and the Cβ resonances of the preceding residue. As the coupling to the own Cα and the Cβ resonances is stronger compared to the preceding Cα and the Cβ resonances, the signal intensities also vary accordingly [11].

In the following we will describe how to obtain a backbone resonance assignment using *CcpNmr Analysis* [12], but similar other programs like *Sparky* or *NMRViewJ* can be used. Although the intensities of the intra- and inter-residual signals differ generally, it can be helpful, especially for a semi-automated sequence assignment, to have in addition to the HNCACB also a HN(CO)CACB, where the amide group resonances of one amino acid residue are correlated only with the Cα and the Cβ resonances of the preceding residue (Fig. 2).

1. Pick the signals in the 2D ^{1}H-^{15}N correlation spectra. They serve as "roots" for the 3D spectra.
2. Based on the roots, display strips of the HNCACB and HN(CO)CACB. Pick the signals corresponding to the Cα and the Cβ resonances, identify which signals belong to the own and preceding residue.
3. Search for the matching Cα and Cβ resonance signals. Connect the sequential spin systems. If no matching signals can be found, it is likely that the following residue is a proline, which due to its missing amide proton does not give signals in the 2D

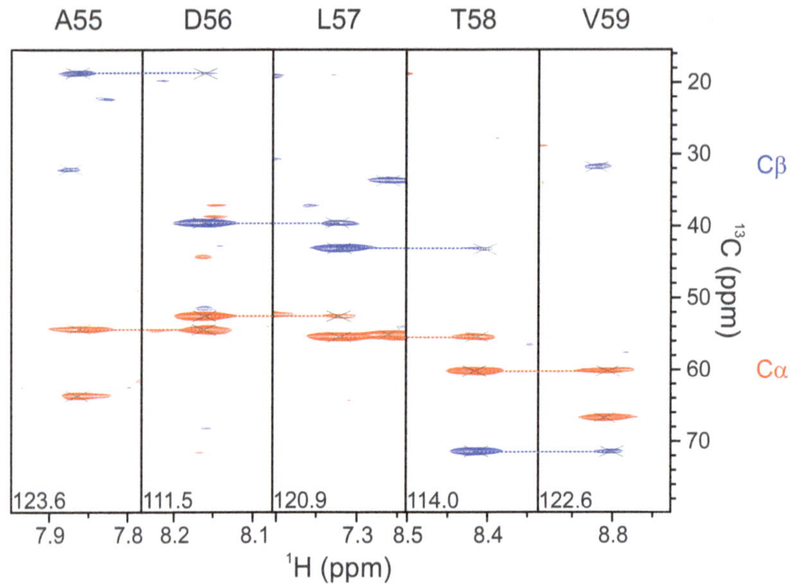

Fig. 2 HNCACB strips of amino acid residues A55–V59 of Atg8. Cα resonance signals are shown in red; Cβ resonance signals are shown in blue. Cβ resonances of alanines are found in the high field region around 18 ppm, whereas Cβ resonances of threonines are found in the low field region around 70 ppm

^{1}H-^{15}N correlation spectra. Only the Cα and the Cβ resonances can be identified in the strip of the following residue.

4. Once a stretch has been connected, link it to the sequence based on unambiguous amino acid types.

   Based on further triple resonance experiments also the resonances of the carbonyls (C′ HNCO), and the Hα and Hβ (HBHA(CO)NH) can be obtained (*see* **Note 8**).

   After the backbone Cα and Hα and often even the Cβ and Hβ chemical shifts have been determined, they can be used as starting points for the side chain assignment. Here, often a HCCH-COSY, together with a HCCH-TOCSY is used. In the HC(C)H-COSY, an HC-group is correlated with the proton(s) of the attached carbon. Start from a Cα/Hα signal to identify the Hβ chemical shifts. Search then the Cβ frequency, where you find again the Hα chemical shift and the Hγ frequencies. Continue the whole side chain along. Side chain assignments of the aromatic residues can be obtained using 2D (HB)CB(CGCD)HD and 2D (HB)CB(CGCDCE)HE experiments as starting points, which correlate the Cβ frequencies with the Hδ and Hε, respectively, of the aromatic ring [13]. Further the NOESY spectra can be used to complete the assignments.

2.2.4 Structural Restraints and Structure Determination

Most information for the structure determination can be gained from the NOESY experiments, which give distance information. Pick the peaks and assign the root resonances if unambiguously possible. Quantify the signal intensities as they are strongly dependent on the respective distances. Based on the complete resonance assignments, the NOE signals are assigned completely. For this, use, for example, *ARIA* [14]. *ARIA* assigns the peaks, calibrates the signal intensities, and calculates an ensemble of conformers. This procedure is repeated in an iterative way several times. The NOE assignments, especially those that are violated in the structure, should be inspected manually (*see* **Note 9**).

Based on the chemical shifts of the HN, N, Hα, Cα, Cβ, and C′, backbone φ/ψ torsion angles and for particular residues also side chain $\chi 1$ torsion angles can be derived. To use *TALOS-N* [15] you have to:

1. Create an input table with the protein sequence and the determined chemical shifts. This can be done easily with the *FormatConverter* included in *CcpNmr* [12].
2. Run TALOS-N.
3. Run RAMA. Inspect the predicted torsion angles. Use predictions classified as "strong" for further structure calculations. Using all derived experimental restraints, determine an ensemble of structures using, e.g., *CNS* [16] (as already used by *ARIA*) or *CYANA* [17] (*see* **Note 10**).

   Examples: For the structure determination of GABARAP [10], 4577 NOE distance restraints were derived in an iterative procedure using manual assignments and *ARIA*. The final structure was determined using *CNS*. The structure of Atg8 [18] was determinate based on 1444 NOE distance restraints and 150 torsion angle restraints using *CYANA*.

2.2.5 Refinement and Validation

1. Similar to X-ray crystallography, the structure ensemble derived by NMR is refined. Here the number and severity of violations of the experimental restraints should be minimized. Depending on the dynamics of the protein, the structures in the ensemble will differ to a varying extent. Often the core region of a protein is well defined, whereas the loop regions show bigger deviations. This is often described by the root-mean-square-distance (RMSD). The structures also have to meet stereochemical criteria, which can be checked using, e.g., *PROCHECK-NMR* [19] or *MolProbity* [20].
2. Deposit the structure in a publicly accessible database, usually in the Protein Data Bank (PDB).

   Example: For GABARAP an ensemble of 15 structures was derived which show a backbone non-hydrogen atom RMSD to the mean structure of 0.049 nm [10]. For Atg8 the

15 structures showed for the backbone heavy atoms of the ubiquitin-like domain an RMSD to the mean structure of 0.065 nm. The N-terminal domain showed a high average local displacement around 0.2 nm due to the high flexibility of that domain [18].

2.3 MD Simulations

In the following it is described how one performs MD simulations for GABARA P using an NMR-derived GABARAP structure (PDB ID 1KOT) as starting conformation and employing the MD software *GROMACS* [21, 22]. However, the protocol given below is applicable to all kinds of proteins. Moreover, any protein structure can be used as starting conformation. Here, care must be taken when the starting structure is not complete and misses certain atoms or residues. For example, if one uses an X-ray structure as starting structure, it usually misses hydrogen atoms. Missing H atoms are, however, no problem for *GROMACS* as they will be added during the setup of the system. In case that heavy atoms should be missing, they can be added using one of several tools that are available, e.g., the "complete.pl" script from the *MMTSB* Tool Set (http://www.mmtsb.org/). In some cases, not only atoms but even residues are missing in PDB structures as they could not be resolved using either X-ray or NMR. For MD simulations, however, these missing residues need to be added before the MD simulation can be started. In this case, one can use *MODELLER* (https://salilab.org/modeller/wiki/Missing%20residues) for adding the heavy atoms or complete missing residues. Once the input structure has been prepared, follow the protocol below for performing an MD simulation. All protocols provided here are based on *GROMACS 5.1.4*, the Amber99SB-ILDN force field [23] and the TIP3P water model [24]. Moreover, it is assumed and recommended that the user works under Linux.

2.3.1 Prepare the Coordinate File Containing the Starting Structure

1. Download the GABARAP structure from the PDB database (1KOT).
   As this is an NMR-derived structure, it contains multiple structural models. Anyone of these models can be used for starting the simulation.
2. Open the "1kot.pdb" file using a protein visualization program (e.g., *VMD*, *PyMOL*, *Chimera*) and save only the first model (aka frame) or split the multi-frame PDB file with grep and awk commands:

```
grep -v 'REMARK' 1kot.pdb > tmp; grep -n 'MODEL\|ENDMDL'
tmp | cut -d: -f 1 | awk '{if(NR%2) printf "sed -n %d,",$1
+1; else printf "%dp tmp > 1kot_%03d.pdb\n", $1-1,NR/2;}'
| bash -sf; rm tmp
```

This creates the files "1kot_001.pdb", . . ., "1kot_015.pdb." As all atoms are present, these PDB files are ready for simulations. We will use "1kot_001.pdb" as an example.

2.3.2 Create a Directory Tree for Running Different Steps in Separate Folders

Create directories for following steps: topology, energy minimization, NVT equilibration, NPT equilibration, and MD production.

```
mkdir 0_topol 1_em 2_nvt 3_npt 4_md
```

For each step, a .mdp file is needed. The mdp file extension stands for molecular dynamics parameters, as the corresponding file contains a list of keywords that are used to set up a simulation. The five .mdp files needed for the current example are provided in the Appendix and are assumed to be located in a directory called "mdp." Thus, the user needs to create this directory:

```
mkdir mdp
```

and copy all .mdp files into this directory.

2.3.3 Create Topology

A *GROMACS* topology file contains information about molecule types and the number of molecules, which will be simulated.

1. Change to the "0_topol" directory to prepare the topology for simulation:

```
cd 0_topol
```

2. Run the "gmx pdb2gmx" command to process the coordinate file and create the topology:

```
echo 2 2 2 2 | gmx pdb2gmx -f ../1kot_001.pdb -o prot.pdb
-p topol.top -ignh -ff amber99sb-ildn -water tip3p -his
```

Explanations:

-f: *GROMACS* reads the coordinate file "1kot_001.pdb," which is located the directory above.

-o & -p: *GROMACS* writes the output files "prot.pdb" and "topol.top."

-ignh: Tells *GROMACS* to ignore the hydrogen atoms in the input file, as hydrogen atoms often have other names in the PDB than in the force fields. New hydrogen atoms will be added by *GROMACS* using the H-atom names of the selected force field.

-ff: Chooses the force field, in the current example Amber99SB-ILDN. If this option is not specified, *GROMACS* offers the user to choose a force field from a list.

-water: Chooses the water model, in the current example TIP3P. If this option is not specified, *GROMACS* offers the user to choose between different water models.

In addition, one could add the option "-inter" allowing the user to interactively choose the protonation states for all titratable residue. To do so, employ another program like *PROPKA* [25] or *H++* [26], which predicts the pK_a values of ionizable groups in proteins based in the 3D structure of the proteins. In case of the GABARAP structure 1KOT, which contains all hydrogen atoms, the protonation states were determined by the NMR assignment. All Glu and Asp residues are deprotonated, while all Lys and Arg are protonated. The most problematic amino acid is the His residues as their pK_a values are very close to the physiological pH. In 1KOT, all His residues are protonated. To pass this information to *GROMACS*, we add the flag "-his" and pass "echo 2 2 2 2," which chooses all His residues as Hip, i.e., in their protonated state. If we do not pass "echo 2 2 2 2," *GROMACS* asks for the protonation state of each His residue interactively. All other residues are assigned to their default states, which corresponds to their state in 1KOT: protonated Lys and Arg and deprotonated Asp and Glu. Please note, if one does not use the "-inter," "-his," "-asp," etc. flags, *GROMACS* always chooses the default states which in many cases can be wrong. Thus, care is required at this step.

When the command 2 has been successfully executed, *GROMACS* prints some messages, including the total charge of the system. For 1KOT it prints: "Total charge 6.000 e."

3. In this step, a simulation box is created. In the current example, we use dodecahedral box where any box edge is at least 1.2 nm away from any protein atom. The protein is centered in the box with the "-c" flag.

```
gmx editconf -f prot.pdb -o prot_box.pdb -c -d 1.2 -bt
dodecahedron
```

4. Now the protein gets solvated so that the water density is ~1000 g/L:

```
gmx solvate -cp prot_box.pdb -o prot_solv.pdb -p topol.top
```

In the current example, 8099 water molecules are added to the system. Please note that the topology file "topol.top" gets updated at this step as the topology of the water molecules is appended to this file that already contained the protein topology from **step 2**.

5. For an MD simulation with periodic boundary conditions (PBCs), which we are going to apply, the charge of the total

system has to be zero. Thus, the charge of +6 needs to be neutralized by adding counterions. For this, one usually adds Cl^- or Na^+ ions in the required number. In the current example, 6 Cl^- ions will thus be added. The following commands need to be executed:

```
gmx grompp -f ../mdp/ions.mdp -c prot_solv.pdb -p topol.
top -o prot_ions.tpr
echo 13 | gmx genion -s prot_ions.tpr -o prot_ions.pdb -p
topol.top -pname NA -nname CL -neutral
```

The first command calls the *GROMACS* preprocessor, *grompp*, which reads a molecular topology file, checks the validity of the file, expands the topology from a molecular description to an atomic description, reads a .mdp file, and translates this into directives for *GROMACS*. The resulting output is a .tpr file, where the file extension stands for portable binary run input. This file contains the starting structure of the simulation, the molecular topology and all the simulation parameters. Because this file is in binary format, it cannot be read with a normal editor. In the current example, "ions.mdp" is read and transformed, together with the coordinate file "prot_solv.pdb" and the topology file "topol.top," to the .tpr file "prot_ions.tpr," which is needed in the second command for adding ions using the "gmx genion" command. Here, the flag "-neutral" makes sure that the system is neutralized with the required number of Na^+ or Cl^- ions. The "echo 13" passed to this command tells *GROMACS* that water molecules should be replaced by ions. If this information is not passed, the user has to provide this information by choosing interactively from a list of options (where option 13 is for "SOL").

After this step, the system's topology has been created, and one can continue with the preparation of the system for the production MD simulation, which includes energy minimization and equilibration MD runs.

2.3.4 Energy Minimization

For the energy minimization of the system, change into the directory for energy minimization:

```
cd ../1_em
```

In the current example, we employ the steepest descent method and perform the minimization until a maximum force of <500 kJ/mol/nm or 2000 minimization steps are reached, depending on what of these two options is reached first. First, "grompp" is called to generate the .tpr file from the .mdp file for minimization:

```
gmx grompp -f ../mdp/em.mdp -c ../0_topol/prot_ions.pdb -p ../
0_topol/topol.top -o prot_em.tpr
```

Afterward the energy minimization can be performed:

```
gmx mdrun -v -deffnm prot_em
```

Here, the option "-v" is for verbose which produces detailed output during the minimization so that the user can follow what happens during this process, and "-deffnm" sets the base name for all input and output files to "prot_em."

2.3.5 NVT Equilibration

The energy minimization is followed by two equilibration runs of the system. For the first equilibration phase, one typically performs a short MD simulation in the NVT ensemble with position restraints on the protein, which allows the solvent molecules and ions to relax into equilibrated positions. In the current example, we run a 100 ps MD simulation under NVT conditions at 298 K using the v-rescale thermostat [27], and with position restraints of 1000 kJ/mol/nm^2 using the constraint method *LINCS* [28]. For doing so, change from the minimization into the NVT equilibration directory:

```
cd ../2_nvt
```

Next, restraints for all protein atoms need to be created:

```
echo 1 | gmx genrestr -f ../1_em/prot_em.gro -o ../0_topol/
posre.itp -fc 1000 1000 1000
```

Here, "echo 1" tells GROMACS to apply restraints on all protein atoms. Without "echo 1" being passed to GROMACS, the user has to choose from a list of atoms or groups to which restraints should be applied to, e.g., Cα atoms, backbone, etc. We restrain the protein structure and save the position restraints file "posre.itp" in the "0_topol" directory. The option "-fc" stands for the force constants which will be applied to the selected atoms in the x, y, and z directions.

Afterward, the MD equilibration can be performed:

```
gmx grompp -f ../mdp/nvt.mdp -c ../1_em/prot_em.gro -p
../0_topol/topol.top -o prot_nvt.tpr
gmx mdrun -v -deffnm prot_nvt
```

2.3.6 NPT Equilibration

The NVT equilibration is succeeded by ten 200 ps MD simulations in the NPT ensemble in which the restraint forces are gradually reduced from 1000 to 10 kJ/mol/nm^2 using the following steps:

1000, 800, 600, 400, 200, 150, 100, 70, 30, and 10 kJ/mol/nm^2. As before, the temperature is kept at 298 K, and the pressure is now set to 1 bar using the Berendsen barostat [29]. For performing the NPT equilibration runs, change into the corresponding directory:

```
cd ../3_npt
```

Then, the first NPT equilibration step with restraints of 1000 kJ/mol/nm^2 is performed:

```
gmx grompp -f ../mdp/npt.mdp -c ../2_nvt/prot_nvt.gro -p
../0_topol/topol.top -t ../2_nvt/prot_nvt.cpt -o prot_npt_1.tpr
```

Here, the option "-t" ensures that the starting coordinates are read from a trajectory, in the current case from the trajectory generated during the NVT equilibration, i.e., "../2_nvt/prot_nvt.cpt." The last frame of this trajectory with coordinates and velocities will be read. The cpt file extension stands for portable checkpoint file, which stores the complete state of the simulation, including the thermostat/barostat variables, random number states, and in case that domain decomposition is used, also the decomposition state is stored. The preprocessor command is followed by the actual MD run:

```
gmx mdrun -v -deffnm prot_npt_1
```

After the first NPT equilibration run, run the other nine equilibration steps using a little shell script. For this, create a list of the remaining force constants:

```
FClist=(800 600 400 200 150 100 70 30 10)
```

followed by

```
for i in {0..8}
do
 # Update force constant in posre.itp file in 0_topol direc-
tory.
 echo 1 | gmx genrestr -f ../1_em/prot_em.gro -o ../0_topol/
posre.itp -fc "${FClist[i]}" "${FClist[i]}" "${FClist[i]}"
 # Run the n-th NPT equilibration step.
 gmx grompp -f ../mdp/npt.mdp -c prot_npt_"$[i+1]".gro -p ../
0_topol/topol.top -t prot_npt_"$[i+1]".cpt -o prot_npt_"$[i
+2]".tpr
 gmx mdrun -v -deffnm prot_npt_"$[i+2]"
done
```

Please note that lines starting with a "#" in the script are comments.

This concludes the equilibration steps.

2.3.7 MD Production

After the energy minimization and equilibration runs, everything is prepared for performing the production MD run. In the current example, a short 10 ns MD simulation in the NPT ensemble will be performed. Usually, production runs are orders of magnitude longer reaching hundreds of nanoseconds if not microseconds. As before, the temperature will be kept at 298 K using the v-rescale thermostat, while the pressure will be regulated with the Parrinello-Rahman barostat [30] with 1 bar as the target pressure. During the production run, no restraints on any atom of the system will be applied. However, all bond lengths will be constrained to their equilibrium values using the *LINCS* method. For later analysis, sufficient snapshots need to be saved during the production MD run. Here, coordinates are saved every 5 ps. For performing the production MD run, following steps need to be performed: changing into the MD production directory,

```
cd ../4_md
```

calling the GROMACS preprocessor,

```
gmx grompp -f ../mdp/md.mdp -c ../3_npt/prot_npt_10.gro -p
../0_topol/topol.top -t ../3_npt/prot_npt_10.cpt -o prot_md.tpr
```

and starting the MD run,

```
gmx mdrun -v -deffnm prot_md
```

2.3.8 MD Analysis

Once the MD production run has been finished, a very important aspect is the analysis of the MD trajectory. The kind of analysis which is being performed depends on the system under study and the research questions that are being addressed. *GROMACS* provides several analysis tools that can directly be applied to a trajectory. If one needs analysis tools, which should not be included in *GROMACS*, one could also use other programs that allow manipulating MD trajectories. One such program is *MDTraj* (http://mdtraj.org/1.8.0/), which is a Python library containing several analysis tools. Both *GROMACS* and *MDTraj* also allow the calculation of NMR observables. For the calculation of S^2 order parameters from MD trajectories, our group also developed a Python program called *MOPS*^{2} (https://github.com/schilli/MOPS). In the current example, we limit ourselves to the calculation of two observables, which are usually always determined from an MD trajectory, the RMSD of the protein motion and the radius of gyration (R_{gyr}) of the protein. Both RMSD and R_{gyr} are often used to check on the stability of a protein during a simulation. RMSD values increasing considerably beyond 5 Å indicate that the protein under study is not stable in the simulation, which also leads to increasing R_{gyr} values. R_{gyr} is also a quantity that is often used to monitor protein (un)folding during a simulation.

For performing the analysis for the current example, first create a subdirectory in the production MD directory where all analysis files will be stored:

```
mkdir analysis
```

Then calculate the RMSD for the backbone of GABARAP. To this end, first align all protein structures saved in the trajectory to a reference structure, which is often the initial structure of the production MD run, and then calculate how much these structures deviate from the reference structure. Typically, one aligns only the backbone of the protein (i.e., excluding the side chains) and then also calculates the RMSD for the backbone. Therefore, we can pass "echo 4 4", where "4" stands for the group of backbone atoms, to *GROMACS*:

```
echo 4 4 | gmx rms -s prot_md.tpr -f prot_md.xtc -tu ns -o
analysis/rmsd.xvg
```

Please note that "gmx rms" does both alignment and RMSD calculation. Each structure from a trajectory (−f) is compared to a reference structure. The reference structure is taken from the structure file provided by the option "-s," and the RMSD is calculated for all structures in the trajectory file given by the option "-f." The output will be saved in the output file "analysis/rmsd.xvg," which will show the time evolution of the RMSD. The extension xvg indicates that this file can be directly viewed with the 2D plotting program *xmgrace*. However, one can also import these files in some other data visualization programs, e.g., *Python*, *MATLAB*, or *R*. The option "-tu" stands for time unit, which is chosen as ns in the current example, i.e., the time axis will be given in ns.

In a similar manner, we can calculate R_{gyr}. Here, we will include all protein atoms ("echo 1") in the calculation:

```
echo 1 | gmx gyrate -s prot_md.tpr -f prot_md.xtc -o analysis/
rg.xvg
```

Please note that "gmx gyrate" does not allow to change the unit of the time axis as done above for "gmx rms" and also in the following "gmx distance" command. Instead, "gmx gyrate" reports time in the default unit used by *GROMACS*, which is ps.

For autophagy-related proteins like GABARAP and Atg8, the motions of the N- and C-termini are of biological relevance. Therefore, it is useful to monitor the distance between the termini during an MD simulation of these proteins. For doing so, we first need to define the necessary index groups

```
gmx make_ndx -f prot_md.tpr -o ../0_topol/index.ndx
```

for which we need to select the Cα atom of the N-terminus (residue 1), creating the group named "C-alpha_&_r_1"

```
3 & ri 1
```

and the same for the C-terminus (residue 119), creating the group named "C-alpha_&_r_119"

```
3 & ri 119
```

These group selections are saved in the file "../0_topol/index.ndx" by typing

```
q
```

and then pressing "Enter." The selections of these groups are based on the predefined group of all Cα atoms (group 3) and its intersection with the group that contains only atoms of residues 1 and 119, respectively. Once these index groups are made, we can calculate the distance between these two atoms:

```
gmx distance -s prot_md.tpr -f prot_md.xtc -tu ns -n
../0_topol/index.ndx -select 'group "C-alpha_&_r_1" plus group
"C-alpha_&_r_119"' -oall analysis/distance.xvg
```

Here, the option "-n" indicates that an extra index group will be used, and "-select" selects the two newly created groups. The option "-oall" ensures that all distances (in the current example only one distance) are plotted as a function of time. If more distances are being calculated, the option "-oav" allows the direct calculation of average distances as function, while "-oxyz" saves the *x*, *y*, and *z* components of the distances. It should be noted that *GROMACS* prints distances in nm—its default unit for lengths—but it is often more convenient to look at atomic distances in Ångstroms.

In Fig. 3 the plots for the RMSD, R_{gyr} and N-C distance as obtained from our current example are shown. Please note that the results from another MD run of the same system are likely to look different as either different random velocities were initially assigned to the atoms or because another computer was used for running the

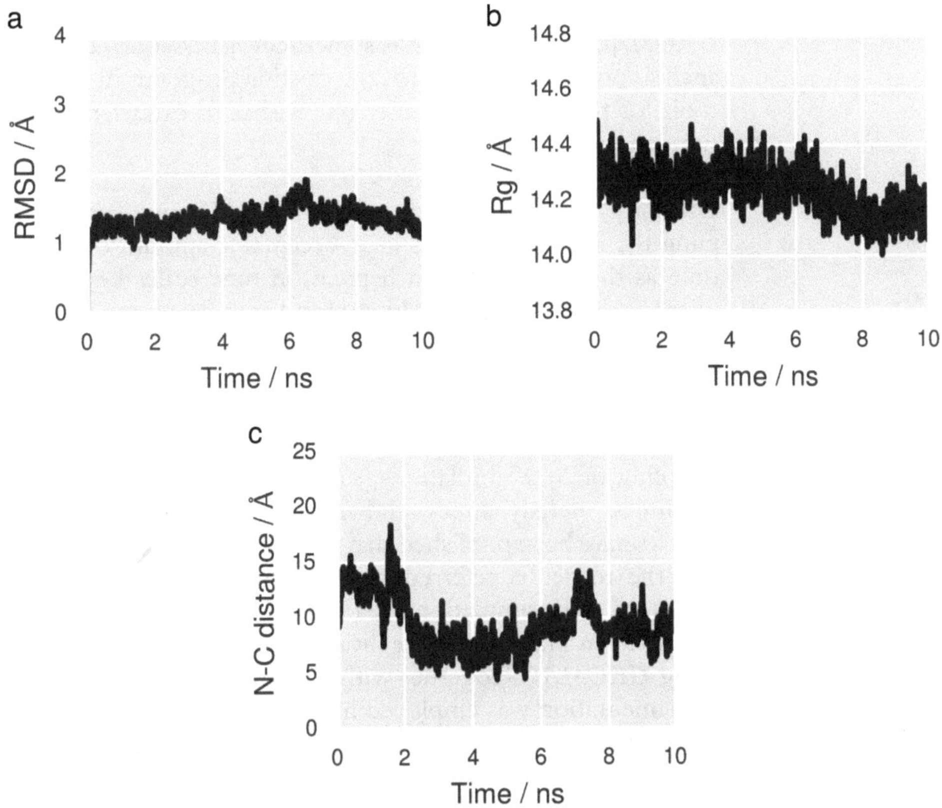

Fig. 3 Analysis of 10 ns MD simulation: (**a**) backbone RMSD, (**b**) protein radius of gyration, and (**c**) distance between the N- and C-termini indicate a stable system

simulation. This does not mean that either of the results is wrong. However, it is necessary to run MD simulations long enough to sample the relevant motions of the system under study. The average values and standard deviations obtained from different MD simulations of the same system modeled with the same force field need to be very similar. If this is not the case, it is clear that the MD simulations have been too short. It is difficult to say prior an MD simulation what the required length is. It depends on what kind of process is being investigated. The larger the system and the larger the motion under study, the longer an MD simulation usually needs to be. A good test to see whether an MD simulation is converging is to calculate averages of the observable for different time blocks of the simulation. For our example above, one could, for instance, calculate the RMSD, R_{gyr}, and N-C distance for the first 5 ns of the simulation and for the whole simulations. If the results are quite different from each other, the simulation needs to be extended. In addition, it also has become common to run not only one MD simulation but multiple copies of the same system, e.g., in triplicate or more, but using different initial velocities in order to obtain statistically significant data.

In addition to the multitude of analysis tools provided by *GROMACS* or *MDTraj*, it is sometimes necessary to develop own analysis programs in case that the existing programs do not provide tools for the physical quantity one wishes to extract from an MD simulation.

2.3.9 HREX-MD Production

Often it is not sufficient to run a simple MD simulation for sampling the relevant motions in an acceptable amount of computing time as these motions can happen on time scales beyond several microseconds. Moreover, biomolecular motions typically involve high-dimensional and rough energy landscapes with many local minima separated by high-energy barriers, which further limit the sampling of the motions of interest. To overcome this limitation, several methods have been developed that help enhancing the sampling of MD simulations, such as replica exchange molecular dynamics, metadynamics, and umbrella sampling. Several recent reviews have been published that explain these different techniques, and the reader is referred to them for further information on enhanced MD simulations [31–34]. Here, we explain how one can perform Hamiltonian replica exchange (HREX) MD [35, 36] using *GROMACS* together with *PLUMED 2.3* [37]. This HREX implementation was employed in our study of the conformational polymorphism of GATE-16 [3]. Here, we will continue to use GABARAP as our model protein. In this example, we will consider 16 replicas which all will be run for 10 ns. Apart from the modifications to the force field in all replicas but the target replica, everything else will be as in standard MD simulations. The exchanges between the replicas are managed by *GROMACS/PLUMED*.

1. The HREX-MD run will be performed in a new directory, which needs to be created first:

```
cd ..
mkdir 5_hrex
cd 5_hrex
```

2. For the different replicas, different topologies need to be produced, which account for the modified Hamiltonians of the replicas. For this we need to write all parameters of the force field into the topology file using the option "-pp hrex.top" of the *GROMACS* preprocessor, so that afterward *PLUMED* scale these parameters as needed:

```
gmx grompp -f ../mdp/md.mdp -c ../3_npt/prot_npt_10.gro -p
../0_topol/topol.top -pp hrex.top -t ../3_npt/prot_npt_10.
cpt -o hrex.tpr
```

3. In HREX-MD, the energy function of parts of the system is being modified. Often it is the protein or parts of the protein under study, for which the Hamiltonian is manipulated, while the energy function of the solvent is kept the same. The parts of the system for which the Hamiltonian is being modified are called the "hot" region even though the temperature is not changed. However, reducing the inter-atom interactions enhances the motion of these atoms in a similar fashion like increasing their temperature. The atoms designated to belong to the hot region are selected by adding an underscore to them in "hrex.top." To do this, open "hrex.top" with the editor *vim*:

```
vim hrex.top
```

switch to visual mode (ctrl + V), and select the whole column containing the atoms. Press "c" to go into change mode and type an underscore (_). Confirm by pressing the "Esc" button and then save and exit with ":x."

For example, following lines

```
...
17  OH  2  SER  OG  17  -0.6546     16  ; qtot 0.5431
18  HO  2  SER  HG  18   0.4275  1.008  ; qtot 0.9706
19  C   2  SER  C   19   0.5973  12.01  ; qtot 1.568
20  O   2  SER  O   20  -0.5679     16  ; qtot 1
; residue  3  MET rtp MET q 0.0
21  N   3  MET  N   21  -0.4157  14.01  ; qtot 0.5843
22  H   3  MET  H   22   0.2719   1.008 ; qtot 0.8562
23  CT  3  MET  CA  23  -0.0237  12.01  ; qtot 0.8325
24  H1  3  MET  HA  24   0.088    1.008 ; qtot 0.9205
...
```

become

```
...
17  OH_  2  SER  OG  17  -0.6546     16 ; qtot 0.5431
18  HO_  2  SER  HG  18   0.4275  1.008 ; qtot 0.9706
19  C_   2  SER   C  19   0.5973 12.01  ; qtot 1.568
20  O_   2  SER   O  20  -0.5679     16 ; qtot 1
; residue 3 MET rtp MET q 0.0
21  N_   3  MET   N  21  -0.4157  14.01 ; qtot 0.5843
22  H_   3  MET   H  22   0.2719   1.008 ; qtot 0.8562
23 CT_   3  MET  CA  23  -0.0237  12.01 ; qtot 0.8325
24 H1_   3  MET  HA  24   0.088    1.008 ; qtot 0.9205
...
```

4. Create a directory for running HREX-MD and change to it:

```
mkdir run_hrex
cd run_hrex
```

5. Now that the hot atoms have been assigned in "hrex.top," the HREX-MD run can be prepared by selecting all parameters. The number of replicas (nrep) needs to be chosen as well as the scaling factors (lambda) for the Hamiltonians of the hot region. In the current example, we set nrep to 16 and the minimum lambda value to 0.65. This corresponds to an effective maximum temperature (tmax) of 457 K, while the unbiased replica (lambda = 1) corresponds to the target replica at the minimum temperature of 298 K. For the other replicas, we assign the lambda values based on a geometric progression of temperatures using a little bash script:

```
nrep=16
tmin=298
tmax=457
list=$(
awk -v n=$nrep \
     -v tmin=$tmin \
     -v tmax=$tmax \
     'BEGIN{for(i=0;i<n;i++){
     t=tmin*exp(i*log(tmax/tmin)/(n-1));
     printf(t);
     if(i<n-1)printf(",");
  }
}'
)
```

6. The temperatures which are saved in "$list" are then used to modify the Hamiltonian of the hot region, which has to be done for each replica using the corresponding temperature. To this end, prepare the .top and .tpr files with biased parameters leading to the modified Hamiltonians:

```
for((i=0;i<nrep;i++))
do
 # Calculate lambda, i.e., the biasing factor.
 lambda=$(echo $list | awk 'BEGIN{FS=",";}{print $1/$'$((i
 +1))';}')
```

```
echo "lambda = $lambda."
# Instruct PLUMED to bias the Hamiltonian in ../hrex.top
using the
# current $lambda and write the biased topology in topol_
$i.top
plumed partial_tempering $lambda < ../hrex.top > topol_
$i.top
# Prepare .tpr files from the .top files.
gmx grompp -maxwarn 1 -f ../../mdp/md.mdp -c
../../3_npt/prot_npt_10.gro -p topol_$i.top -t
../../3_npt/prot_npt_10.cpt -o topol$i.tpr
done
```

The option "-maxwarn 1" is needed for the preprocessor because lambda scales charges, which can cause the system to become charged in some replicas. However, for MD using periodic boundary conditions the system has to be neutral. *GROMACS* solves nonzero charges by adding a compensating background with PBCs.

7. For HREX-MD the file "plumed.dat" is needed, which can contain variables used by *PLUMED*. Furthermore, it can contain information for analysis. For example, we can instruct *PLUMED* to write relevant distances (e.g., the N-C distance), dihedral angles, or other such observables which can later be used for analysis, such as free-energy calculations. However, this is not necessary for HREX-MD, and "plumed.dat" can also stay empty but has to be present. For creating an empty file one can invoke the command "touch":

```
touch plumed.dat
```

In the present example, we leave the file empty.

8. At this stage everything has been prepared so that the HREX-MD simulation can be started. This requires an MPI version of *GROMACS* allowing to run multiple simulations in parallel:

```
gmx_mpi mdrun -v -plumed plumed.dat -multi 16 -hrex -replex
2500 -s topol_.tpr -x traj_.xtc -cpo state_.cpt -c conf-
out_.gro -e ener_.edr -g md_.log
```

At the beginning of an HREX-MD run, the user has to make sure that the exchange rates between the replicas fall in an acceptable range of 0.2–0.7. If the exchange rate is smaller, then the information exchange between the replicas is not sufficient. In this case more replicas need to be added, especially in the region where the exchange rate is too low. An exchange rate above 0.7, while not problematic per se, is a waste of computational resources as the replicas are too narrowly spaced. In such a case, some of the replicas can be removed, and the lambda values of the remaining replicas need to be adapted. The exchange rates can be found at the end of the *.log file. In the current example, we obtained an average exchange probability of about 0.3:

```
Replica exchange statistics
Repl 1999 attempts, 1000 odd, 999 even
Repl average probabilities:
Repl  0  1  2  3  4  5  6  7  8  9  10  11  12  13  14  15
Repl .30 .27 .29 .29 .26 .27 .29 .33 .29 .27 .30 .30 .32 .33 .32
Repl number of exchanges:
Repl  0  1  2  3  4  5  6  7  8  9  10  11  12  13  14  15
Repl 302 254 305 295 263 273 282 327 274 271 283 292 338 334 323
Repl average number of exchanges:
Repl  0  1  2  3  4  5  6  7  8  9  10  11  12  13  14  15
Repl .30 .25 .31 .30 .26 .27 .28 .33 .27 .27 .28 .29 .34 .33 .32
```

9. As for a standard MD run, analysis has to be performed after an HREX-MD analysis. In the current example, we will calculate the RMSD, R_{gyr}, and N-C distance as above but for all replicas. First, create an analysis directory:

```
mkdir analysis
```

and then perform the calculations:

```
for i in {0..15}
do
 echo 4 4 | gmx rms -s topol_$i.tpr -f traj_$i.xtc -tu ns
 -o analysis/rmsd_$i.xvg
 echo 1 | gmx gyrate -s topol_$i.tpr -f traj_$i.xtc -o
 analysis/rg_$i.xvg
 gmx distance -s topol_$i.tpr -f traj_$i.xtc -tu ns -n ../
 ../0_topol/index.ndx -select 'group "C-alpha_&_r_1" plus
 group "C-alpha_&_r_119"' -oall analysis/distance_$i.xvg
done
```

The results for the target replica corresponding to the unbiased Hamiltonian are usually the results of interest, which are presented for the current example in Fig. 4. From enhanced sampling simulations one often also calculates free energies as we did for the distributions of the N-C distances of GATE-16 [3].

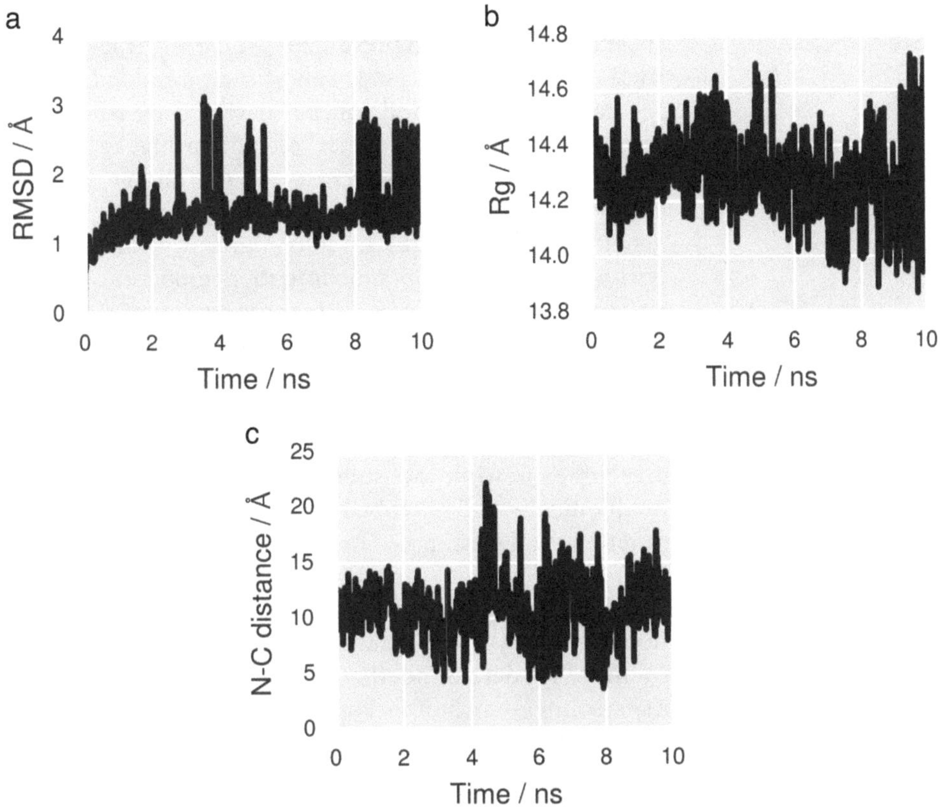

Fig. 4 Analysis of 10 ns HREX-MD simulation with the unperturbed Hamiltonian (i.e., scaling factor of 1): (**a**) backbone RMSD, (**b**) protein radius of gyration, and (**c**) distance between the N- and C-termini indicate that the system samples several different conformational states

3 Notes

1. Sample requirements

   While protein expression and purification for purposes of crystallization follow standard protocols (not covered by this article), a number of important quality criteria need to be considered. First, crystallization of biological macromolecules requires a degree of supersaturation that enables the system to nucleate either de novo or on pre-existing phase boundaries. While this supersaturation is commonly achieved by the addition of precipitants, it obviously requires an appropriate initial concentration of the sample, which depends on the solubility of the molecule of interest and has to be determined experimentally. The concentration of a protein sample is commonly determined from its UV absorbance in the 280 nm region, using extinction coefficients calculated from the known content of aromatic residues (and presumed disulfides). Since these calculated coefficients are strictly valid for an unfolded protein, the accuracy of this method can be increased by performing the measurement in the presence of denaturing agents. Besides concentration, the homogeneity of the sample is of utmost importance for the success of crystallization. The term homogeneity can be thought of as comprising both the chemical purity (absence of contaminants or degradation products, as evidenced by SDS-PAGE or mass spectrometry), the conformational state (accessible by NMR spectroscopy), and the oligomericity (evaluated by size-exclusion chromatography or dynamic light scattering analysis). For phasing by anomalous diffraction methods, selenium is often incorporated during protein expression by substituting methionine by selenomethionine in the microbial growth media; the ratio of incorporation can be estimated by mass spectrometry of the purified sample. In protein preparations destined for crystallization screening, buffers should generally be used at low concentrations (typically 10–20 mM) in order to allow the reservoir solution to determine the final pH. Other low-molecular-mass solutes—including compounds improving solubility or stability of the macromolecule (salts, detergents, etc.) and specific ligands—may be added as desired. It is good practice to keep the solute system as simple as possible.

2. Crystallization and cryoprotection
Conditions for crystal growth must be established anew for each macromolecular entity, which usually requires extensive screening, followed by optimization of initial hits [38]. Keep in mind that seemingly minor modifications to a protein, such as single amino acid exchanges, post-translational modifications, expression tags, etc., may sometimes alter the behavior in crystallization experiments in unpredictable ways and may require returning to the initial screening stage.

Data collection at cryogenic temperatures is intended to reduce radiation damage by essentially limiting the diffusion of reactive radicals [39]. However, treatment with unsuitable cryosolutions may itself induce severe damage to a macromolecular crystal, either by direct interference of the cryoprotectant with crystal integrity or by a mismatch of thermal expansion coefficients between the crystal lattice and the bulk solution. For this reason, conditions for flash cooling must be tuned for each type of crystal, with the aim to optimize diffraction quality.

3. Molecular replacement
Phasing by MR relies on similarities of the structure to be determined with a structure that is already known. Guided by

the experimental data, a search model derived from the known coordinates is positioned within the asymmetric unit of the new crystal type. In case of success, the positioned model can be used to calculate phases for displaying an initial electron density; moreover, it may directly serve as a starting point of structural refinement [40, 41].

The quality of the search model is the most critical factor determining the success of MR. Since the actual structural similarity between the model and the structure to be determined is unknown at this stage, sequence similarity is used as a substitute. This strategy is justified by the observation that during the evolution of proteins, the tertiary structure is usually much more strongly conserved than the amino acid sequence, and sequence similarity thus tends to underestimate structural similarity. In fact, the two parameters are found to be strongly correlated, and a sequence identity of >30%, corresponding to an expected RMSD of <1.5 Å, has been used as a threshold to gauge the chances of success. Of note, this is only a rule of thumb: an MR search may be worth a try even at barely detectable sequence similarity, while it may unexpectedly fail at high similarity, depending on the actual structural deviation. In the presence of sequence differences, various treatments may be applied to the model in order to increase the chances of success. These generally aim at removing atoms which are either absent in the new structure or whose positions are likely to differ from the search model. Examples include truncation of non-conserved side chains to different extents and removal of terminal segments or loops that are presumed to be flexible. If several homologous structures are available, their superposition may reveal polymorphous regions, and an ensemble model in which the superimposed molecules are truncated to the conserved core may be used as an alternative to individual (full-size or truncated) models. Finally, coordinate sets for an MR search may also be generated by template-based modeling, which may include more or less extensive molecular dynamics steps for optimization. While such models naturally feature the correct primary structure of the molecule of interest, their accuracy will be determined by the quality of the template and the details of the procedure. Models created by ab initio methods, i.e., in the absence of a template, are only rarely useful for MR, since they are usually not close enough to the target structure, even if they correctly capture the overall fold. Most software packages in current use for MR apply search strategies based on the Patterson function, thus dividing the six-dimensional problem into three-dimensional rotation and translation searches. Prominent examples include *MOLREP* [42] and *Phaser* [43].

A six-dimensional optimization, using either a brute-force grid search or an evolutionary algorithm, may be considered in cases where other methods fail. The latter is exemplified by EPMR [44]; owing to stochastic sampling of starting conditions, many independent attempts are often required to obtain a solution.

4. Experimental phasing

   If a structure with presumed similarity to the target is unavailable, or if MR fails, experimental phase information is required. This usually implies the introduction of a limited number of marker atoms into the crystal structure. The positions of these atoms can be determined ab initio, and this information together with measured intensities is used to estimate phases for all recorded diffraction signals [45]. Traditionally, marker atoms are electron-rich (heavy) atoms which induce changes in the diffracted intensities when part of the ordered crystal structure, leading to isomorphous differences w.r.t. a native (reference) crystal. In addition, such atoms can display anomalous scattering when exposed to X-rays close to an absorption edge, leading to anomalous differences between pairs of intensities (Friedel mates) that would otherwise be equal. As an alternative to soaking pre-formed crystals, anomalous scatterers may be introduced by replacing methionine by selenomethionine during protein expression. Both isomorphous differences (between datasets) and anomalous differences (within a single dataset) can be exploited to determine the positions of the respective marker atoms. While this is classically accomplished by examining a Patterson map computed with these differences (squared) as coefficients, more sophisticated approaches using dual-space recycling are in routine use today [46]. Once the marker atom coordinates are known, structure factors (with amplitude and phase) can be calculated for this substructure alone, and phases for the protein structure factors are determined by considering the additive relations between the individual complex scattering components of protein and marker atoms. The resulting expectation for the phase angle is expressed as a probability distribution for each diffraction signal; while a single set of differences initially provides symmetric bimodal probability densities, combination of phase information from different sources may reduce this ambiguity. In any case, experimental phasing is complemented by density modification procedures, which can greatly improve the interpretability of an electron density map [47]. This is achieved by enforcing general properties of protein electron densities, such as the difference in density between protein and solvent regions, the similarity in density between copies of the same macromolecule, the characteristic density histogram, etc., without biasing the map toward any specific structural feature.

Using the modified density map, an initial protein model is constructed. In many cases this model will be very incomplete and not yet suitable for refinement, but even a very fragmented set of coordinates may be useful since combination of model-derived phases with the original experimental phases will often yield an improved map which, after another round of density modification, may support extension of the model. This procedure is repeated until a model is obtained that is suitable for refinement against native data. Software solutions implementing the entire pipeline from substructure determination to initial model building include the *SHELX* suite [48] and the *phenix.autosol* wizard [49]. Since the intensity differences commonly used in experimental phasing are often very small, especially in the case of anomalous scattering, obtaining high-quality data is of paramount importance. In practice, this requires a trade-off between conflicting goals: while generally the X-ray dose on the sample should be limited in order to prevent radiation damage (unless radiation-induced alterations are themselves exploited for phasing), the aim to achieve high-redundancy data at a reasonable resolution would seem to favor longer total exposure times and/or higher dose rates. With the advent of hybrid pixel array detectors featuring single-photon sensitivity, noise-free readout, and high frame rates, high-redundancy data collection schemes avoiding excessive X-ray exposure have become feasible.

5. Refinement and validation

   The refinement step aims at improving the current model, i.e., increasing its probability of being correct, conditional on the observed data [50]. Regardless of the method applied for initial phasing, the final cycles of refinement should be carried out with respect to the highest-quality dataset available, which most often represents a native crystal. In addition to global scaling parameters, refinement usually includes—for every atom in the model—the coordinates of the mean position along with a measure of displacement among this position. Depending on the available resolution and thus the number of unique diffraction signals observed, these atomic displacement parameters can be chosen to denote an isotropic or anisotropic coordinate uncertainty, corresponding to one or six model parameters per atom. Modeling rigid-body anisotropic displacement for groups of residues, as implemented in the TLS parameterization requiring only 20 parameters per group, provides a means of incorporating anisotropy at low to medium resolution. Most programs for macromolecular refinement, such as *REFMAC5* [51] or *phenix.refine* [52], are based on maximum-likelihood target functions incorporating sophisticated error models, which are found superior to classical least-squares methods in most application scenarios. It is important to realize that the overall probability of a model being correct is determined not only by its agreement with the measured data but also by its a priori plausibility, i.e., its consistency with expected stereochemical properties. For this reason, stereochemical restraints are usually incorporated into the refinement targets to be optimized, with a relative weight which should be adjusted in each refinement step, depending on the behavior of the system. Unrestrained refinement is useful at very high resolution only, where the data are thought to provide sufficient information to accurately define atomic positions and displacement parameters. The process of structure refinement is carried out iteratively, with automated refinement alternating with careful examination and manual adjustment by the crystallographer. In this way, refinement is effectively intertwined with validation, i.e., evaluation of global as well as local quality indicators, with the aim of spotting potential problems. Similar to the refinement task, criteria for validation need to address both the consistency with measured data and the stereochemical quality of a model. The most important parameter indicating data consistency is the crystallographic residual R and its cross-validation relative R_{free}, while stereochemical validation includes, among many other metrics, the Ramachandran diagram of main-chain torsion angles. Tools for validation of crystallographic structures are often integrated into refinement or visualization packages, but standalone solutions such as *MolProbity* [20] are also available. Note that outliers w.r.t. the statistics-based stereochemical dictionaries may be real in certain cases; in fact, such features often occur in regions which are important for protein function. Since the a priori probability of such features is small, however, they should only be kept in the model if they are strongly supported by the data, i.e., the electron density. In practice, most geometry outliers are generated in regions of weak density and thus represent problems with refinement rather than true properties of the macromolecular structure.

6. Interpretation in a biological context
If applied properly, X-ray crystallography provides macromolecular structures at high precision. It should be kept in mind, however, that the conditions of sample preparation (crystallization in the presence of precipitating agents, starting with a concentrated solution of purified protein) and measurement (usually cryogenic temperature) differ significantly from physiological settings. While structural alterations w.r.t. the native state are generally believed to be small, their relevance will of course depend on the conclusions to be supported by the data.

The process of crystallization, for instance, is known to be potentially selective among conformers present in a solution, thus largely excluding the analysis of conformational ensembles. Similarly, lattice contacts tend to dampen conformational changes induced by chemical or physical means; while this effect can be beneficial in the generation of isomorphous phasing derivatives, it is disadvantageous when evaluating conformational changes induced by ligands or light exposure in pre-formed crystals.

7. NMR sample preparation
   To obtain a high signal-to-noise ratio while recording the NMR data, it is advantageous to use the highest protein concentration possible. Anyhow, the protein concentration is often limited due to aggregation of the protein. Aggregated protein will lead to broad signals, which may be even not detectable. If the protein amount is limited, Shigemi NMR tubes can be used reducing the needed volume from about 600 to 350 μL. The protein should have a high homogeneity as different conformations can led to multiple signals, which impede the data analysis. Nonetheless, these different conformations can be unrevealed by NMR in a quantitative approach giving information about the kinetics and thermodynamics. To properly reference the chemical shifts, it is recommended to prepare a sample with 4,4-dimethyl-4-silapentane-1-sulfonic acid (DSS).

8. Sequence-specific resonance assignments
   Depending on the protein properties (size, dynamics, conformation), it is possible that several resonances cannot be detected, especially in insensitive experiments like the HNCACB. Then it can be helpful to work with the HNCA. If the HNCA does not give unambiguous assignments, an HN (CA)CO or an HNCO+, which correlate the amide group with the C′ of the preceding residue and the own C′, can be helpful.

9. NOE assignments
   Often in the first runs of structure calculation several NOEs are violated. Wrong assignments are the prevailing reason for this. Thus, check again the sequence-specific resonance assignment of the affected atoms. Further artifacts can cause problems. Sometimes very strong signals can show so-called sinc wiggles, when the FID (often in the indirect dimension) was truncated. This can be overcome by applying an appropriate window function during processing.

10. Structure information
    In addition to the distance information from NOEs and the dihedral information from the secondary structure, further restraints can be used for structure determination. The scalar ^{3}J coupling is sensitive to dihedral angles as described by the

Karplus relation. Thus, the backbone dihedral angle φ (defined by the atoms C′-N-Cα-C′) can also be determined by evaluating an HNHA experiment as this is sensitive to the $^3J_{HNH\alpha}$ coupling. Similar experiments are also available for the side chain χ_1 angle ($^3J_{NH\beta}$-HNHB and $^3J_{H\alpha H\beta}$-HAHB(CACO)NH [53, 54]). Further residual dipolar couplings can be used.

Acknowledgments

The authors acknowledge funding by the Deutsche Forschungsgemeinschaft (SFB1208 and SFB974) and the Jürgen Manchot Stiftung.

Appendix

In the following, the five .mdp files needed for the MD simulation using *GROMACS*, which is explained in Subheading 2.3, are provided.

"ions.mdp" for Adding Ions

```
;; ions.mdp
; Run setup
integrator                      = steep
emtol                           = 1000
emstep                          = 0.01
nsteps                          = 2000
; Neighbor search
cutoff-scheme                   = Verlet
pbc                             = xyz
; Electrostatics and vdW
coulombtype                     = PME
pme-order                       = 4
fourierspacing                  = 0.1
rcoulomb                        = 1.2
rvdw                            = 1.2
```

"em.mdp" for Energy Minimization

```
;; em.mdp
; Run setup
integrator                      = steep
emtol                           = 500
emstep                          = 0.001
nsteps                          = 2000
nstxout                         = 100
; Neighbor searching
cutoff-scheme                   = Verlet
nstlist                         = 20
ns-type                         = grid
```

```
pbc                         = xyz
; Electrostatics
coulombtype                 = PME
pme-order                   = 4
fourierspacing              = 0.1
rcoulomb                    = 1.2
; VdW
rvdw                        = 1.2
```

"nvt.mdp" for the Equilibration MD Run in the NVT Ensemble

```
;; nvt.mdp
define                      = -DPOSRES
; Run setup
integrator                  = md
dt                          = 0.002
nsteps                      = 50000
; Output control
nstxout                     = 5000
nstvout                     = 5000
nstfout                     = 5000
nstlog                      = 500
nstenergy                   = 500
; Bonds
constraints                 = all-bonds
constraint-algorithm        = LINCS
lincs-order                 = 4
lincs-iter                  = 1
; Neighbor searching
cutoff-scheme               = Verlet
nstlist                     = 20
ns-type                     = grid
pbc                         = xyz
; Electrostatics
coulombtype                 = PME
pme-order                   = 4
fourierspacing              = 0.1
rcoulomb                    = 1.2
; VdW
rvdw                        = 1.2
; T coupling is on
tcoupl                      = v-rescale
tc-grps                     = Protein Non-Protein
tau-t                       = 0.1 0.1
ref-t                       = 298 298
nsttcouple                  = 10
; P coupling is off
pcoupl                      = no
; Velocity generation
gen-vel                     = yes
gen-temp                    = 298
continuation                = no
```

"npt.mdp" for the Equilibration MD Run in the NPT Ensemble

```
;; npt.mdp
define                          = -DPOSRES
; Run setup
integrator                      = md
dt                              = 0.002
nsteps                          = 100000
; Output control
nstxout                         = 5000
nstvout                         = 5000
nstfout                         = 5000
nstlog                          = 500
nstenergy                       = 500
; Bonds
constraints                     = all-bonds
constraint-algorithm            = LINCS
lincs-order                     = 4
lincs-iter                      = 1
; Neighbor searching
cutoff-scheme                   = Verlet
nstlist                         = 20
ns-type                         = grid
pbc                             = xyz
; Electrostatics
coulombtype                     = PME
pme-order                       = 4
fourierspacing                  = 0.1
rcoulomb                        = 1.2
; VdW
rvdw                            = 1.2
; T coupling is on
tcoupl                          = v-rescale
tc-grps                         = Protein Non-Protein
tau-t                           = 0.1 0.1
ref-t                           = 298 298
nsttcouple                      = 10
; P coupling is on
pcoupl                          = berendsen
pcoupltype                      = isotropic
tau_p                           = 1.0
ref_p                           = 1.0
compressibility                 = 4.5e-5
refcoord_scaling                = com
; Velocity generation
gen-vel                         = no
gen-temp                        = 298
continuation                    = yes
```

"md.mdp" for the Production MD Run

```
;; md.mdp
; Run setup
integrator                  = md
dt                          = 0.002
nsteps                      = 5000000
; Output control
nstxout                     = 0
nstvout                     = 0
nstfout                     = 0
nstlog                      = 2500
nstenergy                   = 2500
nstxout-compressed          = 2500
compressed-x-grps           = System
; Bonds
constraints                 = all-bonds
constraint-algorithm        = LINCS
lincs-order                 = 4
lincs-iter                  = 1
; Neighbor searching
cutoff-scheme               = Verlet
nstlist                     = 20
ns-type                     = grid
pbc                         = xyz
; Electrostatics
coulombtype                 = PME
pme-order                   = 4
fourierspacing              = 0.1
rcoulomb                    = 1.2
; VdW
rvdw                        = 1.2
; T coupling is on
tcoupl                      = v-rescale
tc-grps                     = Protein Non-Protein
tau-t                       = 0.1 0.1
ref-t                       = 298 298
nsttcouple                  = 10
; Pressure coupling is off
pcoupl                      = Parrinello-Rahman
pcoupltype                  = isotropic
tau_p                       = 2.0
ref_p                       = 1.0
compressibility             = 4.5e-5
; Velocity generation
gen-vel                     = no
gen-temp                    = 298
continuation                = yes
```

References

1. Weiergräber OH, Stangler T, Thielmann Y, Mohrlüder J, Wiesehan K, Willbold D (2008) Ligand binding mode of $GABA_A$ receptor-associated protein. J Mol Biol 381:1320–1331
2. Thielmann Y, Weiergräber OH, Mohrlüder J, Willbold D (2009) Structural framework of the GABARAP-calreticulin interface—implications for substrate binding to endoplasmic reticulum chaperones. FEBS J 276:1140–1152
3. Ma P, Schillinger O, Schwarten M, Lecher J, Hartmann R, Stoldt M, Mohrlüder J, Olubiyi O, Strodel B, Willbold D, Weiergräber OH (2015) Conformational polymorphism in autophagy-related protein GATE-16. Biochemistry 54:5469–5479
4. Michel M, Schwarten M, Decker C, Nagel-Steger L, Willbold D, Weiergräber OH (2015) The mammalian autophagy initiator complex contains 2 HORMA domain proteins. Autophagy 11:2300–2308
5. Stangler T, Mayr LM, Dingley AJ, Luge C, Willbold D (2001) Sequence-specific ^{1}H, ^{13}C and ^{15}N resonance assignments of human GABA receptor associated protein. J Biomol NMR 21:183–184
6. Schwarten M, Stoldt M, Mohrlüder J, Willbold D (2009) Sequence-specific ^{1}H, ^{13}C, and ^{15}N resonance assignment of the autophagy-related protein Atg8. Biomol NMR Assign 3:137–139
7. Pervushin K, Riek R, Wider G, Wüthrich K (1997) Attenuated T_2 relaxation by mutual cancellation of dipole-dipole coupling and chemical shift anisotropy indicates an avenue to NMR structures of very large biological macromolecules in solution. Proc Natl Acad Sci U S A 94:12366–12371
8. Schanda P, Van Melckebeke H, Brutscher B (2006) Speeding up three-dimensional protein NMR experiments to a few minutes. J Am Chem Soc 128:9042–9043
9. Solyom Z, Schwarten M, Geist L, Konrat R, Willbold D, Brutscher B (2013) BEST-TROSY experiments for time-efficient sequential resonance assignment of large disordered proteins. J Biomol NMR 55:311–321
10. Stangler T, Mayr LM, Willbold D (2002) Solution structure of human GABA(A) receptor-associated protein GABARAP: implications for biological function and its regulation. J Biol Chem 277:13363–13366
11. Sattler M, Schleucher J, Griesinger C (1999) Heteronuclear multidimensional NMR experiments for the structure determination of proteins in solution employing pulsed field gradients. Prog Nucl Magn Reson Spectrosc 34:93–158
12. Vranken WF, Boucher W, Stevens TJ, Fogh RH, Pajon A, Llinas M, Ulrich EL, Markley JL, Ionides J, Laue ED (2005) The CCPN data model for NMR spectroscopy: development of a software pipeline. Proteins 59:687–696
13. Yamazaki T, Formankay JD, Kay LE (1993) 2-dimensional Nmr experiments for correlating C-13-Beta and H-1-Delta/epsilon chemical-shifts of aromatic residues in C-13-Labeled proteins via scalar couplings. J Am Chem Soc 115:11054–11055
14. Rieping W, Habeck M, Bardiaux B, Bernard A, Malliavin TE, Nilges M (2007) ARIA2: automated NOE assignment and data integration in NMR structure calculation. Bioinformatics 23:381–382
15. Shen Y, Bax A (2013) Protein backbone and sidechain torsion angles predicted from NMR chemical shifts using artificial neural networks. J Biomol NMR 56:227–241
16. Brünger AT, Adams PD, Clore GM, DeLano WL, Gros P, Grosse-Kunstleve RW, Jiang JS, Kuszewski J, Nilges M, Pannu NS, Read RJ, Rice LM, Simonson T, Warren GL (1998) Crystallography & NMR system: a new software suite for macromolecular structure determination. Acta Crystallogr D Biol Crystallogr 54:905–921
17. Güntert P (2004) Automated NMR structure calculation with CYANA. Methods Mol Biol 278:353–378
18. Schwarten M, Stoldt M, Mohrlüder J, Willbold D (2010) Solution structure of Atg8 reveals conformational polymorphism of the N-terminal domain. Biochem Biophys Res Commun 395:426–431
19. Laskowski RA, Rullmannn JA, MacArthur MW, Kaptein R, Thornton JM (1996) AQUA and PROCHECK-NMR: programs for checking the quality of protein structures solved by NMR. J Biomol NMR 8:477–486
20. Chen VB, Arendall WB 3rd, Headd JJ, Keedy DA, Immormino RM, Kapral GJ, Murray LW, Richardson JS, Richardson DC (2010) MolProbity: all-atom structure validation for macromolecular crystallography. Acta Crystallogr D Biol Crystallogr 66:12–21
21. Abraham MJ, Murtola T, Schulz R, Páll S, Smith JC, Hess B, Lindahl E (2015) GROMACS: high performance molecular simulations through multi-level parallelism from

laptops to supercomputers. SoftwareX 1–2:19–25
22. Pronk S, Pall S, Schulz R, Larsson P, Bjelkmar P, Apostolov R, Shirts MR, Smith JC, Kasson PM, van der Spoel D, Hess B, Lindahl E (2013) GROMACS 4.5: a high-throughput and highly parallel open source molecular simulation toolkit. Bioinformatics 29:845–854
23. Lindorff-Larsen K, Piana S, Palmo K, Maragakis P, Klepeis JL, Dror RO, Shaw DE (2010) Improved side-chain torsion potentials for the Amber ff99SB protein force field. Proteins 78:1950–1958
24. Jorgensen WL, Chandrasekhar J, Madura JD, Impey RW, Klein ML (1983) Comparison of simple potential functions for simulating liquid water. J Chem Phys 79:926–935
25. Olsson MH, Søndergaard CR, Rostkowski M, Jensen JH (2011) PROPKA3: consistent treatment of internal and surface residues in empirical pKa predictions. J Chem Theory Comput 7:525–537
26. Anandakrishnan R, Aguilar B, Onufriev AV (2012) H++ 3.0: automating pK prediction and the preparation of biomolecular structures for atomistic molecular modeling and simulations. Nucleic Acids Res 40:W537–W541
27. Bussi G, Donadio D, Parrinello M (2007) Canonical sampling through velocity rescaling. J Chem Phys 126:014101
28. Hess B (2008) P-LINCS: a parallel linear constraint solver for molecular simulation. J Chem Theory Comput 4:116–122
29. Berendsen HJC, Postma JPM, Vangunsteren WF, Dinola A, Haak JR (1984) Molecular-dynamics with coupling to an external bath. J Chem Phys 81:3684–3690
30. Parrinello M, Rahman A (1981) Polymorphic transitions in single crystals: a new molecular dynamics method. J Appl Phys 52:7182–7190
31. Miao YL, McCammon JA (2016) Unconstrained enhanced sampling for free energy calculations of biomolecules: a review. Mol Simul 42:1046–1055
32. Bernardi RC, Melo MCR, Schulten K (2015) Enhanced sampling techniques in molecular dynamics simulations of biological systems. BBA-Gen Subjects 1850:872–877
33. Abrams C, Bussi G (2014) Enhanced sampling in molecular dynamics using metadynamics, replica-exchange, and temperature-acceleration. Entropy 16:163–199
34. Klenin K, Strodel B, Wales DJ, Wenzel W (2011) Modelling proteins: conformational sampling and reconstruction of folding kinetics. Biochim Biophys Acta 1814:977–1000
35. Bussi G (2013) Hamiltonian replica exchange in GROMACS: a flexible implementation. Mol Phys 112:379–384
36. Wang LL, Friesner RA, Berne BJ (2011) Replica exchange with solute scaling: a more efficient version of replica exchange with solute tempering (REST2). J Phys Chem B 115:9431–9438
37. Bonomi M, Branduardi D, Bussi G, Camilloni C, Provasi D, Raiteri P, Donadio D, Marinelli F, Pietrucci F, Broglia RA, Parrinello M (2009) PLUMED: a portable plugin for free-energy calculations with molecular dynamics. Comput Phys Commun 180:1961–1972
38. Moreno A (2017) Advanced methods of protein crystallization. Methods Mol Biol 1607:51–76
39. Garman EF, Owen RL (2006) Cryocooling and radiation damage in macromolecular crystallography. Acta Crystallogr D Biol Crystallogr 62:32–47
40. Scapin G (2013) Molecular replacement then and now. Acta Crystallogr D Biol Crystallogr 69:2266–2275
41. Dodson E (2008) The befores and afters of molecular replacement. Acta Crystallogr D Biol Crystallogr 64:17–24
42. Vagin A, Teplyakov A (1997) MOLREP: an automated program for molecular replacement. J Appl Crystallogr 30:1022–1025
43. McCoy AJ, Grosse-Kunstleve RW, Adams PD, Winn MD, Storoni LC, Read RJ (2007) Phaser crystallographic software. J Appl Crystallogr 40:658–674
44. Kissinger CR, Gehlhaar DK, Fogel DB (1999) Rapid automated molecular replacement by evolutionary search. Acta Crystallogr D Biol Crystallogr 55:484–491
45. Taylor GL (2010) Introduction to phasing. Acta Crystallogr D Biol Crystallogr 66:325–338
46. Grosse-Kunstleve RW, Schneider TR (2007) Substructure determination in isomorphous replacement and anomalous diffraction experiments. Methods Mol Biol 364:197–214
47. Cowtan K (2010) Recent developments in classical density modification. Acta Crystallogr D Biol Crystallogr 66:470–478
48. Sheldrick GM (2008) A short history of SHELX. Acta Crystallogr A 64:112–122
49. Terwilliger TC, Adams PD, Read RJ, Mccoy AJ, Moriarty NW, Grosse-Kunstleve RW, Afonine PV, Zwart PH, Hung LW (2009) Decision-making in structure solution using Bayesian estimates of map quality: the

PHENIX AutoSol wizard. Acta Crystallogr D Biol Crystallogr 65:582–601

50. Tronrud DE (2004) Introduction to macromolecular refinement. Acta Crystallogr D Biol Crystallogr 60:2156–2168
51. Murshudov GN, Vagin AA, Dodson EJ (1997) Refinement of macromolecular structures by the maximum-likelihood method. Acta Crystallogr D Biol Crystallogr 53:240–255
52. Afonine PV, Grosse-Kunstleve RW, Echols N, Headd JJ, Moriarty NW, Mustyakimov M, Terwilliger TC, Urzhumtsev A, Zwart PH, Adams PD (2012) Towards automated crystallographic structure refinement with phenix.refine. Acta Crystallogr D Biol Crystallogr 68:352–367
53. Vajpai N, Gentner M, Huang JR, Blackledge M, Grzesiek S (2010) Side-chain $\chi 1$ conformations in urea-denatured ubiquitin and protein G from (3)J coupling constants and residual dipolar couplings. J Am Chem Soc 132:3196–3203
54. Dux P, Whitehead B, Boelens R, Kaptein R, Vuister GW (1997) Measurement of (15)N-(1)H coupling constants in uniformly (15)N-labeled proteins: application to the photoactive yellow protein. J Biomol NMR 10:301–306

Chapter 3

Structural Studies of Mammalian Autophagy Lipidation Complex

Kazuto Ohashi, Chinatsu Otomo, Zoltan Metlagel, and Takanori Otomo

Abstract

Members of the autophagy-related protein 8 (Atg8) family of ubiquitin-like proteins (ublps), including mammalian LC3 and GABARAP proteins, play crucial roles in autophagosome biogenesis, as well as selective autophagy. Upon induction of autophagy, the autophagic ublps are covalently attached to a phosphatidylethanolamine (PE) molecule of the autophagosomal membrane. This unique lipid conjugation of the autophagic ublps, which is essential for their functions, occurs in a ubiquitination-like reaction cascade consisting of the E1 enzyme ATG7, the E2 ATG3, and the E3 ATG12~ATG5-ATG16L1 complex (~denotes a covalent linkage). These enzymes are structurally unique among those of the canonical ubiquitination cascades, necessitating structural and biochemical studies of these molecules for understanding the molecular mechanisms underlying the lipidation cascade. Here, we will describe methods that were employed in our previous studies (Otomo et al., Nat Struct Mol Biol 20:59–66, 2013; Metlagel et al., Proc Natl Acad Sci U S A 110:18844–18849, 2013; Ohashi and Otomo, Biochem Biophys Res Commun 463:447–452, 2015), including the production of recombinant enzymes, in vitro enzymatic reactions, the crystallization of the E3 complexes, and the NMR-based investigations of E1–E2 and E2–E3 interactions.

Key words LC3, GABARAP, ATG12, ATG5, ATG3, Conjugation, Crystallization, NMR

1 Introduction

Mammals possess seven members of the ATG8 family of ublps: LC3A, LC3B, LC3B2, LC3C, GABARAP, GABARAPL1, and GABARAPL2 (or GATE-16) [1, 2]. These proteins are required for efficient formation of autophagosomes and the degradation of the inner membrane of the autophagosomes in lysosomes [3], as well as selective autophagy via binding to a linear motif termed LIR/AIM in adaptor proteins that directly recruit cargo materials [4–7]. For the Atg8 family ublps to exert their functions, they must be conjugated to a PE molecule contained in the autophagosomal membrane, and this lipidation reaction is stimulated upon induction of autophagy [8, 9].

Nicholas Ktistakis and Oliver Florey (eds.), *Autophagy: Methods and Protocols*, Methods in Molecular Biology, vol. 1880,
https://doi.org/10.1007/978-1-4939-8873-0_3,

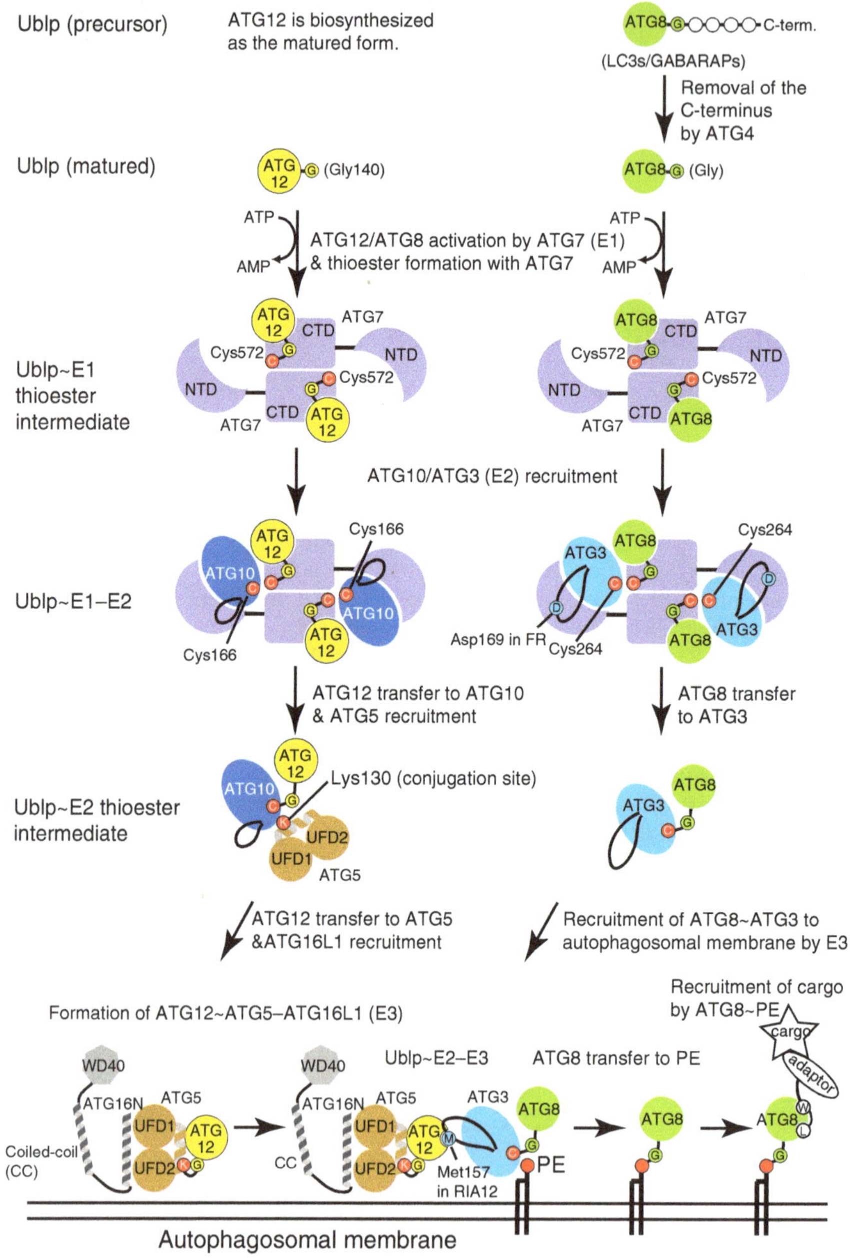

Fig. 1 Ubiquitination-like conjugation systems in autophagy. E1, E2, and E3 enzymes are presented as cartoon models. The residues involved in conjugation reactions are indicated. Asp169 in ATG3^{FR} and Met157 in ATG3^{RIA12}, which are critical for E1–E2 and E2–E3 interactions, respectively, are indicated

The Atg8 family ublps undergo a reaction cascade that resembles the canonical ubiquitination pathway comprised of E1, E2, and E3 enzymes (Fig. 1). Upon translation and folding of the Atg8 family ublps, several residues of their C-terminus are removed by ATG4 protease, such that a glycine residue becomes the new

C-terminal residue [10]. The E1 ATG7 activates this C-terminal glycine using ATP, resulting in the covalent attachment of the glycine residue on its catalytic cysteine through a thioester bond. Then, the catalytic cysteine residue of the E2 enzyme ATG3 replaces that of the E1 ATG7, resulting in the transfer of the Atg8 family ublp onto ATG3. The thioester-bonded Atg8 family ublp is transferred from ATG3 to PE, which is enhanced by the ATG12~ATG5-ATG16L1 E3 complex.

Despite the chemical similarity of the reaction cascades, the structural organizations of the corresponding E1, E2, and E3 enzymes differ considerably from the canonical enzymes. For example, ATG7 forms a gliding bird-shaped dimeric architecture that can interact with two E2s through its two N-terminal domains [11–14]. ATG3 consists of a canonical E2 domain and a unique flexible region (referred to as ATG3^{FR}), which interacts with the N-terminus of the E1 ATG7 as well as ATG12 of the E3 ATG12~ATG5-ATG16L1 [7, 11–16]. ATG3 shuttles between the E1 and the E3 through these mutually exclusive interactions [17–19]. ATG12 is also a ublp and is evolutionarily closest to the Atg8 family within the ublp kingdom [9]. Notably, the interaction between ATG12 and the region of ATG3 interacting with ATG12 (RIA12) is structurally similar to that between Atg8 family ublps and LIR/AIM motifs [15, 19]. ATG12 is conjugated exclusively to a specific lysine of ATG5 by the same E1, ATG7, and specialized E2, ATG10, for the ATG12-ATG5 conjugation [9, 20, 21]. In vivo, this protein-protein conjugation reaction is constitutive, and the resulting ATG12~ATG5 conjugate is stable because there is no de-conjugation enzyme for ATG12, allowing the ATG12~ATG5 conjugate to function as the E3 ligase for the lipidation cascade immediately upon induction of autophagy. The ATG12~ATG5 conjugate promotes the transfer of Atg8 family ublps from ATG3 to PE [22–24], acting as an E2-stimulating unit, despite no structural similarity with any such domains like the RING domain of the canonical ubiquitination pathways. ATG5 is a structural protein that consists of three domains: two ubiquitin-like fold domains and a helical region inserted between the two other domains [22, 25–28]. Conjugation of ATG12 onto the lysine in the helical region of ATG5 creates a single structural platform on which evolutionarily conserved residues of both proteins are assembled in a surface patch, which is important for E3 activity [22]. The surface opposite from the ATG12 conjugation site is also conserved and is the binding site for ATG16L1, a multi-domain structural protein that is responsible for localization of the E3 complex to autophagosome-forming sites [29, 30]. ATG16L1 interacts with ATG5 through an amphipathic helical region in its N-terminus (referred to as ATG16N) [22, 25], which is followed by a coiled-coil dimerization domain [31, 32] and a WD40-repeat β-propeller domain [33]. It is not known whether ATG16L1 is biochemically required for the lipidation reaction. Yeast Atg16 is not required for in vitro lipidation [23]. Nonetheless, ATG16N has been included in the E3 complex used for structural and biochemical studies because ATG16N improves the solubility and stability of ATG5 [15, 22, 25–28].

The lipidation of Atg8 family ublps has served as the central molecular feature of autophagy, underscoring its importance for autophagy as well as autophagy research [34]. Elucidation of the mechanistic details underlying the ATG8 lipidation cascade requires structural studies of the set of those unique enzymes. We will describe procedures that were used in our previous studies for production of all E1, E2, and E3 proteins, their crystallization, NMR analysis, and in vitro enzymatic assays [15, 17, 22].

2 Materials

2.1 Bacterial Protein Expression

1. Plasmids: pCDF Duet-1, pACYC Duet-1, pET Duet-1 (Novagen), pGEX4T3 (GE Healthcare).
2. *Escherichia coli* competent cells: BL21(DE3).
3. Culture media: Luria-Bertani (LB) medium or Terrific Broth (TB) medium.
4. M9 minimal culture medium for stable isotopic labeling: 7 g/L sodium phosphate dibasic, 3 g/L potassium phosphate monobasic, 5 g/L NaCl, 1 g/L ^{15}N-ammonium chloride, 4 g/L D-glucose (or 2 g/L ^{13}C-D-glucose for ^{15}N/^{13}C double labeling), 20 mg/L thiamine, 6.6 μM ferric chloride, 1 mM magnesium sulfate, 50 mM manganese (II) chloride.
5. M9 minimal culture medium for selenomethionine labeling: 7 g/L sodium phosphate dibasic, 3 g/L potassium phosphate monobasic, 5 g/L NaCl, 1 g/L ammonium chloride, 4 g/L D-glucose, 20 mg/L thiamine, 6.6 μM ferric chloride, 1 mM magnesium sulfate, 50 mM manganese (II) chloride.
6. Amino acid mixture: 100 mg/L L-phenylalanine, 100 mg/L L-threonine, 100 mg/L L-lysine, 50 mg/L L-isoleucine, 50 mg/L L-leucine, 50 mg/L L-valine, and 60 mg/L selenomethionine.
7. Antibiotics stocks (1000×): streptomycin (50 mg/mL) in water; chloramphenicol, 34 mg/mL in ethanol; ampicillin 100 mg/mL in 50% ethanol.
8. LB agar plates containing antibiotics.
9. Isopropyl-beta-thiogalactopyranoside (IPTG): 1 M in water.

2.2 Insect Cell Protein Expression

1. DH10B bacterial competent cells (Life Technologies).
2. pFastBacHT plasmid vector (Life Technologies).

3. LB agar plates containing 7 μg/mL gentamicin, 50 μg/mL kanamycin, 10 μg/mL tetracycline, 40 μg/mL IPTG, and 100 μg/mL X-gal.
4. Transfection reagent: X-tremeGENE 9 (Sigma-Aldrich).
5. Insect cell medium: Insect-XPRESS (Lonza).
6. Insect cells: *Spodoptera frugiperda* 9 (Sf9) cells (Expression Systems).
7. Cell Resuspension Solution: 50 mM Tris–HCl pH 7.5, 10 mM EDTA, 100 μg/mL RNase A.
8. Cell Lysis Solution: 0.2 M sodium hydroxide, 1% SDS.
9. Neutralization Solution: 4.1 M guanidine hydrochloride, 0.76 M potassium acetate, 2.1 M glacial acetic acid, final pH 4.2.
10. Isopropanol.
11. 70% ethanol.
12. Nuclease-free water.

2.3 Protein Purification

1. Tobacco etch virus (TEV) protease (2 mg/mL) stored at −80 °C.
2. Phenylmethylsulfonyl fluoride (PMSF): 100 mM in isopropanol.
3. Purification buffer stock solutions: 2 M imidazole pH 7.0, 1 M HEPES pH 7.5, 1 M HEPES pH 7.0, 0.5 M sodium phosphate dibasic (Na_2HPO_4), 0.5 M sodium phosphate monobasic (NaH_2PO_4), 1 M Tris–HCl pH 8.0.
4. 5 M NaCl.
5. 20% (v/v) Triton-X 100.
6. 1 M dithiothreitol (DTT) stored at −20 °C.
7. 0.5 M tris(2-carboxyethyl)phosphine (TCEP) stored at −20 °C.
8. 0.5 M EDTA pH 8.0.
9. Glutathione (reduced form).
10. C3 cell homogenizer (Avestin).
11. Econo-Pac gravity column (Bio-Rad).
12. AKTA purifier chromatography system (GE Healthcare).
13. Ni-NTA Superflow (Qiagen).
14. Glutathione Sepharose 4B (GE Healthcare).
15. Source 15S cation exchange chromatography column (GE Healthcare).
16. Source 15Q anion exchange chromatography column (GE Healthcare).

17. Superdex 200 size exclusion chromatography column (GE Healthcare).
18. Amicon stirred concentrator cell (EMD Millipore).
19. Amicon Ultra centrifugal concentrator (EMD Millipore).

2.4 Biochemical Assays

1. Lipids (Avanti Polar Lipids): 1,2-dioleoyl-sn-glycero-3-phosphocholine (PC), 1,2-dioleoyl-sn-glycero-3-phosphoethanolamine (PE), bovine liver l-α-phosphatidylinositol (blPI).
2. Nitrogen gas cylinder.
3. Water bath.
4. Mini extruder (Avanti Polar Lipids).
5. Polycarbonate filter (Avanti Polar Lipids): 100 μm pore size.
6. Thermal cycler.
7. ATP: 100 mM in water. Adjusted pH to ~7.5.
8. 6 M urea 13.5% acrylamide SDS-PAGE.
9. SDS-PAGE sample buffer (2×): 100 mM Tris–HCl pH 6.8, 2% (w/v) SDS, 20% (w/v) glycerol, 0.02% (w/v) bromophenol blue, and 10% (v/v) 2-mercaptoethanol.
10. Nonreducing SDS-PAGE sample buffer (2×): 100 mM Tris–HCl pH 6.8, 2% (w/v) SDS, 20% (w/v) glycerol and 0.02% (w/v) bromophenol blue.
11. Gel stain: GelCode Blue (Thermo Pierce).
12. 12% acrylamide NuPAGE with MOPS buffer (Life Technologies).
13. Odyssey infrared scanner (LI-COR).

2.5 Crystallization of the ATG12~ATG5-ATG16N Complex

1. Crystallization plates: 24-well Cryschem Plates (Hampton Research).
2. Crystallization stock solutions: 1 M HEPES, pH 7.0–8.0; 1 M BICINE, pH 9.0; 0.5 M MES, pH 6.0–6.75; 1.5 M lithium citrate tribasic; 1 M sodium phosphate dibasic; 3.6 M ammonium sulfate; 50% (w/v) PEG 5000 monoethyl ether; 50% (w/v) PEG 3350; ethylene glycol.

3 Methods

3.1 Bacterial Expression and Purification of ATG16N

To generate the ATG12~ATG5-ATG16N complex, ATG16N is separately expressed and purified for in vitro reconstitution of the E3 complex. ATG16N is an amphipathic helical region spanning ~33 residues from the residue 11–43 in human ATG16L. Such a short peptide can be generated as a fusion protein in bacteria. Here, a method using maltose-binding protein (MBP) is described. MBP is chosen to improve the solubility of the ATG16N, which has a poor solubility on its own.

1. Clone the DNA sequence encoding a human ATG16N fragment (e.g., the residues 11–43) into a pET Duet-based plasmid vector harboring a 6×His-MBP-TEV protease recognition sequence (TEVs) tag.
2. Transform BL21(DE3) cells with the plasmid and spread the cells on an LB-agar plate with 100 μg/mL ampicillin.
3. Incubate the plate at 37 °C for overnight.
4. Pick multiple colonies and inoculate them into 1 L LB medium with ampicillin in a 4 L flask.
5. Shake the flask at 220 rpm at 37 °C.
6. Add IPTG (final 1 mM) to induce protein expression at an OD_{600} of 0.8.
7. After 4 h, centrifuge cells at 4000 × *g* (max) for 15 min, and resuspend cell pellets in the lysis buffer [20 mM Hepes pH 7.5, 300 mM NaCl]. Flash-freeze cells in a 50 mL plastic tube and store at −80 °C until use.
8. Place the plastic tube in a water-filled beaker to thaw cells. Add PMSF (1 mM) slowly to the tube while swirling the cells.
9. Disrupt the cells using Avestin C3 pressure homogenizer.
10. Clarify the cell lysate by centrifugation at 43,667 × *g* (max).
11. Pour the supernatant into an Econo-Pac column, and add imidazole (20 mM), Triton X-100 (0.2% (v/v)), and Ni-NTA agarose beads to the supernatant. Nutate the column for 1 h at 4 °C.
12. Let the column stand and immediately discard the solution.
13. Wash the beads with seven column volumes (CVs) of the buffer [20 mM HEPES pH 7.5, 20 mM imidazole pH 7.0, 1 M NaCl, 0.2% (v/v) Triton-X 100], followed by 7 CVs of the buffer [20 mM Hepes pH 7.5, 20 mM imidazole pH 7.0, 300 mM NaCl]. Elute protein with five CVs of the buffer [300 mM imidazole pH 7.0, 300 mM NaCl].
14. Confirm the elution of the protein by SDS-PAGE.
15. Pool the fractions containing the protein, and dilute the protein solution sixfold with 20 mM sodium phosphate pH 6.8.
16. Inject the diluted solution into a Source 15S column equilibrated with the buffer [20 mM sodium phosphate pH 6.8, 50 mM NaCl]. Elute the protein in a NaCl gradient from 50 to 500 mM.
17. Confirm the protein fractions by SDS-PAGE and pool them.

18. Quantitate the concentration of the protein using absorbance at 280 nm. Flash-freeze the protein in liquid nitrogen in aliquots, and store the aliquots at −80 °C until use.

3.2 Expression and Purification of the ATG12~ATG5 Conjugate and Reconstitution of the ATG12~ATG5-ATG16L1 Complex

The ATG12~ATG5 protein-protein conjugation can be achieved in bacteria by co-expression of ATG12, ATG5, the E1 ATG7, and the E2 ATG10. The in vivo conjugation simplifies the production procedure of the ATG12~ATG5 conjugate, which otherwise would have to be synthesized in an inefficient in vitro reaction. The co-expression of the four proteins can be achieved using a pair of commercial Duet vectors, and this approach has been taken for producing the yeast Atg12~Atg5 conjugate in vivo as well [23].

1. Clone the DNA sequence encoding human ATG12 (residues 52–140) into a multiple cloning site (MCS) of pACYC Duet-1 and human ATG5 (full-length) into the other MCS of the same vector. Insert a six-histidine (6×His) tag, a maltose-binding protein (MBP), and a TEV protease cleavage sequence (TEVs) at the N-terminus of ATG12. Similarly, clone human ATG7 into a MCS of pCDF Duet-1 and human ATG10 into the other MCS (*see* **Note 1**).
2. Transform BL21(DE3) with both plasmids and spread the bacteria on an LB plate containing 50 μg/mL streptomycin and 34 μg/mL chloramphenicol.
3. Incubate the plate at 37 °C for overnight.
4. Pick multiple colonies and inoculate them into 1 L LB or TB medium with streptomycin and chloramphenicol (TB produces about four times as much protein as LB per volume.) (*see* **Note 2**).
5. Shake the flask at 220 rpm at 37 °C.
6. When the OD_{600} reaches 0.8 (LB) or 2.4 (TB), cool the flask on ice water. Then, add IPTG (final 0.2 mM) to induce protein expression.
7. Continue shaking the flask at 18 °C. After 16–20 h, harvest and store cells as described above.
8. Disrupt the cells and purify the 6×His-MBP-TEVs-ATG12~ATG5 by Ni-NTA chromatography as described above (*see* **Notes 3** and **4**).
9. Add DTT (final 1 mM) and EDTA (final 0.5 mM) to the eluate and concentrate the eluted protein in an Amicon pressure cell with a 30 kDa molecular weight (MW) cutoff membrane. Inject the concentrated proteins into a Superdex 200 size exclusion column equilibrated with the buffer [20 mM sodium phosphate pH 6.8, 300 mM NaCl, 1 mM DTT]. Identify the fractions containing proteins by SDS-PAGE and pool them.

10. Add an aliquot of TEV protease and incubate the solution at 4 °C or room temperature to cleave off the 6×His-MBP tag. Confirm the cleavage by SDS-PAGE.
11. Dilute the protein solution threefold with 20 mM sodium phosphate pH 6.8.
12. Inject the diluted solution into a Source 15S column equilibrated with the buffer [20 mM sodium phosphate pH 6.8, 50 mM NaCl, 1 mM DTT]. Elute the protein in a NaCl gradient from 50 to 500 mM (*see* **Note 5**).
13. Confirm the protein fractions by SDS-PAGE and pool them. Quantitate the concentration of the purified ATG12~ATG5 conjugate using absorbance at 280 nm.
14. To reconstitute the ATG12~ATG5-ATG16N complex, add a 1.2-fold molar excess of 6×His-MBP-TEVs-ATG16N to the pooled 6×His-MBP-TEVs-ATG12~ATG5 proteins. Without any further incubation, add an aliquot of TEV protease to the protein mixture solution to remove the 6×His-MBP tag from ATG16N. Incubate the protein solution at 4 °C or R.T. Check the cleavage by SDS-PAGE.
15. Upon complete digestion, dilute the sample solution threefold with 20 mM sodium phosphate pH 6.8 to reduce salt concentration to ~150 mM.
16. Inject the diluted solution into a Source 15S column equilibrated with the buffer [20 mM sodium phosphate pH 6.8, 50 mM NaCl]. Elute the protein with a NaCl gradient from 150 to 500 mM.
17. Confirm the protein fractions by SDS-PAGE. Pool the fractions and concentrate the pooled solution in an Amicon pressure cell with a 30 kDa MW cutoff membrane or using an Amicon Ultra centrifugal concentrator.
18. Inject the concentrated proteins into a Superdex 200 size exclusion column equilibrated with the buffer [10 mM HEPES pH 7.0, 300 mM NaCl, 1 mM DTT].
19. Concentrate the purified protein using an Amicon Ultra concentrator higher than 10 mg/mL. Quantitate the concentration of the protein using absorbance at 280 nm. Flash-freeze the protein in aliquots in liquid nitrogen and store at −80 °C until use.

3.3 Bacterial Expression and Purification of the Full-Length ATG3

1. Clone the DNA sequence of ATG3 (full-length) into a modified pGEX4T3 vector (GE healthcare). Also, insert a TEV between GST and ATG3.
2. Transform BL21(DE3) and spread the bacteria on an LB agar plate containing ampicillin.

3. Incubate the plate at 37 °C for overnight.
4. Inoculate multiple bacterial colonies in LB media, and shake the flask at 220 rpm at 37 °C.
5. Induce protein expression at an OD_{600} of 0.8 by adding IPTG (final 0.25 mM). Decrease the temperature to 25 °C.
6. Next day, harvest cells in the buffer [20 mM Tris–HCl pH 8.0, 150 mM NaCl, 1 mM DTT] and store cells as described above.
7. To purify the protein, thaw and lyse cells, and clarify the lysate as described above.
8. ^Mix the cleared lysate with Glutathione Sepharose beads in an Econo-Pac column. Nutate the column for 1 h at 4 °C.
9. Let the column stand and immediately discard the solution.
10. Wash the beads with 5 CVs of the buffer [20 mM Tris–HCl pH 8.0, 1 M NaCl, 0.1% (v/v) Triton-X100, 1 mM DTT], followed by 7 CVs of the buffer [20 mM Tris–HCl pH 8.0, 150 mM NaCl, 1 mM DTT]. Elute the protein with the buffer [20 mM Tris–HCl pH 8.0, 150 mM NaCl, 1 mM DTT, 15 mM glutathione (reduced form)] (*see* **Note 6**).
11. Confirm the protein fractions by SDS-PAGE. Pool the fractions, and add an aliquot of TEV protease to the pooled solution. Incubate the solution at 4 °C to cleave off the GST tag. Check the cleavage by SDS-PAGE.
12. Upon complete digestion, inject the cleaved ATG3 into a Source 15Q column equilibrated with the buffer [20 mM Tris–HCl pH 8.0, 50 mM NaCl, 1 mM DTT]. Elute the protein in a NaCl gradient from 150 to 400 mM.
13. Confirm the protein fractions by SDS-PAGE and pool them. Concentrate in an Amicon pressure cell with a 10 kDa MW cutoff membrane or using an Amicon Ultra centrifugal concentrator.
14. Inject the concentrated protein into a Superdex 200 size exclusion column equilibrated with the buffer [10 mM Hepes pH 7.5, 150 mM NaCl, 1 mM DTT].
15. Confirm the protein fractions by SDS-PAGE. Concentrate the protein and store at −80 °C.

3.4 Bacterial Expression and Purification of the ATG3 Flexible Region ($ATG3^{FR}$) for NMR Study

1. Clone $ATG3^{FR}$ (residues 88–192) into a pET-Duet-1 plasmid vector harboring a 6×His-MBP-TEV protease recognition sequence (TEVs) tag.
2. Use BL21(DE3) cells to express the protein as described above for ATG16N. Use M9 minimal medium for ^{15}N or $^{15}N/^{13}C$ labeling.
3. Disrupt the cells and purify the 6×His-MBP-TEVs-$ATG3^{FR}$ by Ni-NTA chromatography as described above for ATG16N.

4. Confirm the protein in the eluate by SDS-PAGE. Add an aliquot of TEV protease to cleave off the MBP tag. Add DTT (final 1 mM) and EDTA (final 0.5 mM) to the eluate.
5. Upon complete cleavage, dilute the protein threefold with 20 mM Tris–HCl pH 8.0. Inject the diluted solution into a Source 15Q column equilibrated with the buffer [20 mM Tris–HCl pH 8.0, 50 mM NaCl]. Elute the protein with a NaCl gradient from 50 to 350 mM.
6. Confirm the purity of the protein by SDS-PAGE. Concentrate the protein and store at −80 °C.

3.5 Bacterial Expression and Purification of the ATG3 Region Interacting with ATG12 (ATG3^{RIA12}) and for Crystallography

1. Clone ATG3^{RIA12} (residues 140–170) into a pET-Duet-1 plasmid vector harboring a 6×His-MBP-TEV protease recognition sequence (TEVs) tag.
2. Use BL21(DE3) cells to express the protein as described above for ATG16N. To incorporate selenomethionine into the protein, use M9 minimal culture media for selenomethionine labeling. Add the amino acid mixture containing selenomethionine at 1 h before induction.
3. Purify the ATG3^{RIA12} as described above for ATG^{FR}.
4. To reconstitute the ATG3^{RIA12}-ATG12~ATG5-ATG16N complex, add a 1.2-fold molar excess of ATG3^{RIA12} to the ATG12~ATG5-ATG16N complex. Without any further incubation, inject the protein mixture into an SD200 size exclusion column equilibrated with 10 mM HEPES pH 7.0, 150 mM NaCl, 1 mM DTT.
5. Concentrate the reconstituted complex to 10 mg/mL using a 30 kDa cutoff Amicon Ultra concentrator. Store the protein at −80 °C.

3.6 Baculovirus-Based Insect Cell Expression and Purification of Human ATG7

Although the fact that the ATG12~ATG5 conjugate can be obtained from bacterial cells as described above suggest that human ATG7 is expressed in bacteria, ATG7 cannot be obtained from bacteria due to its poor expression level. To overcome this issue, baculovirus-based insect cell expression was utilized for the production of ATG7.

3.6.1 Preparation of Bacmid DNA

1. Clone ATG7 into pFASTBACHT with an N-terminal His-tag as well as TEVs using an appropriate pair of restriction enzymes.
2. Transform DH10Bac with the generated plasmid and plate the bacteria on an LB-agar containing 7 μg/mL gentamicin, 50 μg/mL kanamycin, 10 μg/mL tetracycline, 40 μg/mL IPTG, and 100 μg/mL X-gal.

3. Inoculate a single, isolated white colony into 2 mL LB medium containing 7 μg/mL gentamicin, 50 μg/mL kanamycin, 10 μg/mL tetracycline. Grow at 37 °C in a shaker overnight.
4. Centrifuge 1.5 mL overnight culture in a 2-mL microcentrifuge tube at 2700 × *g* for 10 min.
5. Remove the supernatant by aspiration, and resuspend the cell pellet in 250 μL Buffer Cell Resuspension Solution.
6. Add 250 μL Cell Lysis Solution, mix by inverting the tube, and incubate at room temperature for 5 min.
7. Add 350 μL Neutralization Solution, mix by inverting the tube and immediately centrifuge at 28,500 × *g* for 10 min.
8. Transfer the supernatant to a 1.5-mL microcentrifuge tube, and centrifuge for another 10 min at 28,500 × *g*.
9. Transfer the supernatant to a new 1.5-mL microcentrifuge tube containing 800 μL isopropanol, mix by inverting tube, and incubate on ice for 5–10 min.
10. Centrifuge at 28,500 × *g* for 15 min.
11. Remove the supernatant by aspiration, add 500 μL of 70% ethanol, and invert the tube a few times to wash the bacmid DNA pellet.
12. Centrifuge at 28,500 × *g* for 15 min.
13. Remove the supernatant by aspiration, and air-dry the pellet.
14. Resuspend the bacmid DNA pellet in 50 to 100 μL nuclease-free water.

3.6.2 Transfection of Sf9 Cells and Generation of Virus Stock

1. Seed 1.5 × 10^6 Sf9 cells in 2 mL medium per well of a six-well plate, and allow to sit for 30 min to 1 h in a 27 °C incubator.
2. Add 6 μL X-tremeGENE 9 to 200 μL medium in a 1.5-mL microcentrifuge tube, and then add 2 μg bacmid DNA, and mix well. Incubate at room temperature for 20 min.
3. Wash the seeded cells once with 1 mL medium, and add 0.8 mL medium per well.
4. Add 200 μL X-tremeGENE 9/DNA mixture to the well. Incubate in a 27 °C incubator.
5. 4–6 h later, replace the transfection mixture with 2 mL fresh medium.
6. After 4 days, collect the supernatant and keep it as P0 virus stock.

3.6.3 Amplification of Baculovirus

1. Prepare 12 mL Sf9 cells at 1–2 × 10^6 cells/mL in a 50 mL flask.
2. Add 1 mL of P0 virus stock and incubate in a 27 °C shaker.
3. After 3–4 days, harvest virus by centrifugation. Keep the supernatant as P1 virus stock.

4. Prepare 50 mL Sf9 cells at 1–2 × 10^6 cells/mL in a 250 mL flask.
5. Add 1 mL of P1 virus stock and incubate in a 27 °C shaker.
6. After 3–4 days, harvest virus by centrifugation. Keep the supernatant as P2 virus stock.
7. Prepare 100 mL Sf9 cells at 1–2 × 10^6 cells/mL in a 500 mL flask.
8. Add 2 mL of P2 virus stock and incubate in a 27 °C shaker.
9. After 3–4 days, harvest virus by centrifugation. Keep the supernatant as P3 virus stock.

3.6.4 Infection for Expression

1. Seed 400 mL of Sf9 cells at 0.8–1 × 10^6 cells/mL in a 2 L flask on the day before infection.
2. Next day, add 40 mL (1/100 volume of the culture) of P3 virus stock to cells.
3. After 2 days, pellet cells by centrifugation at 2000 × *g* for 20 min. Resuspend the cells with 45 mL of 10 mM HEPES pH 7.5, 150 mM NaCl, and 1 mM DTT in a 50 mL tube, and centrifuge the tube at 2000 × *g* for 10 min.
4. Discard the solution and flash-freeze the tube containing cell pellets.
5. Store the tube at −80 °C until use.

3.6.5 Purification of ATG7

1. Thaw, disrupt, and clarify cells as described above.
2. Purify His-ATG7 from the lysate using Ni-NTA chromatography as described above for ATG16N.
3. Dilute the eluate from the Ni-NTA column sixfold with 20 mM Tris–HCl pH 8.0.
4. Inject the diluted protein into a source 15Q column equilibrated with the buffer [20 mM Tris–HCl pH 8.0, 50 mM NaCl, 1 mM DTT]. Elute the protein in a NaCl gradient from 50 to 400 mM.
5. Pool the protein fraction identified by SDS-PAGE and concentrate it using a concentrator.
6. Inject the concentrated protein into an SD200 size exclusion column equilibrated with the buffer [10 mM HEPES pH 7.5, 300 mM NaCl, 1 mM DTT, 10% glycerol].
7. Confirm the protein fractions by SDS-PAGE and store the protein in aliquots at −80 °C.

3.7 Lipidation Assay

The activity of the human ATG12~ATG5-ATG16N E3 complex used for structural studies can be examined by in vitro LC3/GABARAP lipidation assay as described below. The assay enables structure-function studies involving the E3 complex.

1. Prepare liposomes consisting of PC, PE, and blPI as follows.
2. Mix the lipids dissolved in chloroform with a molar ratio of 40% PC, 40% PE, and 20% blPI in a small glass tube.
3. Dry the lipid mixture under a nitrogen gas stream. Place the glass tube under vacuum for 2 h.
4. Add the buffer [20 mM HEPES pH 7.5, 100 mM NaCl] to the tube. Incubate the tube for 1 h at room temperature, followed by three to six times of freeze-thaw cycles. Use liquid nitrogen for freezing and a water bath (42–50 °C) for thawing.
5. Extrude the hydrated lipid mixture through a 100 μm polycarbonate filter (30 times). Keep the liposomes at 4 °C and use within 2 days.
6. For lipidation assay, thaw aliquots of ATG7, ATG3, the ATG12~ATG5-ATG16N complex, and LC3/GABARAP. The concentrations of the aliquots need to be at least 50 μM or higher.
7. Add 1 mM freshly dissolved DTT to ATG7 and ATG3 stocks, and incubate them on ice for 30 min.
8. Mix the proteins (0.5–1 μM ATG7, 1 μM ATG3, 1 μM ATG12~ATG5-ATG16N, and 10 μM LC3/GABARAP (matured from)) and the liposomes (final 1 mM lipid) in 50 mM HEPES pH 7.5, 100 mM NaCl, 0.3 mM DTT (final), and 1 mM $MgCl_2$. Make this mixture in a PCR tube.
9. Place the PCR tube on a thermal cycler set to 37 °C.
10. Add ATP to the tube at final 1 mM.
11. Quench the reaction at a desired time point by adding an equal volume of standard SDS-PAGE sample buffer (2×).
12. Run a 6 M urea 13.5% acrylamide SDS-PAGE gel with the samples (*see* **Note 7**).
13. Stain the gel with GelCode Blue and destain the gel with water.
14. Scan the stained gel on an Odyssey infrared scanner and quantitate the band intensities.

3.8 LC3/GABARAP-ATG3 Thioester Formation Assay

The role of $ATG3^{FR}$ in E1-E2 interaction can be studied by monitoring single turnover LC3/GABARAP~ATG3 thioester formation reactions as described below.

1. Thaw aliquots of ATG7, ATG3, and LC3/GABARAP on ice.
2. Treat ATG7 and ATG3 with 1 mM DTT as described above.
3. First, load ATG7 with LC3/GABARAP as follows: Mix 20 μM ATG7 and 10 μM LC3/GABARAP in 50 mM HEPES pH 7.0, 150 mM NaCl, 0.3 mM DTT, 200 μM $MgCl_2$, and 200 μM ATP in a 1.5 mL plastic tube on ice for 10 min.

4. Second, quench LC3/GABARAP loading on ATG7 and initiate the transfer of LC3/GABARAP to ATG3 as follows: Immediately after the ATG7 loading reaction, dilute the reaction mixture tenfold with a solution containing 4 μM ATG3 in 50 mM HEPES pH 7.0, 150 mM NaCl, 0.5 mM DTT, and 50 mM EDTA. Incubate the diluted mixture on ice for time course.
5. Remove small aliquots of the reaction mixture during a time course experiments, and mix them with an equal volume of nonreducing SDS-PAGE sample buffer (2×). This quenches the reaction. To prevent degradation of the formed thioester conjugates, keep the sample on ice (*see* **Note 8**).
6. Immediately after completion of a time course experiment, load the sample onto 12% acrylamide NuPAGE. Run the gel in MOPS buffer.
7. Stain and scan the gel as described above for visualization and quantification.

3.9 Crystallization of the ATG12~ATG5-ATG16N Complex

1. Thaw an aliquot of the ATG12~ATG5-ATG16N complex (10 mg/mL). Place a 1–2 μL drop of the protein on a well of a 24-well Cryschem sitting-drop plate, and add an equal volume of the reservoir solution [condition I: 0.1 M HEPES, pH 7.5, 0.2 M lithium citrate, 20% (w/v) PEG 3350] or [condition II: 0.1 M BICINE, pH 9.0, 0.2 M sodium phosphate dibasic, 18% (w/v) PEG 3350]. To maximize the chance of getting high-quality crystals, titrate pH as well as PEG concentration. Seal the plate with a transparent tape.
2. Incubate the plate at 20 °C. Crystals appear in ~1–2 days.
3. When crystal growth stops (in ~3–5 days), open the well and exchange the crystal solution with the cryoprotection buffer [condition I: 0.1 M HEPES, pH 7.5, 0.2 M lithium citrate, 21.5% (w/v) PEG 3350, 0.1 M NaCl, 25% (v/v) ethylene glycol] or [condition II: 0.1 M MES, pH 6.5, 0.2 M sodium phosphate dibasic, 19.5% (w/v) PEG 3350, 20% (v/v) ethylene glycol].
4. Pick up crystals with a nylon loop and flash-cool them in liquid nitrogen.
5. Screen crystals and collect diffraction data at a synchrotron facility.

3.10 NMR Resonance Assignment of $ATG3^{FR}$ and Identification of the E1 and E3 Binding Sites

NMR chemical shift perturbation is a powerful method to investigate protein–protein interactions and was used for identifying the E1 and E3 binding sites of $ATG3^{FR}$.

1. Use Millipore Ultra centrifugal concentrator device to exchange the buffer of ^{15}N-labeled $ATG3^{FR}$ into [20 mM

sodium phosphate pH 6.8, 150 mM NaCl, 2 mM DTT, 10% (v/v) D_2O], and place the concentrated sample solution (0.3 mM protein) in an NMR tube.

2. Record ^1H-^{15}N 2D Heteronuclear Single Quantum Coherence (HSQC) spectrum of ^{15}N-labeled $ATG3^{FR}$ at 25 °C (*see* **Note 9**).
3. Add unlabeled ATG7 (E1) or the ATG12~ATG5-ATG16N complex (E3) and record a ^1H-^{15}N HSQC spectrum. To identify the peaks affected by binding, calculate the ratio of each peak of ^{15}N-labeled $ATG3^{FR}$ in the presence to the absence of the unlabeled protein.
4. For assignment, record 3D HNCACB and HN(CO)CACB triple resonance spectra with 1 mM $^{13}C/^{15}N$-labeled $ATG3^{FR}$ (*see* **Note 10**).
5. Perform sequential assignment of the backbone atoms of $ATG3^{FR}$ using the acquired triple resonance spectra.
6. Identify the residues whose ^1H-^{15}N HSQC peak was broadened or shifted upon the addition of E1 or E3.

3.11 Crystallization of the ATG12~ATG5-ATG16L1 Complex Bound to the $ATG3^{RIA12}$

1. Place a 1–2 μL drop of the protein on a well of a 24-well Cryschem sitting-drop plate and add an equal volume of the reservoir solution [0.1 M MES, pH 6.0–6.75, 0.2 M ammonium sulfate, 10–16% (w/v) PEG 5000 monoethyl ether]. To maximize the chance of getting high-quality crystals, titrate pH as well as PEG concentration. Seal the plate with a transparent tape.
2. Incubate the plate at 4 °C. Crystals appear in ~2 days.
3. When crystals stop to grow (in ~2 weeks), open the well and exchange the crystal solution with cryoprotection buffer [0.1 M MES, pH 6.5, 0.2 M ammonium sulfate, 10–16% (w/v) PEG 5000 MME, 0.1 M NaCl, 26% (v/v) ethylene glycol).
4. Pick up crystals with a nylon loop and flash-cool them in liquid nitrogen.
5. Screen crystals and collect diffraction data at a synchrotron facility.

4 Notes

1. The co-expression of ATG7, ATG10, ATG12, and ATG5 can be achieved using other sets of plasmid vectors, such as pET Duet and pRSF Duet. The copy number of these plasmid vectors is different from that of the used ones. Therefore, it is possible that the final expression level is affected by choice. Use

of bacterial strains co-expressing rare tRNAs, such as BL21-CodonPlus (Agilent) and Rosetta (EMD Biosciences) cells, may boost the final yield of the ATG12~ATG5 conjugate. To use these strains, avoid using pACYC, as pACYC is not compatible with these strains.

2. Bacteria can be grown from a single colony or multiple colonies. Inoculate a single colony in a small medium (~3 mL) for overnight pre-culture or multiple colonies in a slightly larger size medium (~10 mL) for short pre-culture (~1 h) in the morning.
3. This co-expression system generates the 6×His-MBP tag without ATG12 more than tenfold and the intact 6×His-MBP-ATG12 protein. Therefore, use a large volume of Ni-NTA beads (50 mL Ni-NTA for 6 L TB culture) to capture all proteins.
4. Amylose affinity chromatography for purification of MBP-fusion proteins can be used in a combination or in place of Ni-NTA chromatography step.
5. The ATG12~ATG5-ATG16N complex elutes from a Source 15S column at a higher salt concentration than the ATG12~ATG5 conjugate. This is because ATG16N is also highly positively charged. The shift of the elution position allows a confirmation of the complex formation.
6. Adjust the pH of the elution buffer containing 15 mM glutathione to ~8.0 using 6 M sodium oxide.
7. 6 M urea is not essential for separating the PE-conjugated LC3/GABARAP from unconjugated proteins in SDS-PAGE. However, it increases the separation between these two forms, facilitating the quantification of the bands on scanned images. Furthermore, ATG3 and the ATG12~ATG5 conjugate migrate separately on 6 M urea gels but not on standard gels, allowing a clear visualization of these proteins on SDS-PAGE. Acrylamides in urea gels are prone to degrade particularly at higher temperatures. Thus, run urea gels at a low voltage or using cold buffers to avoid heating up the gel.
8. The LC3/GABARAP~ATG3 thioester is not stable. Do not freeze the samples to avoid degradation.
9. The unfolded nature of ATG3^{FR} makes the spectral dispersion narrow. Thus, increase the acquisition time of ^{15}N dimension to maximize the resolution.
10. Use of the carbon constant time versions of these experiments maximizes the resolution of the carbon dimension, which would help separate otherwise overlapping peaks of the unfolded ATG3^{FR}. Also record other alternative types of triple resonance experiments, such as HNCO, HNCACO, HNN, and HNCN, to facilitate the assignment.

Acknowledgments

This work was supported by US National Institute of Health grant GM092740 (to T.O.) and Japan Society of Promoting Science (JSPS) fellowship (to K.O.).

References

1. Cappadocia L, Lima CD (2017) Ubiquitin-like protein conjugation: structures, chemistry, and mechanism. Chem Rev. https://doi.org/10.1021/acs.chemrev.6b00737
2. Schulman BA, Harper JW (2009) Ubiquitin-like protein activation by E1 enzymes: the apex for downstream signalling pathways. Nat Rev Mol Cell Biol 10(5):319–331
3. Tsuboyama K, Koyama-Honda I, Sakamaki Y, Koike M, Morishita H, Mizushima N (2016) The ATG conjugation systems are important for degradation of the inner autophagosomal membrane. Science 354(6315):1036–1041. https://doi.org/10.1126/science.aaf6136
4. Ichimura Y, Kumanomidou T, Sou YS, Mizushima T, Ezaki J, Ueno T, Kominami E, Yamane T, Tanaka K, Komatsu M (2008) Structural basis for sorting mechanism of p62 in selective autophagy. J Biol Chem 283 (33):22847–22857
5. Noda NN, Kumeta H, Nakatogawa H, Satoo K, Adachi W, Ishii J, Fujioka Y, Ohsumi Y, Inagaki F (2008) Structural basis of target recognition by Atg8/LC3 during selective autophagy. Genes Cells 13 (12):1211–1218
6. Noda NN, Ohsumi Y, Inagaki F (2010) Atg8-family interacting motif crucial for selective autophagy. FEBS Lett 584(7):1379–1385. https://doi.org/10.1016/j.febslet.2010.01.018
7. Klionsky DJ, Schulman BA (2014) Dynamic regulation of macroautophagy by distinctive ubiquitin-like proteins. Nat Struct Mol Biol 21(4):336–345. https://doi.org/10.1038/nsmb.2787
8. Ichimura Y, Kirisako T, Takao T, Satomi Y, Shimonishi Y, Ishihara N, Mizushima N, Tanida I, Kominami E, Ohsumi M, Noda T, Ohsumi Y (2000) A ubiquitin-like system mediates protein lipidation. Nature 408:488–492
9. Ohsumi Y (2001) Molecular dissection of autophagy: two ubiquitin-like systems. Nat Rev Mol Cell Biol 2:211–216
10. Kirisako T, Ichimura Y, Okada H, Kabeya Y, Mizushima N, Yoshimori T, Ohsumi M, Takao T, Noda T, Ohsumi Y (2000) The reversible modification regulates the membrane-binding state of Apg8/Aut7 essential for autophagy and the cytoplasm to vacuole targeting pathway. J Cell Biol 151(2):263–276
11. Noda NN, Satoo K, Fujioka Y, Kumeta H, Ogura K, Nakatogawa H, Ohsumi Y, Inagaki F (2011) Structural basis of Atg8 activation by a homodimeric E1, Atg7. Mol Cell 44 (3):462–475. https://doi.org/10.1016/j.molcel.2011.08.035
12. Taherbhoy AM, Tait SW, Kaiser SE, Williams AH, Deng A, Nourse A, Hammel M, Kurinov I, Rock CO, Green DR, Schulman BA (2011) Atg8 transfer from Atg7 to Atg3: a distinctive E1-E2 architecture and mechanism in the autophagy pathway. Mol Cell 44 (3):451–461. https://doi.org/10.1016/j.molcel.2011.08.034
13. Yamaguchi M, Matoba K, Sawada R, Fujioka Y, Nakatogawa H, Yamamoto H, Kobashigawa Y, Hoshida H, Akada R, Ohsumi Y, Noda NN, Inagaki F (2012) Noncanonical recognition and UBL loading of distinct E2s by autophagy-essential Atg7. Nat Struct Mol Biol 19(12):1250–1256. https://doi.org/10.1038/nsmb.2451
14. Kaiser SE, Mao K, Taherbhoy AM, Yu S, Olszewski JL, Duda DM, Kurinov I, Deng A, Fenn TD, Klionsky DJ, Schulman BA (2012) Noncanonical E2 recruitment by the autophagy E1 revealed by Atg7-Atg3 and Atg7-Atg10 structures. Nat Struct Mol Biol 19(12):1242–1249. https://doi.org/10.1038/nsmb.2415
15. Metlagel Z, Otomo C, Takaesu G, Otomo T (2013) Structural basis of ATG3 recognition by the autophagic ubiquitin-like protein ATG12. Proc Natl Acad Sci U S A 110 (47):18844–18849. https://doi.org/10.1073/pnas.1314755110
16. Yamada Y, Suzuki NN, Hanada T, Ichimura Y, Kumeta H, Fujioka Y, Ohsumi Y, Inagaki F (2007) The crystal structure of Atg3, an autophagy-related ubiquitin carrier protein (E2) enzyme that mediates Atg8 lipidation. J Biol Chem 282:8036–8043
17. Ohashi K, Otomo T (2015) Identification and characterization of the linear region of ATG3

that interacts with ATG7 in higher eukaryotes. Biochem Biophys Res Commun. https://doi.org/10.1016/j.bbrc.2015.05.107

18. Qiu Y, Hofmann K, Coats JE, Schulman BA, Kaiser SE (2013) Binding to E1 and E3 is mutually exclusive for the human autophagy E2 Atg3. Protein Sci. https://doi.org/10.1002/pro.2381
19. Metlagel Z, Otomo C, Ohashi K, Takaesu G, Otomo T (2014) Structural insights into E2-E3 interaction for LC3 lipidation. Autophagy 10(3):522–523. https://doi.org/10.4161/auto.27594
20. Mizushima N, Sugita H, Yoshimori T, Ohsumi Y (1998) A new protein conjugation system in human. The counterpart of the yeast Apg12p conjugation system essential for autophagy. J Biol Chem 273:33889–33892
21. Mizushima N, Noda T, Yoshimori T, Tanaka Y, Ishii T, George MD, Klionsky DJ, Ohsumi M, Ohsumi Y (1998) A protein conjugation system essential for autophagy. Nature 395 (6700):395–398. https://doi.org/10.1038/26506
22. Otomo C, Metlagel Z, Takaesu G, Otomo T (2013) Structure of the human ATG12~ATG5 conjugate required for LC3 lipidation in autophagy. Nat Struct Mol Biol 20(1):59–66. https://doi.org/10.1038/nsmb.2431
23. Hanada T, Noda NN, Satomi Y, Ichimura Y, Fujioka Y, Takao T, Inagaki F, Ohsumi Y (2007) The Atg12-Atg5 conjugate has a novel E3-like activity for protein lipidation in autophagy. J Biol Chem 282:37298–37302
24. Fujioka Y, Noda NN, Fujii K, Yoshimoto K, Ohsumi Y, Inagaki F (2008) In vitro reconstitution of plant Atg8 and Atg12 conjugation systems essential for autophagy. J Biol Chem 283:1921–1928
25. Matsushita M, Suzuki NN, Obara K, Fujioka Y, Ohsumi Y, Inagaki F (2007) Structure of Atg5. Atg16, a complex essential for autophagy. J Biol Chem 282:6763–6772
26. Kim JH, Hong SB, Lee JK, Han S, Roh KH, Lee KE, Kim YK, Choi EJ, Song HK (2015) Insights into autophagosome maturation revealed by the structures of ATG5 with its interacting partners. Autophagy 11(1):75–87. https://doi.org/10.4161/15548627.2014.984276
27. Kim M, Sandford E, Gatica D, Qiu Y, Liu X, Zheng Y, Schulman BA, Xu J, Semple I, Ro SH, Kim B, Mavioglu RN, Tolun A, Jipa A, Takats S, Karpati M, Li JZ, Yapici Z, Juhasz G, Lee JH, Klionsky DJ, Burmeister M (2016) Mutation in ATG5 reduces autophagy and leads to ataxia with developmental delay. Elife 5. https://doi.org/10.7554/eLife.12245
28. Noda NN, Fujioka Y, Hanada T, Ohsumi Y, Inagaki F (2013) Structure of the Atg12-Atg5 conjugate reveals a platform for stimulating Atg8-PE conjugation. EMBO Rep 14 (2):206–211. https://doi.org/10.1038/embor.2012.208
29. Fujita N, Itoh T, Omori H, Fukuda M, Noda T, Yoshimori T (2008) The Atg16L complex specifies the site of LC3 lipidation for membrane biogenesis in autophagy. Mol Biol Cell 19(5):2092–2100
30. Nishimura T, Kaizuka T, Cadwell K, Sahani MH, Saitoh T, Akira S, Virgin HW, Mizushima N (2013) FIP200 regulates targeting of Atg16L1 to the isolation membrane. EMBO Rep 14(3):284–291. https://doi.org/10.1038/embor.2013.6
31. Fujioka Y, Noda NN, Nakatogawa H, Ohsumi Y, Inagaki F (2010) Dimeric coiled-coil structure of Saccharomyces cerevisiae Atg16 and its functional significance in autophagy. J Biol Chem 285(2):1508–1515. https://doi.org/10.1074/jbc.M109.053520
32. Parkhouse R, Ebong IO, Robinson CV, Monie TP (2013) The N-terminal region of the human autophagy protein ATG16L1 contains a domain that folds into a helical structure consistent with formation of a coiled-coil. PLoS One 8(9):e76237. https://doi.org/10.1371/journal.pone.0076237
33. Bajagic M, Archna A, Busing P, Scrima A (2017) Structure of the WD40-domain of human ATG16L1. Protein Sci. https://doi.org/10.1002/pro.3222
34. Mizushima N, Yoshimori T, Levine B (2010) Methods in mammalian autophagy research. Cell 140(3):313–326. https://doi.org/10.1016/j.cell.2010.01.028

Chapter 4

Structural Studies of Selective Autophagy in Yeast

Akinori Yamasaki, Yasunori Watanabe, and Nobuo N. Noda

Abstract

Budding yeast has been utilized as a model system for studying basic mechanisms of autophagy. The cytoplasm-to-vacuole targeting (Cvt) pathway, which delivers some vacuolar enzymes into the vacuole selectively and constitutively, is one of the most characterized examples of selective autophagy in budding yeast. Here we summarize the methods of X-ray crystallography, NMR, and other biophysical analyses to study the structural basis of the Cvt pathway.

Key words Cvt pathway, X-ray crystallography, NMR, Atg8, Atg19, Ape1

1 Introduction

Budding yeast has been utilized as the most valuable model organism for studying the basic molecular mechanisms of autophagy that includes not only starvation-induced bulk autophagy but also selective types of autophagy [1]. The firstly identified and most extensively characterized type of selective autophagy in budding yeast is the cytoplasm-to-vacuole targeting (Cvt) pathway, in which some vacuolar enzymes such as aminopeptidase I (Ape1) are delivered to the vacuole selectively and constitutively through mechanisms similar to autophagy [2]. The Cvt pathway has been utilized as a model system for studying molecular mechanisms of selective autophagy, through which the first cargo receptor/adaptor in selective autophagy, Atg19, has been identified and the basic concept of selective autophagy, that is, cargo receptors guaranteeing selectivity through interactions with both selective cargoes and autophagy machineries, has been established [3–6]. In the Cvt pathway, a cargo receptor Atg19 binds Atg8, the precursor form of Ape1 (prApe1), and α-mannosidase (Ams1) using the C-terminal WEEL sequence, the coiled-coil (CC) region, and the Ams1 binding domain (ABD), respectively [5, 7, 8]. Atg19 also binds Atg11, a scaffold protein, thereby recruiting autophagy machineries to selective cargoes [9]. Atg34 is a paralog of Atg19 and functions

Nicholas Ktistakis and Oliver Florey (eds.), *Autophagy: Methods and Protocols*, Methods in Molecular Biology, vol. 1880, https://doi.org/10.1007/978-1-4939-8873-0_4, © Springer Science+Business Media, LLC, part of Springer Nature 2019

specifically to the selective autophagy of Ams1 [10]. Another important concept established from the Cvt pathway is that the selective cargo itself functions as a template for autophagosome formation during selective autophagy [11]. In the Cvt pathway, prApe1 forms large aggregates named the Ape1 complex utilizing the self-assembling property of the propeptide, which functions as a template for isolation membrane formation [11–13]. The Ape1 complex is then delivered into the vacuole in a manner dependent absolutely on the autophagic pathway, where prApe1 is processed into a mature form (mApe1) by vacuolar hydrolases [2]. Thus, the Cvt pathway is an ideal model system for studying basic mechanisms of selective autophagy. Recently, the Cvt pathway has also been utilized as a model system for structural biological studies of selective autophagy [13–16]. Here we summarize the methods of X-ray crystallographic and NMR studies as well as in vitro biophysical analyses of the Cvt pathway.

2 Materials

Prepare all solutions using ultrapure water prepared by Milli-Q (Millipore) except for those used for bacterial culture, for which deionized water can be used. We do not add sodium azide to any reagents.

2.1 Sample Preparation

2.1.1 Protein Expression

1. Bacterial expression vectors of glutathione *S*-transferase (GST)-fused proteins (e.g., pGEX-6p-1, GE Healthcare).
2. Bacterial expression vectors of hexa-histidine (6×His)-tagged proteins (e.g., pHT1, which is a modified version of pET28a (+) vector, Novagen, with insertion of sequences coding a 6xHis tag and a TEV protease cleavage site [7]).
3. NEBuilder HiFi DNA Assembly system (New England Biolabs).
4. Thermal cycler.
5. *E. coli* strain BL21 (DE3) for protein expression.
6. 2×YT medium: 16 g/L tryptone, 10 g/L yeast extract, 5 g/L sodium chloride (NaCl).
7. 1000× vitamin mix: 1 mg/mL biotin, 1 mg/mL choline chloride, 1 mg/mL folic acid, 1 mg/mL naiacinamide, 1 mg/mL D-pantothenate, 1 mg/mL pyridoxal, 0.1 mg/mL riboflabine, 1 mg/mL thiamine.
8. M9 medium: 17.1 g/L disodium phosphate dodecahydrate ($Na_2HPO_4 \cdot 12H_2O$), 3.0 g/L monopotassium phosphate (KH_2PO_4), 1.0 g/L [^{15}N]ammonium chloride ($^{15}NH_4Cl$), 0.5 g/L NaCl. Autoclave, and then mix the following: 2 mM

$MgSO_4$, 1 mM $CaCl_2$, 1 μM $FeCl_3$, 0.2% (w/v) D-[^{13}C]glucose, 1× vitamin mix (*see* above; *see* **Note 1**).

9. Minimal medium for the expression of selenomethionine-labeled protein (SeMet medium): 0.50 g/L L-alanine, 0.58 g/L L-arginine hydrochloride, 0.40 g/L L-aspartate, 0.03 g/L L-cystine, 0.67 g/L L-glutamate, 0.33 g/L L-glutamine, 0.54 g/L L-glycine, 0.06 g/L L-histidine, 0.46 g/L L-isoleucine, 0.23 g/L L-leucine, 0.84 g/L L-lysine hydrochloride, 0.13 g/L L-phenylalanine, 0.10 g/L L-proline, 2.08 g/L L-serine, 0.46 g/L L-threonine, 0.17 g/L L-tyrosine, 0.23 g/L L-valine, 0.50 g/L adenine, 0.67 g/L guanosine, 0.17 g/L thymine, 0.50 g/L uracil, 1.50 g/L sodium acetate, 1.50 g/L succinic acid, 0.75 g/L ammonium chloride, 1.08 g/L sodium hydroxide, 10.50 g/L dipotassium phosphate (K_2HPO_4). Autoclave and then mix the following: 0.1% D-glucose, 0.25 g/L $MgSO_4{\cdot}7H_2O$, 4.18 mg/L $FeSO_4{\cdot}7H_2O$, 8.4 μL/L H_2SO_4, 25 mg/L L-selenomethionine, 10 mL/L Kao and Michayluk vitamin solution (Sigma-Aldrich).
10. Isopropyl β-D-1-thiogalactopyranoside (IPTG) for expression induction.
11. Bioshaker (e.g., GBR-200, TAITEC) (*see* **Note 2**).
12. High-speed refrigerated centrifuge (e.g., Model 7000, Kubota).
13. Turbidimeter.

2.1.2 Protein Purification

1. Sonicator (e.g., Sonifier 250, Branson) (*see* **Note 3**).
2. Glutathione-Sepharose 4B (GS4B, GE Healthcare) or GST-accept (Nacalai Tesque).
3. Ni-NTA agarose (QIAGEN).
4. Chromatography column for gravity-flow chromatography (e.g., Econo-Pac columns, Bio-Rad).
5. PreScission protease (GE Healthcare).
6. TEV protease (GE Healthcare).
7. Phosphate buffered saline (PBS).
8. Tris buffer: 20 mM Tris–HCl pH 8.0, 150 mM NaCl.
9. Tris-AS buffer: 20 mM Tris pH 8.0, 500 mM ammonium sulfate.
10. GS4B elution buffer: 50 mM Tris–HCl pH 8.0, 10 mM glutathione (reduced form) (*see* **Note 4**).
11. Ni-NTA wash buffer: 50 mM Tris–HCl pH 8.0, 20 mM imidazone, 500 mM NaCl.
12. Ni-NTA elution buffer: 50 mM Tris–HCl pH 8.0, 200 mM imidazone, 500 mM NaCl.
13. CM-A buffer: 20 mM Tris–HCl pH 8.0.

14. CM-B buffer: 20 mM Tris–HCl pH 8.0, 1 M NaCl.
15. Phosphate buffer: 25 mM sodium phosphate pH 7.0, 100 mM NaCl.
16. HiPrep 26/10 Desalting (GE Healthcare).
17. HiLoad 26/600 Superdex 75 pg (GE Healthcare).
18. HiLoad 26/60 Superdex 200 pg (GE Healthcare).
19. HiTrap CM (GE Healthcare).
20. Liquid chromatography system (e.g., ÄKTA explorer 100, GE Healthcare).
21. Bio-Gel P-6 Desalting Cartridge (Bio-Rad).
22. Centrifugal filters (e.g., Centriprep, Merck-Millipore).

2.2 NMR Measurement and Crystallization

1. NMR spectrometers (e.g., Varian UNITY Inova 600 and 800, Varian).
2. Magnetic susceptibility-matched NMR microtube, 5 mm diameter (Shigemi Inc.).
3. Liquid handling system (e.g., mosquito Crystal, TTP Labtech).
4. Intelli-Plate (Art Robbins) (*see* **Note 5**).
5. Crystal Clear sealing tape (Hampton Research).
6. Crystallization reservoir reagents (summarized in Table 1).
7. Synthesized WEEL peptide (Sigma-Aldrich).
8. Incubator.
9. CryoLoop (Hampton Research).
10. Synchrotron radiation facilities (e.g., SPring-8, Japan, or Photon Factory, Japan).

Table 1
Crystallization conditions

Crystal	Protein concentration	Reservoir
Atg8-Atg19^{WEEL}	13.6 mg/mL of Atg8 0.98 mg/mL of Atg19^{WEEL}	2 M ammonium sulfate and 100 mM sodium citrate (pH 4.2)
mApe1	2.5 mg/mL	30% (v/v) PEG 400, 100 mM magnesium chloride, 1.1 M NaCl and 100 mM Tris–HCl (pH 6.5)
prApe1 L11S	12 mg/mL	50% (v/v) PEG 200, 200 mM magnesium chloride, and 100 mM sodium cacodylate (pH 6.5)
Ape1^{N20}-Atg19^{CC3}	9.3 mg/mL of Ape1^{N20} 13.3 mg/mL of Atg19^{CC3}	30% (w/v) PEG 4000, 200 mM sodium acetate, and 100 mM Tris–HCl (pH 8.5)
Ape1^{N22} P22L	25.3 mg/mL	500 mM magnesium sulfate and 100 mM sodium acetate (pH 4.6)

2.3 In Vitro Experiments

1. Zetasizer Nano S (Malvern).
2. Quartz batch cuvette ZEN2112.
3. J-720W spectropolarimeter (Jasco Co.).
4. Optima XL-I (Beckman Coulter).
5. An50Ti rotor (Beckman Coulter).
6. Charcoal-filled Epon double-sector centerpiece.

3 Methods

Carry out all procedures at room temperature unless otherwise specified.

3.1 Sample Preparation

All proteins can be expressed and purified from *E. coli* BL21 (DE3). Three-dimensional NMR experiments for structural calculation and sequential main chain assignment require uniformly $^{15}N,^{13}C$-labeled protein. The crystal structure determination of the $Ape1^{N20}$-$Atg19^{CC}$ complex and $Ape1^{N22}$ P22L requires selenomethionine-labeled $Atg19^{CC}$ and platinum-derivatized crystals, respectively. Proteins can be purified by affinity-tag-based chromatography followed by either size-exclusion or cation exchange chromatography. Used affinity-tags, columns, and buffer solution for each protein are listed in Table 2.

3.1.1 Protein Expression

1. Insert each gene into pGEX-6p-1 vector (GE healthcare) or pHT1 using NEBuilder HiFi DNA Assembly (New England Biolabs) or traditional cloning techniques using restriction enzymes and ligases (*see* **Note 6**). Add tyrosine residue at the C-terminus of Atg19 CC2 and CC3 in order to make it easy to calculate protein concentration based on A280. Introduce point mutations by PCR-based site-directed mutagenesis.
2. Inoculate 10 mL of 2×YT medium with ampicillin (100 μg/mL) or kanamycin (50 μg/mL) with a single colony of BL21 (DE3) carrying the expression plasmid and grow overnight at 37 °C with reciprocal shaking (*see* **Note 7**).
3. Transfer 10 mL of overnight culture into 1 L of 2×YT, M9 or SeMet medium containing 100 μg/mL ampicillin or 50 μg/mL kanamycin for production of native, uniformly $^{15}N,^{13}C$-labeled or selenomethionine-labeled proteins, respectively.
4. Grow at 37 °C with orbital shaking until cells reach OD_{600} of 0.5–1.0 (approx. 2.5 h) (*see* **Note 8**).
5. Reduce the incubator temperature to 16–25 °C, and induce protein expression by addition of 0.01–0.5 mM IPTG (*see* **Note 9**).

Table 2
List of the plasmids, columns, and buffer used for purification of each protein

Protein	Purpose	Plasmid	Tag	Second column	Buffer
Atg8	Crystal	pHT1	6×His	HiTrap CM	CM-A/CM-B buffer
Atg19 WT	Analysis	pGEX-6p-1	GST	Superdex 200 26/60	Tris buffer
Atg19 CC1 (153–191)	Analysis	pGEX-6P-1	GST	Superdex 75 26/60	Tris buffer
Atg19 CC2 (156–187)	NMR	pGEX-6p-1	GST	Superdex 75 26/60	Tris buffer
Atg19 CC3 (160–187)	Crystal	pGEX-6p-1	GST	Superdex 75 26/60	Tris buffer
Atg19 ABD (254–367)	NMR	pGEX-6p-1	GST	Superdex 75 26/60	Tris buffer
Atg34 ABD (246–358)	NMR	pGEX-6p-1	GST	Superdex 75 26/60	Tris buffer
prApe1 WT	Analysis	pGEX-6p-1	GST	Superdex 200 26/60	Tris buffer
prApe1 L11S	Crystal	pGEX-6p-1	GST	Superdex 200 26/60	Tris buffer
mApe1	Crystal	pGEX-6p-1	GST	Superdex 200 26/60	Tris buffer
Ape1 (1–45) WT	Analysis	pGEX-6p-1	GST	Superdex 75 26/60	Tris buffer
Ape1 (1–45) L11S	Analysis	pGEX-6p-1	GST	Superdex 75 26/60	Tris buffer
Ape1 (1–45) P22L	Analysis	pGEX-6p-1	GST	Superdex 75 26/60	Tris buffer
Ape1 (1–20) WT	Crystal	pGEX-6p-1	GST	Superdex 75 26/60	Tris buffer
Ape1 (1–22) P22L	Crystal	pGEX-6p-1	GST	Superdex 75 26/60	Tris-AS buffer

6. After overnight induction at 16–25 °C with orbital shaking, collect cells by centrifugation at 4000 × *g* for 15 min at 4 °C.
7. Suspend cell pellets in PBS or Tris buffer containing 5 mM DTT, 0.5 mM EDTA, and 0.1 mM PMSF for GST-fused proteins, or in Ni-NTA wash buffer with 0.1 mM PMSF for 6xHis-tagged proteins in a volume of ~30 mL per cell pellet from 1 L culture and store at −80 °C until needed.

3.1.2 Protein Purification

1. After thawing cell pellets, disrupt cells by sonication on ice water (600 × 0.2 s pulses with 0.8 s intervals between pulses) (*see* **Note 10**) and centrifuge at 35,000 × *g* for 40 min at 4 °C.
2. Load the cleared lysate onto an affinity resin column (bed volume of ~5 mL) pre-equilibrated with each buffer, and allow the lysate to flow through.
3. Wash the column two times with five column volumes of PBS for GS4B/GST-accept or Ni-NTA wash buffer for Ni-NTA.
4. Elute the bound protein by adding two to three column volumes of each elution buffer, and collect the eluate.
5. For buffer exchange, load the eluate onto a HiPrep 26/10 Desalting column equilibrated with Tris buffer or Bio-Gel

P-6 Desalting Cartridge equilibrated with PBS using a liquid chromatography system.

6. Collect the peak fractions, add ~0.1 mg of PreScission protease to cleave GST-tag, and incubate overnight at 4 °C. In the case of 6×His-tagged proteins, add ~0.1 mg of TEV protease, and incubate overnight at 25 °C (*see* **Note 11**).
7. To remove PreScission protease, cleaved GST-tag and any non-cleaved protein, load the protein solution onto a new GS4B gravity-flow column (bed volume of ~5 mL) pre-equilibrated with Tris buffer or PBS and collect the flow-through fraction (*see* **Note 12**).
8. Purify the flow-through fraction of the digested GST-fused proteins over a HiLoad 26/600 Superdex 75 pg or Superdex 200 pg gel filtration column in Tris buffer (or phosphate buffer for NMR experiments) using liquid chromatography system, and collect peak fractions. In the case of the digested 6×His-tagged protein, purify using a HiTrap CM cation exchange column using a linear gradient of the CM-A and CM-B buffers using liquid chromatography system (*see* **Note 13**). Analyze the elution fractions by SDS-PAGE and CBB staining to check the purity of eluted protein.

3.2 NMR Experiments

3.2.1 NMR Measurements and Structural Calculation

Atg19 and Atg34 ABDs are monomeric, small (~12 kDa), and stable proteins and thus suitable for structurer determination by NMR. The following protocol describes solution structure determination procedures for Atg19 and Atg34 ABDs [8] and will be applicable to solution structure determination for other proteins like ABDs.

1. Prepare approx. 250 μL of a solution of ^{15}N,^{13}C-labeled protein at 0.7–1.0 mM in phosphate buffer.
2. Add 1 μL of 5 mg/mL 2,2-dimethyl-2-silapentane-5-sulfonate (DSS) for chemical shift referencing, 20 μL of D_2O for field-frequency locking, and 230 μL of the protein solution to a magnetic susceptibility-matched NMR tube.
3. Measure NMR spectra described below at 25 °C. For assignment of the protein backbone resonances, two-dimensional ^{1}H-^{15}N HSQC and three-dimensional HNCA, HN(CO)CA, HNCACB, CBCA(CO)NH, HNCO, HNCAHA, and HBHA (CO)NH spectra. For assignment of the aliphatic side chain atoms, two-dimensional ^{1}H-^{13}C HSQC and three-dimensional HCCH TOCSY spectra. For assignment of the aromatic side chain atoms, two-dimensional ^{1}H-^{13}C HSQC, (Hb)Cb (CgCd)Hd, and (Hb)Cb(CgCdCe)He and three-dimensional HCCH TOCSY spectra. For structural calculation, three-dimensional ^{15}N-edited NOESY and ^{13}C-edited NOESY spectra using a 100 ms mixing time (in case of Atg19 and Atg34

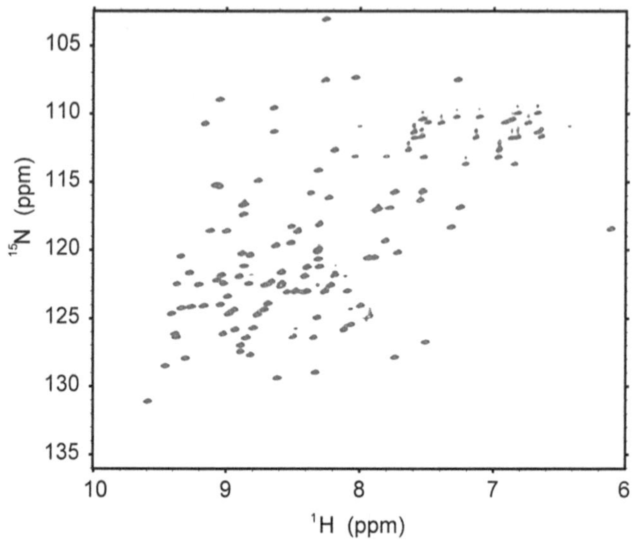

Fig. 1 Example of a typical ^{1}H-^{15}N HSQC spectrum of Atg19 ABD suitable for structure determination by NMR. The signals are well-dispersed and have similar intensities, indicating that this protein forms a rigid structure

ABDs). Spectral processing system (e.g., NMRPipe program [17]) is necessary for data process. Make ^{1}H, ^{13}C, and ^{15}N resonance assignments using NMR assignment and integration program (e.g., Sparky, T. D. Goddard and D. G. Kneller, SPARKY 3, University of California, San Francisco). A typical ^{1}H-^{15}N HSQC spectrum suitable for structure determination by NMR is shown in Fig. 1.

4. Integration of the cross peak intensities on the NOESY spectra and structural calculation can be performed by Sparky and the CANDID/CYANA 2.1 program [18], respectively.

3.2.2 NMR Titration Experiments

The following protocols describe Ape1^{N45}-Atg19^{CC2} interaction analysis procedures using NMR spectroscopy.

1. Prepare approx. 250 μL of a solution of ^{15}N,^{13}C-labeled protein (Ape1^{N45}) at 25 μM in phosphate buffer and another solution of other native protein (Atg19^{CC2}) at 500 μM in phosphate buffer.
2. Using **steps 2** and **3** of Subheading 3.2.1 (*see* above), make Ape1^{N45} backbone resonances assignment.
3. Add 1 μL of 5 mg/mL DSS for chemical shift referencing, 20 μL of D_2O for field-frequency locking, and 230 μL of the uniformly labeled protein solution to a magnetic susceptibility-matched NMR tube, and measure a two-dimensional ^{1}H-^{15}N HSQC spectrum.

4. Remove the NMR tube from NMR spectrometer, and add an appropriate amount of Atg19^{CC2} to the NMR tube. A starting point is 12.5 μM (0.5 molar equivalent of Atg19^{CC2}).
5. Measure a two-dimensional ^{1}H-^{15}N HSQC spectrum again. Plot the ^{1}H-^{15}N HSQC spectrum, and check for peaks that have changed intensity or position after addition of Atg19^{CC2}. In case of Ape1^{N45}-Atg19^{CC2} interaction analysis, some peaks have changed intensity after addition of Atg19^{CC2}.
6. Repeat **steps 3** and **4** twice with addition of more Atg19^{CC2} (1.0 and 2.0 molar equivalent of Atg19^{CC2}).
7. Plot peak intensity ratios of Ape1^{N45} residues in the presence of 2.0 molar equivalent of Atg19^{CC2} to Ape1^{N45} alone, and identify residues of Ape1^{N45} involved in the interaction with Atg19^{CC2}.

3.3 X-ray Crystallography

3.3.1 Crystallization

The following protocol describes crystallization procedures of Atg8-Atg19^{WEEL}, prApe1 L11S, mApe1, Ape1^{N20}-Atg19^{CC3}, and Ape1^{N22} P22L. All crystallization experiments can be performed by the sitting-drop vapor diffusion method at 20 °C. The procedures will be applicable to solution structure determination for other proteins. Protein concentration and reservoir condition for crystallization are listed in Table 1 [13, 19].

1. Prepare ~40 μL of a solution of 9.3 mg/mL native Ape1^{N20} and 13.3 mg/mL selenomethionine-labeled Atg19^{CC3} (in a molar ratio of 1:1) per crystallization screen (96 conditions) to be tested. Crystal Screen, Index HT, PEG RX, Salt RX (Hampton Research), and Wizard Classic 1/2 (Molecular Dimensions) can be used as initial screening reagents.
2. Set up 96-well plates by depositing sitting drops of 0.3 μL protein solution and equal amounts of crystallization screen solutions using a liquid handling system. Incubate at 20 °C.
3. Monitor for crystal growth daily.
4. If initial crystals are obtained, optimize conditions around the original hit solution by varying pH, salt concentration, precipitant concentration, as well as protein concentration. Second screen can be done by either hand or the liquid handling system. An example of obtained crystals is shown in Fig. 2.

3.3.2 Diffraction Data Collection and Structure Determination

Diffraction data collection experiments can be done in synchrotron radiation facilities such as Photon Factory, Japan, or SPring-8, Japan.

1. Soak obtained crystals in the reservoir solution supplemented with 10–30% glycerol as a cryoprotectant. Pick up a crystal using a mounted CryoLoop (Hampton Research), and then immediately place on the goniometer of beamline for flash-

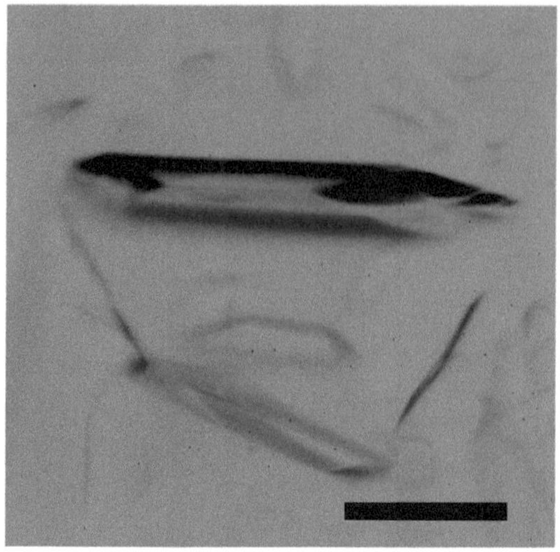

Fig. 2 Example of Ape1^{N22} P22L crystals suitable for structure determination by X-ray crystallography. The bar indicates 0.1 mm

cooling. Keep the crystal in a stream of nitrogen gas at −178 °C during data collection.

2. Collect peak anomalous diffraction data from a selenomethionine-labeled protein crystal on a charge-coupled device (CCD) detector using a synchrotron beamline (wavelength at 0.9792 Å) (*see* **Note 14**). HKL2000 program suite [20] or XDS [21] can be used for indexing, integrating, and scaling the diffraction data.
3. Structure determination can be performed by the molecular replacement method for Atg8-Atg19^{WEEL}, prApe1 L11S, and mApe1 using PDB models of GABARAP (PDB 1GNU), mApe1 (5JGF), and *Borrelia burgdorferi* B31 (PDB 1Y7E), respectively. Structure determination of the Ape1^{N20}-Atg19^{CC3} complex can be performed by the single-wavelength anomalous dispersion method using the peak data of the selenomethionine-labeled crystal using Phenix program suite [22], and that of the propeptide of Ape1 can be performed by the single isomorphous replacement with anomalous scattering (SIRAS) using data from native and platinum-derivatized crystals using HKL2MAP graphical user interface [23] with SHELX C/D/E program [24]. Crystallographic refinement can be performed using REFMAC5 in CCP4, Phenix, or CNS [25] (*see* **Note 15**), and manual model building can be performed using the COOT program [26].

3.4 Dynamic Light Scattering Measurement

Dynamic light scattering can measure the average size of protein particles in solution. It can be applied to detect the size of prApe1 aggregates with and without Atg19^{CC}.

1. Prepare to purify the proteins in 20 mM Tris pH 8.0 and 150 mM NaCl by size-exclusion chromatography.
2. Centrifuge 5 μM of the proteins 15,000 × *g* for 10 min at 4 °C (*see* **Note 16**).
3. Incubate the proteins with or without 1 μg of PreScission protease and 10 μM Atg19^{CC1} for 1 h at room temperature.
4. Apply the samples to a quartz batch cuvette ZEN2112.
5. Measure the samples by Zetasizer Nano S (Malvern) at 25 °C (*see* **Note 17**).
6. Make an average from three measurements.

3.5 Circular Dichroism Spectroscopy

Circular dichroism spectroscopy can measure the approximate ratio of a random coil, helix, and β-strand in a protein in solution. It can be applied to study the conformation of the propeptide of Ape1 which can be random-coiled or α-helical dependent on the buffer condition, point mutation, or complex formation with Atg19^{CC}.

1. Dilute purified proteins to the concentration of A280 = 0.1 in PBS diluted to four times with water (*see* **Note 18**).
2. Load the sample into a quartz cuvette.
3. Measure the samples by J-720W spectropolarimeter (Jasco Co.) at 20 °C between 200 and 250 nm with a bandwidth of 1 nm, a response time of 4 s, and a scanning speed of 50 nm/min.
4. Obtain a spectrum by averaging 16 consecutive scans.

3.6 Analytical Ultracentrifugation

Analytical ultracentrifugation, especially the sedimentation velocity experiments, can measure the shape, molar mass, and size distribution of proteins in solution. It can be applied to detect the molar mass of the Ape1^{N45}-Atg19^{CC2} complex in solution.

1. Dilute protein samples to the concentration of A280 = 0.50 in Tris buffer.
2. Load the sample and Tris buffer to the sample and reference sectors of charcoal-filled Epon double-sector centerpieces, respectively (*see* **Note 19**).
3. Set the centerpieces to an An50Ti rotor, and centrifuge at 50,000 rpm (182,000 × *g* at the center of the sample cell) at 20 °C.
4. Obtain the sedimentation coefficient distribution function, c(s), using the SEDFIT program.
5. Calculate partial specific volumes of proteins based on their amino acid sequences using the Sednterp program.

6. Calculate the solvent density and viscosity using the Sednterp program.
7. Obtain molecular weights of the protein by converting c(s) to c (M) with SEDFIT based on Svedberg equation using the obtained sedimentation coefficients and the diffusion coefficients (*see* **Note 20**).

4 Notes

1. Autoclaving the reagents together will cause reaction between glucose and amino acids and turn into brown. Precipitation will also happen.
2. Bioshaker specifications require up to 200 rpm rotation capability and both heating and cooling capabilities in order to keep the induction temperature.
3. When the efficiency of cell disruption is reduced, polish or change the flat tip at the head of the horn (refer to the manufacturer's manual).
4. Glutathione is easy to be oxidized. Prepare the solution before use, or store the solution at −20 °C or below.
5. Low-profile type of Intelli-Plate is useful for saving space during incubation.
6. When constructing plasmids, it is important to reduce the number of artificial amino acids attached to the protein as small as possible.
7. Alternatively, multiple colonies can be used for inoculation when not wanting to perform colony selection.
8. Cell growth is slower when kanamycin is used as an antibiotic.
9. Optimize the concentration of IPTG in order to maximize the amount of soluble protein.
10. Use ice water instead of just ice in order to increase the cooling efficiency.
11. When in hurry, increase the concentration of the protease.
12. After the application is ended, wash the column with 5 mL or more PBS, and collect it in order to increase the recovery.
13. Increase the percentage of the B buffer from 0 to 40% over 20–30 column volumes for making a gradient.
14. The optimum peak wavelength should be determined by measuring XAFS. When XAFS cannot be measured, use a little shorter wavelength (e.g., 0.978 Å).
15. When using multiple software for refinement, be careful not to change freeR flag in the diffraction dataset.

16. When measuring smaller proteins, increase the concentration in order to obtain enough scattering.
17. Before starting measurement, incubate the cuvette for 60 s or more at 25 °C.
18. High ion strength is not permitted to measure the spectrum at shorter wavelength (~200 nm).
19. As a reference buffer, the same lot of the buffer that is used for protein preparation should be used. It is better to use external dialysis solution of the protein as a reference buffer.
20. When multiple components exist in one sample, the determined molecular weight is less accurate.

Acknowledgments

This work was supported by Japan Society for the Promotion of Sciences KAKENHI [grant numbers 25111001, 25111004 to N. N.N., 17K18339 to A.Y.], CREST, Japan Science and Technology Agency [grant number JPMJCR13M7 to N.N.N.], and The Naito Foundation [to N.N.N.].

References

1. Mizushima N, Yoshimori T, Ohsumi Y (2011) The role of Atg proteins in autophagosome formation. Annu Rev Cell Dev Biol 27:107–132
2. Lynch-Day MA, Klionsky DJ (2010) The Cvt pathway as a model for selective autophagy. FEBS Lett 584:1359–1366
3. Scott SV, Guan J, Hutchins MU, Kim J, Klionsky DJ (2001) Cvt19 is a receptor for the cytoplasm-to-vacuole targeting pathway. Mol Cell 7:1131–1141
4. Leber R, Silles E, Sandoval IV, Mazon MJ (2001) Yol082p, a novel CVT protein involved in the selective targeting of aminopeptidase I to the yeast vacuole. J Biol Chem 276:29210–29217
5. Shintani T, Huang WP, Stromhaug PE, Klionsky DJ (2002) Mechanism of cargo selection in the cytoplasm to vacuole targeting pathway. Dev Cell 3:825–837
6. Suzuki K, Kamada Y, Ohsumi Y (2002) Studies of cargo delivery to the vacuole mediated by autophagosomes in Saccharomyces cerevisiae. Dev Cell 3:815–824
7. Noda NN et al (2008) Structural basis of target recognition by Atg8/LC3 during selective autophagy. Genes Cells 13:1211–1218
8. Watanabe Y, Noda NN, Kumeta H, Suzuki K, Ohsumi Y, Inagaki F (2010) Selective transport of alpha-mannosidase by autophagic pathways: structural basis for cargo recognition by Atg19 and Atg34. J Biol Chem 285:30026–30033
9. Yorimitsu T, Klionsky DJ (2005) Atg11 links cargo to the vesicle-forming machinery in the cytoplasm to vacuole targeting pathway. Mol Biol Cell 16:1593–1605
10. Suzuki K, Kondo C, Morimoto M, Ohsumi Y (2010) Selective transport of alpha-mannosidase by autophagic pathways: identification of a novel receptor, Atg34p. J Biol Chem 285:30019–30025
11. Shintani T, Klionsky DJ (2004) Cargo proteins facilitate the formation of transport vesicles in the cytoplasm to vacuole targeting pathway. J Biol Chem 279:29889–29894
12. Oda MN, Scott SV, Hefner-Gravink A, Caffarelli AD, Klionsky DJ (1996) Identification of a cytoplasm to vacuole targeting determinant in aminopeptidase I. J Cell Biol 132:999–1010
13. Yamasaki A et al (2016) Structural basis for receptor-mediated selective autophagy of aminopeptidase I aggregates. Cell Rep 16:19–27
14. Bertipaglia C et al (2016) Higher-order assemblies of oligomeric cargo receptor complexes

form the membrane scaffold of the Cvt vesicle. EMBO Rep 17:1044–1060

15. Su MY et al (2015) Structure of yeast Ape1 and its role in autophagic vesicle formation. Autophagy 11:1580–1593
16. Yamasaki A, Noda NN (2017) Structural biology of the Cvt pathway. J Mol Biol 429:531–542
17. Delaglio F, Grzesiek S, Vuister GW, Zhu G, Pfeifer J, Bax A (1995) NMRPipe: a multidimensional spectral processing system based on UNIX pipes. J Biomol NMR 6:277–293
18. Guntert P (2004) Automated NMR structure calculation with CYANA. Methods Mol Biol 278:353–378
19. Adachi W, Suzuki NN, Fujioka Y, Suzuki K, Ohsumi Y, Inagaki F (2007) Crystallization of Saccharomyces cerevisiae aminopeptidase 1, the major cargo protein of the Cvt pathway. Acta Crystallogr Sect F Struct Biol Cryst Commun 63:200–203
20. Otwinowski Z, Minor W (1997) Processing of X-ray diffraction data collected in oscillation mode. Methods Enzymol 276:307–326
21. Kabsch W (2010) Xds. Acta Crystallogr D Biol Crystallogr 66:125–132
22. Adams PD et al (2010) PHENIX: a comprehensive python-based system for macromolecular structure solution. Acta Crystallogr D Biol Crystallogr 66:213–221
23. Pape T, Schneider TR (2004) HKL2MAP: a graphical user interface for macromolecular phasing with SHELX programs. J Appl Crystallogr 37:843–844
24. Sheldrick GM (2010) Experimental phasing with SHELXC/D/E: combining chain tracing with density modification. Acta Crystallogr D Biol Crystallogr 66:479–485
25. Brunger AT et al (1998) Crystallography & NMR system: a new software suite for macromolecular structure determination. Acta Crystallogr D Biol Crystallogr 54:905–921
26. Emsley P, Lohkamp B, Scott WG, Cowtan K (2010) Features and development of Coot. Acta Crystallogr D Biol Crystallogr 66:486–501

Chapter 5

Biophysical Studies of LC3 Family Proteins

Javier H. Hervás, Zuriñe Antón, and Alicia Alonso

Abstract

Autophagy is an important cellular process in which cell components are degraded in a controlled way and their building blocks are recycled into new macromolecules. Autophagy starts within a double-membrane container, the autophagosome, itself the result of a number of interconversions of cell membranous elements. In our recent work, we have described reconstituted model systems for the interactions of autophagy proteins with membrane lipid bilayers and for the autophagy protein-mediated vesicle tethering and fusion, with the aim of ultimately reconstituting the autophagosome formation. The present chapter describes in detail (a) the steps required for the preparation of semisynthetic lipid vesicles (liposomes), including giant unilamellar vesicles, (b) ultracentrifugation and fluorescence methods for assaying protein binding to membranes, and (c) procedures for assessing vesicle–vesicle aggregation and fusion. The latter include methods for intervesicular total lipid mixing, mixing of lipids in the vesicle inner monolayers, and aqueous contents mixing.

Key words LC3/GABARAP proteins, Autophagy proteins, Membrane fusion, Liposomes, Fusion assays, Membrane biophysics, Protein–lipid interactions

1 Introduction

Biophysical studies of protein–lipid interactions and processes guided by them may have a huge impact on the understanding of molecular mechanisms of cellular functions. Among the latter, autophagy comprises the formation, elongation, and closure of a double-membrane structure called autophagosome in which unused or damaged cell materials are engulfed for degradation. During this event, a large variety of proteins are involved, some of them being in contact with the lipid matrix of the newly formed structure. This is the case of the LC3/GABARAP protein family that has been described to contribute to membrane tethering and fusion events along the autophagosome formation process and subsequent fusion with lysosomes. Moreover, lipid geometry and

Javier H. Hervás and Zuriñe Antón contributed equally to this work.

Nicholas Ktistakis and Oliver Florey (eds.), *Autophagy: Methods and Protocols*, Methods in Molecular Biology, vol. 1880, https://doi.org/10.1007/978-1-4939-8873-0_5, © Springer Science+Business Media, LLC, part of Springer Nature 2019

membrane curvature also modulate the fusion process induced by these proteins [1]. However, the origin of the membrane material and the precise mechanism of the fusion process itself are still unclear and objects of debate [2–4].

Moreover, the phospholipid cardiolipin (CL) has been proposed to play a role in selective mitochondrial autophagy (mitophagy). Indeed during rotenone-induced mitophagy, CL externalization to the outer mitochondrial membrane can act as a signal for LC3 binding [5]. The latter would underpin both mitochondrial recognition and autophagosome formation, ultimately leading to removal of damaged mitochondria. A comprehensive in vitro study of LC3B, GABARAP, and GABARAPL2 interaction with CL-enriched membranes shows that LC3B interacts preferentially with CL over other di-anionic lipids, that the membrane binding capacity of LC3B resides in its N-terminus, and that GABARAP, but not GABARAPL2, interacts with CL in vitro [6].

A variety of biophysical assays for the study of protein–lipid interactions and characterization of the membrane fusion process induced by the LC3/GABARAP proteins will be explained in detail in this chapter.

1.1 Model Membrane Systems for the Study of Mammalian Autophagy

For exploring how proteins of the LC3/GABARAP family interact with and perform their function in a membrane environment, model lipid membranes are used in vitro. This technique allows the creation of membranes with a defined lipid composition characteristic of specific organelles or membrane regions. The use of lipid vesicles is essential when studying the effect of perturbing agents, autophagy proteins in this case, on membrane architecture, or the impact of different membrane characteristics on protein activity. Lipid vesicles, depending on their size, can be classified into small unilamellar vesicles (SUVs), large unilamellar vesicles (LUVs) (*see* Subheading 3.1), or giant unilamellar vesicles (GUVs). Of note, GUVs' size, in the order of 5–100 micrometers, allows performing direct microscopy on individual vesicles (*see* Subheading 3.2).

1.2 In Vitro Protein–Lipid Interaction Studies of ATG Proteins

Liposome flotation assays are widely applied in studying the affinity and lipid specificity of protein binding to membranes [7, 8]. In the liposome float-up experiment, the vesicles and bound proteins are enriched by density gradient centrifugation (Subheading 3.4). Moreover, the results obtained from protein–lipid overlay (PLO) assays provide a good qualitative indication of the relative affinity of different proteins for the lipid(s) with which they interact (*see* Subheading 3.5). The main advantage of the PLO assay over the other binding assays is that it allows the simultaneous analysis of the lipid binding specificities of many protein samples for a range of naturally occurring lipids.

In addition, approaches based on fluorescence spectroscopy are also useful for studying many aspects of lipid–protein interactions, due to their intrinsic sensitivity, suitable timescale, and minimal perturbation effects [9]. Extrinsic probes can be used for studying protein–lipid interactions, meaning that the lipid and/or protein is conjugated to a fluorescent dye. IANBD, a thiol-specific iodoacetamide derivative of 7-nitrobenz-2-oxa-1,3-diazol-4-yl (NBD), was chosen as one of the fluorescent probes in our studies with Atg8 human orthologues (*see* Subheading 3.6). Since the protein is labeled with a unique NBD moiety at a single location, this approach also provides an opportunity to characterize the structural changes that occur at specific sites of the protein when associated with the membrane.

1.3 Insights into Membrane Fusion Induced by LC3/GABARAP Family Proteins

Vesicle tethering and fusion induced by GABARAP and GABARAPL2 in the context of autophagosome elongation can be systematically studied by a combination of aggregation, lipid mixing, aqueous contents mixing, and leakage assays. Fluorescence is one of the most sensitive spectroscopic techniques that allows real-time characterization of events occurring in membranes such as fusion, aggregation, or permeation. It has been applied previously to study the fusogenic activity associated with hydrolytic enzymes sphingomyelinase (SMase) and phospholipase C (PLC) [10].

The first step in membrane fusion is to bring membranes together in close proximity, in order to remove the water between the two membranes. This process is called membrane tethering or vesicle aggregation, and it can be easily monitored through the increase in suspension turbidity (Subheading 3.7). Once this step has been accomplished, the membrane is destabilized and undergoes first membrane hemifusion of the outer monolayers and then the formation of an aqueous fusion pore by membrane hemifusion of the inner monolayers. These two events can be measured by fluorescence resonance energy transfer (FRET), a distance-dependent interaction between the electronic excited states of two dye molecules in which excitation is transferred from a donor (NBD) molecule to an acceptor (rhodamine, Rho) molecule without emission of a photon. In order to study vesicle–vesicle fusion, lipid mixing between vesicles can be monitored using a FRET pair of fluorescently labeled lipids, a methodology developed by Struck et al. [11]. 7-Nitrobenz-2-oxa-1,3-diazol-4-yl (NBD) is used as the energy donor and rhodamine (Rho) as the energy acceptor. Each of the two fluorophores is coupled to the free amino group of a phosphatidylethanolamine to provide an analogue that can be incorporated into a lipid vesicle bilayer. When both fluorescent lipids are in LUVs at the appropriate surface densities (ratio of fluorescent lipid to total lipid), efficient energy transfer is observed. When such vesicles are fused with a population of unlabeled vesicles by the addition of GABARAP or GABARAPL2, the two probes mix with the other lipids present, getting diluted in the resulting new membrane. Thus, mixing reduces the surface density of the fluorophores, and the energy transferred between molecules decreases, and the fusion can be measured as an increase of NBD (donor) fluorescence (*see* Subheadings 3.8 and 3.9).

In the vesicle–vesicle fusion process, a pore connecting the vesicle inner spaces is opened, and therefore intervesicular mixing of aqueous contents should occur. This can be studied by using the ANTS/DPX fluorescence quenching couple (Subheading 3.10). It should be noted that the extent of contents mixing corresponds roughly with that of inner monolayer lipid mixing. The protein-induced fusion process could be also studied in terms of the concomitant vesicle leakage. During membrane fusion, proteins cannot always maintain the vesicle's aqueous contents separated from the aqueous environment. This could be addressed by a leakage assay in which the ability of particular molecules to permeabilize the lipid bilayer is tested by following the externalization of the well-known low-molecular-weight ANTS fluorescent molecule and its quencher DPX (*see* Subheading 3.11). This approach was initially developed in 1985 to measure vesicle fusion [12]. Most biological fusion events occur in such a way that vesicle contents do not leak out. This is also the case in GABARAP- and GABARAPL2-induced fusion processes.

2 Materials

2.1 Components for Lipid Model Vesicle Preparation

1. Egg phosphatidylcholine (PC), 1,2-dioleoyl-sn-glycero-3-phosphoethanolamine (DOPE), bovine liver phosphatidylinositol (PI), heart bovine cardiolipin (CL), brain phosphatidylserine (PS), egg phosphatidylglycerol (PG), egg phosphatidic acid (PA), 1,2-dioleoyl-sn-glycero-3-phosphoethanolamine-N-(lissamine rhodamine B sulfonyl) (Rho-PE), ovine wool cholesterol (Chol), and 1,2-dihexadecanoyl-sn-glycero-3-phosphocholine (DPPC).
2. Chloroform/methanol solution (2:1 v/v).
3. Liposomes hydration buffer:
   (a) PreScission buffer [50 mM Tris–HCl (pH 7.5), 150 mM NaCl, 1 mM EDTA] with freshly added 1 mM DTT.
   (b) UBL buffer: 50 mM Tris–HCl, pH 7.5, 150 mM NaCl, and 1 mM MgCl2, supplemented with 1 mM DTT.
4. 0.05 and 0.1 μm polycarbonate filters for vesicle extrusion.
5. High-pressure extruder [LIPEX Liposome Extrusion System (Transferra Nanosciences, Burnaby, Canada)] with gas flow.
6. Platinum electrodes in a specially designed Teflon chamber (Industrias Técnicas ITC, Bilbao; *see* Fig. 1).

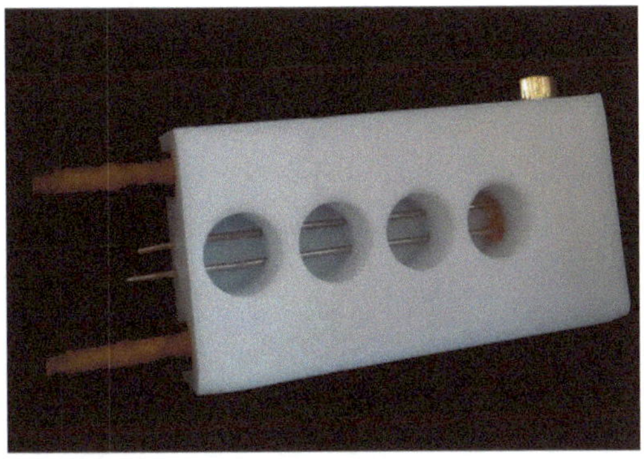

Fig. 1 Homemade chamber and PTFE circular wells for the formation of GUV in solution

7. 300 mM sucrose solution.
8. Glass coverslips.
9. Vacuum grease.
10. Electric field generator.
11. Eight-well microscopy chambers.
12. Fatty acid free BSA (10 mg/mL) in distilled water.
13. Confocal microscope (Nikon D-ECLIPSE C1, Nikon, Melville, NY).

2.2 Components for Phospholipid Assay

1. 1 mM NaH_2PO_4 standard solution.
2. 20-mL glass tubes.
3. Hamilton microsyringes for volumes lower than 10 μL of phospholipids in chloroform/methanol.
4. 60% perchloric acid ($HClO_4$).
5. Ammonium heptamolybdate solution [$(NH_4)_6Mo_7O_{24}.4H_2O$]. 22 g of ammonium heptamolybdate dissolved in 143 mL of H_2SO_4 and 857 mL of distilled water.
6. 10% (v/v) ascorbic acid.
7. Heating block (205 °C).
8. Spectroscopy cuvettes.
9. Spectrophotometer.

2.3 Components for Equilibrium Sucrose Gradient Centrifugation

1. 3 mM liposome solutions [PC:DOPE (80:20 mol ratio) and PC:DOPE:CL (50:20:30 mol ratio)].
2. PreScission buffer: 50 mM Tris–HCl, pH 7.5, 150 mM NaCl, and 1 mM EDTA with freshly added 1 mM DTT.
3. 2.4 M sucrose stock solution prepared in PreScission buffer.
4. 0.8 M and 0.5 M sucrose solutions.

5. 1-mL polycarbonate tubes (Beckman Coulter).
6. Beckman Optima TLX benchtop ultracentrifuge with a TLA-120.2 rotor (Beckman Coulter).
7. 500-μL Hamilton syringe.
8. 6× loading buffer.
9. SDS-polyacrylamide gels.
10. Anti-LC3 monoclonal antibody (Medical & Biological Laboratories Co. Ltd., M186-3), anti-ATG3 polyclonal antibody [Santa Cruz Biotechnology, Inc. (Dallas, TX), sc-70139], anti-GABARAP monoclonal antibody (Medical & Biological Laboratories Co. Ltd., M135-3), and anti-GABARAPL2 polyclonal antibody (Medical & Biological Laboratories Co. Ltd., PM038).
11. Goat anti-mouse IgG-HRP secondary antibody [Santa Cruz Biotechnology, Inc. (Dallas, TX), sc-2005] and goat anti-rabbit IgG-HRP secondary antibody [Santa Cruz Biotechnology, Inc. (Dallas, TX), sc-2004].

2.4 Components for Protein–Lipid Overlay Assay

1. Methanol/chloroform/water (2:1:0.8 v/v/v) solution.
2. Hybond-C nitrocellulose membrane (GE Healthcare).
3. Fat-free powder milk.
4. Phosphate-buffered saline (PBS).
5. PreScission buffer: 50 mM Tris–HCl, pH 7.5, 150 mM NaCl, and 1 mM EDTA with freshly added 1 mM DTT.
6. Anti-LC3 monoclonal antibody (Medical & Biological Laboratories Co. Ltd., M186-3).
7. Goat anti-mouse IgG-HRP secondary antibody [Santa Cruz Biotechnology, Inc. (Dallas, TX), sc-2005].
8. Supersignal West Pico Chemiluminescent Substrate (Thermo Fisher Scientific).
9. Amersham Hyperfilm ECL (GE Healthcare, UK) and a Curix 60 processor (Agfa, Belgium).

2.5 Components for NBD Fluorescence Spectroscopy

1. 1 mM TCEP in PreScission buffer (50 mM Tris–HCl, 150 mM NaCl, 1 mM EDTA, pH 7.5).
2. PD-10 desalting columns (GE Healthcare).
3. Labeling buffer (10 mM HEPES-HCl, pH 7.3, 100 mM KCl, 1 mM EDTA).
4. IANBD (Molecular Probes, Eugene, OR, USA), dissolved in DMSO.
5. 3 mM liposomes: PC:DOPE (80:20 mol ratio), and PC:DOPE:CL (50:20:30 mol ratio).

6. 1-cm path length quartz cuvette.
7. QuantaMaster spectrofluorometer with thermostatic control and constant stirring [Photon Technology Instruments (PTI)-Horiba].

2.6 Components for Vesicle Aggregation Measurements

1. 3 mM LUV or SUV solutions [PC:DOPE (45:55 mol ratio), PC:DOPE:bl-PI (35:55:10 mol ratio), and PC:DOPE:CL (35:55:10 mol ratio)].
2. UBL buffer (50 mM Tris–HCl, pH 7.5, 150 mM NaCl, and 1 mM $MgCl_2$) supplemented with 1 mM DTT.
3. 100 mM ATP stock in 50 mM Tris–HCl, pH 6.8.
4. 1-mL quartz cuvette.
5. Uvikon 922 spectrophotometer (Kontron instruments, Gro-β-Zimmern, Germany).

2.7 Components for Total and Inner Lipid Mixing Assay

1. Two LUV populations of PC:DOPE (45:55 mol ratio), PC: DOPE:bl-PI (35:55:10 mol ratio), and PC:DOPE:CL (35:55:10 mol ratio):
   (a) Population 1: with no fluorescence probes.
   (b) Population 2: with 1.5 mol% NBD-PE and 1.5 mol% Rho-PE.
2. Reaction buffer: 50 mM Tris–HCl, pH 7.5, 150 mM NaCl, and 1 mM $MgCl_2$, supplemented with 1 mM DTT.
3. 100 mM ATP stock in 50 mM Tris–HCl, pH 6.8.
4. 1-mL quartz cuvette.
5. QuantaMaster spectrofluorometer with thermostatic control and constant stirring (PTI-Horiba).
6. 1 M *n*-octyl-β-D-glucopyranoside (OG) stock.
7. 100 mM stock solution of sodium dithionite, prepared in reaction buffer.
8. PD-10 desalting columns (GE Healthcare).

2.8 Components for Vesicle Aqueous Contents Mixing Assay

1. Three LUV populations of PC:DOPE (45:55 mol ratio), PC: DOPE:bl-PI (35:55:10 mol ratio), and/or PC:DOPE:CL (35:55:10 mol ratio):
   (a) Population 1: with ANTS encapsulated.
   (b) Population 2: with DPX encapsulated.
   (c) Population 3: with ANTS and DPX encapsulated.
2. Liposome hydration buffers:
   (a) Population 1: 25 mM ANTS, 40 mM NaCl, and 50 mM Tris–HCl, pH 7.5.

(b) Population 2: 90 mM DPX and 50 mM Tris–HCl, pH 7.5.

(c) Population 3: 12.5 mM ANTS, 45 mM DPX, 20 mM NaCl, and 50 mM Tris–HCl, pH 7.5.

3. PD-10 desalting columns (GE Healthcare).
4. Isosmotic buffer solution: 50 mM Tris–HCl, 150 mM NaCl, 1 mM $MgCl_2$, and 1 mM DTT, pH 7.5.
5. 100 mM ATP stock in 50 mM Tris–HCl, pH 6.8.
6. 1-mL quartz cuvette.
7. QuantaMaster spectrofluorometer with thermostatic control and constant stirring (PTI-Horiba).

2.9 Components for Vesicle Contents Efflux Measurements

1. Liposome hydration buffer: 20 mM ANTS, 70 mM DPX, 1 mM $MgCl_2$, 40 mM NaCl, and 50 mM Tris–HCl, pH 7.5.
2. 3 mM LUV population [PC:DOPE (45:55 mol ratio), PC: DOPE:bl-PI (35:55:10 mol ratio), and PC:DOPE:CL (35:55:10 mol ratio)].
3. PD-10 desalting columns (GE Healthcare).
4. Isosmotic buffer solution: 50 mM Tris–HCl, 150 mM NaCl, 1 mM $MgCl_2$, and 1 mM DTT, pH 7.5.
5. 100 mM ATP stock in 50 mM Tris–HCl, pH 6.8.
6. 1-mL quartz cuvette.
7. QuantaMaster spectrofluorometer with thermostatic control and constant stirring (PTI-Horiba).
8. 1 mM Triton X-100.

3 Methods

3.1 Membrane Lipid Model Vesicle Preparation (MLVs, LUVs, and SUVs)

1. Prepare pure lipid stocks by dissolving lipids in powder in chloroform/methanol (2:1 v/v) to the desired concentration.
2. Pipette the desired amount of lipid from the stock in organic solution into a glass test tube.
3. Evaporate the organic solvent under a nitrogen gas flow.
4. To completely remove any traces of organic solvent, place the sample into a high-vacuum desiccator for 2 h to obtain a solvent-free dried lipid film at the bottom of the test tube.
5. Hydrate the lipid by pipetting the desired amount of an aqueous solution at a temperature above the lipid main phase transition temperature, and shake the suspension by vigorous vortexing for lipid detachment from the bottom of the test tube (*see* **Note 1**).

This method is often enough for the preparation of multilamellar vesicles (MLVs). MLVs are spontaneously formed, containing usually seven to ten concentric bilayers, each of them separated by a thin water layer. Vesicles are heterogeneous in size, ranging from 100 to 5000 nm.

6. If smaller lipid vesicles (<500 nm) are required, the MLV suspension must be treated by extrusion to obtain large unilamellar vesicles (LUVs) of 80–400 nm or sonication for small unilamellar vesicles (SUVs) (<80 nm).
7. For LUV subject the MLV suspension to 10 freeze-thaw cycles in order to reduce the number of lamellae in the vesicles. This is achieved by placing the vesicles in liquid nitrogen for 1 min and then 3 min in a water bath at a temperature above that of the lipid with the highest main transition temperature. Then vortex the suspension before starting the next cycle. Repeat this process ten times.
8. Extrude mechanically the MLV suspensions [LIPEX Liposome Extrusion System (Transferra Nanosciences, Burnaby, Canada)] ten times through polycarbonate porous filters of the desired diameter (usually between 0.1 and 0.4 μm), using a high-pressure extruder (nitrogen gas flow) [13]. The extruder can be connected to a water bath in order to maintain the vesicles at a high temperature if necessary. An LUV population homogeneous in size is obtained.
9. For SUV preparation, sonicate MLV suspensions by either bath sonication or better by probe tip sonication. Immerse the probe in the MLV suspension (MSE Soniprep 150, MSE, Crawley, UK) for 20 min (10 s on/10 s off) on ice. In this way, large MLVs are broken down into SUVs.
10. Centrifuge the SUVs preparation at 14,100 × *g*, 10 min to remove aggregates and traces of the sonicator tip.
11. SUV generation can be directly observed as the vesicle solution becomes transparent. Due to their small diameter (smaller than 80 nm), SUVs present a high curvature stress, which induces a lipid enrichment in the external monolayer as compared to the inner one [14].
12. Finally, in all cases check vesicle size using dynamic light scattering (DLS). LUVs usually have a homogeneous average diameter between 100 and 500 nm, depending on the filter, and SUVs are less than 80 nm.
13. Assay the lipid phosphorus concentration of the vesicle suspension using the Fiske–Subbarow method (*see* Subheading 3.3), because some lipid can be lost in the process.

3.2 Membrane Lipid Model Vesicle Preparation (GUVs)

1. As in the case of LUVs, prepare pure lipid stocks by diluting the appropriate lipids in chloroform/methanol (2:1 v/v).
2. Starting from the pure stock, prepare GUV stocks containing the desired mix of lipids at a final lipid concentration of 2 mM. Add the desired percentage of a fluorescent dye, i.e., Rho-PE (approximately 0.5%), to the lipid stock to allow the visualization of vesicles in a confocal fluorescence microscope.
3. Add 4× 1 μL of the appropriate GUV stock solution onto the surface of different platinum electrodes in a suitable chamber (Industrias Técnicas ITC, Bilbao; *see* Fig. 1) containing four holes with two platinum electrodes each. Spread uniformly the volume along the platinum wires.
4. Introduce the chamber into a high-vacuum desiccator for 2 h to remove any remaining solvent traces, and then equilibrate for 15 min at the desired temperature by an incorporated water bath.
5. Next, to enable vesicle visualization, cover the platinum wires with 400 μL of a 300 mM (high-density) sucrose solution previously equilibrated at the desired temperature. Thus, the sucrose solution will be encapsulated in the vesicles so that when these are transferred to the chambers with an isosmotic solution, they will sediment at the bottom of the chamber so they can be imaged.
6. Cover the chamber holes with a glass slide in place with vacuum grease to avoid evaporation during electroformation.
7. Connect the platinum electrodes to a generator and apply the desired electric field (870 mV) for approximately 2 h. Vesicles are then generated enclosing sucrose. Temperature during preparation must be always higher than that of the lipid with the highest main phase transition temperature. After that, decrease the applied frequency (10 Hz) for 1 h and 30 min to induce vesicle detachment from the electrodes.
8. Disconnect the electric field and water bath and leave the vesicles 30 min for equilibration.
9. Then transfer 80 μL of the vesicle suspension to a microscopy chamber, which has been pretreated with fatty acid free BSA (10 mg/mL) for 45 min and contains 320 μL of an isosmotic aqueous solution or buffer without sucrose (*see* **Notes 2** and **3**). In this way, vesicles sediment due to the higher density of the enclosed sucrose, and this allows direct fluorescent confocal microscopy measurements to be performed (Fig. 2).

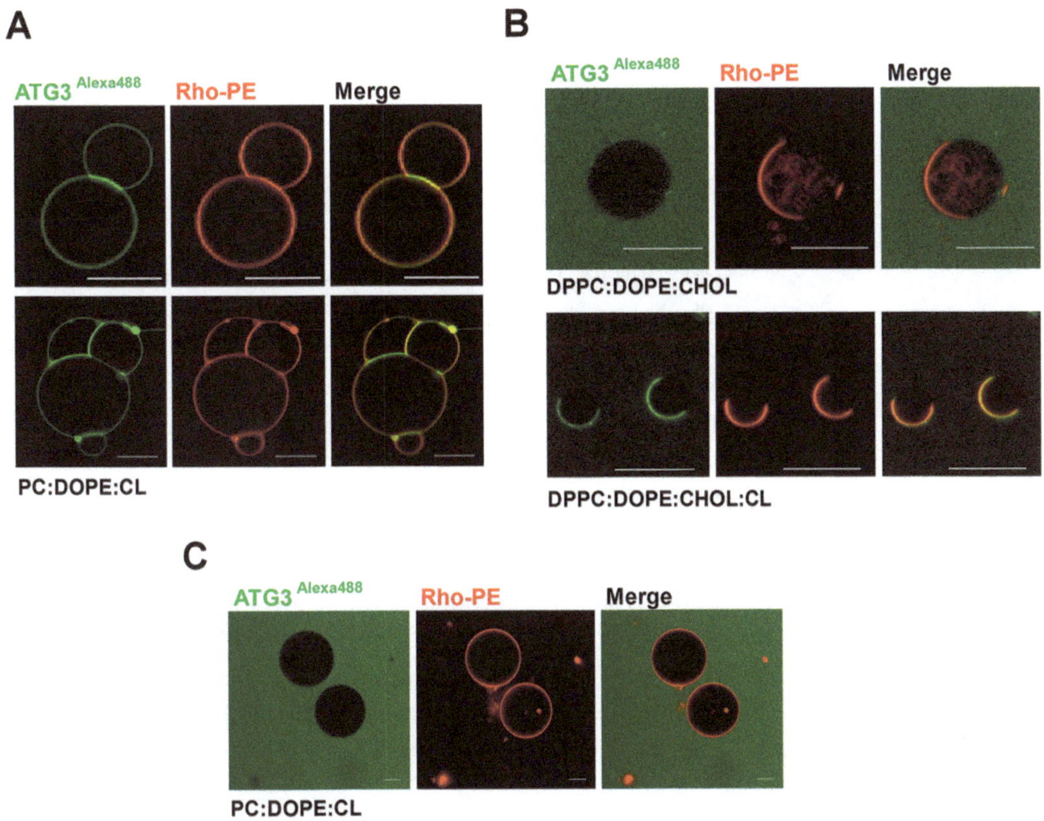

Fig. 2 Interaction of ATG3 with cell-sized lipid vesicles, GUVs. (**a**) ATG3 preferential localization in GUV contact sites. Membranes labeled with Rho-PE and ATG3 labeled with Alexa 488. Merge: yellow indicates colocalization of both probes. (**b**) ATG3 binding to GUV with segregated domains. Two different lipid domains are clearly seen, one of them containing CL and labeled with Rho-PE. (**c**) ATG3 K9D/K11D does not interact with PC:DOPE: CL (60:30:10 mol ratio) GUV. Lipid compositions are given in each panel. Scale bars: 10 μm [17]

3.3 Phospholipid Assay (Fiske–Subbarow Method)

Briefly, this approach consists of hydrolyzing the phospholipids until the lipid phosphate group is free to interact with specific reagents that will color the solution in a concentration-dependent manner.

1. A phosphorus calibration curve must be set up from which the exact lipid sample concentration will be determined. Pipette 0, 25, 50, 75, and 100 nmol phosphorus, respectively, into duplicate separate test tubes from a 1 mM NaH_2PO_4 standard solution.
2. Pipette the study sample (liposomes) into separate tubes (at least triplicate) to contain about 50 nmol lipid phosphorus, which will be in the center of the calibration curve. When measuring volumes in organic solution, use Hamilton microsyringes.
3. Add 500 μL of 60% perchloric acid ($HClO_4$) to each tube.

4. Vortex the tubes vigorously and place them in a heating block at 205 °C for 45 min. In this way phospholipid hydrolysis is achieved, yielding free inorganic phosphate.
5. Collect the tubes, cool down to room temperature, and pipette the following solutions:
   (a) 4 mL 1× ammonium heptamolybdate solution [$(NH_4)_6Mo_7O_{24}\cdot 4H_2O$].
   (b) 500 μL 10% (v/v) ascorbic acid (added while vortexing).

   The inorganic phosphate reacts with molybdate, which subsequently reacts with the ascorbic acid giving a yellow-colored solution.
6. Place the tubes into a boiling water bath and leave for 6 min. During that period, the solution color is blueshifted depending on the amount of phosphorus.
7. Finally, cool the tubes in water and measure absorbance at 812 nm for each of the samples. In our case, an Ultrospec 500 pro spectrophotometer from Amersham Biosciences (Piscataway, NJ, USA) was used for this purpose.
8. To obtain the sample phospholipid concentration, plot the standard absorbance against the phosphorus concentration and fit to a straight line. Use the slope of the curve and the sample absorbance to find out the sample concentration.

3.4 Equilibrium Sucrose Gradient Centrifugation of Liposomes

1. Incubate LC3B, GABARAP, GABARAPL2, or ATG3 (10 μM) with liposomes (3 mM) [containing 0.05 mol percent Rho-PE for detection] in a final volume of 300 μL at the desired temperature (*see* **Note 4**).
2. After that, adjust the sample to 1.4 M sucrose using a 2.4 M sucrose stock solution prepared in PreScission buffer (125 μL of sample + 175 μL of 2.4 M sucrose) and load it (300 μL) at the bottom of a discontinuous sucrose gradient overlaid with a 0.8 M medium layer (400 μL) and an 0.5 M upper layer (300 μL) sucrose (*see* **Note 5**).
3. Equilibrate the tubes and centrifuge the sample at 356,000 × *g* for 3 h at 4 °C in a Beckman Optima TLX benchtop ultracentrifuge, using a TLA-120.2 rotor (Beckman Coulter). During ultracentrifugation, due to their lower density, liposomes and liposome-bound proteins float up to the top fraction of the gradient whereas free protein remains at the bottom (*see* Fig. 3 and **Note 6**).
4. Finally, take 250-μL aliquots starting from the bottom (a total of 4 samples) and add 50 μL 6× loading buffer. Then analyze the top fraction (T), containing liposomes as indicated by the rhodamine fluorescence, and the bottom fraction (B), containing the unbound protein, by Coomassie blue-stained SDS-PAGE and Western blotting (*see* Fig. 4 and **Note 7**).

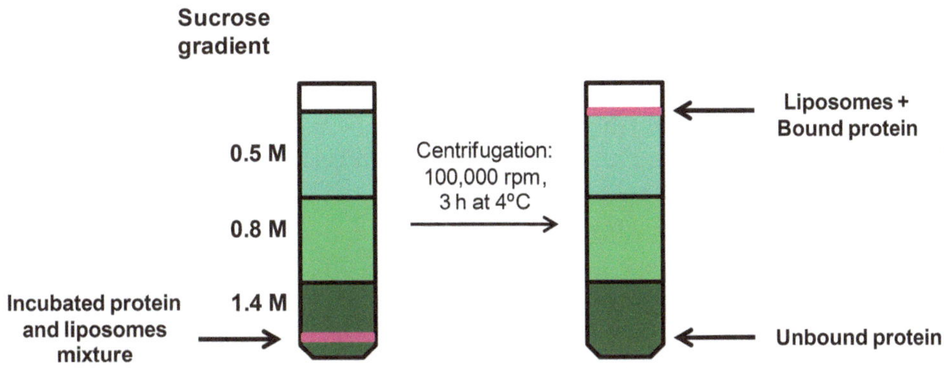

Fig. 3 Schematic representation of an equilibrium sucrose gradient centrifugation assay. First, a sucrose gradient is generated and the sample is located at the bottom. After centrifugation, liposomes and liposome-bound protein float up while the unbound protein remains at the bottom

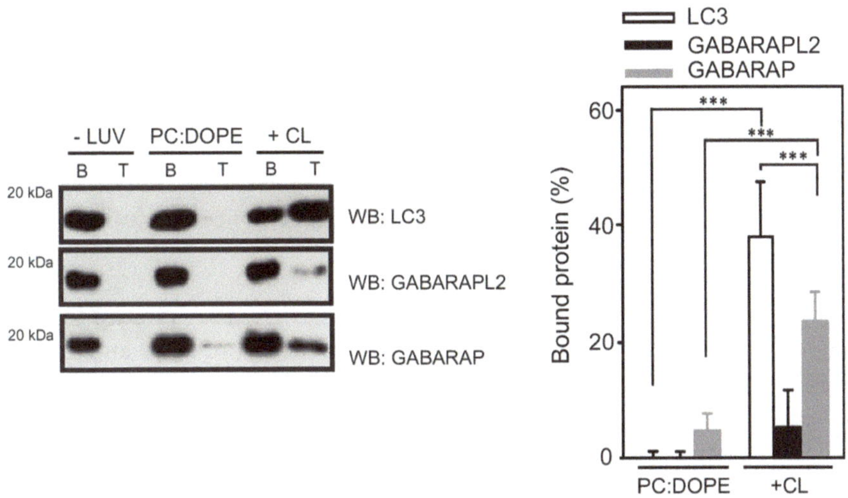

Fig. 4 Different CL-interacting properties among human Atg8 orthologues. Each orthologue (10 μM) was incubated with 3 mM LUVs composed of PC:DOPE (80:20 mol ratio) or PC:DOPE:CL (50:20:30 mol ratio) followed by flotation of the liposomes by gradient centrifugation. Bound protein fraction was quantified by SDS-PAGE/immunoblot analysis using anti-LC3, GABARAPL2, and GABARAP antibodies. Molecular masses are shown in kDa on the left-hand side. Data shown as mean ± SD ($n = 3$); $P < 0.001$ (Reproduced from ref. [6] with permission from Taylor and Francis)

3.5 Protein–Lipid Overlay Assay (Lipid Dot-Blot)

1. Dissolve stock solutions of different lipids in methanol/chloroform/water (2:1:0.8, v/v/v) at different final concentrations. If required, prepare serial 1:3 dilutions from the stock, ranging from 0 to 3000 pmol lipid.
2. Spot lipids (0.5 μL) onto Hybond-C nitrocellulose membrane (GE Healthcare) and leave for 1 h at room temperature to let the lipid dry (*see* **Note 8**).

3. To detect the protein by immunoblotting, first block the nitrocellulose with 5% fat-free milk in PBS for 1 h at room temperature to avoid nonspecific interactions.
4. Further incubate the membrane for 1 h with the protein (e.g., 0.5–0.75 μM LC3B) in PreScission buffer at 37 °C (*see* **Note 9**).
5. Wash the nitrocellulose 4 times, 10 min each, with PBS, and soak in 5% fat-free milk in PBS with the primary antibody at 1:1000 dilution overnight at 4 °C (*see* **Note 10**).
6. Wash the nitrocellulose 4 times with PBS and soak in 5% fat-free milk in PBS with a HRP-conjugated antibody (Santa Cruz Biotechnology, Dallas, TX, USA) at a 1:5000 dilution for 1 h at room temperature.
7. After washing with PBS 4 times 10 min each at room temperature to eliminate the unbound secondary antibody, detect the protein by chemiluminescence. Thus, the primary results obtained from the PLO assay are in the form of a standard immunoblot (Fig. 5). Develop the blot on a Curix 60 processor (Agfa, Belgium) using Amersham Hyperfilm ECL (GE Healthcare, UK).
8. Digitalize protein bands and measure and interpret integrated densities in a semiquantitative manner using Quantity one software (Bio-Rad).

3.6 NBD Fluorescence Spectroscopy

1. Site-directed mutagenesis is used to remove all endogenous cysteine residues present in the protein and to introduce a new Cys residue at the desired position in the generated Cys-free protein.
2. Labeling:
   (a) Concentrate cysteine-substituted proteins (LC3B^{Q15C}, LC3B^{S101C}, GABARAP^{S16C}, and GABARAP^{S88C} mutants) and GABARAPL2^{C15} (containing an endogenous C15 cysteine) to 2–3 mg/mL, and incubate them with 1 mM TCEP in PreScission buffer for 30 min at room temperature.
   (b) Remove TCEP by passing the sample through a PD-10 desalting column (GE Healthcare), previously equilibrated with the labeling buffer (10 mM HEPES, pH 7.3, 100 mM KCl, 1 mM EDTA).
   (c) Add IANBD (Molecular Probes, Eugene, OR, USA), dissolved in DMSO, in a small volume (15–25 μL) to the protein solution to give a final protein/probe molar ratio of 1:10 (NBD stock solution = 30 mM), and incubate overnight at 4 °C with gentle stirring protecting samples from the light (*see* **Note 11**).

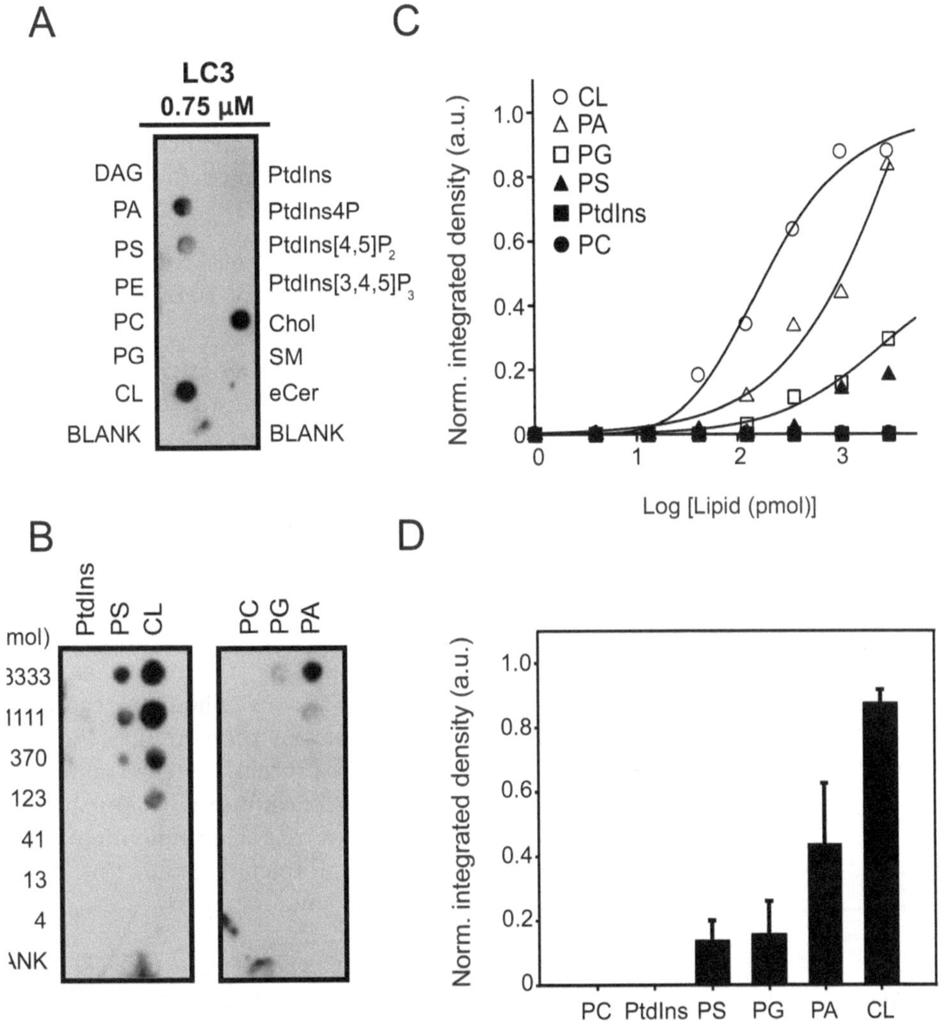

Fig. 5 LC3B binds CL directly and preferentially over other anionic lipids. LC3B immunoreactivity is shown from a representative protein/lipid dot-blot experiment with some of the most common glycerolipids and sphingolipids present in mammalian cells (1 nmol of each lipid) (**a**) and with increasing amounts of some of those lipids (**b**). (**c**) Integrated LC3B immunoreactivity was measured at each spot and normalized for each lipid, and the data were fitted with a sigmoidal equation of the type $f = (y_0 + a \times x^b)/(c^b + x^b)$, from which an apparent Hill coefficient (b) of 5.3 ± 1.2 for CL binding could be estimated ($R^2 = 0.991$). (**d**) Quantitative assessment of bound LC3B (0.75 μM) with 1.11 nmol of each lipid. LC3B immunoreactivity was measured at each spot by immunoblotting using anti-LC3 antibody and densitometry. Data shown as mean ± SEM ($n = 3$). a.u., arbitrary units. Norm., normalized (Reproduced from ref. [6] with permission from Taylor and Francis)

(d) Centrifuge the sample for 10 min at 14,000 × *g* and load it onto a PD-10 column previously equilibrated with labeling buffer to separate the conjugate from unreacted labeling reagent.

(e) For determining the degree of labeling, the following equation is used:

% Degree of labeling: $[(Abs_{NBD}/\varepsilon_{NBD})/(Abs_{prot}/\varepsilon_{prot})] \times 100$
where: Abs_{NBD} = NBD absorbance at 488 nm
Abs_{prot} = Protein absorbance at 280 nm
ε_{NBD} = NBD molar extinction coefficient
ε_{prot} = Protein molar extinction coefficient

3. NBD fluorescence measurements:
   (a) Incubate the sample components (protein or protein plus liposomes at the desired ratio) for 10 min at 37 °C prior to measurements.
   (b) Add the sample to a 1-cm path length cuvette, and record NBD spectra in a QuantaMaster spectrofluorometer (PTI-Horiba) with thermostatic control and constant stirring under the following conditions:

λ_{ex} = 465 nm	Band-pass = 5 nm
λ_{em} = 520–650 nm	Cutoff filter at 515 nm
T = 37 °C	Rate = 1 nm/s

When NBD moves from an aqueous milieu (i.e., the surface of a soluble protein) to a hydrophobic environment (i.e., the nonpolar interior of the membrane or a protein), its emission intensity increases while its wavelength of maximum emission shifts to the blue (lower wavelengths). In all cases, background intensities from samples lacking protein are subtracted from the intensities measured in protein-containing samples (Fig. 6).

3.7 Vesicle Aggregation Measurements

1. Prepare LUVs or SUVs at 3 mM lipid concentration as described in Subheading 3.1 in 50 mM Tris–HCl, 150 mM NaCl, 1 mM $MgCl_2$, and 1 mM DTT, pH 7.5 buffer.
2. Add 600 μL of preheated reaction buffer with 0.4 mM of liposomes, ATG7 (1 μM), ATG3 (1 μM), and the indicated Atg8 human homologs (10 μM) in a 1 mL quartz cuvette, and place it in the corresponding holder of a Uvikon 922 - spectrophotometer (Kontron instruments, Groß-Zimmern, Germany).
3. Start measuring the absorbance at 400 nm for 5 min with continuous stirring at 37 °C.
4. Add 5 mM of ATP to the cuvette in order to start the reaction, and continue measuring absorbance at 400 nm for the desired time or until signal stabilization. Lipid aggregation is monitored as an increase in turbidity (absorbance at 400 nm) of the sample (*see* Figs. 7 and 8).

Measures of liposomes and protein alone are required in order to know the contribution of both components to the increase in absorbance.

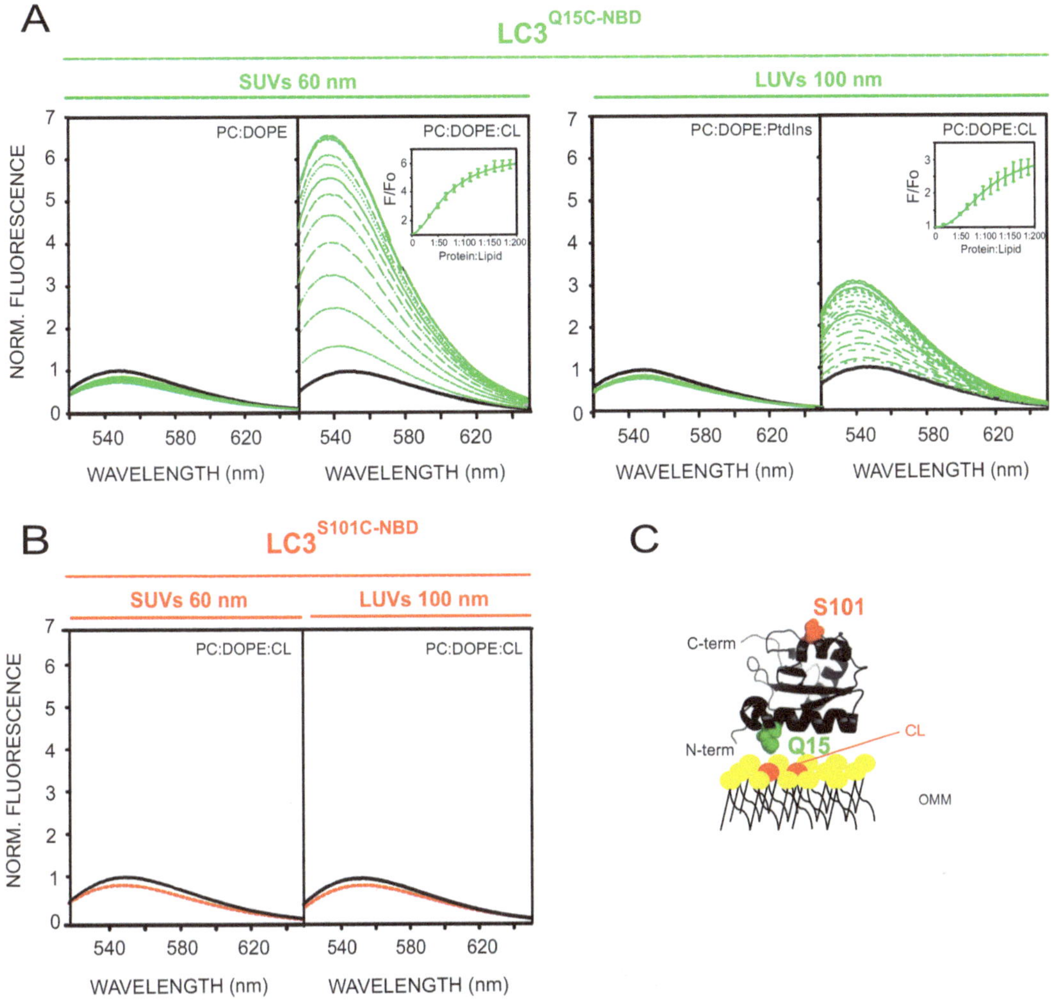

Fig. 6 LC3B C-terminus remains exposed to the hydrophilic environment after protein binding to CL-enriched membranes. Representative NBD fluorescence emission spectra of (**a**) LC3B$^{Q15C\text{-}NBD}$ (1 mM) in the absence or presence of increasing amounts of liposomes containing either PC:DOPE (80:20 mol ratio), PC:DOPE:CL, or PC:DOPE:PtdIns (50:20:30 mol ratio) and (**b**) LC3B$^{S101C\text{-}NBD}$ (1 mM) in the absence or presence of PC:DOPE:CL liposomes. In each case, fluorescence was normalized to the peak intensity of the protein spectrum in the absence of liposomes. (**c**) Structural model generated with PyMOL depicting the two LC3B residues that were individually mutated to cysteine obtaining single-cysteine LC3B mutants. The environmentally sensitive fluorophore NBD was used to label each of these single-cysteine residues. PDB: 1UGM. *OMM* outer mitochondrial membrane. *Norm.*, normalized (Reproduced from ref. [6] with permission from Taylor and Francis)

3.8 Total Lipid Mixing Assay

1. Prepare two LUV populations with the desired lipid composition (*see* **Note 12**), population 1 with no fluorescent probes and population 2 containing 1.5 mol% NBD-PE and 1.5 mol% Rho-PE.
2. Determine the lipid concentration with the lipid phosphorous assay (*see* Subheading 3.3).

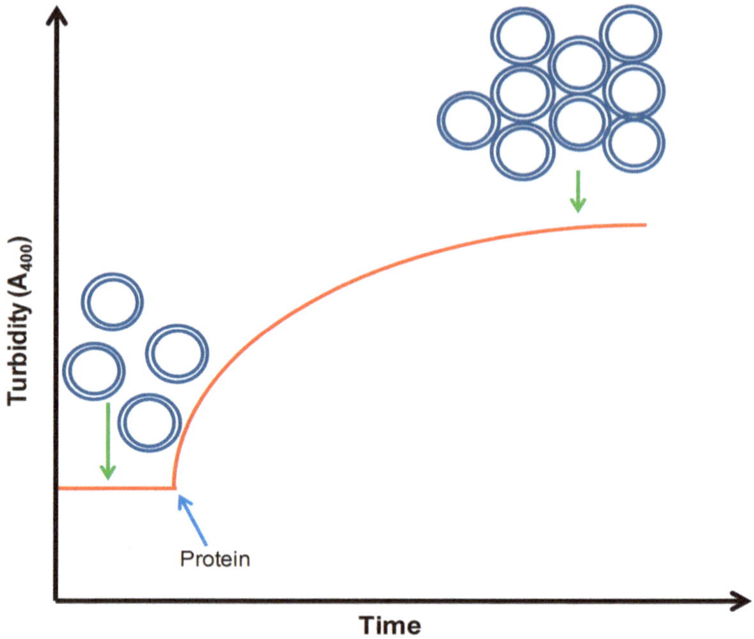

Fig. 7 Schematic overview of a time course of an aggregation assay. Initially vesicles scarcely scatter light. Upon protein addition, aggregation occurs and turbidity (light scattering) increases in parallel

3. Add in a 1 mL quartz cuvette: 1 mL of preheated reaction buffer (*see* **Note 13**) containing the two populations of liposomes mixed in a 9:1 (population 1/population 2) ratio to a 0.4 mM total concentration, ATG7 (1 μM), ATG3 (1 μM), and the indicated Atg8 human homologs (10 μM). Place the cuvette in the holder of a QuantaMaster spectrofluorometer (PTI-Horiba).
4. Start recording the fluorescence at 530 nm with an excitation wavelength of 465 nm for 5 min with continuous stirring at 37 °C. Use the following conditions:

   $\lambda_{ex} = 465$ nm

   $\lambda_{em} = 530$ nm

   Interference filter at 515 nm
5. After 5-min stabilization, add 5 mM of ATP in the quartz cuvette in order to start the reaction and continue to monitor NBD fluorescence at 530 nm (*see* Fig. 9).
6. When NBD fluorescence signal stabilizes, add 30 mM *n*-octyl-β-D-glucopyranoside (OG) detergent in order to obtain the 100% fusion value for subsequent calculations.
7. Calculate the amount of lipid mixing using the following equation:

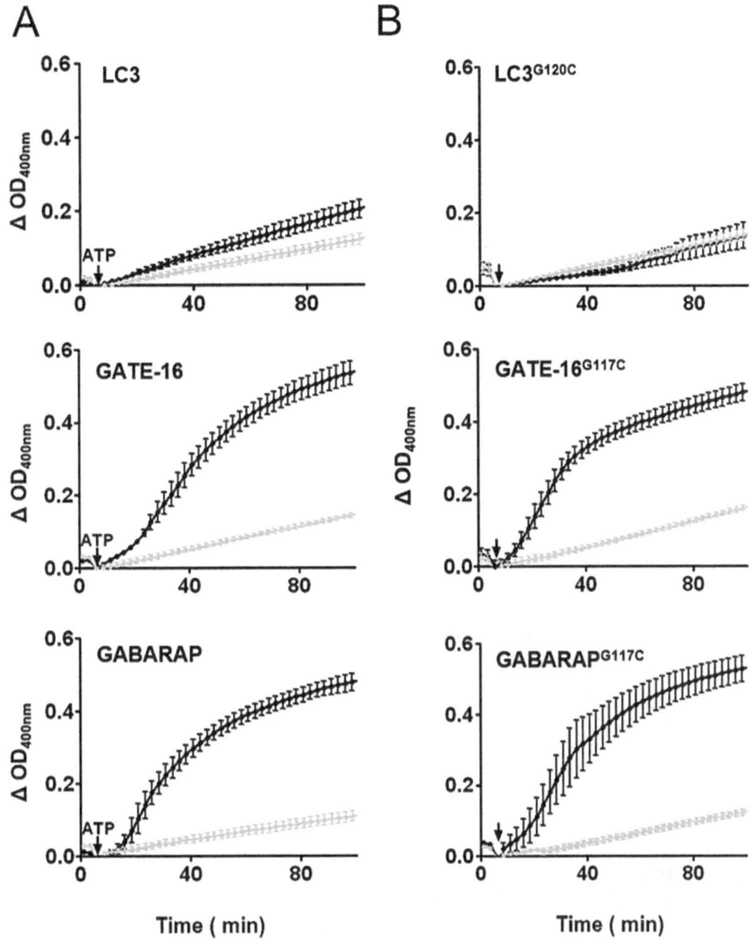

Fig. 8 Membrane tethering activities by human autophagic proteins. (**a**) Enzymatic conjugation reactions (maleimide-free) were performed followed by changes in absorbance at 400 nm, 37 °C. Liposomes were composed of PC/DOPE/bl-PI (35/55/10 mol ratio). For a negative control, ATG7 was excluded from the reaction (gray trace). (**b**) Chemical conjugation to PEmal was performed followed by measurements of changes in the absorbance at 400 nm, 37 °C. Liposomes were composed of PC/DOPE/PEmal/bl-PI (35/40/15/10 mol ratio). For negative controls, liposomes lacking PEmal were used (gray traces) (Reproduced from ref. [1] with permission from Elsevier)

$$\text{Lipid mixing (\%)} = [(F - F_0)/(F_{100} - F_0)] \times 100$$

where F is the fluorescence at equilibrium, F_0 is the fluorescence of vesicles at time zero, and F_{100} is the maximum fusion (fluorescence after OG addition).

3.9 Inner Lipid Mixing Assay

1. Prepare two LUV populations as detailed in Subheading 3.8.
2. Prepare a fresh 100 mM stock solution of sodium dithionite in the same buffer as the vesicles (*see* **Note 14**).

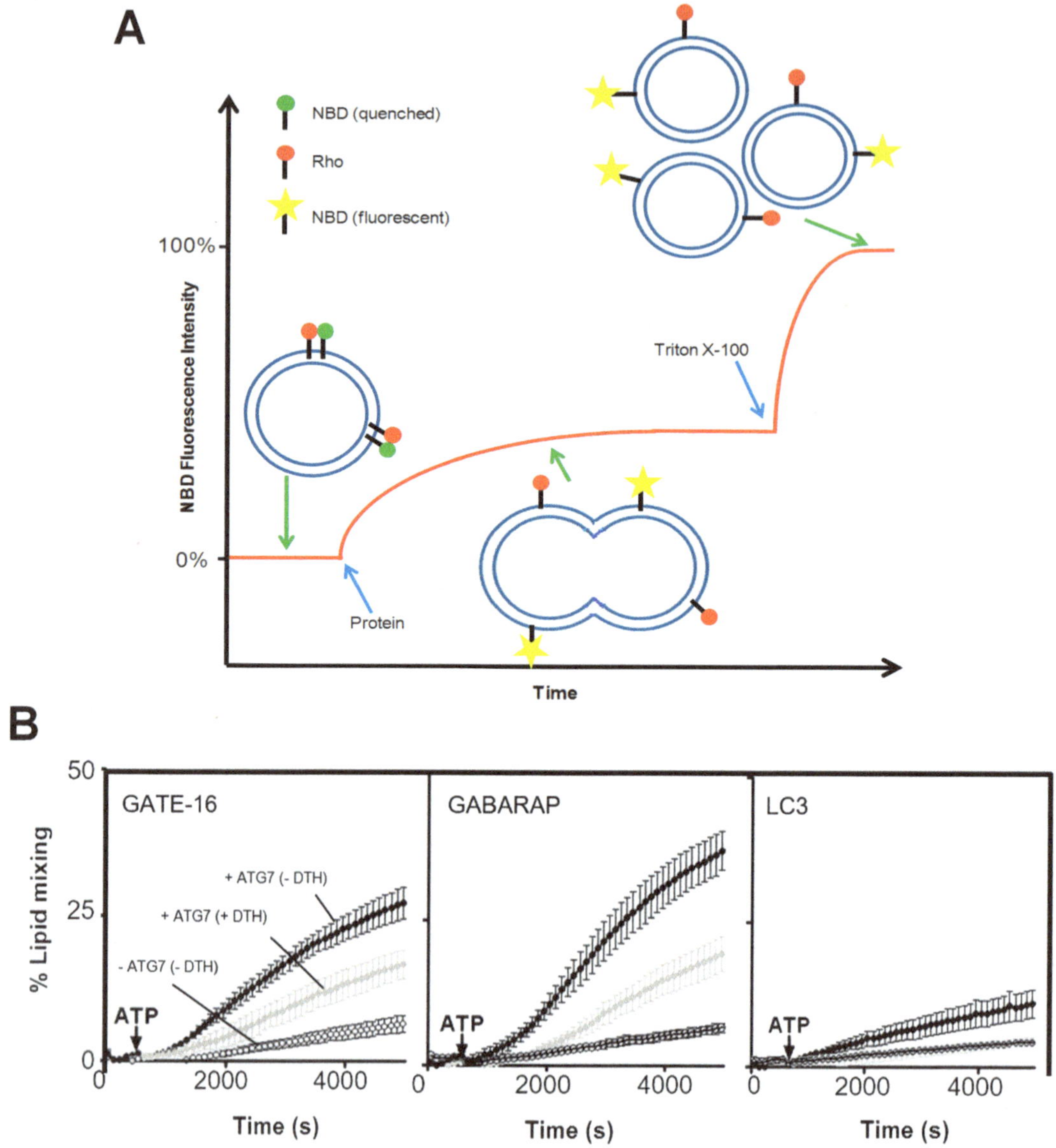

Fig. 9 (**a**) Schematic overview of an intervesicular lipid mixing assay. Initially there are two populations of vesicles, one with and the second without the FRET pair. Upon protein addition, when fusion occurs lipids are mixed and rhodamine cannot quench NBD. Triton X-100 is added to achieve vesicle solubilization; under these conditions 100% lipid mixing value is obtained. (**b**) Membrane fusion induced by enzymatically conjugated LC3/GABARAP-PE. Time courses of total (black circles) and inner lipid (black circles) mixing induced by ATG7 (1 μM), ATG3 (1 μM) and the indicated ATG8 human homologs (10 μM) in liposomes (0.4 mM) composed of PC/DOPE/bl-PI (35/55/10 mol ratio). Lipid mixing was monitored by the NBD/rhodamine lipid dilution assay. For inner monolayer lipid mixing, NBD + rho-liposomes were pretreated with the appropriate amounts of sodium dithionite to quench NBD fluorescence of the outer leaflet. Unlabeled and (NBD + Rho) labeled liposomes (9:1) were mixed and incubated for 5 min before ATP addition. The extent of lipid mixing was quantified on a percentage basis according to the following equation: $(F_t - F_0/F_{100} - F_0) \times 100$ where F_t is the measured fluorescence of protein-treated LUVs at time t, F_0 is the initial fluorescence of the LUV suspension before protein addition, and F_{100} is the fluorescence value after complete disruption of LUVs by addition of 30 mM β-octylglucoside. Each curve represents the average and SE of three independent experiments. For negative controls, ATG7 was excluded from the reactions (open circles) (Reproduced from ref. [1] with permission from Elsevier)

3. Place population 2 in a 1-ml quartz cuvette and start monitoring fluorescence signal at 530 nm with an excitation wavelength of 465 nm.
4. Add small amounts (1–3 μL) of sodium dithionite solution to the sample until the fluorescence is quenched to half of initial value (*see* **Note 15**).
5. After that, recover the vesicle suspension and immediately pass it through a Sephadex G-25 column to remove dithionite.
6. Determine the lipid concentration with the lipid phosphorous assay (*see* Subheading 3.3).
7. Add in a 1-mL quartz cuvette: 1 mL of preheated reaction buffer containing the two populations of liposomes mixed in a 9:1 (population 1/population 2) ratio to a 0.4 mM total concentration, ATG7 (1 μM), ATG3 (1 μM), and the indicated Atg8 human homologs (10 μM). Place the cuvette in the corresponding holder of a QuantaMaster spectrofluorometer (PTI-Horiba).
8. Start recording the fluorescence at 530 nm with an excitation wavelength of 465 nm for 5 min with continuous stirring at 37 °C. Use the following conditions:

   $\lambda_{ex} = 465$ nm

   $\lambda_{em} = 530$ nm

   Interference filter at 515 nm
9. After 5-min stabilization, add 5 mM of ATP in the quartz cuvette in order to start the reaction and continue to monitor NBD fluorescence at 530 nm (Fig. 9).
10. When NBD fluorescence signal stabilizes, add 30 mM *n*-octyl-β-D-glucopyranoside (OG) detergent in order to obtain the 100% fusion value for subsequent calculations.
11. Calculate the amount of lipid mixing using the following equation:

$$\text{Lipid mixing (\%)} = [(F - F_0)/(F_{100} - F_0)] \times 100$$

where F is the fluorescence at equilibrium, F_0 is the fluorescence of vesicles at time zero, and F_{100} is the maximum fusion (fluorescence after OG addition).

3.10 Vesicle Aqueous Contents Mixing Assay

1. Prepare three LUV populations with the desired lipid composition (*see* **Note 12**), population 1 containing encapsulated ANTS (25 mM ANTS, 40 mM NaCl, and 50 mM Tris–HCl, pH 7.5), population 2 containing DPX (90 mM DPX and 50 mM Tris–HCl, pH 7.5), and population 3 containing both dyes (12.5 mM ANTS, 45 mM DPX, 20 mM NaCl, and 50 mM Tris–HCl, pH 7.5).

2. Pass the vesicle suspensions through a Sephadex G-25 column to remove non-entrapped ANTS and DPX. A previously adjusted isosmotic buffer solution is used for this process: 50 mM Tris–HCl, 150 mM NaCl, 1 mM $MgCl_2$, and 1 mM DTT, pH 7.5.
3. Determine the lipid concentration with the lipid phosphorous assay (*see* Subheading 3.3).
4. Add in a 1-mL quartz cuvette: 1 mL of preheated reaction buffer containing the two populations of liposomes mixed in a 1:1 (population 1/population 2) ratio to a 0.4 mM total concentration, ATG7 (1 μM), ATG3 (1 μM), and the indicated Atg8 human homologs (10 μM). Place the cuvette in the holder of a QuantaMaster spectrofluorometer (PTI-Horiba).
5. Start recording the fluorescence at 530 nm with an excitation wavelength of 360 nm (slits, 2 nm) for 5 min with continuous stirring at 37 °C. Place a 470 nm cutoff filter between the sample and the emission monochromator to avoid scattering interference.
6. After 5-min stabilization, add 5 mM of ATP in the quartz cuvette in order to start the reaction and continue to monitor ANTS fluorescence at 530 nm (*see* Fig. 10).
7. When ANTS fluorescence signal stabilizes, stop recording.
8. To obtain the 100% mixing of contents value for subsequent calculations, 0.4 mM liposomes corresponding to population 3 are added to a quartz cuvette in a final volume of 1 mL of preheated reaction buffer. Place the cuvette in the holder and monitor ANTS fluorescence at 530 nm at 37 °C.
9. Calculate the amount of intervesicular aqueous contents mixing using the following equation:

$$\text{Contents mixing (\%)} = [(F - F_0)/(F_{100} - F_0)] \times 100$$

where F is the fluorescence at equilibrium, F_0 is the fluorescence of vesicles at time zero, and F_{100} is the maximum fusion (fluorescence of vesicles containing co-encapsulated ANTS and DPX).

3.11 Vesicle Contents Efflux Measurements (Leakage Assay)

1. Prepare 100 nm LUVs as previously described in Subheading 3.1, in the following buffer containing both ANTS and DPX molecules: 20 mM ANTS, 70 mM DPX, 1 mM $MgCl_2$, 40 mM NaCl, and 50 mM Tris–HCl, pH 7.5.

   A high DPX/ANTS ratio is used to ensure complete quenching inside vesicles. Samples are covered with aluminum foil to protect fluorescent molecules from bleaching.
2. Pass the vesicle suspension through a Sephadex G-25 column to remove non-entrapped ANTS and DPX. A previously adjusted isosmotic buffer solution is used for this process:

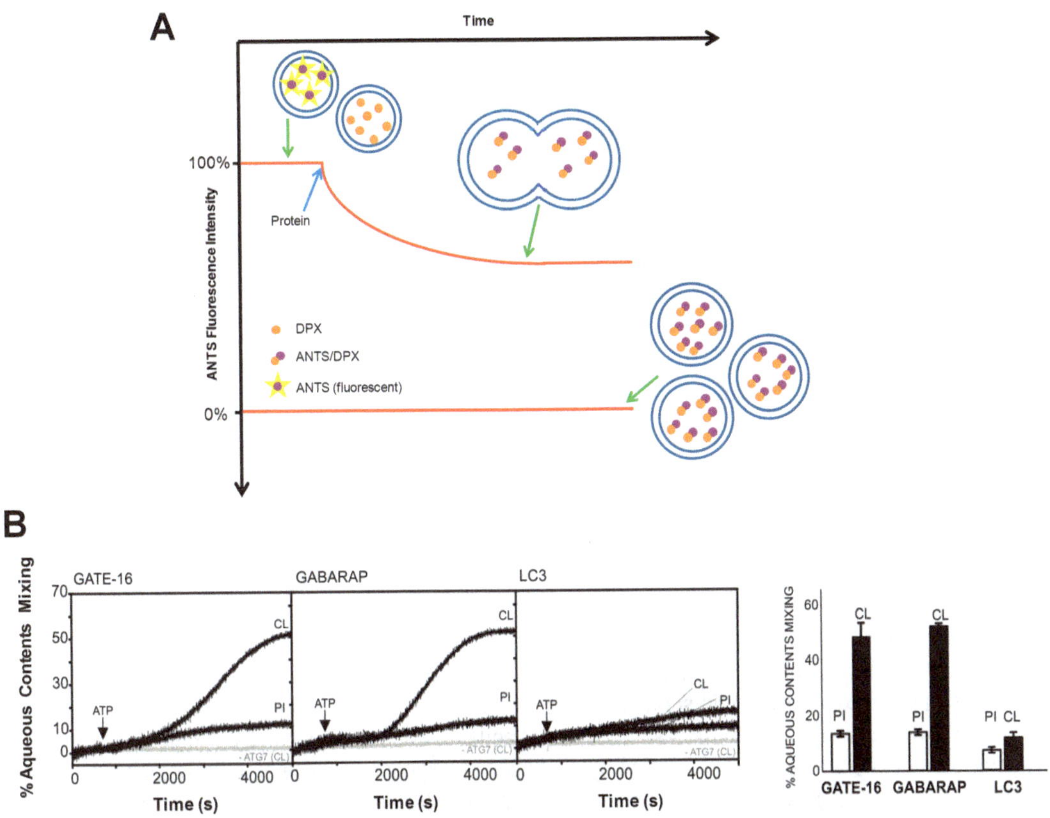

Fig. 10 (**a**) Schematic overview of an intervesicular aqueous contents mixing assay. Initially there are two populations of vesicles, one with ANTS and the second one with DPX. Upon protein addition, when complete fusion occurs, the aqueous contents of the vesicles are mixed, and DPX can quench ANTS. 100% fusion value is obtained measuring the fluorescence of a vesicle population containing both molecules in the same proportion. (**b**) Effect of lipids modifying the membrane intrinsic curvature on membrane fusion induced by LC3/GABARAP-PE conjugates. Representative time courses (left panels) and comparison at $t = 2000$ s (right panel) of aqueous contents mixing induced by ATG7 (1 μM), ATG3 (1 μM), and the indicated ATG8 human homologs (10 μM) in liposomes (0.4 mM) with the basic lipid composition PC/DOPE (35/55 mol ratio) to which either bl-PI (+10 mol ratio) or CL (+10 mol ratio) was added. ATG7 was excluded from the reaction in control experiments (gray traces). Liposomes and proteins were incubated for 5 min before ATP addition. Each curve represents the average and SE of three independent experiments (Reproduced from ref. [1] with permission from Elsevier)

50 mM Tris–HCl, 150 mM NaCl, 1 mM $MgCl_2$, and 1 mM DTT, pH 7.5.

3. Measure the lipid concentration as previously described (*see* Subheading 3.3).
4. Add 1 mL of preheated reaction buffer with 0.4 mM liposomes, ATG7 (1 μM), ATG3 (1 μM), and the indicated Atg8 human homologs (10 μM) in a 1-mL quartz cuvette, and place it in the corresponding holder of a QuantaMaster spectrofluorometer (PTI-Horiba).

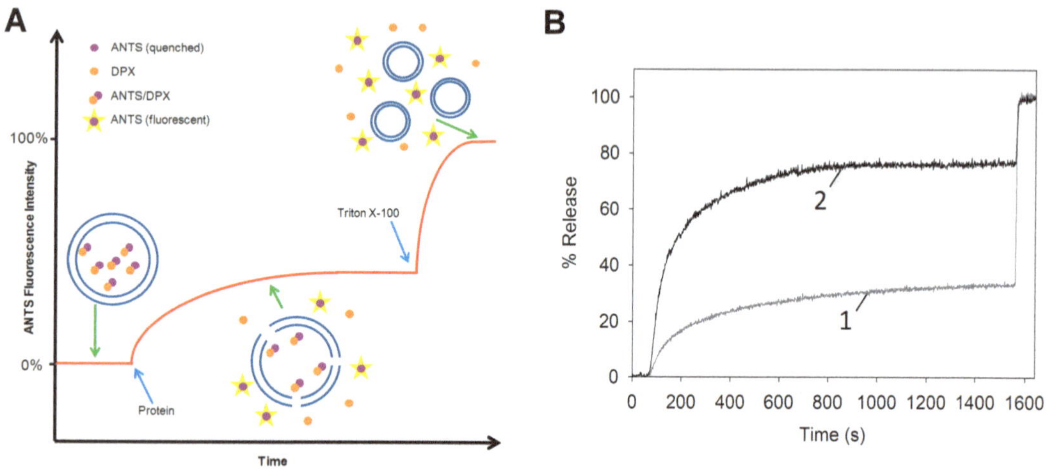

Fig. 11 (**a**) Schematic overview of the ANTS/DPX leakage assay. By entrapping both ANTS and DPX inside the vesicles, their close proximity allows DPX to interact with ANTS quenching its fluorescence. Upon incorporation of the permeabilizing agent (protein in this case) into the membrane, vesicle-enclosed ANTS and DPX are externalized through membrane defects. Both molecules are diluted into the external medium, and they hardly interact, inducing an increase of ANTS fluorescence. Thus, by following ANTS fluorescence vesicle bilayer, permeabilization or solubilization can be followed. Triton X-100 is generally added for complete vesicle solubilization into mixed vesicle lipid/surfactant/Triton X-100 micelles that give the 100% leakage value (i.e., maximum ANTS fluorescence). (**b**) Release of vesicular aqueous contents induced by C16 ceramide and C16 dihydroceramide. Ceramides were generated by the action of sphingomyelinase on LUVs composed of *1* SM:PE:PC:Ch (1:1:1:1) (initial slope: 0.25 ± 0.030) and *2* dhSM:PE:PC:Ch (1:1:1:1) (initial slope: 0.51 ± 0.041) ($n = 3$; P [Student's t-test] $= 0.007$). *SM* sphingomyelin, *dhSM* dihydrosphingomyelin, *PE* phosphatidylethanolamine, *PC* phosphatidylcholine, *Ch* cholesterol (Reproduced from ref. [18] with permission from Taylor and Francis)

5. Start measuring the fluorescence at 520 nm with an excitation wavelength of 355 nm for 5 min with continuous stirring at 37 °C. Use the following conditions:

   $\lambda_{ex} = 355$ nm.

   $\lambda_{em} = 520$ nm.

   Interference filter at 450 nm

6. Add 5 mM of ATP in the quartz cuvette in order to start the reaction and follow vesicle bilayer permeabilization or solubilization following ANTS fluorescence (Fig. 11).

7. When ANTS fluorescence signal stabilizes, add 1 mM Triton X-100 in order to obtain the 100% release value for subsequent calculations.

8. Finally, calculate the amount of leakage using the following equation:

$$\%\ \text{Leakage} = [(F - F_0)/(F_{100} - F_0)] \times 100$$

   where F is the fluorescence at equilibrium, F_0 is the fluorescence at time zero, and F_{100} is the maximum leakage (fluorescence after Triton X-100 addition).

4 Notes

1. By increasing temperature, the lipid is hydrated faster, and when preparing vesicles with a mixture of lipids, a more homogeneous lipid composition through the vesicles can be expected by maintaining every lipid in the fluid phase state.
2. Vesicle generation under electric fields is strongly dependent on the lipid composition, on the aqueous solution's ionic strength and pH, and on the electric voltage and frequency conditions [15]. In particular, only very low ionic strengths can be used in the buffers to prepare GUVs under this procedure. However, the approach has considerably evolved, and nowadays even the generation of giant vesicles from cell membranes in physiological salt solutions is possible [16].
3. Cut a pipette tip diagonally to take the GUV solution so that vesicles are not broken, and place them carefully drop by drop on top of the aqueous solution.
4. In general, small or large unilamellar vesicles are the best model membrane systems for the study of protein–lipid interactions.
5. When creating the sucrose gradient, try to add the different sucrose layers to the side of the tube wall to avoid mixing of the layers.
6. Liposome flotation assays are not sensitive to protein oligomerization or aggregation events.
7. When analyzing gradient fractions by Coomassie staining or Western blotting for the first time, all fractions should be analyzed in order to see whether some protein has not floated up to the upper fraction.
8. In order to spot the lipids in the correct place, equidistant one from each other and straight, small marks can be made with a pencil and a ruler on the edge of the nitrocellulose membrane.
9. Optimize the protein concentration for each experiment taking into account the final protein/lipid ratio.
10. Incubating the nitrocellulose membrane in 15-mL tubes avoids using relatively big volumes of protein. Only 5-mL total volume is needed when using these tubes.
11. The number of commercially available dyes is enormous, and choosing the most suitable one for each experiment is a very important step. Many of them, e.g., NBD, are very sensitive to the presence of water, while others, like fluorescein, are more sensitive to changes in the pH or the electrostatic environment.
12. The lipid composition of vesicles constitutes also a key factor when designing the experiments; of particular importance is to consider the use of cone-shaped or inverted cone-shaped

phospholipids which might be present in regions of the membrane where these events take place.

13. The osmolarity of the buffer solution should be adjusted so that vesicles do not suffer distortion and release their content.
14. Keep sodium dithionite solution in the dark at 4 °C until use.
15. **Steps 3** and **4** should be done as quickly as possible because some dithionite can penetrate the vesicles.

Acknowledgments

The authors thank Ms. Araceli Marcos for her technical support. This work was supported in part by grants from FEDER/Spanish Ministry of Economy (BFU 2011-28566, BFU 2015-66306-P) and the Basque Government (IT838-13). J.H.H. and Z.A. were predoctoral fellows supported by the University of the Basque Country; J.H.H. is a postdoctoral fellow supported by the University of the Basque Country, and Z.A. is a postdoctoral research associate supported by BrisSynBio at the University of Bristol.

References

1. Landajuela A, Hervás JH, Antón Z et al (2016) Lipid geometry and bilayer curvature modulate LC3/GABARAP-mediated model autophagosomal elongation. Biophys J 110:411–422
2. Lamb CA, Yoshimori T, Tooze SA (2013) The autophagosome: origins unknown, biogenesis complex. Nat Rev Mol Cell Biol 14:759–774
3. Tooze SA, Yoshimori T (2010) The origin of the autophagosomal membrane. Nat Cell Biol 12:831–835
4. Ktistakis NT, Tooze SA (2016) Digesting the expanding mechanisms of autophagy. Trends Cell Biol 26:624–635
5. Chu CT, Ji J, Dagda RK et al (2013) Cardiolipin externalization to the outer mitochondrial membrane acts as an elimination signal for mitophagy in neuronal cells. Nat Cell Biol 15:1197–1205
6. Antón Z, Landajuela A, Hervás JH et al (2016) Human Atg8-cardiolipin interactions in mitophagy: specific properties of LC3B, GABARAPL2 and GABARAP. Autophagy 12:2386–2403
7. Landeta O, Landajuela A, Gil D et al (2011) Reconstitution of proapoptotic BAK function in liposomes reveals a dual role for mitochondrial lipids in the BAK-driven membrane permeabilization process. J Biol Chem 286:8213–8230
8. Yethon JA, Epand RF, Leber B et al (2003) Interaction with a membrane surface triggers a reversible conformational change in Bax normally associated with induction of apoptosis. J Biol Chem 278:48935–48941
9. Johnson AE (2005) Fluorescence approaches for determining protein conformations, interactions and mechanisms at membranes. Traffic 6:1078–1092
10. Goñi FM, Villar AV, Nieva JL et al (2003) Interaction of phospholipases C and sphingomyelinase with liposomes. Methods Enzymol 372:3–19
11. Struck DK, Hoekstra D, Pagano RE (1981) Use of resonance energy transfer to monitor membrane fusion. Biochemistry 20:4093–4099
12. Ellens H, Bentz J, Szoka FC (1985) H+- and Ca2+-induced fusion and destabilization of liposomes. Biochemistry 24:3099–3106
13. Mayer LD, Hope MJ, Cullis PR (1986) Vesicles of variable sizes produced by a rapid extrusion procedure. Biochim Biophys Acta Biomembr 858:161–168
14. Szoka F, Papahadjopoulos D (1980) Comparative properties and methods of preparation of lipid vesicles (liposomes). Annu Rev Biophys Bioeng 9:467–508

15. Bagatolli LA (2003) Thermotropic behavior of lipid mixtures studied at the level of single vesicles: giant unilamellar vesicles and two-photon excitation fluorescence microscopy. Methods Enzymol 367:233–253
16. Montes L-R, Alonso A, Goñi FM et al (2007) Giant unilamellar vesicles electroformed from native membranes and organic lipid mixtures under physiological conditions. Biophys J 93:3548–3554
17. Hervás JH, Landajuela A, Antón Z et al (2017) Human ATG3 binding to lipid bilayers: role of lipid geometry, and electric charge. Sci Rep 7:15614
18. Hernández-Tiedra S, Fabriàs G, Dávila D et al (2016) Dihydroceramide accumulation mediates cytotoxic autophagy of cancer cells via autolysosome destabilization. Autophagy 12:2213–2229

Chapter 6

Reconstituting Autophagy Initiation from Purified Components

Peter Mayrhofer and Thomas Wollert

Abstract

The hallmark of macroautophagy is the de novo generation of a membrane structure that collects cytoplasmic material and delivers it to lysosomes for degradation. The nucleation of this precursor membrane, termed phagophore, involves the coordinated assembly of the Atg1-kinase complex and the recruitment of Atg9 vesicles. The latter represents one important membrane source in order to produce phagophores in vivo. We explain how the process of phagophore nucleation can be reconstituted from purified components in vitro. We describe the assembly of the ~500 kDa pentameric Atg1-kinase complex from its purified subunits. We also explain how Atg9-donor vesicles are generated in vitro to study the interaction of Atg9 and Atg1-kinase complexes by floatation experiments.

Key words Autophagy, Atg9, Atg1-kinase complex, In vitro reconstitution, Membrane tethering

1 Introduction

Macroautophagy (autophagy in the following) is a major degradation pathway that sequesters cytoplasmic material in de novo formed autophagosomes to deliver it to lysosomes for degradation [1]. The pathway is highly conserved from yeast to humans [2]. A coordinated activity of more than 40 autophagy-related (Atg) proteins promotes in conjugation with canonical membrane trafficking factors the biogenesis of autophagosomes. Most of these factors are organized into five complexes or macromolecular assemblies [3]. The sequential recruitment of these complexes to the phagophore assembly site (PAS) is required to form autophagosomes [4].

Atg17 is a scaffolding protein and integral part of the Atg1-kinase complex, which is essential for initiation of nonselective autophagy in yeast. This S-shaped homodimer is the earliest Atg protein to be recruited to the PAS upon induction of autophagy by starvation [4, 5]. Atg17 coordinates the recruitment of Atg9 vesicles that provide membranes in order to nucleate the phagophore

Nicholas Ktistakis and Oliver Florey (eds.), *Autophagy: Methods and Protocols*, Methods in Molecular Biology, vol. 1880,
https://doi.org/10.1007/978-1-4939-8873-0_6,

[6]. However, Atg17 is constitutively accompanied by Atg31 and Atg29, which have been suggested to regulate Atg17 function [7].

Upon induction of autophagy by starvation, the target of rapamycin complex 1 (TORC1) is inactivated. This complex phosphorylates amongst other targets the Atg1-kinase complex component Atg13. TORC1 deactivation results in a rapid dephosphorylation of Atg13 [8]. This stimulates the assembly of the pentameric Atg1-kinase complex consisting of the kinase Atg1, Atg13, and the constitutive Atg17-Atg31-Atg29 subcomplex [9, 10]. Dephosphorylated Atg13 directly interacts with Atg17 to facilitate complex formation [11], but both Atg13 and Atg1 are dispensable for the recruitment of Atg9 to the PAS [12]. The nucleation of phagophores from Atg9 vesicles requires, on the other hand, Atg1 and Atg13.

Through our in vitro reconstitution approach, we revealed the molecular function of the Atg1-kinase complex components and how they cooperate to recruit Atg9 vesicles. The six-transmembrane helix protein Atg9 was expressed in *E. coli*, purified as detergent-solubilized protein and reconstituted in small unilamellar vesicles. These vesicles had a size of ~30 nm, which is equivalent to the size of Atg9 vesicles in vivo [13]. Moreover, we found that dimeric Atg17 binds two Atg9 vesicles and therefore functions as membrane tether to selectively bring Atg9 vesicles in juxtaposition to allow them to fuse in a presumably SNARE-mediated fashion [14]. Interestingly, the constitutive Atg17-Atg31-Atg29 complex displayed only ~50% of the Atg9-binding capacity of Atg17. We could demonstrate that the Atg31-Atg29 regulatory subcomplex masks at least one of the two Atg9-binding sites per constitutive Atg17 dimer. This prevented Atg17 from tethering Atg9 vesicles in the absence of stimulating factors, although inactivated Atg17 is still able to interact with Atg9. This observation also explains why Atg17-Atg31-Atg29 is sufficient to recruit Atg9 vesicles to the PAS.

Starvation or other cues lead to dephosphorylation of Atg13 and permit assembly of the pentameric Atg1-kinase complex. We could demonstrate that Atg1-Atg13 releases the autoinhibition in Atg17-Atg31-Atg29 upon complex formation, allowing the Atg1-kinase complex to bind and tether Atg9 vesicles as Atg17 itself does [15]. Our study thus revealed how Atg17 function is controlled by the subunits of the Atg1-kinase complex to restrict its activity under non-induced conditions but activate it during starvation [16, 17].

2 Materials

2.1 Cell Lines

1. *E. coli* Rosetta (DE3).
2. *E. coli* DH10-MultiBac.

3. High Five insect cells (BTI-TN-5B1-4).
4. Sf9 insect cells.
5. LB medium.

2.2 Plasmids

1. pET28a(+)-Atg9core: Atg9 amino acids 281–779 with a C-terminal Strep-His6-tag. A PreScission cleavage site was inserted between Atg9 and the tag.
2. pST39-Atg17: Atg17 was cloned as N-terminal His6-tag fusion protein that could be cleaved through a PreScission protease cleavage site in between Atg17 and the His6-tag accompanied by a C-terminal myc-tag.
3. pST39-Atg29-Atg31: Atg29 and Atg31 were cloned into the polycistronic pST39 vector [18]. Atg29 contained an N-terminal His6-tag preceded by a PreScission protease cleavage site.
4. pCoofy-Atg1: Atg1 was cloned into a modified pFastBac-derived pCoofy vector (pCoofy29) with an N-terminal MBP-tag preceded by a PreScission protease cleavage site and a C-terminal His6-tag which was preceded by a TEV protease cleavage site [19].
5. pCoofy37-Atg13: Atg13 was cloned into the pCoofy37 vector containing an N-terminal MBP-tag preceded by a PreScission protease cleavage site and a C-terminal His6-tag. A TEV cleavage site was present in between Atg13 and the His6-tag.

2.3 Chemicals

1. LDAO (*N,N*-Dimethyldodecylamine *N*-oxide).
2. Lipids: cholesterol; 1-palmitoyl-2-oleoyl-sn-glycero-3-phosphocholine (POPC); 1-palmitoyl-2-oleoyl-sn-glycero-3-phosphoethanolamine (POPE); 1-palmitoyl-2-oleoyl-sn-glycero-3-phospho-L-serine (POPS); 1,2-dipalmitoyl-sn-glycero-3-phosphoethanolamine-*N*-(lissamine rhodamine B sulfonyl) (rhodamine-PE).
3. Kanamycin.
4. Ampicillin.
5. Chloramphenicol.
6. Isopropyl β-D-1-thiogalactopyranoside (IPTG).
7. PMSF.

2.4 Buffer Compositions

2.4.1 Atg9core Purification

1. Atg9 lysis buffer: 50 mM Tris/HCl pH 8.0, 500 mM KCl, 10% glycerol.
2. Atg9 solubilization buffer: 25 mM Tris/HCl pH 8.0, 500 mM KCl, 10% glycerol.
3. Atg9 HisTrap buffer A: 25 mM Tris/HCl pH 8.0, 500 mM KCl, 10% glycerol, 40 mM imidazole, 0.1% LDAO.

4. Atg9 HisTrap buffer B: 25 mM Tris/HCl pH 8.0, 200 mM KCl, 10% glycerol, 500 mM imidazole, 0.1% LDAO.
5. Atg9 SEC buffer: 25 mM Tris/HCl pH 8.0, 150 mM NaCl, 0.05% LDAO.

2.4.2 Atg17 Purification

1. Atg17 lysis buffer: 100 mM Tris/HCl pH 7.4, 300 mM NaCl, 10% glycerol, 10 mM imidazole, 5 mM β-mercaptoethanol.
2. Atg17 HisTrap buffer A: 25 mM Tris/HCl pH 7.4, 300 mM NaCl, 10% glycerol, 20 mM imidazole, 5 mM β-mercaptoethanol.
3. Atg17 HisTrap buffer B: 25 mM Tris/HCl pH 7.4, 300 mM NaCl, 10% glycerol, 500 mM imidazole, 5 mM β-mercaptoethanol.
4. Atg17 SEC buffer: 25 mM Tris/HCl pH 8.0, 150 mM NaCl.

2.4.3 Atg31-Atg29 Purification

1. Atg29/31 lysis buffer: 100 mM Tris/HCl pH 8.0, 300 mM NaCl, 10% glycerol, 5 mM imidazole, 5 mM β-mercaptoethanol.
2. Atg29/31 wash buffer: 25 mM Tris/HCl pH 7.2, 300 mM NaCl, 10% glycerol, 5 mM imidazole, 5 mM β-mercaptoethanol.
3. Atg29/31 elution buffer: 25 mM Tris/HCl pH 7.2, 300 mM NaCl, 10% glycerol, 250 mM imidazole, 5 mM β-mercaptoethanol.
4. Atg29/31 SEC buffer: 25 mM Tris/HCl pH 7.2, 275 mM NaCl.

2.4.4 Atg1 Purification

1. Atg1 lysis buffer: 100 mM Tris/HCl pH 7.4, 300 mM NaCl, 20 mM imidazole, 10% glycerol, 1% Triton X-100, 5 mM β-mercaptoethanol.
2. Atg1 HisTrap buffer A: 25 mM Tris/HCl pH 7.4, 300 mM NaCl, 20 mM imidazole, 10% glycerol, 0.1% Triton X-100, 5 mM β-mercaptoethanol.
3. Atg1 HisTrap buffer B: 25 mM Tris/HCl pH 7.4, 300 mM NaCl, 500 mM imidazole, 10% glycerol, 0.1% Triton X-100, 5 mM β-mercaptoethanol.
4. Atg1 SEC buffer: 25 mM Tris/HCl pH 7.4, 300 mM NaCl, 10% glycerol.

2.4.5 Atg13 Purification

1. Atg13 lysis buffer: 100 mM Tris/HCl pH 7.4, 300 mM NaCl, 20 mM imidazole, 5% glycerol, 5 mM β-mercaptoethanol, 0.2% Tween-20.

2. Atg13 HisTrap buffer A: 25 mM Tris/HCl pH 7.4, 300 mM NaCl, 20 mM imidazole, 5% glycerol, 5 mM β-mercaptoethanol, 0.05% Tween-20.
3. Atg13 HisTrap buffer B: 25 mM Tris/HCl pH 7.4, 300 mM NaCl, 500 mM imidazole, 5% glycerol, 5 mM β-mercaptoethanol, 0.05% Tween-20.
4. Atg13 MonoQ dilution buffer: (25 mM Tris/HCl pH 7.4, 5% glycerol, 10 mM DTT).
5. Atg13 MonoQ buffer A: 25 mM Tris/HCl pH 7.4, 125 mM NaCl, 5% glycerol, 10 mM DTT.
6. Atg13 MonoQ buffer B: 25 mM Tris/HCl pH 7.4, 1 M NaCl, 5% glycerol, 10 mM DTT.

2.4.6 Floatation Assay

1. Liposome buffer: 25 mM HEPES/NaOH pH 7.0, 100 mM NaCl.
2. Histodenz buffer: 25 mM HEPES/NaOH pH 7.0, 100 mM NaCl, 80% Histodenz.

3 Methods

3.1 Atg9^{core} Purification

Atg9 is a membrane protein that contains six transmembrane helices. Consequently, the protein is co-translationally inserted into the plasma membrane of its expression host (*E. coli*). Thus, Atg9 was purified by extracting it from whole membrane fractions as explained below. Care must been taken regarding the used detergent, which should solubilize the target proteins efficiently without interfering with the structure of the protein. For Atg9, which comprises >70% helical secondary structure elements, circular dichroic spectroscopy was performed to evaluate the secondary structure content of each purified batch. Moreover, functional studies required Atg9 to be embedded within its native membrane environment. Detergent-solubilized Atg9 was therefore reconstituted in liposomes as described below and proper protein folding was again confirmed by circular dichroism.

3.1.1 Atg9^{core} Expression

1. Inoculate 35 mL LB medium containing 35 μg/mL kanamycin and 20 μg/mL chloramphenicol with *E. coli* Rosetta (DE3) transformed with pET28a(+)-Atg9core, and incubate at 37 °C overnight.
2. Add 30 mL overnight culture to 3 L TB medium supplemented with kanamycin and chloramphenicol, and grow the culture at 37 °C.
3. At an OD600 of 0.7, move the culture to an incubator set at 18 °C, and cool it down for approx. 1 h. Induce at an OD600

of 1.2 with a final concentration of 0.3 mM IPTG, and express for 18 h.

3.1.2 Membrane Fraction Preparation and Solubilization

1. Spin the culture for 15 min at 4500 × *g*, and resuspend the cell pellet in 200 mL Atg9 lysis buffer with freshly added PMSF (1 mM final concentration), 500 μL Protease Inhibitor Cocktail, and 20 μL benzonase.
2. Pass the cell suspension through a cell disruption system (*see* **Note 1**), and centrifuge for 10 min at 24,000 × *g* and 4 °C.
3. Collect the supernatant, and centrifuge again 10 min at 24,000 × *g* and 4 °C.
4. Collect the supernatant, and ultracentrifuge for 1 h at 150,000 × *g* (*see* **Note 2**).
5. To wash, resuspend the membrane pellet in 100 mL Atg9 lysis buffer supplemented with 100 μL Protease Inhibitor Cocktail using a douncer.
6. Ultracentrifuge again for 1 h at 150,000 × *g*. Weight the pellet, and resuspend it with 80 mL solubilization buffer (supplemented with 0.1% Protease Inhibitor Cocktail) per mg membrane.
7. Add LDAO to reach 0.5% (w/V) final concentration.
8. Ultracentrifuge for 30 min at 150,000 × *g* to remove insoluble particles.

3.1.3 Affinity and Size-Exclusion Chromatography

1. Add imidazole to a final concentration of 20 mM to the supernatant, and load it on a 5 mL HisTrap FF crude column attached to an Aekta system (1.5 mL/min). Collect the flow-through, and load it once again.
2. Wash with 120 mL Atg9 HisTrap buffer A (2 mL/min), and perform a step elution with 100% Atg9 HisTrap buffer B (1 mL/min).
3. Collect all peak fractions, and concentrate to 5 mL. Centrifuge for 10 min full speed in a tabletop centrifuge, and load on a HiLoad 16/600 Superdex 200 gel filtration column.
4. Run with Atg9 SEC buffer, and collect 1.5 mL fractions.
5. Monomeric Atg9 shows a peak at 72 mL, which was used for Atg9-proteoliposome preparation.

3.2 Atg17 Purification

Atg17 is a scaffold protein that structurally organizes the Atg1-kinase complex by its direct interaction with the two regulatory subcomplexes Atg1-Atg13 through Atg13 and Atg31-Atg29 through Atg31. It forms S-shaped dimers in solution and possesses Atg9-binding sites. Atg17 was expressed in *E. coli* and purified by affinity chromatography using its cleavable N-terminal His-tag and gel filtration (Fig. 1, *left lane*).

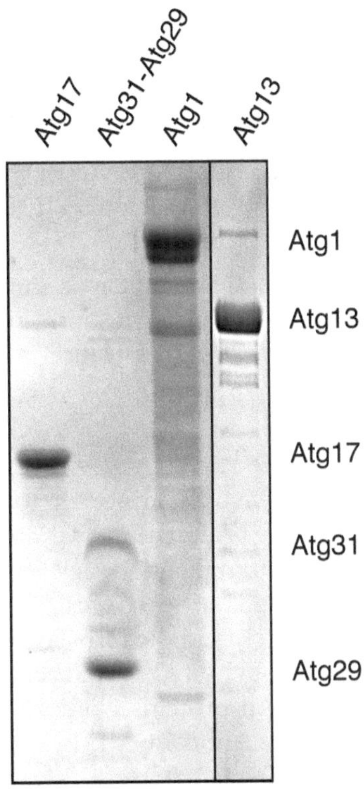

Fig. 1 Purified subunits of the Atg1-kinase complex. The SDS-polyacrylamidgel shows proteins of the Atg1-kinase complex as indicated after purification. Atg31 is very weekly stained by Coomassie. The double band of Atg1 corresponds to phosphorylated (lower) and non-phosphorylated (upper) species

3.2.1 Atg17 Expression

1. Inoculate 35 mL LB medium containing 100 μg/mL ampicillin and 20 μg/mL chloramphenicol with *E. coli* Rosetta (DE3) transformed with pST39-Atg17, and incubate at 37 °C overnight.
2. Add 30 mL overnight culture to 3 L LB medium supplemented with ampicillin and chloramphenicol, and grow the culture at 37 °C.
3. At an OD600 of 0.5, move the culture to an incubator set at 18 °C. Induce at an OD600 of 0.8 with a final concentration of 0.3 mM IPTG, and express for 18 h.

3.2.2 Affinity and Size-Exclusion Chromatography

1. Spin the culture for 15 min at 4500 × *g*, and resuspend the cell pellet in 80 mL Atg17 lysis buffer with freshly added PMSF (1 mM final concentration), 250 μL Protease Inhibitor Cocktail, and 10 μL benzonase.
2. Lyse the cells by sonication, and centrifuge for 1 h at 50,000 × *g* and 4 °C.

3. Load on a 1 mL HisTrap FF crude column attached to an Aekta system (1.5 mL/min).
4. Wash with 30 mL Atg17 HisTrap buffer A, and perform a step elution with 100% Atg17 HisTrap buffer B.
5. Add DTT and EDTA to a final concentration of 1 mM each, as well as 15 μL His-PreScission. Incubate at room temperature for 1 h.
6. Centrifuge the sample for 10 min full speed in a tabletop centrifuge, and load it on a HiLoad 16/600 Superdex 200 gel filtration column.
7. Run with Atg17 SEC buffer. Atg17 elutes at around 64 mL.

3.3 Atg31-Atg29 Purification

The Atg31-Atg29 subcomplex constitutively interacts with Atg17 in vivo [7]. In order to investigate how this subcomplex regulates Atg17 function, we compared Atg17 activity in terms of Atg9-binding and vesicle tethering in the presence and absence of Atg31-Atg29. This required the subcomplex to be produced separately and the ternary complex to be reconstituted from Atg17 and Atg31-Atg29. We therefore produced Atg31-Atg29 by expressing both proteins from a polycistronic vector in *E.coli* and purifying the subcomplex by affinity chromatography using N-terminal His-tagged Atg29, cleavage of the affinity tag by PreScission protease digestion followed by size-exclusion chromatography (Fig. 1, *second left lane*).

3.3.1 Atg31-Atg29 Expression

1. Inoculate 35 mL LB medium containing 100 μg/mL ampicillin and 20 μg/mL chloramphenicol with *E. coli* Rosetta (DE3) transformed with pST39Atg29-Atg31, and incubate at 37 °C overnight.
2. Add 30 mL overnight culture to 3 L LB medium supplemented with ampicillin and chloramphenicol, and grow the culture at 37 °C.
3. At an OD600 of 0.5, move the culture to an incubator set at 18 °C. Induce at an OD600 of 0.8 with a final concentration of 0.3 mM IPTG, and express for 18 h.

3.3.2 Affinity and Size-Exclusion Chromatography

1. Spin the culture for 15 min at 4500 × *g*, and resuspend the cell pellet in 80 mL Atg29/31 lysis buffer with freshly added PMSF (1 mM final concentration), 250 μL Protease Inhibitor Cocktail, and 10 μL benzonase.
2. Lyse the cells by sonication, and centrifuge for 1 h at 50,000 × *g* and 4 °C.
3. Equilibrate 1 mL Ni-NTA bead slurry in Atg29/31 lysis buffer, and incubate with supernatant for 1 h at 4 °C under constant mixing.

4. Transfer the suspension to a gravity flow column, remove the lysate, and wash the beads 5× with 30 mL Atg29/31 wash buffer.
5. Elute with 5 mL Atg29/31 elution buffer. Add DTT and EDTA to a final concentration of 1 mM each, as well as 50 μL His-PreScission. Incubate at room temperature for 1 h, and keep at 4 °C overnight (*see* **Note 3**).
6. Centrifuge the sample for 10 min full speed in a tabletop centrifuge, and load it on a HiLoad 16/600 Superdex 200 gel filtration column.
7. Run with Atg29/31 SEC buffer. The complex elutes at around 64 mL.

3.4 Atg1 Purification

Atg1 exerts a kinase-independent function during autophagy initiation and a kinase-dependent regulatory function during later stages of autophagosome biogenesis. Its activity is regulated by phosphorylation of threonine 226 and serine 230 [20]. We expressed Atg1 in High Five cells (insect cells) to preserve phosphorylation of both regulatory sites. Purified Atg1 shows two distinct bands in SDS-polyacrylamid-gelelectrophoresis with the lower band corresponding to the phosphorylated species (Fig. 1, *third left line*). This ensured that Atg1 was expressed and purified in its active form.

3.4.1 Atg1 Expression

Atg1 was expressed in baculovirus-infected High Five cells as described previously [21].

1. Transform the plasmid into chemocompetent DH10-MultiBac cells, and select for successful integration into the bacmid DNA by blue/white screening.
2. Make a bacmid DNA preparation from a 2 mL overnight culture.
3. Transfect Sf9 cells (one well of a six-well plate) with the bacmid DNA.
4. Three days post transfection, harvest the medium, and add it to a 25 mL Sf9 suspension culture (0.5×10^6 cells/mL).
5. On the day of proliferation arrest (dpa), add medium if necessary to keep cell number at around 10^6 cells/mL, and culture the cells for 2 more days.
6. On dpa two, harvest the medium, and use this virus stock to infect a large culture of High Five cells (*see* **Note 4**). Ideally, cells should double once before proliferation arrests. The correct amount of virus stock needed can be tested first with a small culture.
7. Harvest the cells at dpa two.

3.4.2 Affinity and Size-Exclusion Chromatography

1. Resuspend cell pellet from 2 L High Five culture in 60 mL lysis buffer supplemented with 500 μL Protease Inhibitor Cocktail, PMSF to a final concentration of 1 mM, and 20 μL benzonase.
2. Disrupt the cells using a douncer, and centrifuge the lysate for 1 h at 50,000 × *g* and 4 °C (*see* **Note 5**).
3. Load the supernatant on a 5 mL HisTrap FF crude column attached to an Aekta system (1.0 mL/min).
4. Wash with 100 mL Atg1 HisTrap buffer A, and perform a step elution with 100% Atg1 HisTrap buffer B.
5. Add DTT and EDTA to a final concentration of 1 mM each, as well as 20 μL His-PreScission and 50 μL His-TEV. Incubate at room temperature for 1 h to cleave off the tags.
6. Centrifuge the sample for 10 min full speed in a tabletop centrifuge, and load it on a HiLoad 16/600 Superdex 200 gel filtration column.
7. Run with Atg1 SEC buffer. The complex elutes at around 65 mL.

3.5 Atg13 Purification

The physical interaction of Atg13 and Atg17 is regulated by phosphorylation, and partial dephosphorylation of Atg13 is required for both proteins to interact. Expression of Atg13 in *E.coli* resulted in non-phosphorylated protein which readily interacted with Atg17 and Atg1 to facilitate assembly of the pentameric complex. Interestingly, addition of ATP to the reconstituted pentameric complex resulted in Atg13-phosphorylation at its regulatory phosphorylation sites and consequently abolished its interaction with Atg17, mimicking the situation in non-induced cells [15]. Its interaction with Atg1 was, however, retained and not sensitive with respect to the phosphorylation state, arguing that both proteins form a constitutive complex in the absence of autophagy in vivo.

3.5.1 Atg13 Expression

1. Inoculate 65 mL LB medium containing 35 μg/mL kanamycin and 20 μg/mL chloramphenicol with *E. coli* Rosetta (DE3) transformed with pCoofy37-Atg13, and incubate at 37 °C overnight.
2. Add 60 mL overnight culture to 6 L TB medium supplemented with kanamycin and chloramphenicol, and grow the culture at 37 °C.
3. Cool the culture to 25 °C prior to inducing with 0.3 mM IPTG at an OD600 of 0.5. Continue expression for 3 h.

3.5.2 Affinity and Ion Exchange Chromatography

1. Spin the culture for 15 min at 4500 × *g*, and resuspend the cell pellet in 100 mL Atg13 lysis buffer supplemented with freshly added PMSF (1 mM final concentration), 500 μL Protease Inhibitor Cocktail, and 10 μL benzonase.

2. Lyse the cells by sonication, and centrifuge for 1 h at 50,000 × *g* and 4 °C.
3. Load on a 5 mL HisTrap FF crude column attached to an Aekta system (1.5 mL/min).
4. Wash with 50 mL Atg13 HisTrap buffer A, and elute with a 0.3%/mL gradient of Atg13 HisTrap buffer B.
5. Add DTT and EDTA to a final concentration of 1 mM each to the pooled elution fractions, as well as 10 μL His-PreScission and 20 μL His-TEV. Incubate at room temperature for 1 h.
6. Dilute with Atg13 MonoQ dilution buffer to a final NaCl concentration of 125 mM.
7. Load the sample on a 1 mL HiTrap Q HP column, and elute with a gradient of 0.32%/mL Atg13 MonoQ buffer B (*see* **Note 6**).
8. To concentrate, pool the Atg13 fractions, and dilute the sample with Atg13 MonoQ dilution buffer to a final NaCl concentration of 125 mM. Bind again to the 1 mL HiTrap Q HP column, and elute with Atg13 MonoQ buffer with 300 mM NaCl (*see* **Note 7**).

3.6 Generation of Atg9 Proteoliposomes

A key step in reconstituting the interactions of Atg9 with the Atg1-kinase complex involves generation of Atg9 vesicles. Atg9 is a transmembrane protein, and meaningful experiments cannot be performed with the detergent-solubilized protein since detergents interfere with protein folding as well as protein-protein interactions. The first step to incorporate Atg9 into membranes requires the detergent-solubilized Atg9 solution to be added to detergent-lipid suspensions. The detergent is solubilizing lipids and Atg9 by forming micelles that shelter the hydrophobic moieties from the aqueous solution. An intrinsic property of all detergents is the concentration at which micelles are being formed, termed critical micelle concentration (CMC). In order to allow solubilized lipids to form membrane bilayers, the detergent concentration must be reduced to values well below the CMC. This can be done by either slow withdrawal of detergent through dialysis or by rapid dilution of the protein-lipid-detergent suspension with buffer. For Atg9, the rapid dilution approach was used. The procedure of generating Atg9-proteoliposomes is explained in detail below.

1. Prepare a mix of the following lipids in chloroform, with the given molar percentage: cholesterol (20%), POPC (60%), POPS (10%), POPE (10%), lissamine-rhodamine-PE (0.1%).
2. Dry the lipids under a gentle nitrogen stream, and completely evaporate the chloroform by incubating under vacuum for 1 h.
3. Resuspend the lipid film by adding 1 mL Atg9 SEC buffer per mg lipids, and vortex for 1 min.

4. Add LDAO to a final concentration of 0.15%, and vortex until the suspension turns clear.
5. Add Atg9 to the solubilized lipids in a protein-lipid molar ratio of 1:200. Incubate for 30 min at room temperature (*see* **Note 8**).
6. Quickly dilute the mixture with Atg9 SEC buffer below the critical micelle concentration of LDAO. Incubate for 30 min at room temperature.
7. Ultracentrifuge the sample for 30 min at 150,000 × *g*. Remove the buffer completely, and resuspend the liposomes in 1 mL liposome buffer per mg lipids.
8. Sonicate shortly to break down larger aggregates.
9. For protein-free liposomes used in control experiments, resuspend the lipid film directly in liposome buffer.

3.7 Co-floatation of Atg Proteins with Atg9-PLs

The interaction of Atg1-kinase complex subunits and Atg9-PLs was assessed using co-floatation experiments. This approach uses the different densities of membranes and water. Membrane vesicles containing Atg9 (Atg9-PLs) or lacking Atg9 (SUVs, control reactions) were mixed with the potential interaction partner. After incubation, a density gradient centrifugation is performed. Therefore, Histodenz is added to the protein mixture to reach a final concentration of 40% and overlay by less-dense layers as described below. During ultracentrifugation, liposomes are migrating to the interface between 30% Histodenz and buffer, whereas unbound protein remains in the bottom fraction (40% Histodenz), and protein aggregates are sedimented. This allows proteins which interact with membranes or Atg9 to be recovered from the top fractions that contain Atg9-PLs or vesicles lacking Atg9 (SUVs).

1. Mix purified proteins with 250 μL Atg9 proteoliposomes (or liposomes without Atg9 for control experiments) at a molar ratio of proteins-Atg9 of 1:6. Add liposome buffer to a final volume of 300 μL.
2. Sonicate the sample using a low power output and short pulses until the sample turns clear (*see* **Note 9**).
3. Mix with the same volume of Histodenz buffer to reach a final concentration of 40% Histodenz. Transfer to an ultracentrifuge tube.
4. Set up the density gradient by overlying with 30% Histodenz (prepared by diluting the stock solution with liposome buffer) and liposome buffer as top layer.
5. Ultracentrifuge for 1 h at 165,000 × *g*.
6. Collect the liposomes from the interface in between liposome buffer and 30% Histodenz layers.

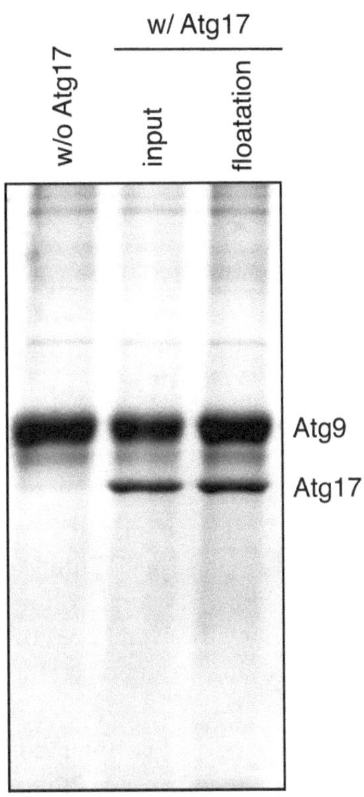

Fig. 2 Floatation experiment of Atg9-proteoliposomes (PLs) and Atg17. The first lane shows Atg9-PLs used to set up the experiment. Input corresponds to the mixture of Atg9-PLs and Atg17 before floatation, and floatation shows the sample recovered from the top floating fraction that contains Atg9-PLs. Atg17 is recovered from the Atg9-PLs containing floating fraction, indicating that both proteins are physically interacting

The recovered liposome fraction is analyzed by SDS-PAGE to determine which of the added proteins are co-floating. An example is shown in Fig. 2.

4 Notes

1. Cell lysis is a critical step during Atg9 production. Good yields were only obtained when *E. coli* cells were lysed using high-pressure homogenizer systems (e.g., Avestin Emulsiflex-C3). Sonication is not recommended as it induces protein aggregation.
2. Be careful when transferring the supernatant; the pellet might be very loose. If larger amounts of supernatant are transferred to the ultracentrifugation tube, the final amount of detergent needed will be significantly higher.

3. Due to the presence of detergent, the activity of the protease for cleaving off the His-tag is low and the cleavage reaction therefore inefficient. Thus, a higher amount of PreScission Protease and a longer incubation times are recommended.
4. Culturing High Five cells requires a critical concentration of L-glutamine to be maintained. It is therefore recommended to supplement the medium with L-glutamine prior to use even if the medium already contains L-glutamine.
5. Take care when transferring the supernatant; the pellet might be quite loose. The supernatant itself is probably not entirely cleared and appears cloudy. This, however, does not affect binding of the protein to the HisTrap column so that the supernatant can still be applied. Alternatively, the centrifugation step can be repeated with the precleared supernatant.
6. It is sufficient to run the salt gradient to about 250 mM NaCl as Atg13 will elute very early. Afterward, regenerate the columns with several column volumes of MonoQ buffer B, and wash with several column volumes MonoQ buffer A before proceeding to the next step.
7. For Atg13, concentration with concentrators based on ultrafiltration is not recommended. Atg13 binds to both, PES and regenerated cellulose filters, which leads to loss of protein.
8. To avoid co-sedimentation of possible Atg9 aggregates with the proteoliposomes, Atg9 can be pre-ultracentrifuged for 10 min at 150,000 × *g* before adding it to the lipids.
9. If the pulses are too strong, the proteoliposomes can easily be damaged or destroyed. The optimal settings vary between different devices, but it is recommended to start with.

References

1. Reggiori F, Klionsky DJ (2005) Autophagosomes: biogenesis from scratch? Curr Opin Cell Biol 17(4):415–422
2. Yang Z, Klionsky DJ (2010) Eaten alive: a history of macroautophagy. Nat Cell Biol 12 (9):814–822. https://doi.org/10.1038/ncb0910-814
3. Feng Y, He D, Yao Z, Klionsky DJ (2014) The machinery of macroautophagy. Cell Res 24 (1):24–41. https://doi.org/10.1038/cr.2013.168
4. Suzuki K, Kubota Y, Sekito T, Ohsumi Y (2007) Hierarchy of Atg proteins in pre-autophagosomal structure organization. Genes Cells 12(2):209–218. https://doi.org/10.1111/j.1365-2443.2007.01050.x
5. Kawamata T, Kamada Y, Kabeya Y, Sekito T, Ohsumi Y (2008) Organization of the pre-autophagosomal structure responsible for autophagosome formation. Mol Biol Cell 19 (5):2039–2050. https://doi.org/10.1091/mbc.E07-10-1048
6. Sekito T, Kawamata T, Ichikawa R, Suzuki K, Ohsumi Y (2009) Atg17 recruits Atg9 to organize the pre-autophagosomal structure. Genes Cells 14(5):525–538. https://doi.org/10.1111/j.1365-2443.2009.01299.x
7. Kabeya Y, Noda NN, Fujioka Y, Suzuki K, Inagaki F, Ohsumi Y (2009) Characterization of the Atg17-Atg29-Atg31 complex specifically required for starvation- induced autophagy in Saccharomyces cerevisiae. Biochem Biophys Res Commun 389(4):612–615. https://doi.org/10.1016/j.bbrc.2009.09.034
8. Kamada Y, Yoshino K-I, Kondo C, Kawamata T, Oshiro N, Yonezawa K, Ohsumi

Y (2010) Tor directly controls the Atg1 kinase complex to regulate autophagy. Mol Cell Biol 30(4):1049–1058. https://doi.org/10.1128/MCB.01344-09

9. Kabeya Y, Kamada Y, Baba M, Takikawa H, Sasaki M, Ohsumi Y (2005) Atg17 functions in cooperation with Atg1 and Atg13 in yeast autophagy. Mol Biol Cell 16(5):2544–2553. https://doi.org/10.1091/mbc.E04-08-0669
10. Kamada Y (2010) Prime-numbered Atg proteins act at the primary step in autophagy: unphosphorylatable Atg13 can induce autophagy without TOR inactivation. Autophagy 6(3):415–416
11. Fujioka Y, Suzuki SW, Yamamoto H, Kondo-Kakuta C, Kimura Y, Hirano H et al (2014) Structural basis of starvation-induced assembly of the autophagy initiation complex. Nat Struct Mol Biol 21(6):513–521. https://doi.org/10.1038/nsmb.2822
12. Mari M, Griffith J, Rieter E, Krishnappa L, Klionsky DJ, Reggiori F (2010) An Atg9-containing compartment that functions in the early steps of autophagosome biogenesis. J Cell Biol 190(6):1005–1022. https://doi.org/10.1083/jcb.200912089
13. Yamamoto H, Kakuta S, Watanabe TM, Kitamura A, Sekito T, Kondo-Kakuta C et al (2012) Atg9 vesicles are an important membrane source during early steps of autophagosome formation. J Cell Biol 198(2):219–233. https://doi.org/10.1083/jcb.201202061
14. Nair U, Jotwani A, Geng J, Gammoh N, Richerson D, Yen WL et al (2011) SNARE proteins are required for macroautophagy. Cell 146(2):290–302. https://doi.org/10.1016/j.cell.2011.06.022
15. Rao Y, Perna MG, Hofmann B, Beier V, Wollert T (2016b) The Atg1-kinase complex tethers Atg9-vesicles to initiate autophagy. Nat Commun 7:10338. https://doi.org/10.1038/ncomms10338
16. Matscheko N, Mayrhofer P, Wollert T (2017) Passing membranes to autophagy: unconventional membrane tethering by Atg17. Autophagy 13(3):629–630. https://doi.org/10.1080/15548627.2016.1276678
17. Rao Y, Matscheko N, Wollert T (2016a) Autophagy in the test tube in vitro reconstitution of aspects of autophagosome biogenesis. FEBS J 283(11):2034–2043. https://doi.org/10.1111/febs.13661
18. Tan S (2001) A modular polycistronic expression system for overexpressing protein complexes in Escherichia coli. Protein Expr Purif 21(1):224–234. https://doi.org/10.1006/prep.2000.1363
19. Scholz J, Besir H, Strasser C, Suppmann S (2013) A new method to customize protein expression vectors for fast, efficient and background free parallel cloning. BMC Biotechnol 13(1):12. https://doi.org/10.1186/1472-6750-13-12
20. Kijanska M, Dohnal I, Reiter W, Kaspar S, Stoffel I, Ammerer G et al (2010) Activation of Atg1 kinase in autophagy by regulated phosphorylation. Autophagy 6(8):1168–1178. https://doi.org/10.4161/auto.6.8.13849
21. Bieniossek C, Richmond TJ, Berger I (2001) MultiBac: multigene baculovirus-based eukaryotic protein complex production, vol 21. John Wiley & Sons, Inc., Hoboken, NJ, pp 5.20.1–5.20.26. https://doi.org/10.1002/0471140864.ps0520s51

Chapter 7

Cell-Free Reconstitution of Autophagic Membrane Formation

Min Zhang and Liang Ge

Abstract

Autophagy is a catabolic pathway for bulk turnover of cytoplasmic components through the lysosome. Completion of autophagy requires a sophisticated membrane remodeling process. The early steps involve autophagic membrane precursor generation from the intracellular membranes. The intricate protein-membrane interactions underlying autophagic membrane precursor generation have been a focus of attention but yet poorly defined. Here, we summarize the procedure of a cell-free system we have established to dissect the molecular mechanism of early autophagic membrane generation.

Key words Autophagy, Autophagosome, Cell-free reconstitution, LC3 lipidation

1 Introduction

Autophagy is a bulk degradation process for the cell to cope with stress and maintain homeostasis. Autophagy decay leads to the accumulation of protein aggregates and damaged organelles which is associated with aging and the related diseases, such as neurodegeneration [1, 2]. Biogenesis of double-membrane autophagosomes is a central step of autophagy, which requires significant membrane remodeling of the intracellular membrane compartments. In starvation-induced autophagy, a hierarchical action of the autophagy-related gene (ATG) proteins, together with non-ATG proteins, occurs at special sites on the intracellular membrane to facilitate the membrane remodeling processes [3–5]. It has been proposed that sophisticated protein-membrane interactions govern the initiation and maturation of the autophagosome, as well as rejuvenation of the lysosome after autophagy completion [3–6]. The underlying molecular events in each step of autophagy have been unclear.

Here we focus our attention on the initiation of autophagic membrane formation. There have been two major questions: (1) which intracellular compartment contributes membrane to

Nicholas Ktistakis and Oliver Florey (eds.), *Autophagy: Methods and Protocols*, Methods in Molecular Biology, vol. 1880, https://doi.org/10.1007/978-1-4939-8873-0_7, © Springer Science+Business Media, LLC, part of Springer Nature 2019

the autophagosome (the membrane source) and (2) how does this compartment generate the autophagic membrane? In the initial step of autophagic membrane formation, the membrane sources and the autophagic membrane precursors lack a distinct feature like a cup-shaped phagophore or a double-membrane autophagosome. Hence, it has been difficult to clarify the early membrane remodeling events based on traditional cell imaging. We developed a cell-free approach to reconstitute an early step of autophagic membrane formation. This assay is based on analyzing LC3 lipidation, a key molecular event of autophagosome biogenesis [7–9]. The cell-free assay allows us to combine a membrane fractionation approach to pinpoint an intracellular compartment, the ER-Golgi intermediate compartment (ERGIC), as a membrane source for the autophagosome [8, 9]. Starting from this point, we have subsequently identified a membrane remodeling pathway involving the ER-exit sites (ERES), the ERGIC, and the COPII machinery, which is essential for early autophagic membrane generation [8–12].

2 Materials

We prepare all buffers using ultrapure water (18 MΩ-cm at 25 °C) and analytical grade reagents. Storage of reagents and buffers is listed specifically below.

1. B88 buffer: 20 mM HEPES-KOH, pH 7.2, 250 mM sorbitol, 150 mM potassium acetate, and 5 mM magnesium acetate (store at 4 °C) (*see* **Note 1**).
2. B88 lysis buffer: B88 buffer with protease inhibitors (Roche), phosphatase inhibitors (Roche), and 0.3 mM DTT (store at −20 °C).
3. PBS: 137 mM sodium chloride, 2.7 mM potassium chloride, 10 mM disodium phosphate, and 1.8 mM monopotassium phosphate; adjust pH to 7.2 (store at 4 °C).
4. EBSS: Earle's balanced salt solution (Thermo) (store at room temperature).
5. DMEM: Dulbecco's Modified Eagle's Medium (store at 4 °C) and FBS (Fetal Bovine Serum) (store at −20 °C).
6. 10×ATP regeneration system: 400 mM creatine phosphate, 2 mg/mL creatine phosphokinase, and 10 mM ATP in B88 (provides ATP during cell-free reaction) (store at −80 °C).
7. HB1 buffer: 20 mM HEPES-KOH, pH 7.2, 400 mM sucrose, and 1 mM EDTA (store at 4 °C).
8. HB1 lysis buffer: HB1 buffer with protease inhibitors (Roche), phosphatase inhibitors (Roche), and 0.3 mM DTT (store at −20 °C).

9. 1.25 M, 1.1 M, and 0.25 M sucrose buffers (Golgi isolation kit, Sigma) (store at 4 °C).
10. 50% Optiprep solution: 50% Optiprep (Sigma, 60% stock solution), 20 mM Tricine-KOH, pH 7.4, 42 mM sucrose, and 1 mM EDTA (store at 4 °C).
11. Optiprep dilution buffer: 20 mM Tricine-KOH, pH 7.4, 250 mM sucrose, and 1 mM EDTA (store at 4 °C).
12. GTP 10 mM stock (activates GTP-binding proteins) (store at −80 °C).
13. 0.05% Trypsin-EDTA 1× (Gibco) (store at 4 °C for short term and − 20 °C for long term).
14. Phospholipids C Reagent [Wako, for measuring phosphatidylcholine (PC) level] (store at 4 °C; after making the reaction solutions, aliquot and freeze at −80 °C).
15. Trypan blue (store at room temperature).
16. Centrifuge tubes (1.5, 15, and 50 mL).
17. 22G needles and syringes (1, 3, and 5 mL).
18. Glass slides and coverslips.
19. Student microscope (to check trypan blue stain).
20. Refrigerated benchtop swinging bucket centrifuge for 15 and 50 mL conical tubes (e.g., Sorvall ST 16R, Thermo).
21. Benchtop ultracentrifuge and rotors (for differential centrifugation and sucrose gradient ultracentrifugation, e.g., Optima Max-XP, Beckman; rotors: TLA100.3, TLS-55).
22. Ultracentrifuge and rotors (for Optiprep ultracentrifugation, e.g., Optima XE-90, Beckman; rotor: SW-55).
23. Microplate spectrophotometer (to measure protein and phospholipid concentration with 96-well plates, e.g., xMark™ Microplate Absorbance Spectrophotometer, Bio-Rad).
24. Ultracentrifuge tubes (11 × 34 mm ultra-clear and polycarbonate, 13 × 51 mm ultra-clear and polycarbonate).
25. Refrigerated bench top microfuge (for centrifuging 1.5 mL Eppendorf tubes to collect total donor membrane, e.g., Eppendorf Centrifuge 5430R).

3 Methods

3.1 Cell-Free Lipidation with Total Cellular Membrane

This section describes the basic procedure of the cell-free lipidation assay (Fig. 1a). Major steps include total cellular membrane collection, cytosol collection, His-T7-LC3 purification, and the cell-free reaction. Below we describe the procedures for preparing the

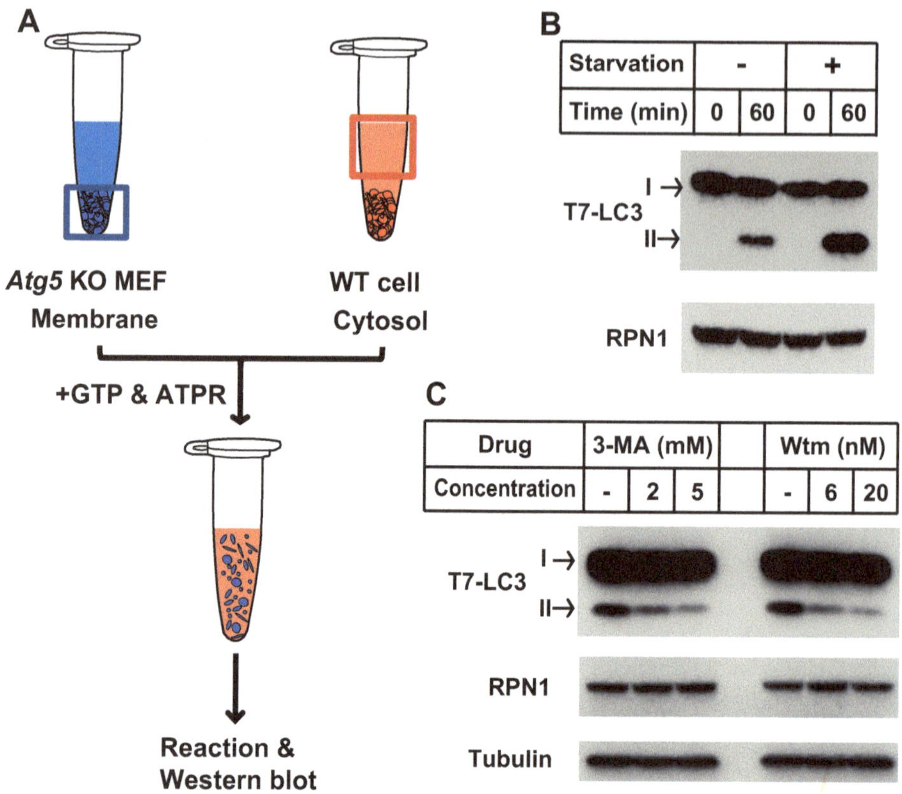

Fig. 1 Cell-free LC3 lipidation assay (modified from ref. [8]). (**a**) Diagram showing cell-free reconstitution of LC3 lipidation. Membranes from Atg5 knockout (KO) MEFs were incubated with cytosol from wild-type (WT) cells plus GTP and an ATP regeneration system (ATPR). Then SDS-PAGE and immunoblot were performed to detect the generation of lipidated LC3. (**b**) LC3 lipidation reaction showing the enhanced activity of cytosol and membrane collected from starved cells. HEK293T and Atg5 KO MEF cells were either untreated or starved for 90 min. The in vitro lipidation reaction was performed by incubating T7-LC3 with HEK293T cytosols and Atg5 KO MEF membranes with indicated treatments for the indicated times followed by SDS-PAGE and immunoblot. The band with higher mobility is the lipidated band (LC3-II). RPN1, ribophorin1. (**c**) LC3 lipidation reaction showing that the phosphatidylinositol-3 kinase inhibitors, 3-methyladenine (3-MA) and wortmannin (wtm), inhibit the reaction in a dose-dependent manner

membrane and the cytosol and setting up the reaction. We skip the His-T7-LC3 purification because it is a standard His-tag protein purification using nickel columns.

3.2 Total Cellular Membrane Preparation

1. Culture Atg5 knockout (KO) mouse embryonic fibroblast (MEF) cells [13] in DMEM plus 10% FBS for 2–3 days to confluence (*see* **Note 2**).
2. Remove the medium and wash the cells once with PBS. Trypsinize the cells with 1 mL/dish 1× trypsin. Then add 5 mL DMEM plus 10% FBS to quench the action of trypsin. Transfer the cell suspension to a 15 mL conical tube, and centrifuge at 600 × *g* for 5 min at 4 °C to collect the cells. Remove the

supernatant and suspend the cell pellet with 5 mL ice-cold PBS. Repeat the centrifugation.

3. Discard the supernatant, suspend the cell pellet with B88 lysis buffer (use a volume about three times the cell pellet volume), and pass ten times through a 22G needle with a syringe. Place 3 μL of the lysate on a glass slide and stain with trypan blue. Check the fraction of cells that are trypan blue positive. If 80–90% of cells are positive, proceed to the next step. If less than 80% of cells are positive, homogenize the cells again, with a check of cell lysis every five strokes until 80–90% of the cells are lysed. Do not exceed more than 90% lysis.
4. Transfer the lysate to an ultracentrifuge tube and centrifuge at 100,000 × *g* for 30 min. Discard the supernatant, and resuspend the total membrane pellet with minimum amount of B88 lysis buffer. Take 1 μL of membrane suspension to measure the PC concentration with a microplate spectrophotometer. Adjust the membrane concentration to 0.6 μg/μL of PC (*see* **Note 3**).

3.3 Cytosol Preparation

1. Culture human HEK293T (human embryonic kidney) cells in DMEM plus 10% FBS for 2–3 days to confluence. Incubate the cells with DMEM plus 10% FBS (nutrient-rich) or EBSS (starvation) for 1 h (*see* **Note 4**).
2. Scrape the cells into the medium and transfer the cell suspension to a 15 mL conical tube. Centrifuge at 600 × *g* for 5 min at 4 °C to collect the cells. Remove the supernatant and suspend the cell pellet with 5 mL ice-cold PBS (*see* **Note 5**). Repeat the centrifugation.
3. Discard the supernatant, suspend the cell pellet with an equal volume of B88 lysis buffer, and pass ten times through a 22G needle with a syringe. Place 3 μL of the lysate on a glass slide and stain with trypan blue. Check the fraction of cells that are trypan blue positive. If 80–90% of cells are positive, proceed to the next step. If less than 80% of cells are positive, homogenize the cells again, with a check of cell lysis every five strokes until 80–90% of the cells are lysed. Do not exceed more than 90% lysis.
4. Transfer the lysate to an ultracentrifuge tube and centrifuge at 100,000 × *g* for 30 min. Collect the supernatant and repeat the centrifugation if the supernatant is not clear. The clear supernatant is the cytosol ready for use in the reaction (*see* **Note 6**). Measure the protein concentration with the Bradford assay. Aliquot and store at −80 °C.

3.4 Cell-Free Lipidation Reaction

For each reaction, mix the following components on ice: cytosol (2 mg/mL final concentration), 3 μL 10×ATP regeneration system, GTP (0.15 mM final concentration), 0.2 μg T7-LC3, membrane (0.2 mg/mL PC content final concentration), and B88 to make a total of 30 μL in a low retention 1.5 mL microfuge tube.

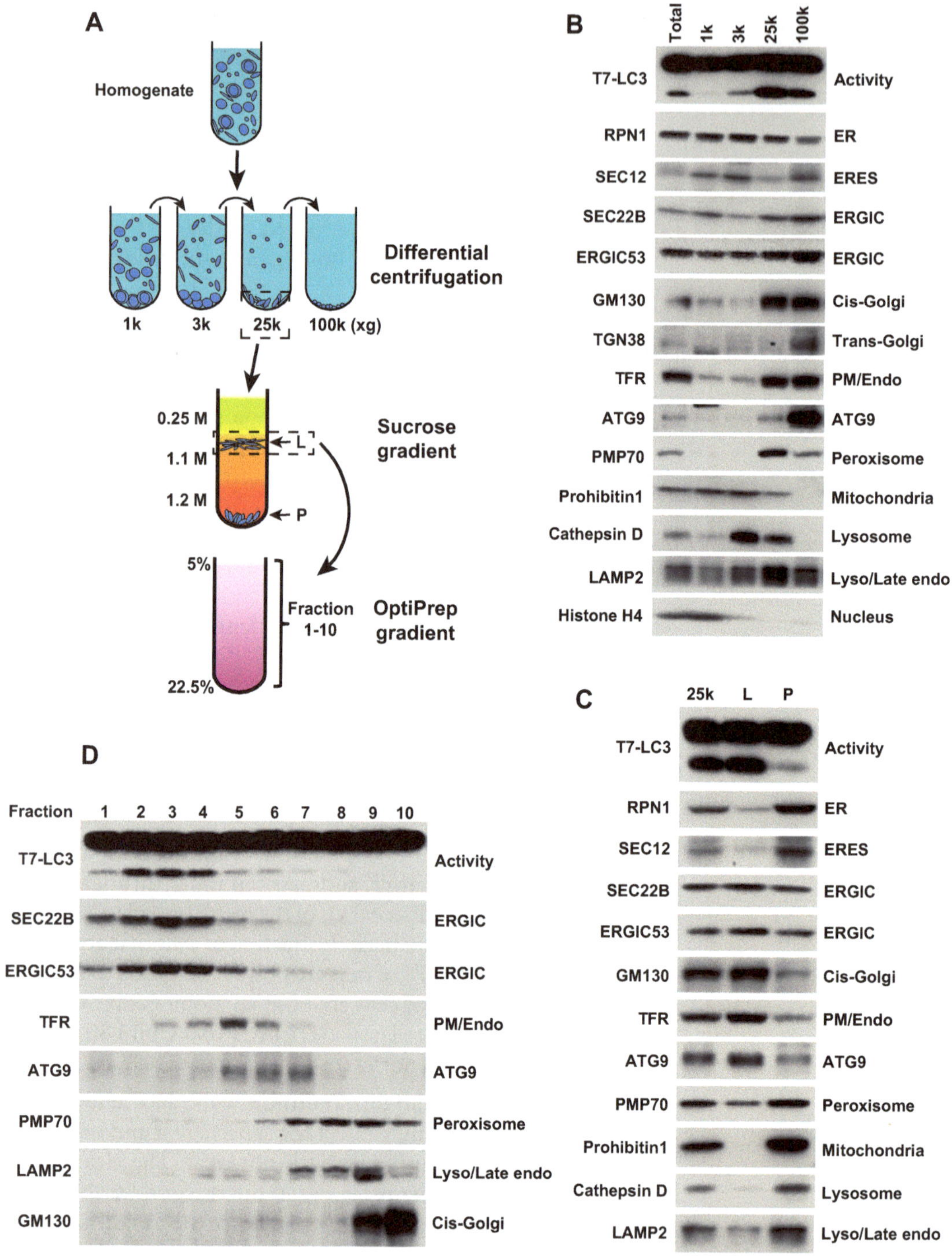

Fig. 2 Three-step membrane fractionation and cell-free lipidation (modified from Ref. [8]). (**a**) Diagram showing the three-step membrane fractionation. Atg5 KO MEFs were homogenized, and the lysates were subjected to differential centrifugations with indicated g forces. The ability of each fraction to trigger T7-LC3 lipidation was examined. The 25,000 × *g* (25k) pellet, which had the most activity (*see* **b**), was selected, and a sucrose gradient ultracentrifugation was performed to separate the 25k pellet to L (Light) and P (Pellet)

Gently tap the tube to mix and incubate at 30 °C for 30–90 min. Quench the reaction by adding SDS-PAGE loading buffer to make a total of 80 μL. Load 12 μL of the sample for immunoblotting with a T7 antibody (*see* **Note 7**) (Fig. 1b, c).

3.5 Three-Step Membrane Fractionation and Cell-Free Lipidation

This section describes the procedure of the cell-free lipidation assay using the membranes from a three-step membrane fractionation to look for the most active membrane fraction (Fig. 2a).

Major steps include the three-step membrane fractionation and the cell-free reaction.

3.5.1 Three-Step Membrane Fractionation

1. Culture Atg5 KO MEF (ten 15 cm dishes) in DMEM plus 10% FBS for 2–3 days to confluence.
2. Remove the medium and wash the cells once with PBS. Trypsinize the cells with 1 mL/dish 1× trypsin. Then add 5 mL DMEM plus 10% FBS to quench the action of trypsin. Transfer the cell suspension to 15 mL conical tubes and centrifuge at 600 × *g* for 5 min at 4 °C to collect the cells. Remove the supernatant and suspend the cell pellet with 5 mL ice-cold PBS. Repeat the centrifugation.
3. Discard the supernatant, suspend the cell pellet with HB1 lysis buffer (use a volume about three times the cell pellet volume), and pass ten times through a 22G needle with a syringe.
4. Pipet 3 μL of the lysate on a glass slide and stain with trypan blue. Check the fraction of cells that are trypan blue positive. If 80–90% of cells are positive, proceed to the next step. If less than 80% of cells are positive, homogenize the cells again, with a check of cell lysis every five strokes until 80–90% of the cells are lysed. Do not exceed more than 90% lysis.
5. Centrifuge the cell lysate at 1000 × *g* in a swinging bucket rotor for 10 min. Transfer the supernatant to a new 15 mL centrifuge tube. Suspend the pellet with HB1 lysis buffer (use a volume about three times the pellet volume) and centrifuge for 10 min at 1000 × *g*. Keep the pellet (1k membrane) and combine the supernatant from the two centrifugations (*see* **Note 8**).
6. Centrifuge the supernatant at 3000 × *g* in a swinging bucket rotor for 10 min. Keep the pellet (3k membrane), and transfer

Fig. 2 (continued) fractions. The L fraction, which contained the majority of the activity to promote T7-LC3 lipidation (*see* **c**), was further resolved on an Optiprep gradient after which ten fractions from the top were collected and the lipidation activity was examined in each. (**b–d**) The differential centrifugation (**b**), sucrose gradient (**c**), and Optiprep gradient (**d**). The total PC of each fraction was adjusted to a concentration of 0.6 μg/μL. The T7-LC3 lipidation activity of each fraction was tested, and immunoblot was performed to examine the generation of lipidated T7-LC3 as well as the distribution of the indicated membrane markers. *PM* plasma membrane, *Endo* endosome, *lyso* lysosome the supernatant to a 13 × 51 mm polycarbonate ultracentrifuge tube.

7. Ultracentrifuge the supernatant at 25,000 × *g* in a TLA100.3 rotor for 20 min. Keep the pellet (25k membrane), and transfer the supernatant to 13 × 51 mm polycarbonate ultracentrifuge tubes.
8. Ultracentrifuge the supernatant at 100,000 × *g* in a TLA100.3 rotor for 30 min. Keep the pellet (100k membrane) and discard the supernatant.
9. Suspend the 25k membrane with 0.75 mL 1.25 M sucrose buffer. Pass the membrane suspension through a 22G needle ten times to fully disperse the membrane pellet (*see* **Note 9**).
10. Transfer the suspension into an 11 × 34 mm ultra-clear ultracentrifuge tube. Layer on top of the membrane suspension with 0.5 mL 1.1 M sucrose buffer and 0.5 mL 0.25 M sucrose. Ultracentrifuge at 120,000 × *g* for 2 h in a TLS-55 rotor.
11. Collect the membrane layer located at the boundary of 0.25/ 1.1 M sucrose (L fraction) as well as the pellet (P fraction). Sometimes the L fraction layer is broad. In that case, collect all floating membranes.
12. Dilute the L fraction three times with HB1 buffer in a 13 × 51 mm polycarbonate ultracentrifuge tubes and ultracentrifuge at 100,000 × *g* in a TLA100.3 rotor for 40 min (*see* **Note 10**).
13. Make 22.5%, 19%, 16%, 12%, 8%, and 5% Optiprep solutions by diluting the 50% Optiprep with Optiprep dilution buffer. Suspend the L fraction membrane pellet with 1 mL 19% Optiprep gradient by passing through a 22G needle ten times. In a 13 × 51 mm ultra-clear ultracentrifuge tube, layer Optiprep solutions of 0.5 mL 22.5%, 1 mL 19% (with the membrane), 0.9 mL 16%, 0.9 mL 12%, 1 mL 8%, 0.5 mL 5%, and 0.2 mL HB1 buffer. Ultracentrifuge at 150,000 × *g* in a SW-55 rotor for 3 h.
14. Collect 10 fractions, 0.5 mL each, from the top to the bottom. Dilute with 2 mL B88 buffer in a 13 × 51 mm polycarbonate centrifuge tube and ultracentrifuge for 40 min at 100,000 × *g*. Keep the membrane pellets (#1–10).

3.5.2 Lipidation Reaction

Suspend all membrane pellets (3k, 25k, 100k, L, P, and ten fractions of Optiprep gradient) with B88 buffer, and centrifuge to collect the membrane to exchange the buffer to B88 (for 3k, 25k, and 100k, just use the same speed as indicated. For L, P, and ten fractions of Optiprep gradient, use 100,000 × *g* to pellet the membrane). Suspend the membrane pellets with minimum amount of B88 lysis buffer, measure PC concentration, and adjust to 0.6 μg/μL of PC. Perform the LC3 lipidation reaction with the membranes using the protocol indicated above (Fig. 2b–d).

3.6 Small Autophagosomal Precursor Formation Assay

The cell-free small autophagosomal precursor formation assay includes collecting donor membranes (from total cellular membrane or membrane fractionation), cytosols, a cell-free small vesicle formation assay, and a cell-free LC3 lipidation assay to determine the amount of autophagosomal precursors generated in the small vesicle fraction (Fig. 3a).

3.7 Donor Membrane Prepared from Total Cellular Membrane

Generally, we perform the assay with the donor membrane from total cellular membrane. Collect and lyse Atg5 KO MEF in B88 lysis buffer according to Subheading 3.1. Centrifuge the lysate at 20,000 × *g* for 20 min in a refrigerated microfuge. Discard the supernatant and carefully suspend the pellet with B88. Repeat the centrifugation, and carefully suspend the donor membrane pellet with B88 lysis buffer. Adjust concentration to 1 μg/μL PC.

3.8 Donor Membrane Prepared from Membrane Fractionation

The purpose is to find the membrane fraction active for generating autophagosomal precursors. The donor membranes include the 3k fraction, the P fraction, and four fractions from the Optiprep gradient (*see* **Note 11**) (Fig. 3b).

3k fraction: Collect 3k membrane fraction according to Subheading 3.5.

P fraction: Collect P membrane fraction according to Subheading 3.5, except that after suspending the pellet with B88 buffer; centrifuge the membrane at 25,000 × *g* for 40 min. Keep the pellet and suspend it in B88 lysis buffer.

Four fractions from Optiprep gradient: Perform the Optiprep gradient and collect ten fractions according to Subheading 3.5. Combine fractions to four fractions F1 (1&2), F2 (3&4), F3 (5&6), and F4 (7–10). Dilute the combined fractions with B88 buffer (at least three times the volume of the Optiprep fraction), and collect the membrane at 25,000 × *g* for 40 min. Suspend in B88 lysis buffer.

Adjust all membrane fractions to a concentration of 1 μg/μ L PC.

3.9 Cytosol Preparation

Prepare Atg5 KO MEF cytosol and HEK293T cytosol according to Subheading 3.2 (*see* **Note 12**). Cell-free small vesicle formation assay (*see* **Note 13**).

1. In a low-retention 1.5 mL Eppendorf tube, make a similar cell-free reaction containing the Atg5 KO MEF cytosol (2 μg/μL), 1×ATP regeneration system, GTP (0.15 mM), and donor membrane (0.2 μg/μL PC). Add B88 buffer to bring the total volume to 50 μL. Gently tap to mix and incubate the mixture at 30 °C for 1 h.

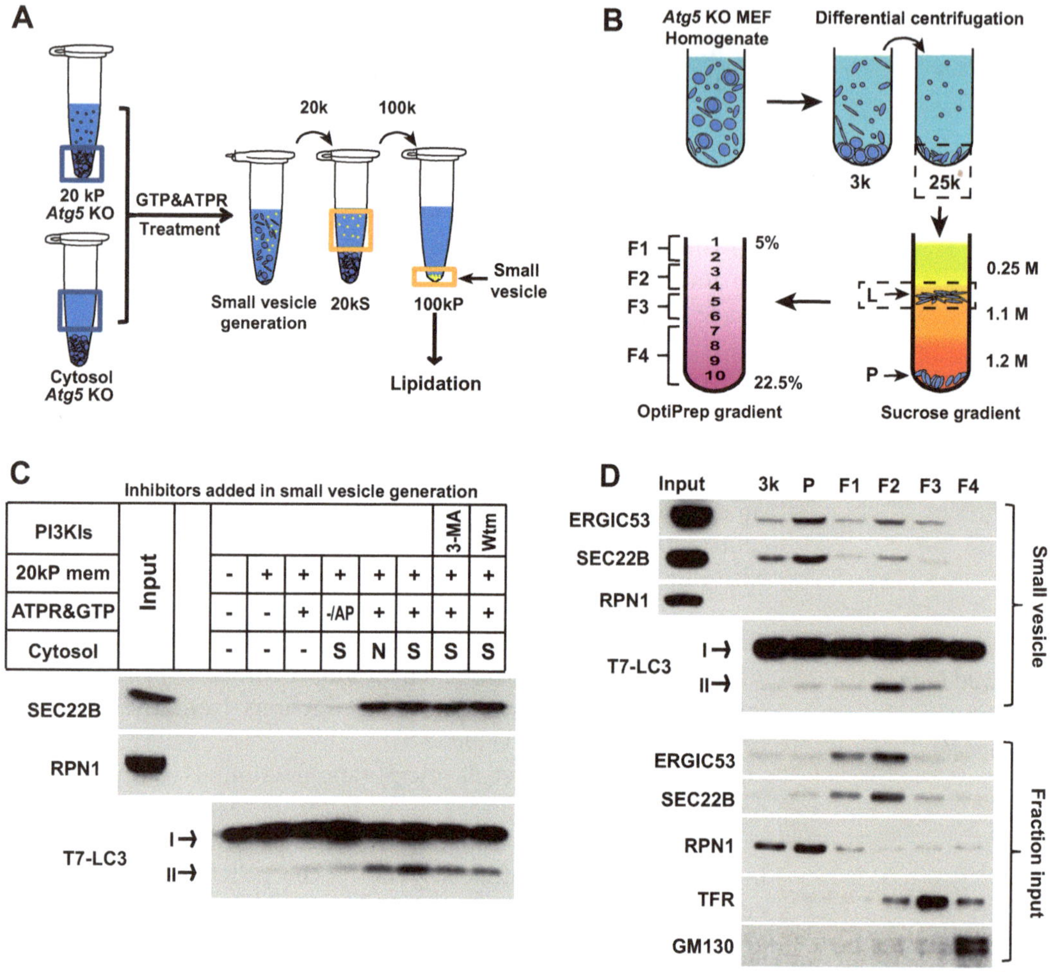

Fig. 3 Cell-free autophagosomal precursor formation assay (modified from Ref. [10]). (**a**) Diagram showing the cell-free small autophagosomal precursor generation assay with total membrane as the donor membrane. The medium-speed membrane pellet (20k $\times$ *g*, 20KP) from Atg5 KO MEFs was harvested and incubated with cytosol from Atg5 KO MEFs, GTP and an ATP regeneration system (ATPR), and in the absence or presence of drug treatments. The slowly sedimenting membranes generated were separated from the donor membrane by a medium-speed centrifugation (20k $\times$ *g*) and collected by high-speed sedimentation (100k $\times$ *g*) of the supernatant fraction (20kS). Then a lipidation assay with HEK293T cytosol was performed to analyze the competency of the small membranes (100kP) to induce LC3 lipidation. (**b**) Diagram showing the cell-free small autophagosomal precursor generation assay with fractionated membranes. Briefly, Atg5 KO MEF membranes were subjected to differential centrifugation to collect 3k $\times$ *g* and 25k $\times$ *g* pellet fractions. Sucrose gradient ultracentrifugation was performed to separate the L and P fraction after which the L was further separated by Optiprep gradient ultracentrifugation. Ten fractions from top to bottom were collected and combined to four as indicated (F1–F4). These membrane fractions were then employed as donor membranes for the cell-free small autophagosomal precursor generation assay. (**c**) A cell-free small autophagosomal precursor generation assay using the total cellular membrane with the indicated conditions. Small vesicles were collected to determine activity in the LC3 lipidation reaction. The result shows the dependence of cytosol, membrane, and ATPR&GTP for generating small autophagosomal precursors. Also, cytosol from starved cells stimulates the production of autophagosomal precursors, and PI3K inhibitors attenuate the process. SEC22B is a control for COPII vesicle

2. To remove the donor membrane, centrifuge the reaction mixture at 20,000 × *g* (total cellular membrane) or 25,000 × *g* (membrane fractions) for 20 min.
3. To collect the small vesicles generated, transfer 35 μL supernatant to a new 1.5 mL Eppendorf tube and ultracentrifuge at 100,000 × *g* for 30 min in a TLA100.3 rotor with adapters or a TLA-55 rotor (*see* **Note 14**). Discard the supernatant and keep the pellet. Sometimes the pellet is hard to see, but it is there.

3.10 Analyzing the Generation of Small Autophagosomal Precursors via LC3 Lipidation

Make a master mix of HEK293T cytosol (2 μg/μL), 1× ATP regeneration system, GTP (0.15 mM), and T7-LC3 (10 ng/mL) in B88. Suspend the small membrane pellet generated in the cell-free small vesicle formation assay with 15 μL master mix by pipetting. Incubate the mixture at 30 °C for 1 h. Ultracentrifuge the mixture at 100,000 × *g* for 30 min. Remove 12 μL supernatant carefully (*see* **Note 15**). Add 60 μL 1× SDS-PAGE loading buffer to the pellet plus the 3 μL supernatant. Determine the LC3 lipidation level using a 12 μL sample by immunoblot (Fig. 3c, d).

4 Notes

1. The pH of the B88 buffer is crucial. Lower pH leads to low lipidation efficiency, and higher pH leads to nonspecific lipidation [14].
2. Use cells with passages less than 20. Membranes from passages higher than 20 are less active in triggering LC3 lipidation and are prone to fragmentation, making it difficult to isolate them in the membrane fractionation described in Subheading 3.2.
3. The pellet in this step is hard and difficult to resuspend. Try incubating the pellet in B88 lysis buffer on ice for 30 min to 1 h, after which the pellet will soften. We do not recommend vortexing. Suspend the pellet by pipetting. PC is the most abundant lipid in the membrane. Therefore, we use PC as a standard to normalize membrane amounts in the following experiments.
4. In addition to starvation, sometimes we also treat the cells with drugs, such as rapamycin and Torin-1, to activate the cytosolic

Fig. 3 (continued) formation. 3-MA, 5 mM; Wortmannin, 20 nM; S, starved Atg5 KO MEF cytosol; N, untreated Atg5 KO MEF cytosol; −/AP, in the absence of ATPR&GTP and in the presence of Apyrase (AP). (**d**) A cell-free small autophagosomal precursor generation assay performed with the 3k, P, F1–F4 fractions (upper four panels: small vesicle). SDS-PAGE and immunoblot with indicated antibodies were carried out to probe the organelle enrichment of each fraction (lower five panels: fraction input). The result shows that the ERGIC is the membrane mostly active in generating the small autophagosomal precursors for LC3 lipidation. SEC22B is a control for COPII vesicle formation. *TFR* transferrin receptor components involved in autophagy. Of the cell lines we have tested, cytosol from HEK293T cells is most active. In addition to HEK293T cells, cytosol from COS7 and wild-type MEF cells are also active in triggering LC3 lipidation.

5. When suspending the cell pellets with PBS, we suggest tapping the pellets or gently pipetting. We do not recommend strong pipetting or vortexing, because this will break the cells and lead to low yield of cytosol.
6. Usually, if a large number of cells are harvested, a lipid layer will appear on the top after centrifugation. Try to avoid the lipid layer when collecting the supernatant. We usually collect the pure supernatant located between the cell-debris pellet and the lipid layer, sacrificing the cytosol that is near either the top or the bottom. If the lipid layer is too broad to avoid, try collecting the supernatant with some lipid contamination and repeat centrifuging the cytosol several times until you can obtain a clear supernatant of cytosol. The lipid layer usually becomes thinner after each round of centrifugation.
7. You should see a big band (unlipidated T7-LC3) migrating around 17kd and a faster migrating thinner band (lipidated T7-LC3) around 3–5 mm lower than the fat band on the membrane. We suggest running a 12.5%, 15%, or 8–16% gradient to better separate the lipidated LC3 from the unlipidated LC3. Sometimes the unlipidated band signal is so strong that it covers the lipidated signal. In this case, we recommend adding a step of 20,000 $\times$ *g* centrifugation after the reaction. Then remove 3/4 of the supernatant, and add SDS-PAGE loading buffer to the pellet and remaining supernatant. The lipidated LC3 should be fully recovered in the pellet. By removing the supernatant, we only reduce the level of unlipidated LC3.
8. If many cells are used, the pellet after the 1000 $\times$ *g* spin is difficult to see because of turbidity due to the high concentration of remaining membranes in the supernatant. Try to look at the pellet against a monochromatic background. Dark blue is usually a good choice. Carefully transfer the supernatant. If the pellet is not visible, try increasing the speed to 1200–1500 $\times$ *g*. To increase recovery, we suggest washing the pellet with lysis buffer and repeating the centrifugation.
9. We performed the cell-free lipidation assay with the four membrane fractions. The 25k membrane is the active membrane (Fig. 2b). Therefore, we further separated the 25k with the sucrose gradient indicated below.
10. We have performed the cell-free lipidation assay with both of these membrane fractions. The L fraction is the active membrane (Fig. 2c). Therefore, we further separated the L fraction with the indicated Optiprep gradient.

11. We do not collect the 100k fraction because membranes in this fraction are usually small vesicles.
12. LC3 lipidation has been shown to regulate autophagic vesicle tethering and fusion [15–17]. Therefore, we use cytosol from Atg5KO cells to perform the small vesicle formation assay to avoid the complication of LC3 lipidation affecting the size of the autophagosomal precursors generated in the cell-free assay.
13. This protocol is basically similar to the COPII vesicle formation assay as described previously [18–20].
14. Pay attention to avoiding donor membrane contamination. After the reaction and the medium-speed centrifugation, carefully transfer the supernatant (2/3 of supernatant), and avoid even a tiny amount of the pellet.
15. Because there is not much membrane in the small membrane fraction, the lipidated LC3 level is low. By removing the supernatant, the amount of unlipidated LC3 is reduced, which prevents the strong signal of the unlipidated LC3 from obscuring the lipidated LC3 signal.

Acknowledgments

L.G. was previously supported by the NIH Pathway to Independence Award (K99/R00) National Institute of General Medical Sciences (Grant Number: 1K99GM114397-02) and is now supported by the State Key Laboratory of Membrane Biology, China; the National Natural Science Foundation of China (31741082), the Tsinghua-Peking Center for Life Sciences, and the National Young Thousand Talents Program. L.G. is deeply grateful for the training provided by Dr. Randy Schekman at the University of California, Berkeley. We thank Bob Lesch at University of California Berkeley for critical reading and helpful editing of the manuscript.

References

1. Levine B, Kroemer G (2008) Autophagy in the pathogenesis of disease. Cell 132(1):27–42. https://doi.org/10.1016/j.cell.2007.12.018
2. Mizushima N, Levine B, Cuervo AM, Klionsky DJ (2008) Autophagy fights disease through cellular self- digestion. Nature 451 (7182):1069–1075. https://doi.org/10.1038/nature06639
3. Burman C, Ktistakis NT (2010) Autophagosome formation in mammalian cells. Semin Immunopathol 32(4):397–413. https://doi.org/10.1007/s00281-010-0222-z
4. Feng Y, He D, Yao Z, Klionsky DJ (2014) The machinery of macroautophagy. Cell Res 24 (1):24–41. https://doi.org/10.1038/cr.2013.168
5. Ge L, Baskaran S, Schekman R, Hurley JH (2014) The protein-vesicle network of autophagy. Curr Opin Cell Biol 29C:18–24. https://doi.org/10.1016/j.ceb.2014.02.005
6. Chen Y, Yu L (2017) Recent progress in autophagic lysosome reformation. Traffic 18 (6):358–361. https://doi.org/10.1111/tra.12484

7. Kabeya Y, Mizushima N, Ueno T, Yamamoto A, Kirisako T, Noda T, Kominami E, Ohsumi Y, Yoshimori T (2000) LC3, a mammalian homologue of yeast Apg8p, is localized in autophagosome membranes after processing. EMBO J 19 (21):5720–5728. https://doi.org/10.1093/emboj/19.21.5720
8. Ge L, Melville D, Zhang M, Schekman R (2013) The ER-Golgi intermediate compartment is a key membrane source for the LC3 lipidation step of autophagosome biogenesis. elife 2:e00947. https://doi.org/10.7554/eLife.0094700947
9. Ge L, Schekman R (2013) The ER-Golgi intermediate compartment feeds the phagophore membrane. Autophagy 10(1):170–172
10. Ge L, Zhang M, Schekman R (2014) Phosphatidylinositol 3-kinase and COPII generate LC3 lipidation vesicles from the ER-Golgi intermediate compartment. Elife 3:e04135. https://doi.org/10.7554/eLife.04135
11. Ge L, Wilz L, Schekman R (2015) Biogenesis of autophagosomal precursors for LC3 lipidation from the ER-Golgi intermediate compartment. Autophagy 11(12):2372–2374. https://doi.org/10.1080/15548627.2015.1105422
12. Ge L, Zhang M, Kenny SJ, Liu D, Maeda M, Saito K, Mathur A, Xu K, Schekman R (2017) Remodeling of ER-exit sites initiates a membrane supply pathway for autophagosome biogenesis. EMBO Rep 18:1586–1603. https://doi.org/10.15252/embr.201744559
13. Mizushima N, Yamamoto A, Hatano M, Kobayashi Y, Kabeya Y, Suzuki K, Tokuhisa T, Ohsumi Y, Yoshimori T (2001) Dissection of autophagosome formation using Apg5-deficient mouse embryonic stem cells. J Cell Biol 152(4):657–668
14. Oh-oka K, Nakatogawa H, Ohsumi Y (2008) Physiological pH and acidic phospholipids contribute to substrate specificity in lipidation of Atg8. J Biol Chem 283(32):21847–21852. https://doi.org/10.1074/jbc.M801836200
15. Nakatogawa H, Ichimura Y, Ohsumi Y (2007) Atg8, a ubiquitin-like protein required for autophagosome formation, mediates membrane tethering and hemifusion. Cell 130 (1):165–178. https://doi.org/10.1016/j.cell.2007.05.021
16. Weidberg H, Shpilka T, Shvets E, Abada A, Shimron F, Elazar Z (2011) LC3 and GATE-16 N termini mediate membrane fusion processes required for autophagosome biogenesis. Dev Cell 20(4):444–454. https://doi.org/10.1016/j.devcel.2011.02.006
17. Landajuela A, Hervas JH, Anton Z, Montes LR, Gil D, Valle M, Rodriguez JF, Goni FM, Alonso A (2016) Lipid geometry and bilayer curvature modulate LC3/GABARAP-mediated model autophagosomal elongation. Biophys J 110(2):411–422. https://doi.org/10.1016/j.bpj.2015.11.3524
18. Schindler AJ, Schekman R (2009) In vitro reconstitution of ER-stress induced ATF6 transport in COPII vesicles. Proc Natl Acad Sci U S A 106(42):17775–17780. https://doi.org/10.1073/pnas.0910342106
19. Lam SK, Yoda N, Schekman R (2010) A vesicle carrier that mediates peroxisome protein traffic from the endoplasmic reticulum. Proc Natl Acad Sci U S A 107(50):21523–21528. https://doi.org/10.1073/pnas.1013397107
20. Merte J, Jensen D, Wright K, Sarsfield S, Wang Y, Schekman R, Ginty DD (2010) Sec24b selectively sorts Vangl2 to regulate planar cell polarity during neural tube closure. Nat Cell Biol 12(1):41–46.; sup pp 41–48. https://doi.org/10.1038/ncb2002

Chapter 8

Use of Peptide Arrays for Identification and Characterization of LIR Motifs

Mads Skytte Rasmussen, Åsa Birna Birgisdottir, and Terje Johansen

Abstract

The mammalian ATG8 proteins (LC3A-C/GABARAP, GABARAPL1, and GABARAPL2) are small ubiquitin-like proteins critically involved in macroautophagy. Their processed C-termini are posttranslationally conjugated to a phosphatidylethanolamine moiety, enabling their insertion into the lipid bilayers of both the inner and outer membranes of the forming autophagosomes. The ATG8s bind a diverse selection of proteins including cargo receptors for selective autophagy, members of the core autophagy machinery, and other proteins involved in formation, transport, and maturation (fusion to lysosomes) of autophagosomes. Protein binding to the ATG8s is in most cases mediated by short, conserved sequence motifs known as *LC3-interacting regions* (LIRs). Here, we present a protocol for identifying putative LIR motifs in a whole protein sequence using peptide arrays generated by SPOT synthesis on nitrocellulose membranes. The use of two-dimensional peptide arrays allows for further identification of specific residues critical for LIR binding.

Key words Autophagy, Atg8, GABARAP, LC3, LIR, Peptide array

1 Introduction

Autophagy processes, whereby cytoplasmic content is degraded in the lysosome, consist of at least three different main pathways including macroautophagy, microautophagy, and chaperone-mediated autophagy [1]. Macroautophagy (for simplicity hereafter referred to as autophagy) can be subdivided into selective autophagy or unspecific, bulk autophagy. Selective autophagy allows for targeted degradation of large substrates (damaged or surplus mitochondria, peroxisomes or ER, and protein aggregates as well as invading bacteria and viral capsids) but also of specific signaling proteins, as well as recycling of metabolic substrates (lipids and glycogen) [2–5]. Selective autophagy receptors are able to simultaneously bind the cargo and ATG8 family proteins attached to the inner membrane of the forming autophagosome [2, 6]. Cargo recognition is often mediated through specific ubiquitin tags on the

Nicholas Ktistakis and Oliver Florey (eds.), *Autophagy: Methods and Protocols*, Methods in Molecular Biology, vol. 1880, https://doi.org/10.1007/978-1-4939-8873-0_8,

substrates that are recognized by the SLR (sequestosome 1-like receptors) family of cargo receptors which includes p62/SQSTM1(sequestosome 1) [6], NBR1 [7], OPTN [8], NDP52 [9], Tax1BP1 [10], and yeast Cue5 [11]. NDP52 is also indirectly recruited to cargo by recognizing Galectin-8, a cytosolic lectin, acting as a danger receptor that detects bacterial invasion by binding host glycans exposed on damaged pathogen-containing vacuoles or sterile damage to lysosomal or endosomal membranes [12]. Some autophagy receptors may directly recognize cargo as exemplified by TRIM family E3 ligases acting as cargo receptors for viral capsids and recruitment hubs for autophagy components [13]. Direct recognition is also mediated by Atg19 and Atg34, acting in the biosynthetic cytoplasm to vacuole targeting (Cvt) pathway in yeast, which mechanistically is similar to autophagy [14, 15]. NCOA4 directly targets surplus ferritin for autophagic degradation [16]. Autophagy receptors may also be present on the cargo itself because they reside in the outer membranes of organelles like NIX/BNIP3L [17], FUNDC1 [18], and FKBP8 [19], in mitochondria, and FAM134B in ER [20].

All cargo receptors identified so far contain an *LC3-interacting region* (LIR) [21], also referred to as AIM (*Atg8-family interacting motif*) in yeast [22]. Canonical LIR sequences consist of a short linear motif of up to 13 amino acids with a 5-amino acid-long core sequence following the consensus Θ-X-X-Γ—, where Θ is an aromatic amino acid (W/F/Y), Γ is a hydrophobic amino acid (L/I/V), and X can be any amino acid [21]. The LIR allows for tight association of the cargo receptor with the ATG8s (LC3s/GABARAPs) through interaction with two hydrophobic pockets on the ATG8s [21, 22]. The ATG8s are small ubiquitin-like proteins covalently attached to the lipid phosphatidylethanolamine (PE) in the autophagosomal membrane and function as anchoring points for the cargo receptors at the autophagosome [23]. Thus, the LIR is crucial for the ability of the cargo receptors to guide their target to degradation by autophagy. In addition to autophagy receptors, ATG8 proteins also bind a diverse selection of LIR-containing proteins including members of the core autophagy machinery [24], proteins involved in transport of autophagosomes [25–27], cross talk with endocytosis [28], and fusion of autophagosomes with lysosomes [29]. The ability of ATG8 proteins to scaffold proteins and protein complexes on membranes via LIR-dependent interactions is not limited to autophagy but also observed in spatially restricted cellular signaling [30].

Canonical LIR sequences can be predicted by use of iLIR, a web-based application [31], or by manually searching the sequence of a protein for LIR-like motifs. However, LIR-like sequences can often be identified in most proteins and require experimental testing for validation. Relying only on linear sequence motifs resembling LIR-like motifs might also result in false negatives as a growing number of proteins have been shown to utilize noncanonical LIRs such as NDP52 [32], Bcl-2 [33], Atg12 [34], and UBA5 [35]. These noncanonical LIRs may lack the conserved aromatic residue, like the C-LIR of NDP52 [32], or the conserved hydrophobic residue, like the LIR of Bcl-2 [33]. In UBA5, the spacing between the conserved aromatic and hydrophobic residue is only one residue, and another hydrophobic residue spaced three amino acids from the aromatic residue is also important for binding [35]. In ATG12 the aromatic and hydrophobic residues are spaced far apart in the primary sequence but brought together in a three-dimensional binding epitope [34]. Furthermore, we are as of yet not able to predict the precise determinants for selective binding to the different ATG8s, as some only bind GABARAPs while others only bind LC3s and some bind all ATG8s with high affinity [21, 36].

In this chapter, we provide a protocol for mapping LIRs in ATG8-interacting proteins by use of peptide arrays produced by SPOT synthesis on nitrocellulose membranes. Consisting of small linear peptides up to about 20 residues, the peptide array is well suited to probe for possible LIRs in the target protein as well as revealing the importance of single residues of the LIR motif for ATG8 binding. To search a protein for candidate LIR motifs, an array of 20-mer peptides covering the entire sequence of the target protein is synthesized and immobilized on a cellulose membrane using SPOT synthesis [37, 38]. For routine identification of LIR motifs, the membrane is populated with 20-mer peptides starting at the N-terminal end of the protein sequence moving the overlapping peptides three residues toward the C-terminal end in each spot relative to the peptide in the previous spot. The membrane is then probed with recombinant GST or GST-LC3s/GABARAPs after which binding is detected with an anti-GST antibody (Fig. 1). Care should be taken in choosing an appropriate ATG8, as several LIR-containing proteins only bind strongly to either GABARAPs [24, 36, 39] or LC3A/B [18, 26]. Once a LIR motif has been identified, the individual LIRs can also be probed for specificity against the different ATG8s by performing single spot analysis of 20-mers containing the LIR motif followed by separate probing with either GABARAP or LC3B [40] (Fig. 2). Two-dimensional peptide arrays consisting of an 18-mer containing the LIR motif can be used to probe the residue requirements of the LIR motifs by substituting each position of the sequence with all of the 19 alternative amino acids [26] (Fig. 3). The results obtained by peptide arrays should always be confirmed with other methods, such as GST pulldowns or immunoprecipitations with the full-length proteins with WT and mutated LIR sequences, or other in vitro and/or in vivo protein-protein interaction assays.

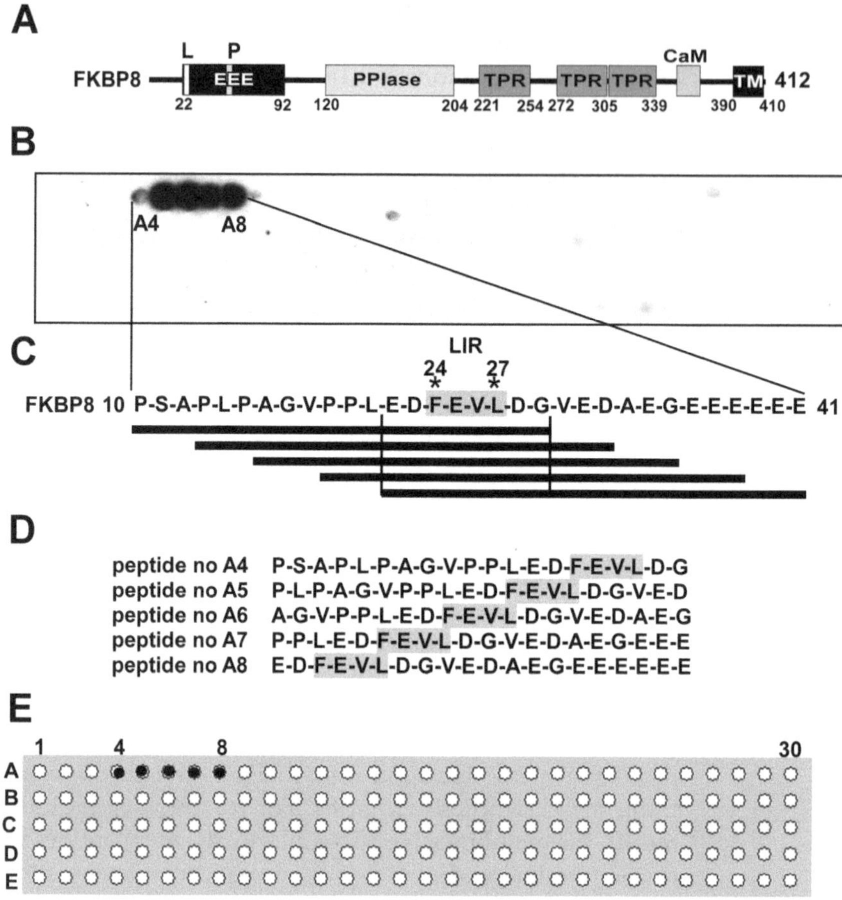

Fig. 1 Mapping of a single LIR motif in FKBP8 using a peptide array of 20-mer peptides covering full-length FKBP8 of 412 amino acids. The array was probed with 1 μg/mL of GST-GABARAP for 2 h, and binding to GST-GABARAP detected with anti-GST antibody using chemiluminescence. (**a**) Schematic drawing of the domain architecture of human FKBP8. (**b**) Image of the developed film of a 20,3 peptide array containing 20-mer peptides spotted on a 540 grid membrane with 18 rows of 30 spots in the horizontal direction. Each peptide was shifted three amino acids relative to the previous peptide. Only the part of the membrane containing peptides derived from the sequence of FKBP8 is shown. Five positive spots are seen in position 4–8 in the array. (**c**) The sequence of the region covered by five peptides corresponding to the positive spots in (**b**). The identified consensus LIR motif is indicated in gray and the relevant amino acid positions are shown. The extent of each peptide is indicated with horizontal bars below the sequence. (**d**) The sequences of the five GABARAP-interacting peptides are shown. (**e**) Schematic drawing of the membrane with the position of spotted peptides shown and the positive spots indicated as black dots. *L* LIR motif, *EEE* glutamate-rich domain, *P* proline-rich sequence, *PPIase* peptidyl-prolyl cis-trans isomerase domain, *TPR* tetratricopeptide repeats, *CaM* calmodulin-binding domain, and *TM* transmembrane domain (Adapted from Ref. [19])

2 Materials

2.1 Synthesis of Peptide Arrays

Synthesis of peptide arrays on cellulose membranes was performed using a MultiPep automated peptide synthesizer (INTAVIS Bioanalytical Instruments AG, Germany). Membranes were ordered

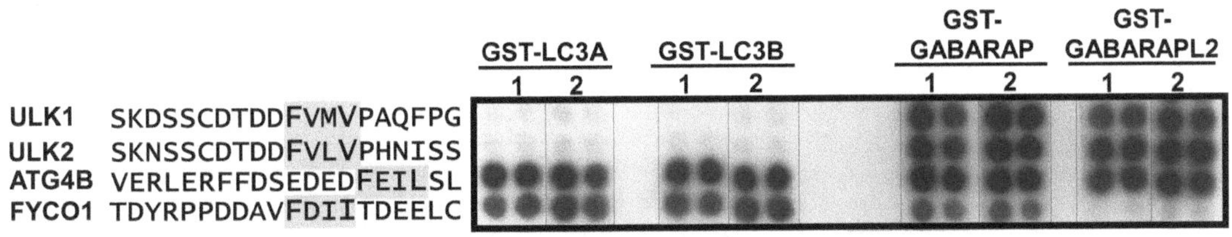

Fig. 2 Peptide arrays with 20-mer peptides containing identified LIR motifs from ULK1, ULK2, ATG4B, and FYCO1 (synthesized in duplicates marked 1 and 2) were examined for their binding preference toward GST-tagged ATG8 homologues LC3A, LC3B, GABARAP, and GABARAPL2. The peptide arrays were probed with 1 μg/mL of the different GST-ATG8 fusion proteins for 2 h, and binding was detected with anti-GST antibodies (Adapted from Ref. [40])

from Ola Rumohr Blingsmo at the Centre for Molecular Medicine Norway, NCMM-Administration and Core Facilities (NCMM ADMIN), Faculty of Medicine, University of Oslo. Design of the arrays was done using MultiPep peptide synthesis control software (INTAVIS AG, Germany) version 1.5.06.

2.2 Production and Purification of GST-Atg8/LC3/GABARAPs

1. GST expression vector, e.g., pDest15 (Invitrogen, 11802014).
2. *Escherichia coli* suitable for protein production, e.g., BL21 STAR (DE3) pLysS cells (Thermo Fisher, C602003).
3. 2× YT medium: [16% Bacto-tryptone (w/v), 10% yeast extract (w/v), 5% NaCl (w/v)] with ampicillin (100 μg/mL).
4. Suitable centrifuge, e.g., JLA 16.250 fixed angle rotor (Beckman Coulter).
5. Lysis buffer: 50 mM Tris–HCl pH 8.0, 250 mM NaCl, 1 mM DTT, 0.35 mg/mL lysozyme.
6. Sonicator, e.g., VCX 130 Sonic Vibra-Cell ultrasonic processor.
7. GST-binding column, e.g., GSTrap™ FF 1 mL; GE Healthcare.
8. Elution buffer: 1× PBS pH 8.0, 1 mM reduced glutathione; the buffer is pH adjusted, sterile filtered, and degassed before use.
9. Chromatography system, e.g., ÄKTAprime chromatography system; GE Healthcare.
10. Bradford protein assay or similar.

2.3 Peptide Overlay Assay

1. Ponceau S Staining Solution: 0.1%(w/v) Ponceau S (Sigma-Aldrich, # P3504) in 5%(v/v) acetic acid.
2. TBS-T buffer: 10 mM Tris–HCl pH 7.4, 150 mM NaCl, and 0.1% Tween-20.
3. HRP-conjugated GST-antibody clone RPN1236; GE Healthcare.

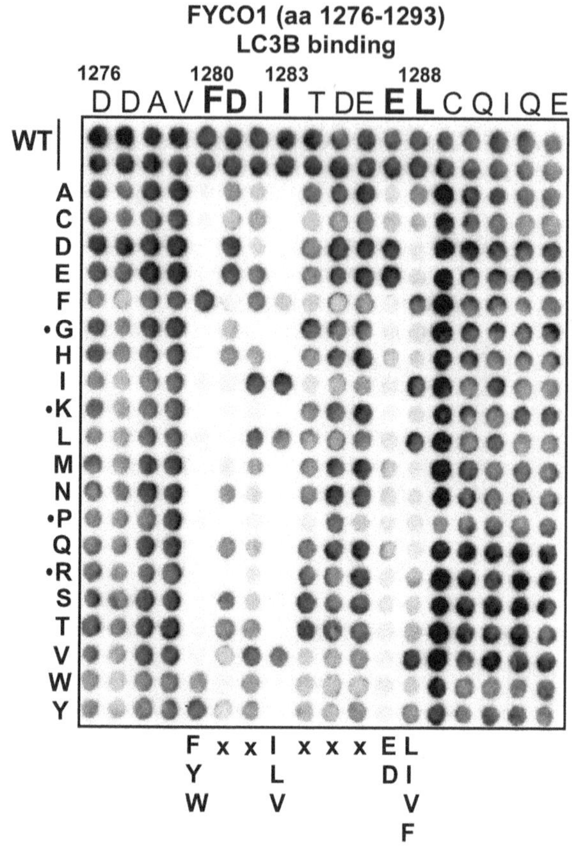

Fig. 3 A two-dimensional peptide array scan analyzing the effects of single amino acid substitutions at all positions of the indicated 18-mer peptides from FYCO1 (amino acids 1276–1293). Each position of the 18-mer peptides was replaced with all 20 amino acids. The peptide array was probed with 1 μg/mL GST-LC3B for 2 h, and binding to GST-LC3B was detected with anti-GST antibodies. This analysis allowed the identification of the residues EL C-terminal to the core LIR motif as crucial for binding and showed that the functional LIR in this case is likely to be at least 13 amino acids. At the aromatic position binding in HP1, only F/W/Y are tolerated. At the hydrophobic position binding in HP2, only I/L/V are tolerated. Basic residues K and R and secondary structure disrupting G and P residues are not tolerated in the core LIR (Adapted from Ref. [26])

4. Western blotting Luminol Reagent kit (SC-2048, Santa Cruz Biotechnology).
5. Chemiluminescence capturing device, e.g., charge-coupled device (CCD) camera or X-ray film exposure.
6. Regeneration buffer for stripping of membrane: 62.5 mM Tris–HCl pH 6.7, 2% SDS, and 70 μL 2-mercaptoethanol per 10 mL buffer.

3 Methods

3.1 Production and Purification of GST-Atg8/LC3/GABARAPs

1. Subclone cDNAs encoding Atg8/LC3/GABARAPs in-frame with an N-terminal GST tag in an IPTG-inducible expression vector. Use the vector to transform a suitable strain of *Escherichia coli*, i.e., BL21 STAR (DE3) pLysS cells (Thermo Fisher).
2. Pick a single colony of transformed *E. coli*, transfer into a sterile tube with 5 mL of 2× YT medium, and incubate overnight at 37 °C on a horizontal shaker (250 rpm).
3. Dilute the overnight culture 1:20 to a final volume of 100 mL of 2× YT medium with ampicillin (100 μg/mL), and grow further under same conditions until the cells reach OD_{600} of 0.9 (about 2–3 h).
4. Induce expression of the target gene by adding IPTG to a final concentration of 0.5 mM. Allow bacteria to grow for another 3–4 h at room temperature on a horizontal shaker.
5. After induction, transfer the bacterial culture into a sterile 250-mL plastic flask and centrifuge at 3000 × *g* at 4 °C for 10 min.
6. Remove the supernatant completely and resuspend the recovered pellet in 4 mL ice-cold lysis buffer (*see* **Note 1**).
7. Keep the lysate on ice for 20 min and then add Triton X-100 to 1% (v/v).
8. Distribute the lysate into three tubes (about 1.5 mL per tube) (*see* **Note 2**).
9. Sonicate on ice (four times at 40% amplitude and 20 kHz) (*see* **Note 3**).
10. Transfer to a microfuge tube and centrifuge at 15,000 × *g* at 4 °C for 10 min.
11. Filter the supernatant by passing through a 0.45-μm sterile filter, and apply on a GST-binding column.
12. Elute proteins from the column with elution buffer at the speed of 1 mL/min using a suitable chromatography system.
13. Use the Bradford protein assay to estimate protein concentration in the fractions where the protein is eluted.

3.2 Set-Up of Peptide Arrays for 20,3 or 18,1 Scanning and 2D Substitution Analyses

1. Before doing a peptide array scan to identify putative LIR motifs in a protein, it is advised to do GST pulldown assays or immunoprecipitations to check for a possible direct interaction between the protein of interest and ATG8 family proteins (*see* **Note 4**). The iLIR prediction server can also be used to look for consensus LIR motifs in the protein.

2. We have usually ordered the synthesis of two 540 grid membranes (30 × 18 peptides SPOT synthesized) to allow the search or scan for motifs binding to ATG8s in multiple proteins as this is the most cost-effective way. This allows the analysis of on average about two to five proteins per membrane (four to ten proteins in total) depending on the length of each protein.
3. Each protein sequence is scanned in a 20,3 manner meaning that the membranes contain overlapping 20-mer peptides that are shifted with three amino acids per peptide starting at the N-terminal end of the protein. This usually also allows for quite accurate determination of the extent of the sequence motif required for efficient binding. If higher resolution is needed, an 18,1 array can be used where the overlapping 18-mer peptides are shifted with only one amino acid per spot.
4. For 2D arrays, we use 18-mer peptides and spot two rows of WT sequence at the top and then peptides with single substitutions of every amino acid at one position at the time from the N-terminal end of the 18-mer to the C-terminal end (*see* Fig. 3). We locate the core LIR motif (W/F/Y-x-x-L/I/V) in the middle of the peptide so we can look for residues both N- and C-terminal to the core that may influence binding. However, some LIRs are located at the C-terminus of the protein in question, like human ATG4B or yeast Atg19. We then, of course, locate the LIR at the end of the peptide as in its native protein.

3.3 Peptide Overlay Assay

1. Stain the peptide array membrane with Ponceau S for 1 min.
2. Wash several times with water to allow visualization of peptide spots.
3. Dry the membrane overnight.
4. On the day of the assay, activate the membrane by incubating in 100% methanol for 10 s.
5. Wash two times, at 5 min each, in TBS-T buffer.
6. Block nonspecific labeling by incubating the membrane in TBS-T containing 5% (w/v) nonfat dry milk powder for 1 h at room temperature.
7. To test binding to GST-Atg8/LC3/GABARAP, incubate the membrane with the purified GST-Atg8/LC3/GABARAP protein at 1–2 μg/mL concentration in TBS-T buffer (the volume is adjusted according to the size of the membrane) for 2 h at room temperature (*see* **Note 4**).
8. After binding, wash the membrane five times in TBS-T.
9. Incubate overnight at 4 °C with HRP-conjugated anti-GST at 1:5000 dilutions in 5% dry milk.
10. Wash the membrane five times in TBS-T.

3.4 Detection of Atg8/LC3/GABARAP Binding by Chemiluminescence

1. Mix equal amounts of Luminol Reagent Solution A and Solution B (use 0.125 mL/cm^2 total volume) by inversion in a screw cap vial.
2. Pour off the TBS-T wash solution from the membrane. Add the mixed Luminol Reagent Solutions to the membrane (peptide spot side facing up). Incubate for 1 min at room temperature.
3. Lift the membrane with tweezers to drain excess Luminol Reagent and wrap the membrane (peptide spot side facing up) tightly in a single layer of plastic wrap.
4. Carefully place the plastic-wrapped membrane (peptide spot side facing up) in a film cassette. In a darkroom, expose the membrane to film for 1 min and develop according to standard protocols for autoradiogram. Exposure time may be altered as necessary.
5. Alternatively, place the plastic-wrapped membrane in a chemiluminescence capturing device, e.g., charge-coupled device (CCD) camera for exposure and signal detection using the appropriate settings (*see* **Note 5**).

3.5 Analysis of Peptide Arrays

1. The 20,3 or 18,1 scans are read off by scoring the peptide spots with a clear signal above background as positive. This allows the identification of positive peptides and the extent of the positive region to be mapped. In a 20,3 scan a putative LIR usually appears as 4-6 consecutive strong spots (*see* Fig. 1) (*see* **Notes 6** and **7**).
2. For the 2D arrays, the analysis of residues important for binding (*see* Fig. 3) is either done in a qualitative manner noting the loss of signal with certain substitutions, or the arrays can be analyzed by quantifying signal intensity from the chemiluminescent detection imaging device.

3.6 Stripping of the Membrane

1. In the event that no LIR is identified and one wants to reprobe the membrane with another ATG8 protein, the membrane can be stripped and reprobed. If the membrane is dried out, it must be regenerated in methanol for some seconds.
2. Wash three times in water for 10 min each at room temperature.
3. Warm the regeneration buffer in a petri dish to 50 °C and incubate the membrane for 10 min. Repeat twice.
4. Incubate three times in 10× TBS-T, for 10 min each, at room temperature.
5. Incubate three times in TBS-T, for 10 min each, at room temperature.

6. To check the success of the stripping procedure, block the membrane in 5% milk in TBS-T for 2 h at room temperature and incubate with HRP-conjugated anti-GST following the developing procedure as described above.
7. To reprobe the membrane with a new protein, wash the membrane five times, 4 min each, in TBS-T buffer, and block the membrane in 5% milk in TBS-T buffer for 2 h at room temperature. Then incubate the membrane with the new GST-ATG8 protein following the peptide array procedure as described above.

4 Notes

1. Although the buffer itself can be stored at RT, lysozyme and DTT must be added prior to use.
2. Alternatively transfer to three cryotubes and store at −70 °C. Allow for slow thawing of the *E. coli* lysates.
3. It is recommended to purify the GST-Atg8/LC3/GABARAP the same day or the day before the overlay assay. The purified protein can be kept at 4 °C for brief storage.
4. We have successfully used GST fusions of yeast (*S. cerevisiae*) Atg8, *Drosophila melanogaster* Atg8a, human LC3A, LC3B, LC3C, GABARAP, and GABARAPL1 to probe peptide arrays to identify LIR motifs. For human proteins, our first choice is to use GST-GABARAP or GST-GABARAPL1 to probe the arrays since they bind to most LIR-containing proteins and are therefore suitable for the first screening. However, if GST pulldowns of in vitro translated protein, or protein from cell extracts, or immunoprecipitations from cells, show a very strong preference for LC3A or LC3B, we use GST-LC3A or GST-LC3B for the peptide array.
5. The chemiluminescence method is a highly sensitive detection method and usually works well for the detection of LIR motifs within minutes with the described protocol. However, be aware to the appearance of clear spots with an outer black ring around them. Such spots usually indicate that all the luminescence substrate has been used up (due to high secondary antibody binding). New substrate should then be added to the membrane followed by immediate detection with shorter detection time. Alternatively wash the membrane extensively in TBS-T and reapply chemiluminescence substrate.
6. Care should be taken when probing sequences enriched in hydrophobic regions containing multiple W or F residues such as WD40 or KELCH repeat proteins, as these regions might produce false positives if spaced in such a way that allows

for docking into the HP1 and HP2 hydrophobic pockets of ATG8s. Knowledge of the structure of the protein in question, or (if a structure does not exist) knowledge about protein disorder, in the identified region might help to remove false negatives, as LIRs usually lie in intrinsically unstructured regions of the protein. Caution is advised when probing sequences containing cysteine residues, as those peptides might form disulfide bridges thus causing the spot to produce a false negative. *See* also **Note 2** in [37].

7. Before doing a 20,3 peptide array scan, we routinely profile binding to a candidate human LIR-containing protein by GST pulldown assays using in vitro translated, ^{35}S-labeled candidate protein and recombinant, bead-bound GST fusions of LC3A, LC3B, LC3C, GABARAP, GABARAPL1, and GABARAPL2 [41].

Acknowledgments

We are extremely grateful to Ola Rumohr Blingsmo at the Centre for Molecular Medicine Norway, NCMM-Administration and Core Facilities (NCMM ADMIN), Faculty of Medicine, University of Oslo, for advice and synthesizing the peptide arrays. The technical assistance of Gry Evjen is greatly appreciated. This work was funded by grants from the FRIBIO and FRIBIOMED programs of the Norwegian Research Council (grant numbers 196898 and 214448) and the Norwegian Cancer Society (grant number 71043-PR-2006-0320) to T.J.

References

1. Mizushima N, Komatsu M (2011) Autophagy: renovation of cells and tissues. Cell 147:728–741
2. Johansen T, Lamark T (2011) Selective autophagy mediated by autophagic adapter proteins. Autophagy 7:279–296
3. Khaminets A, Behl C, Dikic I (2016) Ubiquitin-dependent and independent signals in selective autophagy. Trends Cell Biol 26:6–16
4. Rogov V, Dotsch V, Johansen T, Kirkin V (2014) Interactions between autophagy receptors and ubiquitin-like proteins form the molecular basis for selective autophagy. Mol Cell 53:167–178
5. Stolz A, Ernst A, Dikic I (2014) Cargo recognition and trafficking in selective autophagy. Nat Cell Biol 16:495–501
6. Pankiv S, Clausen TH, Lamark T, Brech A, Bruun JA, Outzen H, Overvatn A, Bjorkoy G, Johansen T (2007) p62/SQSTM1 binds directly to Atg8/LC3 to facilitate degradation of ubiquitinated protein aggregates by autophagy. J Biol Chem 282:24131–24145
7. Kirkin V, Lamark T, Sou YS, Bjorkoy G, Nunn JL, Bruun JA, Shvets E, Mcewan DG, Clausen TH, Wild P, Bilusic I, Theurillat JP, Overvatn A, Ishii T, Elazar Z, Komatsu M, Dikic I, Johansen T (2009) A role for NBR1 in autophagosomal degradation of ubiquitinated substrates. Mol Cell 33:505–516
8. Wild P, Farhan H, Mcewan DG, Wagner S, Rogov VV, Brady NR, Richter B, Korac J, Waidmann O, Choudhary C, Dotsch V, Bumann D, Dikic I (2011) Phosphorylation of the autophagy receptor optineurin restricts Salmonella growth. Science 333:228–233

9. Thurston TL, Ryzhakov G, Bloor S, Von Muhlinen N, Randow F (2009) The TBK1 adaptor and autophagy receptor NDP52 restricts the proliferation of ubiquitin-coated bacteria. Nat Immunol 10:1215–1221
10. Newman AC, Scholefield CL, Kemp AJ, Newman M, Mciver EG, Kamal A, Wilkinson S (2012) TBK1 kinase addiction in lung cancer cells is mediated via autophagy of Tax1bp1/Ndp52 and non-canonical NF-kappaB signalling. PLoS One 7:e50672
11. Lu K, Psakhye I, Jentsch S (2014) Autophagic clearance of polyQ proteins mediated by ubiquitin-Atg8 adaptors of the conserved CUET protein family. Cell 158:549–563
12. Thurston TL, Wandel MP, Von Muhlinen N, Foeglein A, Randow F (2012) Galectin 8 targets damaged vesicles for autophagy to defend cells against bacterial invasion. Nature 482:414–418
13. Mandell MA, Jain A, Arko-Mensah J, Chauhan S, Kimura T, Dinkins C, Silvestri G, Munch J, Kirchhoff F, Simonsen A, Wei Y, Levine B, Johansen T, Deretic V (2014) TRIM proteins regulate autophagy and can target autophagic substrates by direct recognition. Dev Cell 30:394–409
14. Lynch-Day MA, Klionsky DJ (2010) The Cvt pathway as a model for selective autophagy. FEBS Lett 584:1359–1366
15. Suzuki K, Kondo C, Morimoto M, Ohsumi Y (2010) Selective transport of alpha-mannosidase by autophagic pathways: identification of a novel receptor, Atg34p. J Biol Chem 285:30019–30025
16. Mancias JD, Pontano Vaites L, Nissim S, Biancur DE, Kim AJ, Wang X, Liu Y, Goessling W, Kimmelman AC, Harper JW (2015) Ferritinophagy via NCOA4 is required for erythropoiesis and is regulated by iron dependent HERC2-mediated proteolysis. Elife 4. https://doi.org/10.7554/eLife.10308
17. Novak I, Kirkin V, Mcewan DG, Zhang J, Wild P, Rozenknop A, Rogov V, Lohr F, Popovic D, Occhipinti A, Reichert AS, Terzic J, Dotsch V, Ney PA, Dikic I (2010) Nix is a selective autophagy receptor for mitochondrial clearance. EMBO Rep 11:45–51
18. Liu L, Feng D, Chen G, Chen M, Zheng Q, Song P, Ma Q, Zhu C, Wang R, Qi W, Huang L, Xue P, Li B, Wang X, Jin H, Wang J, Yang F, Liu P, Zhu Y, Sui S, Chen Q (2012) Mitochondrial outer-membrane protein FUNDC1 mediates hypoxia-induced mitophagy in mammalian cells. Nat Cell Biol 14:177–185
19. Bhujabal Z, Birgisdottir AB, Sjottem E, Brenne HB, Overvatn A, Habisov S, Kirkin V, Lamark T, Johansen T (2017) FKBP8 recruits LC3A to mediate Parkin-independent mitophagy. EMBO Rep 18:947–961
20. Khaminets A, Heinrich T, Mari M, Grumati P, Huebner AK, Akutsu M, Liebmann L, Stolz A, Nietzsche S, Koch N, Mauthe M, Katona I, Qualmann B, Weis J, Reggiori F, Kurth I, Hubner CA, Dikic I (2015) Regulation of endoplasmic reticulum turnover by selective autophagy. Nature 522:354–358
21. Birgisdottir AB, Lamark T, Johansen T (2013) The LIR motif - crucial for selective autophagy. J Cell Sci 126:3237–3247
22. Noda NN, Ohsumi Y, Inagaki F (2010) Atg8-family interacting motif crucial for selective autophagy. FEBS Lett 584:1379–1385
23. Mizushima N, Yoshimori T, Ohsumi Y (2011) The role of Atg proteins in autophagosome formation. Annu Rev Cell Dev Biol 27:107–132
24. Alemu EA, Lamark T, Torgersen KM, Birgisdottir AB, Larsen KB, Jain A, Olsvik H, Overvatn A, Kirkin V, Johansen T (2012) ATG8 family proteins act as scaffolds for assembly of the ULK complex: sequence requirements for LC3-interacting region (LIR) motifs. J Biol Chem 287:39275–39290
25. Fu MM, Holzbaur EL (2014) Integrated regulation of motor-driven organelle transport by scaffolding proteins. Trends Cell Biol 24:564–574
26. Olsvik HL, Lamark T, Takagi K, Larsen KB, Evjen G, Overvatn A, Mizushima T, Johansen T (2015) FYCO1 contains a C-terminally extended, LC3A/B-preferring LC3-interacting region (LIR) motif required for efficient maturation of autophagosomes during basal autophagy. J Biol Chem 290:29361–29374
27. Pankiv S, Alemu EA, Brech A, Bruun JA, Lamark T, Overvatn A, Bjorkoy G, Johansen T (2010) FYCO1 is a Rab7 effector that binds to LC3 and PI3P to mediate microtubule plus end-directed vesicle transport. J Cell Biol 188:253–269
28. Popovic D, Akutsu M, Novak I, Harper JW, Behrends C, Dikic I (2012) Rab GTPase-activating proteins in autophagy: regulation of endocytic and autophagy pathways by direct binding to human ATG8 modifiers. Mol Cell Biol 32:1733–1744
29. Mcewan DG, Popovic D, Gubas A, Terawaki S, Suzuki H, Stadel D, Coxon FP, Miranda De Stegmann D, Bhogaraju S, Maddi K, Kirchof A, Gatti E, Helfrich MH, Wakatsuki S, Behrends C, Pierre P, Dikic I (2015) PLEKHM1 regulates autophagosome-lysosome fusion through HOPS complex

and LC3/GABARAP proteins. Mol Cell 57:39–54

30. Genau HM, Huber J, Baschieri F, Akutsu M, Dotsch V, Farhan H, Rogov V, Behrends C (2015) CUL3-KBTBD6/KBTBD7 ubiquitin ligase cooperates with GABARAP proteins to spatially restrict TIAM1-RAC1 Signaling. Mol Cell 57:995–1010
31. Kalvari I, Tsompanis S, Mulakkal NC, Osgood R, Johansen T, Nezis IP, Promponas VJ (2014) iLIR: a web resource for prediction of Atg8-family interacting proteins. Autophagy 10:913–925
32. Von Muhlinen N, Akutsu M, Ravenhill BJ, Foeglein A, Bloor S, Rutherford TJ, Freund SM, Komander D, Randow F (2012) LC3C, bound selectively by a noncanonical LIR motif in NDP52, is required for antibacterial autophagy. Mol Cell 48:329–342
33. Ma P, Schwarten M, Schneider L, Boeske A, Henke N, Lisak D, Weber S, Mohrluder J, Stoldt M, Strodel B, Methner A, Hoffmann S, Weiergraber OH, Willbold D (2013) Interaction of Bcl-2 with the autophagy-related GABAA receptor-associated protein (GABARAP): biophysical characterization and functional implications. J Biol Chem 288:37204–37215
34. Kaufmann A, Beier V, Franquelim HG, Wollert T (2014) Molecular mechanism of autophagic membrane-scaffold assembly and disassembly. Cell 156:469–481
35. Habisov S, Huber J, Ichimura Y, Akutsu M, Rogova N, Loehr F, Mcewan DG, Johansen T, Dikic I, Doetsch V, Komatsu M, Rogov VV, Kirkin V (2016) Structural and functional analysis of a novel interaction motif within UFM1-activating enzyme 5 (UBA5) required for binding to ubiquitin-like proteins and ufmylation. J Biol Chem 291:9025–9041
36. Rogov VV, Stolz A, Ravichandran AC, Rios-Szwed DO, Suzuki H, Kniss A, Lohr F, Wakatsuki S, Dotsch V, Dikic I, Dobson RC, Mcewan DG (2017) Structural and functional analysis of the GABARAP interaction motif (GIM). EMBO Rep 18:1382–1396
37. Frank R, Overwin H (1996) SPOT synthesis. Epitope analysis with arrays of synthetic peptides prepared on cellulose membranes. Methods Mol Biol 66:149–169
38. Kramer A, Schneider-Mergener J (1998) Synthesis and screening of peptide libraries on continuous cellulose membrane supports. Methods Mol Biol 87:25–39
39. Lystad AH, Ichimura Y, Takagi K, Yang Y, Pankiv S, Kanegae Y, Kageyama S, Suzuki M, Saito I, Mizushima T, Komatsu M, Simonsen A (2014) Structural determinants in GABARAP required for the selective binding and recruitment of ALFY to LC3B-positive structures. EMBO Rep 15:557–565
40. Skytte Rasmussen M, Mouilleron S, Kumar Shrestha B, Wirth M, Lee R, Bowitz Larsen K, Abudu Princely Y, O'reilly N, Sjottem E, Tooze SA, Lamark T, Johansen T (2017) ATG4B contains a C-terminal LIR motif important for binding and efficient cleavage of mammalian orthologs of yeast Atg8. Autophagy 13:834–853
41. Johansen T, Birgisdottir AB, Huber J, Kniss A, Dotsch V, Kirkin V, Rogov VV (2017) Methods for studying interactions between Atg8/LC3/GABARAP and LIR-containing proteins. Methods Enzymol 587:143–169

Chapter 9

Studying Autophagic Lysosome Reformation in Cells and by an In Vitro Reconstitution System

Yang Chen, Qian Peter Su, and Li Yu

Abstract

Autophagic lysosome reformation (ALR) is the terminal step of autophagy. ALR functions to recycle lysosomal membranes and maintain lysosome homeostasis. Maintaining a functional lysosome pool is critical for generating autolysosomes, in which cellular components are degraded and turned over during autophagy. This unit describes methods to visualize ALR in cells. In addition, this unit provides detailed protocols to establish in vitro systems which can be used to reconstitute ALR as well as to reconstitute mitochondrial tubulation/network formation, another process that is driven by motor proteins.

Key words Autophagy, Autophagic lysosome reformation, Autolysosome, In vitro reconstitution, Dynamic tubulation, Mitochondrial network formation

1 Introduction

Autophagic lysosome reformation (ALR) is the terminal step of autophagy [1]. During ALR, reformation tubules are extruded from autolysosomes, and nascent lysosomes are generated. Thus, ALR marks the termination step of autophagy and is important for maintaining lysosome homeostasis and ensuring the turnover of autolysosomal contents [1, 2]. ALR defects result in accumulation of enlarged lysosomes at a cellular level and a series of diseases, including neurological disorders [3–7]. Previous studies have identified the key players in ALR at a molecular level through proteomics combined with large-scale RNAi screening [8]. Using purified autolysosomes as the membrane source, and in vitro polymerized microtubules together with the purified motor protein KIF5B to provide the driving force, it is possible to reconstitute membrane tubulation during ALR in vitro [9]. The same system can be applied to study the effect of vesicle size on movement and membrane deformation [10]. In addition, using purified mitochondria as membrane sources, the in vitro system can be applied to reconstitute the dynamic tubulation and fusion of mitochondria, which is a

Nicholas Ktistakis and Oliver Florey (eds.), *Autophagy: Methods and Protocols*, Methods in Molecular Biology, vol. 1880, https://doi.org/10.1007/978-1-4939-8873-0_9, © Springer Science+Business Media, LLC, part of Springer Nature 2019

recently discovered mechanism for forming mitochondrial networks [11].

2 Materials

All the solutions and buffers should be prepared using analytical grade reagents and dissolved in ddH_2O (MilliQ, Millipore, 18.2 MΩ-cm at 25 °C), followed by filtration with 0.2 μm syringe filters to remove all the impurities. Prepare and store all reagents at room temperature (unless indicated otherwise). Diligently follow all waste disposal regulations when disposing of waste materials. We do not add sodium azide to reagents.

2.1 Observation of ALR in Cells

1. Culture medium: DMEM/high glucose supplemented with 10% fetal bovine serum (FBS), penicillin, streptomycin, and GlutaMAX-ITM.
2. Starvation medium: DMEM medium (serum- and glutamate-free).
3. Phosphate-buffered saline (PBS).
4. Amaxa Cell Line Nucleofector kit.
5. Plasmid: LAMP1-mcherry.

2.2 In Vitro Reconstitution System

2.2.1 Protein Purification

1. Wash buffer: 20 mM Tris/HCl, pH 7.5, 500 mM NaCl, 30 mM imidazole.
2. Elution buffer: 50 mM Tris/HCl, pH 8.0, 150 mM NaCl, 250 mM imidazole.
3. Storage buffer: 50 mM HEPES-KOH, pH 7.4, 300 mM NaCl, 1 mM $MgCl_2$, 10% (w/v) sucrose, 50 mM ATP (*see* **Note 1**). Store at 4 °C.
4. Sf9 cells.
5. Medium for insect cells.
6. Ni Sepharose 6 fast flow.

2.2.2 Organelle Isolation

Autolysosome Isolation from Cultured Cells

1. Starvation medium: DMEM (serum- and glutamate-free).
2. Phosphate-buffered saline (PBS).
3. Lysosome isolation kit (Cat. No LYSISO1; Sigma-Aldrich).
4. Protease inhibitor cocktail.

Isolation of Mitochondria from Rat Liver

1. Phosphate-buffered saline (PBS): 135 mM NaCl, 4.7 mM KCl, 10 mM Na_2HPO_4, 2 mM NaH_2PO_4.
2. H-S buffer: 10 mM HEPES-KOH, pH 7.4, 320 mM sucrose, 5 mM $MgSO_4$, 1 mM EGTA, 1 mM DTT, and protease inhibitors (*see* **Note 2**). Store at 4 °C.

3. Dounce homogenizer.
4. Optima MAX-XP ultracentrifuge.
5. 5 mL ultracentrifuge tubes.

2.2.3 In Vitro Reconstitution Assay

1. Coverslips (No.1, 0.17 μm, 24 × 50 mm) and slides (24 × 60 mm).
2. Double-sided tape (*see* **Note 3**).
3. Ultrasonic cleaner.
4. Acetone.
5. 1 M KOH solution.
6. HTS tubulin powder (*see* **Note 3**). Store at −20 °C before dissolving in buffer.
7. HiLyte 647-labeled tubulin powder (*see* **Note 3**). Store at −20 °C before dissolving in buffer.
8. General tubulin buffer (*see* **Note 3**). Store at 4 °C.
9. Tubulin glycerol buffer (*see* **Note 3**). Store at 4 °C.
10. Motility assay buffer (MAB)/BRB80: 80 mM PIPES-KOH, pH 6.8, 1 mM $MgCl_2$, 1 mM EGTA. Store at 4 °C for up to 1 month.
11. 3 mg/mL Casein solution in MAB. Aliquot and store at −80 °C for long-term storage; once taken out, store at 4 °C for up to 1 month (*see* **Note 4**).
12. GLOX solution: 60 mg/mL glucose oxidase, 6 mg/mL catalase in PBS with 40% (v/v) glycerol. Divide into aliquots of 200 μL and store at −20 °C for up to 2 months; once thawed, store at 4 °C for up to 1 week (*see* **Note 5**).
13. ATP working solution (with an ATP regeneration system and an oxygen scavenger system): 10 mM phosphocreatine (PC), 300 μg/mL creatine kinase (CK), 20 μM Taxol, 10 mM DTT, 0.1 mg/mL casein, 2.5% glucose, 100× GLOX, and 20 μM ATP in MAB (*see* **Note 6**).
14. Syringe and syringe filter (0.2 μm).
15. Inverted microscope with 561 nm and 640 nm lasers, high NA objective, TIRF illumination mode, EMCCD, or sCMOS.
16. CMDiI for membrane staining.

3 Methods

3.1 Visualization of ALR in Cells (See Note 7)

1. Transfect cells with LAMP1-mcherry plasmid and plate in a glass-bottom dish. Wait for at least 14 h for cells to settle down and to express the proteins from the exogenous plasmid.

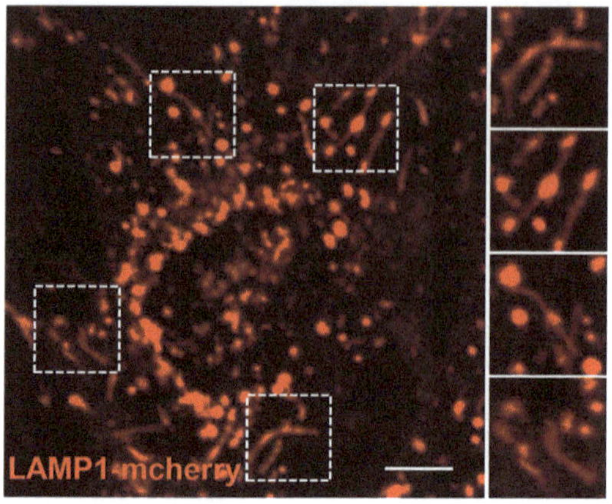

Fig. 1 Autophagic lysosome reformation in NRK cells. After NRK cells are transfected with LAMP1-mcherry and starved for 6 h, reformation tubules are labeled red. Scale bar, 5 μm. Right panels show enlarged images of reformation tubules (boxed areas in the main image)

2. Change the culture medium to starvation medium after washing the cells twice with PBS.
3. Incubate cells in a tissue culture incubator at 37 °C supplemented with 5% CO_2 for 4–8 h.
4. Use the live-cell imaging system on a confocal microscope (F1000, Olympus) to visualize ALR at 4–8 h post-starvation. The pinhole of the confocal microscope should be adjusted to 80–120 μm (Fig. 1).
5. Autolysosomes and protrusion tubules are positive for LAMP1 (*see* **Note 8**).

3.2 In Vitro Reconstitution System

3.2.1 Motor Protein Purification

1. pFastBac DUAL was used as the vector for baculovirus expression. This vector does not contain a tag for purification, so a histidine (×6) tag was introduced before the first codon of full-length KIF5B. The histidine-tagged KIF5B coding sequence was cloned between the BamH I/Xba I sites of the pFastBac DUAL plasmid.
2. Full-length KIF5B protein was expressed using the Bac-to-Bac expression system. Sf9 cells were grown in Lonza media to a density of ~2×10^6 cells/mL and incubated with virus containing the full-length KIF5B construct. After 60 h, cells were collected, lysed by freeze-thaw cycles, and centrifuged at 4 °C.
3. The soluble fraction was bound in batches to Ni-NTA agarose, and the resin was washed with wash buffer. His6-tagged KIF5B was eluted with elution buffer. The eluate was concentrated and stored in storage buffer at −80 °C [9] (*see* **Note 9**).

3.2.2 Organelle Purification

Autolysosome Purification from Cultured Cells

1. Six dishes (15 cm) of cultured NRK cells were first starved for 4 h using starvation medium. The cells were trypsinized from the culture dish and washed with PBS twice.
2. The cell pellet was collected, and autolysosomes were purified according to the protocol in the lysosome purification kit. Briefly, the cell pellet was homogenized with a Dounce homogenizer in 1000 μL extraction buffer supplemented with protease inhibitor cocktail and centrifuged at 1000 × *g* for 10 min at 4 °C to remove the unbroken cells and nuclei. The supernatant of the previous spin was centrifuged at 20,000 × *g* for 20 min at 4 °C to collect the membranes. The pellet was resuspended in 400 μL extraction buffer supplemented with 253 μL of 60% OptiPrep with 137 μL dilution buffer. Next, an OptiPrep density gradient was loaded into a 5 mL ultracentrifuge tube. The density and volume from bottom to top is 27% 0.8 mL, 22.5% 1 mL, 19% 0.8 mL (sample), 16% 1 mL, 12.5% 0.9 mL, and 8% 0.3 mL. The tube was centrifuged at 150,000 × *g* for 4 h at 4 °C. The top 1 mL of the gradient fraction contains the best purity of autolysosomes as confirmed by Western blot and immunofluorescence analysis of the autolysosome markers LC3 and LAMP1 [9] (*see* **Note 10**).
3. Collect the first two fractions from the top (500 μL for each fraction) and add 1 mL PBS to each fraction to dilute the OptiPrep. Centrifuge the sample at 20,000 × *g* for 20 min at 4 °C.
4. Recover the autolysosomes in the pellet by resuspending in a buffer that is compatible with the following experiment.

Mitochondrial Purification

1. Rat liver (2 g) or 20 dishes (15 cm) of cultured cells were washed with PBS several times, homogenized with a Dounce homogenizer in 4 mL of H-S buffer and centrifuged at 1000 × *g* for 10 min at 4 °C.
2. The pellet was discarded, and the supernatant was centrifuged at 11,400 × *g* for 20 min at 4 °C. The pellet was resuspended in 500 μL H-S buffer and applied to the top of a HEPES-buffered sucrose step density gradient (0.3 mL of 2.3 M, 1.7 mL of 1.7 M, and 1.5 mL of 1 M sucrose). The gradient was then centrifuged at 100,000 × *g* for 30 min at 4 °C.
3. 500 μL of the 1 M/1.7 M interface was recovered, diluted threefold with H-S buffer, and pelleted by centrifuging for 10 min at 11,400 × *g*. This wash step was repeated three times.

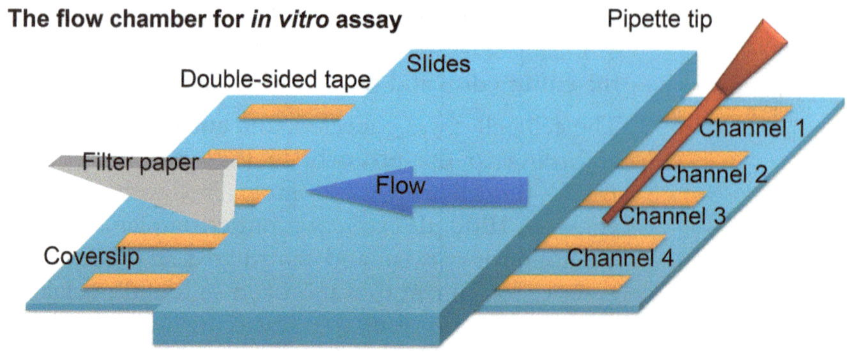

Fig. 2 The flow chamber for the in vitro assay

3.2.3 In Vitro Reconstitution Assay

Flow Chamber Assembly

1. The coverslips were cleaned in a staining jar by sonicating in acetone for 30 min, then in 1 M KOH for 30 min. The coverslips were then washed with ddH_2O and stored in ddH_2O to keep the surface hydrophilic.
2. The flow chamber (shown in Fig. 2, containing four individual channels) used for the in vitro reconstitution assay was homemade from cleaned coverslips and slides and assembled just before the experiment. Strips of double-sided tape were used to stick the coverslip and slide together and to isolate the flow channels. Folded filter papers were used to absorb and change the solution in the chamber.

Preparation of Polymerized Microtubule Filaments

1. Dissolve the dark and/or HiLyte 647 tubulin powder in general tubulin buffer and tubulin glycerol buffer with GTP as described in the manufacturer's manual to give a final concentration of 4 mg/mL protein. Aliquot the tubulin solution and store in the −80 °C freezer after quick freezing in liquid nitrogen, protecting from light.
2. Mix the dark tubulin and dye-labeled tubulin at a molar ratio of 25:1. Add 1 mM GTP and 20 μM taxol to the tubulin solution. Incubate and polymerize the tubulin mix in a 37 °C water bath for 30 min followed by 30 min centrifugation at 20,000 × *g* to remove the tubulin dimers or oligomers. Resuspend the pellet of microtubule filaments in MAB containing 20 μM taxol and keep the tube in a 37 °C water bath for at least 1 day before using (*see* **Note 11**).
3. Check the length of the microtubule filaments by direct visualization with 640 nm TIRF illumination on a microscope.

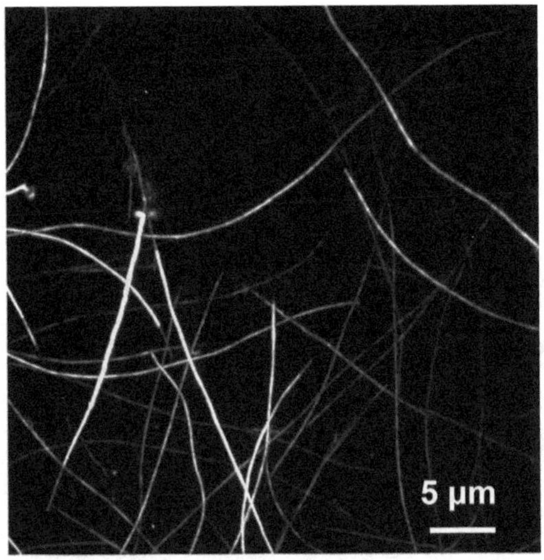

Fig. 3 The maximum intensity projection of the tracking trajectory (white lines) of the microtubule filaments. The image was obtained every 500 ms for 60 s on a Nikon TIRF microscope under 640 nm laser illumination

Gliding Assays to Confirm the Activity of Purified Motor Proteins

1. The gliding assay chambers were prepared by using hydrophilic coverslips. 15 μL of ~1 mg/mL full-length KIF5B was incubated in flow chamber channels for 5 min.
2. The kinesin-coated coverslips were blocked with 3 mg/mL casein for 5 min, and then 40 nM of HyLite 647-labeled microtubules was introduced into the flow chamber.
3. An ATP solution with an ATP regeneration system and an oxygen scavenger system was subsequently added to the chamber.
4. Images were recorded every 500 ms with TIRF illumination mode under 640 nm laser excitation (Fig. 3).

In Vitro Reconstitution of ALR and Mitochondrial Tubulation

1. The chambers used in the reconstitution assay were prepared as for the gliding assay. The channels were coated with 15 μL 10 μg/mL anti-tubulin antibody for 5 min and then blocked with 60 μL 3 mg/mL casein for 5 min. Next, 60 μL 1 mg/mL microtubule filaments were immobilized on the coverslip surface. The binding ratio and density of the microtubule filaments were checked with 640 nm laser TIRF illumination.
2. The autolysosomes or mitochondria were incubated with CMDiI in MAB for 5 min in a 37 °C water bath. The vesicles were then centrifuged and washed with MAB.
3. Full-length KIF5B (80 nM) was incubated with ~0.4 mg/mL CMDiI-labeled autolysosomes or mitochondria for 10 min on ice. The motor-coated autolysosomes or mitochondria were introduced into the microtubule-coated flow chambers for 10 min.

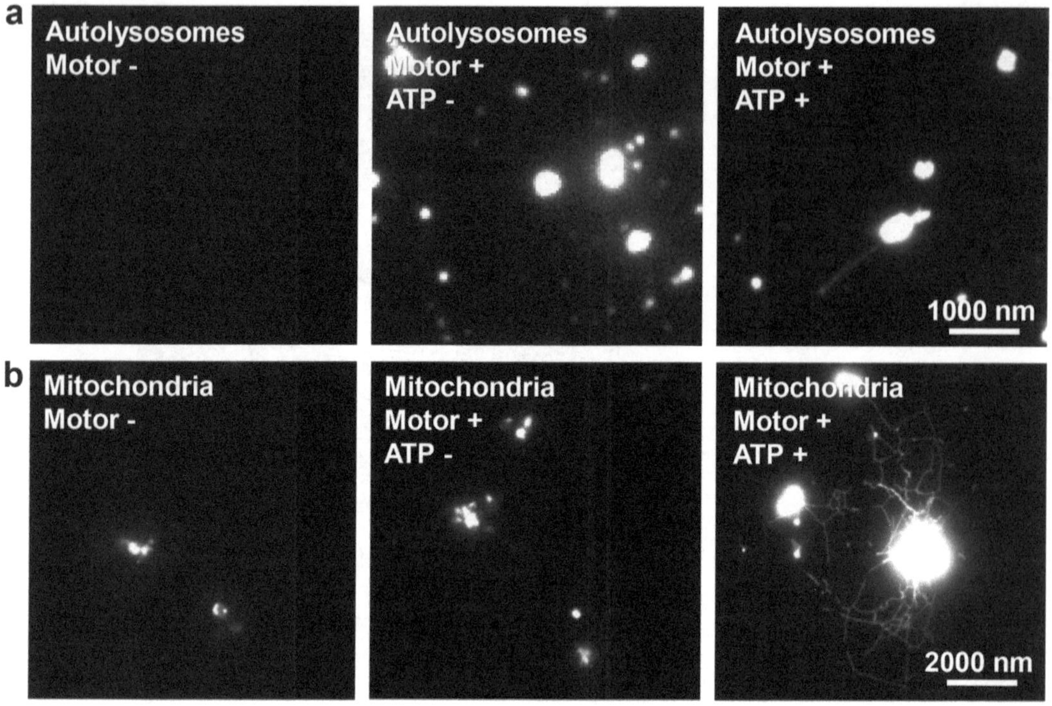

Fig. 4 In vitro reconstitution assays for autolysosomes and mitochondria. (**a**) In vitro reconstituted tubulation process driven by 80 nM KIF5B for CMDiI-labeled autolysosomes in the presence of 20 μM ATP. (**b**) In vitro reconstituted tubulation and network formation driven by 80 nM KIF5B for CMDiI-labeled mitochondria in the presence of 20 μM ATP

4. 60 μL of ATP solution with an ATP regeneration system and an oxygen scavenger system were added to the chambers. The formation of tubes and/or networks was visualized using a Nikon TIRF microscope under 561 nm laser illumination (*see* **Note 12**) (Fig. 4).

4 Notes

1. When preparing the storage buffer for KIF5B, ATP is added freshly before use.
2. When preparing the H-S buffer for mitochondria purification, protease inhibitors are added freshly before use.
3. When assembling the flow chamber for the in vitro reconstitution, tape of 200 μm thickness is recommended. We recommend using the double-sided tape from 3 M (Cat. No. 200MP). We also recommend using the HTS tubulin, HiLyte 647 tubulin, general tubulin buffer, and tubulin glycerol buffer from Cytoskeleton Corporation.

4. When preparing the casein solution in MAB, the pH value will be reduced after the casein powder is dissolved. Adjust the pH back to 6.8 with KOH.
5. When preparing the GLOX solution, weigh and dissolve 60 mg glucose oxidase and 6 mg catalase powder in 500 μL PBS on a vortex shaker, and avoid bubbles. Centrifuge the yellow solution at 20,000 × *g* for 1 min and discard the dark-colored pellet. Finally, add 500 μL 80% glycerol and aliquot the solution.
6. Prepare the ATP working solution freshly before each experiment and keep it on ice. Immediately before adding the solution to the chamber, add the ATP and GLOX to warm the ATP solution.
7. All media and buffers used in the starvation procedures should be pre-warmed to 37 °C.
8. In NRK cells, reformation tubules appeared at 4–8 h post-starvation. Approximately 60% of cells show reformation tubules. The number and size of reformation tubules vary according to the cell status and cell density and also among different cell types.
9. Add ATP to the storage buffer freshly before usage. To avoid multiple freeze-thaw cycles, we recommend making protein aliquots of no more than 5 μL per tube (every experiment needs ~0.4 μL kinesin solution). Store the protein aliquots in the −80 °C freezer after quickly freezing them in liquid nitrogen.
10. When homogenizing cells, use trypan blue solution to monitor the extent of homogenization. The homogenization is completed when 80–85% of cells are broken. Too little homogenizing will result in a low yield of membranes. Excessive homogenizing will possibly disrupt the autolysosome membranes, and the recovered autolysosomes will contain lots of membrane fragments.
11. Pre-warm the buffers and centrifuge before dealing with microtubules, because low temperatures cause rapid depolymerization of microtubule filaments. Cut the pipette tips when resuspending the microtubule pellet to avoid shearing forces. Sometimes the microtubule filaments are not long enough (<5 μm) for the in vitro assay. If so, just leave the resuspended solution in a 37 °C water bath and the filaments will anneal again.
12. ALR tubulation is visible on about 30–40% of the autolysosomes attached to the bottom of the chamber. The percentage of mitochondrial tubulation is about 70–80% of attached mitochondria.

References

1. Yu L et al (2010) Termination of autophagy and reformation of lysosomes regulated by mTOR. Nature 465(7300):942–946
2. Chen Y, Yu L (2017) Recent progress in autophagic lysosome reformation. Traffic 18 (6):358–361
3. Rong Y et al (2011) Spinster is required for autophagic lysosome reformation and mTOR reactivation following starvation. Proc Natl Acad Sci U S A 108(19):7826–7831
4. Schulze RJ et al (2013) Lipid droplet breakdown requires dynamin 2 for vesiculation of autolysosomal tubules in hepatocytes. J Cell Biol 203(2):315–326
5. Magalhaes J et al (2016) Autophagic lysosome reformation dysfunction in glucocerebrosidase deficient cells: relevance to Parkinson disease. Hum Mol Genet 25(16):3432–3445
6. Chang J, Lee S, Blackstone C (2014) Spastic paraplegia proteins spastizin and spatacsin mediate autophagic lysosome reformation. J Clin Invest 124(12):5249–5262
7. Varga RE et al (2015) In vivo evidence for lysosome depletion and impaired autophagic clearance in hereditary spastic paraplegia type SPG11. PLoS Genet 11(8):e1005454
8. Rong Y et al (2012) Clathrin and phosphatidylinositol-4,5-bisphosphate regulate autophagic lysosome reformation. Nat Cell Biol 14(9):924–934
9. Du W et al (2016) Kinesin 1 drives autolysosome tubulation. Dev Cell 37(4):326–336
10. Su QP et al (2016) Vesicle size regulates nanotube formation in the cell. Sci Rep 6:24002
11. Wang C et al (2015) Dynamic tubulation of mitochondria drives mitochondrial network formation. Cell Res 25(10):1108–1120

Chapter 10

Formation of Autophagosomes Coincides with Relaxation of Membrane Curvature

Jaime Agudo-Canalejo and Roland L. Knorr

Abstract

Autophagy is an intracellular degradation process that employs complex membrane dynamics to isolate and break down cellular components. However, many unanswered questions remain concerning remodeling of autophagic membranes. Here, we focus on the advantages of theoretical modeling to study the formation of autophagosomes and to understand the origin of autophagosomal membranes. Starting from the well-defined geometry of final autophagosomes, we ask the question of how these organelles can be formed by combining various pre-autophagosomal membranes such as vesicles, membrane tubules, or sheets. We analyze the geometric constraints of autophagosome formation by taking the area of the precursor membranes and their internal volume into account. Our results suggest that vesicle fusion contributes little to the formation of autophagosomes. In the second part, we quantify the curvature of the precursors and report that the formation of autophagosomes is associated with a strong relaxation of membrane curvature energy. This effect we find for a wide range of membrane asymmetries. It is especially strong for small distances between both autophagosomal membranes, as observed in vivo. We quantify the membrane bending energies of all precursors by considering membrane asymmetries. We propose that the generation and supply of pre-autophagosomal membranes is one limiting step for autophagosome formation.

Key words Reverse autophagy, Membrane remodeling, Fusion, Scission, Bending energy, Membrane curvature, Theory modeling, Retrograde autophagy

1 Introduction

During the intracellular process of macroautophagy (hereafter autophagy), a membrane-bound organelle, the autophagosome, is generated de novo. The remodeling of the autophagic membrane during the life cycle of the organelle is a complex multistep process involving changes in the morphology and topology of the autophagic membranes [1]. Autophagosomes grow as flattened double-membrane sheets, the phagophores, which then bend into spherical double-membrane organelles, the autophagosomes. Canonical mechanisms of membrane bending such as protein coats [2] were assumed to mediate the bending step [3]. However, using theoretical modeling, we previously showed that bending of phagophores

Nicholas Ktistakis and Oliver Florey (eds.), *Autophagy: Methods and Protocols*, Methods in Molecular Biology, vol. 1880, https://doi.org/10.1007/978-1-4939-8873-0_10,

has to be understood in terms of two coupled morphological changes. Bending of the flat sections of the sheet competes with the removal of the strongly curved edge of the phagophore. We showed that the energy profile of the bending process strongly depends on the relative size of the phagophore. The edge removal and bending of the phagophore are energetically favored for typical sizes of phagophores/autophagosomes in vivo [4]. This suggests that phagophores require mechanisms to prevent premature bending rather than a machinery to drive their bending.

Nevertheless, the origin and remodeling of early autophagic membranes is still a matter of debate [5]. Atg9 and COPII vesicles as well as membrane tubules seem to play central roles during autophagy initiation [6]. In mammals, autophagosomes can start, for example, at contact sites of organelles [7] or as PI(3)P-positive domains of the endoplasmic reticulum (ER) known as the omegasome [8]. In itself, the ER is a combination of tubular and sheet-like membranes [9]. In general, membrane vesicles, tubules, or sheets and combinations thereof (including more complex shapes) could contribute membrane material to form autophagosomes (*see* Fig. 1). In contrast to complicated, pre-autophagosomal mechanisms and potential membrane geometries, the final autophagosomes are rather simple and well-defined in terms of geometry: two spherical vesicles stuck within each other. Starting from the final shape of an autophagosome, we calculate the size and number of pre-autophagosomal vesicles and the size of membrane tubules and sheets required to match the geometric constraints of the autophagosome. We consider the combinations of precursors

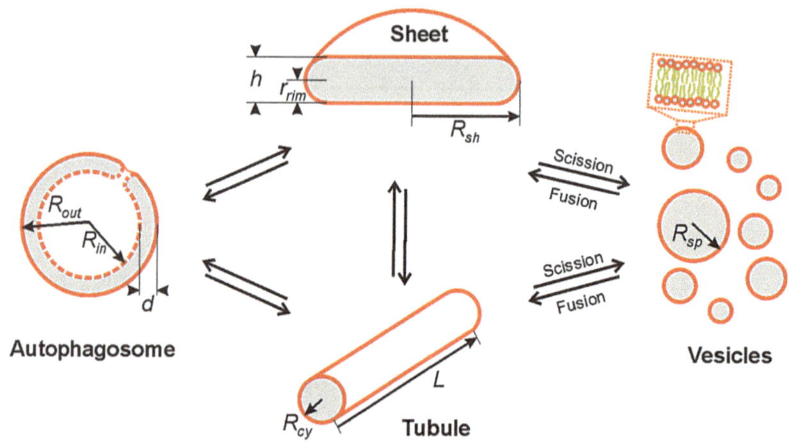

Fig. 1 Major shapes of membranes during autophagy and their geometric parameters. In principle, autophagosomes can be generated from several membranous precursors such as sheets, tubules, fused vesicles, and mixtures of various sources. Two-dimensional views show membranes in red as indicated. Dashed lines and gray areas refer to membranes and enclosed volumes which are visible only in sections. Included are the parameters which define each of the shapes and are used in this chapter

such as vesicles, tubules, and sheets. Moreover, we study the energetics of autophagosome formation.

This chapter is structured as follows: in Subheading 2, we describe the parameters required to calculate the bending energy and show where the model can be simplified. In Subheading 3 we describe the geometry and energetics of common membrane shapes and consider the formation of autophagosomes under area and volume constraints present. In Subheading 4 we compute the energetics of autophagosome formation.

2 Materials (Theory)

All biological membranes are in a fluid state. Because of their small thickness, of the order of 5 nm, they can be considered as quasi-two-dimensional fluid materials. The physical properties of the membrane differ from other quasi-two-dimensional fluids we might be familiar with in our macroscopic world. Soap bubbles, for example, conserve their internal volume but not their surface area. However, both the enclosed volume and the area of biological membranes are conserved. Thus, it is difficult to predict the behavior of membranes using intuition based on our macroscopic experience with two- or three-dimensional fluids. Here we describe modeling of membranes as a method to understand autophagic membrane dynamics. The shape of fluid membranes is primarily governed by their bending energy [10]. This mesoscopic description has been corroborated by a detailed and quantitative comparison between experimentally observed and theoretically calculated shapes [11, 12].

2.1 The Bending Energy of the Membrane

The bending energy of a membrane is given by

$$E_{\mathrm{be}} = 2\kappa \int \mathrm{d}A\,(M - m)^2 + \kappa_{\mathrm{G}} \int \mathrm{d}A\,K \tag{1}$$

1. The integration runs over the whole *surface area* A of the membrane or vesicle.
2. The *curvature* of any surface can be described locally by two perpendicular arcs. The inverse radii of these arcs are the two principal curvatures, C_1 and C_2, which characterize the local shape of a membrane. The principal curvatures can be positive or negative if the membrane bulges toward the exterior or interior compartment of a vesicle, respectively. The arithmetic mean of both curvatures defines the mean curvature $M = (C_1 + C_2)/2$, whereas their product defines the Gaussian curvature $K = C_1 C_2$.
3. The bending energy of any material is governed by its *bending rigidity* (or elastic modulus) κ, and by its *Gaussian curvature*

modulus κ_G, both of which have units of energy. Fluid bilayers are very flexible, and their bending rigidity is on the order of $\kappa = 10 - 20\ k_B T \sim 10^{-19}$ J, where k_B is the Boltzmann constant and T the temperature. The Gaussian curvature modulus is typically negative with $\kappa_G \approx -\kappa$, and we will use this value throughout this work unless otherwise noted.

4. If the two leaflets of the bilayer membrane differ in their molecular composition, the membrane can develop a certain preferred or *spontaneous curvature m*. The composition of a membrane can change, e.g., by binding of small molecules or proteins or by their modification such as through phosphorylation.
5. Due to the Gauss-Bonnet theorem, the second integral in (1) does not depend on the specific shape of the membrane; it only depends on the membrane topology. For any closed surface with spherical topology, such as a sphere, a tubule closed at both ends, a pancake-like membrane sheet, or an autophagosome (prior to the scission of the neck), this term evaluates simply to $4\pi\kappa_G$.

2.2 Simplifications of the Model

Depending on the actual aim of the model, a number of simplifications can be included. Here we introduce some generalizations we apply later in our reductionist model, mostly related to the interaction of proteins with the pre-autophagosomal membranes.

1. Binding and unbinding of any molecules or structural changes of these molecules alter the spontaneous curvature m. The formation of the autophagosome is achieved by the concerted actions of Atg proteins. Eighteen of the core proteins are grouped into six distinct complexes, and most of these core complexes and their interactions are conserved, and their interaction with the membrane can be restricted in space and time [13]. For example, in vitro we showed that conjugating the ubiquitin-like protein Atg8 to phospholipids changes m [14]. However, it is unknown how m changes dynamically during the process of autophagy. It will be challenging to quantify such values and moreover, these values may differ between various types of autophagy. Thus, throughout most of this chapter, we do not take m into account. When we do, we will consider a homogeneous value of m throughout the membrane that could arise, for example, from a global asymmetry due to the different compositions of the aqueous compartments inside and outside the closed membranes. We will however not consider the action of proteins that might localize only to specific parts of the membrane, leading to varying values of m throughout the membrane.
2. High surface densities of proteins can influence the elastic moduli κ and κ_G of the membrane. Atg8-PE, for example, has

been shown to induce membrane tubulation and rigidification [15]. This can however be attributed to unphysiological high surface densities of Atg8-PE [14]. As protein coats were not found on autophagosomes thus far [4], the assumption of constant elastic moduli seems reasonable.

3 Methods: Formation of Autophagosomes Under Consideration of Geometric Constraints

3.1 Geometry and Energetics of Common Membrane Shapes

1. The geometry of a spherical *vesicle* is defined by its radius R_{sp} (*see* Fig. 1). Its surface area is given by $A = 4\pi R_{sp}^2$ and its enclosed volume by $V = 4\pi R_{sp}^3/3$. Its mean curvature is given by $M = 1/R_{sp}$ everywhere, with which we can calculate its bending energy to be $E_{be} = 8\pi\kappa(1 - mR_{sp})^2 + 4\pi\kappa_G$. Interestingly, in the absence of spontaneous curvature (with $m = 0$), the bending energy of a sphere becomes independent of its radius. Throughout this chapter, vesicles are considered spherical.
2. A cylindrical *membrane tubule* is defined by its radius R_{cy}, as well as by its length L (*see* Fig. 1). Assuming that the tubule is capped at both ends, its surface area will be given by $A = 2\pi R_{cy}L + 4\pi R_{cy}^2$ and its volume by $V = \pi R_{cy}^2 L + 4\pi R_{cy}^3/3$. In both cases, the second term corresponding to the area and volume of the two caps becomes negligible if the tubule is very long, i.e., if $L \gg R_{cy}$. The mean curvature along the cylindrical part is $M = 1/2R_{cy}$, and it is $M = 1/R_{cy}$ at the two end caps. With these, we can calculate the bending energy of a tubule to be $E_{be} = \pi\kappa\frac{L}{R_{cy}}(1 - 2mR_{cy})^2 + 8\pi\kappa(1 - mR_{cy})^2 + 4\pi\kappa_G$.
3. A flat *phagophore*, a pancake-like, double-*membrane sheet*, can be defined by its long radius R_{sh} and by its rim radius r_{rim} (*see* Fig. 1). Such a definition is equivalent to constructing the pancake as two flat single-membrane disks connected to each other by a toroidal rim. Because the equations for the area, volume, and mean curvature of a toroidal segment are rather complicated, it is useful to consider only the case of a very thin sheet, i.e., of $r_{rim} \ll R_{sh}$. In this case, the toroidal segment can be approximated as being half a cylinder of length $L = 2\pi R_{sh}$, and the relevant equations become simpler. The area of the sheet then becomes $A \approx 2\pi R_{sh}^2 + 2\pi^2 R_{sh} r_{rim}$ and the enclosed volume $V \approx 2\pi R_{sh}^2 r_{rim} + \pi^2 R_{sh} r_{rim}^2$. The flat disks of the pancake have $M = 0$, while the rim has $M \approx 1/2r_{rim}$, leading to a bending energy $E_{be} \approx 4\pi\kappa m^2 R_{sh}^2 + \pi^2\kappa\frac{R_{sh}}{r_{rim}}(1 - 2mr_{rim})^2 + 4\pi\kappa_G$. Furthermore, we note that, for sheets that are large but thin, the calculation of their area and volume is particularly simple, even in cases when their shape is not pancake-like. In this case, a sheet (or a collection of sheets) can be defined by its projected area A_{sh}

(corresponding to πR_{sh}^2 for a pancake-like sheet) and its thickness h (corresponding to $2r_{rim}$ in the case of a pancake-like sheet). The total area of the sheet is then $A \approx 2A_{sh}$ and its volume is $V \approx A_{sh}h$.

4. Finally, the shape of an *autophagosome* consists of an inner autophagic vesicle, a sphere of radius R_{in} which is nested inside a slightly larger outer autophagic vesicle, and a sphere of outer radius R_{out} (*see* Fig. 1). Alternatively, we could also define the shape of an autophagosome by its outer radius R_{out} and the spacing between the inner and outer spheres $d \equiv R_{out} - R_{in}$. The area of the autophagosome will be $A = 4\pi(R_{out}^2 + R_{in}^2)$ and its lumen or enclosed volume $V = 4\pi(R_{out}^3 - R_{in}^3)/3$. We note that, by enclosed volume, we refer to the volume within the bounds of the continuous membrane that forms the autophagosome prior to scission of the autophagosomal neck, and not the volume captured within the inner vesicle. The mean curvature of the inner segment of the autophagosome is $M = -1/R_{in}$ and is negative because it bulges toward the interior, whereas the mean curvature of the outer segment is $M = 1/R_{out}$. With these, the bending energy of an autophagosome can be calculated as $E_{be} = 8\pi\kappa(1 - mR_{out})^2 + 8\pi\kappa(1 + mR_{in})^2 + 4\pi\kappa_G$. Notice that, in the absence of spontaneous curvature (with $m = 0$), the bending energy of the autophagosome is simply $E_{be} = 16\pi\kappa + 4\pi\kappa_G$, which is independent of the size of the autophagosome or the relative size of its inner and outer radii.
5. *Typical dimensions in cells of the membrane shapes considered here*: (1) R_{sp} of vesicles is in the range of 20–50 nm for Atg9 and COPII vesicles, respectively [16, 17]. (2) R_{cy} of ER tubules is about 15 nm in yeast and 30 nm in mammals [9]. (3) The thickness h of ER sheets is about 50 nm [18]. (4) The size of autophagosomes can vary strongly. They can span more than one order of magnitude, from few hundred nanometers to several micrometers [19–21]. The relation between the size of autophagosomes and spacing between their inner and outer autophagic vesicles is not well characterized. The spacing is in the order of tens of nanometers [22, 23]. Throughout most of this chapter, we consider a *typical autophagosome* of 1 μm diameter (implying an outer radius R_{out} = 500 nm) and intermembrane spacing $d = 20$ nm (implying an inner radius $R_{in} = 480$ nm).

3.2 Formation of Autophagosomes from Various Pre-autophagosomal Membranes

1. *Vesicles*

   Suppose that we observe an autophagosome with an outer radius R_{out} and inner radius R_{in}. How many spherical vesicles, of what size, would need to fuse in order to generate such an autophagosome? Because successful membrane fusion will conserve both area and enclosed volume, we need to find the number and size of vesicles that will have the same area and volume as the final autophagosome. The conservation of area gives us the condition $A = 4\pi(R_{out}^2 + R_{in}^2) = N4\pi R_{sp}^2$, where N is the number of vesicles, whereas the conservation of volume gives the condition $V = 4\pi(R_{out}^3 - R_{in}^3)/3 = N4\pi R_{sp}^3/3$. Combining both conditions and solving for N and R_{sp}, we obtain

$$N = \frac{\left(R_{out}^2 + R_{in}^2\right)^3}{\left(R_{out}^3 - R_{in}^3\right)^2} \tag{2}$$

$$R_{sp} = \frac{R_{out}^3 - R_{in}^3}{R_{out}^2 + R_{in}^2} \tag{3}$$

This highlights a perhaps unintuitive fact: there is only one possible vesicle size that, when combined in a specific number, will lead to an autophagosome of given size and shape (i.e., given spacing between outer and inner membranes). One might have thought that it would be possible to create the same autophagosome from a smaller number of larger vesicles, or from a larger number of smaller vesicles, but this is not the case, due to the fact that both the area and the enclosed volume of the autophagosome must be taken into account. One could, of course, use combinations of vesicles of different sizes, but as long as the difference in size between the vesicles involved is small, the typical vesicle size should be given by Eq. (3). For our example autophagosome of 1 μm diameter and $d = 20$ nm, Eqs. (2) and (3) predict that $N = 534$ vesicles of radius $R_{sp} = 30$ nm must have fused (*see* Fig. 2). In Fig. 3, we plot the number and radius of the vesicles required for the generation of an autophagosome with three different outer radii and for varying intermembrane spacing d of the autophagosome.

The size of the vesicles required is well within the biologically reasonable range (*see* **Note 5** of Subheading 3.1). In contrast, hundreds of vesicles have not been reported in the context of autophagosome formation so far. This is especially important as autophagosomes are characterized by a small intermembrane spacing. To obtain autophagosomes from vesicles exclusively in this regime, hundreds of vesicles are required to fuse with each other (*see* Fig. 3a). This is valid for a large range of autophagosomal sizes. Thus, the absence of excessive numbers of vesicles suggests that vesicles might contribute a minor fraction of the final autophagosomal membrane.

Equations (2) and (3) become particularly simple when we consider the biologically relevant case of very small intermembrane spacing of the autophagosome, i.e., $d \ll R_{out}$. In this case, a series expansion of (2) and (3) in powers of d gives

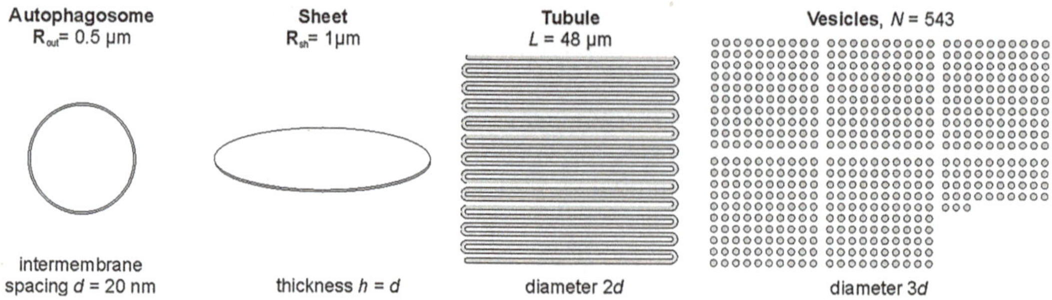

Fig. 2 Geometries of autophagosomal membranes. Precursor membranes are drawn to scale and match the membrane area and internal volume of our standard autophagosome with $R_{out} = 500$ nm and intermembrane spacing $d = 20$ nm. Characteristic length scales and vesicle numbers are given. Diameters of the sheet's rim, the tubule, and the vesicles are given relative to the intermembrane spacing of the autophagosome d

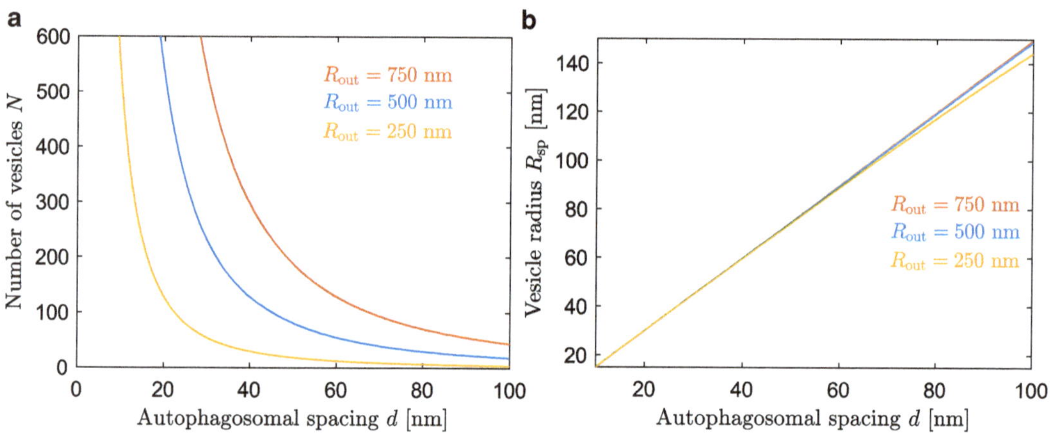

Fig. 3 Generation of an autophagosome from fusion of spherical vesicles. (**a**) Number of vesicles and (**b**) radius of the vesicles required for the generation of autophagosomes with outer radii $R_{out} = 750$ nm, 500 nm, and 250 nm (i.e., diameters 1.5 μm, 1 μm, and 500 nm), as a function of the intermembrane spacing d of the autophagosome. These are calculated from Eqs. (2) and (3) in the text. For smaller values of the spacing d, a larger number of smaller vesicles is required

$R_{sp} \approx (3/2)d$ and $N \approx (8/9)(R_{out}/d)^2$. In this limit, therefore, the diameter of the fusing vesicles must be three times the intermembrane spacing of the final autophagosome, and the required number of vesicles grows quadratically with the ratio of outer radius to intermembrane spacing of the autophagosome. Importantly, this limit of $d \ll R_{out}$ is the typical case in vivo.

2. Membrane Tubules

   It is also possible that the membrane used to generate autophagosomes has its origin in membrane tubules such as those in the ER. What is then the total length and diameter of tubules needed to generate a particular autophagosome? Assuming that the tubules involved are long and thin (which, as we will see a posteriori, corresponds to autophagosomes with narrow intermembrane spacing), we can neglect the area and volume of the tubule end caps, and the conservation of area and volume during formation of the autophagosome gives us the conditions $A = 4\pi(R_{out}^2 + R_{in}^2) = 2\pi R_{cy} L$ and $V = 4\pi(R_{out}^3 - R_{in}^3)/3 = \pi R_{cy}^2 L$. We note that these two conditions are valid independently of whether we consider a single tubule of length L or several separate shorter tubules that fuse, as long as their total length adds up to L. Taking both conditions together and solving for L and R_{cy}, we find

$$L = 3\frac{\left(R_{out}^2 + R_{in}^2\right)^2}{R_{out}^3 - R_{in}^3} \tag{4}$$

$$R_{cy} = \frac{2}{3}\frac{R_{out}^3 - R_{in}^3}{R_{out}^2 + R_{in}^2} \tag{5}$$

As in the case of spherical vesicles, there is only one specific combination of total length and radius of tubules that can give rise to an autophagosome of given geometry. To generate our typical autophagosome of 1 μm diameter and $d = 20$ nm, Eqs. (4) and (5) predict that a total length $L = 48$ μm of tubules of radius $R_{cy} = 20$ nm is required (*see* Fig. 2).

Interestingly, comparing Eqs. (3) and (5) we see that the required radius of the cylindrical tubules involved is two thirds smaller than the radius required for the vesicles. In particular, if we once again consider the case of the very small intermembrane spacing of the autophagosome ($d \ll R_{out}$), Eqs. (4) and (5) simplify to $L \approx 4R_{out}^2/d$ and $R_{cy} \approx d$. Therefore, the diameter of the cylindrical tubules is generally twice the intermembrane spacing, and the total length of tubules required grows quadratically with the size of the autophagosome (if the intermembrane spacing is kept constant).

3. Membrane sheets

The ER is also composed of membrane sheets, and it is possible that these participate in the generation of autophagosomes. As before, the conservation of membrane area and enclosed volume during autophagosome formation can be used to calculate the total amount and the thickness of the sheets involved in the process. Assuming that the sheets involved are large but thin, area and volume conservation gives the conditions $A = 4\pi(R_{out}^2 + R_{in}^2) = 2A_{sh}$ and $V = 4\pi(R_{out}^3 - R_{in}^3)/3 = A_{sh}h$. As in the case of the tubules, these conditions are valid independently of whether we consider a large sheet of projected area A_{sh} or several smaller sheets that fuse, as long as their total projected area adds up to A_{sh}. Taking both conditions together and solving for A_{sh} and h, we obtain

$$A_{\mathrm{sh}} = 2\pi\left({R_{\mathrm{out}}}^2 + {R_{\mathrm{in}}}^2\right) \tag{6}$$

$$h = \frac{2}{3}\frac{{R_{\mathrm{out}}}^3 - {R_{\mathrm{in}}}^3}{{R_{\mathrm{out}}}^2 + {R_{\mathrm{in}}}^2} \tag{7}$$

For our typical autophagosome of 1μm diameter and $d = 20$ nm, Eqs. (6) and (7) predict that a total projected area $A_{\mathrm{sh}} = 3.0\ \mu\mathrm{m}^2$ of sheets of thickness $h = 20$ nm is required. Such a total projected area would be equivalent to a single pancake-like sheet with diameter $2\sqrt{A_{\mathrm{sh}}/\pi} = 2.0\ \mu m$ (*see* Fig. 2).

As in the case of spheres and tubules, Eqs. (6) and (7) become particularly simple for the case of an autophagosome with very small intermembrane spacing ($d \ll R_{\mathrm{out}}$). In this case, they become $A_{\mathrm{sh}} \approx 4\pi {R_{\mathrm{out}}}^2$ and $h \approx d$. The required thickness of the sheets is therefore approximately equal to the intermembrane spacing of the final autophagosome, and the total projected area of the sheets required grows quadratically with the size of the autophagosome.

4. Mixtures of Vesicles, Tubules, and Sheets
In short, we have found above that an autophagosome with intermembrane spacing d may be formed from many vesicles of diameter $3d$, from one or multiple long tubules of diameter $2d$, or from one or multiple membrane sheets of thickness d (*see* Fig. 2). Moreover, the autophagosome could be formed from any mixture of such vesicles, tubules, and sheets: as long as their total area equals the total area of the autophagosome, the enclosed volume will automatically have the correct value.

When the autophagosome is generated from mixtures of these three components, it also becomes possible for the diameter/thickness of the components to be different from the values calculated above. However, if some of the components are smaller than the values calculated above, the rest of the components must be larger. As an example, it is impossible to generate an autophagosome of intermembrane spacing d from a mixture of vesicles of a diameter smaller than $3d$ and tubules of a diameter smaller than $2d$. If the vesicle diameter is smaller than $3d$, the tubule diameter will need to be larger than $2d$, and vice versa.

5. *In cells*, it seems rather unreasonable to assume that the membrane source of autophagosomes is controlled precisely to account for the area-volume constraint. In the case that one of the membrane sources was too large, excess luminal volume (relative to the membrane area) may accumulate during the formation of autophagosomes. Because autophagosomes are known for a small luminal volume [22], this raises the question of how can the cell correct for such events.

The membrane of a developing autophagosome can be continuous with other organelles such as the sheets of the ER [24–26]. This implies that the lumen of the sheets and lumen of the autophagosome may be continuous as well and might be used to transfer luminal volumes between ER and autophagosomes. Interestingly, the luminal spacing of ER sheets is significantly larger than that of the autophagosome connected to it [25, 26]. This suggests that autophagosomes are either more likely to be grown from tubular ER than sheet-like ER and/or that a mechanism exists which supplies membranes with a disproportionately large area-to-volume ratio.

3.3 Method: Energetics of Autophagosome Formation

We have shown so far that, because membrane area and enclosed volume are conserved during autophagosome formation, there are important constraints on the number, shape, and dimensions of the precursor membranes (vesicles, sheets, tubules) participating in the process. A second important question concerns the energetics of the process: under which conditions is autophagosome formation favorable energetically? Is the bending energy of the system reduced or increased during autophagosome formation?

3.3.1 Autophagosome-Like Shapes Can Form by Fusing Three Vesicles

1. Straightforward considerations can be made to show that autophagosome formation is generally an energetically favorable process. In ref. 12 the energy of closed vesicles of different shapes was studied in detail, as a function of their volume-to-area ratio $v \equiv V/[(4\pi/3)(A/4\pi)^{3/2}]$, which can range between 0 and 1, with $v = 1$ corresponding to a sphere. For the case of membranes with zero spontaneous curvature, it was found that tubule-like (prolate) shapes have lowest energy in the range $1 > v > 0.652$, sheet-like (oblate) shapes have lowest energy in the range $0.652 > v > 0.592$, and autophagosome-like (stomatocyte) shapes have lowest energy in the range $0.592 > v > 0$. A straightforward but important consequence of this is the following. When two, three, or four spherical vesicles fuse, the volume-to-area ratio of the resulting post-fusion vesicle is 0.707, 0.577, and 0.5, respectively. The fusion of even larger numbers of vesicles results in even lower volume-to-area ratio of the final vesicle. Taking these values together with the ranges for lowest energy shapes described above, one concludes that fusion of three or more vesicles is sufficient for autophagosome-like shapes to be energetically preferred over sheet-like or tubule-like shapes. This was recently confirmed by computer simulations [27]. Note: The volume-to-area ratio of our typical autophagosome of 1 μm diameter and $d = 20$ nm is $v = 0.043$.

To fully make sure that the lowest possible energy corresponds to the autophagosome-like shape, we should also compare this energy to the total energy of the individual vesicles before fusion. As described in Subheading 3.1, in the absence of spontaneous curvature, the energy of N spherical vesicles is $E_{ve} = N[8\pi\kappa + 4\pi\kappa_G]$. The energy of the resulting autophagosome after fusion, on the other hand, is $E_{ap} = 16\pi\kappa + 4\pi\kappa_G$. The condition that the autophagosome shape has lower or equal energy, i.e., that $E_{ap} \leq E_{ve}$, is then equivalent to

$$N \geq \frac{4\kappa + \kappa_G}{2\kappa + \kappa_G} \tag{8}$$

For a typical membrane, we have $\kappa_G \approx -\kappa$, which results in $N \geq 3$. Therefore, fusion of three or more vesicles into an autophagosome-like shape can decrease the energy of the system indeed. If the Gaussian curvature modulus κ_G were smaller or larger than $-\kappa$, this minimal number of vesicles would increase or decrease, respectively.

3.3.2 Relaxation of Total Bending Energy During Autophagosome Formation

The calculations described above show that, already for autophagosomes formed from as little as three vesicles, the formation process can be energetically favorable. One may wonder then, how large is the relaxation of bending energy during the formation of a realistic autophagosome for which, as we saw above, hundreds of vesicles (or a comparable amount of membrane in the form of tubules or sheets) must fuse?

1. In Subheading 3.2, we obtained the number and radius of vesicles, length and radius of tubules, or size and thickness of membrane sheets, that are needed to generate an autophagosome of given size and intermembrane spacing. It is straightforward to combine these results with the expressions for the bending energy of each of these shapes that we derived in Subheading 3.1 and in this way be able to compare the total bending energy of each of these four configurations.
2. In Fig. 4a, we plot the bending energy of each of these configurations for the particular case of zero spontaneous curvature of the membrane, for an autophagosome with outer radius $R_{out} = 500$ nm, as a function of the intermembrane spacing d of the autophagosome. We find that, for biologically reasonable values of the intermembrane spacing, the ordering of the different configurations from highest to lowest energy is long tubule, set of vesicles, membrane sheet, and autophagosome. The total relaxation of bending energy when a tubule becomes an autophagosome can be extremely large, of the order of $10^3\ \kappa \approx 10^4\ k_BT \approx 10^{-16}$ J. This energy is larger for smaller intermembrane spacings of the autophagosome.

3.3.3 Pathways of Autophagosome Formation

Autophagosomes form by a complex cascade of membrane remodeling events [1]. In general, such remodeling processes can be described as a combination of changes in membrane morphology and topology [28–30]. Morphological transformations arise from continuous and smooth changes of the membrane curvature and

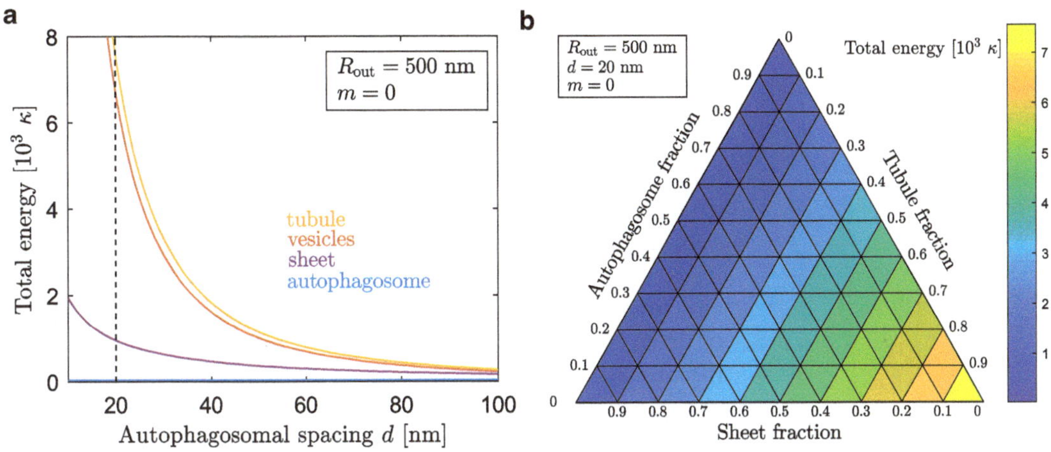

Fig. 4 Bending energy relaxation during autophagosome formation in the absence of spontaneous curvature. (**a**) Bending energy of an autophagosome with outer radius $R_{\rm out} = 500$ nm (blue), as well as the corresponding single sheet (purple), set of N vesicles (red, Fig. 3a), and single tubule (yellow) that have the same total area and enclosed volume as the autophagosome. Within a realistic range of autophagosomal spacing (10 – 100 nm), we find that the autophagosome has lowest energy, followed by the sheet, the vesicles, and the tubule, in this order. The vertical dashed line corresponds to the particular case discussed throughout the chapter, with intermembrane spacing $d = 20$ nm. (**b**) Bending energy of a sheet, a long tubule, and an autophagosome, such that the total membrane area and enclosed volume of the mixture corresponds to an autophagosome with outer radius $R_{\rm out} = 500$ nm and intermembrane spacing $d = 20$ nm. The top, bottom left, and bottom right corners of the triangle therefore correspond to the special cases of only autophagosome, only sheet, and only tubule. The energy values at these corners are thus identical to the values of the blue, purple, and yellow lines in (**a**) at $d = 20$ nm

shape, whereas topological transformations of the membrane involve intermediate states in which the membranes have to deviate strongly from their usual bilayer structure.

1. *Topological transformations*, membrane fusion or scission, are required to change the shape of the membrane between vesicles and other shapes such as tubules (*see* Fig. 1). The total energy of the tubules and vesicles is in a similar range (*see* Fig. 4a). Alterations in membrane topology include rather complicated energy barriers with non-bilayer transition states. Despite the fact that vesicles are known to be involved in autophagosome formation [6], their total number required in vivo seems to be much lower than estimated ($N > 500$) (*see* **Note 1** in Subheading 3.2). For these reasons, this transition will not be considered in what follows.
2. *Morphological transformations* could continuously transform a long tubule into a large sheet or an autophagosome without requiring membrane fusion/fission. This transition can in principle occur in a single discrete step, involving full-scale shape transformation of the tubule into a sheet via intermediate

paddle-like shapes of lower symmetry, which represent a sizable energy barrier [27, 31, 32]. Alternatively, one could consider a process that begins after the nucleation (or fusion) of a small sheet or autophagosomal structure at one end of a long tubule. This small sheet or autophagosomal structure can then continuously grow into a large sheet or mature autophagosome by exchanging area and volume with the tubule, which must then continuously shrink as a consequence. To explore such pathways, we consider the energy of the system when the total area and volume (which correspond to a "mature" autophagosome of given size and shape) is partitioned into a single, continuous membrane. Various membrane fractions represent a sheet of changing size, a tubule of varying length, and a "growing" autophagosome. The radius of the vesicles and the tubule is given by Eqs. (3) and (5). In Fig. 4b, we plot the energy of such a mixture for the particular case of a target autophagosome with $R_{out} = 500$ nm and $d = 20$ nm. The corners of the ternary diagram correspond to the target autophagosome (top corner), the case with only a sheet (bottom left corner), and the long tubule only (bottom right corner). The lines directly connecting the corners of the diagram represent mixtures between two of the shapes. As we had seen before, the tubule has the highest energy, whereas the "mature" autophagosome has the lowest energy. All pathways going through the inside of the ternary diagram are more complex and involve the simultaneous coexistence of three membrane shapes: sheet, tubule, and autophagosome.

3. Finally, let us briefly consider the *effect of spontaneous curvature* on the energetics of autophagosome formation. For zero spontaneous curvature, we found the ordering of tubule, vesicles, sheet, and autophagosome from larger to lower energy. Would this ordering change for nonzero spontaneous curvature? In Fig. 5, we plot the energies of the four configurations as a function of spontaneous curvature, for the particular case of an autophagosome with an outer radius $R_{out} = 500$ nm and intermembrane spacing $d = 20$ nm. Indeed, we find that this ordering can change, and in particular vesicles and tubules become more energetically favorable for sufficiently high spontaneous curvature. This was to be expected, given that, for example, proteins inducing positive spontaneous curvature stabilize tubular structures in the ER [33]. Nevertheless, we find that autophagosomes are always energetically more stable than membrane sheets of the corresponding size, which is in agreement with previous work considering the bending of phagophores [4]. Moreover, we show that the ordering of tubule, vesicles, sheet, and autophagosome from larger to lower energy

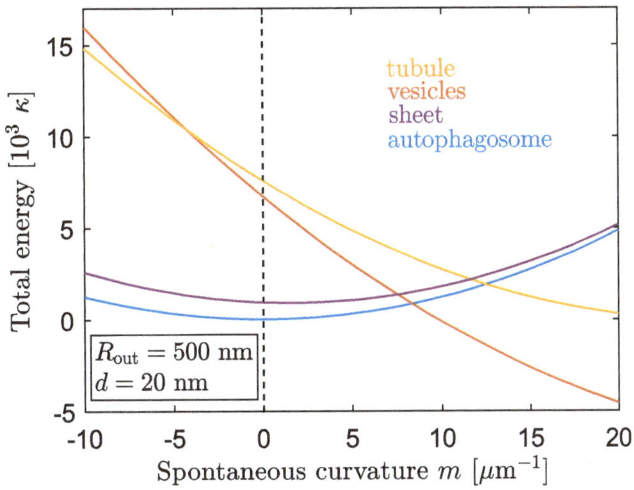

Fig. 5 Bending energy relaxation during autophagosome formation as a function of spontaneous curvature. The bending energy of an autophagosome with outer radius $R_{out} = 500$ nm and intermembrane spacing $d = 20$ nm (blue), as well as the bending energies of the corresponding single sheet (purple), set of $N = 534$ vesicles (red), and single tubule (yellow) that have the same total area and enclosed volume as the autophagosome are plotted as a function of spontaneous curvature. Spontaneous curvature can strongly affect the energy of the different configurations, but the ordering of tubule, vesicles, sheet, and autophagosome from highest to lowest bending energy remains valid in a wide range of spontaneous curvatures above and below $m = 0$ (vertical dashed line). Formation of autophagosomes from sheet-like membranes, such as phagophores, is always energetically favorable

survives for a wide range of positive and negative spontaneous curvatures around $m = 0$.

Acknowledgments

We thank Reinhard Lipowsky (MPI of Colloids and Interfaces) for stimulating discussions and institutional and financial support.

References

1. Knorr RL, Mizushima N, Dimova R (2017) Fusion and scission of membranes: ubiquitous topological transformations in cells. Traffic 18:758–761
2. McMahon HT, Gallop JL (2005) Membrane curvature and mechanisms of dynamic cell membrane remodelling. Nature 438:590–596
3. Xie Z et al (2009) Indirect estimation of the area density of Atg8 on the phagophore. Autophagy 5:217–220
4. Knorr RL, Dimova R, Lipowsky R (2012) Curvature of double-membrane organelles generated by changes in membrane size and composition. PLoS One 7:e32753
5. Lamb CA, Yoshimori T, Tooze SA (2013) The autophagosome: origins unknown, biogenesis complex. Nat Rev Mol Cell Biol 14:nrm3696
6. Hurley JH, Young LN (2017) Mechanisms of autophagy initiation. Annu Rev Biochem 86:225–244

7. Hamasaki M et al (2013) Autophagosomes form at ER-mitochondria contact sites. Nature 495:389–393
8. Axe EL et al (2008) Autophagosome formation from membrane compartments enriched in phosphatidylinositol 3-phosphate and dynamically connected to the endoplasmic reticulum. J Cell Biol 182:685–701
9. Shibata Y, Hu J, Kozlov MM, Rapoport TA (2009) Mechanisms shaping the membranes of cellular organelles. Annu Rev Cell Dev Biol 25:329–354
10. Helfrich W (1973) Elastic properties of lipid bilayers: theory and possible experiments. Z Naturforsch C 28:693–703
11. Lipowsky R (1991) The conformation of membranes. Nature 349:475–481
12. Seifert U, Berndl K, Lipowsky R (1991) Shape transformations of vesicles: phase diagram for spontaneous- curvature and bilayer-coupling models. Phys Rev A 44:1182–1202
13. Noda NN, Inagaki F (2015) Mechanisms of autophagy. Annu Rev Biophys 44:101–122
14. Knorr RL et al (2014) Membrane morphology is actively transformed by covalent binding of the protein Atg8 to PE-lipids. PLoS One 9: e115357
15. Kaufmann A, Beier V, Franquelim HG, Wollert T (2014) Molecular mechanism of autophagic membrane-scaffold assembly and disassembly. Cell 156:469–481
16. Stagg SM et al (2008) Structural basis for cargo regulation of COPII coat assembly. Cell 134:474–484
17. Yamamoto H et al (2012) Atg9 vesicles are an important membrane source during early steps of autophagosome formation. J Cell Biol 198:219–233
18. Shibata Y et al (2010) Mechanisms determining the morphology of the peripheral ER. Cell 143:774–788
19. Jin M, Klionsky DJ (2014) Regulation of autophagy: modulation of the size and number of autophagosomes. FEBS Lett 588:2457–2463
20. Kovács AL, Réz G, Pálfia Z, Kovács J (2000) Autophagy in the epithelial cells of murine seminal vesicle in vitro. Formation of large sheets of nascent isolation membranes, sequestration of the nucleus and inhibition by wortmannin and 3-ethyladenine. Cell Tissue Res 302:253–261
21. Xie Z, Nair U, Klionsky DJ (2008) Atg8 controls phagophore expansion during autophagosome formation. Mol Biol Cell 19:3290–3298
22. Biazik J, Vihinen H, Anwar T, Jokitalo E, Eskelinen E-L (2015) The versatile electron microscope: an ultrastructural overview of autophagy. Methods 75:44–53
23. Biazik J, Ylä-Anttila P, Vihinen H, Jokitalo E, Eskelinen E-L (2015) Ultrastructural relationship of the phagophore with surrounding organelles. Autophagy 11:439–451
24. Bars RL, Marion J, Borgne RL, Satiat-Jeunemaitre B, Bianchi MW (2014) ATG5 defines a phagophore domain connected to the endoplasmic reticulum during autophagosome formation in plants. Nat Commun 5:4121
25. Yamaguchi A et al (2009) A subdomain of the endoplasmic reticulum forms a cradle for autophagosome formation. Nat Cell Biol 11:1433
26. Ylä-Anttila P, Vihinen H, Jokitalo E, Eskelinen E-L (2009) 3D tomography reveals connections between the phagophore and endoplasmic reticulum. Autophagy 5:1180–1185
27. Bahrami AH, Lin MG, Ren X, Hurley JH, Hummer G (2017) Scaffolding the cup-shaped double membrane in autophagy. PLoS Comput Biol 13:e1005817
28. Knorr RL, Lipowsky R, Dimova R (2015) Autophagosome closure requires membrane scission. Autophagy 11:2134–2137
29. Lipowsky R (2014) Remodeling of membrane compartments: some consequences of membrane fluidity. Biol Chem 395:253–274
30. Kozlov MM, McMahon HT, Chernomordik LV (2010) Protein-driven membrane stresses in fusion and fission. Trends Biochem Sci 35:699–706
31. Heinrich V, Svetina S, Žekš B (1993) Nonaxisymmetric vesicle shapes in a generalized bilayer-couple model and the transition between oblate and prolate axisymmetric shapes. Phys Rev E 48:3112–3123
32. Bahrami AH, Hummer G (2017) Formation and stability of lipid membrane nanotubes. ACS Nano 11:9558–9565
33. Shemesh T et al (2014) A model for the generation and interconversion of ER morphologies. Proc Natl Acad Sci 111:E5243–E5251

Chapter 11

Studies of Receptor-Atg8 Interactions During Selective Autophagy

Christine Abert and Sascha Martens

Abstract

Autophagy research frequently requires the determination of protein-protein interactions. The experimental system described in this chapter allows a simple, versatile, and quantitative in vitro analysis of interactions between recombinant cargo receptor and Atg8 proteins by fluorescence microscopy. The assay can be easily modified to study other protein-protein interactions. The purified autophagy receptor is recruited to affinity resins via a suitable tag and then added to fluorescently labeled ATG8 in solution. The relative strength of the interaction can be assessed by determination of the fluorescence intensity on the surface of the bead at an equilibrium binding state. Thereby different interaction partners can be quantitatively compared, and weak or interactions with high off rates can be detected and quantified.

Key words Autophagy, Selective autophagy, Cytoplasm-to-vacuole targeting pathway, Cargo receptor, Atg8, Atg19, Protein-protein interaction, In vitro reconstitution, Fluorescence microscopy, Quantification

1 Introduction

Autophagy receptors confer selectivity to the autophagic process due to their ability to link the cargo to the nascent autophagosomal membrane. Apart from cargo binding, the interaction of the receptor with the small ubiquitin-like ATG8 proteins that decorate the autophagic membrane [1, 2] is crucial for efficient cargo degradation by autophagy [3–6]. These interactions that often rely on multiple low affinity binding sites can be studied in vitro employing a microscopy-based system using purified proteins and affinity resins [5, 7]. The recombinant cargo receptor fused to an affinity tag is bound to specific resins, e.g., via a GST-tag to glutathione-coated sepharose [8], and subsequently added to fluorescently labeled or tagged ATG8 proteins in solution. The interactions are subsequently assessed by fluorescence microscopy. As opposed to classical pull-down experiments where the beads are usually washed, binding events can be observed at equilibrium allowing

Nicholas Ktistakis and Oliver Florey (eds.), *Autophagy: Methods and Protocols*, Methods in Molecular Biology, vol. 1880, https://doi.org/10.1007/978-1-4939-8873-0_11,

for detection of weak interactions or interactions with high off rates. The fluorescence intensity on the bead surface can be easily measured and quantified, and thus relative differences in binding strength can be accurately determined. The system relies on high-quality purified proteins, but once these are available, the experimental setup is simple and highly versatile. For instance, binding regions within the receptor can be identified by comparing the binding of wild-type and mutant receptors. In addition, other factors such as secondary binding partners or competitors could be added. Titration experiments can determine binding constants by assessing the signal increase depending on the concentration of the binding partner. This versatility makes the system a powerful tool to reconstitute interactions during selective autophagy in a controlled in vitro environment.

2 Materials

1. Affinity resin, here beads coated with reduced glutathione (Glutathione Sepharose 4B with an average diameter of 90 μm, GE Healthcare, Catalogue number: 17075601).
2. Reaction buffer: 25 mM HEPES pH 7.5, 150 mM NaCl, 1 mM DTT, filtered through 0.2 μm membrane.
3. Purified recombinant proteins, affinity or fluorescently tagged.
4. Fluorescence microscope, here a spinning disc confocal microscope (Visitron). 20× and 63× objectives (LD Achroplan 20×/0.4 Corr; Plan-Apochromat 63×/1.4 Oil DIC, Carl Zeiss).
5. 384-well microscopy plates.
6. SDS-PAGE equipment and if needed Western blotting equipment and specific antibodies.

3 Methods

3.1 Binding Reaction

1. To prepare the purified proteins for the subsequent assay, thaw them on ice and centrifuge to remove any aggregates that may interfere with the experiment (*see* **Notes 1** and **2**). All proteins should be kept on ice throughout the experiment, unless it is known that other temperatures are required to maintain their stability. Determine their protein concentration by UV- or Bradford-based spectrophotometry and confirm by SDS-PAGE and Coomassie staining if possible. Determine the needed protein amounts for each experiment with a final reaction volume of 20 μL: a typical concentration of 20 μM is used for the affinity tagged bait-protein, in this case the

autophagy receptor Atg19, to saturate the affinity resin. The saturation can be checked by SDS-PAGE by loading the supernatant after the binding reaction. If the GST-tagged receptor is also found in the supernatant, its concentration exceeded the beads binding capacity, i.e., the beads are saturated. The protein concentration of the binding partner, here fluorescently labeled Atg8, typically amounts to 1–5 μM (*see* **Note 3**).

2. Prepare 10 μL of the affinity resin (here glutathione immobilized on sepharose; *see* **Note 4**) per planned experimental reaction. Equilibrate the resin in the reaction buffer by adding at least 10× more buffer than the volume of beads, gently resuspend the beads, and centrifuge at 4 °C for 1–3 min at 15,000 rcf until the beads are settled. Repeat this **step 2** times. Resuspend the beads 1:1 in reaction buffer, and aliquot 20 μL of this slurry into separate reaction tubes. Spin shortly and remove buffer to obtain 10 μL equilibrated beads per reaction.

3. To coat the beads with the bait-protein (here with the glutathione-S-transferase (GST)-tagged Atg19 autophagy receptor), add first the amount of reaction buffer required for a final volume of 20 μL to the equilibrated beads. Then add the bait-protein solution to a final volume of 20 μL, and mix carefully by pipetting up and down without introducing any air. Incubate for 45–60 min at 4 °C with gentle mixing to prevent the beads from settling, e.g., on a rotating wheel at 10 rpm.

4. Remove unbound receptor molecules by addition of 200 μL reaction buffer, centrifugation at 4 °C for 3 min at 15,000 rcf, and careful removal of as much supernatant as possible without disturbing the settled beads. Repeat this **step 2** times. Remove all supernatant and resuspend 1:1 in buffer (10 μL). Take a sample of 5 μL beads slurry for SDS-PAGE to confirm binding of the receptor to the affinity beads (*see* **Notes 5** and **6**).

5. For assessment of Atg8 binding to the receptor on the beads, aliquot 10 μL of a solution of Atg8 fluorescently labeled with meGFP (monomeric enhanced green fluorescent protein; *see* **Note 7**) at the desired concentration, e.g., 1 μM directly into the wells of a 384-well plate. Add 0.5–1 μL of the receptor-coated beads to the Atg8 solution. Incubate for 30 min at room temperature (*see* **Notes 1** and **8** and Fig. 1).

6. Image with a 20× objective (or 63× objective for weak signals). Image the focus plane of several beads (*see* **Note 9** and Fig. 2).

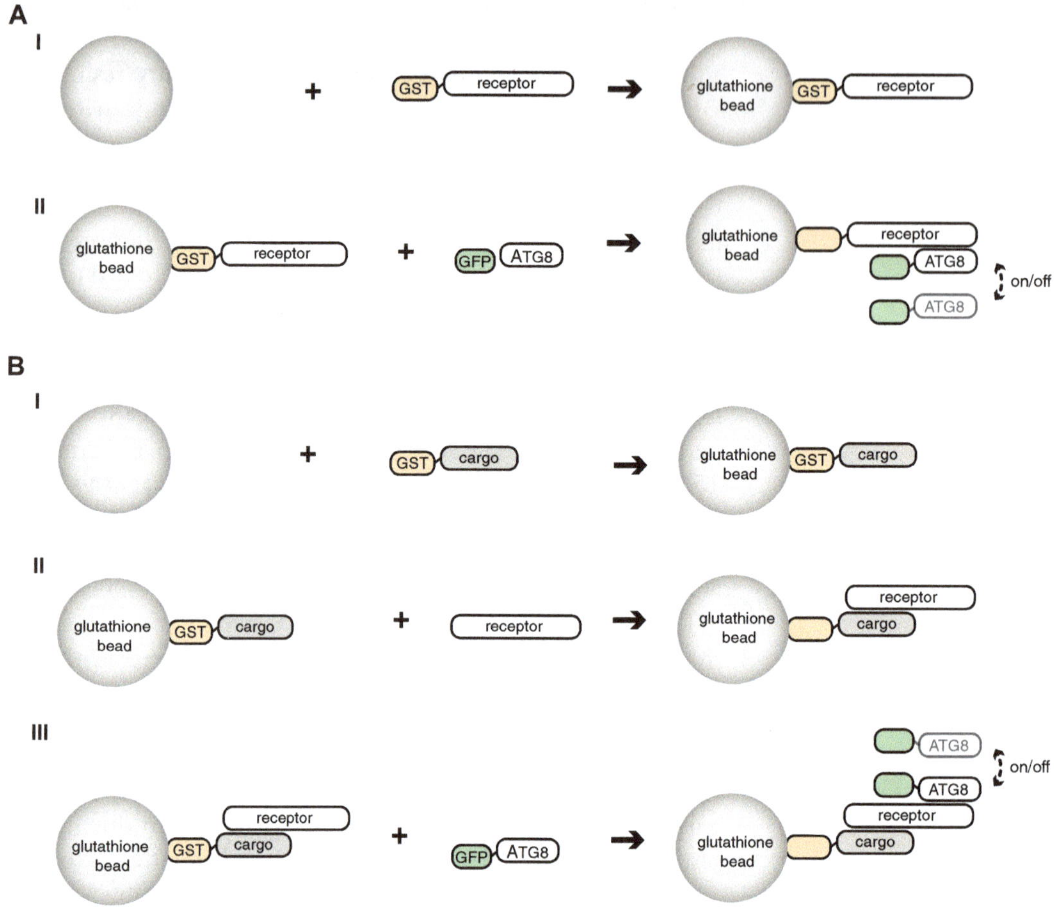

Fig. 1 Steps of the interaction assay. (**a**) In the first step, the receptor is recruited to beads (Glutathione Sepharose) via a tag (GST). Then, the fluorescently labeled interaction partner, i.e., meGFP-Atg8, is added. Potential binding is assessed at an equilibrium binding state. (**b**) This setup can be extended to reconstitute further aspects of selective autophagy. Affinity tagged cargo proteins bind to the receptor in solution, which in turn recruits labeled meGFP-Atg8 to the bead surface

3.2 Quantification of the Relative Binding Strength

1. Determine the intensity of the meGFP-Atg8 signal on the bead surface by randomly drawing a straight line across the bead. The "plot profile" function in ImageJ will yield the two maxima of the meGFP-Atg8 signal at the surface of the bead (*see* **Note 10** and Fig. 3).
2. Determine the background signal by measuring the average signal intensity of an area of the picture that does not contain any beads (*see* Fig. 3). Subtract this value from the maximum intensity of the beads of the same picture.
3. Quantify at least three images per experiment, and include the same number of beads in each quantification. Do not quantify broken or overlapping beads.
4. Compare the signal strength of multiple experiments by averaging the maximal signals on beads for each experiment.

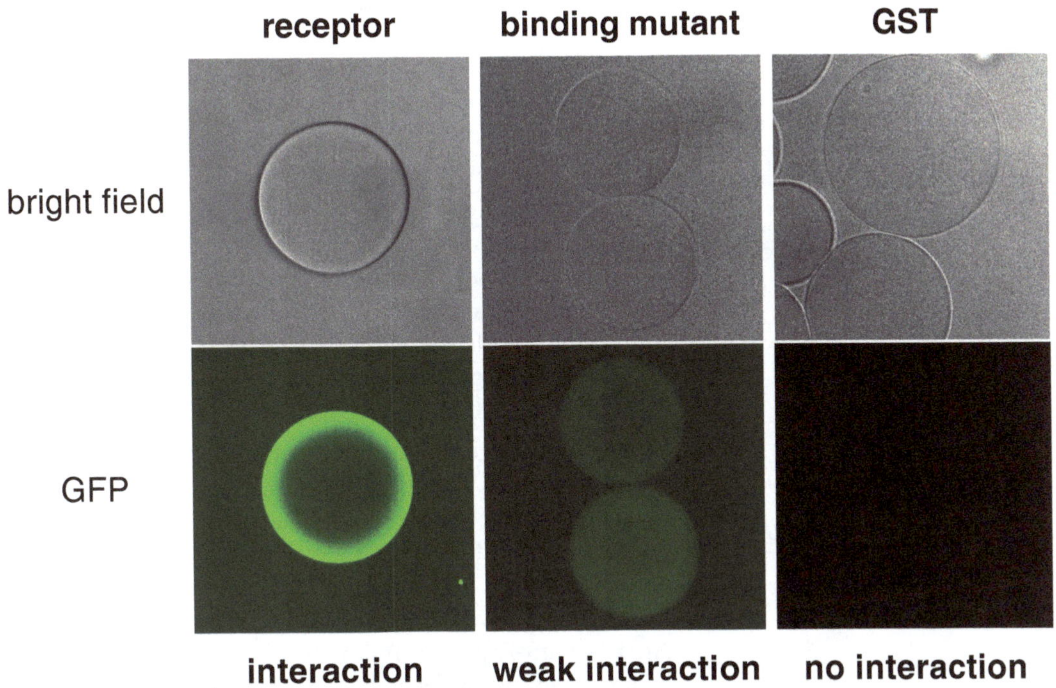

Fig. 2 Typical results of the microscopy-based interaction assay. The affinity tagged receptor, in this case the *S. cerevisiae* receptor Atg19 with a GST-tag, binds meGFP-tagged *S. cerevisiae* Atg8. This is reflected by the green signal on the surface of the beads ("receptor"). Using an Atg8 binding deficient mutant of Atg19 results in a decreased fluorescence signal on the surface of the beads due to weaker Atg19-Atg8 interaction ("binding mutant"). As a control, GST alone was bound to the beads and incubated with meGFP-Atg8. No unspecific protein binding of Atg8 to GST can be observed ("GST"). Bright-field images of the glutathione beads are shown above as reference. Scale bar, 100 mm

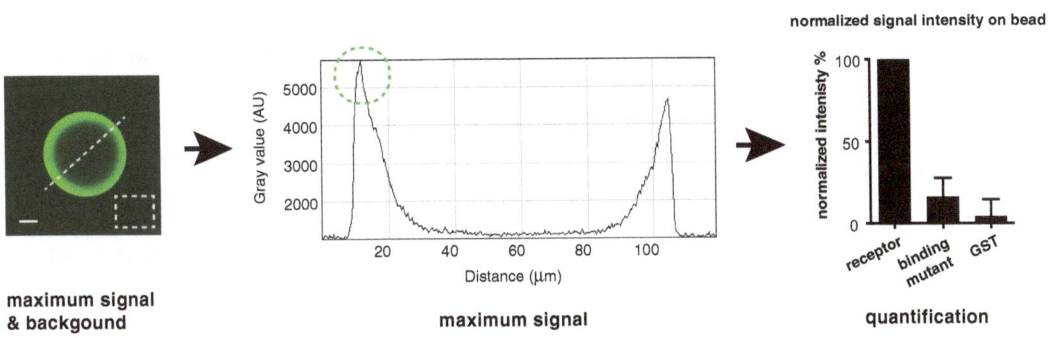

Fig. 3 Quantification. The relative binding strength can be determined by quantifying the signal of the fluorescent prey on the surface of the beads. The maximum signal intensity is measured along the diameter of a bead in focus. The background intensity of an area without bead is also measured. This value is subtracted from the maximum intensity on the bead. The signal intensities of meGFP-Atg8 binding to wild-type Atg19 and various mutants/binding partners, etc. can be accurately compared

4 Notes

1. The experimental setup described here can be readily modified or extended. The following two examples will help to illustrate the versatility of the system:
   (a) The cargo-binding activity of the receptor can be taken into account by coating the beads with the purified cargo (e.g., *S. cerevisiae* propeptide of prApe1) via a specific tag. The untagged receptor can then be recruited to the cargo-mimicking beads via direct cargo-receptor interaction, and meGFP-Atg8 binding can be monitored on the beads (*see* Fig. 1b).
   (b) To probe for potential additive or competitive effects, two binding partners can be added simultaneously or sequentially to the receptor-coated beads. An example is the competitive binding of Atg8 and the Atg12~Atg5-Atg16 complex to the receptor Atg19 [9]. Using prApe1-cargo beads coated with the receptor, it was found the Atg12~Atg5-Atg16 complex, an E3-like enzyme involved in autophagy, is displaced by Atg8 from the cargo receptor coupled beads.
2. The quality of purified proteins is crucial to obtain conclusive results [4, 7, 9]. Expression and purification of proteins of interest may have to be optimized in case unexpected/varying results are observed. Before the experiment, the protein solution should be centrifuged at 15,000 rcf, 4 °C for at least 10 min in a tabletop centrifuge to remove aggregates that might interfere with the experiment. After centrifugation, the protein concentration should be determined.
3. The protein concentrations for each experiment may have to be adapted. For example, larger proteins may coat affinity beads already at lower concentrations and smaller proteins only at higher concentrations. A strong/weak interaction of an autophagy receptor with ATG8 may require lower/higher amounts of labeled ATG8 to obtain an optimal signal during microscopy. It is important not to use too high a concentration of the fluorescent binding partner, because this may result in oversaturation of the signal, which consequently cannot be quantified reliably.
4. Affinity beads may be chosen according to the protein tag. Here sepharose beads coated with reduced glutathione were used. In any case, the chosen resin has to be tested for suitability including ease of handling, unspecific binding, and autofluorescence. GSH Sepharose (GE Healthcare) or GFP-/RFP-Trap beads (ChromoTek) bind with a high affinity to the respective tags and usually do not interact unspecifically with

proteins. Nickel nitrilotriacetic acid beads (Ni-NTA, Qiagen), which bind polyhistidine-tagged proteins, have a higher autofluorescence and the propensity to bind unspecifically and thus yield false-positive results.

5. If the affinity tagged receptor has no fluorescence tag allowing for direct visualization of the protein by fluorescence microscopy, binding should be confirmed by SDS-PAGE and Coomassie staining or Western blotting with specific antibodies.
6. If desired, a classical, gel-based pull-down assay can be easily done in parallel. To this end use an initial volume of 20 μL of receptor-coated beads. Split into two aliquots of 10 μL, and add 10 μL Atg8 solution of 1 μM to 10 μL beads (use the other 10 μL for the microscopy-based assay). Incubate for 1 h with gentle mixing at 4 °C. Take an input sample for SDS-PAGE, and wash 3× with 200 μL buffer with 3 min centrifugations at 15,000 rcf at 4 °C. Add SDS-PAGE sample buffer to the beads slurry. Detect by Coomassie staining or Western blotting with specific antibodies after SDS-PAGE.
7. Here an optimized version of the green fluorescent protein of *Aequoria victoria*, meGFP, is used. The protein contains the A206K mutation to render it monomeric [10]. Dimerization of the GFP-tag could lead to an enhanced signal on the beads and thus overestimation of the binding capacity of the proteins of interest. The fluorescence intensity of the mGFP moiety is enhanced by the two F64L and S65T mutations [11].
8. Include controls for unspecific binding of labeled meGFP-Atg8 to the resin (meGFP-Atg8 and beads without Atg19), unspecific protein binding of meGFP-Atg8 (meGFP-Atg8 and GST-coated beads), unspecific protein binding of the GST-tagged Atg19 receptor (GST-Atg19 coated beads and meGFP), and autofluorescence (receptor-coated beads without meGFP-ATG8).
9. If possible take stack pictures throughout the height of a bead to ensure inclusion of the focus plane, e.g., take ca. 10 stack pictures with a spacing of 3 μm at a 20× magnification when using GSH beads with an average diameter of 90 μm. If multiple experiments will be compared to each other and/or quantified, all pictures have to be taken with the same settings of exposure time, laser intensity, and objective. The laser power has to be adapted to the signal intensity to avoid oversaturation of the signal, which would make comparison of signal intensities impossible. Imaging should be done quickly, and repeated imaging of the same beads should be avoided.
10. In this case the ImageJ [12] software was used for quantification (https://imagej.nih.gov/ij/download.html).

Acknowledgments

We thank Justyna Sawa-Makarska and Verena Baumann for comments on the manuscript. Christine Abert is supported by a Doc fellowship of the Austrian Academy of Sciences (ÖAW).

References

1. Ichimura Y, Kirisako T, Takao T, Satomi Y, Shimonishi Y, Ishihara N, Mizushima N, Tanida I, Kominami E, Ohsumi M, Noda T, Ohsumi Y (2000) A ubiquitin-like system mediates protein lipidation. Nature 408 (6811):488–492. https://doi.org/10.1038/35044114
2. Shpilka T, Weidberg H, Pietrokovski S, Elazar Z (2011) Atg8: an autophagy-related ubiquitin-like protein family. Genome Biol 12 (7):226. https://doi.org/10.1186/gb-2011-12-7-226
3. Rogov V, Dotsch V, Johansen T, Kirkin V (2014) Interactions between autophagy receptors and ubiquitin-like proteins form the molecular basis for selective autophagy. Mol Cell 53(2):167–178. https://doi.org/10.1016/j.molcel.2013.12.014
4. Sawa-Makarska J, Abert C, Romanov J, Zens B, Ibiricu I, Martens S (2014) Cargo binding to Atg19 unmasks additional Atg8 binding sites to mediate membrane-cargo apposition during selective autophagy. Nat Cell Biol 16(5):425–433. https://doi.org/10.1038/ncb2935
5. Wurzer B, Zaffagnini G, Fracchiolla D, Turco E, Abert C, Romanov J, Martens S (2015) Oligomerization of p62 allows for selection of ubiquitinated cargo and isolation membrane during selective autophagy. Elife 4:e08941. https://doi.org/10.7554/eLife.08941
6. Zaffagnini G, Martens S (2016) Mechanisms of selective autophagy. J Mol Biol 428(9 Pt A):1714–1724. https://doi.org/10.1016/j.jmb.2016.02.004
7. Abert C, Kontaxis G, Martens S (2016) Accessory interaction motifs in the Atg19 cargo receptor enable strong binding to the clustered ubiquitin-related Atg8 protein. J Biol Chem 291(36):18799–18808. https://doi.org/10.1074/jbc.M116.736892
8. Smith DB, Johnson KS (1988) Single-step purification of polypeptides expressed in Escherichia coli as fusions with glutathione S-transferase. Gene 67(1):31–40
9. Fracchiolla D, Sawa-Makarska J, Zens B, Ruiter A, Zaffagnini G, Brezovich A, Romanov J, Runggatscher K, Kraft C, Zagrovic B, Martens S (2016) Mechanism of cargo-directed Atg8 conjugation during selective autophagy. Elife 5:e18544. https://doi.org/10.7554/eLife.18544
10. von Stetten D, Noirclerc-Savoye M, Goedhart J, Gadella TW Jr, Royant A (2012) Structure of a fluorescent protein from Aequorea victoria bearing the obligate-monomer mutation A206K. Acta Crystallogr Sect F Struct Biol Cryst Commun 68 (Pt 8):878–882. https://doi.org/10.1107/S1744309112028667
11. Cinelli RA, Ferrari A, Pellegrini V, Tyagi M, Giacca M, Beltram F (2000) The enhanced green fluorescent protein as a tool for the analysis of protein dynamics and localization: local fluorescence study at the single-molecule level. Photochem Photobiol 71(6):771–776
12. Schneider CA, Rasband WS, Eliceiri KW (2012) NIH image to ImageJ: 25 years of image analysis. Nat Methods 9(7):671–675

Part II

Imaging Autophagy in Tissue Culture

Chapter 12

Correlative Light and Electron Microscopy of Autophagosomes

Sigurdur Gudmundsson, Jenny Kahlhofer, Nastassia Baylac, Katri Kallio, and Eeva-Liisa Eskelinen

Abstract

Live-cell imaging has been widely used to study autophagosome biogenesis and maturation. When combined with correlative electron microscopy, this approach can be extended to reveal ultrastructural details in three dimensions. The resolution of electron microscopy is needed when membrane contact sites and tubular connections between organelles are studied.

Key words Live-cell imaging, Phagophore, Autophagosome, Serial sectioning, Electron tomography

1 Introduction

Live-cell imaging is a powerful method to reveal protein and organelle dynamics. However, in fluorescence microscopy one only sees objects that are fluorescent. On the contrary, in transmission electron microscopy (EM), all membrane-bound organelles and cytoskeletal filaments are visible and in most cases can be identified by morphology. When combined in correlative light and electron microscopy (CLEM), live-cell imaging and transmission electron microscopy can reveal information that cannot be obtained using any other method. In this article, we describe a protocol that we have successfully used to combine live- and fixed-cell fluorescence microscopy and transmission electron microscopy.

2 Materials

1. Gridded glass-bottom cell culture dishes (MatTek P35G-1.5-14-CGRD). Before using the dishes, make sure the grid markings are clearly visible under a phase contrast microscope with a 10× or 20× objective (*see* **Note 1**).

Nicholas Ktistakis and Oliver Florey (eds.), *Autophagy: Methods and Protocols*, Methods in Molecular Biology, vol. 1880, https://doi.org/10.1007/978-1-4939-8873-0_12,

2. Polylysine in water, 0.1 mg/mL (Sigma P4707).
3. Starvation medium (recipe from the Nicholas Ktistakis laboratory, Babraham Institute, Cambridge, UK).

   Wash a large beaker well with double-distilled or Milli-Q water (ddH_2O).

   Fill the beaker with 600 mL ddH_2O and place on a magnetic stirrer.

   Add the following salts and solutions to the beaker, and stir until everything has dissolved.

   NaCl, 8.81816 g

   1 M $CaCl_2$, 1 mL

   1 M $MgCl_2$, 1 mL

   Glucose, 0.99 g

   HEPES, 4.766 g

   Top up to approximately 950 mL and adjust pH to 7.4 with 8 M NaOH. Add 10 g bovine serum albumin, stir until it has dissolved, and then adjust the volume to 1000 mL. Take to a laminar flow cabinet; sterile filter into new bottles or tubes. Aliquot and store at −20 °C.
4. Glutaraldehyde, 25% in water, electron microscopy grade (Sigma-Aldrich G5882 or Electron Microscopy Sciences 16200).
5. 0.2 M HEPES buffer, pH 7.4.
6. Osmium tetroxide, 4% water solution (Electron Microscopy Sciences 19,140).
7. Uranyl acetate (Electron Microscopy Sciences 22,400).
8. Ethanol, absolute.
9. BEEM Embedding Capsules, Size 00, 8mm I.D. (Electron Microscopy Sciences 70000-B).
10. Embedding resin, such as TAAB embedding resin, T028.
11. Sodium cacodylate buffer, 0.1 M, pH 7.4 (Sodium cacodylate trihydrate, Electron Microscopy Sciences 12300).
12. Single-slot copper grids (Electron Microscopy Sciences GS2×1, G2010-Cu), coated with Formvar.
13. Lead citrate, Electron Microscopy Sciences 17800.

3 Methods

3.1 Polylysine Coating of MatTek Dishes and Seeding the Cells

In case you have problems with cells not adhering tight enough to the dishes, you can try polylysine coating. Otherwise omit this step.

1. Coat 35 mm gridded MatTek dishes with polylysine (polylysine in water, 0.1 mg/mL).

Add 250–500 μL of the polylysine solution to the dishes (enough to cover the glass bottom).
Incubate for 30 min at room temperature.

Remove the polylysine and wash once with sterile cell culture grade phosphate-buffered saline (PBS).

2. Seed cells of interest on MatTek dishes so that they will be at approximately 50% confluent on the day of the experiment. Having a relatively low confluency will greatly help locating the same cells in light and electron microscopy.

3.2 Live-Cell Imaging, Confocal Microscopy, and Fixation

1. Perform your experimental treatment, e.g., to induce autophagy. Serum- and amino acid-free medium can be used in case starvation-induced autophagy is of interest. Since we have observed that in Earle's Balanced Salt Solution (EBSS) the cytoplasm contracts, we use a starvation medium made by recipe obtained from the Nicholas Ktistakis laboratory (*see* Materials, Subheading 2), for 60 min.

   (In case you do not need live-cell imaging, you can omit this step and fix the cells directly with 4% paraformaldehyde.)

2. Live-cell imaging can be performed using an imaging system equipped for this purpose, e.g., it is equipped with a chamber to keep the temperature and gaseous atmosphere in correct readings. Preferably, both wide-field microscopy and spinning disc confocal imaging should also be available. For wide-field live imaging, we use a 63× water immersion objective and keep the cells at +37 °C and 5% CO_2.

3. Locate cell(s) of interest and image live over a desired, preferably relatively short, period of time.

4. To fix the cells under the microscope, remove most of the medium without touching the dish, and then add warm (+37 °C) 4% paraformaldehyde in 0.2 M HEPES (pH 7.4) as fixative. Incubate for 10 min, and then remove the fixative and add 0.2 M HEPES buffer.

5. After the fixation the cell(s) should be imaged again, e.g., using spinning disc confocal, to acquire a Z stack of optical sections (Fig. 1c, d). The imaging after fixation is important for correlation since the location of the fluorescent structures may slightly change during fixation. Phase contrast or DIC images also need to be taken in order to image the shapes of the cells of interest, as well as the neighboring cells (Fig. 1a, b).

6. Next reduce the magnification to 20× phase contrast or DIC objective, and take low-magnification images of the cells and the grid coordinates on the dish, covering at least one whole grid square, preferable more (Fig. 1a). You may need to take several images and stich them together to get a large enough

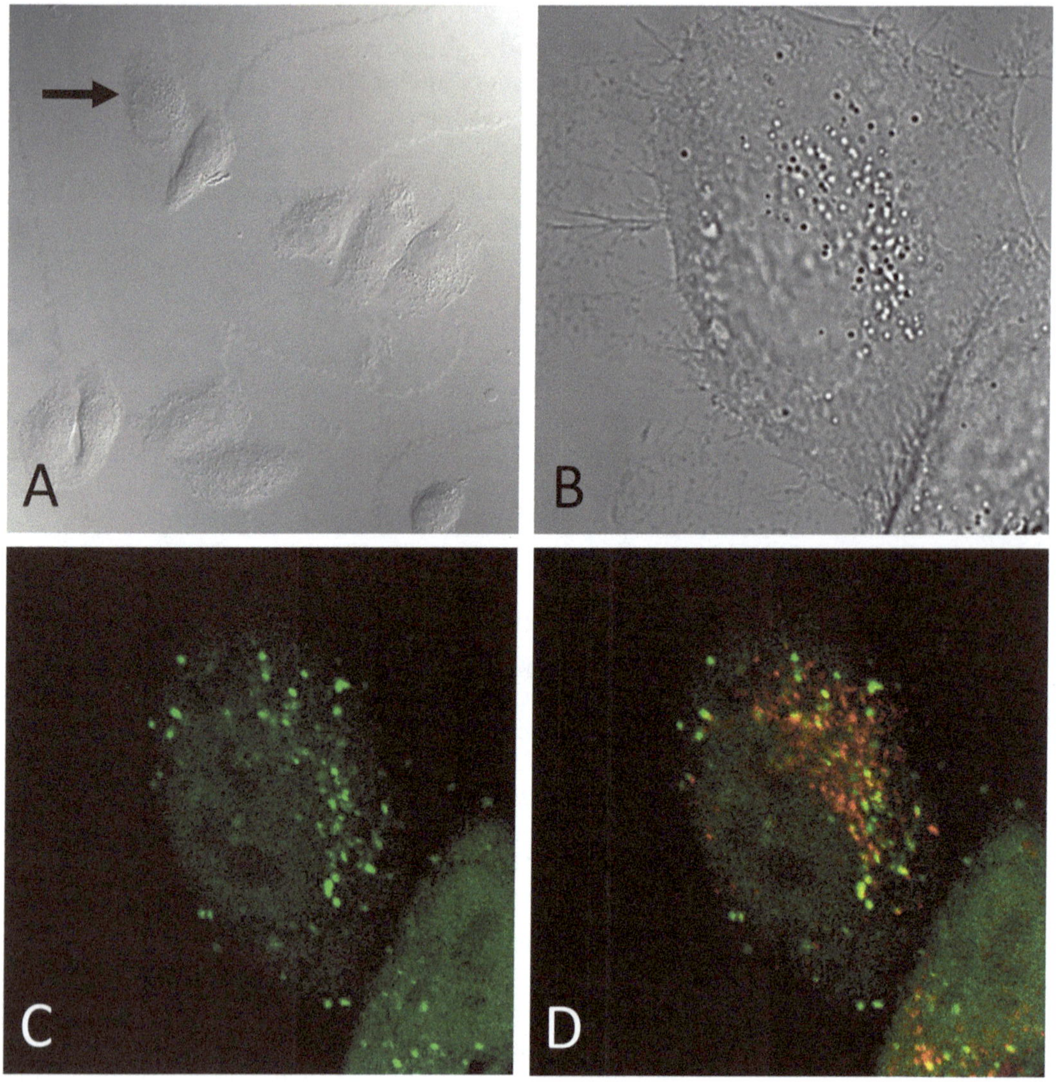

Fig. 1 Phase contrast and fluorescence images of HeLa cells stably expressing the autophagy marker mRFP-GFP-LC3. This probe can be used to monitor autophagosome acidification, since GFP fades in an acidic environment, while mRFP is more acid resistant. The cells were incubated in starvation medium for 1 h. (**a**) Phase contrast image showing part of letter C on the MatTek grid. The arrow indicates the cell of interest, shown at higher magnification in the other panels. (**b**) Phase contrast image of the cell of interest. (**c**) A confocal optical section showing the GFP fluorescence. (**d**) A confocal optical section showing an overlay of the GFP and mRFP fluorescence

field of view. These images will be used to locate the cell(s) of interest after embedding for electron microscopy (*see* **Note 2**).

7. Finally, fix the cells again in a fume hood: 2% glutaraldehyde in 0.2 M HEPES, pH 7.4, 30–120 min at room temperature. Then you can store the cells in 0.2 M HEPES buffer at +4 °C for some days.

3.3 EM Preparation

1. Before you start this step, check the cells once more with a phase contrast microscope. Mark the outside bottom of the dish with a marker pen or marker objective, so that you will know where to place the BEEM capsule with resin after dehydration (*see* **Note 3**).
2. Wash the cells twice for 3 min with 0.1 M sodium cacodylate buffer, pH 7.4.
3. Postfix the cells in osmium tetroxide at room temperature for 1 h. We use 1% OsO4, 0.1 M cacodylate buffer, containing 15 mg/mL freshly added $K_4[Fe(CN)_6]$ to reduce the osmium [1, 2]. This adds contrast to phagophore and autophagosome membranes.
4. Wash cells twice for 3 min with sodium cacodylate buffer.
5. Wash three times for 5 min with ddH_2O.
6. Incubate in 1% uranyl acetate in ddH_2O, +4 °C, 1 h. Uranyl acetate staining enhances the contrast.
7. Wash the cells three times with ddH_2O.
8. Remove the lids of the BEEM capsules and fill them to the top with embedding resin mixture, prepared according to the manufacturer's instructions.
9. Dehydrate the cells: once with 50%, 70%, and 96% and twice with 100% ethanol, each step for 3 min at room temperature.
10. Drain ethanol completely but quickly and immediately drop one drop of resin to cover the cells. The cells must not dry at any step during the embedding, and this step is the most critical (*see* **Note 4**).
11. Place a BEEM capsule filled with resin upside down on top of the cells, so that it is standing on top of your marking (*see* first step in 3.3.). Make sure there are no air bubbles on top of the cells. In case you see bubbles, make sure they lift up and float on top of the resin. Incubate at room temperature for 2 h.
12. Polymerize the resin at +60 °C for 14–24 h.
13. After taking the samples from the oven, detach them from the MatTek dish immediately, before they cool down. By having a firm grip on the BEEM capsule, dip the outside bottom of the dish in liquid nitrogen and then tap the MatTek dish softly. It should easily come off the capsule, leaving the cells and an imprint of the grid at the bottom of the resin block.

3.4 Trimming the Pyramid and Thin Sectioning

1. Fix the BEEM capsule in a holder, for instance, the block holder of an ultramicrotome. The MatTek grid should be facing up. Observe the block surface and grid under a stereo microscope. Trim the block surface either by hand or by using a trimming device, so that the block face will contain the cells of

interest. The block face should be small enough to produce sections that fit onto an electron microscopy grid. A suitable block face is smaller than the size of one square in the MatTek grid. We use a diamond knife to cut either 60-nm serial sections for conventional transmission electron microscopy or 230-nm serial sections for electron tomography [2].

2. Stain the 60-nm sections with 0.5% uranyl acetate for 30 min and then with 3% led citrate for 1 min. The tomography sections are used unstained.

3.5 Correlating Light and Electron Microscopy Images

Tracing the same organelles in the fluorescence image and EM sections can be difficult, particularly if the fluorescent organelles are small and numerous. We have used morphological markers like lipid droplets, which are visible in both phase contrast and EM images, to help this process (Figs. 2 and 3). In this case we use an overlay of fluorescence and phase contrast images as a starting point for the correlation with the EM images. Accuracy of the correlation can be improved by using fluorescently tagged markers localizing to morphologically easy-to-identify organelles such as mitochondria,

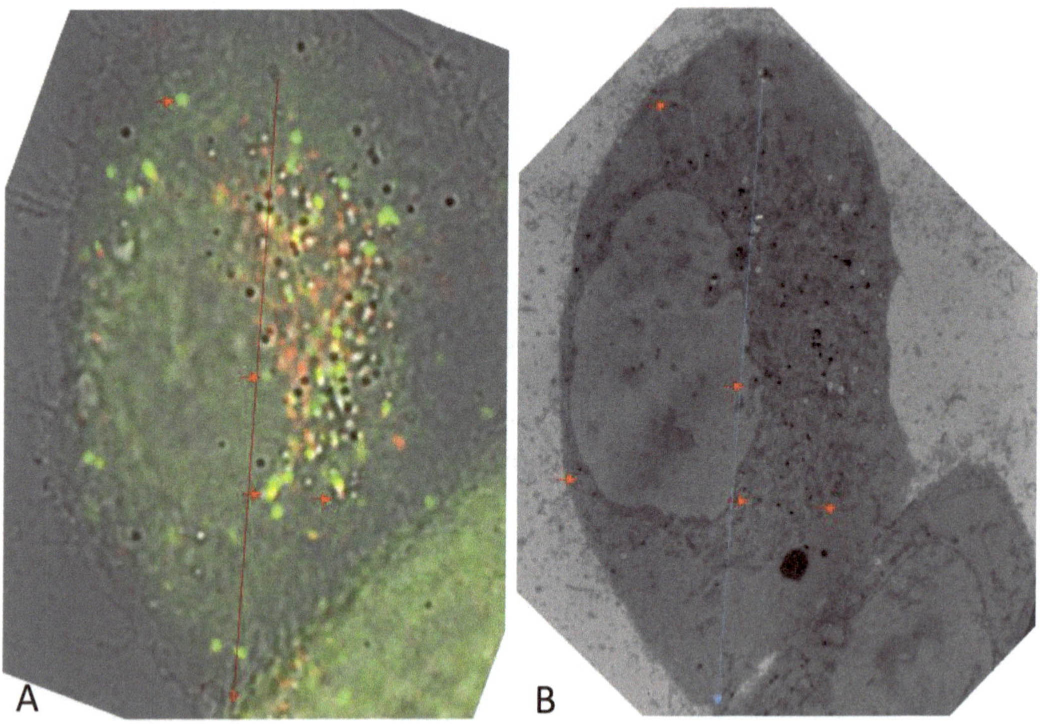

Fig. 2 Manual correlation of fluorescence signals to the EM image for the cell shown in Fig. 1. The blue and red bars in panels (**a**) and (**b**) were drawn using features visible in both the phase contrast and EM images, i.e., from the two lipid droplets at the upper end of the bar to the cell edge at the lower end of the bar. The location of some of the fluorescent spots was then traced in the EM image (red arrows), using their location in relation to the bar as a guideline. A higher magnification of the same cell is presented in Fig. 3

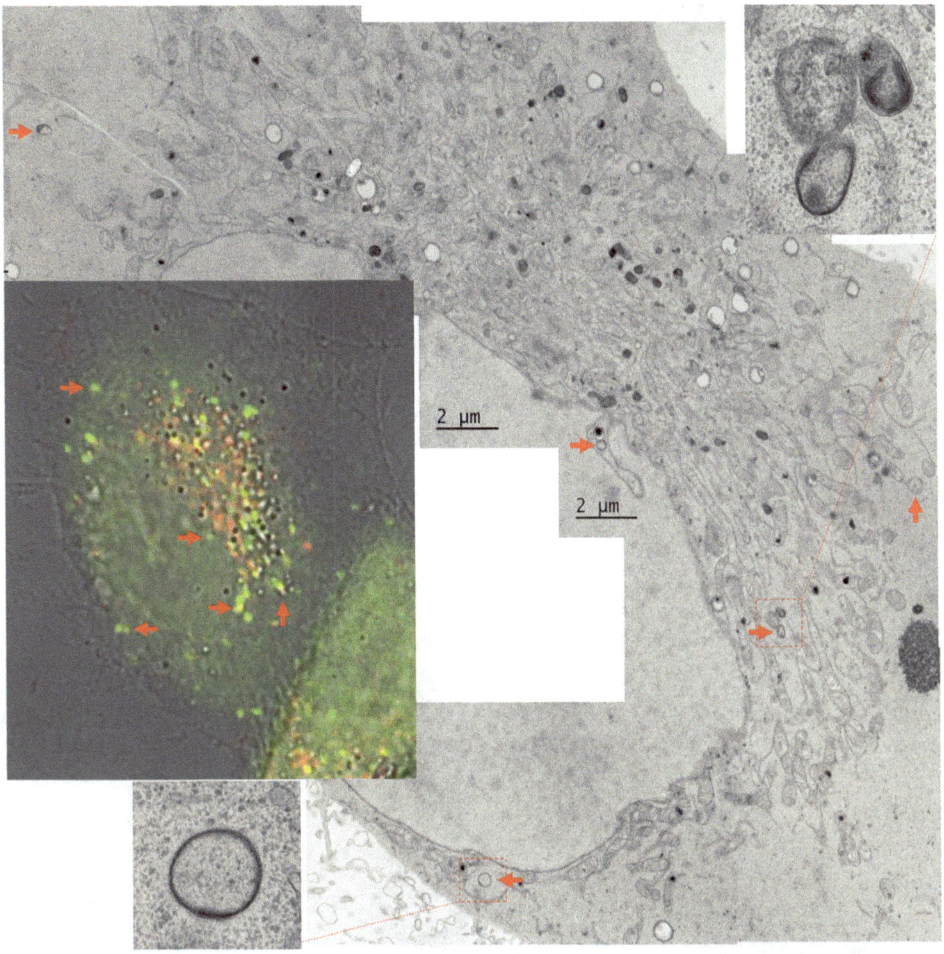

Fig. 3 Multiple EM images were taken at 2000× primary magnification of the cell shown in Figs. 1 and 2. The images were stitched together to produce a large field of view at high resolution. The red arrows indicate the structures most likely corresponding to the green fluorescent spots on the fluorescence insert. These structures have the morphology of autophagosomes, as demonstrated by the two boxed structures shown at higher magnification in the inserts

nucleus, or plasma membrane. Another option is to use the mini singlet oxygen generator (MiniSOG) tag. MiniSOG is a small fluorescent protein engineered to produce singlet oxygen that can catalyze the polymerization of diaminobenzidine into an osmiophilic, electron-dense reaction product [3]. A third option is to feed or coat the cells with beads that are both fluorescent and electron dense. Fluorescently labelled gold particles (FluoroNanogold) [4] and quantum dots [5] have been used for this purpose (*see* **Note 5**).

It is also possible to use software to help in correlating the fluorescence and EM images. Figure 4 is an example where we used the TrakEM2 plugin in ImageJ/Fiji software [6, 7] to correlate the fluorescence and EM images.

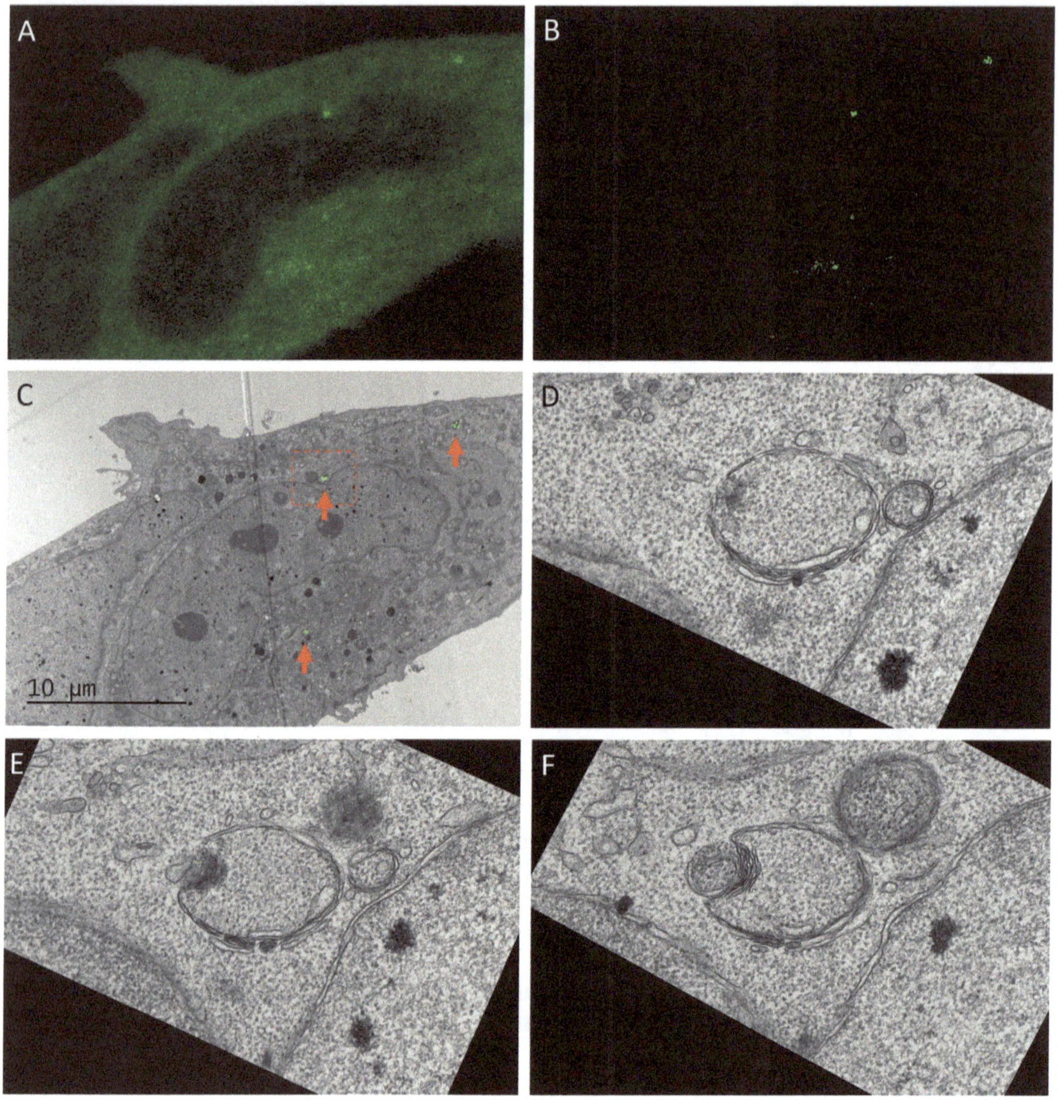

Fig. 4 An example of image correlation performed with the TrakEM2 plugin. HEK293 cells stably expressing ATG13 were incubated in starvation medium for approximately 20 min. Panels (**a**) and (**b**) show an optical section of the ATG13 signals before (**a**) and after (**b**) image processing to reduce the diffuse background. (**c**) A correlated image showing an overlay of the processed fluorescence image (panel **b**) and an EM image. (**d–f**) Serial sections showing two autophagosomes located in the boxed area in panel (**c**)

3.5.1 Correlation of a Single-Fluorescence Image to an EM Image Using the TrakEM2 Plugin in ImageJ Software

Detailed instructions for the TrakEM2 can be found online (http://imagej.net/TrakEM2_tutorials, https://imagej.net/TrakEM2). The workflow explained below can be expanded to correlate a stack of images of different types such as phase contrast, EM serial sections, and fluorescence images.

1. Make sure that both images are the same size. Open both images in ImageJ and note the size of the larger image in the upper left corner. Open the scale menu (image>scale). Change

the values (width and height in pixels) for the smaller image to correspond the larger one. Change the image type of the fluorescence image to RBG (image > type > RBG), and then save both images in a new folder containing nothing but the images that are to be correlated. Close the two images.

2. Open a new TrakEM2 document (File>new>trackem2 (blank)), and select where you would like the project (intermediate files) to be saved. Next drag the folder containing the two images to the Trakem2 window to add the data as a stack. Use the default values for slide separation but select "Resize canvas to fit stack."
3. Unlink all the images under the patches tab on the left side of the TrakEM2 window.
4. Switch to the layers tab and scroll to your fluorescence image, right click the image, and select nonlinear transformation (transform > nonlinear transformation).

   You are now not able to scroll through your images while transforming them. To switch between views, you must go to the layers tab and use the sliders to switch between views.
5. Add the points that you use to manually transform the fluorescence image. By adding one point, you can move your image around the workspace, adding a second point will then allow you to scale and rotate the image, and by adding even more points you can correlate your fluorescence image more precisely to the EM image.
6. Set the first point by holding down shift and left clicking on the point, and then switch to the EM image by going to the layer menu on the left and dragging the slider for the EM layer to the right. Now drag the point that you just placed to the location you think corresponds the same location on the EM image.
7. To add additional points, return the fluorescence image using the slider. For best results add multiple points distributed over the area or cell you wish to correlate.

   Note that you can apply the transformation and restart the transformation process with the partially transformed dataset at any time.
8. Export the aligned images out of TrakEM2 by right clicking the image (Export > Make a flat image) in the menu that appears. Select 8-bit grayscale and export to show. This will produce a stack of two images.
9. Split the stack into two images (Image > Stacks > Stacks to Images), and then merge the two images (Image > Color > Merge Channels) in the menu that appears: select the appropriate color for the fluorescence image and gray for the EM image.
10. Finally save the new correlated image.

4 Notes

1. We have had some problems with MatTek dishes (P35G-1.5-14-CGRD). One problem has been that in some dishes, the grid is on the wrong side of the dish and thus will not be visible on top of the Epon block. Another problem we have faced is that in part of the dishes, the grid lines and markings are blurred and poorly visible. Therefore, before using the dishes, please check that the letters are correctly oriented (not mirror images) and clearly visible. The correct appearance of the grid can be checked on this web page: https://www.mattek.com/store/p35g-1-5-14-cgrd/.
2. For electron microscopy, cells located on top of the MatTek grid markings (numbers, letters, and grid lines) are less optimal since they are easily lost during sectioning. Therefore, look for cells located outside any grid markings.
3. Perform the EM embedding in a fume hood, using gloves and protective clothing. The reagents are either toxic or allergenic or both. Make sure you dispose of both liquid and other waste properly; they should not end up in the sink or normal waste basket.
4. The cells should not dry out at any point during the handling and sample preparation. This means you should replace the buffer or ethanol quickly after removing the previous solution from the cells. Take extreme care during and after dehydration since alcohol evaporates very quickly.
5. Note also that from one dish, you can only use one area equal to one grid square for EM. If you aim to image more than one cell from one dish, the cells need to be close enough to each other to fit into one EM thin section.

Acknowledgments

Authors' laboratory is supported by the Academy of Finland and Magnus Ehrnrooth Foundation. Live-cell imaging and confocal microscopy were carried out in the Light Microscopy Unit at the Institute of Biotechnology, University of Helsinki. We thank the Electron Microscopy Unit at the Institute of Biotechnology, University of Helsinki, for technical help in thin sectioning and for the possibility to use an electron microscope. HeLa cells stably expressing mRFP-GFP-LC3 were a kind gift from Tamotsu Yoshimori, University of Osaka, Japan, and HEK293 cells stably expressing ATG13 were a kind gift from Nicholas Ktistakis, Babraham Institute, Cambridge, UK.

References

1. Biazik JM, Vihinen H, Anwar T, Jokitalo E, Eskelinen EL (2015) The versatile electron microscope: an ultrastructural overview of autophagy. Methods 75:44–53. https://doi.org/10.1016/j.ymeth.2014.11.013
2. Yla-Anttila P, Vihinen H, Jokitalo E, Eskelinen EL (2009) Monitoring autophagy by electron microscopy in Mammalian cells. Methods Enzymol 452:143–164
3. Souslova EA, Mironova KE, Deyev SM (2017) Applications of genetically encoded photosensitizer miniSOG: from correlative light electron microscopy to immunophotosensitizing. J Biophotonics 10(3):338–352. https://doi.org/10.1002/jbio.201600120
4. Takizawa T, Powell RD, Hainfeld JF, Robinson JM (2015) FluoroNanogold: an important probe for correlative microscopy. J Chem Biol 8(4):129–142. https://doi.org/10.1007/s12154-015-0145-1
5. Killingsworth MC, Bobryshev YV (2016) Correlative light- and electron microscopy using quantum dot nanoparticles. J Vis Exp. https://doi.org/10.3791/54307
6. Schindelin J, Arganda-Carreras I, Frise E, Kaynig V, Longair M, Pietzsch T et al (2012) Fiji: an open-source platform for biological-image analysis. Nat Methods 9(7):676–682. https://doi.org/10.1038/nmeth.2019
7. Cardona A, Saalfeld S, Schindelin J, Arganda-Carreras I, Preibisch S, Longair M et al (2012) TrakEM2 software for neural circuit reconstruction. PLoS One 7(6). https://doi.org/10.1371/journal.pone.0038011

Chapter 13

Improved Electron Microscopy Fixation Methods for Tracking Autophagy-Associated Membranes in Cultured Mammalian Cells

Ritsuko Arai and Satoshi Waguri

Abstract

Autophagy-related organelles, including omegasomes, isolation membranes (or phagophores), autophagosomes, and autolysosomes, are characterized by dynamic changes in lipid membranes including morphology as well as their associated proteins. Therefore, it is critical to define and track membranous elements for identification and detailed morphological analyses of these organelles. However, it is often difficult to clearly observe these organelles with good morphology in conventional electron microscopy (EM), thus hampering 3D analyses and correlative light-electron microscopy (CLEM). Here, we focus on describing fixation procedures using (1) ferrocyanide-reduced osmium for CLEM and (2) aldehyde/OsO_4 mixture for detecting omegasome structures and isolation membrane-associated tubules (IMATs). These methods can be easily applied to cultured mammalian cells for conventional and cutting-edge EM analyses, leading to a better understanding of ultrastructural details in autophagosome formation.

Key words Ferrocyanide-reduced osmium fixation, CLEM, Aldehyde/OsO_4 mixture, Autophagosome, Omegasome, Isolation membrane-associated tubule (IMAT)

1 Introduction

Autophagy-related structures in mammalian cells, such as isolation membranes (IM) (or phagophores), autophagosomes, and autolysosomes, have been well documented by electron microscopy (EM) [1]. With the discovery of several autophagy-related molecules and advanced imaging techniques that utilize fluorescent protein-tagged indicators, it is evident that we have just begun to understand the mechanisms of the biogenesis of IMs and autophagosomes. In the case of endoplasmic reticulum (ER)-derived, starvation-induced autophagy, which has been well characterized in this field, the site of initial nucleation represented by the presence of the ULK1 complex is reported to coincide with tubulovesicular structures in the vicinity of the ER [2]. In the subsequent elongation and closing steps, electron tomography demonstrated the

Nicholas Ktistakis and Oliver Florey (eds.), *Autophagy: Methods and Protocols*, Methods in Molecular Biology, vol. 1880,
https://doi.org/10.1007/978-1-4939-8873-0_13,

presence of narrow tubular extensions connecting the ER and IM [3, 4]. A live imaging study of a phosphatidylinositol 3-phosphate (PI3P)-binding molecule, DFCP1, revealed the presence of the omegasome, an autophagosome precursor structure [5], which was later demonstrated to possess a cluster of tubular structures, IMATs (IM-associated tubules), that connect the ER and IM [6].

Intracellular localization of fluorescently tagged proteins observed by fluorescence microscopy can be linked to EM-level structures using correlative light-electron microscopy (CLEM). Once embedded in resin, the fine structures can be subjected to 3D analysis using techniques such as electron tomography and focused ion beam scanning electron microscopy (FIB-SEM). One of the key conditions for these analyses is the ability to unambiguously detect membranes with high contrast. However, conventional preparation processes for EM may result in the morphological deteriorations and/or destaining of the membranes of autophagy-related structures, making it difficult to follow each membrane in 3D and even in 2D. Therefore, some improvements such as those described below are required.

1.1 Ferrocyanide-Reduced Osmium Fixation for CLEM

Since the ferrocyanide-reduced osmium method was first reported [7], it has been applied to several types of mammalian cells and organelles including autophagosomes [8–11]. The protocol uses a solution of 1–2% osmium tetroxide (OsO_4) containing 1.5% tetrapotassium ferrocyanide as a reducing agent after conventional primary aldehyde fixation. This method results in specific increases in the densities of isolation membranes and autophagosomes as well as the narrowing of the space between the double membrane structures (Fig. 1). It should be noted that this method is used for the discovery of connections between the IM and ER [3, 4] and of multiple contacts of IM with other organelles, such as the ER, ER exit site, late endosomes/lysosomes, Golgi complex, and mitochondria [12]. The rationale for the increased osmium deposition by the presence of reducing agents appears to be complex and has been hypothesized to be involved in different levels of oxidation and water solubility of Os intermediates [13, 14]. Importantly, this method can be applied to CLEM in cells expressing fluorescently tagged proteins; thus the present protocol also includes handling procedures of cells for acquiring images at light microscopic levels.

1.2 Fixation with Aldehyde/OsO_4 Mixture for Detecting Omegasome Structure, IMAT

Another fixation method employs a mixture of aldehyde and OsO_4, which had been only temporarily used during the 1960s and 1970s [15, 16]. We have modified this method for the detection of ER-IM intermediate structures, IMATs [6]. Because of high contrast of IMATs and IMs, this method is suitable for electron tomography but is not applicable for CLEM. We further optimized the conditions of this method and found that fixation at relatively higher temperatures (up to 30 °C) improves IMAT morphology

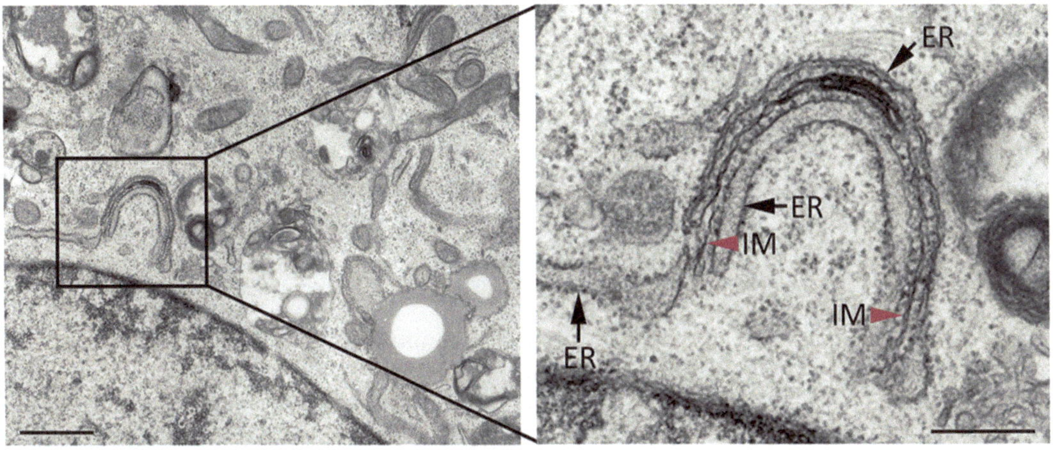

Fig. 1 Isolation membrane by ferrocyanide-reduced osmium fixation. *Atg3*$^{-/-}$ mouse embryonic fibroblasts (MEF) expressing GFP-DFCP1 (double FYVE domain-containing protein 1) were cultured in Hank's balanced salt solution (HBSS) for 120 min (nutrient-deprived conditions) and then processed for EM by the ferrocyanide-reduced osmium method. Isolation membrane (IM) is sandwiched by two endoplasmic reticulum (ER) profiles that possess ribosomes. Note that the electron density of IM is higher than that of other organelles and that the outer and inner membranes of the IM are so close that the cleft between them is not so apparent. *Bars*, 0.5 μm (*left*) and 0.2 μm (*right*). Also refer to Fig. 3 for CLEM

[17]. Unfortunately, the molecular mechanism underlying this stainability is not well understood, and morphological preservation varies between regions.

2 Materials

Prepare all solutions using ultrapure water (e.g., Milli-Q® water) and analytical or EM grade reagents. Prepare and store all solutions at room temperature unless otherwise indicated.

2.1 Light Microscopy for CLEM

1. 0.2 M phosphate buffer (PB): Mix 400 mL of 0.2 M $Na_2HPO_4 \cdot 12H_2O$ with 100 mL of 0.2 M $NaH_2PO_4 \cdot 2H_2O$, pH 7.4.
2. 10% paraformaldehyde (PFA): Add 2.0 g of PFA to 15.0 mL of 70 °C ultrapure water. Add one drop of 1 N NaOH to the suspension using a transfer pipette (3 mL), and mix at 70 °C until dissolved. Dilute to 20.0 mL with ultrapure water and filter through a filter paper. Store at 4 °C. Use this stock within a few days.
3. 25% glutaraldehyde (GA): Store at 4 °C.
4. Fixative: 2% PFA-0.1–2% GA in 0.1 M PB, pH 7.4. Mix 2.0 mL of 10% PFA, 0.04–0.8 mL of 25% GA, 5.0 mL of 0.2 M PB, pH 7.4. Dilute to 10.0 mL with ultrapure water. Use immediately (*see* **Note 1**).

5. Round coverslips (13 mm in diameter) with engraved grid and block labels.
6. Glass bottom cell culture dish of ~3.0 cm in diameter.
7. Confocal microscope.

2.2 Ferrocyanide-Reduced Osmium Fixation

1. Fixative: 2% PFA-2% GA in 0.1 M PB, pH 7.4. Mix 2.0 mL of 10% PFA, 0.4–0.8 mL of 25% GA, 5.0 mL of 0.2 M PB, pH 7.4. Dilute to 10.0 mL with ultrapure water. Use immediately.
2. 4% OsO_4: Split a glass ampule containing 1.0 g of OsO_4, and place into a wide-necked brown glass bottle with a ground-glass stopper. Add 25.0 mL of ultrapure water, mix, and leave at room temperature overnight to completely dissolve. Handle in a fume hood with care. Store at 4 °C.
3. Post-fixative: 1% OsO_4–1.5% tetrapotassium ferrocyanide in 0.1 M PB (pH 7.4). Mix 0.5 mL of 0.2 M PB (pH 7.4), 0.25 mL of ultrapure water, and 0.015 g of tetrapotassium ferrocyanide. After complete dissolution, add 0.25 mL of 4% OsO_4. Use immediately.

2.3 Fixation with Aldehyde/OsO_4 Mixture

1. 20% PFA: Add 4.0 g of PFA to 12.0 mL of 70 °C ultrapure water. Add two drops of 1 N NaOH to the suspension using a transfer pipette (3 mL), and mix at 70 °C until dissolved. Dilute to 20.0 mL with ultrapure water and filter through a filter paper. Store at 4 °C. Use this stock within a few days.
2. Fixative: 2% PFA-2% GA-2% OsO_4 in 0.1 M sodium cacodylate buffer (pH 7.4). Mix 2.5 mL of 0.4 M sodium cacodylate buffer, 0.8 mL of 25% GA, 1.0 mL of 20% PFA, and 0.7 mL of ultrapure water. Pre-warm this aldehyde solution (5.0 mL in total) and 5.0 mL of 4% OsO_4 in a 30 °C water bath, and mix them just before use.
3. Round coverslip.

2.4 Embedding, Sectioning, and Image Processing

1. 50, 70, 80, and 95% ethanol.
2. 100% ethanol: Dehydrate 99.5% ethanol with molecular sieves.
3. Epon 812 resin mixture: Mix 30.14 g of MNA, 54.82 g of Epon 812, and 15.04 g of DDSA in a disposable plastic beaker using a stir bar for 10 min. Add 1.5 mL of DMP-30 and mix for 30 min and then degas under vacuum (*see* **Note 2**).
4. Saturated uranyl acetate: Dissolve 1.0 g of uranyl acetate in 50.0 mL of ultrapure water. Store at 4 °C. Use supernatant after centrifugation each time.
5. 0.25–0.30% lead citrate: Dissolve 25–30 mg of lead citrate in 10.0 mL of ultrapure water by adding 2 drops of 10 N NaOH using a transfer pipette (3 mL). Filter with a 0.22 μm filter. Use within 1 month.

6. Copper grid of 200 or 300 mesh: Apply ion sputtering to make the surface hydrophilic.
7. Copper grid of single slit: Cover with formvar membrane and apply ion sputtering.
8. Plastic lid of cell culture plate (e.g., 4-well plate).
9. Vacuum pump.
10. Ultramicrotome.
11. Ion-sputtering equipment.
12. 80 °C oven.

3 Methods

Refer to the flowchart in Fig. 2 for an overview and comparison of the two fixation methods.

3.1 Light Microscopy for CLEM

1. Place round coverslips in a multi-well plate so that the engraved side faces upward (*see* **Note 3**).
2. Grow mammalian culture cells expressing fluorescently tagged marker proteins on the coverslips.
3. Fix the cells with 2% PFA-0.1–2% GA in 0.1 M PB (pH 7.4) for 15 min (*see* **Note 1**).
4. Rinse three times with 0.1 M PB (pH 7.4).
5. Transfer the coverslip to a 3 cm glass bottom dish with 0.1 M PB so that the cell side faces downward. Observe with a confocal laser scanning microscope equipped with a 60× or 100× objective lens. Acquire both fluorescence and bright field images (Fig. **3A**) along the z-axis that covers a whole cell. Record the block number for each acquisition (*see* **Note 4**).
6. Carefully return the coverslip to the multi-well plate that contains fixative (*see* Subheading 3.2, **step 1** below). The cell side of the coverslip should face upward.

3.2 Ferrocyanide-Reduced Osmium Fixation

1. Fix the cells with 2% PFA-2% GA in 0.1 M PB (pH 7.4) for 10 min at room temperature and then 50 min at 4 °C (*see* **Note 5**).
2. Rinse three times for 15 min each in 0.1 M PB (pH 7.4).
3. Post-fix with 1% OsO_4-1.5% tetrapotassium ferrocyanide in 0.1 M PB (pH 7.4) for 60 min at room temperature in the dark (*see* **Note 6**).
4. Rinse twice for 5 min each in 0.1 M PB (pH 7.4).

3.3 Fixation with Aldehyde/OsO_4 Mixture

1. Place round coverslips of ~12 mm in diameter in a multi-well plate.
2. Grow mammalian cells on the coverslips.

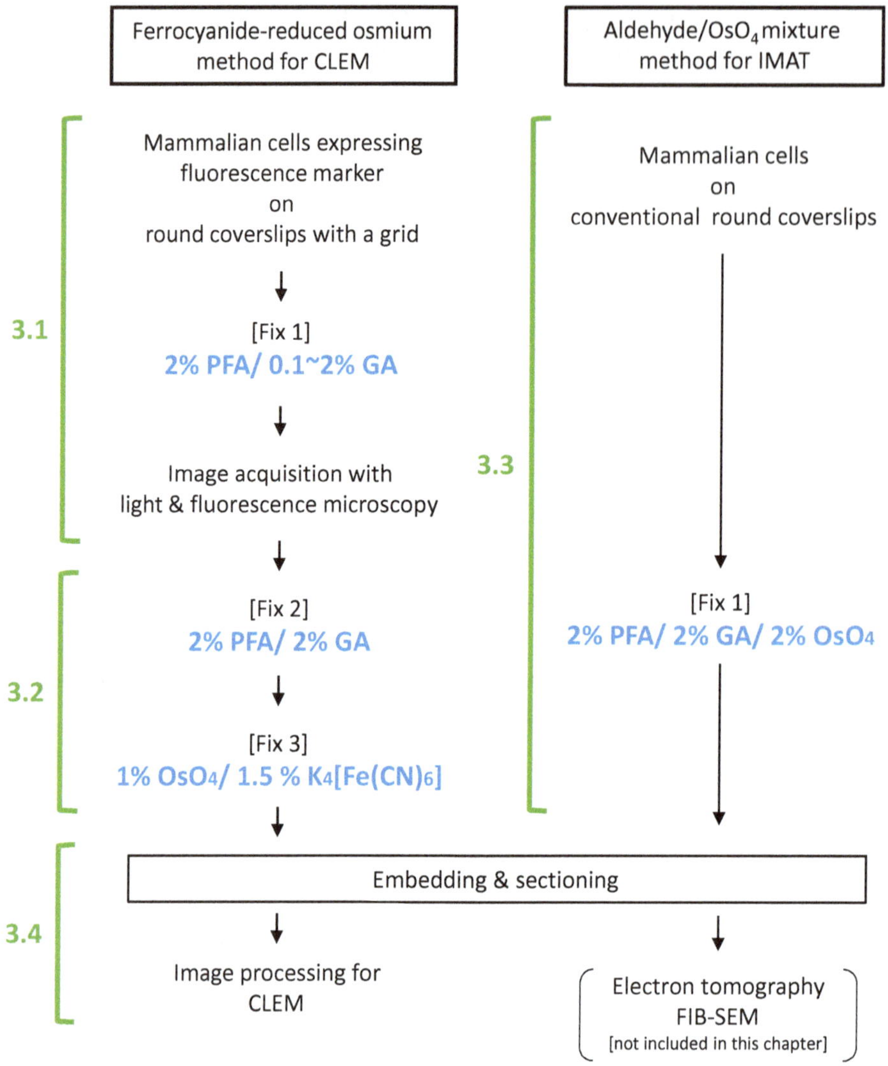

Fig. 2 Flowchart of the two fixation protocols in this chapter. Corresponding section numbers are indicated on the left of two protocols

3. Fix the cells with pre-warmed 2% PFA-2% OsO_4 in 0.1 M sodium cacodylate buffer (pH 7.4) for 60 min at 20–30 °C (*see* **Note 7**).
4. Rinse three times for 5 min each in sodium cacodylate buffer (pH 7.4).

3.4 Embedding, Sectioning, and Image Processing

1. Immerse coverslips in 50, 70, 80, and 95% ethanol sequentially for 10 min each.
2. Immerse coverslips in 100% ethanol three times for 20 min each (*see* **Note 8**).

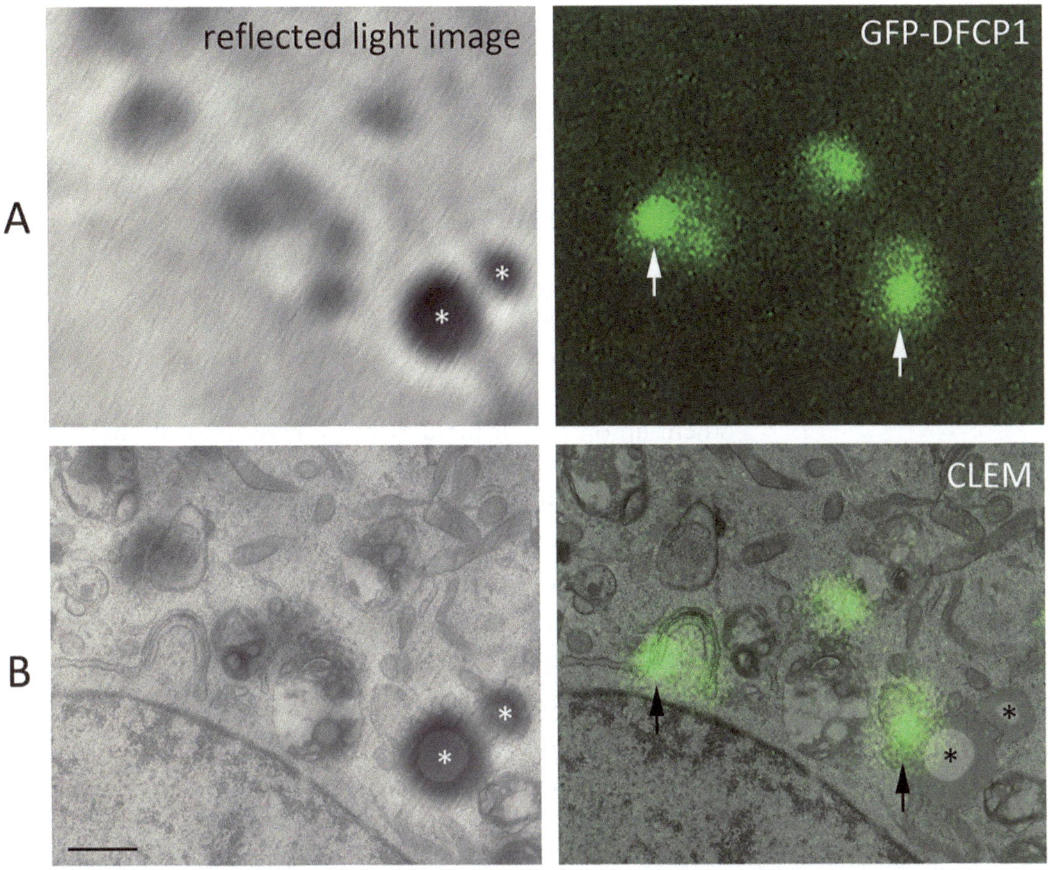

Fig. 3 An example of CLEM data. The same cells described in Fig. 1 were processed for CLEM analyses. (**A**) Reflected light images (*left*) and fluorescence image for GFP (*right*) were acquired by a confocal scanning microscope. Positions of these two images should be aligned during image acquisition. (**B**) Resulting CLEM pictures after positional adjustment between images of EM and of (**A**) that were given transparency using software. In the *left* column, positional alignment was performed according to the position of lipid droplets (*asterisk*). Note that in the *right* column, two intense signals for GFP-DFCP1 (*arrows*) are localized near the edge of IM. The *left* structure of IMs is magnified and shown in Fig. 1. *Bar*, 0.5 μm

3. Immerse coverslips in propylene oxide twice for 10 min each for resin infiltration (*see* **Note 9**).
4. Immerse coverslips in a mixture of Epon 812 resin and propylene oxide (1:1) for 60 min.
5. Immerse coverslips in Epon 812 resin overnight under weak negative pressure generated with a vacuum pump.
6. Place the coverslips into a shallow plastic container such as the lid of a 4-well plate with the cell side facing upward. Pour the resin mixture into the container.
7. Heat in an oven at 80 °C for 2 days for polymerization of the Epon 812 resin.

8. Carefully strip a region of the resin block from the coverslip while heating on a hot plate at around 130 °C. It is important to keep the resin soft enough to be easily stripped (*see* **Note 10**).
9. Excise a small block from the stripped resin disk, and attach onto the polymerized resin stand with instant glue. In the case of CLEM, refer to the grid number and geographic features that were recorded during Subheading 3.1, **step 5**, for identification and trimming of the region of interest on the cell-containing surface of the resin. It may be helpful to take a photo of the trimmed resin block for confirmation.
10. Prepare ultrathin sections with ~70 nm thickness, and pick them up on copper mesh grids for conventional observation. For CLEM, make serial sections and pick up on single-slit grids with formvar membrane (*see* **Note 11**).
11. Stain the sections for 5 min with uranyl acetate.
12. Rinse with distilled water.
13. Stain the sections with lead citrate for 5 min in a glass petri dish with solid NaOH.
14. Rinse with distilled water.
15. Dry the sections and observe with an EM.
16. For CLEM analysis, both images of bright field and fluorescence are compared and aligned with that of EM by using Photoshop (Adobe Systems Inc.). Figure 3 shows an example of prepared CLEM data. Images of bright field and fluorescent signals that are processed to gain some transparency are layered on the EM images, and relative positions of the layers are adjusted according to marker organelles, lipid droplets in this case.

4 Notes

1. Prior optimization of GA concentration (e.g., 0.1–2%) is required so that the fixatives still retain the fluorescent signal while ensuring morphological preservation during EM.
2. Epon resin can be stored at −30 °C in a disposable syringe. Note that the frozen resin with the syringe should be warmed to room temperature before use to prevent dew condensation.
3. In the case of using Matsunami #GC1310 coverslips, the block number should be read from the back side through a stereomicroscope.
4. In the case of using Matsunami #GC1310 coverslips, grid lines and numbers are observed in green when the optical settings of the microscope are for GFP, FITC, or Alexa488.

Avoid selecting cells that straddle the grid lines. When multiple cells (e.g., from 4 to 9 cells) are selected per coverslip, keep a distance of more than 2 mm between the cells, so that each trimmed block contains a single selected cell during the later sectioning step.

5. Cells can be stored for a few days in a refrigerator in this fixative. The fixed cells are processed and embedded in a resin with the coverslip, and ultrathin sections will be cut in parallel to the plane of the coverslip. Therefore, two images of EM and fluorescence can be compared and aligned by PC.

6. OsO_4 should be protected from direct exposure to light during storage and use. OsO_4 is also volatile and a powerful oxidizing agent; thus it is highly toxic. Careful handling and disposal in a fume hood are required.

7. Keeping temperatures between 20 and 30 °C during fixation is critical. When cells were fixed at lower temperatures (4 and 15 °C), the contrast of IMAT in EM pictures decreased (Fig. 4) [17].

8. It is required to move the coverslip to glass dishes at this step, because the propylene oxide that is used in the next step is corrosive for plasticware.

9. Propylene oxide is so volatile that the solution may dry up if it takes longer to transfer the coverslips.

10. An ultrasonic cutter for handicraft (e.g., ZO-41 [Nisshin EM, Co., Ltd.]) is very useful for this process. Make a groove along the hem of the coverslip using the cutter blade, and then strip the resin disk from the cover slip on the hot plate by picking it up at the interface. The resin disk will be detached

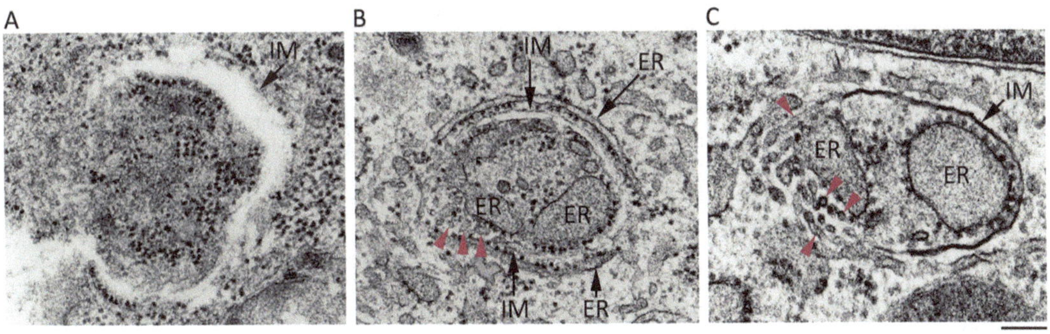

Fig. 4 Temperature-dependent preservation of IMAT structure in the fixation protocol using aldehyde/OsO_4 mixture. *Atg3*$^{-/-}$ MEFs were cultured in HBSS to expose nutrient-deprived conditions for 120 min and fixed with aldehyde/OsO_4 mixture at 4 °C (**A**), 15 °C (**B**), and 30 °C (**C**) for 10 min and then for 50 min at 4 °C. Note that electron densities of IM and IMATs (*arrowheads*) are higher for the fixation at higher temperature. Bar, 0.2 μm. These figures are reproduced and modified from a previous publication [17] with permission from the Japanese Society of Electron Microscopy Technology for Medicine and Biology

spontaneously. Stripping from grid-engraved coverslips should be done with utmost care, because small masses of resin containing cells may remain on the coverslip due to sticky attachment to the coverslip. The embedded cells and, in the case of CLEM, block number will be observed on the surface of the stripped resin by a stereomicroscope.

11. For serial sectioning, the size of sections is recommended to be slightly wider than the slit width so that the sections are better affixed to the grid. This gives stable EM observation against the electron beam.

Acknowledgments

We thank all the members in our department for helpful discussions. This work was supported by JSPS KAKENHI Grant numbers 24390048 and 15H04670 (to S. Waguri).

References

1. Eskelinen EL, Reggiori F, Baba M, Kovacs AL, Seglen PO (2011) Seeing is believing: the impact of electron microscopy on autophagy research. Autophagy 7(9):935–956
2. Karanasios E, Walker SA, Okkenhaug H, Manifava M, Hummel E, Zimmermann H, Ahmed Q, Domart MC, Collinson L, Ktistakis NT (2016) Autophagy initiation by ULK complex assembly on ER tubulovesicular regions marked by ATG9 vesicles. Nat Commun 7:12420. https://doi.org/10.1038/ncomms12420
3. Hayashi-Nishino M, Fujita N, Noda T, Yamaguchi A, Yoshimori T, Yamamoto A (2009) A subdomain of the endoplasmic reticulum forms a cradle for autophagosome formation. Nat Cell Biol 11(12):1433–1437. https://doi.org/10.1038/ncb1991
4. Yla-Anttila P, Vihinen H, Jokitalo E, Eskelinen EL (2009) 3D tomography reveals connections between the phagophore and endoplasmic reticulum. Autophagy 5(8):1180–1185
5. Axe EL, Walker SA, Manifava M, Chandra P, Roderick HL, Habermann A, Griffiths G, Ktistakis NT (2008) Autophagosome formation from membrane compartments enriched in phosphatidylinositol 3-phosphate and dynamically connected to the endoplasmic reticulum. J Cell Biol 182(4):685–701. https://doi.org/10.1083/jcb.200803137
6. Uemura T, Yamamoto M, Kametaka A, Sou YS, Yabashi A, Yamada A, Annoh H, Kametaka S, Komatsu M, Waguri S (2014) A cluster of thin tubular structures mediates transformation of the endoplasmic reticulum to autophagic isolation membrane. Mol Cell Biol 34(9):1695–1706. https://doi.org/10.1128/MCB.01327-13
7. Karnovsky MJ (1971) Use of ferrocyanide-reduced osmium in electron microscopy. In: Proc 14th Annual Meeting Amer Soc Cell Biol
8. Hoshino Y, Shannon WA, Seligman AM (1976) A study on ferrocyanide-reduced osmium tetroxide as a stain and cytochemical agent. Acta Histochem Cytochem 9:125–136
9. Langford LA, Coggeshall RE (1980) The use of potassium ferricyanide in neural fixation. Anat Rec 197(3):297–303. https://doi.org/10.1002/ar.1091970304
10. Schnepf E, Hausmann K, Herth W (1982) The osmium tetroxide-potassium ferrocyanide (OsFeCN) staining technique for electron microscopy: a critical evaluation using ciliates, algae, mosses, and higher plants. Histochemistry 76(2):261–271
11. Yla-Anttila P, Vihinen H, Jokitalo E, Eskelinen EL (2009) Monitoring autophagy by electron microscopy in Mammalian cells. Methods Enzymol 452:143–164. https://doi.org/10.1016/S0076-6879(08)03610-0
12. Biazik J, Yla-Anttila P, Vihinen H, Jokitalo E, Eskelinen EL (2015) Ultrastructural relationship of the phagophore with surrounding organelles. Autophagy 11(3):439–451. https://doi.org/10.1080/15548627.2015.1017178

13. Hua Y, Laserstein P, Helmstaedter M (2015) Large-volume en-bloc staining for electron microscopy-based connectomics. Nat Commun 6:7923. https://doi.org/10.1038/ncomms8923
14. White DL, Mazurkiewicz JE, Barrnett RJ (1979) A chemical mechanism for tissue staining by osmium tetroxide-ferrocyanide mixtures. J Histochem Cytochem 27 (7):1084–1091. https://doi.org/10.1177/27.7.89155
15. Audrey MG (1975) Fixation, dehydration and embedding of biological specimens. In: Audrey MG (ed) Practical methods in electron microscopy, vol 3. North-Holland Publishing Company, Amsterdam
16. Hayat M (1981) Fixation for electron microscopy. Academic Press, New York
17. Yabashi A, Uemura T, Waguri S (2014) Optimal temperature in applying a fixative mixture of aldehyde and osmium tetroxide to the observation of isolation membrane-associated tubules. J Electr Microsc Technol Med Biol 28(1):9–11

Chapter 14

Three-Color Simultaneous Live Imaging of Autophagy-Related Structures

Hiroyuki Ueda, Ouin Kunitaki, and Maho Hamasaki

Abstract

Simultaneous live cell imaging of multiple proteins helps to analyze mobility and interactions among proteins over time. Since autophagosomes depend on other organelles for their formation, it is necessary to observe this process with multiple fluorsphores to mark multiple organelles and the autophagosomes. To do so, we set up three cameras on one microscope to be able to acquire three colors at the same time. Here we describe the setup using a Yokogawa spinning disk confocal microscope (CSU-W1) with Andor TuCam system attaching 3 × Zyla 4.2 CMOS cameras (Andor) and detail the method for acquiring live images.

Key words Live cell imaging, Three cameras

1 Introduction

Since the very first autophagic-related genes (termed Atg now, but originally called Apg) were identified in 1993, there are now close to 40 Atgs shown to be involved in formation of autophagosomes. Characterization of Atg proteins were performed, and some were shown to form complexes, like the Ulkl complex, but most were found to be soluble proteins existing in the cytosol. Orchestration of Atg proteins is required to form autophagosomes; therefore, to observe how the formation process is taking place, imaging of multiple Atg proteins is required. Live cell imaging of two proteins with different fluorsphores using a single camera creates a delay during the filter changes and even more so if z-stacks are taken. Adding one more channel causes more delay. To characterize how proteins interact or co-localize with each other, simultaneous live cell imaging of the proteins is better. Here we demonstrate how to perform live cell imaging on three proteins. To do so, we set up the three-camera system on one microscope to simultaneously observe three proteins (Fig. 1).

Nicholas Ktistakis and Oliver Florey (eds.), *Autophagy: Methods and Protocols*, Methods in Molecular Biology, vol. 1880, https://doi.org/10.1007/978-1-4939-8873-0_14,

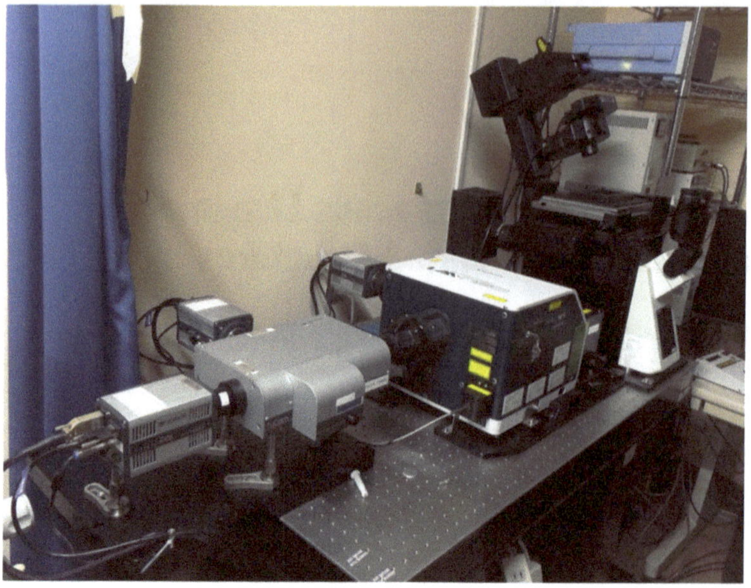

Fig. 1 Setup of the microscope: three Andor CMOS cameras are attached to an Olympus IX83 microscope equipped with a Yokogawa CSU-W1 spinning disk system

2 Materials

2.1 Cell Culture

Dulbecco's Modified Eagle's Medium – high glucose (Sigma, D6429).

Dulbecco's Modified Eagle's Medium – phenol red-free (Sigma, D5921).

Fetal bovine serum (Gibco, 10270), heat inactivated (56 °C, 30 min).

Earle's Balanced Salt Solution (Sigma, E3024).

L-Glutamine solution (Sigma, G7513-100 mL).

Trypsin/EDTA (Sigma, T4174-100 mL).

PBS (Wako, 163-03545, 197-02865, 169-04245, 191-01665).

2.2 Imaging Equipment

Glass bottom plate (Greiner, 627860, 627870).

TetraSpeck microspheres (Thermo Fisher, T7280).

2.3 Microscope

Here is an outline of the microscope setup we used.

IX83 fully motorized and automated inverted microscope (Olympus).

Stage top incubator (Tokai hit).

Yokogawa spinning disk unit (CSU-W1).

TuCam (two-camera adapter) (Andor).

Zyla 4.2 CMOS camera (Andor).

ALC laser combiner (Andor): 445 nm, 515 nm, 488 nm, 561 nm.

Filter sets: CSUW1 – 2nd dichroic mirror 514LP (Split C/Y), 561LP (Split G/R).

CSUW1: Emission filter for 2nd camera port 475/28(CFP), 617/73(RFP).

TuCam: Dichroic mirror 580LP.

TuCam: Emission filter 617/73 (RFP), 543/22 (YFP) (SemLock) (Fig. 2a–c).

Imaging software: iQ (Andor).

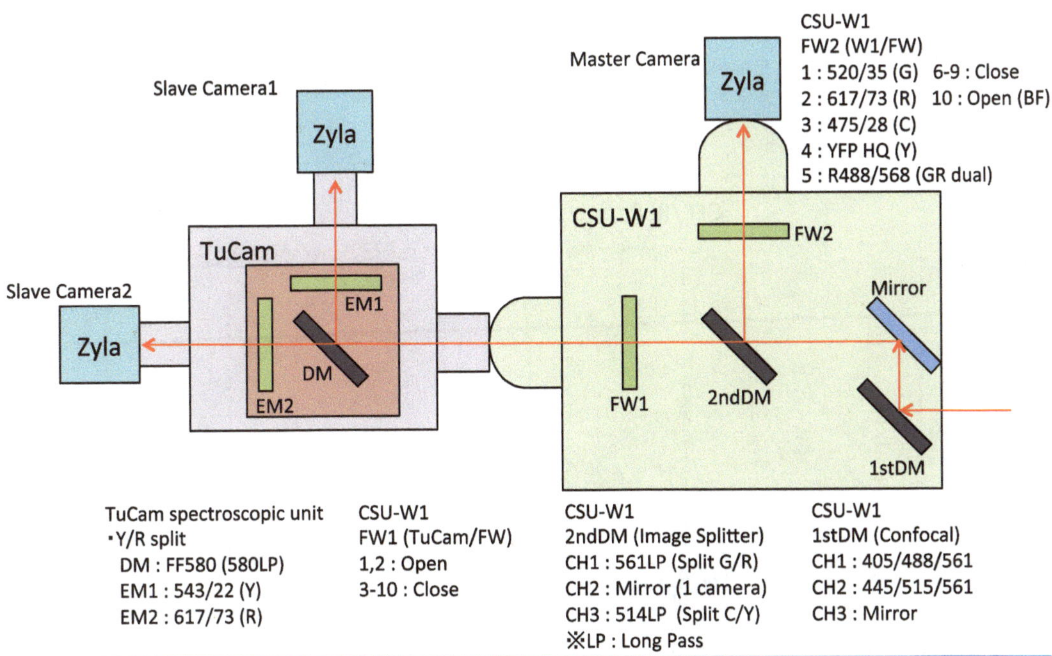

Imaging method	Slave2	Slave1	Master	TuCam	FW1	FW2	2ndDM	1stDM
1 camera	-	-	○	-	-	各EM	2: Mirror	1: 405/488/561 or 2: 445/514/561
GFP/RFP	R	-	G	Y/R	1: Open	1: 520/35	1: 561LP	1: 405/488/561
CFP/YFP/RFP	R	Y	C	Y/R	1: Open	3: 482/35	3: 514LP	2: 445/514/561

Fig. 2 Setup of our microscope, cameras, and filters

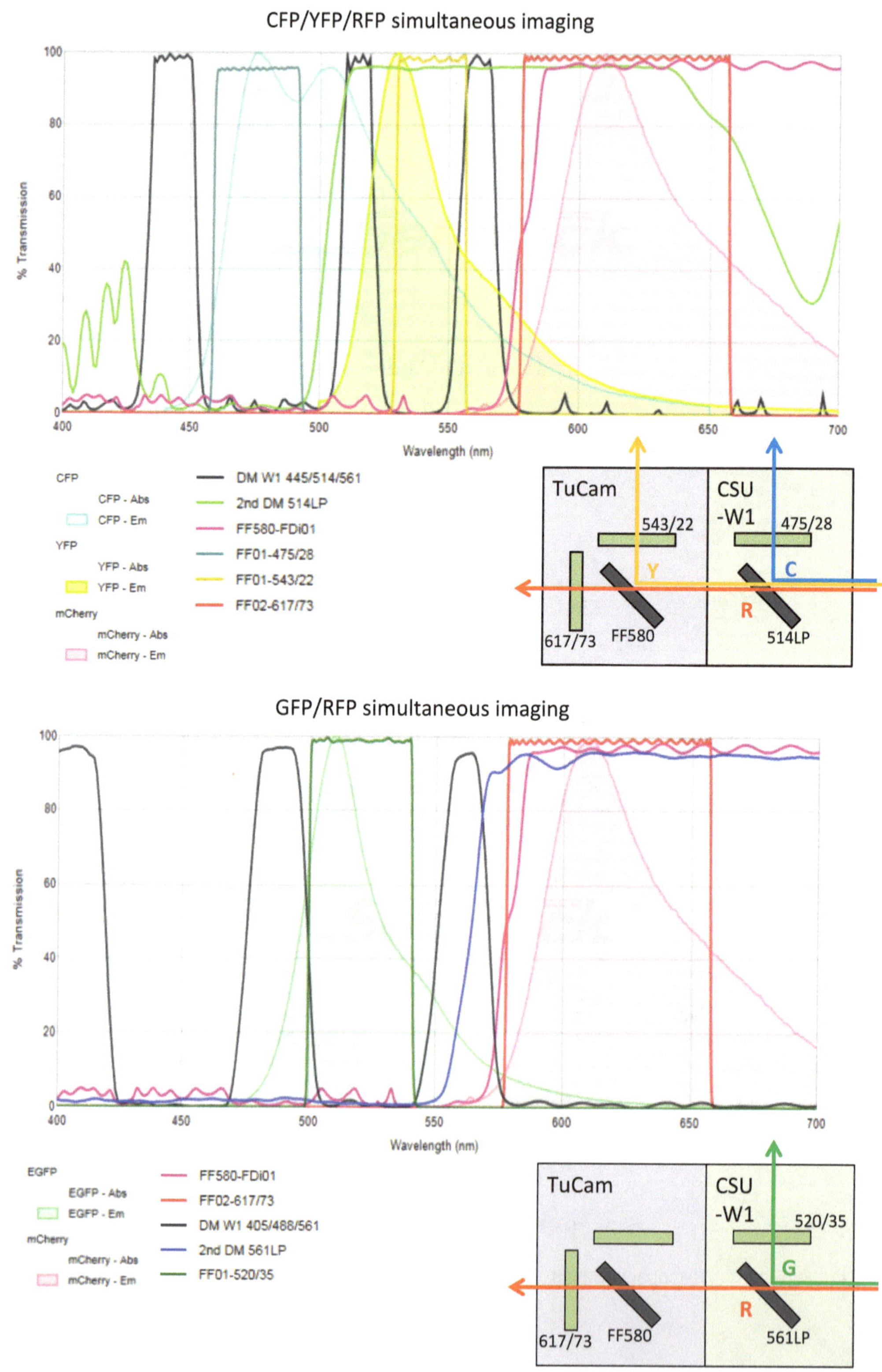

Fig. 2 (continued)

3 Methods

3.1 PC Setup

Two PCs are needed to control three cameras.

1st PC: Master camera, microscope, CSU, Andor Laser Combiner (ALC).

Piezo Z stage.

2nd PC: two slave cameras.

3.2 Aligning Three Cameras Using iQ Software (Andor)

This part is a critical step for three-color simultaneous live cell imaging. Alignment of the three cameras needed to be precise in order to examine co-localization of each protein. Here is an example of the adjustment steps employed using the iQ software (Andor).

1. Three cameras have to be aligned. First, roughly align cameras using CSU spinning disk. Stop the CSU-W1 scanning spinning disk.
2. Acquire pinhole image using slave camera 2.
3. Using camera align on iQ software, the image taken at **step 2** can be overlaid on the live image of pinhole taken by the master camera.
4. Using this overlay image, align the master camera to slave camera 2 by using *xy* adjuster (Fig. 3).
5. Repeat above to align slave camera 2 and slave camera 1.
6. The chromatic aberration of each channel can be adjusted by the focus function of TuCam. Focus of the master camera can be adjusted to slave cameras 1 and 2.

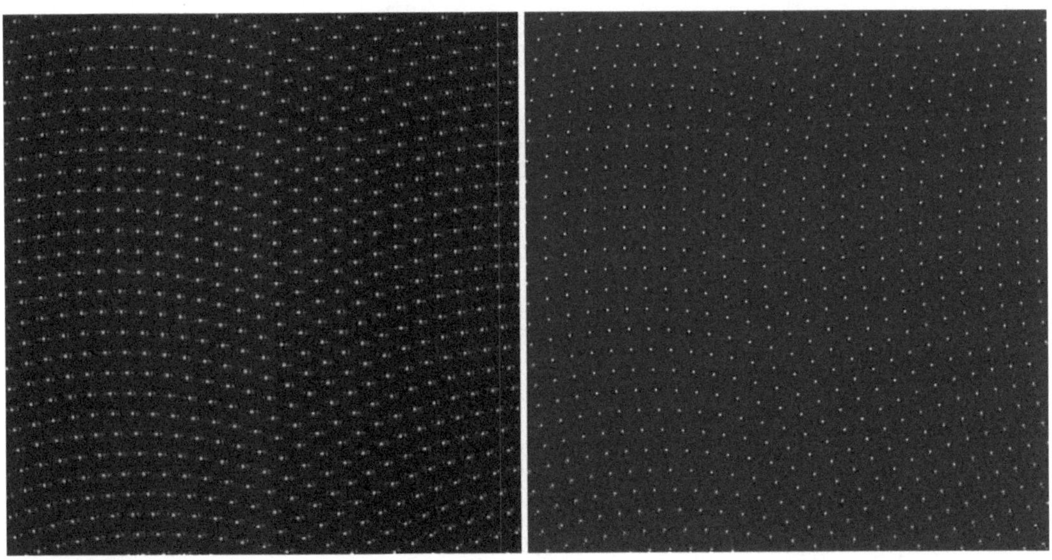

Fig. 3 Example of rough alignment using CSU spinning disks. Left: two channels are not matched and need alignment. Right: aligned image

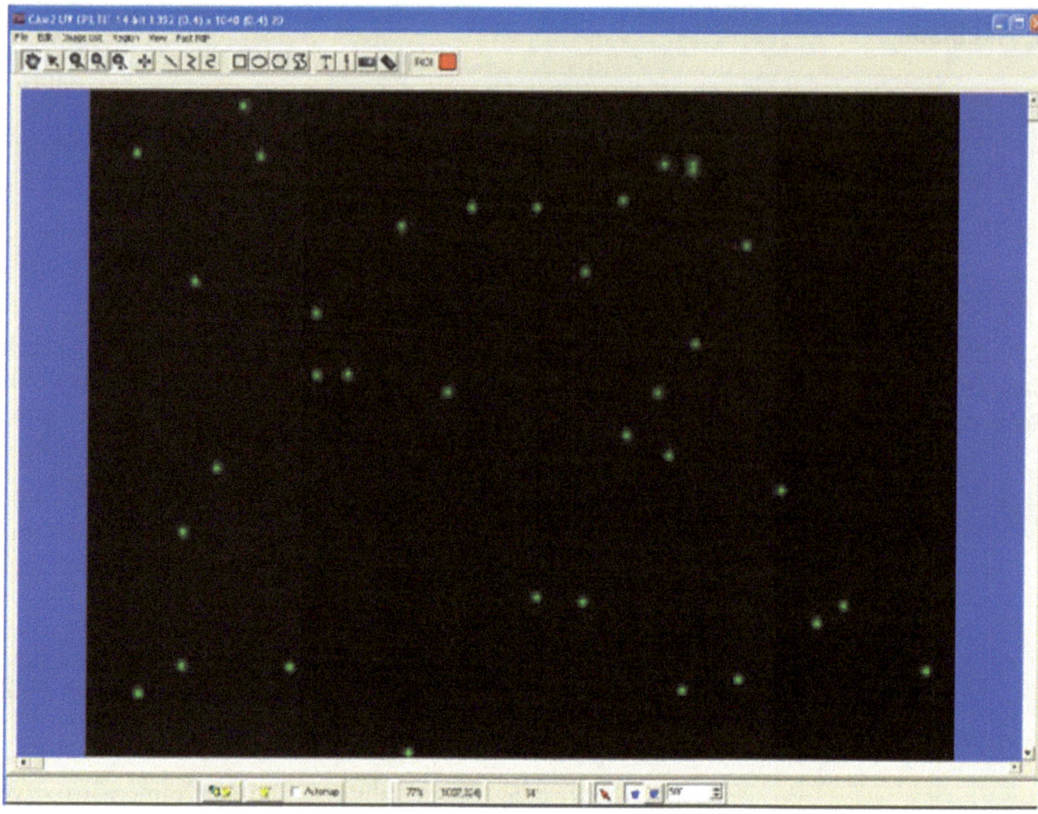

Fig. 4 Example image of TetraSpeck microspheres (Image taken from TuCam_User_Guide by Andor)

7. Repeat **steps 2–5** using TetraSpeck beads (Thermo Fisher) as an imaging target for precise alignments (Fig. 4).

Physical synchronization of three cameras can be done as follows: the fire cable (output) of the master camera is diverged into two and connected to external trigger cables (input) of the two slave cameras.

3.3 Plating Cells and Changing Medium (see Note 1)

1. Plate cells on glass bottom plate 24 h prior to observation using phenol red-free medium. Bigger and flatter cells (such as COS7 or U2Os) work best. If transfection is needed prior to imaging, make sure not to exceed 40–50% of confluence on the day of observation so that cells can stretch nicely.
2. Turn on the temperature and CO_2 of the microscope stage incubator an hour prior to using.
3. Place cell plate in the microscope stage incubator.
4. Use PBS to rinse out nutrient-rich medium. Make sure to use the same or a larger amount of PBS compared to the nutrient-rich medium to ensure medium is rinsed from the bottom and the wall of the plate. Repeat two times.
5. Change medium to starvation medium (EBSS, etc.) carefully using a dropper.

3.4 Acquire Live Cell Images

1. Set each channel's exposure time. Exposure time of slave cameras cannot be longer than the master camera.
2. Set the camera acquisition parameters (exposure time, duration, intervals, *z*-stacks, repeat time, etc.) of the master camera.
3. The number of total images has to be entered into the control software for the two slave cameras (# of time points × # of *z*-stacks).
4. To synchronize all cameras, click external trigger on both slave cameras (iQ software: camera → exposure → external trigger). With this step, the two cameras are now waiting for the master camera to run.
5. Click Run on the master camera to start acquiring live cell images
(*see* **Notes 2** and **3**).

3.5 Image Analysis

Image analysis is required as it is advisable to set exposure time to a minimum in order to avoid leakage of channels and bleaching (*see* **Note 2**). There are so many ways to process images, so here we show an example of how we processed the image for Atg5 dots (Fig. 5) using the Fiji distribution of ImageJ [1].

1. The acquired image data are first background-subtracted. If necessary, the images are corrected for the fluorescence bleaching over time [2].
2. Dot signals can be enhanced by processing data using a median filter to attenuate noise and an unsharp filter is applied to increase their contrast (Fig. 5).

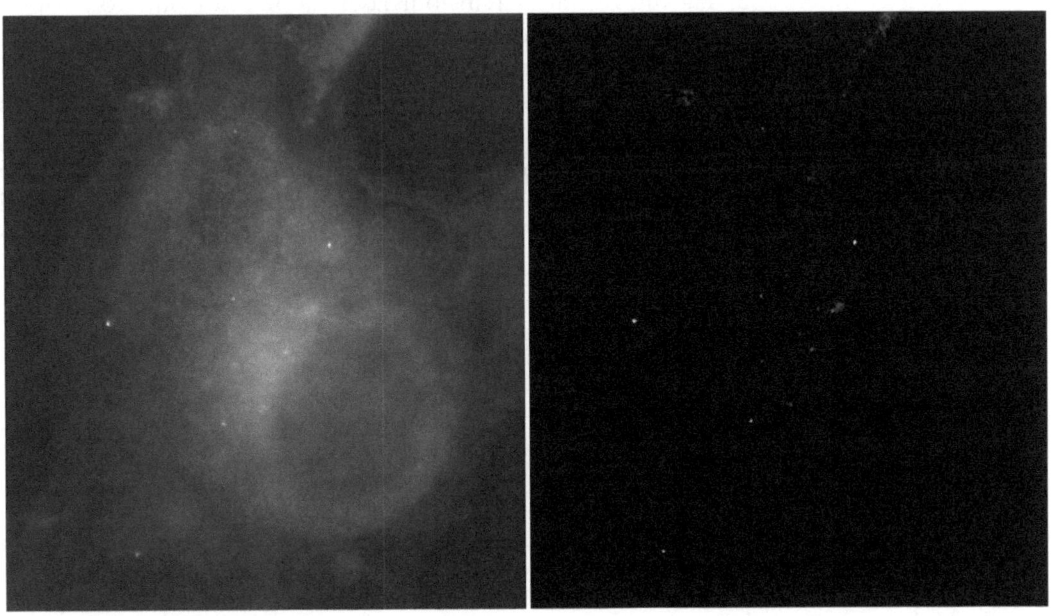

Fig. 5 Before (left) and after (right) image of data processing by Fiji on Atg5

3. Dots are binarized using the Otsu algorithm.
4. The co-localization of signals is evaluated by computing the logical conjunction (AND) of all three binarized channels. Dots present after this operation are determined to be "co-localized."
5. For measurement of distances between a dot signal and a larger structure, e.g., endoplasmic reticulum, the nearest distance from the dot to the edge of the binarized structures is taken as the distance between these two signals.

4 Notes

1. Culturing properly is a critical step to achieve reproducible images. Since autophagy is induced by starvation, accessibility to nutrients is important. Therefore, how evenly cells are plated throughout the dish so as to obtain consistent confluency in each experiment and keeping the number of passages low and consistent would produce more reproducible results.
2. For three-color imaging, it is important to minimize the bleed-through of each channel. Best is to observe different shaped organelles in each channels (e.g., ER, mitochondria, and autophagosomes. Each has different shapes so that easy to detect bleed-through). Also, we set the exposure time to as short as possible to try to avoid bleaching so that we can acquire many sequential images (longer duration with frequent intervals, etc.). To image three fluoresphore using CFP/YFP/RFP, set the brightest of all three to CFP.
3. Having in our system four lasers allows not only to perform three-color simultaneous live cell imaging but also dual-color simultaneous live cell imaging (Fig. 2, GFP/RFP simultaneous imaging). Also, if movement of proteins is not fast enough to facilitate simultaneous live cell imaging, one-camera imaging can also be performed using the Yokogawa CSU-W1. It can acquire confocal images at video rate. The CSU-W1 confocal system can acquire images with wider and better signal to noise than the CSU-X1.

References

1. Schneider CA, Rasband WS, Eliceiri KW (2012) NIH Image to ImageJ: 25 years of image analysis. Nat Methods 9:671–675
2. Miura K, Rueden C, Hiner M, Schindelin J, Riefdort J (2014) ImageJ Plugin CorrectBleach V2.0.2. Zenodo. https://doi.org/10.5281/zenodo.30769

Chapter 15

Correlative Live-Cell Imaging and Super-Resolution Microscopy of Autophagy

Eleftherios Karanasios

Abstract

Correlative live-cell imaging and super-resolution microscopy of autophagy was developed to combine the temporal resolution of time-lapse fluorescence microscopy with the spatial resolution of super-resolution microscopy. HEK293 cells that express recombinant proteins of interest fused to fluorescent tags are imaged live to capture the formation of autophagosomes, fixed on stage to "snap-freeze" these structures, stained with appropriate antibodies, relocated, and imaged at super resolution by direct stochastic optical reconstruction microscopy. This chapter provides an easy-to-follow protocol along with practical tips and background information to help set up and perform an experiment.

Key words Autophagy, Autophagosome, Endoplasmic reticulum, Membrane trafficking, Time-lapse fluorescence microscopy, Super-resolution microscopy, dSTORM, HEK293

1 Introduction

Autophagy is a quality control pathway that engulfs cytoplasmic substrates and delivers them for degradation to lysosomes [1]. The formation of the functional unit of autophagy, the autophagosome, is very dynamic, lasting less than 10 min from initiation to completion. To dissect such a dynamic cellular phenomenon, several studies have exploited the power of time-lapse fluorescence microscopy and have been pivotal to our current understanding of both the origin of the autophagosome membrane [2] and of the interplay between the modules of the autophagic machinery [3, 4]. Though these studies have visualized the process in unprecedented temporal resolution, the spatial resolution has not been as impressive. The main reason behind this problem is the diffraction limit of light, which allows fluorescent structures to be laterally resolved only if they are 0.3 μM apart (or axially resolved if they are 0.5 μM apart). However, autophagosomes induced by amino acid starvation in mammalian cells (the main model used so far in these studies) are 0.5–1 μM in diameter [5], which means that their precursor

Nicholas Ktistakis and Oliver Florey (eds.), *Autophagy: Methods and Protocols*, Methods in Molecular Biology, vol. 1880, https://doi.org/10.1007/978-1-4939-8873-0_15,

structures should be far smaller. For instance, the membrane compartment of ATG9, which interacts with initiating autophagosomes, consists of vesicles of 30–35 nm in diameter [6].

Correlative live-cell imaging and super-resolution microscopy was developed in order to upgrade the spatial resolution of time-lapse fluorescence microscopy experiments by up to one order of magnitude. This chapter provides an easy-to-follow protocol that combines time-lapse fluorescence microscopy with direct stochastic optical reconstruction microscopy (dSTORM). HEK293 cells expressing the previously reported combination of GFP-ATG13 (stably) and mCherry-ER (transiently) will be used as example [3]. The cells are seeded out on gridded coverslips, imaged live by time-lapse fluorescence microscopy, fixed on stage, stained with antibodies against the fluorescent proteins, relocated on the gridded coverslips, and finally imaged by dSTORM. For more information on the technical side of the microscopy as well as a protocol that includes a step of structured illumination microscopy (SIM), readers are referred to an excellent extended protocol published by Walker and colleagues [7].

2 Materials

1. HEK293 cells (ATCC no. CRL-1573) at low passage number.
2. The plasmid pmCherry-dgk1 used for the expression of a fluorescent ER reporter was a gift of S. Siniossoglou (Cambridge Institute for Medical Research, Cambridge, UK). A great variety of plasmids for the expression of fluorescent ER and other organelle reporters in mammalian cells are available from Addgene (www.addgene.org).
3. X-tremeGENE 9 transfection reagent (Roche Cat# 06365787001).
4. Cell culture medium: DMEM (Invitrogen) supplemented with 10% FBS (Gibco).
5. Phosphate buffered saline (PBS)—1×.
   (a) 4.3 mM Na_2HPO_4.
   (b) 1.4 mM KH_2PO_4.
   (c) 137 mM NaCl.
   (d) 2.7 mM KCl.
   (e) It may also be prepared as 10× stock. Stable for several months at room temperature.
6. Amino acid starvation medium.
   (a) 20 mM HEPES-NaOH (pH 7.4).
   (b) 140 mM NaCl.

   (c) 1 mM $CaCl_2$.
   (d) 1 mM $MgCl_2$.
   (e) 5 mM glucose.
   (f) 1% (w/v) BSA.
   (g) Stable for months at 4 °C.

7. 3.7% formaldehyde solution (*see* **Note 1**).
   (a) Preheat 3 L of H_2O at 60 °C.
   (b) Add 18.5 g paraformaldehyde powder (Sigma, Cat. P-6148), and stir to dissolve.
   (c) Slowly, add 8 M NaOH to the solution to help paraformaldehyde to dissolve.
   (d) Let the solution to cool down below 30 °C.
   (e) In the meantime, prepare 1 L 1 M HEPES, pH 7.4.
   (f) Add to the formaldehyde solution 1 L of 1 M HEPES-NaOH (pH 7.4) and 1 L of double-distilled water and stir. Stable for several months at 4 °C when protected from light.
8. Blocking buffer (*see* recipe below).
   (a) Dulbecco Modified Eagle Medium (DMEM).
   (b) 10 mM HEPES (pH 7.4).
   (c) 0.1% (w/v) sodium azide (*see* **Note 2**).
   (d) Stable for several months at 4 °C.
9. NETgel (*see* recipe below).
   (a) 150 mM NaCl.
   (b) 5 mM EDTA.
   (c) 50 mM Tris-HCl, pH 7.4.
   (d) 0.05% (v/v) NP-40.
   (e) 0.25% (w/v) gelatin (from bovine skin; Sigma, Cat. no. G-6650).
   (f) 0.02% (w/v) sodium azide.
   (g) Stable for several months at room temperature. To prepare, make up the gelatin solution separately. In order to help the gelatin dissolve, preheat the water to 55 °C before adding the gelatin. Leave the gelatin solution to cool down before combining with the remaining ingredients.
10. Permeabilization buffer.
   (a) NETgel containing 0.25% (v/v) NP-40. Stable for several months at room temperature.

11. Primary antibodies: The immunostaining protocol described below was developed using custom-made antibodies that were obtained as gifts; therefore, no specific recommendations will be made. All major antibody manufacturers though distribute good working antibodies against GFP and mCherry. Alternatively, the endoplasmic reticulum (or other organelles of interest) can be stained using antibodies against endogenous proteins of these organelles.
12. Secondary antibodies: Goat anti-rabbit and goat anti-mouse secondary antibodies conjugated to Alexa Fluor 647 (Thermo Fisher Scientific) or CF 568 (Biotium).
13. 35 mm dishes with a gridded glass coverslip base (MatTek, Cat# P35G-1.5-14-CGRD-D.S).
14. 6-well plate.
15. Rocker.
16. dSTORM imaging buffer (*see* **Note 3**).
    (a) 1 μg/mL catalase (Sigma Cat# C100).
    (b) 0.8 μM TCEP (Sigma Cat# C4706).
    (c) 40 mg/mL glucose (Sigma Cat# G7528).
    (d) 50 μg/mL glucose oxidase (Sigma Cat# G2133)
    (e) 12.5% glycerine (Sigma Cat# G2289).
    (f) 1.25 mM KCl (Sigma Cat# P9541).
    (g) 1 mM Tris-HCl (Sigma Cat# 93352).
    (h) 100 mM MEA-HCl 100 mM (Sigma Cat# M6500).
    (i) A 20× stock solution of the catalase, TCEP, KCl, Tris-HCl, and glucose oxidase can be prepared in PBS, aliquoted, and stored at −20 °C. Glucose and glycerine can also be combined in a separate 2.5× stock solution in PBS, aliquoted and stored at −20 °C. Both stocks are stable for more than a year. On the other hand, the MEA-HCl solution should be made fresh on the day of the experiment. The buffers should be kept on ice and combined to 1× STORM buffer just before imaging.
17. 1 mM Tris-HCl (Sigma Cat# 93352).
18. 100 mM MEA-HCl 100 mM (Sigma Cat# M6500).

3 Methods

3.1 Cell Preparation

1. Plate 0.4×10^6 low passage number of HEK293 cells stably expressing GFP-ATG13 per well of 6-well plate in Dulbecco Modified Eagle Medium (DMEM) (*see* **Note 4**).
2. Incubate the cells overnight at 37 °C—5% CO_2.

3.2 Cell Transfection

1. Prepare the transfection complex mix according to the manufacturer's protocol. This will contain 100 μL Opti-MEM I reduced serum medium, 2 μL X-tremeGENE 9 DNA transfection reagent per 1 μg of plasmid DNA, and 0.2–0.5 μg of plasmid DNA for the expression of mCherry-dgk1 [3] (*see* **Note 5**). Mix gently and incubate for 15 min at room temperature.
2. Add the transfection complex to cells by pipetting.
3. Incubate the cells overnight at 37 °C—5% CO_2.

3.3 Cell Plating on Gridded Dishes

1. Next day, trypsinize and count the cells from each well.
2. Plate $<0.3 \times 10^6$ cells per each gridded 35 mm dish (*see* **Note 6**).
3. Incubate cells overnight at 37 °C—5% CO_2.

3.4 Induction of Autophagy Through Amino Acid Starvation of Cells

1. Take one 35 mm gridded dish from the incubator to the imaging room (*see* **Note** 7).
2. Aspirate DMEM and wash with amino acid starvation medium 3 times.
3. Add 2 mL of starvation medium and set the timer ON.

3.5 Selection of Cells to Image

1. Select cells with a clear definition of the strands of the endoplasmic reticulum. The ER strands will be better defined nearer the periphery of the cell. Also, select cells that are producing more GFP-ATG13 particles. If the microscope has a motorized stage, it will be helpful to mark the position of several cells before selecting one or more.
2. Start recording the video once a sufficient number of GFP-ATG13 particles are forming (*see* **Note 8**).

3.6 Time-Lapse Fluorescence Microscopy (See Note 9)

1. Use a high-magnification lens (preferably 100× 1.4 NA).
2. Adjust the intensity of excitation light, aiming to balance a good signal-to-noise with the minimum exposure to light possible, in order to prevent photobleaching and phototoxicity.
3. Set the camera to minimum exposure possible: typically 100–500 ms, 2 × 2 binning and 100 gain (*see* **Note 10**).
4. Set the image acquisition interval at 10 s.
5. Remove the lid from the gridded dish (*see* **Note 11**) and start capturing. Continue until one or more structures of interest appear on the screen.

3.7 On-stage Fixation of Cells

1. Fix the cells by adding 2 mL of 3.7% formaldehyde to the gridded disk (*see* **Note 12**).

2. Refocus and capture images at both the GFP and mCherry imaging channels and at differential interference contrast (DIC).
3. Switch to a lower-power lens (typically 10×), and capture a DIC picture of the location of the imaged cell(s) on the gridded coverslip (*see* **Note 13**).
4. Mark the dish at 6 o'clock in order to help set its orientation on the second microscope.

3.8 Immunostaining (See Note 14)

1. As soon as possible after the on-stage fixation, aspirate the medium-formaldehyde mix from the gridded dish, and replace with 2 mL of formaldehyde solution. Incubate at room temperature for a total time of 15 min protected from light.
2. Aspirate the formaldehyde solution and wash with 2 mL of blocking buffer. Incubate the samples in 2 mL of blocking buffer for at least 15 min at room temperature (*see* **Note 15**).
3. Aspirate the blocking buffer and wash twice with 2 mL of NETgel.
4. Add 2 mL of permeabilization buffer, and incubate at room temperature for 10 min (*see* **Note 16**).
5. Aspirate the permeabilization buffer, and wash twice with 2 mL of NETgel. Add 2 mL of NETgel, and incubate at room temperature for at least 30 min (*see* **Note 17**).
6. Prepare the solution with the appropriate working concentration of the primary antibodies using NETgel.
7. Aspirate the NETgel, add 30 μL of the antibody solution into the middle of the coverslip, and incubate at room temperature for at least 30 min (*see* **Note 18**).
8. Wash the coverslip with 2 mL of NETgel, and incubate at room temperature for 5 min rocking. Repeat twice.
9. Prepare the solution with the secondary antibodies using NETgel.
10. Aspirate the NETgel from each well, add 30 μL of the secondary antibody solution into the middle of the coverslip, and incubate at room temperature for at least 30 min.
11. Wash the samples with 2 mL of PBS and incubate at room temperature for 5 min rocking. Repeat 6 times.
12. Store the samples in PBS at 4 °C and protected from light (*see* **Note 19**).

3.9 Prepare the Microscope for dSTORM Imaging (See Note 20)

1. Set the camera relay lens to position 1 (*see* **Note 21**).
2. Ensure that the objective lens is clean and the correction collar is set correctly.

3. Insert the 2×-magnifying lens into the illumination path in the TIRF condenser, and check the laser beam exiting the objective for focusing and centering.

3.10 dSTORM Imaging (See Note 22)

1. Prepare the dSTORM imaging buffer (*see* Subheading 2).
2. Aspirate the PBS and add 500 μL of the dSTORM imaging buffer.
3. Place the dish onto the microscope stage using the orientation mark on the dish to correct its position.
4. Use a lower-power lens (10× or 20×) and bright-field illumination to visualize the etchings on the grid using the microscope camera.
5. Use the reference images from the **step 3** to identify the cell of interest and position the cell to the center of the field of view.
6. Switch to the 100× objective, and focus on the sample fluorescence, observing through the microscope camera under appropriate epifluorescence illumination.
7. Focus on the sample and apply the dynamic focus lock (*see* **Note 23**).
8. Correct the illumination angle on the TIRF condenser (*see* **Note 24**).
9. Image acquisition settings. Select the MultiColor Continuous acquisition mode. In our system, the iXon 897 camera is set to acquire images of 256 × 256 pixels (0.16 μM/px) at a rate of 55 frames per sec. Set the preamplifier gain set to 2.4 and the EM gain at 100. Collect more than 15,000 images (*see* **Note 25**).
10. Increase the laser transmission power to maximum and start the image acquisition (*see* **Note 26**).
11. Image reconstruction. Use the Nikon system built-in software to reconstruct the single-molecule localization maps from the captured frames. We typically apply the default parameters, and select for automatic correction of any lateral drift and of the registration between the different color channels (requires calibration with fluorescent beads). To obtain visually similar images to conventional light microscopy, use the Gaussian display setting in which the more precisely localized fluorophores are shown brighter and smaller.

4 Notes

1. Heat and hydroxide ions break the paraformaldehyde polymers to monomeric formaldehyde, which is the form that will cross-link the proteins in the samples.
2. Sodium azide is added to the buffers in order to prevent the growth of microorganisms. Handle with care always wearing gloves.
3. The suggested dSTORM buffer has been evaluated for the combination of Alexa Fluor 647 and CF 568 fluorophores. Different buffer compositions may be required if the secondary antibodies are conjugated to different fluorophores.
4. Aim for a confluency of 80% on the day of the transfection.
5. For better results titrate DNA to achieve optimal expression levels, balancing good signal-to-noise with structure preservation. For instance, overexpression of some ER reporters shifts the equilibrium from ER strands to ER sheets.
6. Splitting the cells on the following day of the transfection allows achieving the optimal confluency for both high transfection efficiency (70–90%) and imaging (less than 70%). A confluency of less than 70% on the day of imaging ensures that cells will be proliferating in a log growth phase and will be allowed enough space to spread out. This will be particularly important in order to image the strands of the endoplasmic reticulum. We have also found that cells with extremely high expression levels of the ER reporter are cleared out at replating (presumably because they are not healthy enough to readhere).
7. The bicarbonate buffering system of DMEM requires high CO_2 concentration that cannot be provided by ambient air. If the microscope chamber has no CO_2 connection, it will be important to avoid prolonged incubation of the cells in DMEM medium that has not been supplemented with a buffering agent in order to prevent changes in pH.
8. Once autophagy is induced, GFP-ATG13 particles will constantly form and collapse (they typically last for 3–4 min [3]). After approximately 30 min of amino acid starvation, the cells with stronger autophagy response will contain more than 10 particles at one time.
9. An appropriate system for time-lapse fluorescence microscopy will typically include an inverted microscope frame, a light source, mirrors and filters specific for the fluorescent proteins of interest, a high-quality objective lens (63× or preferably 100×), a CCD/sCMOS camera, and a heated incubation chamber connected to CO_2. The protocol described herein was developed using an Olympus IX81 Cell^R imaging system

(now replaced by IX83 CellSens), with a Solent Scientific full enclosure chamber, a 100× 1.4 N.A. objective (Olympus), a Polychrome V monochromator (TILL Photonics), and a Hamamatsu Orca ER CCD camera. GFP and mCherry were excited at 488 and 585 nm, respectively, with 15 nm bandwidth and 20% transmission. Bandpass filters at 505–545 nm and 605–680 nm were used to collect fluorescent light from GFP and mCherry, respectively. For multichannel imaging, the delay in capturing between channels is an important consideration since it may produce artifacts that could be wrongly interpreted as dynamic changes. A solution to this problem is the use of multicamera adapters.

10. We use the 2 × 2 binning on the camera to achieve shorter exposure times and faster image acquisition and avoid phototoxicity and photobleaching without compromising the maximum resolution of the objective, as the combination with the 100× lens gives pixels of ~130 μm in *x*/*y* that meet the Nyquist sampling criterion (https://en.wikipedia.org/wiki/Nyquist%E2%80%93Shannon_sampling_theorem). We also focus on regions near the periphery of the cells because the depth of the cytoplasm is near the axial resolution limit of conventional light microscopy, thus avoiding the necessity for optical sectioning and further reducing light exposure.
11. The lid is removed in order to allow access to apply the fixative on stage without disturbing the position of the dish.
12. Formaldehyde should be pre-warmed at room temperature for 1 h. Cells are very efficiently fixed with a final concentration of 1.85% formaldehyde (any motion observed through the camera is paused within a few seconds). This will "freeze" the underlying structures captured in the live cells. Once the dish is out of the stage, discard the mix of amino acid starvation medium and formaldehyde, and replace with 3.7% formaldehyde.
13. It will be important to ensure that the etched numbers and letters are clearly depicted in order to make possible to relocate the cells in the second microscope.
14. Immunostaining is performed on the bench, always protecting the samples from light during the incubation steps.
15. The NH_3 chains of the amino acids in the DMEM will cross-link any remaining formaldehyde, preventing nonspecific cross-linking of the primary and secondary antibodies to subcellular structures that would create high background.
16. Care should be taken to avoid exceeding the suggested time, as the detergent in the permeabilization buffer will extract the lipids from the cellular membranes damaging the underlying structures.

17. The NETgel contains gelatin, which will block nonspecific binding sites, preventing cross-reactivity with the antibodies. Samples can be left in NETgel up to 2 h at room temperature or overnight at 4 °C.
18. 30 μL of antibody solution will be enough because the gridded coverslip is attached to a hole at the bottom of a 35 mm dish and the surface tension will keep the solution within the area of the glass. The lid of the dish should be kept on to prevent the sample from drying.
19. The samples are stable for at least 1 week.
20. The details provided below apply to the Nikon Ti-E system and 2D dSTORM.
21. The STORM position corresponds to an astigmatic lens that is used for 3D STORM and will not be discussed here.
22. The microscope should be placed in an isolation chamber or in a room without temperature fluctuations. Once the temperature of the system has stabilized at around 29 °C, the objective correction collar position should be adjusted to prevent spherical aberration. In order to do this, measure the point spread function (PSF) using TetraSpeck 0.1 μm fluorescent beads (Thermo Fisher Scientific, Cat# T7279) in a 35 mm dish with a gridded coverslip overlaid with the dSTORM buffer. Use wide-field epi-illumination to capture a 3 μm deep *z*-stack of individual beads adhered to the cover glass at 0.2 μm steps. Use the PSFj software [8] to measure the PSF, and adjust the correction collar to minimal lateral and axial PSF values. PSF values are affected by unnecessary optical elements in the light path (e.g., DIC prism, dirty objective and sample holder, etc.). If despite any adjustments made the measured values are not close to the theoretical ones, an alternative objective should be tested.
23. Because of the high precision of the relocation, any drifts of the objective from the sample during the long acquisition times will introduce errors in the relative positions of adjacent molecules that are blinking and thus detected in different frames. If such a feature is not available, then minimizing the room temperature fluctuations should also minimize drift. A drift correction option is also provided by the Nikon analysis software at the image reconstruction stage.
24. Reaching the correct TIRF angle will cause a sudden and dramatic increase in the intensity of the emission.
25. The total number of frames required to capture enough blinks that will be used to reconstruct the localization of the protein of interest will depend on a number of factors, including the abundance of the protein, the quality of the staining, and of

course the quality and frequency of blinking. For the combination of our imaging system and our proteins of interest, typically a total of 15,000–20,000 images are required to reach this number. As the rate of blinking drops with time, a camera with faster acquisition rate can exploit the initial window of more efficient blinking and will collect a sufficient number of blinks in a fraction of this number of frames.

26. Lasers of sufficient power are necessary for dSTORM experiments. For instance, on our system and with the buffer described above, the 300 mW 647 nm laser works efficiently with various fluorophores (e.g., Alexa Fluor 647, Cy5, etc.), while the 150 mW 561 nm laser allows imaging only with some fluorophores (e.g., CF 555, CF 568); however the 50 mW 488 argon laser does not have sufficient power for dSTORM experiments. As the laser power increases, the sample will initially shine very bright because all the fluorophores transition through a near simultaneous excitation-emission state before reaching the dark state. From this moment on, only the occasional blinks of individual fluorophores fluorescing and then returning to the dark state should be observed. The efficiency of this transition is determined by the combination of the light reaching the sample and the imaging buffer. Care should be taken to allow as much light reaching the sample as possible through adjusting several parameters, including the TIRF angle, the condenser focusing/centering, the presence of 2× lens in TIRF condenser, and additional power settings for the individual lasers. Poor buffer quality on the other hand may prevent blinking despite the fluorophores entering the dark state. Finally, the frequency of the blinks gradually drops as more fluorophores are bleached. In order to boost the blinking frequency of the sample, simultaneous illumination with low power 405 nm light can be applied. However, above a certain power that needs to be determined empirically, the application of 405 nm light will also increase the number of fluorophores that return to the normal excitation-emission state, creating a background signal from constantly fluorescing fluorophore clusters that will introduce artifacts at the image reconstruction stage.

References

1. Mizushima N, Komatsu M (2011) Autophagy: renovation of cells and tissues. Cell 147:728–741. https://doi.org/10.1016/j.cell.2011.10.026
2. Karanasios E, Ktistakis NT (2014) Live-cell imaging for the assessment of the dynamics of autophagosome formation: focus on early steps. Methods (San Diego, Calif) 75:54–60. https://doi.org/10.1016/j.ymeth.2014.12.007
3. Karanasios E, Stapleton E, Manifava M et al (2013) Dynamic association of the ULK1 complex with omegasomes during autophagy induction. J Cell Sci 126:5224–5238. https://doi.org/10.1242/jcs.132415

4. Koyama-Honda I, Itakura E, Fujiwara TK, Mizushima N (2013) Temporal analysis of recruitment of mammalian ATG proteins to the autophagosome formation site. Autophagy 9:1491–1499. https://doi.org/10.4161/auto.25529
5. Mizushima N, Klionsky DJ (2007) Protein turnover via autophagy: implications for metabolism. Annu Rev Nutr 27:19–40. https://doi.org/10.1146/annurev.nutr.27.061406.093749
6. Yamamoto H, Kakuta S, Watanabe TM et al (2012) Atg9 vesicles are an important membrane source during early steps of autophagosome formation. J Cell Biol 198:219–233. https://doi.org/10.1083/jcb.201202061
7. Walker SA, Karanasios E, Ktistakis NT (2017) Correlative live cell and super resolution imaging of autophagosome formation. Methods Enzymol 587:1–20
8. Theer P, Mongis C, Knop M (2014) PSFj: know your fluorescence microscope. Nat Methods 11:981–982

Chapter 16

Methods for Imaging Autophagosome Dynamics in Primary Neurons

Audrey Dong, Vineet Vinay Kulkarni, and Sandra Maday

Abstract

Autophagy is an essential degradative pathway that maintains neuronal homeostasis and prevents axon degeneration. However, the mechanisms of autophagy in neurons are only beginning to be understood. To address this fundamental gap in knowledge, we have established several key methodologies for live-cell imaging and quantitative analysis of autophagy in primary hippocampal neurons. Using these methods, we have defined compartment-specific dynamics of autophagy in real-time under basal versus stress conditions. For example, we have characterized autophagosome biogenesis in the distal axon and subsequent retrograde transport to the soma for degradation. Autophagosomes are also generated locally within the soma. In contrast to the axon, the majority of autophagosomes in dendrites are stationary, while some exhibit bidirectional movement. These studies establish an initial road map for autophagosome dynamics in each compartment of the neuron and set the stage for a more detailed understanding of neuronal autophagy in stress and disease.

Key words Autophagy, Hippocampal neurons, Autophagosome, Transport, Biogenesis, Axon, Dendrite

1 Introduction

Autophagy is a lysosomal degradation pathway that eliminates damaged or aged cellular components from the cytoplasm [1–8]. In this process, proteotoxins are packaged into an autophagosome and targeted for destruction by fusion with degradative lysosomes. Post-mitotic cells such as neurons are particularly dependent on autophagy to maintain cellular homeostasis [1–3]. Most neurons in the brain are born during embryogenesis and must survive a lifetime [9]. Consequently, to support their long-term viability and functionality, neurons require robust quality-control mechanisms. In fact, loss of autophagy is sufficient to induce axon degeneration and neuron death in mice [10–13]. Further, mutations in

Audrey Dong and Vineet Vinay Kulkarni are co-first authors.

Nicholas Ktistakis and Oliver Florey (eds.), *Autophagy: Methods and Protocols*, Methods in Molecular Biology, vol. 1880, https://doi.org/10.1007/978-1-4939-8873-0_16,

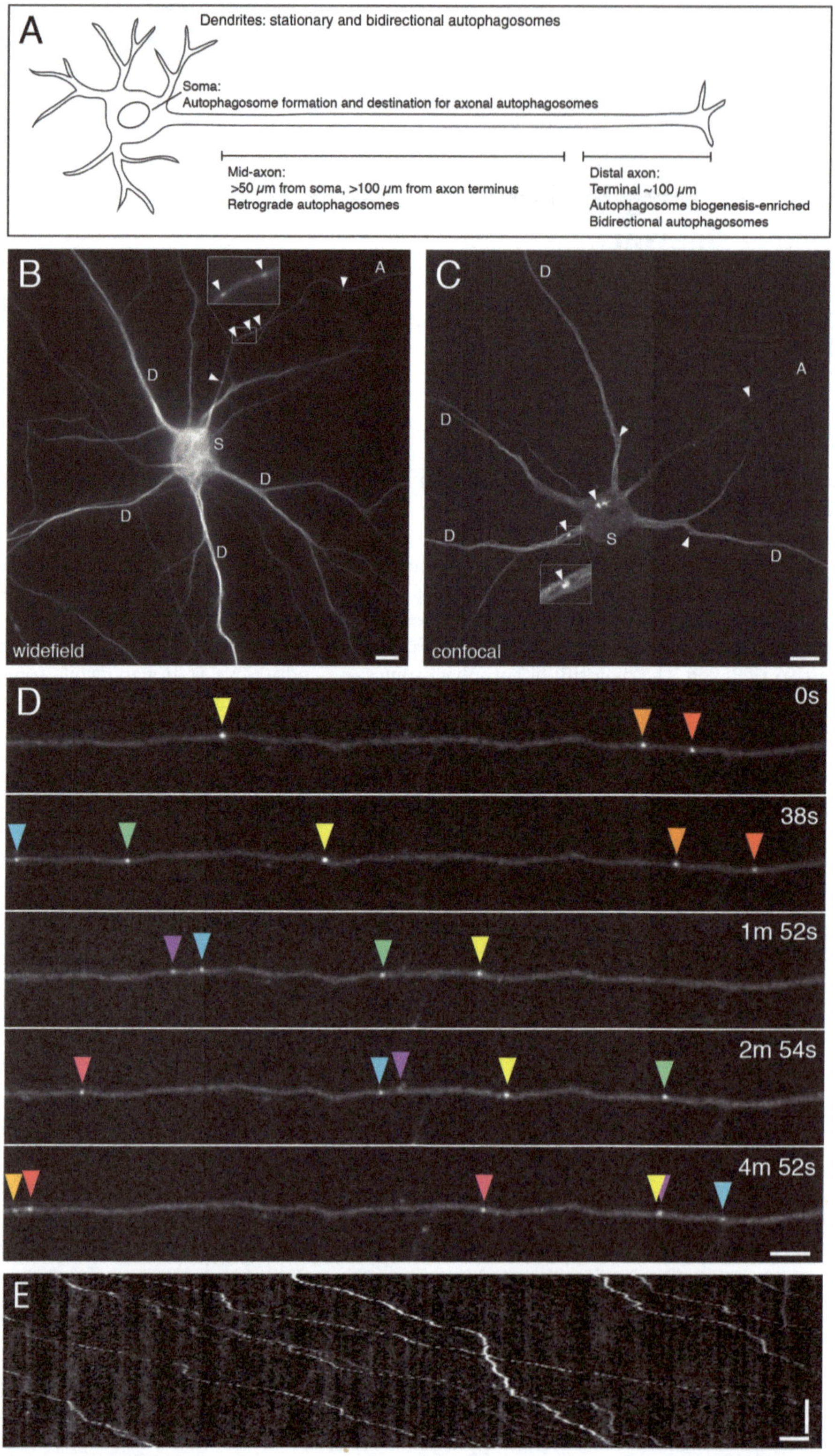

Fig. 1 GFP-LC3 distribution and dynamics in primary hippocampal neurons. (**a**) Schematic of compartment-specific autophagosome dynamics in primary neurons. (**b**) Widefield image of GFP-LC3 in a hippocampal

autophagy genes have been linked to the progression of neurodegenerative disorders in humans [1, 14]. Despite evidence that autophagy is a key regulator of neuronal homeostasis, very little is known about the mechanisms driving this process in neurons.

Here we describe a protocol for live-cell imaging and quantitative analysis of autophagy in primary hippocampal neurons. Autophagy is readily followed using GFP-tagged LC3. LC3 is a well-characterized marker for the autophagosome [15, 16]. LC3 has a cytosolic and lipidated population, and the lipidated form is associated with autophagosomes, appearing as discrete puncta in the neuron (Fig. 1). Our live-cell analyses of GFP-LC3-positive autophagosomes have revealed striking features of autophagosome dynamics in each compartment of the neuron [4–6]. Axonal autophagosomes are preferentially generated in the distal end of the axon and undergo robust retrograde transport toward the soma [4–6]. Autophagosomes in dendrites, however, are predominantly stationary or undergo bidirectional movement [5, 6], likely owing to the mixed polarity of the underlying microtubule network [17, 18]. Once in the soma, autophagosomes are confined within the somatodendritic domain [6], which likely facilitates cargo degradation by promoting fusion with proteolytically active lysosomes that are enriched in this region [19, 20]. Thus, live-cell imaging has revealed that constitutive autophagy is a highly compartmentalized pathway in neurons.

To measure autophagosome dynamics in neurons, we isolate hippocampal neurons from GFP-LC3 transgenic mouse embryos at embryonic day 15.5 [16]. The embryos are separated by genotype based on GFP fluorescence (GFP-LC3 transgenic vs. non-transgenic littermates), and hippocampal neurons are prepared independently from each genotype. At the final stage of plating, GFP-LC3 transgenic neurons are combined with non-transgenic neurons at a 1:20 dilution. This dilution of GFP-LC3 transgenic neurons enables resolution of single-cell dynamics and directionality of axonal processes, while maintaining a dense environment

Fig. 1 (continued) neuron grown 8 DIV in glial-conditioned maintenance media. Image was taken with a 63× objective. Arrowheads denote autophagosomes in the axon (A). Inset, magnified view of axonal autophagosomes. *D*, dendrite, *S*, soma. Bar, 10 μm. (**c**) Confocal image of GFP-LC3 in a hippocampal neuron grown 9 DIV in glial-conditioned maintenance media. Image was taken with a spinning disk confocal microscope using a 100× objective. Arrowheads denote autophagosomes in the axon (A), dendrites (D), and soma (S). Inset, magnified view of an autophagosome in a dendrite. Bar, 10 μm. (**d**) Time series of GFP-LC3-positive autophagosomes in the axon undergoing retrograde transport; retrograde direction is toward the right. Each colored arrowhead represents an individual autophagosome. Images were acquired using a spinning disk microscope and 63× objective on a hippocampal neuron grown 8 DIV in glial-conditioned maintenance media. Bar, 5 μm. (**e**) Corresponding kymograph to the movie in (**d**) showing autophagosome retrograde transport in the axon; retrograde direction is toward the right. Each diagonal line represents a single autophagosome moving toward the soma. Vertical bar, 1 min. Horizontal bar, 5 μm that is conducive for the neurons to form synaptic connections. Neurons are imaged at 8–10 DIV, at a developmental time point when axon-dendrite polarity is established. For older neurons (>2 weeks), neurons are cultured in glial-conditioned maintenance media to promote long-term viability. Imaging of GFP-LC3 can be performed on a widefield fluorescence microscope or a spinning disk confocal microscope; imaging conditions for both microscopes are described. The spinning disk confocal microscope provides enhanced resolution of autophagosome dynamics within regions of greater depth, such as the soma and dendrites (Figure 1b, c), and also facilitates the tracking of autophagosome biogenesis events. The protocols described here can be used to quantitate autophagy levels and dynamics in primary neurons in real time in response to various modalities of stimuli and stress, thus enabling a more detailed understanding of the mechanisms and regulation of neuronal autophagy in health and disease.

2 Materials

2.1 Primary Mouse Hippocampal Culture

1. GFP-LC3 transgenic mice, strain name B6.Cg-Tg (CAG-EGFP/LC3)53Nmi/NmiRbrc, obtained from RIKEN BioResource Center [16] (*see* **Note 1**).
2. Neuronal Maintenance Media: Neurobasal media supplemented with 2% B-27, 37.5 mM NaCl, 33 mM glucose, 2 mM GlutaMAX, 100 U/mL penicillin, and 100 μg/mL streptomycin. Do not filter the B-27. Add the B-27 after the media have been filtered through an 0.2 μm filter unit.
3. Neuronal Attachment Media: Minimal Essential Media supplemented with 10% heat inactivated horse serum, 33 mM glucose, 1 mM pyruvic acid, and 37.5 mM NaCl.
4. Trypsin solution: 2.5% (w/v) trypsin.
5. Hanks' Balanced Salt Solution (HBSS): 1× HBSS buffered with 10 mM HEPES, pH 7.0.
6. Borate Buffer: 0.1 M borate buffer, pH 8.5 (prepared from boric acid and sodium tetraborate).
7. Poly-L-Lysine solution: 2 mg/mL poly-L-lysine in borate buffer (prepared from poly-L-lysine, molecular weight 30,000–70,000 kDa, Sigma cat. no. P2636).
8. Cytosine–D-arabinofuranoside hydrochloride (Ara-C): 1 mM Ara-C.
9. Acid-washed glass coverslips: 25 mm Deckgläser coverslips (Carolina Biosciences cat. no. 633037).

2.2 Primary Mouse Glial Culture

1. Glial Media: DMEM supplemented with 10% heat inactivated Fetal Bovine Serum, 2 mM GlutaMAX, 100 U/mL penicillin, and 100 μg/mL streptomycin.
2. Trypsin solution: 2.5% (w/v) trypsin.
3. Hanks' Balanced Salt Solution (HBSS): 1× HBSS buffered with 10 mM HEPES, pH 7.0.
4. Cell strainer, 40 μm pores (Falcon, cat. no. 352340).
5. Dulbecco's phosphate buffered saline (DPBS), pH 7.1.
6. Trypsin-EDTA solution: 0.05% trypsin-EDTA (1×) in DPBS.

2.3 Live-Cell Imaging of GFP-LC3 in Primary Hippocampal Neurons

1. Imaging chamber: Chamlide CMB magnetic chamber for imaging 25 mm round coverslips (Live Cell Instrument, cat. no. CM-B25–1).
2. Neuronal imaging media: Hibernate E (BrainBits) supplemented with 2% B-27 and 2 mM GlutaMAX.
3. Inverted widefield fluorescence microscope: Leica DMi8 inverted fluorescence microscope equipped with an sCMOS camera (Hamamatsu Orca Flash 4.0 V2+). GFP-LC3 is imaged with 63×/1.40 NA or 100×/1.40 NA Plan-Apochromat oil immersion objectives and FITC filter cube. For live-cell imaging, the microscope has Leica Adaptive Focus Control to maintain the focal plane during image acquisition, a climate-controlled chamber to maintain 37 °C, and is mounted on an air table to buffer against vibration. Images are acquired with Leica LAS-AF software.
4. Spinning disk confocal microscope: Leica DMi8 inverted widefield microscope with motorized stage equipped with a Yokogawa W1 spinning disk confocal scanning head, VisiScope Homogenizer for uniform laser illumination, and Photometrics Prime 95B sCMOS camera with 95% quantum efficiency. Imaging GFP-LC3 is performed with 63×/1.40 NA or 100×/1.40 NA Plan-Apochromat oil immersion objectives, 488 nm/200 mW solid-state laser line, and 525/50 nm emission filter. For live-cell imaging, the microscope has Leica Adaptive Focus Control to maintain the focal plane during image acquisition, a climate-controlled chamber to maintain 37 °C, and is mounted on an air table to buffer against vibration. Images are acquired with VisiView software.

2.4 Analysis of Autophagosome Dynamics

1. FIJI with Multiple Kymograph Plugin.
2. Ilastik.
3. Microsoft Excel.
4. GraphPad Prism.

3 Methods

3.1 Primary Mouse Hippocampal Culture

1. Prepare acid-washed glass coverslips:
   - (a) Wash glass coverslips in 0.25% (v/v) acetic acid overnight at room temperature, rocking.
   - (b) Rinse the coverslips 3 times in 100% EtOH.
   - (c) Store coverslips in 100% EtOH at 4 °C until needed.
2. The day before neuron plating, coat the coverslips with poly-L-lysine (PLL):
   - (a) Dilute 2 mg/mL PLL to 0.5 mg/mL in borate buffer.
   - (b) Add 200 μL 0.5 mg/mL PLL to each 25 mm glass coverslip, and spread evenly over the coverslip surface.
   - (c) Incubate overnight at 37 °C.
   - (d) Remove the PLL and rinse twice with sterile water.
   - (e) Remove the water and add attachment media and incubate in 37 °C tissue culture incubator during the dissection (typically ~1–2 h).
3. Dissect and plate hippocampal neurons:
   - (a) Timed pregnancies are set up in advance between a GFP-LC3 transgenic mouse and C57BL6 mouse. We typically use male mice to supply the GFP-LC3 transgene since they do not get sacrificed for the neuronal dissection.
   - (b) Euthanize the pregnant dam on gestation day 15.5, dissect out the uterus and place in a sterile 10 cm cell culture dish with cold HBSS.
   - (c) Remove the embryos from the uterus, decapitate, and remove the brains. Collect the brains in a sterile 35 mm cell culture dish with cold HBSS and store on ice.
   - (d) Using a fluorescence dissecting microscope, separate the GFP-LC3 transgenic brains from the non-transgenic littermate brains and collect each genotype in a 35 mm cell culture dish with HBSS and store on ice. On average, we obtain ~4 GFP-LC3 transgenic embryos and ~4 non-transgenic embryos.
   - (e) Remove the meninges from the cerebral hemispheres.
   - (f) Dissect out the hippocampus and collect the hippocampi of each genotype in separate 35 mm cell culture dishes containing cold HBSS and store on ice. The neurons from each genotype will be prepared separately and combined in the final step of plating.

(g) In a biological safety cabinet for tissue culture, transfer the hippocampi to 4.5 mL HBSS pre-warmed to 37 °C. Add 0.5 mL 2.5% (w/v) trypsin solution and incubate 10 min at 37 °C.

(h) After the incubation, the hippocampi will have settled to the bottom of the tube. Carefully remove the trypsin solution without disturbing the tissue.

(i) Add 10 mL attachment media and incubate at room temperature long enough for the hippocampi to settle to the bottom of the tube. Remove the attachment media and wash twice more with attachment media.

(j) Add 1 mL attachment media and dissociate the hippocampi by triturating (pipetting up and down) through a Pasteur pipette. First, using a Pasteur pipet fire-polished to round the edges, triturate five times round trip to break up large pieces of tissue. Second, using a Pasteur pipet fire-polished to decrease the bore size by 1/3, triturate 8–10 times round-trip to achieve a homogeneous cell suspension. Minimize bubbles during the trituration.

(k) Allow any remaining undissolved tissue fragments to settle and transfer the supernatant (hippocampal cell suspension) to a new tube.

(l) Determine the cell density using a hemocytometer.

(m) For plating neurons onto coverslips, 8 coverslips are placed into a 10 cm cell culture dish and a total of 750,000 cells are plated per 10 cm (~12,000 cells/ cm^2). To resolve single-cell dynamics of autophagosomes, GFP-LC3 transgenic neurons are diluted with non-transgenic neurons at 1:20 (*see* **Note 2**).

(n) Incubate neurons at 37 °C in a 5% CO_2 incubator.

(o) After the cells attach to the coverslip (typically within ~4 h of plating), transfer coverslips to 6-well plates containing pre-equilibrated maintenance media and return to the 37 °C, 5% CO_2 incubator.

(p) Every 3–4 days, replace ~25% of the media with pre-equilibrated maintenance media. The first feed is supplemented with Ara-C (1 μM final concentration). No differences have been noted in autophagy with Ara-C addition.

(q) For studies performed within 8–10 DIV, neurons are cultured in neuronal maintenance media. If studies require neurons >2 weeks in culture, neurons are grown in glial-conditioned neuronal maintenance media (as described below).

3.2 Primary Mouse Glial Culture

1. Dissect and plate mouse glia (*see* **Note 3**):
   (a) Euthanize 4–5 wild-type mouse pups at P0–P1 and remove the brains. Collect the brains in a 10 cm cell culture dish with cold HBSS and store on ice.
   (b) Using a dissecting microscope, remove the meninges from the cerebral hemispheres.
   (c) Dissect out the cortex and collect the cortices in a 35 mm cell culture dish containing cold HBSS and store on ice.
   (d) In a biological safety cabinet for tissue culture, transfer the cortices to 4.5 mL HBSS pre-warmed to 37 °C. Add 0.5 mL 2.5% (w/v) trypsin solution and incubate 10 min at 37 °C.
   (e) After the incubation, the cortices will have settled to the bottom of the tube. Carefully remove the trypsin solution without disturbing the tissue.
   (f) Add 10 mL glial media and incubate at room temperature long enough for the cortices to settle to the bottom of the tube. Remove the glial media and wash twice more with glial media.
   (g) Add 1 mL glial media and dissociate the cortices by trituration. First, triturate through a 5 mL pipet to break up large pieces of tissue. Then triturate with a p1000 pipet until you achieve a homogenous cell suspension.
   (h) Add 4 mL glial media, pass cells through a cell strainer, and collect in a 50 mL conical tube. Add an additional 5 mL glial media to strained cells.
   (i) Determine the cell density using a hemocytometer.
   (j) Plate approximately three million glia per 10 cm dish in glial media and incubate in a 37 °C, 5% CO_2 incubator.
   (k) The next day, replace media with fresh glial media and return to 37 °C, 5% CO_2 incubator.
   (l) Cells will appear sparse for the first few days in culture. The glia will be 80–90% confluent within approximately 10 days in culture. Every 3–4 days, perform a full media change.
2. When the glia are approximately 80% confluent, passage the glia into new 10 cm cell culture dishes:
   (a) Aspirate and discard the glial media.
   (b) Wash cells once with DPBS.
   (c) Add 3 mL trypsin-EDTA solution.
   (d) Incubate at 37 °C, 5% CO_2 until cells are detached (~5 min).

(e) Add 7 mL glial media to cells to deactivate the trypsin.

(f) Spin 2000 rpms (644 × *g*) for 2 min in a swing-out rotor.

(g) Discard the supernatant.

(h) Resuspend the cell pellet in 10 mL glial media.

(i) Determine the cell density using a hemocytometer.

(j) Plate 2–three million cells per 10 cm cell culture dish in glial media and incubate at 37 °C, 5% CO_2.

3. Generate glial-conditioned neuronal maintenance media for neuronal culture:

 (a) The day before neuron plating, replace glial media on glia with neuronal maintenance media.

 (b) The day of neuron plating, transfer coverslips with neurons from attachment media to 6 well trays containing glial-conditioned neuronal maintenance media. Replace maintenance media on glia.

 (c) Use glial-conditioned neuronal maintenance media as a source of media for subsequent feedings of the neurons. After feeding neurons, replace neuronal maintenance media on glia for the next round of feeding.

3.3 Live-Cell Imaging of GFP-LC3 in Primary Hippocampal Neurons

1. Equilibrate environmental chamber on the microscope to 37 °C for at least 1 h prior to imaging.

2. Place coverslip in Chamlide chamber and add 1 mL neuronal imaging media, pre-warmed to 37 °C.

3. Select a GFP-LC3-positive neuron that has a well-defined cell soma, axon, and dendrites, following criteria from Kaech and Banker [21]. At 8–10 DIV, dendrites should be ~100–300 μm in length, tapered, and branched. The neuron should have a single axon that extends well beyond the field of view, with a uniformly thin width along its length. Avoid neurons that are not well developed or axons that appear to have distensions.

4. Note the location of the movie within the neuron. We have observed very different patterns of motility in different regions of the neuron (Fig. 1a). Autophagosome biogenesis is enriched in the distal axon (defined as the terminal 100 μm of the axon) [4, 5] and in the soma [6]. Autophagosomes exhibit bidirectional motility in the distal axon and transition to robust retrograde transport in the mid-axon (defined as >50 μm from the soma and >100 μm from the axon terminus) (Fig. 1a, d, e) [4, 5]. In dendrites, autophagosomes are primarily stationary, while some exhibit bidirectional movement [5, 6].

5. Note the distribution of LC3 in the neuron. LC3 exists in a cytosolic form, microtubule-bound form, and lipidated form; the lipidated form associates with autophagosomes and appears as discrete puncta within the neuron (Fig. 1b–e).

Autophagosomes are identified by GFP-LC3-positive punctate structures. As these structures become larger (~800 nm), they appear as ring-shaped structures.

6. Using a 63× or 100× objective, image GFP-LC3 dynamics over time using the following guidelines.
   (a) Acquire images at a rate of 1–2 s per frame for at least 5 min. Longer imaging durations are required to measure events of autophagosome biogenesis [4, 5]. Faster frame rates may be required to track the dynamics of bidirectional autophagosomes in dendrites. For movies within the axon, it is important to note the direction of the cell body by obtaining tiled still images from the location of the axon movie back to the soma to document direction.
   (b) Adjust exposure time and light intensity to minimize bleaching of the GFP-LC3 signal. Typically, we are using the following conditions:
      - Widefield microscope: 100 ms exposure, 10% fluorescence intensity manager (intensity of LED lamp).
      - Spinning Disk Confocal microscope: 200 ms exposure, 50–80% laser power.
   (c) If available, use the Adaptive Focus Control to maintain the focal plane throughout image acquisition.
   (d) If available, use an air table to minimize vibrations that contribute to image drift.
7. Due to the increased depth of the soma, total GFP-LC3 in the soma is quantitated with a *Z*-stack of the entire cell body, acquired at 0.1 μm sections using a confocal spinning disk microscope. *Z*-stacks can also be obtained with a widefield microscope and processed with deconvolution.
8. Coverslips are imaged for a maximum of ~2 h. Beyond this point, cell health will decline.

3.4 Analysis of Autophagosome Dynamics (See Note 4)

1. Open microscope image files in FIJI using the Bio-Formats Importer and quantify the following metrics in the axon:
   (a) Autophagosome density:
      - In the first frame of the image series, count the number of GFP-LC3-positive puncta (autophagosomes) along the axon. Play the movie to determine the number of puncta, as autophagosomes often collide during transport and may appear as a single larger structure in one frame but actually consist of several individual puncta that resolve as they move.
      - Determine the length of the axon by tracing the axon of interest with a line and measuring the length (in μm).

- Normalize the number of autophagosomes by the axon length and present as No. of Autophagosomes/100 μm.

(b) Autophagosome flux through a midpoint:

- Draw a line bisecting the axon.
- Count the number of GFP-LC3-positive puncta that cross the midpoint during the movie.
- Normalize the number of crossing autophagosomes by the duration of the movie and present as No. of Autophagosomes/min.

(c) Direction of motility (percent retrograde vs. anterograde vs. stationary/bidirectional):

- Using the Multiple Kymograph Plugin, generate a kymograph of the axon using a line width of 3 (Fig. 1e). Note the retrograde and anterograde direction of the kymograph.
- Referencing the movie and kymograph together, bin autophagosomes into the following categories: (a) retrograde (autophagosomes that move a net distance of ≥5 μm in the retrograde direction within the 5 min imaging window), (b) anterograde (autophagosomes that move a net distance of ≥5 μm in the anterograde direction within the 5 min imaging window), and (c) stationary/bidirectional (autophagosomes that do not undergo processive motility and are confined within a net distance of <5 μm within the 5 min imaging window). Eliminate edge effects due to processive autophagosomes that enter the field of view at the end of the movie and would need to be binned as stationary. In this case, truncate the region of the axon used for quantitation to eliminate these edges.
- Determine the percentage that each category represents based on the total number of autophagosomes recorded.

(d) Each of these metrics is determined for a single axon from a single neuron. Obtain measurements from at least 25–30 individual neurons from 3 independent experiments.

(e) Process data using excel. Graph data and perform statistical analyses in Prism.

2. Similar metrics can be obtained in dendrites, although individual autophagosomes are more difficult to track persistently for an entire 5 min due to the higher background of GFP-LC3 signal in dendrites. Further, the bidirectional nature of autophagosomes in dendrites requires higher time resolution to

facilitate tracking and quantification of individual autophagosome dynamics. In dendrites, the populations of stationary versus bidirectional autophagosomes can be distinguished using the following criteria: stationary autophagosomes exhibit a displacement of <5 μm in 5 minutes, and bidirectional autophagosomes exhibit sustained movement of 5 μm in both anterograde and retrograde directions, irrespective of displacement, during the imaging window [5].

3. Quantitate GFP-LC3 area in the soma:
   (a) Using FIJI, generate a maximum projection of the Z-stack generated from the soma.
   (b) Identify and segment GFP-LC3-positive autophagosomes using ilastik. Since GFP-LC3 has a cytosolic and microtubule-bound component, it can be challenging to reliably segment GFP-LC3-positive autophagosomes away from these other signals using global thresholding techniques based on fluorescence intensity or size. Ilastik uses a machine learning-based approach using multiple criteria defined by the user to successfully identify and distinguish autophagosomes from other GFP-LC3-positive signals in the cell.
   (c) Import the segmented images generated in ilastik into FIJI and measure the area occupied by GFP-LC3-positive structures using the "analyze particles function".
   (d) Normalize the total area of GFP-LC3-positive autophagosomes for a given cell body by the total area of the corresponding cell body.

   See **Notes 5–7**.

4 Notes

1. The use of the GFP-LC3 transgenic mice to observe autophagosome dynamics in neurons is preferred over transient transfection of GFP-LC3, as the transgenic mice provide less variation in expression level. In addition, transfection of GFP-LC3 can also result in high GFP-LC3 concentration which can make individual autophagic events difficult to resolve. If transient transfection methods are used, optimize transfection conditions to minimize high expression of GFP-LC3.
2. For experiments requiring higher neuronal densities or longer time in culture, we recommend diluting GFP-LC3 transgenic neurons more than 1:20, such as 1:30–1:40, to resolve single-cell dynamics and axon direction.
3. The procedure for isolation and culture of glia enriches primarily for astrocytes and not oligodendrocytes or microglia.

4. Events of autophagosome biogenesis or formation can be measured in the terminal 100 μm of the axon or in the soma. These events are marked by the appearance of GFP-LC3-positive puncta that grow progressively into ring-shaped structures ~800 nm in diameter. Rates of autophagosome formation can be measured as the number of growth events occurring within a unit of time. To measure the kinetics of individual autophagosome development, Fluorescence Recovery after Photobleaching (FRAP) can be used [4].
5. Other autophagy markers that label nascent autophagosomes, such as mCherry-tagged Atg5 or Atg13, can be expressed in hippocampal neurons using Lipofectamine 2000 [5]. Expression of organelle markers for maturation state (e.g., mCherry-LAMP1 or mCherry-GFP-LC3 in non-transgenic neurons) can distinguish immature versus mature and degradative autophagic compartments [4, 6]. Fluorescently-labeled proteins should be validated and expressed at low levels for live-cell imaging.
6. To measure specific characteristics of autophagosome motility such as instantaneous velocities, time paused, and frequency or duration of directional reversals, tracking programs in Metamorph [4] or FIJI can be applied.
7. The GFP-LC3 imaging protocol described here can be performed in other neuronal cell types (e.g., dorsal root ganglion neurons, cortical neurons, motor neurons) isolated from GFP-LC3 transgenic mice. We have observed similar autophagosome dynamics across the aforementioned neuronal subtypes [5]. We have found that mechanistic studies of autophagosome biogenesis are facilitated in dorsal root ganglion neurons that have more flat, fan-like growth cones [4].

Acknowledgments

This work was funded by NIH grants K99NS082619 and R00NS082619, the McCabe Fund Fellow Award, the University of Pennsylvania Alzheimer's Disease Core Center, the Intellectual and Developmental Disabilities Research Center at the Children's Hospital of Philadelphia and the University of Pennsylvania, and The Philadelphia Foundation to S.M.

References

1. Maday S (2016) Mechanisms of neuronal homeostasis: autophagy in the axon. Brain Res 1649(Pt B):143–150. https://doi.org/10.1016/j.brainres.2016.03.047
2. Yamamoto A, Yue Z (2014) Autophagy and its normal and pathogenic states in the brain. Annu Rev Neurosci 37:55–78. https://doi.org/10.1146/annurev-neuro-071013-014149
3. Ariosa AR, Klionsky DJ (2016) Autophagy core machinery: overcoming spatial barriers in neurons. J Mol Med (Berl) 94(11):1217–1227. https://doi.org/10.1007/s00109-016-1461-9

4. Maday S, Wallace KE, Holzbaur EL (2012) Autophagosomes initiate distally and mature during transport toward the cell soma in primary neurons. J Cell Biol 196(4):407–417. https://doi.org/10.1083/jcb.201106120
5. Maday S, Holzbaur EL (2014) Autophagosome biogenesis in primary neurons follows an ordered and spatially regulated pathway. Dev Cell 30(1):71–85. https://doi.org/10.1016/j.devcel.2014.06.001
6. Maday S, Holzbaur EL (2016) Compartment-specific regulation of autophagy in primary neurons. J Neurosci 36(22):5933–5945. https://doi.org/10.1523/JNEUROSCI.4401-15.2016
7. Weidberg H, Shvets E, Elazar Z (2011) Biogenesis and cargo selectivity of autophagosomes. Annu Rev Biochem 80:125–156. https://doi.org/10.1146/annurev-biochem-052709-094552
8. Mizushima N, Yoshimori T, Ohsumi Y (2011) The role of Atg proteins in autophagosome formation. Annu Rev Cell Dev Biol 27:107–132. https://doi.org/10.1146/annurev-cellbio-092910-154005
9. Spalding KL, Bhardwaj RD, Buchholz BA, Druid H, Frisen J (2005) Retrospective birth dating of cells in humans. Cell 122 (1):133–143. https://doi.org/10.1016/j.cell.2005.04.028
10. Hara T, Nakamura K, Matsui M, Yamamoto A, Nakahara Y, Suzuki-Migishima R, Yokoyama M, Mishima K, Saito I, Okano H, Mizushima N (2006) Suppression of basal autophagy in neural cells causes neurodegenerative disease in mice. Nature 441 (7095):885–889. https://doi.org/10.1038/nature04724
11. Komatsu M, Waguri S, Chiba T, Murata S, Iwata J, Tanida I, Ueno T, Koike M, Uchiyama Y, Kominami E, Tanaka K (2006) Loss of autophagy in the central nervous system causes neurodegeneration in mice. Nature 441(7095):880–884. https://doi.org/10.1038/nature04723
12. Komatsu M, Wang QJ, Holstein GR, Friedrich VL Jr, Iwata J, Kominami E, Chait BT, Tanaka K, Yue Z (2007) Essential role for autophagy protein Atg7 in the maintenance of axonal homeostasis and the prevention of axonal degeneration. Proc Natl Acad Sci U S A 104 (36):14489–14494. https://doi.org/10.1073/pnas.0701311104
13. Nishiyama J, Miura E, Mizushima N, Watanabe M, Yuzaki M (2007) Aberrant membranes and double-membrane structures accumulate in the axons of Atg5-null Purkinje cells before neuronal death. Autophagy 3 (6):591–596
14. Kim M, Sandford E, Gatica D, Qiu Y, Liu X, Zheng Y, Schulman BA, Xu J, Semple I, Ro SH, Kim B, Mavioglu RN, Tolun A, Jipa A, Takats S, Karpati M, Li JZ, Yapici Z, Juhasz G, Lee JH, Klionsky DJ, Burmeister M (2016) Mutation in ATG5 reduces autophagy and leads to ataxia with developmental delay. Elife 5. https://doi.org/10.7554/eLife.12245
15. Kabeya Y, Mizushima N, Ueno T, Yamamoto A, Kirisako T, Noda T, Kominami E, Ohsumi Y, Yoshimori T (2000) LC3, a mammalian homologue of yeast Apg8p, is localized in autophagosome membranes after processing. EMBO J 19(21):5720–5728. https://doi.org/10.1093/emboj/19.21.5720
16. Mizushima N, Yamamoto A, Matsui M, Yoshimori T, Ohsumi Y (2004) In vivo analysis of autophagy in response to nutrient starvation using transgenic mice expressing a fluorescent autophagosome marker. Mol Biol Cell 15 (3):1101–1111. https://doi.org/10.1091/mbc.E03-09-0704
17. Baas PW, Deitch JS, Black MM, Banker GA (1988) Polarity orientation of microtubules in hippocampal neurons: uniformity in the axon and nonuniformity in the dendrite. Proc Natl Acad Sci U S A 85(21):8335–8339
18. Kleele T, Marinkovic P, Williams PR, Stern S, Weigand EE, Engerer P, Naumann R, Hartmann J, Karl RM, Bradke F, Bishop D, Herms J, Konnerth A, Kerschensteiner M, Godinho L, Misgeld T (2014) An assay to image neuronal microtubule dynamics in mice. Nat Commun 5:4827. https://doi.org/10.1038/ncomms5827
19. Gowrishankar S, Yuan P, Wu Y, Schrag M, Paradise S, Grutzendler J, De Camilli P, Ferguson SM (2015) Massive accumulation of luminal protease-deficient axonal lysosomes at Alzheimer's disease amyloid plaques. Proc Natl Acad Sci U S A 112(28):E3699–E3708. https://doi.org/10.1073/pnas.1510329112
20. Lee S, Sato Y, Nixon RA (2011) Lysosomal proteolysis inhibition selectively disrupts axonal transport of degradative organelles and causes an Alzheimer's-like axonal dystrophy. J Neurosci 31(21):7817–7830. https://doi.org/10.1523/JNEUROSCI.6412-10.2011
21. Kaech S, Banker G (2006) Culturing hippocampal neurons. Nat Protoc 1(5):2406–2415. https://doi.org/10.1038/nprot.2006.356

Chapter 17

Imaging Autophagy in hiPSC-Derived Midbrain Dopaminergic Neuronal Cultures for Parkinson's Disease Research

Petros Stathakos, Natalia Jimenez-Moreno, Lucy Crompton, Paul Nistor, Maeve A. Caldwell, and Jon D. Lane

Abstract

To appreciate the positive or negative impact of autophagy during the initiation and progression of human diseases, the isolation or de novo generation of appropriate cell types is required to support focused in vitro assays. In human neurodegenerative diseases such as Parkinson's disease (PD), specific subsets of acutely sensitive neurons become susceptible to stress-associated operational decline and eventual cell death, emphasizing the need for functional studies in those vulnerable groups of neurons. In PD, a class of dopaminergic neurons in the ventral midbrain (mDANs) is affected. To study these, human-induced pluripotent stem cells (hiPSCs) have emerged as a valuable tool, as they enable the establishment and study of mDAN biology in vitro. In this chapter, we describe a stepwise protocol for the generation of mDANs from hiPSCs using a monolayer culture system. We then outline how imaging-based autophagy assessment methodologies can be applied to these neurons, thereby providing a detailed account of the application of imaging-based autophagy assays to human iPSC-derived mDANs.

Key words Autophagy, hiPSC, Stem cells, Dopaminergic neurons, Cell culture, Immunofluorescence, Parkinson's disease

Abbreviations

BDNF	Brain derived neurotrophic factor
BSA	Bovine serum albumin
cAMP	Cyclic adenosine monophosphate
DAPT	*N*-[*N*-(3,5-difluorophenacetyl)-L-alanyl]-*S*-phenylglycine t-butyl ester

Maeve A. Caldwell and Jon D. Lane contributed equally to this work.

Nicholas Ktistakis and Oliver Florey (eds.), *Autophagy: Methods and Protocols*, Methods in Molecular Biology, vol. 1880, https://doi.org/10.1007/978-1-4939-8873-0_17, © Springer Science+Business Media, LLC, part of Springer Nature 2019

DMEM	Dulbecco's modified Eagle's medium
DMSO	Dimethyl sulfoxide
GDNF	Glial-derived neurotrophic factor
LDN193189	4-[6-[4-(1-piperazinyl)phenyl]pyrazolo[1,5-a]pyrimidin-3-yl]-quinoline hydrochloride
NEAA	Nonessential amino acids
Pen/Strep	Penicillin/Streptomycin
SB431542	4-(5-benzol[1,3]dioxol-5-yl-4-pyrldin-2-yl-1H-imidazol-2-yl)-benzamide hydrate
SHH	Sonic hedgehog

1 Introduction

Human-induced pluripotent stem cell (hiPSC) technology is making major contributions to our understanding of the causes of human disease and is a promising source of material for replacement therapies. By differentiating hiPSC into specific, disease-affected cell types, researchers are able to investigate in vitro the molecular pathways that drive human diseases in the appropriate cell/tissue type. This includes neurodegenerative diseases—e.g., Alzheimer's disease (AD) and Parkinson's disease (PD)—where autophagy deregulation is a significant factor [1, 2]. For human neurodegenerative diseases, key questions center on why particular subsets of neurons are susceptible to degeneration, either in patients genetically predisposed to the disease or in those whose condition is regarded as idiopathic. Thus, there is a need to understand disease processes in precisely those cells that are susceptible in vivo. In PD, a distinct class of ventral midbrain dopaminergic neurons (mDANs) in the substantia nigra pars compacta (SNpc) is lost. It is thought that the atypical morphology and distinctive physiology of these neurons force mDANs to operate at or close to their energy demand/supply threshold in vivo [3]. This predisposes mDANs to declining cytoplasmic quality control pathways during aging and stress. Here, we describe a protocol for the generation of mDAN cultures from hiPSCs and methods to study autophagy in these cells using imaging. The methods can be applied equally to healthy patient-derived hiPSCs and hiPSCs from PD patients for comparative studies of underlying genetic defects and to genome-edited hiPSC lines. We are also able to use a modified version of the protocol described here for the production of cortical neurons. Altering differentiation parameters enables researchers to push hiPSC neuralization toward other neuronal subtypes for studies of different neurodegenerative diseases, and the reader is advised to consult focused reports for alternative neuralization conditions (reviewed in [1]). This chapter will be divided into three separate sections with their own descriptions of materials and methods due to the complexity of the subject matter. Section 2 deals with culturing iPSCs, Sect. 3 with differentiation protocols, and Sect. 4 with microscopy.

2 Culturing Human-Induced Pluripotent Stem Cells (hiPSCs) (*See* Table 1)

Preface and general considerations

Successful neural differentiation depends to a great extent on the quality of the human-induced pluripotent stem cells (hiPSCs) that are used. Thus, their growth and expansion must be carried out meticulously and very carefully monitored before neuralization can be attempted (*see* Fig. 1). In our lab, hiPSCs are maintained in fully defined xeno-free and feeder-free complete Essential 8 (E8) Medium [4] (*see* **Note 1**).

2.1 Materials

Table 1
List of reagents used for the Essential 8 culturing system

Reagents	Provider	Catalog number
Essential 8 medium/E8 supplement	Thermo Fisher Scientific	A1517001
Vitronectin	Thermo Fisher Scientific	A1413301
EDTA	Thermo Fisher Scientific	15575-020
DPBS	Thermo Fisher Scientific	14190-144
Y-27632 dihydrochloride	Tocris	1254
RevitaCell	Thermo Fisher Scientific	A26445
CryoStor CS10	Stem Cell Technologies	07930

EDTA ethylenediaminetetraacetic acid, *DPBS* Dulbecco's phosphate-buffered saline

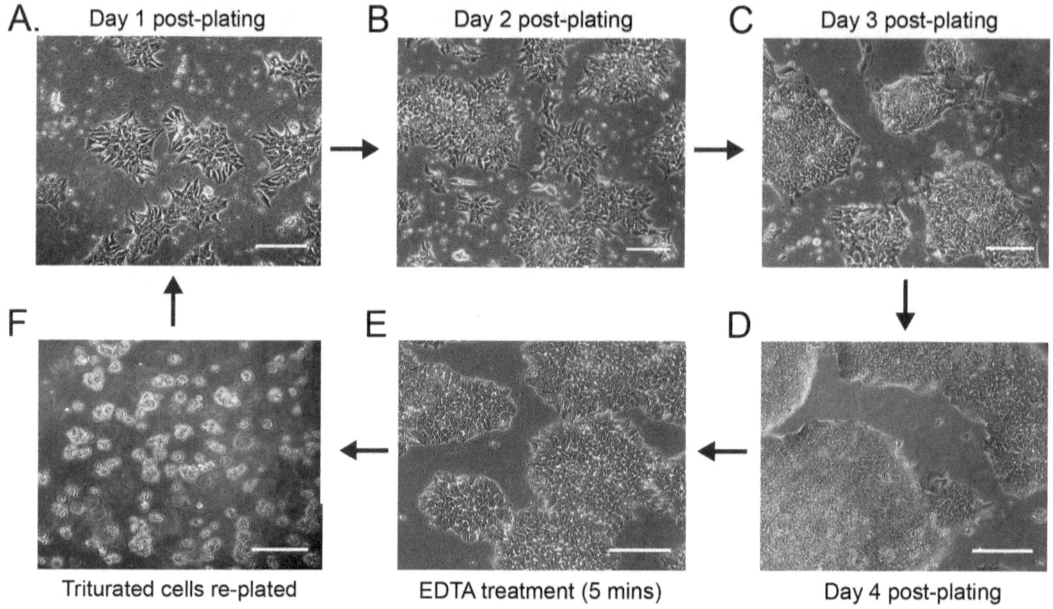

Fig. 1 Feeder-free hiPSC colony morphology during a complete cycle of seeding, proliferation, enzymatic detachment, and reseeding. (**a**) One day after initial plating. The colonies start forming, but cells within remain loosely connected and the colonies have non-uniform margins. (**b**) Day 2; the colonies have expanded and have started to adopt a tighter density, still with uneven margins. (**c**) Day 3; tighter colonies of cells are seen with smoother margins. (**d**) Day 4: colonies have become significantly larger, with more densely packed cells and with smooth margins. The colonies are now ready to be passaged. (**e**) Chemical passaging with EDTA; 5 min after EDTA application, the cells in the colonies start to separate. They are detached by gently forcing with a 1 mL pipette tip. (**f**) This image shows colonies that have detached, have been triturated into small clusters using a 1 mL pipette tip, and have been passaged on to new vitronectin-coated plates. The following day, the cells will once again look like those of image (**a**). Scale bars: 200 μm

2.2 Preparation of Reagents

2.2.1 Complete Essential 8 Medium

Complete E8 medium consists of 500 mL E8 basal medium (store at 2–8 °C) and 10 mL E8 supplement (50×; store at −5 to −30 °C; thaw at room temperature, NOT at 37 °C). The manufacturer suggests adding 10 mL E8 supplement in 490 mL of E8 basal medium to make 500 mL of complete E8 medium; however, the complete E8 can be stored for only up to 2 weeks at 2–8 °C. For longer storage time, we recommend decanting the E8 basal medium into 49 mL aliquots (store at 2–8 °C) and the E8 supplement into 1 mL aliquots (refreeze at −5 to −30 °C). Before use, mix one aliquot of each to make 50 mL of complete E8. We store the unmixed reagents for up to 3 months without any noticeable loss of activity.

2.2.2 Vitronectin

Vitronectin is supplied at 500 μg/mL. Upon receipt, thaw at room temperature, then decant into 20 μL aliquots, refreeze, and store at −80 °C. Prior to use, thaw and dilute vitronectin 1:100 in DPBS (e.g., 20 μL vitronectin in 1980 μL DPBS) and use immediately. Do not refreeze.

2.2.3 EDTA

Dilute 50 μL of 0.5 M EDTA to 50 mL of DPBS (final concentration of 0.5 mM), filter sterilize, and store in room temperature for up to 6 months.

2.3 Plating hiPSC

2.3.1 Vitronectin Plate Coating

1. Thaw and dilute vitronectin in DPBS as described in Subheading 2.2.2.
2. Apply diluted vitronectin to cell culture dishes at a final concentration of ~0.5 μg/cm^2 (e.g., 500 μL/well of a 12-well plate; 1 mL/well of a 6-well plate; 2 mL in a 60-mm-diameter dish).
3. Coat the dishes for at least 1 h at room temperature. If not to be used immediately, Parafilm-shield the vitronectin-containing dishes to prevent evaporation and keep them for up to 7 days at 2–8 °C. Before use, ensure that the plates have not dried out, and allow them to reach room temperature (about 20 min; *see* **Note 2**).

2.3.2 Thawing hiPSCs

1. While waiting for the dishes to be coated, warm an appropriate volume of complete E8 medium at room temperature for approximately 30 min.
2. After coating is completed, remove a Cryovial of hiPSCs from liquid nitrogen and thaw in a 37 °C water bath for around 1 min.
3. Before being completely thawed, add 1 mL of warm complete E8 medium to the cells.
4. Gently transfer the cells into a 15 mL Falcon tube containing 3 mL of complete E8 medium, and centrifuge (100 × *g* for 3 min at room temperature).
5. Aspirate the medium, and gently resuspend the pelleted cells in 1 mL of fresh complete E8 medium. It is important to not over-triturate the cells, as they must be plated in the form of small aggregates/clusters (and not as single cells) in order to thrive.
6. Aspirate the vitronectin solution and plate the cell suspension on to the coated dishes at a density of 4–5 × 10^4 cells/cm^2 in complete E8 medium supplemented with 1× RevitaCell (1:100 or use 10 μM Y-27632) for the first 24 h to improve cell survival (final complete E8 medium volumes: 1 mL/well for 12-well plates; 2 mL/well for 6-well plates; 4 mL/60 mm dish).
7. Gently agitate the plates to ensure equal cell distribution and incubate at 37 °C, 5% CO_2.
8. Change media daily. No additional RevitaCell is required.
9. Do not remove any cells undergoing spontaneous differentiation during cell expansion; these will be eliminated though consecutive passaging.

2.4 Passaging hiPSCs

hiPSCs cells need to be passaged when they reach approximately 80% density, or 4–5 days after initial plating (*see* Fig. 1d and **Note 3**):

1. Coat culture dishes with vitronectin for 1 h (*see* Subheading 2.3.1), and warm complete E8 medium.
2. When coating is complete, aspirate spent media from hiPSC cultures and wash once with 0.5 mM EDTA or PBS (500 μL/well of a 12-well plate; 1 mL/well of a 6-well plate; 2 mL/60 mm diameter dish) to remove growth medium residues.
3. Replace with fresh 0.5 mM EDTA (500 μL/well of a 12-well plate; 1 mL/well of a 6-well plate; 2 mL/60 mm diameter dish) and incubate for 5–8 min at room temperature or 3–5 min at 37 °C.
4. When colonies start loosening and rounding up, the cells are ready to be harvested (Fig. 1e). The next two steps must be carried out as quickly as possible.
5. Aspirate EDTA and wash once with pre-warmed complete E8 medium (*see* **Note 4**).
6. Immediately aspirate the media as it neutralizes the EDTA very quickly, and gently harvest the colonies by liquid flow using either a 1 mL or a 5 mL serological pipette.
7. Collect the dislodged cells into 15 mL falcon tubes (*see* **Note 5**).
8. Triturate the cell suspension up to 3 times to break the colonies into smaller clusters (Fig. 1f). Do not over-triturate as very small clusters and single cells do not survive passaging.
9. Aspirate vitronectin from the newly coated dishes.
10. Passage hiPSC cell suspension at a dilution of 1:4 (approximately 4–5 × 10^4 cells/cm^2) into complete E8 medium to a final volume of 1 mL/well of 12-well plates, 2 mL/well of 6-well plates, or 4 mL/60 mm diameter dish.
11. Gently shake the plates for equal cell distribution of cells and incubate at 37 °C, 5% CO_2. If the next day the cell density is too high or too low, adjust future passaging ratios accordingly.
12. Feed daily until the cells are ready to be passaged again.

    See **Notes 6** and 7.

2.5 hiPSC Cryopreservation

1. Prepare a cryo 1 °C freezing container (e.g., Mr Frosty™ from ThermoFisher Scientific). "Mr Frosty" is an isopropanol-filled Cryovial container that upon freezing has the property of gradually decreasing temperature at a rate of 1 °C/min.
2. Harvest the hiPSC as described in Subheading 2.4, **steps 1–7** and collect them in 15 mL falcon tubes.
3. Count the cells using a hemocytometer.

4. Centrifuge at 100 × *g* for 3 min at room temperature.
5. Aspirate spent media.
6. Gently resuspend cell pellet in CryoStor CS10 (~1 × 10^6 cells per mL).
7. Mix very gently with the tip of 1 mL pipette (triturate up to three times) and distribute into Cryovials.
8. Immediately place the Cryovials into "Mr Frosty" (room temperature) and freeze at −80 °C overnight.
9. Transfer the frozen Cryovials into liquid nitrogen for long-term storage.

See **Note 8**.

3 hiPSC Differentiation into Midbrain Dopaminergic Neurons (mDANs), *See* Tables 2 and 3

Preface and general considerations.

In order to produce hiPSC-derived neurons that are suitable for the investigation of intracellular degradation pathway function and dysfunction in PD-related research, we have optimized strategies from the literature [5–7] to develop a differentiation protocol that generates sufficient numbers of uniformly-dispersed mDANs to perform whole-cell (soma and neurites) autophagy imaging assays. At a glance, hiPSCs acquire neural identity by blocking transforming growth factor β (TGFβ) pathways [8] via inhibition of SMAD signaling [9], while the midbrain dopaminergic fate is induced via activation of sonic hedgehog (SHH) and WNT signaling cascades that determine dorsal/ventral and anterior/posterior patterning of neural progenitors [10].

3.1 Materials

Table 2
List of reagents for neural differentiation

Reagent	Provider, Cat. Num.	Preparation and storage
Accutase	Thermo F.S., A11105-01	Store at 2–8 °C
Ascorbic acid	Sigma, A5960	0.2 M stock solution in dH_2O. Store 50 μL aliquots at −20 °C
β-mercaptoethanol	Thermo F.S., 31350-010	50 mM. Store at 2–8 °C
B27	Thermo F.S., 17504-044	Store 500 μL aliquots at −20 °C

(continued)

Table 2
(continued)

Reagent	Provider, Cat. Num.	Preparation and storage
Human recombinant BDNF	R&D Systems, 248-BD or PeproTech 450-02	Make 20 μg/mL stock solution in PBS, supplemented with 0.1% BSA. Store 50 μL aliquots at −20 °C
Recombinant human GDNF	R&D Systems, 212-GD or PeproTech 450-10	Make 20 μg/mL stock solution in PBS, supplemented with 0.1% BSA. Store 50 μL aliquots at −20 °C
cAMP dibutyryl (db-cAMP)	Sigma, D0627	500 mM stock solution in dH_2O. Store 50 μL aliquots at −20 °C
CHIR99021	Axon Medchem, 1386	600 mM stock solution in DMSO. Store 50 μL aliquots at −20 °C
CryoStor CS10	Stem Cell Tech., 07930	Store at 2–8 °C
DAPT	Tocris, 2634	10 mM stock solution in DMSO. Store 50 μL aliquots at −20 °C
DMEM/F-12 + Glutamax	Thermo F.S., 31331-028	Store at 2–8 °C
Glutamax	Thermo F.S., 35050-038	Store at 2–8 °C
Human insulin	Sigma, I9278	10 mg/mL. Store at 2–8 °C
Laminin	Sigma, L2020	Store 50 μL aliquots at −20 °C
LDN193189 HCl	Sigma, SML0559	1 mM stock solution in dH_2O. Store 5–10 μL aliquots at −80 °C
N2	Thermo F.S., 17502-048	Store 250 μL aliquots in −20 °C
NEAA	Thermo F.S., 11140-035	Store at 2–8 °C
Neurobasal	Thermo F.S., 21103-049	Store at 2–8 °C
Pen/Strep	Sigma, P4333	Store at 2–8 °C
Polyornithine	Sigma, P4957	Store at 2–8 °C
SB431542	Tocris, 1614	10 mM stock solution in of DMSO Store 50 μL aliquots at −80 °C
Recombinant human SHH C24II	R&D Systems, 1845SH025	100 μg/mL stock solution in PBS, supplemented with 0.1% BSA. Store 20 μL aliquots at −20 °C
Y-27632	Tocris, 1254 or Axon, 1683	10 mM stock solution in dH_2O (or DMSO). Store 50 μL aliquots at −20°

3.2 Basic Neuralization Medium (N2B27) Recipe

Table 3
N2B27 (basic neuralization) medium recipe

Reagents	Concentration	Example for 50 mL N2B27
DMEM/F-12 + Glutamax	48%	24 mL
Neurobasal	48%	24 mL
Glutamax	0.5%	250 μL
NEAA	1%	500 μL
PenStrep	1%	500 μL
N2	0.5%	250 μL
B27	1%	500 μL
Insulin	5 μg/mL	25 μL
β-mercaptoethanol	75 μM	75 μL

DMEM Dulbecco's Modified Eagle's medium, *NEAA* nonessential amino acids, *Pen/Strep* Penicillin/streptomycin

3.3 Plate and Coverslip Preparation

3.3.1 Plate Coating

1. Add polyornithine (500 μL/well of a 12-well plate, 1 mL/well of a 6-well plate or 2 mL/6 cm diameter dish). If using tissue culture coated plates, proceed directly to **step 4** and omit the polyornithine coating step.
2. Incubate for 2 h at 37 °C.
3. Remove polyornithine. Do not wash.
4. Add laminin diluted 1:100 in DPBS (10 μg/mL) (500 μL/well of a 12-well plate, 1 mL/well of a 6-well plate or 2 mL/60 mm diameter dish).
5. Incubate for 2 h at 37 °C.
6. If not used immediately, Parafilm-shield the polyornithine- or laminin-containing plates to prevent evaporation. These can be kept for up to 7 days at 2–8 °C. Before use ensure that the plates have not dried out and incubate for 15–20 min at 37 °C.

   See **Note 9**.

3.3.2 Coverslip Coating

1. Transfer dry-autoclaved borosilicate glass coverslips (No.1 thickness; 13 mm diameter) in 4- or 24-well plates using sterile tweezers.
2. Add water or DPBS to the empty spaces between the wells to ensure humidity is maintained.

3. Add a 70 μL drop of polyornithine in the center of each coverslip. Incubate for 2 h at 37 °C. If the droplet breaks, flood the well with another 150 μL of polyornithine and/or laminin (Step dependent). Press down the coverslip with a 200 μL pipette tip.
4. Remove polyornithine and without washing, add a 70 μL drop of DPBS diluted laminin (dilution: 1:100, 10 μg/mL) to each coverslip.
5. Incubate for 2 h at 37 °C.
6. Use the same protocol when treating live-cell imaging dishes (*see* Subheading 4.2).

   See **Note 10**.

3.4 mDAN Differentiation Protocol (Fig. 2)

The efficiency of neural differentiation depends on the quality of the hiPSC culture. Consistency in hiPSC maintenance and expansion leads to reproducible neural differentiations. We recommend comparing/contrasting experimental conditions and conducting biological repeats with hiPSCs of similar passage number, otherwise epigenetic changes in hiPSCs must also be considered [11]. Newly reprogrammed hiPSCs must be checked for karyotypic abnormalities prior to using them for neural differentiation or functional studies (*see* **Note 11**).

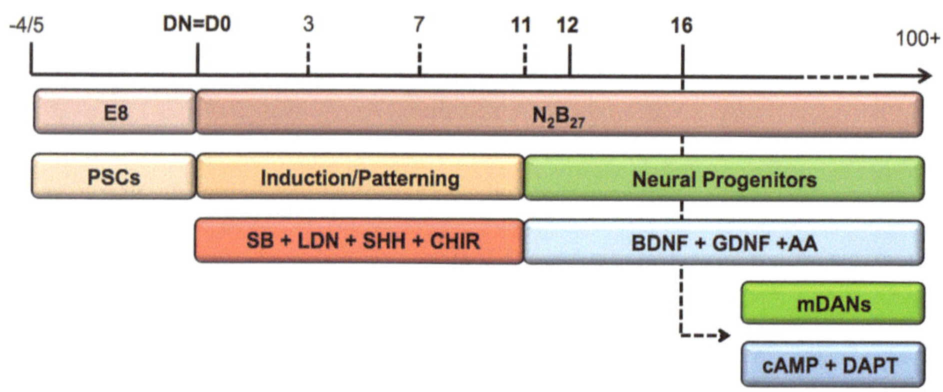

Fig. 2 Adapted monolayer protocol for the differentiation of hiPSCs into mDANs. hiPSCs are cultured in E8 essential medium until reaching 70% confluency (days −4 to 0). Neuralization starts at day 0 when the media are exchanged for N2B27 supplemented with neural fate inducing (SB and LDN) and patterning (SHH and CHIR) factors. Days 3 and 7 are the typical passaging days. At day 9 all of the above factors are removed, and the cells are maintained for another 2 days in N2B27. At this point, media is replaced with N2B27 supplemented with neurotrophic factors (BDNF, GDNF and AA), and the cells are passaged again at day 12 into N2B27 with neurotrophic factors. From that point the cells are passaged once per week or when their confluency reaches the 80–85%. For terminal differentiation, from day 15 and onward, the cells are administrated with db-cAMP and DAPT

3.4.1 hiPSC mDAN Neuralization Procedure

1. Passage hiPSCs as described in Subheading 2.4, but at **step 8**, triturate the cell suspension five times to produce smaller clusters of cells. This action produces higher numbers of smaller colonies, which we find results in higher proportions of mDANs in the final culture.
2. Depending on the needs of the particular hiPSC line, after up to ~4 days in complete E8 medium, cells are transferred to N2B27 media. Remove spent E8 medium and wash twice with pre-warmed N2B27 (500 μL/well of a 12-well plate, 1 mL/well of a 6-well plate or 2 mL/60 mm diameter dish).
3. Aspirate and replace with N2B27 supplemented with 100 nM LDN, 10 μM SB, 200–500 ng/mL SHH, and 0.6–1.0 μM CHIR (1 mL/well of 12-well plates, 2 mL/well of 6-well plates or 4 mL/6 cm diameter dish) and incubate at 37 °C, 5% CO_2. This is neuralization day 0 (DN = D0). Change media daily.
4. At D3 the cells will be passaged for the first time after the neuralization was initiated. At least 4 h before splitting the cells, prepare the plates as described in Subheading 3.3.1 (plating on to coverslips will be required after D16 for imaging studies).
5. Following laminin coating, remove spent media from the cells to be passaged and wash once with pre-warmed Accutase or DBPS (500 μL/well of a 12-well plate, 1 mL/well of a 6-well plate, or 2 mL/60-mm-diameter dish).
6. Aspirate and replace with fresh Accutase (500 μL/well of a 12-well plate, 1 mL/well of a 6-well plate, or 2 mL/60-mm-diameter dish) and incubate for 5 min at 37 °C, 5% CO_2.
7. Gently force the cells away from the plate surface by agitating the Accutase once only by drawing up and down using a 1 mL pipette tip. If the cells are easily detached, continue to the next step. If not, incubate for another 5 min at 37 °C, 5% CO_2 and repeat the process.
8. Add an equal volume of N2B27 and gently agitate the cells 1–3 times using a 1 mL pipette tip.
9. Collect the cell suspension into a 15 mL falcon tube.
10. Add fresh N2B27 in the dish and repeat **steps 8** and **9** to collect any remaining cells.
11. Repeat **step 10** until all of the cells have been harvested.
12. Add additional N2B27 to the Falcon tube until reaching a N2B27: Accutase volume ratio of 3:1.
13. Centrifuge at 100 × g/3 min at room temperature to wash out the Accutase.

14. Aspirate spent media and break up the pellet in 1 mL fresh N2B27 by gently triturating for up to 5 times using a 1 mL pipette tip.
15. Remove laminin from the coated plates, and plate the cells at a density of approximately $4 \times 10^4/cm^2$ (passaging ratio of 1:2 to 1:5) in N2B27 supplemented with 100 nM LDN, 10 μM SB, 200–500 ng/mL SHH, and 0.6–1.0 μM CHIR and 10 μM Y-27632 (1 mL/well for 12-well plates, 2 mL/well for 6-well plates or 4 mL/60 mm diameter dish). Incubate at 37 °C, 5% CO_2. Change media daily without further addition of Y-27632.
16. At D7 the cells will need to be passaged for a second time. Repeat **steps 4–15**.
17. At D11, the neural inducing (SB and LDN) and patterning (SHH and CHIR) factors need to be removed from the neuralization media. Without washing, aspirate media and add N2B27 supplemented with 20 ng/mL BDNF, 20 ng/mL GDNF and 0.2 mM ascorbic acid (1.5 mL/well of 12-well plates, 3 mL/well of 6-well plates or 6 mL/6 cm diameter dish). Incubate at 37 °C, 5% CO_2. If the next day the cells occupy the entire surface of the dish, move to the next step. If not, allow them to form a complete monolayer before proceeding to the next step, changing the media daily.
18. At D12 (or when a full monolayer is formed), passage the cells following **steps 4–15** in the appropriate amount of N2B27 supplemented with 20 ng/mL BDNF, 20 ng/mL GDNF, 0.2 mM ascorbic acid and 10 μM Y-27632, and incubate at 37 °C, 5% CO_2 on laminin coated plates and incubate at 37 °C, 5% CO_2. Change the media every second day or sooner if needed.
19. At D16 (or when a full monolayer is formed again) passage the cells for the fourth time (*see* **step 18**). From this passage onward, cells can be plated on coverslips if required (*see* Subheading 3.4.3).
20. Keep expanding and passaging the cells as described in **step 18** until D50, but not more than once per week. Change media when needed (judged by color change), at least once every 3 days.

See **Note 12**.

3.4.2 Maintaining a Neural Culture Beyond D50

1. To split the cells after D50, repeat **steps 5** and **6** of Subheading 3.4.1, and proceed to the next step.
2. Gently, remove Accutase and wash once with fresh N2B27. Do not use an aspirator.

3. Remove the wash media and harvest the cells by dislodging them using a 1 mL pipette tip and 1–2 mL of fresh N2B27 supplemented with 20 ng/mL BDNF, 20 ng/mL GDNF, 0.2 mM ascorbic acid, 1: 1000 laminin, and 10 μM Y-27632. Collect them in a 15 mL falcon tube, but *do not* pellet by centrifugation.
4. Triturate the cells gently up to 7 times with a 1 mL pipette tip and plate them in laminin-coated plates (*see* Subheading 3.3.1) at a density of approximately $10^5/cm^2$.
5. Passage cells after they have established a full monolayer, but not more frequently than every 10 days (and every 15 days after neurons have reached D100). Change media when needed or at least every 4 days with N2B27 supplemented with 20 ng/mL BDNF, 20 ng/mL GDNF, and 0.2 mM ascorbic acid, adjusting the medium volume accordingly. Add 1:1000 laminin every second feeding. Adhering to this methodology, we have kept the neurons in culture for more than a year.

3.4.3 Passaging on Coverslips: Terminal Differentiation

For imaging-based investigations, we recommend plating neuralized cells on coverslips from D16 onward. However, cells can be plated on coverslips earlier should analysis of early differentiation parameters be required.

1. Coat coverslips with polyornithine and laminin as described in Subheading 3.3.2.
2. Harvest the neural progenitors as described in Subheading 3.4.1, **step 19**. Collect in a 15 mL Falcon tube, centrifuge at 100 × *g*/3 min at room temperature, and resuspend the pellet in a maximum volume of 2 mL of fresh N2B27 supplemented with 20 ng/mL BDNF, 20 ng/mL GDNF, 0.2 mM ascorbic acid, and 10 μM Y-27632 to initiate terminal differentiation.
3. Gently, triturate up to 5 times with a 1 mL pipette tip.
4. Remove laminin droplets from the pre-coated coverslips and replace with 100 μL droplet of N2B27 supplemented with 20 ng/mL BDNF, 20 ng/mL GDNF, 0.2 mM ascorbic acid, 500 μM db-cAMP, 10 μM DAPT, and 10 μM Y-27632.
5. Add 5–10,000 cells per droplet (approximately 5000 cells/cm^2). This amount of cells is usually contained within 10–50 μL of cell suspension, depending on how much media was used to resuspend the pellet in **step 2**.
6. With care so as to not disturbed the droplets, incubate at 37 °C, 5% CO_2 for 1 h to allow cells to settle and attach.
7. Flood the wells by topping them up with 350 μL of N2B27 supplemented with 20 ng/mL BDNF, 20 ng/mL GDNF, 0.2 mM ascorbic acid, 500 μM db-cAMP, 10 μM DAPT, and 10 μM Y-27632.

8. Before incubating at 37 °C, 5% CO_2, ensure that the coverslips are not floating after flooding by gently pushing them down with a pipette tip.
9. Feed the cells at 3-day intervals by removing 250 μL spent medium and replacing with 350 μL fresh N2B27 supplemented with 20 ng/mL BDNF, 20 ng/mL GDNF, 0.2 mM ascorbic acid, 500 μM db-cAMP, and 10 μM DAPT. Thereafter, continue the 3-day feeding regime, adding 1:1000 laminin every second feeding (no additional Y-27632 is required.)
10. After 30 days on coverslips, completely remove DAPT from the feeding media and feed once every 5 days. In this way, we have kept neurons on the same coverslips for more than 3 months.

    See **Note 13**.

3.5 Cryopreservation and Thawing of Neural Progenitors

We cryopreserve neural progenitors from differentiation D6 until ~D60. Beyond D60, progenitors can be frozen, but cell survival (after thawing) is diminished. We suggest (after D60) freezing 3× more progenitors than is recommended at earlier time points (outlined below).

3.5.1 Cryopreservation

1. At differentiation D16 prepare a "Mr Frosty" and follow the process as described in Subheading 2.5.
2. After centrifuging and removing the spent media and Accutase, resuspend the pellet in CryoStor CS10 (~1×10^6 cells per mL) triturating up to three times.
3. Distribute in Cryovials and follow the same process as described in Subheading 2.5.

3.5.2 Thawing (See Note 14)

1. Prepare laminin coated plated as described in Subheading 3.3.1.
2. Thaw cell by placing Cryovials in a 37 °C water bath for up to 1 min.
3. Before they are completely thawed, add 1 mL of warm N2B27 per vial.
4. When completely thawed, add another 2 mL of N2B27 to dilute the CryoStor.
5. Centrifuge (100 × *g*/3 min at room temperature) to wash out the freezing medium.
6. Aspirate the medium and gently resuspend the cells in 1 mL of fresh N2B27, triturating up to 5 times.
7. Remove laminin from the pre-coated plates and add the appropriate amount of N2B27 (1 mL/well for 12-well plates, 2 mL/well for 6-well plates, or 4 mL/60-mm-diameter dishes) supplemented with 20 ng/mL BDNF, 20 ng/mL GDNF,

0.2 mM ascorbic acid, and 10 μM Y-27632. After this initial step, inclusion of Y-27632 is not recommended.

8. Add cell suspension to plates and gently agitate for equal cell distribution. Incubate at 37 °C, in 5% CO_2. Replace with fresh N2B27 supplemented with 20 ng/mL BDNF, 20 ng/mL GDNF, and 0.2 mM ascorbic acid the following day if necessary.
9. Feed the cells when needed but at least once every 3 days with N2B27 supplemented with 20 ng/mL BDNF, 20 ng/mL GDNF, and 0.2 mM ascorbic acid.

4 Imaging-Based Methods to Monitor Autophagy in mDAN Cultures

Preface and general considerations.

In the following sections, we outline approaches to image (macro) autophagy in hiPSC-derived mDANs. In theory, imaging approaches that are used to study autophagy in common laboratory cell-line models can be transferred to these cells, but caution is needed when working with such fragile cells. In addition, certain typical cell fixation and staining techniques are less successful when using hiPSC-derived mDANs, so pilot studies are always necessary to find the best conditions. Researchers wishing to stimulate autophagy in hiPSC-derived neuronal cultures need to test whether cells are responsive to the inducer. In our hands, amino acid starvation is less effective than pharmacological mTOR inhibition (using, e.g., AZD8055). In all experiments, it is essential that autophagic flux is taken into account, since an increase in autophagosomes might indicate upregulated autophagy and/or reduced lysosomal clearance.

4.1 Fixed-Cell Imaging for Autophagy Analysis in mDAN-Enriched hiPSC Neuronal Cultures

In autophagy research, image-based quantitation of autophagosomes and autophagosome assembly sites provides researchers with a rapid and direct means for measuring autophagic activity [12]. Careful flux analysis is particularly important during static, fixed-cell autophagy analysis, since the location of autophagosome assembly with respect to the positioning of the lysosomal pool might mean that impaired autophagosome retrograde trafficking will alter flux kinetics (for discussion, *see* [1]). For fixed-cell neuronal imaging, it is important to confirm that cells analyzed are of the correct neuronal classification (e.g., for mDANs, this is best done by counterstaining with anti-tyrosine hydroxylase (TH) antibodies and for midbrain markers, e.g., FOXA2 (*see* Fig. 3a)). To analyze autophagy, we use anti-WIPI2 antibodies to detect assembly sites (*see* [1]). In our hands, anti-LC3 antibodies are less than

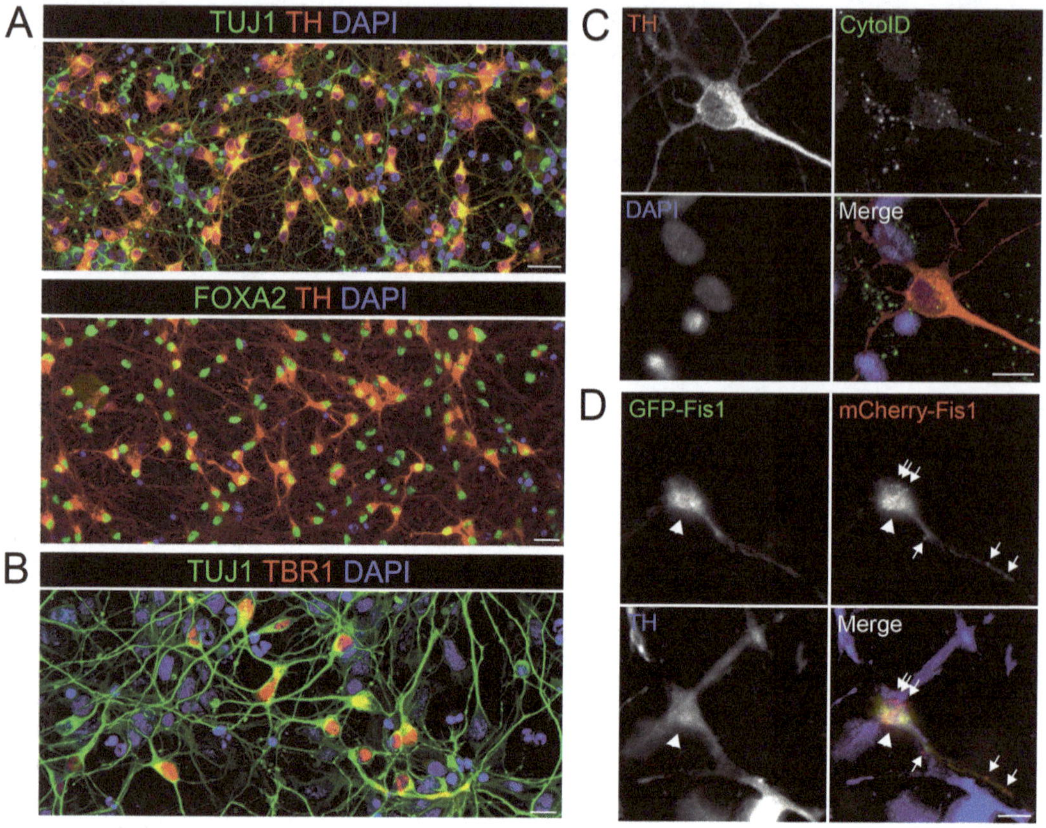

Fig. 3 Examples images of hiPSC neuronal cultures and autophagy/mitophagy assessments. (**a**) Examples of hiPSC-derived mDAN cultures stained with anti-TH antibodies in combination with anti-βIII tubulin (TUJ1; top) and FOXA2 (bottom) antibodies. Bar = 20 μm. (**b**) Example field of hiPSC-derived cortical neurons stained with anti-βIII tubulin (TUJ1) and anti-TBR1. Bar = 20 μm. (**c**) Example image of a TH-positive mDAN labelled using CytoID. Cells were treated for 4 h with 1 μm AZD8055, labelled with CytoID, then fixed and stained with anti-TH antibodies. Bar = 10 μm. (**d**) Mitophagy assessment using the tandem mCherry-GFP-FIS1 reporter [14]. hiPSC-derived neuronal cultures were transduced with lentivirus expressing mCherry-GFP-FIS1 and imaged live before fixing and labeling with anti-TH antibodies. Arrows show red puncta indicative of mitochondria delivered to lysosomes. Arrowhead indicates the position of the soma. Bar = 10 μm

satisfactory, because it is often hard to distinguish cell debris from "true" autophagosomes in these cultures. This is particularly an issue when analyzing autophagosomes within neurites—visualization with anti-LC3 antibodies is more straightforward when observations are restricted to the soma. For a rapid and reliable method to visualize autophagosomes and lysosomes, we co-stain with CytoID (Fig. 3c) and LysoTracker (*see* [1]). For a general summary of immunofluorescence techniques as applied to standard cell-line models, *see* [13].

4.1.1 Materials

Table 4
Reagents for autophagy induction and flux assessment hiPSC neuronal cultures

Reagent	Provider, Cat. Num.	Preparation and storage	Working concentration and incubation length
AZD8055	Stratech, S1555	1 mM stock solution in DMSO. Store at −20 °C	1 μM for 2–4 h
Rapamycin	Calbiochem, 53123-88-9	1 mM stock solution in DMSO. Store at −20 °C	1 μM for 2–4 h
Bafilomycin	Enzo, CM110-0100	20 μM stock solution in DMSO. Store at −20 °C	20 nM for at least 1 h
Leupeptin	Sigma, L2884	20 mM stock solution in DMSO. Store at −20 °C	20 μM for 6 h *or* 10 μM overnight

4.1.2 Methanol Fixation for Immunofluorescence Microscopy (Cells on Coverslips, See Note 15)

1. Wash neurons rapidly, but carefully ×2 with PBS, and replace the final wash with −20 °C methanol. Incubate cells for 5 min. We store methanol at −20 °C, and remove immediately prior to cell fixation. Fixation is carried out on the bench.
2. Wash fixed cells ×2 with PBS (~5 min each wash) and proceed directly to primary antibody stage (no blocking is required).
3. Incubate coverslips with primary antibodies diluted in PBS for 2 h (room temperature) or overnight at 4 °C (depending on the antibody; *see* Table 5).
4. Wash coverslips ×3 with PBS and proceed directly to secondary antibody stage.
5. Incubate with appropriate secondary antibodies at room temperature for 1 h.
6. Wash ×3 with PBS (5 min each).
7. Remove coverslips and then mount by inverting on to a 5 μL drop of Mowiol placed on a standard glass slide. Press down carefully and leave to set for minimum 6 h.

4.1.3 Paraformaldehyde Fixation for Immunofluorescence Microscopy (Cells on Coverslips)

1. Wash neurons rapidly but carefully with PBS.
2. Fix in 4% PFA for 20 min (enough to comfortably immerse the coverslip).
3. Permeabilization and blocking is achieved by incubating neurons with blocking solution containing 5% BSA and 0.3% Triton X-100 in PBS for 30–60 min at room temperature (*see* **Note 16**).

Table 5
Antibodies and associated reagents for fluorescence analysis of hiPSC neuronal cultures

Reagent	Provider, Cat. Num.	Fixation	Working dilution and incubation length
CytoID	Enzo, ENZ-51031	PFA	2 μL/mL for 30 min in live cells
LysoTracker Red	Thermo F.S., L7528	PFA	0.5 μM for 20 min in live cells
LysoTracker deep red	Thermo F.S., L12492	PFA	0.5 μM for 20 min in live cells
MitoTracker Red	Thermo F.S., M7512	PFA	0.5 μM for 20 min in live cells
MitoTracker Green	Thermo F.S., M7514	PFA	0.5 μM for 20 min in live cells
MitoSOX	Thermo F.S., M36008	PFA	5 μM for 20 min in live cells
Hoechst 33342	Thermo F.S., 62249	PFA/methanol	1 μg/mL for 5 min in live cells
DAPI	Sigma D9542	PFA/methanol	0.1 μg/mL for 45 min with fixed cells
Anti-TH (rabbit)	Millipore, AB152	PFA/methanol	1:300. Incubate for 2 h-to-overnight
Anti-TH (mouse)	Santa Cruz, sc-25269	PFA/methanol	1:100. Incubate for 2 h-to-overnight
Anti-βIII tubulin (TUJ1)	Biolegend, 801201	PFA/methanol	1:1000. Incubate for 2 h-to-overnight
Anti-FOXA2	Santa Cruz, sc-101060	PFA/methanol	1:100. Incubate for 2 h-to-overnight
Anti-LMX1A	Millipore, AB10533	PFA/methanol	1:1000. Incubate for 2 h-to-overnight
Anti-LMX1B	Proteintech, 18278	PFA fixation	1:250. Incubate for 2 h-to-overnight
Anti-TBR1	Abcam, AB31940	PFA fixation	1:500. Incubate for 2 h-to-overnight
Anti-WIPI2 (mouse monoclonal)	Bio-Rad, MCA5780GA	Methanol	1:400 incubate for 1 h
Anti-LC3B (rabbit polycolonal)	Sigma, L7543	Methanol	1:400. Incubate for 1 h
Anti-P62 (mouse monoclonal)	Abnova, H00008878	PFA/methanol	1:400. Incubate for 1 h
Anti-LAMP1	DSHB, H4A3–5	PFA and saponin	1:400. Incubate for 2 h-to-overnight
Donkey Anti-Rabbit Alexa Fluor 488	Thermo F.S., A21206		1:400. Incubate for 1 h
Donkey Anti-Rabbit Alexa Fluor 568	Thermo F.S., A10042		1:400. Incubate for 1 h

(continued)

Table 5
(continued)

Reagent	Provider, Cat. Num.	Fixation	Working dilution and incubation length
Donkey Anti-Mouse Alexa Fluor 488	Thermo F.S., A21202		1:400. Incubate for 1 h
Donkey Anti-Mouse Alexa Fluor 488	Thermo F.S., A10037		1:400. Incubate for 1 h
Donkey Anti-Rabbit Alexa Fluor 647	Thermo F.S., A31573		1:400. Incubate for 1 h
Donkey Anti-Mouse Alexa Fluor 647	Thermo F.S., A31571		1:400. Incubate for 1 h

Table 6
Reagents for fixation, permeabilization, blocking, and mounting of hiPSC neuronal cultures

Reagent	Provider, Cat. Num.	Preparation and storage	Working concentration
Paraformaldehyde (PFA)	Thermo F.S., 28908	1 ampule content diluted in 30 mL PBS. Store at 2–8 °C	4%
Bovine Serum Albumin (BSA)	Sigma, A9418	BSA solutions are prepared in PBS. Store at 2–8 °C	5% or 2.5%
Triton X-100	Sigma, X100	Triton X100 is diluted in PBS. Store at room temperature	0.3% or 0.15%
Saponin	Sigma, S4521	Prepare 10% stock solution in PBS. Store at 2–8 °C for up to 4 weeks	0.1% for blocking and antibody solutions
Donkey serum	Sigma, D9663	Store aliquots at −20 °C	10%
Mowiol 40–88	Sigma, 324590	Store in 1 mL aliquots at −20 °C. Defrost, add 25 mg DABCO and store at 4 °C	N/A
DABCO	Sigma, D2522	Store at 2–8 °C	25 mg/mL
Live-cell imaging dishes	MatTek, P35G-0.170-14-C	N/A	N/A

4. Incubate coverslips with primary antibodies diluted in 50% blocking solution (2.5% BSA and 0.15% Triton X-100 in PBS) for 2 h (room temperature) or overnight at 4 °C (depending on the antibody; *see* Table 5).
5. Wash coverslips ×3 with PBS and proceed directly to secondary antibody stage.

6. Incubate with appropriate secondary antibodies diluted in 50% blocking solution at room temperature for 1 h.
7. Wash ×3 with PBS (5 min each).
8. Mount by inverting on to a 5 μL drop of Mowiol placed on a standard glass slide. Press down carefully and leave to set for minimum 6 h (*see* **Note 17**).

4.2 Live-Cell Imaging of Autophagy Dynamics in mDAN Cultures

Neuronal architecture means that more careful consideration of the location of autophagosome assembly and of the maturation-coupled transportation kinetics of the autophagosome pool is often desirable. Live-cell imaging provides a direct means to measure these parameters, and a number of vital stains (Table 5) are available. We also use lentiviruses to express common autophagy reporters (e.g., GFP-LC3B (*see* [1]); mCherry-GFP-LC3B; ATG5-GFP (*see* [1]); mCherry-GFP-FIS1 (Fig. 3d) [14]) (*see* **Note 18**).

1. Prepare live-cell imaging dishes by coating as described in Subheading 3.3.2.
2. Plate cells as described in Subheading 3.4.3.
3. For lentiviral transduction, infect cells for 3 days and then replace media with N2B27 supplemented with 20 ng/mL BDNF, 20 ng/mL GDNF, 0.2 mM ascorbic acid, 500 μM db-cAMP, and 10 μM DAPT. Incubate cells for another 2–4 days (*see* **Note 19**).
4. For staining with fluorescent probes, wash cells with PBS and then treat cells with the appropriate probes diluted in N2B27 supplemented with 20 ng/mL BDNF, 20 ng/mL GDNF, 0.2 mM ascorbic acid, 500 μM db-cAMP, and 10 μM DAPT for the indicated times at 37 °C (*see* Table 5). Wash cells twice with PBS and then add fresh N2B27 supplemented with 20 ng/mL BDNF, 20 ng/mL GDNF, 0.2 mM ascorbic acid, 500 μM db-cAMP, and 10 μM DAPT (*see* **Note 20**).

See (**Note 21**).

5 Summary

The ability to monitor autophagy in disease-specific cell types derived from hiPSCs provides researchers with the capacity to determine how autophagy contributes to neurodegenerative disease onset and progression and the means to carry out functional screens in appropriate cell-line models [1, 2]. We have presented a protocol for the rapid generation of human mDANs from hiPSCs for PD-based research and have outlined considerations for imaging-based autophagy analysis. In theory, any imaging-based autophagy assessment approach should be applicable; however, the unique architecture and fragility of hiPSC-derived neuronal cultures dictate that additional care is needed in the selection and application of assessment methodologies when using these cells.

6 Notes

1. General considerations: We describe using E8 medium for feeder-free hiPSC culturing. Other feeder-free systems are available, although we have not carried out direct comparisons. This is an indicative method for culturing hiPSCs. Minor adjustments may be required depending on the cell line and the age of the cells (early or late passage). All procedures must be carried out under sterile conditions. Cells or culture surfaces must not be allowed to dry out at any point. It is highly recommended to review the suppliers' webpages for product and method updates.
2. Vitronectin does not need to be washed off the tissue culture surface prior to plating cells.
3. Cells must be split after a maximum 5 days, even if they have not reached the desired density. On the contrary, if over confluent, cells should be split earlier; however, we do not recommend passaging hiPSCs within 48 h of plating.
4. The manufacturer does not suggest washing after the EDTA treatment (**step 5**); however, we have noticed that a quick wash with complete E8 medium increases cell survival.
5. During harvesting (**steps 6** and 7), do not over-triturate, do not scrape, and do not intensively force any cells that have not detached under gentle liquid flow.
6. RevitaCell or Y-27632 is not required in established lines; however, if the cells do not recover after 2–3 consecutive passages, add 1× RevitaCell (1:100) or 10 μM Y-27632 on the day of passage only.
7. If the cells do not detach readily from the dish, repeat **steps 2–7**, but incubate with EDTA for longer. On the contrary, if the cells detach during EDTA incubation, do not aspirate but neutralize the EDTA with 3× volume of complete E8 medium. Gently collect all the cells from the dish with a 5 mL serological pipette into a 15 mL falcon tube, centrifuge (100 × g for 3 min at room temperature), resuspend the pellet in fresh complete E8 medium, and progress to **step 9**.
8. General considerations. hiPSCs are very sensitive to temperature fluctuations. Use dry ice to transfer the Cryovials between different locations. Cells can also be frozen in complete E8 medium supplemented with 10% dimethyl sulfoxide (DMSO); however, we have noticed decreased cell survival

after thawing by this approach. Refill "Mr Frosty" with isopropanol every 3 uses. We routinely keep frozen hiPSCs at −80 °C for up to 2 weeks or in liquid nitrogen for >6 months without affecting cell survival after thawing.

9. We reuse polyornithine and laminin solutions twice (i.e., three uses in total), without having noticed any loss of their activity.
10. Coverslips can be preprepared and stored at 2–8 °C for up to 3 days, after carefully Parafilm-shielding the plates without disturbing the droplets. Before use, ensure that the droplets have not dried out and incubate at 37 °C for 15–20 min.
11. General considerations. Do not start neuralizations with freshly thawed hiPSCs. Instead, allow them to recover and settle for at least two consecutive passages, and ensure that the hiPSCs look similar to those shown in Fig. 1. The concentrations of SHH and CHIR are crucial factors in obtaining cultures with high proportions of mDANs. Concentration ranges of 200–500 ng/mL for SHH and 0.6–1.0 μM CHIR are typically stated in publications [6, 7, 10]. The protocols outlined below are designed to generate high numbers of mDANs. It is possible to generate cortical neurons following the same protocols described here but in the absence of SHH and CHIR [15]. This can provide a useful comparison when examining mDAN biology.
12. Do not let the cells dry at any stage—during feeding, leave a small amount of spent media covering the cells and add the fresh media on top. If the day following plating cells is sparse or patchy, add 1:1000 volume of laminin. Cell death after passaging is expected; however, if extensive, replace with media supplemented with 10 μM Y-27632 and 1:1000 laminin the day after passaging. The suggested cell densities are indicative. Until familiarizing with the process and the particular properties of the cell line, we recommend optimizing by plating the cells at several different densities.
13. Do not use aspirators to remove spent media. The suggested cell densities are indicative. To familiarize with the process and the cell line behavior, we recommend optimizing by plating the cells in several different densities. If any droplet breaks while plating, flood the well with another 350 μL media and add the double amount of cells. To ensure equal cell distribution after adding the cells in the droplet, very gently triturate both together once on the coverslip with a 200 μL pipette tip (one up/down). Cells can stay in droplets overnight; however, if adopting this approach, ensure that the empty spaces between the wells are filled with sterile water or DPBS to increase the humidity inside the plate. Flood the wells first thing in the next morning (*see* **step 7**).

14. Do not recover the cells directly on to coverslips. Allow them to settle on coated cell culture dishes and readapt for one passage.
15. Preservation of neuronal fine structure is difficult with methanol, so care must be taken to ensure that coverslips do not dry out when transferring from aqueous to solvent and vice versa. Equally, coverslips should not be too wet so as to avoid mixing disruption when transferring to methanol.
16. For lysosomal staining, 0.1% saponin should be used as permeabilization reagent in place of Triton X-100 to maintain lysosomal structure. Alternative blocking solutions comprising 10% normal donkey serum (for anti-TH staining), or 0.2% BSA with 0.2% Triton X-100, have also been tested with similar results.
17. We store Mowiol in 1 mL aliquots at −20 °C. As needed, thaw and add 25 mg DABCO antifade and store for up to 6 weeks at 4 °C.
18. Common promoters for expression of transgenes do not work with all hiPSC-derived neuronal types (e.g., the CMV promoter is silenced in midbrain neuronal cultures). For promoters, we use synapsin, EF1a, or spleen focus forming virus (SFFV) [16].
19. For lentiviral transduction, neural progenitors are plated in live imaging dishes or on coverslips. Cells are infected for 3d, and then media is replaced with complete N2B27 and cells incubated for another 2–4 days. These long incubation periods are needed to allow sufficient transgene expression.
20. After a live imaging experiment, it is advisable to ensure that cells examined are confirmed mDANs. To do this, we stain cells post-imaging with anti-TH antibodies after PFA fixation as described in Subheading 4.1.3 (Fig. 3d). Using an automated microscope setup and image acquisition software such as MetaMorph (Universal Imaging) can be of assistance as it is possible to record field positions to allow multipoint time-lapse imaging with retrospective TH labeling.
21. Like primary neurons, hiPSC-derived neuronal cultures are fragile and are more prone to illumination damage than standard, undifferentiated laboratory cell lines. Neuronal architecture dictates that the cell body (soma) may be some mm distance from growing neurite tips. This affects analysis of autophagosome assembly and intensity analyses. Successful hiPSC-derived neuronal cultures are complex, with multiple neurites intersecting one another on the plane of the coverslip. This makes it difficult to associate cell bodies with peripheral neuronal structures.

Acknowledgments

This work is supported by a Parkinson's UK project grant (G1402), a Wellcome Trust Ph.D. studentship awarded to NJM through the Dynamic Cell Biology program (grant number 083474), and a Medical Research Council Ph.D. studentship (to PS).

References

1. Jimenez-Moreno N et al (2017) Induced pluripotent stem cell neuronal models for the study of autophagy pathways in human neurodegenerative disease. Cell 6(3):E24
2. Jungverdorben J, Till A, Brustle O (2017) Induced pluripotent stem cell-based modeling of neurodegenerative diseases: a focus on autophagy. J Mol Med (Berl) 95(7):705–718
3. Surmeier DJ, Obeso JA, Halliday GM (2017) Selective neuronal vulnerability in Parkinson disease. Nat Rev Neurosci 18(2):101–113
4. Technologies L (2015) Culturing pluripotent stem cells (PSCs) in essential 8TM medium
5. Jaeger I et al (2011) Temporally controlled modulation of FGF/ERK signaling directs midbrain dopaminergic neural progenitor fate in mouse and human pluripotent stem cells. Development 138(20):4363–4374
6. Kirkeby A et al (2012) Generation of regionally specified neural progenitors and functional neurons from human embryonic stem cells under defined conditions. Cell Rep 1 (6):703–714
7. Kriks S et al (2011) Dopamine neurons derived from human ES cells efficiently engraft in animal models of Parkinson's disease. Nature 480 (7378):547–551
8. Ozair MZ, Kintner C, Brivanlou AH (2013) Neural induction and early patterning in vertebrates. Wiley Interdiscip Rev Dev Biol 2 (4):479–498
9. Chambers SM et al (2009) Highly efficient neural conversion of human ES and iPS cells by dual inhibition of SMAD signaling. Nat Biotechnol 27(3):275–280
10. Arenas E, Denham M, Villaescusa JC (2015) How to make a midbrain dopaminergic neuron. Development 142(11):1918–1936
11. Sullivan GJ et al (2010) Induced pluripotent stem cells: epigenetic memories and practical implications. Mol Hum Reprod 16 (12):880–885
12. Klionsky DJ et al (2016) Guidelines for the use and interpretation of assays for monitoring autophagy (3rd edition). Autophagy 12 (1):1–222
13. Allan VJ (ed) (2000) Protein localization by fluorescence microscopy. A Practical Approach. Oxford University Press, Oxford
14. Allen GF et al (2013) Loss of iron triggers PINK1/Parkin-independent mitophagy. EMBO Rep 14(12):1127–1135
15. Nistor PA et al (2015) Long-term culture of pluripotent stem-cell-derived human neurons on diamond—a substrate for neurodegeneration research and therapy. Biomaterials 61:139–149
16. Betin VM et al (2012) A cryptic mitochondrial targeting motif in Atg4D links caspase cleavage with mitochondrial import and oxidative stress. Autophagy 8(4):664–676

Chapter 18

Correlative Light and Electron Microscopy to Analyze LC3 Proteins in *Caenorhabditis elegans* Embryo

Céline Largeau and Renaud Legouis

Abstract

In this chapter, we present a protocol to perform correlative light and electron microscopy (CLEM) on *Caenorhabditis elegans* embryos. We use a specific fixation method which preserves both the GFP fluorescence and the structural integrity of the samples. Thin sections are first analyzed by light microscopy to detect GFP-tagged proteins, then by transmission electron microscopy (TEM) to characterize the ultrastructural anatomy of cells. The superimposition of light and electron images allows to determine the subcellular localization of the fluorescent protein. We have used this method to characterize the roles of autophagy in the phagocytosis of apoptotic cells in *C. elegans* embryos. We analyzed in apoptotic cell and phagocytic cell the localization of the two homologs of LC3/GABARAP proteins, namely, LGG-1 and LGG-2.

Key words Green fluorescent protein, LC3-associated phagocytosis, High-pressure freezing, Freeze substitution, LGG-1, LGG-2, GMA resin

1 Introduction

Macroautophagy is a vesicular process involved in lysosome-mediated degradation and recycling of numerous cellular materials [1]. Autophagy-inducing signals trigger the nucleation of a flat membranous structure, the phagophore, which elongates and sequesters autophagy cargoes and then closes to generate a double-membrane vesicle, called autophagosome. The autophagosome finally fuses with the lysosome to promote the degradation of engulfed cytoplasmic compounds (Fig. 1).

Apoptosis is an essential process involved in the elimination of excessive or damaged cells during tissue remodeling and homeostasis [2]. Cells undergoing apoptosis are efficiently recognized and eliminated through a process called phagocytosis, which is performed by professional and nonprofessional cells. Phagocytosis is a dynamic multistep process, starting with the recognition of "eat-me" signals, exposed on the surface of apoptotic cells, by specific

Nicholas Ktistakis and Oliver Florey (eds.), *Autophagy: Methods and Protocols*, Methods in Molecular Biology, vol. 1880, https://doi.org/10.1007/978-1-4939-8873-0_18,

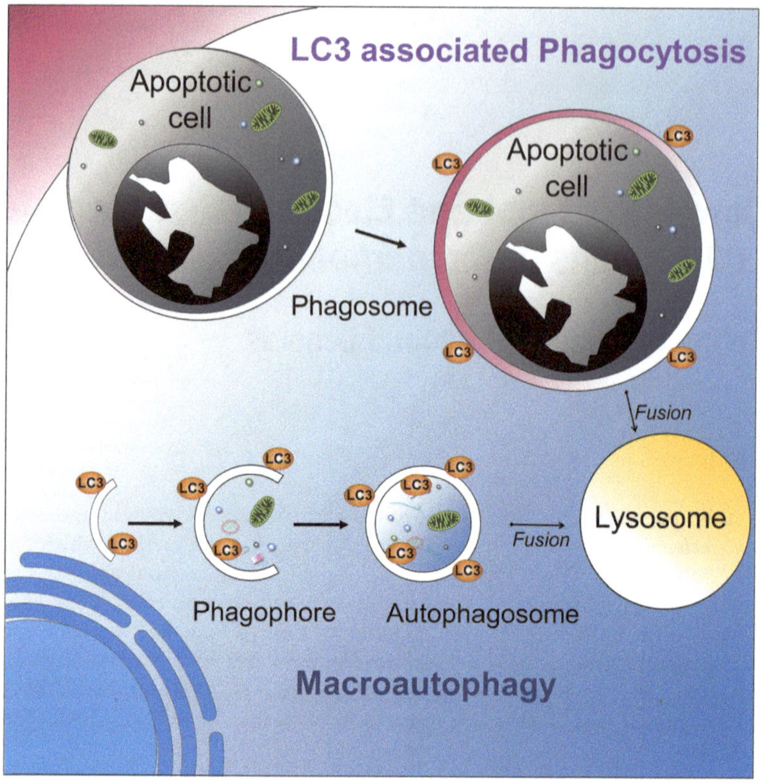

Fig. 1 Macroautophagy and phagocytosis are lysosomally mediated degradative pathways. During LC3-associated phagocytosis, an apoptotic cell is engulfed in a single-membrane vesicle, the phagosome. During macroautophagy, an isolation membrane elongates, sequesters cytoplasmic compounds, and closes to form a double-membrane vesicle, the autophagosome. Both processes involve the recruitment of the ubiquitin-like protein LC3 which allows the fusion to the lysosome and the degradation

cell-surface receptors leading to the internalization of apoptotic cells by phagocytic cells. The engulfing vesicle containing apoptotic cells, also called phagosome, undergoes a maturation process, which results in the degradation of the dead cells. Recent studies have shown that autophagy is involved in optimal clearance of apoptotic cells. In phagocytic cells, it has been shown that the autophagic protein LC3 rapidly decorates the phagosome formed upon engulfment of an apoptotic cell, facilitating the degradation of the internalized cell (Fig. 1). This process, termed LC3-associated phagocytosis (LAP), also occurs during phagocytosis of pathogens. Although LAP shares common components with classical autophagy, it is a distinct process, since LC3 is not associated with a double-membrane autophagosome, but recruited and lipidated to a single-membrane phagosome [3–8].

To address the specific functions of each of the two subfamilies, LC3 and GABARAP, during phagocytosis of apoptotic cells, we used *Caenorhabditis elegans* which possesses only one homolog for each family, LGG-1/GABARAP and LGG-2/LC3 [9]. Using correlative light and electron microscopy (CLEM) [10], we analyzed GFP::LGG-1 and GFP::LGG-2 localization during apoptosis in the embryo and showed that LGG-1 is enriched in vesicles inside the apoptotic corpses, while LGG-2 is enriched in phagocytic cells. We report here the detailed protocol used to perform CLEM on *C. elegans* embryo, which is subdivided into four main steps: (1) cryofixation by high-pressure freezing to preserve structures, (2) freeze substitution and inclusion in GMA resin to preserve GFP fluorescence, (3) ultramicrotomy and successive observations of the sections by light and electron microscopies, and (4) correlation of fluorescent and electronic images.

This protocol allows to perform an easy and rapid correlation between the electron and the light microscopy, based on the dual imaging of thin sections. The CLEM pictures have a lower contrast than classical transmission electron microscopy, but this approach circumvents the difficulty of the retrospective location of the regions of interest, which necessitates very specific approaches [11].

2 Materials

2.1 *Caenorhabditis elegans* Strains and Culture

The conditions for culturing nematode strains are very classical, and they will not be detailed here (*see* Ref. [12]).

1. Strains: RD286 (*plgg-1::GFP::LGG-1* in Ced-1 context); ZH2333 (*pCed-1::GFP::LGG-2*).
2. Nematode Growth Medium (NGM) plate: 500 mL of water with 1.5 g of NaCl, 1.5 g of bactopeptone, and 0.5 mL of cholesterol (5 mg/mL) autoclaved and supplemented with 0.5 mL $CaCl_2$ (1 mol/L), 0.5 mL $MgSO_4$ (1 mol/L), 10 mL KH_2PO_4 (1 mol/L), and 1.65 mL K_2HPO_4 (1 mol/L).
3. *Escherichia coli* OP50 strain (available at the Caenorhabditis Genetics Center https://cgc.umn.edu/). NGM plates are seeded with 100 μL of OP50 culture in exponential growth phase.

2.2 Buffers and Equipments for Cryofixation

1. Cryofixation is performed in automatic mode with HPF apparatus EM PACT2 (Leica Microsystem) using liquid nitrogen for cryogenic conditions and methyl cyclohexane to generate pressure.
2. M9 buffer: 1 L of water with 6 g of Na_2HPO_4, 3 g of KH_2PO_4, and 5 g of NaCl, autoclaved and aliquoted by 200 mL. Aliquots are supplemented with 200 μL of 1 M $MgSO_4$.
3. M9 buffer containing 20% bovine serum albumin (BSA), store at 4 °C after complete dissolution.

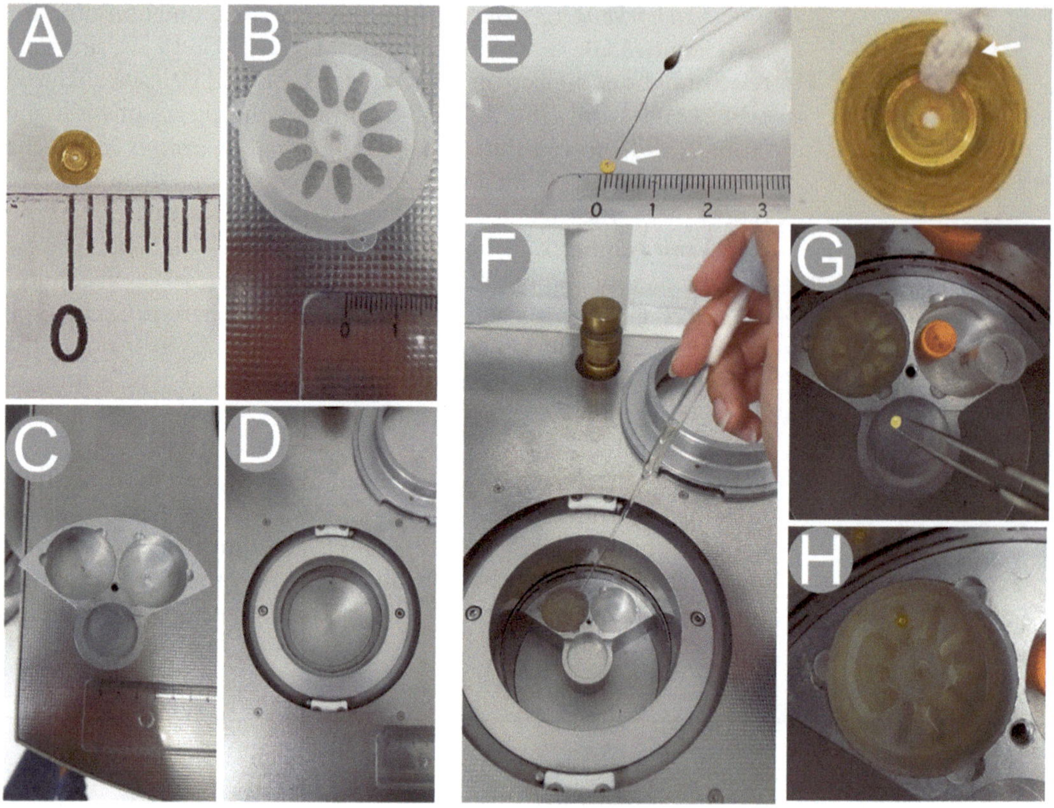

Fig. 2 Cryofixation and substitution materials. (**a**) 200 μm deep flat carrier. (**b**) Specimen processing metal block. (**c**) Flat embedding insert. (**d**) AFS incubating chamber. (**e**) Embryos are transferred in the flat carrier with a "worm pick" (white arrow). (**f**) A Pasteur pipette is used for loading and changing the buffers inside the flat embedding insert. (**g**, **h**) A 200 μm deep flat carrier is transferred into the flat embedding insert with tweezers. Ruler graduations are cms

4. 1% phosphatidylcholine (Sigma-Aldrich) pre-coated 200 μm deep flat carriers (Leica Microsystem) (Fig. 2a).
5. 4% low gelling agarose in 1 mL aliquots in small glass tubes, store at 4 °C.

2.3 Buffers and Equipment for Freeze Substitution and Resin Inclusion

1. Freeze substitution is performed with automated freeze-substitution system (AFS2) with integrated binocular lens and incubating chamber (Leica Microsystem) (Fig. 2d).
2. Pasteur pipettes.
3. Flat embedding insert (sample container containing a reagent tray and a continuous flow ring) (Leica Microsystem) (Fig. 2b).
4. Aclar sheet (Leica Microsystem).
5. Glass bottle of 10 mL.
6. Tweezers and "worm pick" with platinum wire (Fig. 2e, g).

7. Specimen processing metal block (Leica Microsystem) (Fig. 2c).
8. Freeze substitution buffer: 0.1% $KMnO_4$; 5% water; 95% anhydride acetone; 0.01% osmium.
9. 95% acetone in water.
10. 95% ethanol in water.
11. 5% uranyl acetate in ethanol 100% is vortexed and filtered (0.2 μm).
12. 0.1% uranyl acetate in 95% acetone.
13. GMA acrylic resin: 68,1 mL glycol methacrylate supplemented with 30 mL of butyl methacrylate, 3 mL of H_2O and 600 mg of benzoyl peroxide, mixed during 1 h at 4 °C and store at −20 °C.
14. 30% GMA in 95% ethanol.
15. 70% GMA in 95% ethanol.
16. 0,32% *N*,*N*-dimethyl-*p*-toluidine (accelerator polymerization agent) (Merck) in GMA acrylic resin. Prepare extemporaneously.

2.4 Ultramicrotomy

1. Ultramicrotomy is performed with an ultramicrotome EM UC7 (Leica Microsystems).
2. Razor blade.
3. Glass knife.
4. Diamond knife for histology (Diatome, Histo, 2 mm, 45°).
5. Toluidine blue: 1% toluidine blue and 1% borate sodium in water filtered through 0.2 μm filters.
6. Standard light microscope.
7. 200-mesh copper formvar carbon-coated finder grid.

2.5 Equipment for Light Microscopy

We used interchangeably the following upright or inverted microscopes.

1. Inverted microscope: AxioObserver Z1 microscope (Zeiss) equipped with Evolve EMCCD camera (Roper Scientific).
2. Upright microscope: Axioskop 2 Plus microscope (Zeiss) equipped with Nomarski optics coupled to a camera (CoolSNAP, Roper Scientific).
3. 35 mm cell view (cell culture dish with glass bottom) (Greiner Bio-One).
4. Glass slides and coverslips.

2.6 Equipment for Transmission Electron Microscopy

Transmission electron microscopy is performed with a JEOL 1400 operating at 80 kV coupled with a Gatan 11 Mpixels SC1000 Orius CCD camera.

2.7 Correlation Analysis

Treatment and superimposition of light and electron microscopy pictures are performed with ImageJ, PowerPoint, and Photoshop software.

3 Methods

3.1 Cryofixation

The current protocol is designed for the analysis of embryos, but it can be used for larvae and adults upon adaptation of the nematode culture [13].

1. Put 60 adult worms on an NGM plate seeded with OP50. Remove them after 1 h at 20 °C.
2. Six hours later, about 200 embryos are transferred with the worm pick, into 1% phosphatidylcholine pre-coated 200 μm deep flat carrier (*see* **Note 1**) containing 5 μL of M9 buffer with 20% BSA (Fig. 2e). Eventually, before freezing, add a drop of 4% low gelling agarose to seal the embryo and the cups (*see* **Note 2**).
3. 200 μm deep flat carriers containing embryos are put in high-pressure freezing (HPF) apparatus and are cryofixed according to protocol outlined in the manual using automatic mode.
4. Flat carriers are transferred in cryogenics tubes and stored in liquid nitrogen container, where they can be kept for several months. During all this process, the carrier should be maintained in liquid nitrogen.

3.2 Freeze Substitution and Resin Inclusion

Freeze substitution process has been adapted from previously published protocols in *C. elegans* or plants [14–16].

1. Day 1, the specimen processing metal block containing the flat embedding insert is put in the incubating chamber of AFS preset at −90 °C. Using a Pasteur pipette fill delicately the flat embedding insert with about 1.5 mL of freeze substitution buffer (Fig. 2f). All subsequent steps of removal or change of buffer are performed similarly with the same volume.
2. After precooling at −90 °C of the AFS for 2 h, remove flat carrier from cryogenic tubes using tweezers, and put them in the specimen processing metal block without defrosting the samples (*see* **Note 3**) (Fig. 2g).
3. Transfer a single flat carrier from cryogenic tubes in each compartment of the flat embedding insert (Fig. 2h).

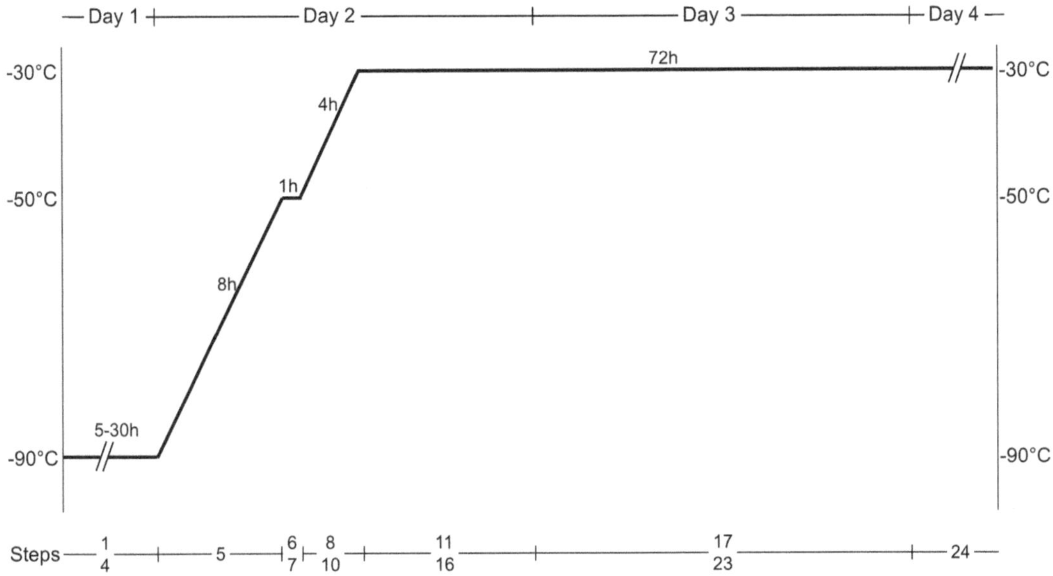

Fig. 3 Graphical summary of the freeze substitution process. The cryosubstitution program occurs between −90 °C and −30 °C. Temperature slopes are 5 °C per hour. Steps are detailed in Subheading 3.2

4. Close AFS and start the AFS program (Fig. 3). Samples are kept at −90 °C for at least 5 h and until 30 h if convenient.
5. A temperature slope of 5 °C by hour during 8 h is applied, to reach −50 °C (Fig. 3).
6. On day 2, prepare 95% acetone and put it in the incubating chamber of AFS for allowing temperature equilibrium at least 15 min before using.
7. Remove the freeze substitution medium from the flat embedding insert and add 95% acetone. Incubate for 10 min and repeat the operation four times. AFS is maintained at −50 °C during 1 h (Fig. 3).
8. Prepare the 5% uranyl acetate and the 0.1% uranyl acetate in acetone and put the last one in the AFS (−50 °C).
9. Remove the 95% acetone buffer and add the 0.1% uranyl acetate in acetone buffer.
10. A temperature slope of 5 °C by hour during 4 h is applied, to reach −30 °C and program AFS to maintain the temperature at −30 °C during the next 72 h (Fig. 3) but proceed to **steps 11** after 1 h, when the temperature has reached −45 °C.
11. Remove acetone with 0.1% uranyl acetate and add 95% acetone anhydride. Incubate for 10 min. Repeat the operation four times and keep in the AFS incubating room for 2 h until the temperature reaches −30 °C.

12. Prepare the 95% ethanol and put it in the incubating chamber when AFS is at −30 °C. Remove 95% acetone and add 95% ethanol for 15 min. Repeat four times.
13. Prepare 30% and 70% GMA in 95% ethanol and store in a − 20 °C freezer until use.
14. After the washes, remove 95% ethanol from the flat embedding insert and add 30% GMA for 2 h.
15. Remove 30% GMA and add 70% GMA for 3 h.
16. Remove 70% GMA and 100% GMA overnight.
17. On day 3, remove 100% GMA and add fresh one for 2 h and 30 min.
18. Remove 100% GMA and add fresh one for 2 h and 30 min again.
19. Using integrated binocular lens and curved metallic spike and tweezers, remove sample from the deep flat carrier and remove the deep flat carrier from the flat embedding insert. Remove very carefully 100% GMA and add fresh one for 15 min.
20. Prepare 100% GMA with accelerator.
21. Put an aclar ring on the flat embedding insert.
22. Remove 100% GMA and add 100% GMA with accelerator.
23. Remove immediately 100% GMA with accelerator and add one more time without overflow of each compartment and incubate overnight.
24. On day 4, switch off the AFS, take off the flat embedding insert, and unmold the blocks of resin that have polymerized in each compartment. Conserve the blocks at −20 °C in black tubes in a vacuum-sealed freezer bag.

3.3 Ultramicrotomy

The thin sectioning used is standard (*see* Ref. [11] and the Leica Microsystems EM UC7 user manual), and the main steps are indicated below without details (*see* **Note 4**).

1. Make a pyramid in the block around the sample with razor blade.
2. Smooth pyramid surface with a glass knife (speed 10, feed 500 nm).
3. Using a histological diamond knife make semi-thin sections (speed 5, feed 200 nm).
4. Collect semi-thin sections on a glass slide and stain them with toluidine blue.
5. Look at stained semi-thin sections under a classical microscope.

6. If the sample is correct, make semi-thin sections slowly (speed 2 feed 200 nm) and collect them on 200-mesh copper formvar carbon-coated finder grid. Keep the grid in darkness and at 4 °C until microscopic observations (*see* **Note 5**).

3.4 Fluorescent and Electron Microscopy

3.4.1 Fluorescent Microscopy Using an Upright Microscope (Fig. 4a)

1. Put the grid on a glass slide with the semi-thin sections on the top side.
2. Add 10 to 20 μL of water on the grid and add a coverslip (*see* **Note 6**).
3. After bright field and fluorescent observations remove the coverslip and collect gently the grid with tweezers (*see* **Note 7**).

3.4.2 Fluorescent Microscopy Using Inverted Microscope (Fig. 4b)

1. Take a 35 mm cell view.
2. In the middle of the glass bottom put a drop of water (5 μL).
3. Return the grid on the water drop using tweezers. The water should just make a link between grid and glass slide, and the grid should not move (*see* **Note 8**).
4. Take bright field and fluorescent images.
5. To remove the grid after observation, gently add 1 mL of water in the dish and use tweezers.

3.4.3 Electron Microscopy

1. Sections are observed with a JEOL 1400 TEM instrument operating at 80 kV, using a standard procedure.
2. The anticontamination device is filled with liquid nitrogen, the high tension is switched on and increased to 80 kV, and the condenser aperture inserted.

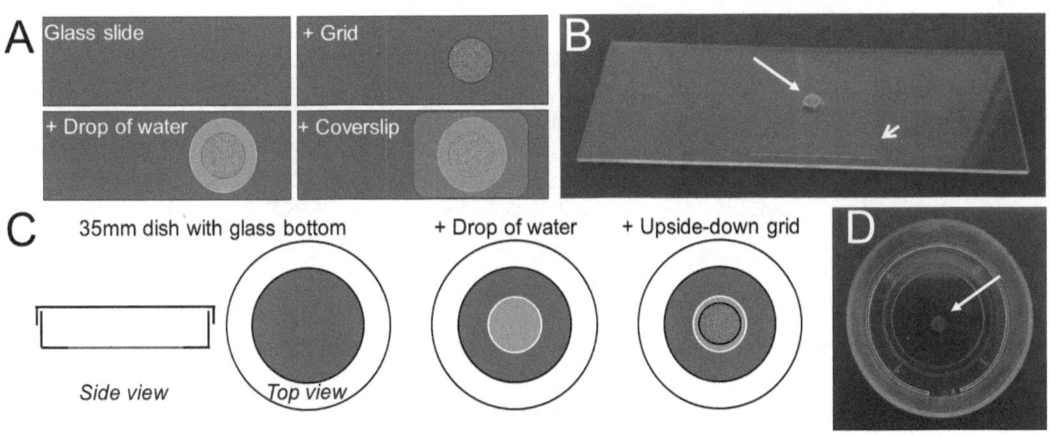

Fig. 4 Mounting grids for light microscopy. (**a**, **b**) When using an upright light microscope, the grid is deposited, upright, on a glass slide and successively covered with a drop of water and a coverslip. (**c**, **d**) When using an inverted microscope, the grid is deposited upside down on a dish with glass bottom above a drop of water. The long arrows point to the grid and the short to the coverslip

3. Grids are mounted on the specimen holder and inserted in the microscope. The filament is then turned on.
4. For each grid, the areas of interest are localized using the letter and the number identified during light microscopy on 200-mesh copper formvar carbon-coated finder grid.
5. Areas of interest are first imaged at low magnification and then analyzed at higher magnification.

3.5 Correlation Between Fluorescent and Electronic Images

1. Perform composite images of bright field and fluorescent images with NIH ImageJ Software (https://imagej.nih.gov/ij/) (Fig. 5a).
2. Import the composite image in a PowerPoint file. Outline the area of interest manually (*see* **Note 9**) and relate this to the transmission electron microscopy images in order to make both of them the same size (Fig. 5b).
3. Open the composite image, of bright field and fluorescent images, in Photoshop (64 Bit).
4. Select all of the image or use a selection tool to choose the area of interest of the image.
5. Copy the image or the selected area, and create a new file with a transparent background.

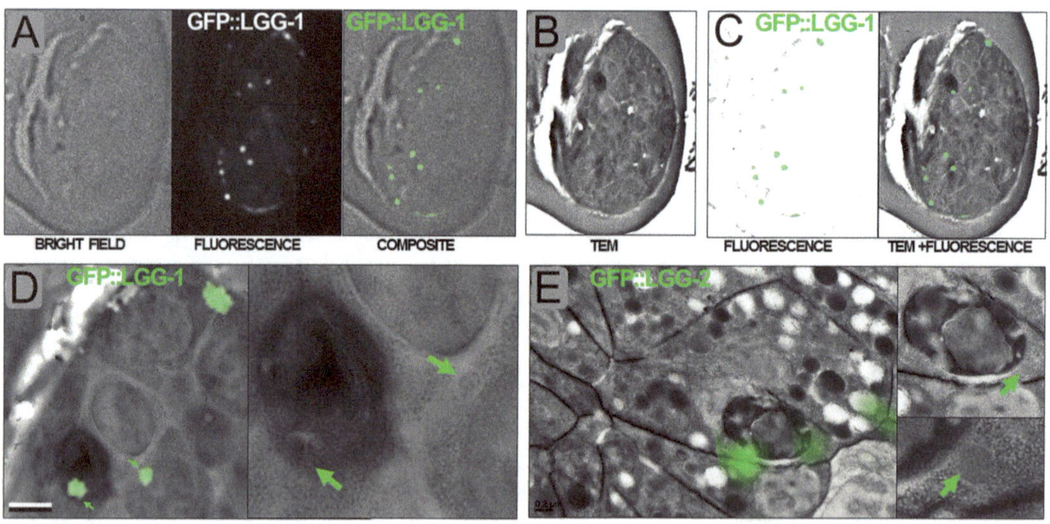

Fig. 5 Correlation between light and electron pictures. (**a**) Bright field and fluorescent images are merged using ImageJ software to obtain a composite image. (**b**) The corresponding transmission electron microscopy image is cropped and sized to fit the light images. (**c**) Using Photoshop software, a thresholding and removal of background noise deletion is performed on the fluorescent image, which is then superimposed with the TEM picture. (**d**, **e**) During our study by CLEM, we were able the see that both GFP::LGG-1 (**d**) and GFP::LGG-2 (**e**) puncta could be associated with vesicular structures of an approximate diameter of 200 μm. Apoptotic corpses appear as dark, small, and round cells rendering the identification of the number of membranes of the GFP-positive vesicles more difficult (green arrows)

6. Paste the image in the new file, select the magic wand tool and click on to select the background of the image. Then suppress the background using the delete command to obtain an image with a transparent background.
7. Save the file in Portable Network Graphics (PNG) format and manually superimpose using PowerPoint, the PNG image, and the composite image to obtain images with identical size. Then, superimpose in a similar way the corrected-size PNG image and the electron microscopy image (Fig. 5c) (*see* **Note 10**).
8. Do the same thing on higher magnification electron microscopy images to correlate the ultrastructure and the fluorescent signals (Fig. 5d, e).

4 Notes

1. Under a hood, put the 200 μm deep flat carriers on the bottom of the crystallizer containing 1% phosphatidylcholine in chloroform, and wait for 5 min. Still under the hood, remove the carriers and transfer them in an empty crystallizer and wait 5 min. Store them at room temperature in a close box.
2. Heat a tube of agarose low gelling 4% with a Bunsen burner until melting of agarose, and then keep in a heating block at 50 °C. In the heating block, also prepare a glass tube containing water. After putting the embryos in the 200 μm deep flat carriers, preheat a Pasteur pipette by pipetting up and down 50 °C water. Immediately pipet few microliters of agarose low gelling and quickly pour a small drop on the 200 μm deep flat carriers containing the embryos.
3. During all the manipulation, always stay in the cold compartment of the AFS, using the least possible light in the room; put a cache on the porthole when you don't manipulate in the AFS and check that there is always enough liquid nitrogen in the AFS.
4. To limit photobleaching of GFP, set the light of the ultramicrotome and the room at minimum level.
5. It is not necessary to contrast grid with uranyl acetate and lead citrate because there is enough contrast with freeze substitution medium and the integrity of the section is better conserved.
6. Add a sufficient volume of water to avoid the grid to be compressed but not too much or it could move during the observation.

7. Sections on the grids are fragile so the fluorescent observations have to be performed quickly.
8. Center the grid as much as possible or the correct focusing of the image could be difficult.
9. To outline the area of interest manually, on the composite image, use different landmarks like an outline of the embryo, an outline of the shell, or even a piece of dirt.
10. A free open-source software called eC-CLEM, which could facilitate the correlation process, has been recently released [17]. It is available as a plugin in the Icy platform.

Acknowledgments

The authors would like to thank the *C. elegans* community and the Caenorhabditis Genetics Center which is funded by the NIH National Center for Research Resources (NCRR) for *C. elegans* strains. We are grateful to Claire Boulogne for critical reading of the manuscript and helpful discussions. The present work has benefited from the expertise of Béatrice Satiat-Jeunemaitre, Cynthia Gillet, Jessica Marion and Claire Boulogne and the core facilities of Imagerie-Gif (http://www.i2bc.paris-saclay.fr) member of IBiSA (http://www.ibisa.net), supported by "France-BioImaging" (ANR-10-INBS-04-01), and the Labex "Saclay Plant Science" (ANR-11-IDEX-0003-02). This work was supported by the Agence Nationale de la Recherche (project EAT, ANR-12-BSV2-018) and the Association pour la Recherche contre le Cancer (SFI20111203826).

References

1. Ohsumi Y (2014) Historical landmarks of autophagy research. Cell Res 24(1):9–23
2. Metzstein MM, Stanfield GM, Horvitz HR (1998) Genetics of programmed cell death in *C. elegans*: past, present and future. Trends Genet 14(10):410–416
3. Henault J, Martinez J, Riggs JM, Tian J, Mehta P, Clarke L, Sasai M, Latz E, Brinkmann MM, Iwasaki A et al (2012) Noncanonical autophagy is required for type I interferon secretion in response to DNA-immune complexes. Immunity 37:986–997
4. Martinez J, Almendinger J, Oberst A, Ness R, Dillon CP, Fitzgerald P, Hengartner MO, Green DR (2011) Microtubule-associated protein 1 light chain 3 alpha (LC3)-associated phagocytosis is required for the efficient clearance of dead cells. Proc Natl Acad Sci U S A 108:17396–17401
5. Sanjuan MA, Dillon CP, Tait SWG, Moshiach S, Dorsey F, Connell S, Komatsu M, Tanaka K, Cleveland JL, Withoff S et al (2007) Toll-like receptor signalling in macrophages links the autophagy pathway to phagocytosis. Nature 450:1253–1257
6. Florey O, Kim SE, Sandoval CP, Haynes CM, Overholtzer M (2011) Autophagy machinery mediates macroendocytic processing and entotic cell death by targeting single membranes. Nat Cell Biol 13:1335–1343
7. Kim JY, Zhao H, Martinez J, Doggett TA, Kolesnikov AV, Tang PH, Ablonczy Z, Chan CC, Zhou Z, Green DR et al (2013) Noncanonical autophagy promotes the visual cycle. Cell 154:365–376
8. Martinez J, Malireddi RKS, Lu Q, Cunha LD, Pelletier S, Gingras S, Orchard R, Guan JL, Tan H, Peng J et al (2015) Molecular

characterization of LC3-associated phagocytosis reveals distinct roles for Rubicon, NOX2 and autophagy proteins. Nat Cell Biol 17:893–906

9. Manil-Ségalen M, Lefebvre C, Jenzer C, Trichet M, Boulogne C, Satiat-Jeunemaitre B, Legouis R (2014) The *C. elegans* LC3 acts downstream of GABARAP to degrade autophagosomes by interacting with the HOPS subunit VPS39. Dev Cell 28(1):43–55
10. Jenzer C, Simionato E, Legouis R (2015) Tools and methods to analyze autophagy in C. elegans. Methods 75:162–171
11. Kolotuev I, Schwab Y, Labouesse M (2009) A precise and rapid mapping protocol for correlative light and electron microscopy of small invertebrate organisms. Biol Cell 102 (2):121–132
12. Brenner S (1974) The genetics of Caenorhabditis elegans. Genetics 77:71–94
13. Lefebvre C, Largeau C, Michelet X, Fourrage C, Maniere X, Matic I, Legouis R, Culetto E (2016) The ESCRT-II proteins are involved in shaping the sarcoplasmic reticulum in *C. elegans*. J Cell Sci 129(7):1490–1499
14. Watanabe S, Punge A, Hollopeter G, Willig KI, Hobson RJ, Davis MW, Hell SW, Jorgensen EM (2011) Protein localization in electron micrographs using fluorescence nanoscopy. Nat Methods 8(1):80–84
15. Watanabe S, Richards J, Hollopeter G, Hobson RJ, Davis WM, Jorgensen EM (2012) Nano-fEM: protein localization using photo-activated localization microscopy and electron microscopy. J Vis Exp 70:e3995
16. Marion J, Le Bars R, Satiat-Jeunemaitre B, Boulogne C (2017) Optimizing CLEM protocols for plants cells: GMA embedding and cryosections as alternatives for preservation of GFP fluorescence in Arabidopsis roots. J Struct Biol 198(3):196–202
17. Paul-Gilloteaux P, Heiligenstein X, Belle M, Domart MC, Larijani B, Collinson L, Raposo G, Salamero J (2017) eC-CLEM: flexible multidimensional registration software for correlative microscopies. Nat Methods 14 (2):102–103

Chapter 19

Imaging Noncanonical Autophagy and LC3-Associated Phagocytosis in Cultured Cells

Elise Jacquin, Katherine Fletcher, and Oliver Florey

Abstract

Monitoring of ATG8 proteins by western blotting and immunofluorescence microscopy are the most common methods to monitor the autophagy pathway. However, it has recently been shown that ATG8 proteins can be lipidated to non-autophagosome, single-membrane compartments through a noncanonical autophagy pathway. This is commonly found to occur during macro-endocytic processes such as phagocytosis, where it has been termed LC3-associated phagocytosis, and upon lysosomotropic drug treatment. Therefore, care is required when interpreting data based on ATG8 in order to conclude whether a signal relates to the canonical or noncanonical pathway. Here we provide methods to monitor noncanonical autophagy through fluorescence microscopy.

Key words Noncanonical autophagy, LAP, Phagocytosis, ATG8, LC3

1 Introduction

During canonical autophagy, two ubiquitin-like conjugation systems act to target the lipidation of ATG8 family proteins to the lipid phosphatidylethanolamine on forming double-membrane autophagosome structures [1, 2]. This is a tightly regulated process which aids in the elongation of autophagosome membranes [3], in cargo selection [4], and ultimately in the fusion with lysosomes [5]. The lipidation of ATG8s has been a hallmark of autophagy since its discovery. ATG8 lipidation can be monitored using western blotting, where the conjugated form (LC3-II) runs faster than the unconjugated form (LC3-I) on SDS-PAGE gels. Alternatively, a number of immunofluorescence microscopy approaches have been utilized to monitor the movement of LC3 from the cytosol to membrane autophagosome structures. Both antibody immunofluorescence of endogenous proteins and the use of fluorescently tagged proteins are commonly used to track the dynamics of the autophagy pathway.

Nicholas Ktistakis and Oliver Florey (eds.), *Autophagy: Methods and Protocols*, Methods in Molecular Biology, vol. 1880, https://doi.org/10.1007/978-1-4939-8873-0_19,

However, recent work has shown that ATG8 lipidation is not restricted to canonical autophagy. A noncanonical autophagy pathway has been identified where ATG8 proteins are lipidated to preformed single-membrane structures of the endolysosomal system [6–8]. This is independent of canonical autophagy and does not require the ULK1/2-FIP200-ATG13 initiation complex or inputs from mTOR signalling. The noncanonical autophagy pathway does not activate in response to nutrient starvation but rather functions during a number of macroscale endocytic engulfment processes. These include phagocytosis of pathogens or dead cells where it is termed LC3-associated phagocytosis (LAP) [9–12], macropinocytosis [12], and the live cell engulfment process entosis [12]. The common feature of these processes is that ATG8 proteins are transiently recruited and lipidated to the single-membrane compartments formed following the engulfment event. The function of such ATG8 lipidation in this context remains unclear. However, there is evidence that noncanonical autophagy can modulate the degradative capacity of the targeted compartment and its fusion with lysosomes. This has important implications for many immune responses where inhibition of noncanonical autophagy has been shown to impair pathogen killing and clearance and antigen presentation [9–11, 13–17].

Interestingly, work from our group has shown that a number of pharmacological drugs commonly used to inhibit the canonical autophagy pathway also have a parallel activity in activating noncanonical autophagy [18, 19]. Lysosomotropic drugs such as chloroquine and ionophores such as monensin both raise the pH of lysosomes and block canonical autophagic flux, thus increasing levels of lipidated ATG8 in cells. However, they also induce an ionic and osmotic imbalance within the endolysosomal compartment, which drives direct lipidation of ATG8s onto the lysosome membranes, independently of canonical autophagy [18, 19]. As such, use and interpretation of ATG8 in monitoring autophagy changes upon use of these drugs is complicated. However, we have recently demonstrated a strategy to distinguish between canonical and noncanonical autophagy through the targeting of the C-terminal domain of ATG16L1, which governs the membrane site of ATG8 lipidation [17].

In light of the recent discovery of the noncanonical autophagy pathway, we cannot assume that ATG8 lipidation is always derived from canonical autophagy. In this chapter we will describe microscopy methods to study both drug-induced and physiological noncanonical autophagy processes.

2 Materials

2.1 Imaging of Monensin-Driven Noncanonical Autophagy

1. HCT116 cell line expressing GFP-LC3.
2. HCT116 culture media: McCoy's 5A, 10% FBS, penicillin, and streptomycin.
3. Monensin.
4. Methanol, stored in −20 °C.
5. 3 μm latex beads.
6. Phosphate-buffered saline (PBS).
7. Blocking buffer: PBS + 5% w/v BSA.
8. 12-well plates.
9. 18 mm glass coverslips.
10. Glass slides.
11. 4',6-diamidino-2-phenylindole (DAPI).
12. Anti-fade mounting solution.
13. Anti-human LAMP1, (e.g., BD Transduction, #611043).
14. Confocal microscope.

2.2 Live Imaging of LAP in Cultured Cells

1. J774.1A murine macrophage cell line expressing GFP-LC3.
2. 35 mm glass bottom dishes.
3. DMEM culture media, +10% FBS, penicillin, and streptomycin.
4. 3 μm latex beads.
5. 0.1 M sodium borate.
6. Human IgG.
7. Interferon gamma (IFNγ).
8. Inverted spinning disc microscope with temperature and CO_2-controlled environment chamber.

2.3 Isolation of Murine Bone Marrow-Derived Macrophage

1. 16-gauge sterile needle.
2. Sterile 0.5 and 1.5 mL Eppendorf.
3. Sterilized or ethanol-washed tweezers and scalpels.
4. 10 cm tissue culture dishes.
5. 12-well plates.
6. 15 and 50 mL Falcon tubes.
7. 70% ethanol.
8. PBS.
9. 1× PBS + 2% w/v FBS, sterile.

10. 1× red blood cell lysis buffer: 10× stock (to dilute 1:10 in sterile Milli-Q H_2O): 8.99 g NH_4Cl, 1 g KHCO3, 37 mg Na_2 EDTA (or 200 μL of a 0.5 M Na_2 EDTA pH 8 solution), adjust pH to 7.3, and complete with 100 mL Milli-Q H_2O.
11. Complete RPMI: RPMI 1640, 10% FBS, 1% Pen/Strep, 50 μM tissue culture grade 2-mercaptoethanol.
12. BMDM differentiation medium: complete RPMI supplemented with 10–30 ng/mL M-CSF and 50 ng/mL Fungizone (amphotericin B).
13. M-CSF.
14. Amphotericin B, 1 mL aliquots stored at −20 °C, aliquot in use can be stored at 4 °C.

2.4 Imaging LAP in Primary BMDM

1. Sterile glass coverslips.
2. Ethanol-washed tweezers.
3. Tissue culture 12-well plates.
4. Glass slides.
5. Poly-L-lysine solution (0.1 mg/mL in tissue culture grade H_2O).
6. Complete RPMI: RPMI 1640, 10% FBS, 1% Pen/Strep, 50 μM tissue culture grade 2-mercaptoethanol.
7. Zymosan A from *Saccharomyces cerevisiae*.
8. PBS.
9. Cell scrapers.
10. Methanol, stored in −20.
11. Blocking buffer: PBS + 5% BSA.
12. Zymosan.
13. 12-well plate.
14. 18 mm glass coverslips.
15. Glass slides.
16. Anti-fade mounting solution.
17. Primary antibodies: anti-LC3A/B (e.g., Cell Signalling #4108), anti-LAMP1 (e.g., BD Transduction #553792).
18. Secondary antibodies.
19. PBS.
20. Confocal microscope.

3 Methods

3.1 Monensin-Driven Latex Bead Assay

1. Place 18 mm coverslips in a 12-well plate.
2. Add 8×10^4 cells and fill with 1 mL media. Leave cells overnight (*see* **Note 1**).
3. The next day, take 200 μL bead solution into a 1.5 mL Eppendorf. Add 1 mL PBS. Centrifuge for 5 min at 2350 × *g*. Aspirate and repeat. Final resuspension of beads in 10 mL PBS.
4. Add 15 μL of washed bead solution. Leave cells for 4 h.
5. Wash cells ×3 in PBS to remove unengulfed beads.
6. Add 100 uM monensin and incubate for 45 min (*see* **Note 2**).
7. Wash cells ×1 in PBS. And then add ice-cold methanol. Place cells in −20 °C freezer for 4 min (*see* **Note 3**).
8. Wash cells in PBS and add blocking buffer at room temperature for 1 h.
9. On a piece of parafilm, add an 80 μl drop of primary anti-LAMP1 antibody and invert cell-containing coverslip over the top. Leave cells in a humidified chamber at 4 °C overnight.
10. Place the coverslips back in a 12-well plate. Wash cells ×3 in PBS and add 0.5 mL secondary antibody in blocking buffer.
11. Wash cells ×3 in PBS. In the last wash, add DAPI at 1 μg/mL for at least 5 min.
12. Mount the coverslips on glass slides with anti-fade solution.
13. Image samples using a confocal microscope. Look for LAMP1-positive bead-containing phagosomes and check for GFP-LC3 recruitment. Count 100–200 LAMP1-positive phagosomes per condition. Monensin should drive the recruitment of GFP-LC3 to bead phagosomes (Fig. 1).

3.2 Generation of IgG-Coated Latex Beads

1. Take 0.5 mL latex beads into a 1.5 mL Eppendorf tube and add 1 mL 0.1 M sodium borate. Vortex the tube.
2. Wash beads three times by pelleting beads in a benchtop centrifuge at 2300 × *g* for 5 min. Aspirate liquid and resuspend pellet in 1 mL 0.1 M sodium borate.
3. At final wash resuspend in 1 mL 0.1 M sodium borate and add 600 μg human IgG. Incubate overnight at 4 °C rotating end over end.
4. Wash beads three times as above and finally resuspend in 1 mL PBS (*see* **Note 4**).

3.3 Live Cell Imaging of IgG-Coated Bead LAP

1. Plate 2× 105 J774.A1 cells expressing GFP-LC3 onto a 35 mm glass bottom dish in 2 mL DMEM complete media in the presence of 200 U/ml IFNγ. Incubate for 2 days.

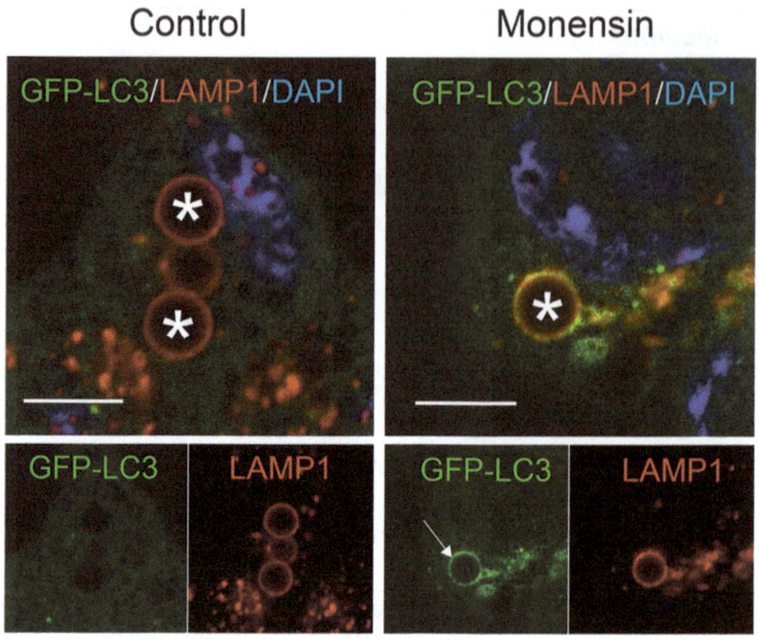

Fig. 1 Monensin induced GFP-LC3 recruitment to latex bead-containing phagosomes. HCT116 cells expressing GFP-LC3 were incubated with plain latex beads for 4 h before the addition of 100 μM monensin for 40 min. Images show confocal sections of cells following staining of LAMP1 (red) and GFP-LC3 (green). Asterisks denote engulfed latex beads. Note GFP-LC3 recruitment to LAMP1-positive phagosomes upon monensin treatment (arrow). Scale bar = 5 μm

2. Set dish up on a spinning disc confocal microscope with a temperature and CO_2-controlled environment chamber.
3. Use ×63 or ×100 objective lenses with DIC capabilities.
4. Add 15 μL of IgG bead solution to cells and leave for 20 min.
5. Under the microscope, locate cells in close association with bead particles or those that have already taken up some particles. Acquire *z*-stacks every 1 min for up to 40 min (*see* **Note 5**).

3.4 Mouse Bone Marrow Cell Isolation

1. Sacrifice mice and apply 70% ethanol on the fur.
2. Dissect the leg bones (2 tibias and 2 femurs/mouse) and transfer them into a petri dish containing sterile 1× PBS + 2% FBS on ice.
3. Incubate for approximately 15 min to soften the tissues.
4. The following steps can be performed in a tissue culture hood using sterilized instruments.
5. Using a 16-gauge sterile needle, pierce the bottom of a sterile 0.5 mL Eppendorf and cut the lid off. Place this cut tube in a 1.5 mL sterile Eppendorf containing 100 μL of 1× PBS + 2% FBS. Prepare one "tube system" per mouse (four bones).

6. On the top of a sterile petri dish, carefully remove the muscles and tendons from the bones using scalpel and tweezers. Transfer the bones into a tissue culture plate containing cold 1× PBS + 2% FBS.
7. Following extraction, wash bones by dipping them into 70% ethanol.
8. Cut the extremities of the bones using scalpel and tweezers and place the bones vertically in the tube system. Close the 1.5 mL Eppendorf lid.
9. Spin on a benchtop centrifuge set at 15900 × *g*. Wait until the speed is reached and spin for 20–30 s before stopping. The bone marrow should have come off the bones and into the PBS through the 0.5 mL tube (red pellet visible in the PBS). The empty bones can be discarded with the 0.5 mL Eppendorf (*see* **Note 6**).
10. Resuspend cells by pipetting up and down and transfer them into 50 mL tubes.
11. Wash by adding 12 mL of complete RPMI media and spinning for 5 min at 620 × *g*.
12. Resuspend the pellet in 1 mL of red blood cell lysis buffer and incubate at room temperature for 1–2 min (gently agitate).
13. Stop red blood cell lysis by adding 12 mL of complete RPMI media. Centrifuge cells for 5 min at 620 × *g*.
14. Discard supernatant and resuspend cells in 1 mL of complete RPMI media.
15. Add BMDM differentiation medium to cells. Culture cells from one mouse into a T75 flask containing 25 mL of medium and cells from two mice into a T175 flask containing 50 mL of medium. After 3 days, add 25 mL or 50 mL of fresh BMDM differentiation medium, respectively, and incubate for 3–6 more days.

3.5 Imaging LAP in Primary Cells

3.5.1 Glass Coverslip Coating

1. Under a tissue culture hood, add one sterile 18 mm glass coverslip per well of a 12-well plate.
2. Add 1 mL of Poly-L-lysine per well and incubate at 37 °C for 10 min.
3. Retrieve the Poly-L-lysine back in the tube, and aspirate the remaining liquid, and allow to air-dry.
4. Wash two or three times with 1 mL of tissue culture grade H_2O and one more time with complete RPMI media.

3.5.2 Cell Seeding

1. Gently harvest BMDM with a sterile scraper and collect them into a 50 mL Falcon tube.
2. Centrifuge the cell suspension at 620 × *g* for 5 min.

3. Resuspend cells in 2 mL of complete RPMI media using a p1000 tip and complete with 8 mL of complete RPMI.
4. Count cells and seed 100,000 cells per well of a 12-well plate in which an 18 mm coverslip has been added, in 1 mL of complete RPMI.

3.5.3 Cell Feeding with Zymosan

1. The next day (or 2 days after), check cell adherence and morphology and gently refresh culture media.
2. Prepare fresh 5 mg/mL zymosan solution in 1× PBS. Vortex and homogenize the solution by passing through a small needle using a 5 mL syringe. 1 mg of zymosan contains approximately 2×10^7 particles. This stock solution concentration is 10×10^7 particles/mL.
3. Incubate the tissue culture plate on ice for about 10 min.
4. Add 25 μg/mL zymosan (5 μL of zymosan stock solution per well) to the cells (zymosan/cell ratio = 5:1) on ice.
5. Incubate at 37 °C to allow cells to engulf zymosan particles (*see* **Note 7**).
6. Wash cells once with PBS and fix for immunostaining as in Subheading 3.1, **steps 7–13**.

4 Notes

1. To get a dispersed plating of cells, place drop of cells in the middle of the coverslip and then add media on top. Make sure to push down the coverslip with a pipette tip to remove any air bubbles under it.
2. Other drugs of choice can be added here to test for noncanonical autophagy. A dose response should be carried out as different cell lines may be differentially sensitive. Alternatively, different genetic backgrounds in cell lines can be used to test for their role in noncanonical autophagy.
3. We find that methanol fixation best retains membrane GFP-LC3 signal, but immunofluorescence is also compatible with other fixatives such as paraformaldehyde.
4. If antibody immunofluorescence is to be used with IgG-coated beads, non-specific binding needs to be blocked by incubating beads in PBS + 5% BSA for 1 h at room temperature. Otherwise primary and secondary antibodies will non-specifically bind the beads.
5. High levels of laser light during image acquisition can suppress phagocytosis. Not all cells will display the same engulfment capacity. Since phagosomes in cell lines are rarely observed before 15 min of incubation, 30 min (or longer) is recommended, but this may need to be optimized.

6. As an alternative to the centrifugation methods to isolate the bone marrow, you may flush the bone marrow cells out of the bones by injecting sterile PBS.
7. Primary cells phagocytose faster than cell lines. Optimization will be required to work out the best time to visualize and quantify LC3 recruitment to zymosan phagosomes.

References

1. Mizushima N et al (1998) A protein conjugation system essential for autophagy. Nature 395(6700):395–398
2. Feng Y et al (2014) The machinery of macroautophagy. Cell Res 24:24–41
3. Weidberg H et al (2010) LC3 and GATE-16/GABARAP subfamilies are both essential yet act differently in autophagosome biogenesis. EMBO J 29(11):1792–1802
4. Lamark T, Svenning S, Johansen T (2017) Regulation of selective autophagy: the p62/SQSTM1 paradigm. Essays Biochem 61 (6):609–624
5. Nguyen TN et al (2016) Atg8 family LC3/GABARAP proteins are crucial for autophagosome-lysosome fusion but not autophagosome formation during PINK1/Parkin mitophagy and starvation. J Cell Biol 215(6):857–874
6. Virgin HW, Levine B (2009) Autophagy genes in immunity. Nat Immunol 10(5):461–470
7. Sprenkeler EG, Gresnigt MS, van de Veerdonk FL (2016) LC3-associated phagocytosis: a crucial mechanism for antifungal host defence against Aspergillus fumigatus. Cell Microbiol 18(9):1208–1216
8. Florey O, Overholtzer M (2012) Autophagy proteins in macroendocytic engulfment. Trends Cell Biol 22(7):374–380
9. Sanjuan MA et al (2007) Toll-like receptor signalling in macrophages links the autophagy pathway to phagocytosis. Nature 450 (7173):1253–1257
10. Martinez J et al (2015) Molecular characterization of LC3-associated phagocytosis reveals distinct roles for Rubicon, NOX2 and autophagy proteins. Nat Cell Biol 17(7):893–906
11. Martinez J et al (2011) Microtubule-associated protein 1 light chain 3 alpha (LC3)-associated phagocytosis is required for the efficient clearance of dead cells. Proc Natl Acad Sci U S A 108(42):17396–17401
12. Florey O et al (2011) Autophagy machinery mediates macroendocytic processing and entotic cell death by targeting single membranes. Nat Cell Biol 13(11):1335–1343
13. Martinez J et al (2016) Noncanonical autophagy inhibits the autoinflammatory, lupus-like response to dying cells. Nature 533 (7601):115–119
14. Ma J et al (2012) Dectin-1-triggered recruitment of light chain 3 protein to phagosomes facilitates major histocompatibility complex class II presentation of fungal-derived antigens. J Biol Chem 287(41):34149–34156
15. Lee HK et al (2010) In vivo requirement for Atg5 in antigen presentation by dendritic cells. Immunity 32(2):227–239
16. Henault J et al (2012) Noncanonical autophagy is required for type I interferon secretion in response to DNA-immune complexes. Immunity 37(6):986–997
17. Fletcher K et al (2018) The WD40 domain of ATG16L1 is required for its non-canonical role in lipidation of LC3 at single membranes. EMBO J 37(4):e97840
18. Jacquin E et al (2017) Pharmacological modulators of autophagy activate a parallel noncanonical pathway driving unconventional LC3 lipidation. Autophagy 13(5):854–867
19. Florey O et al (2015) V-ATPase and osmotic imbalances activate endolysosomal LC3 lipidation. Autophagy 11(1):88–99

Part III

Assays to Measure (and Screen For) Autophagy and Related Processes

Chapter 20

Measurement of Bulk Autophagy by a Cargo Sequestration Assay

Nikolai Engedal, Morten Luhr, Paula Szalai, and Per O. Seglen

Abstract

The nonselective bulk sequestration of cytoplasm and its subsequent delivery to lysosomes for degradation was originally defined as *autophagy* or *macroautophagy*. However, both terms are now increasingly being applied in a generic sense to encompass the many recently described mechanisms for *selective* sequestration and degradation of individual cellular elements. We will therefore use the term *bulk autophagy* to denote the non-exclusive and largely nonselective process.

Bulk autophagy can be measured directly by a cargo sequestration assay, using a cargo marker representative of total cytoplasm. The assay described here measures the sequestration and accumulation of the ubiquitous cytosolic protein lactate dehydrogenase (LDH) in the sedimentable autophagic vacuoles of cells incubated with an inhibitor of intravacuolar degradation such as bafilomycin or leupeptin. Electrodisruption of the plasma membrane followed by centrifugal sedimentation of the "cell corpses" (which retain their organelles in an intact state, bound to the cytoskeleton) is a convenient, efficient, and reproducible way to separate the small fraction of sequestered, sedimentable LDH from the major pool of cytosolic LDH.

Key words Autophagy, Bulk autophagy, Bafilomycin, Cargo assay, Cell corpse, Cell culture, Electrodisruption, Lactate dehydrogenase, Lysosome, Phagophore, Sequestration

1 Introduction

Cargo assays of actual bulk-autophagic activity were introduced in the early 1980s [1–4], based on the sequestration of soluble cargo markers, initially distributed throughout the cytoplasm, into sedimentable autophagic vacuoles. The first markers were membrane-impermeant radiolabeled sugars such as [^{14}C]sucrose and [^{3}H]raffinose, which were introduced into isolated rat hepatocytes by reversible electropermeabilization of the plasma membrane prior to the autophagy measurements. Later, the ubiquitous cytosolic enzyme lactate dehydrogenase (LDH) was used as a cargo probe [5, 6], obviating the need for electropermeabilization, but necessitating treatment of the cells with an inhibitor of intravacuolar degradation such as leupeptin [5, 6] or bafilomycin A1 [7–9] to

Nicholas Ktistakis and Oliver Florey (eds.), *Autophagy: Methods and Protocols*, Methods in Molecular Biology, vol. 1880, https://doi.org/10.1007/978-1-4939-8873-0_20,

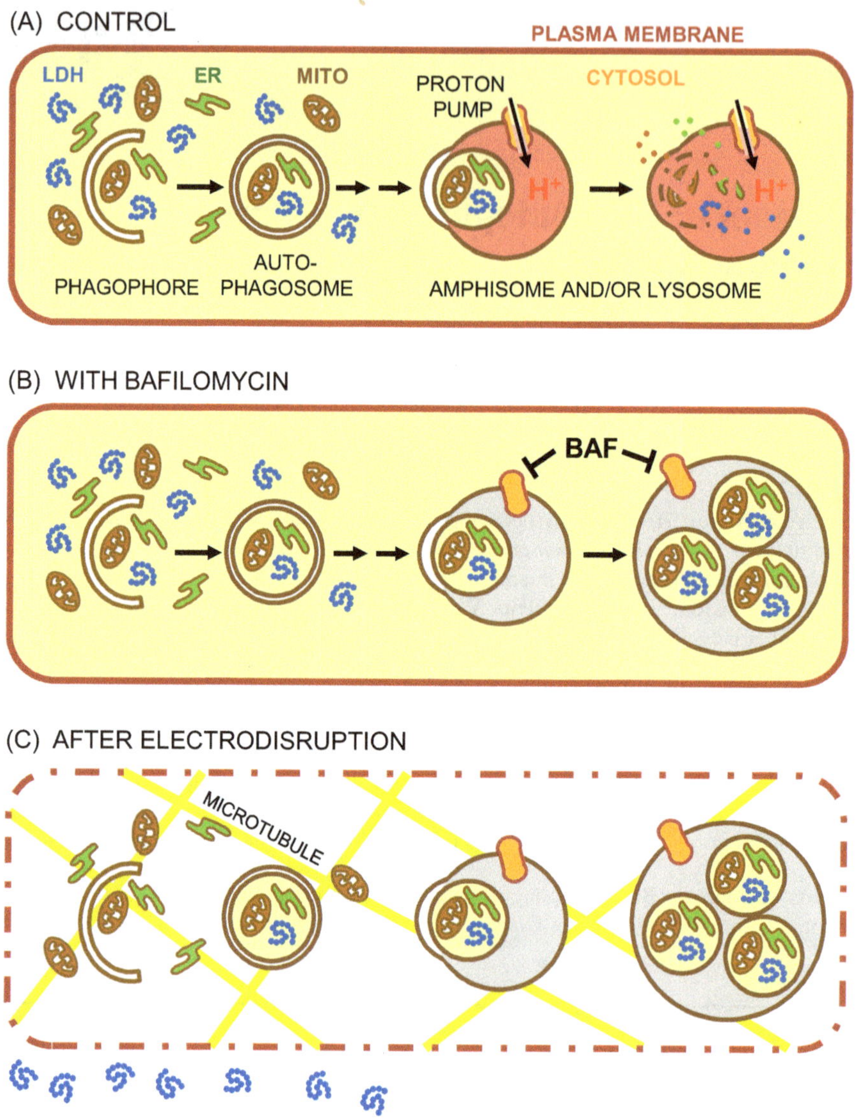

Fig. 1 Schematic representation of bulk-autophagic sequestration of cytoplasm and inhibition of its intravacuolar degradation by bafilomycin A1. (**a**) In untreated cells, cytoplasm (including the cargo marker LDH) is sequestered by phagophores and enclosed in autophagosomes which fuse with acidic autophagic vacuoles (amphisomes and/or lysosomes) to have the cytoplasm degraded. (**b**) Bafilomycin (BAF) inhibits the vacuolar proton pump (v-ATPase), thereby arresting the low-pH-dependent intravacuolar degradation without stopping the autophagic influx, thus causing accumulation of sequestered cytoplasm (including LDH). (**c**) Electrodisruption destroys the plasma membrane, causing cytosol (with LDH) to leak out from the "cell corpses," while organelles (including LDH-containing autophagic vacuoles) remain intact, trapped by the cytoskeletal meshwork, and can be sedimented

prevent LDH degradation during the sequestration period (Fig. 1). The largely nonselective nature of bulk autophagy during amino acid starvation has been well documented by the identical sequestration rates measured for cytosolic enzymes with widely different half-lives [5] as well as by the identical subcellular compositions of purified autophagosomes and total cytoplasm [10].

As a rapid way of separating the small fraction of autophagically sequestered probe from the total cytoplasmic pool, the method of *electrodisruption* has been found to be convenient [1, 5]. By submitting cells suspended in a nonionic medium (isotonic sucrose) to a strong electric pulse, the plasma membrane can be completely disrupted, while intracellular organelles (including autophagic vacuoles) remain intact (Fig. 1). The organelles are enmeshed by the cytoskeleton and can be cleanly separated from the leaked-out cytosol by centrifugal sedimentation of the coherent "cell corpses" [5–9].

For measurements of bulk autophagy in isolated rat hepatocytes by the LDH cargo assay, leupeptin has routinely been used to prevent LDH degradation [5, 6], apparently acting both as a direct protease inhibitor and as a suppressant of the fusion between lysosomes and prelysosomal amphisomes [11, 12]. The hepatocyte assay, which is still recommended for the study of bulk autophagy in this cell type, has been described in detail previously [5, 6], and will, therefore, not be further considered here.

However, in a variety of cultured cell lines investigated, we have found leupeptin to be relatively inefficient as an inhibitor of intravacuolar LDH degradation, possibly reflecting its poor cellular uptake. In contrast, bafilomycin A1, a proton pump inhibitor [13], prevents the degradation of sequestered LDH in all cell lines examined [7–9, 14]. In the routine cargo assay for bulk autophagy measurements in cultured cells described below, we have, accordingly, replaced leupeptin with bafilomycin as an inhibitor of LDH degradation. Other inhibitors of intravacuolar degradation such as concanamycin A or ammonia (10 mM NH_4Cl) may also be used.

2 Materials

2.1 Chemicals

1. Accumax (Innovative Cell Technologies).
2. Bafilomycin A1.
3. Imidazole.
4. Pyruvate.
5. NADH.

2.2 Buffers and Solutions

1. Earle's balanced salt solution (EBSS, with 0.1% glucose).
2. Phosphate-buffered saline (PBS).
3. Fetal bovine serum (10% FBS).
4. Bovine serum albumin (BSA).

5. Triton X-405.
6. Tween-20.
7. Phosphate-buffered sucrose: 100 mM sodium monophosphate, 2 mM EDTA, 2 mM DTT, 1.75% sucrose, pH 7.5.
8. Resuspension buffer: 50 mM sodium monophosphate, 1 mM EDTA, 1 mM DTT.
9. Pyruvate-imidazole buffer: 0.75 mM pyruvate, 65 mM imidazole, pH 7.5.
10. NADH-imidazole buffer: 1.8 mM NADH, 65 mM imidazole, pH 7.5.

2.3 Labware

1. 12-well plates.
2. 0.4 cm electroporation cuvette.
3. 1.5 and 2 mL microcentrifuge tubes.

2.4 Instruments

1. Electroporator (e.g., BTX Harvard EMC 630).
2. Multianalyzer (MaxMat PL-II, Erba Diagnostics).

3 Method

1. Seed cells onto plates of any format and let them grow to the desired confluency in the medium chosen (*see* **Note 1**).
2. For experimental incubations, aspirate the medium, rinse the cells once with the medium or buffer to be used in the experiment, and add fresh medium/buffer with the additives required in the experiment (*see* **Note 2**).
3. Incubate the cells (triplicate samples for each condition tested is recommended) for the desired length of time (e.g., 3–4 h) in the presence of bafilomycin A1 (100–200 nM) (*see* **Note 3**). Always include 0-h samples (in fresh medium/buffer only) to define the background of sedimented LDH.
4. At the end of the experiment, aspirate the medium, rinse the cells once with 500 μL PBS, and incubate them at 37 °C with 200 μL Accumax until adherent cells have detached. Add 800 μL PBS/5% BSA and gently pipette the cells to obtain a near single-cell suspension, which is transferred to a microcentrifuge tube and kept on ice (*see* **Note 4**).
5. Centrifuge the cells at ~500 × *g* for 5 min at 4 °C.
6. Aspirate the supernatant, add 400 μL ice-cold 10% sucrose to the pellet, and gently resuspend the cells to a near single-cell suspension (*see* **Note 5**).
7. Transfer the cell suspension to a 0.4-cm electroporation cuvette and discharge a single electric pulse of ~8 ms duration

(800 V, 25 μF, and 400 Ω settings in the BTX electroporator). This electric shock will cause a uniform and selective plasma membrane *electrodisruption*, which converts the cells to permeable *cell corpses* from which cytosolic LDH leaks out (Fig. 1). However, autophagic vacuoles with sequestered LDH remain intact in situ and thus sediment readily along with the cell corpses (*see* **Note 6**).

8. Transfer the cell disruptate (400 μL) to a microcentrifuge tube containing 400 μL ice-cold phosphate-buffered sucrose. Mix by pipetting and store on ice.
9. Repeat **steps** 7 and **8** with each sample (one at a time).
10. To obtain a measure of the total cellular LDH content ("total LDH") in the sample, pipette 100 μL of the diluted disruptate into a microcentrifuge tube and freeze-thaw it once at −80 °C (storage at −80 °C overnight may be convenient).
11. To measure sedimentable LDH, transfer 500 μL of the diluted disruptate to a microcentrifuge tube containing 900 μL ice-cold "resuspension buffer" with 0.5% BSA and 0.01% Tween-20. Mix briefly by pipetting and store on ice.
12. Centrifuge the mixture from the previous step at 18,000 × *g* for 45 min at 4 °C (*see* **Note 7**).
13. Aspirate the supernatant and freeze-thaw the pellet ("sedimented LDH") at −80 °C (*see* **Note 8**).
14. Thaw all frozen-stored samples on ice.
15. Dilute the "total LDH" sample (from **step 10**) with 300 μL ice-cold resuspension buffer containing 1.33% Triton X-405 and resuspend until the solution is homogeneous.
16. Add 400 μL ice-cold resuspension buffer containing 1% Triton X-405 to the "sedimented LDH" pellet (from **step 13**) and resuspend to homogeneity.
17. Centrifuge all samples at 18,000 × *g* for 5 min at 4 °C (to sediment debris).
18. Measure LDH activity in the "sedimented LDH" and "total LDH" samples spectrophotometrically as the decline in NADH absorbance at 340 nm (*see* **Note 9**).
19. Calculate LDH values back to the total cell sample, taking dilutions and sampling into account. Subtract the 0-h "sedimentable LDH" values from the final "sedimentable LDH" values and divide by each sample's "total LDH"/100 to obtain the percentage of cellular LDH rendered sedimentable (i.e., sequestered) during the incubation period. Bulk-autophagic sequestration rates can be conveniently expressed as %/h (*see* **Note 10**).

4 Notes

1. 12-well plates with a growth surface of ~3.8 cm^2 per well (and harvested at ~70% confluency) will usually yield enough cells for the cargo assay. The culture medium should be chosen to suit the cell line and experimental conditions used.
2. To measure ***maximal*** bulk-autophagic activity (i.e., ***capacity***), the cells may be washed and incubated in an amino acid-free buffer such as EBSS (with glucose). Make sure to use bicarbonate-buffered solutions when cells are incubated in a CO_2 incubator, otherwise not.
3. Bafilomycin addition defines the start of the experiment, when intravacuolar LDH degradation is stopped and autophagically sequestered LDH begins to accumulate.
4. Accumax contains DNase, which degrades "sticky" DNA and helps to maintain single-cell suspensions.
5. The supernatant should be aspirated thoroughly by suction to leave the pellet as dry as possible. Most cultured cells will form a well-packed pellet with little carry-over of medium and minimal cell damage, thus with insignificant transfer of ions to the subsequently added nonionic sucrose solution. However, if the cell disruption (*see* **step** 7) is incomplete (<99%), the cells should be washed once in ice-cold 10% sucrose.
6. While standard homogenization and subcellular fractionation methods can, in principle, be used to separate cytosol from the formed elements of the cell, electrodisruption is both an extremely simple and an exceptionally mild procedure which preserves the structure of intracellular organelles perfectly [5].
7. This step may need to be adjusted to suit the cell type or cell line used. Too harsh centrifugation may result in too packed pellets that are difficult too solubilize, whereas too weak centrifugation may cause incomplete cell corpse sedimentation or loose pellets with excessive carry-over of cytosolic LDH. Centrifugation at 18,000 × *g* for 45 min at 4 °C has been shown to be suitable for most cell lines tested with the current protocol.
8. The supernatant should be aspirated thoroughly by suction to leave the pellet as dry as possible.
9. In principle, quantified immunoblots can also be used in this assay, but enzyme activity measurements are more accurate. An autoanalyzer/multianalyzer is very convenient for this purpose; we routinely use the MaxMat instrument. We previously used the accompanying LDH assay kit (RM LADH0126V from Erba Diagnostics), but we now make our own assay solution by mixing four parts of cold pyruvate-imidazole buffer with one part of cold NADH-imidazole buffer to obtain an imidazole working solution containing 0.6 mM pyruvate and

0.36 mM NADH. We found this NADH concentration (twice as high as in the Erba kit) to be necessary for adequate LDH measurements.

10. Bulk-autophagic sequestration rates, measured under amino acid starvation conditions, vary widely between different cell types, with approximate values of, e.g., Huh7, 0.2; U2OS, 0.6; PC3, 0.9; LNCaP, 1.7; and primary rat hepatocytes, 3–5%/h.

Acknowledgments

This work has been generously supported by grants from the Norwegian Cancer Society, the Research Council of Norway, the University of Oslo, the Nansen Foundation, the Blix Family Fund, and Astri and Birger Torsted's Legacy.

Disclosure: No conflict of interest was disclosed.

References

1. Gordon PB, Seglen PO (1982) Autophagic sequestration of [^{14}C]sucrose, introduced into isolated rat hepatocytes by electropermeabilization. Exp Cell Res 142:1–14
2. Seglen PO, Gordon PB (1984) Amino acid control of autophagic sequestration and protein degradation in isolated rat hepatocytes. J Cell Biol 99:435–444
3. Seglen PO, Gordon PB, Tolleshaug H, Høyvik H (1986) Use of [^{3}H]raffinose as a specific probe of autophagic sequestration. Exp Cell Res 162:273–277
4. Høyvik H, Gordon PB, Seglen PO (1986) Use of a hydrolysable probe, [^{14}C]lactose, to distinguish between pre-lysosomal and lysosomal steps in the autophagic pathway. Exp Cell Res 166:1–14
5. Kopitz J, Kisen GØ, Gordon PB, Bohley P, Seglen PO (1990) Non-selective autophagy of cytosolic enzymes in isolated rat hepatocytes. J Cell Biol 111:941–953
6. Seglen PO, Øverbye A, Sætre F (2009) Sequestration assays for mammalian autophagy. Methods Enzymol 452:63–83
7. Engedal N, Torgersen ML, Guldvik IJ, Barfeld SJ, Bakula D, Sætre F, Hagen LK, Patterson JB, Proikas-Cezanne T, Seglen PO, Simonsen A, Mills IG (2013) Modulation of intracellular calcium homeostasis blocks autophagosome formation. Autophagy 9:1475–1490
8. Seglen PO, Luhr M, Mills IG, Sætre F, Szalai P, Engedal N (2015) Macroautophagic cargo sequestration assays. Methods 75:25–36
9. Luhr M, Szalai P, Sætre F, Gerner L, Seglen PO, Engedal N (2017) A simple cargo sequestration assay for quantitative measurement of nonselective autophagy in cultured cells. Methods Enzymol 587:351–364
10. Strømhaug PE, Berg TO, Fengsrud M, Seglen PO (1998) Purification and characterization of autophagosomes from rat hepatocytes. Biochem J 335:217–224
11. Berg TO, Fengsrud M, Strømhaug PE, Berg T, Seglen PO (1998) Isolation and characterization of rat liver amphisomes: Evidence for fusion of autophagosomes with both early and late endosomes. J Biol Chem 273:21883–21892
12. Øverbye A, Sætre F, Hagen LK, Johansen HT, Seglen PO (2011) Autophagic activity measured in whole rat hepatocytes as the accumulation of a novel BHMT fragment (p10) generated in amphisomes by the asparaginyl proteinase, legumain. Autophagy 7:1011–1027
13. Klionsky DJ, Elazar Z, Seglen PO, Rubinsztein DC (2008) Does bafilomycin A_1 block the fusion of autophagosomes with lysosomes? Autophagy 4:849–850
14. Szalai P, Hagen LK, Sætre F, Luhr M, Sponheim M, Øverbye A, Mills IG, Seglen PO, Engedal N (2015) Autophagic bulk sequestration of cytosolic cargo is independent of LC3, but requires GABARAPs. Exp Cell Res 333:21–38

Chapter 21

Methods to Detect Loss of Lysosomal Membrane Integrity

Sonja Aits

Abstract

Loss of lysosomal membrane integrity, often referred to as lysosomal membrane permeabilization (LMP), occurs in many instances of cell death either as an initiating or as an amplifying event. Currently, the best method for detecting LMP is the galectin puncta formation assay which can be used for a broad range of sample types, both fixed and live, is easy to perform, and highly sensitive. This method, which is similar to the widely used LC3 puncta formation assay for autophagy, is based on the translocation of galectins to damaged lysosomes resulting in a change from uniform to punctate staining pattern. Here, we provide protocols for the galectin puncta formation assay in fixed and live cells and for an alternative assay based on fluorescent dextran release from damaged lysosomes, which can be performed in parallel.

Key words Lysosome, Lysosomal membrane permeabilization, Cell death, Immunocytochemistry, Galectin, EGFP-galectin-3, Lysosome-associated membrane protein, Dextran, Microscopy

1 Introduction

Lysosomes are the central degradative organelle of the cell which receives cargo from processes such as autophagy, endocytosis, and phagocytosis. The material is then broken down by over 50 lysosomal hydrolases and released for reuse or antigen presentation [1–3]. In case of lysosomal membrane permeabilization (LMP), lysosomal hydrolases such as cathepsin proteases leak into the cytosol where they can cause cell death through their direct action or by activating secondary apoptosis or other death pathways [2, 4–7]. LMP can result from direct damage to the lysosomes, e.g., by lysosome-targeted drugs or pathogens [4]. Alternatively, LMP occurs as a death-amplifying secondary event in all major forms of cell death, including apoptosis, creating positive feedback loops [4, 8]. LMP thus contributes to cell death induced by a broad range of stimuli, including anticancer drugs, oxidative stress, death-receptor activation, and pathogens, and seems to play a role in many diseases including neurodegeneration, stroke, liver disease, and infections [4, 5].

Nicholas Ktistakis and Oliver Florey (eds.), *Autophagy: Methods and Protocols*, Methods in Molecular Biology, vol. 1880, https://doi.org/10.1007/978-1-4939-8873-0_21,

In the past, the study of lysosomal involvement in cell death has been difficult because available methods for detecting LMP had poor sensitivity or were prone to artifacts. Permeabilized lysosomes often release only part of their content, and although this is sufficient to cause cell death, limited damage is usually impossible to detect by immunostaining for lysosomal enzymes or label-free electron microscopy [9–11]. Monitoring LMP by the release of preloaded fluorescent dextran or by electron microscopy of BSA-gold has higher sensitivity [10–12]. However, the loading of lysosomes with such molecules may disturb lysosomal function and thereby lead to artifacts. Assays that measure lysosomal enzyme activity in the cytosol are even more prone to artifacts because alterations in lysosomal membrane composition affect the fractionation procedure [11]. Equally problematic is the assessment of lysosomal integrity with lysosomal dyes such as acridine orange or LysoTracker because their staining is altered by changes in lysosomal pH, which can occur independently of LMP [13]. Caution is also advised when using protease inhibitors to determine which pathway is responsible for cell death because blockage of one pathway can lead to activation of another and because many inhibitors lack specificity (e.g. the widely used "caspase inhibitor" zVAD-fmk also inhibits lysosomal cathepsin proteases [14]).

The advent of the galectin puncta formation assay, described in the protocols below (adapted from [10, 11]), has overcome many of these difficulties. This method is based on the observation that several galectins (galectin-1, galectin-3, galectin-8, and galectin-9 and possibly additional galectins), which are small sugar-binding proteins, accumulate on permeabilized lysosomes because they gain access to the glycocalyx that covers the inside of the lysosomal membrane [10, 15]. Subsequently, galectins promote the removal of the damaged lysosomes by autophagy (lysophagy) [15–20]. By expressing fluorescent galectin or by immunostaining, the galectin accumulation can be visualized as a change from uniform to punctate staining pattern (similar to LC3 during autophagy) with puncta representing permeabilized lysosomes [10].

The biggest advantage of the galectin translocation assay is that it enables the detection of individual permeabilized lysosomes for the first time. It is also more sensitive than other methods for detecting LMP, easy to perform, and suitable for both fixed (Method A, described in section 3.1, Fig. 1) and live cells (Method B, described in section 3.2, Fig. 2), tissues, and whole model organisms [10]. Furthermore, it is well suited for automated image analysis and high-throughput microscopy screening [10, 21].

Galectins also accumulate on organelles related to lysosomes such as permeabilized phago- and endosomes [15, 16, 22–26]. Consequently, it is best to determine the identity of observed galectin puncta, which can easily be done by co-staining for lysosomal proteins (Method A, Fig. 1) [10]. If a second LMP

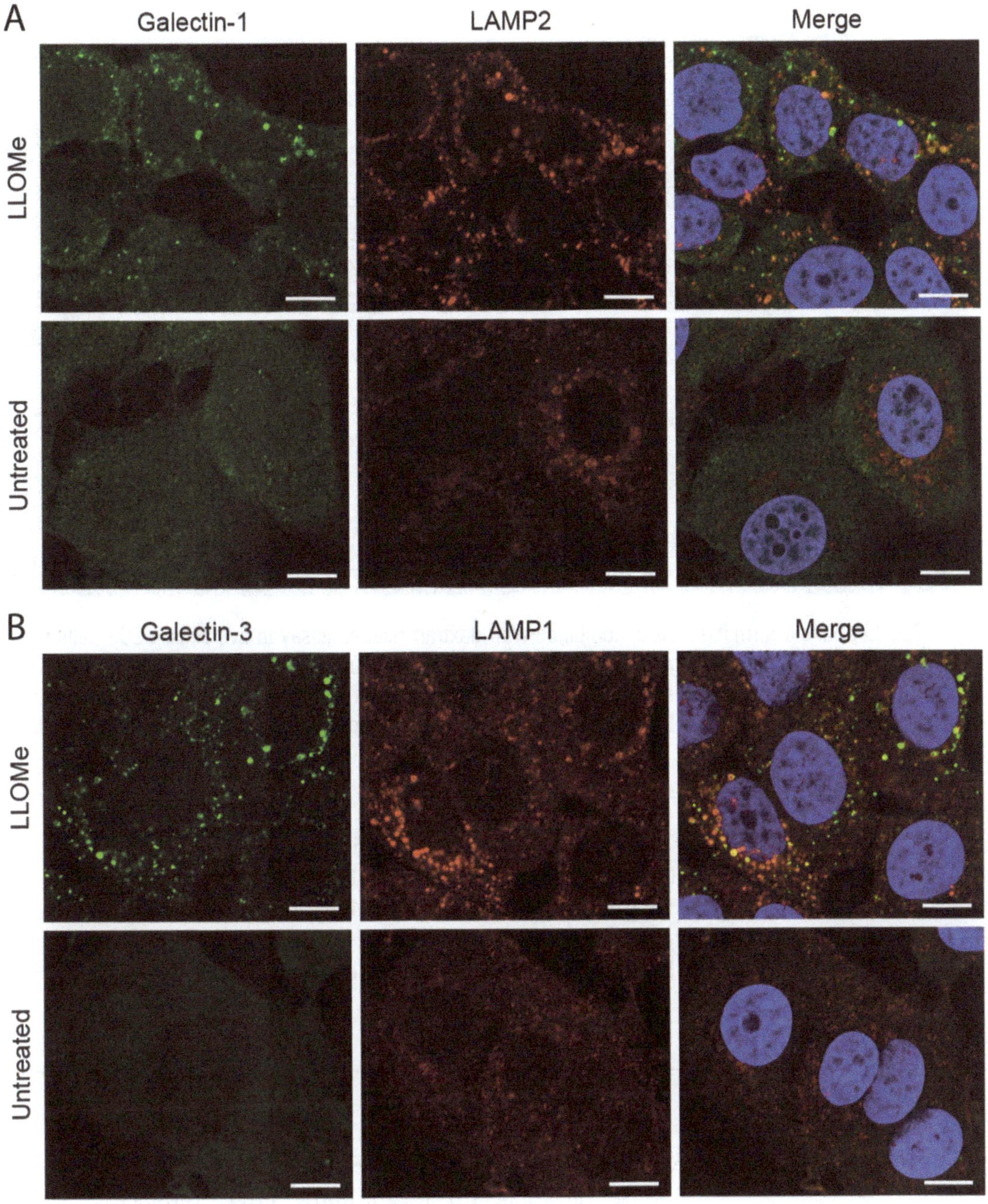

Fig. 1 Galectin puncta formation assay with co-staining for lysosomal markers in fixed cells. MCF7 cells were treated with 2 mM LLOMe for 2 h, fixed and co-stained with rabbit anti-galectin-1 and mouse anti-LAMP2 antibodies (**A**) or mouse anti-galectin-3 and rabbit anti-LAMP1 antibodies (**B**). Nuclei were stained with Hoechst 33342 (shown in merged image only). Size bars 10 μm

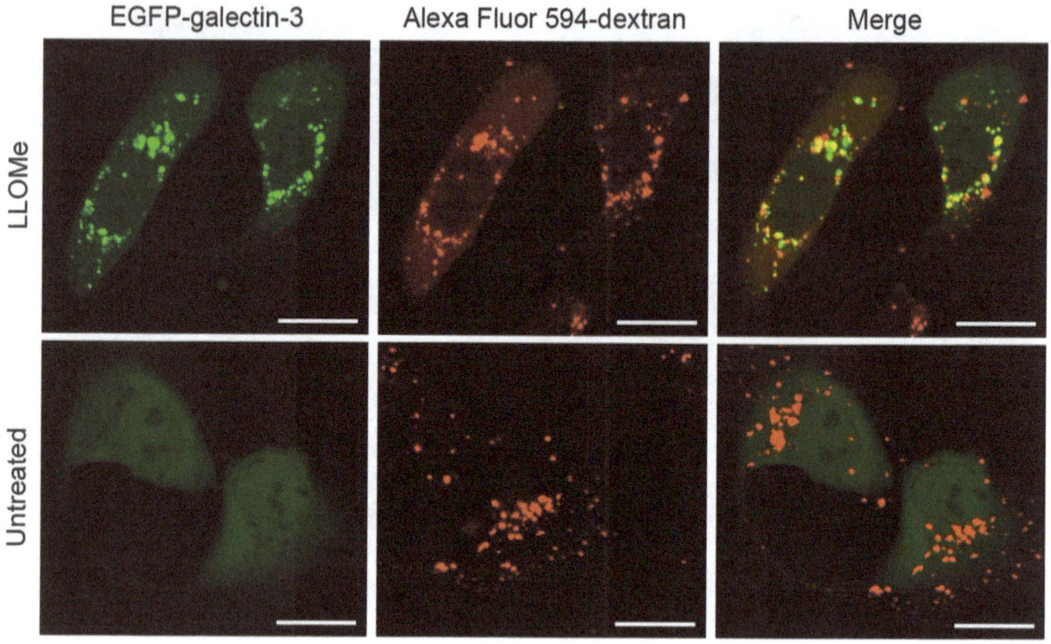

Fig. 2 Galectin puncta formation assay combined with dextran release assay in live cells. U2OS cells were transiently transfected with EGFP-galectin-3 and loaded with Alexa Fluor 594 dextran, 10,000 MW overnight. After a 2 h chase period to allow dextran to reach the lysosomes, untreated cells (LMP negative control) and cells treated with 2 mM LLOMe for 2 h (LMP positive control) were imaged in a confocal microscope. Size bars 20 μm

assay is desired to confirm key findings, the dextran release assay is a good choice especially since it can be performed in parallel with the galectin puncta formation assay (Method B, Fig. 2) [10, 11]. Alternatively, galectin staining can be combined with cathepsin staining, albeit keeping in mind the lower sensitivity of the latter for detecting LMP [10, 11, 27].

2 Materials

2.1 Method A: Galectin Puncta Formation Assay with Co-staining for Lysosomal Markers in Fixed Cells

2.1.1 Antibodies (See Note 1)

1. Rabbit anti-galectin-1 antibody (Abcam, cat. no. ab25138).
2. Mouse anti-galectin-3 antibody (BD Biosciences, cat. no. 556904).
3. Rabbit anti-LAMP1 antibody (Abcam, cat. no. ab24170).
4. Mouse anti-LAMP2 antibody (H4B4, concentrate, Developmental Studies Hybridoma Bank).
5. Alexa 488 and 594-donkey anti-mouse antibody (Life Technologies, cat. no. A21202 and A21203).
6. Alexa 488 and 594-donkey anti-rabbit antibody (Life Technologies, cat. no. A21206 and A21207).

2.1.2 Other Reagents and Materials

1. Cells grown on microscopy chamber slides with detachable chambers or on glass coverslips placed into wells of 24-well plate (*see* **Note 2**).
2. 1 M H-Leu-Leu-OMe hydrochloride (LLOMe, PubChem CID: 15598024) in ultrafiltered (Mili-Q) water (store at −80 °C; *see* **Note 3**).
3. 4% Formaldehyde, freshly diluted in PBS from 37% stock.
4. PBS.
5. 50 mM ammonium chloride in PBS (store at 4 °C).
6. Goat serum.
7. Buffer 1: 1% BSA, 0.3% Triton X-100 in PBS (store at −20 °C).
8. Buffer 2: 0.25% BSA, 0.1% Triton X-100 in PBS (store at 4 °C).
9. Buffer 3: 0.05% Tween-20 in PBS (store at 4 °C).
10. 25 mg/mL Hoechst 33342 in ultrafiltered (Mili-Q) water (store at −20 °C in the dark).
11. Mounting medium.
12. Rectangular coverslips (when using chamber slides) or microscopy glass slides (when using coverslips).

2.1.3 Equipment

1. Vacuum suction.
2. Only if using coverslips: large Petri dish with bottom covered in Parafilm.
3. Plastic box lined with moist tissues (sufficiently large to hold the samples).
4. Fine tweezers (bent models work best).
5. Widefield fluorescent or confocal microscope with 20× or higher magnification objective and blue, red, and green channel.

2.2 Method B: Galectin Puncta Formation Assay Combined with Dextran Release Assay in Live Cells

1. Cells Transiently or Stably Expressing EGFP-Galectin-3 and Matching Cells Without It (*see* **Note 10**).
2. Microscopy chamber slides.
3. 5 mg/mL Alexa Fluor 594-dextran 10,000 MW (anionic, fixable) in PBS (store at 4 °C, *see* **Note 11**).
4. Phenol red-free medium with required additives (serum, etc.).
5. 25 mg/mL Hoechst 33342 in ultrafiltered (Mili-Q) water (store at −20 °C in the dark).
6. Inverted widefield fluorescent or confocal microscope with 20× or higher magnification objective and blue, red, and green channel, preferably with heating and CO_2.

3 Methods

3.1 Galectin Puncta Formation Assay with Co-staining for Lysosomal Markers in Fixed Cells

Cells should not be allowed to dry out at any step. Use sufficient volumes so that cells are always fully covered, and complete liquid removal and addition of next solution for one sample before going to the next. Liquid can be removed with vacuum suction placing the tip to the edge of the sample. All steps are performed at room temperature unless indicated otherwise.

1. Treat cells as desired, preparing two replicates per condition. Include the following control samples: two replicates with untreated cells (LMP negative control) and eight replicates with cells treated with 2 mM LLOMe for 2 h (two for LMP positive control and six for antibody controls) (*see* **Note 3**).
2. Remove medium and add 4% formaldehyde fixative solution (*see* **Note 4**). Incubate for 10 min.
3. Remove fixative and add PBS to wash cells (*see* **Note 5**).
4. Remove PBS and add ammonium chloride solution. Incubate for 10 min to quench remaining aldehyde. Only for coverslips: to minimize the required volume for this and the subsequent steps, transfer coverslips to a Parafilm-covered Petri dish with cells facing up. 90 μL solution can then be carefully added onto each coverslip to form a drop which is held in place by the liquid-repelling Parafilm (*see* **Note 6**).
5. Remove ammonium chloride and wash cells twice in PBS.
6. Remove PBS and add Buffer 1 with 5% goat serum. Incubate for 20 min to block and permeabilize.
7. Prepare primary antibody mixtures in Buffer 1 according to Table 1 (*see* **Note 7**).
8. Remove blocking solution and add appropriate antibody mixtures. Transfer chamber slides or Parafilm-covered Petri dish with coverslips to plastic box lined with moist tissue to prevent samples from drying out and incubate overnight at 4 °C.
9. Remove antibody solutions and wash cells 3 × 5 min in Buffer 2.
10. Centrifuge secondary antibodies 2 min at 20000 × *g* to pellet fluorescent particles which could be mistaken for galectin dots. Take liquid from the top of the supernatant, and prepare secondary antibody mixtures in Buffer 2 according to Table 1 (*see* **Note 7**).
11. Remove wash solution from cells and add appropriate antibody mixtures. Incubate for 1 h in the dark.
12. Remove antibody solutions, and wash cells 3 × 5 min in Buffer 3 and once briefly in PBS.
13. Dilute Hoechst 33342 1:5000 in PBS.

Table 1
Antibody mixtures

Sample	Primary antibody mixture	Secondary antibody mixture
Samples of interest, LMP negative control, LMP positive control	Replicate 1: Rabbit anti-galectin-1 (1 μg/mL) + mouse anti-LAMP2 (0.9 μg/mL) Replicate 2: Mouse anti-galectin-3 (5–10 μg/mL) + rabbit anti-LAMP1 (1 μg/mL)	Replicate 1: Alexa 488 anti-rabbit (2 μg/mL) + Alexa 594 anti-mouse (2 μg/mL) Replicate 2: Alexa 488 anti-mouse (2 μg/mL) + Alexa 594 anti-rabbit (2 μg/mL)
No primary antibody (autofluorescence) control	Replicates 1 and 2: Only Buffer	Replicate 1: Alexa 488 anti-rabbit (2 μg/mL) + Alexa 594 anti-mouse (2 μg/mL) Replicate 2: Alexa 488 anti-mouse (2 μg/mL) + Alexa 594 anti-rabbit (2 μg/mL)
Single antibody control—Galectin-1	Rabbit anti-galectin-1 (1 μg/mL)	Alexa 488 anti-rabbit (2 μg/mL)
Single antibody control—Galectin-3	Mouse anti-galectin-3 (5–10 μg/mL)	Alexa 488 anti-mouse (2 μg/mL)
Single antibody control—LAMP1	Rabbit anti-LAMP1 (1 μg/mL)	Alexa 594 anti-rabbit (2 μg/mL)
Single antibody control—LAMP2	Mouse anti-LAMP2 (0.9 μg/mL)	Alexa 594 anti-mouse (2 μg/mL)

14. Remove wash solution from cells and add Hoechst solution. Incubate for 10 min to stain nuclei (*see* **Note 8**).
15. Remove Hoechst solution and wash cells twice in PBS.
16. If using chamber slides, remove detachable chambers and glue from one slide at a time, add small drops of mounting medium, and cover with rectangular coverslip. If using coverslips, mount one at a time: add a small drop of mounting medium (~7 μL) onto a glass slide, pick up coverslip with tweezers, carefully dip the edge of the coverslip on tissue to remove excess liquid, and then place it on the mounting medium drop with the cell side downward.
17. Let the mounting medium solidify for 16–24 h in the dark. Afterwards, store slides at −20 °C in the dark.
18. Set imaging parameters with the help of the antibody control samples, and acquire images in the blue (nuclei), red (lysosomes), and green (galectin) channel with a fluorescence or confocal microscope. To avoid bias, acquire images in predetermined positions or select fields while looking only at the blue channel.
19. To determine the extent of LMP in each sample, quantify the number and/or combined area of galectin puncta per cell and the percentage of cells scoring positive for punctate galectin staining (*see* **Note 9**). In addition, verify that galectin puncta correspond to LAMP-labelled lysosomes.

3.2 Method B: Galectin Puncta Formation Assay Combined with Dextran Release Assay in Live Cells

If live imaging is desired or endogenous galectin levels are too low for method A, the galectin puncta formation assay can be performed by stably or transiently expressing fluorescently labelled galectin in the cells [10, 15, 20]. This can even be combined with the dextran release assay as described here in live cells to confirm LMP with two assays (Fig. 2) [10].

All incubations should be performed at 37 °C in 5% CO_2 atmosphere.

1. Plate cells expressing EGFP-galectin-3 into chamber slides including four extra samples: two samples for microscope setup and one each as LMP positive and negative control (*see* **Note 12**). In addition, plate one sample of same cells without the fluorescent construct (unstained control). Incubate overnight so cells can adhere.
2. Add dextran at a final concentration of 100 μg/mL to the medium of all samples except the unstained control. Incubate for 18 h (*see* **Note 13**).
3. Remove dextran-containing medium, and wash cells twice with phenol red-free medium (*see* **Note 14**).

4. Remove wash medium and add fresh medium. Incubate for 2 h (chase period) to allow dextran to reach the lysosomal compartment.
5. 1 h after the start of the chase period, add 2 mM LLOMe to one of the microscope setup samples. Leave the second microscope setup sample untreated.
6. 30 min before the end of the chase period, add Hoechst 33342 at a final concentration of 8 μg/mL to the medium of all samples except the unstained control (*see* **Note 15**).
7. Set up imaging parameters for the blue (nuclei), red (dextran), and green (galectin) channel in an inverted fluorescence or confocal microscope using the LLOMe-treated and untreated microscope setup samples and the unstained control sample (*see* **Note 16**).
8. Acquire images of all experimental samples including LMP positive and negative control before treatment (*see* **Note 17**).
9. Treat cells as desired. The LMP negative control sample should be left untreated and the LMP positive control sample treated with 2 mM LLOMe (*see* Method A and **Note 18**).
10. Acquire images or time-lapse series of the treated cells (*see* **Note 19**). To avoid bias, acquire images in predetermined positions, or select fields while looking only at the blue channel.
11. To determine the extent of LMP in each sample, quantify the number and/or combined area of galectin puncta per cell, the percentage of cells scoring positive for punctate galectin pattern, and the percentage of cells with released dextran (*see* Method A and **Note 20**).

4 Notes

1. The antibodies can be replaced with different antibodies binding the same proteins, e.g. rat or rabbit anti-galectin-3 antibodies [10, 16, 28].
2. Plate sufficient cells so that confluency in control samples at the time of fixation is approximately 80%.
3. LLOMe stock solution can be added directly to the medium of the samples or prediluted in water. Stock solution can be stored at −20 °C for some weeks and freeze-thawed several times without significant loss of activity. In the lysosomes LLOMe is converted to a detergent by cathepsin C causing rapid damage to the lysosomal membrane [29–31]. Treatment time and concentration of LLOMe may need to be adjusted depending on the cell type. Choose treatment conditions in which most lysosomes (LAMP positive puncta) are marked with galectins

to reveal the maximal response. In general, LLOMe treatment times in the LMP positive control should not exceed 6 h because LLOMe may kill the cells and because galectin puncta in cells treated with sublethal concentrations disappear over time [10]. In cells with low cathepsin C expression, LLOMe should be replaced with another lysosome-damaging agent, e.g., terfenadine, sphingosine, or photosensitizers [10, 17, 32–34].

4. Galectin staining can also be performed after fixation with ice-cold methanol for 2 min [10]. In this case, **step 4** can be left out.

5. If a break in the procedure is desired, wells can be filled with PBS after this step and samples stored at 4 °C for several weeks. Evaporation should be prevented (e.g. with Parafilm or cling wrap).

6. When covering the Petri dish with Parafilm, ensure that it is completely flat. Instead of a Petri dish, a glass tile or other flat surface that is small enough to fit into the plastic box can also be used. Transfer coverslips quickly and one at a time with tweezers, and add the ammonium chloride solution before transferring the next coverslip so that cells do not dry out. Keep 2 cm distance between coverslips and remove any liquid between them to keep liquid drops in place and avoid solutions from one coverslip running into the next.

7. Antibody dilutions may have to be adjusted depending on the cell type and microscope. If galectin expression levels in the cells are unknown, initial experiments should be performed with both galectin-1 and galectin-3 as expression differs significantly (*see* [10] for comparison of expression in different tissues). In subsequent experiments, it is sufficient to use the galectin with which the assay performed best. In previous studies, galectin-1 antibodies have been used for human fibroblasts, A549, HCT11, HeLa, MDA-PATC53, NCI-H1299 and U2OS cells, and galectin-3 antibodies for astrocytes, murine embryonic fibroblasts, HeLa, HepG2, JIMT-1, MCF7, MDA-PATC53, THP-1, and U2OS cells [10, 16, 35–40]. Once the identity of the observed galectin puncta is confirmed as lysosomes, LAMP antibodies can be omitted in subsequent experiments.

8. Hoechst 33342 can be substituted with another nuclear dye, e.g. DAPI.

9. LAMPs mark the lysosomal membrane, and lysosomes appear as small puncta or rings. The LAMP co-staining is used to verify that galectin puncta indeed correspond to lysosomes, which should be done at some point in each study but not necessarily in all individual experiments. In the LMP negative control

sample, galectin staining should be uniform with no or very few puncta. In the LMP positive control sample, galectin staining should be punctate with most LAMP-labelled lysosomes also marked by galectin (Fig. 1). Frequently, the cytosol remains stained as diffuse background as not all galectin translocates to the permeabilized lysosomes. If galectin puncta are not localized to lysosomes, they may represent damaged endosomes or phagosomes which can be verified using appropriate markers [10, 15, 22, 23, 25, 26]. If few or no galectin puncta are observed in the LMP positive control sample, increase LLOMe concentration, try an alternative LMP-inducing drug (*see* **Note 3**), and verify that galectins are detected correctly using galectin siRNA as described [10]. Generally, galectin puncta should be quantified in at least 50 cells per sample by manual counting or image analysis software (e.g., CellProfiler, freely available from cellprofiler.org). As lysosomes often aggregate in the perinuclear area, it can be difficult to discern individual lysosomes, especially at 20× magnification. In this case, it may be more suitable to quantify the combined area of all galectin puncta per cell rather than the number of galectin puncta [21]. To help reveal heterogeneity in the cell population, the percentage of cells showing an LMP phenotype should also be quantified. A threshold for what constitutes LMP positivity (e.g. >2 galectin puncta) can be set by examining the LMP negative control sample in which most cells should score as negative. To determine the extent to which galectins have translocated to the LAMP-stained lysosomes, co-localization coefficients should be determined after setting thresholds with the single antibody controls [10].

10. EGFP-galectin-3 can be replaced with constructs containing different fluorescent tags or galectins (galectin-1, galectin-8, and galectin-9) [10, 15]. However, tags may interfere slightly with galectin-1 translocation due to its small size (14 kDa) [10]. The fluorescent galectin construct can be expressed transiently or stably using standard methods. Transfection induces some LMP, and experiments should be performed after this has been cleared (usually after 48 h). To monitor the removal of permeabilized lysosomes by lysophagy, a tandem mRFP-GFP-galectin-3 construct can be used [20, 28].

11. Dextran is taken up by the cells and accumulates in the lysosomes. Minor LMP may only release 10 kDa dextran from the lysosomes but not the larger cathepsins. Different dextran sizes can be used to determine size limitations in the observed LMP [11, 41]. If no second LMP assay is required, EGFP-galectin-3 cells should be plated in phenol red-free medium, and **steps 2–4** can be left out.

12. Plate sufficient cells so that confluency in the control samples at the end of imaging is approximately 80%. The unstained sample and each one of the LMP positive and negative control samples is used to set up the imaging parameters in **step** 7.
13. Ensure dextran solution is well mixed by pipetting up and down to obtain equal loading between samples. Dextran concentration and incubation period may need to be adjusted depending on the uptake efficiency of the cells and microscope sensitivity. 50–200 μg/mL and 16–24 h are generally suitable. Shorter incubation (e.g., 4 h) results in reduced staining but might be sufficient in some cases. Dextran can also be added immediately after cell plating to save time with incubation taking place, while cells adhere overnight.
14. From this step onward, phenol red-free medium (preferably same formulation as standard culturing medium) should be used as phenol red can increase background fluorescence.
15. Hoechst 33342 is added only if labelling of the nuclei is desired. For imaging over long periods, Hoechst 33342 should be used at the lowest possible concentration to avoid toxicity, e.g., at a final concentration of 0.01 μg/mL [42]. With such low concentration, maximal staining may only be achieved after 2 h incubation.
16. Extra samples are used to set up the imaging parameters (exposure time, autofocus, etc.) because repeated focusing and exposure cause cell stress. In subsequent experiments in which the microscope settings are reused, these samples and **steps 5** and 7 can be omitted.
17. This is to verify equal dextran loading across samples and record the state of the cells before treatment so that changes observed afterwards can be compared to this.
18. If images are to be recorded at one fixed endpoint, 2 h is usually a suitable time point for the LLOMe-treated LMP positive control (*see* Method A). If time-lapse imaging is to be performed, imaging of the LMP positive control can be started immediately after LLOMe addition as galectin puncta usually become visible within a few minutes [10].
19. If toxicity from repeated autofocusing is observed during time-lapse imaging, it can be beneficial to autofocus in the bright-field channel instead of a fluorescent channel. If a microscope with heating and CO_2 control is unavailable, imaging can usually be performed up to 10 min before the cells become stressed. In this case medium should be pre-warmed, and samples which are not yet being imaged should remain in the incubator. If several samples are taken out at the same time because they are on the same slide, include two LMP negative control samples, one imaged at the beginning and one at the

end, to verify that the lysosomes have not become damaged during the duration of the imaging. Alternatively, if a time-course is not needed, cells can be treated as desired and then fixed as described in Method A before imaging. However, fixation alters dextran staining in some cases, and this should be excluded in a small pilot experiment comparing the same sample before and after fixation.

20. In healthy cells, dextran should be confined to the lysosomes which is reflected in a punctate staining pattern without background (Fig. 2). Cells with cytosolic dextran staining should be scored as positive for LMP even if the release is incomplete. Minor release can be difficult to detect, and galectin puncta usually become visible before dextran release.

Acknowledgments

The author is supported by the Danish Cancer Society, the Swedish Research Council, the Swedish Brain Foundation, the Längmanska Cultural Fund, the Einar Willumsen Foundation, the Segerfalk Foundation, the Sigurd and Elsa Golje's Memorial Foundation and the Thora and Viggo Groves Memorial Fund.

References

1. Luzio JP, Pryor PR, Bright NA (2007) Lysosomes: fusion and function. Nat Rev Mol Cell Biol 8(8):622–632. https://doi.org/10.1038/nrm2217
2. Mrschtik M, Ryan KM (2015) Lysosomal proteins in cell death and autophagy. FEBS J 282 (10):1858–1870. https://doi.org/10.1111/febs.13253
3. Schroder BA, Wrocklage C, Hasilik A, Saftig P (2010) The proteome of lysosomes. Proteomics 10(22):4053–4076. https://doi.org/10.1002/pmic.201000196
4. Aits S, Jäättelä M (2013) Lysosomal cell death at a glance. J Cell Sci 126(Pt 9):1905–1912. https://doi.org/10.1242/jcs.091181
5. Gomez-Sintes R, Ledesma MD, Boya P (2016) Lysosomal cell death mechanisms in aging. Ageing Res Rev 32:150–168. https://doi.org/10.1016/j.arr.2016.02.009
6. Repnik U, Stoka V, Turk V, Turk B (2012) Lysosomes and lysosomal cathepsins in cell death. Biochim Biophys Acta 1824(1):22–33. https://doi.org/10.1016/j.bbapap.2011.08.016
7. Appelqvist H, Waster P, Kagedal K, Ollinger K (2013) The lysosome: from waste bag to potential therapeutic target. J Mol Cell Biol 5 (4):214–226. https://doi.org/10.1093/jmcb/mjt022
8. Vanden Berghe T, Vanlangenakker N, Parthoens E, Deckers W, Devos M, Festjens N, Guerin CJ, Brunk UT, Declercq W, Vandenabeele P (2010) Necroptosis, necrosis and secondary necrosis converge on similar cellular disintegration features. Cell Death Differ 17(6):922–930. https://doi.org/10.1038/cdd.2009.184
9. Brunk UT, Ericsson JL (1972) Cytochemical evidence for the leakage of acid phosphatase through ultrastructurally intact lysosomal membranes. Histochem J 4(6):479–491
10. Aits S, Kricker J, Liu B, Ellegaard AM, Hamalisto S, Tvingsholm S, Corcelle-Termeau E, Hogh S, Farkas T, Holm Jonassen A, Gromova I, Mortensen M, Jaattela M (2015) Sensitive detection of lysosomal membrane permeabilization by lysosomal galectin puncta assay. Autophagy 11 (8):1408–1424. https://doi.org/10.1080/15548627.2015.1063871
11. Aits S, Jaattela M, Nylandsted J (2015) Methods for the quantification of lysosomal membrane permeabilization: a hallmark of

lysosomal cell death. Methods Cell Biol 126:261–285. https://doi.org/10.1016/bs.mcb.2014.10.032

12. Repnik U, Cesen MH, Turk B (2016, 2016) Studying lysosomal membrane permeabilization by analyzing the release of preloaded bsa-gold particles into the cytosol. Cold Spring Harb Protoc (6). https://doi.org/10.1101/pdb.prot087122
13. Repnik U, Cesen MH, Turk B (2016, 2016) The use of lysosomotropic dyes to exclude lysosomal membrane permeabilization. Cold Spring Harb Protoc (5). https://doi.org/10.1101/pdb.prot087106
14. Schotte P, Declercq W, Van Huffel S, Vandenabeele P, Beyaert R (1999) Non-specific effects of methyl ketone peptide inhibitors of caspases. FEBS Lett 442 (1):117–121
15. Thurston TL, Wandel MP, von Muhlinen N, Foeglein A, Randow F (2012) Galectin 8 targets damaged vesicles for autophagy to defend cells against bacterial invasion. Nature 482 (7385):414–418. https://doi.org/10.1038/nature10744
16. Chauhan S, Kumar S, Jain A, Ponpuak M, Mudd MH, Kimura T, Choi SW, Peters R, Mandell M, Bruun JA, Johansen T, Deretic V (2016) TRIMs and galectins globally cooperate and TRIM16 and galectin-3 co-direct autophagy in endomembrane damage homeostasis. Dev Cell 39(1):13–27. https://doi.org/10.1016/j.devcel.2016.08.003
17. Hung YH, Chen LM, Yang JY, Yang WY (2013) Spatiotemporally controlled induction of autophagy-mediated lysosome turnover. Nat Commun 4:2111. https://doi.org/10.1038/ncomms3111
18. Kim BW, Hong SB, Kim JH, Kwon DH, Song HK (2013) Structural basis for recognition of autophagic receptor NDP52 by the sugar receptor galectin-8. Nat Commun 4:1613. https://doi.org/10.1038/ncomms2606
19. Li S, Wandel MP, Li F, Liu Z, He C, Wu J, Shi Y, Randow F (2013) Sterical hindrance promotes selectivity of the autophagy cargo receptor NDP52 for the danger receptor galectin-8 in antibacterial autophagy. Sci Signal 6(261):ra9. https://doi.org/10.1126/scisignal.2003730
20. Maejima I, Takahashi A, Omori H, Kimura T, Takabatake Y, Saitoh T, Yamamoto A, Hamasaki M, Noda T, Isaka Y, Yoshimori T (2013) Autophagy sequesters damaged lysosomes to control lysosomal biogenesis and kidney injury. EMBO J 32(17):2336–2347. https://doi.org/10.1038/emboj.2013.171
21. Pagliero RJ, D'Astolfo DS, Lelieveld D, Pratiwi RD, Aits S, Jaattela M, Martin NI, Klumperman J, Egan DA (2016) Discovery of small molecules that induce Lysosomal cell death in cancer cell lines using an image-based screening platform. Assay Drug Dev Technol 14(8):489–510. https://doi.org/10.1089/adt.2016.727
22. Paz I, Sachse M, Dupont N, Mounier J, Cederfur C, Enninga J, Leffler H, Poirier F, Prevost MC, Lafont F, Sansonetti P (2010) Galectin-3, a marker for vacuole lysis by invasive pathogens. Cell Microbiol 12 (4):530–544. https://doi.org/10.1111/j.1462-5822.2009.01415.x
23. Chen X, Khambu B, Zhang H, Gao W, Li M, Yoshimori T, Yin XM (2014) Autophagy induced by calcium phosphate precipitates targets damaged endosomes. J Biol Chem 289 (16):11162–11174. https://doi.org/10.1074/jbc.M113.531855
24. Dupont N, Lacas-Gervais S, Bertout J, Paz I, Freche B, Van Nhieu GT, van der Goot FG, Sansonetti PJ, Lafont F (2009) Shigella phagocytic vacuolar membrane remnants participate in the cellular response to pathogen invasion and are regulated by autophagy. Cell Host Microbe 6(2):137–149. https://doi.org/10.1016/j.chom.2009.07.005
25. Maier O, Marvin SA, Wodrich H, Campbell EM, Wiethoff CM (2012) Spatiotemporal dynamics of adenovirus membrane rupture and endosomal escape. J Virol 86 (19):10821–10828. https://doi.org/10.1128/JVI.01428-12
26. Ray K, Bobard A, Danckaert A, Paz-Haftel I, Clair C, Ehsani S, Tang C, Sansonetti P, Tran GV, Enninga J (2010) Tracking the dynamic interplay between bacterial and host factors during pathogen-induced vacuole rupture in real time. Cell Microbiol 12(4):545–556. https://doi.org/10.1111/j.1462-5822.2010.01428.x
27. Giraldo AMV, Ollinger K, Loitto V (2017) Microscopic analysis of lysosomal membrane permeabilization. Methods Mol Biol 1594:73–92. https://doi.org/10.1007/978-1-4939-6934-0_5
28. Hasegawa J, Maejima I, Iwamoto R, Yoshimori T (2015) Selective autophagy: lysophagy. Methods 75:128–132. https://doi.org/10.1016/j.ymeth.2014.12.014
29. Uchimoto T, Nohara H, Kamehara R, Iwamura M, Watanabe N, Kobayashi Y (1999) Mechanism of apoptosis induced by a lysosomotropic agent, L-Leucyl-L-Leucine methyl ester. Apoptosis 4(5):357–362

30. Thiele DL, Lipsky PE (1985) Regulation of cellular function by products of lysosomal enzyme activity: elimination of human natural killer cells by a dipeptide methyl ester generated from L-leucine methyl ester by monocytes or polymorphonuclear leukocytes. Proc Natl Acad Sci U S A 82(8):2468–2472
31. Thiele DL, Lipsky PE (1990) Mechanism of L-leucyl-L-leucine methyl ester-mediated killing of cytotoxic lymphocytes: dependence on a lysosomal thiol protease, dipeptidyl peptidase I, that is enriched in these cells. Proc Natl Acad Sci U S A 87(1):83–87
32. Kagedal K, Zhao M, Svensson I, Brunk UT (2001) Sphingosine-induced apoptosis is dependent on lysosomal proteases. Biochem J 359(Pt 2):335–343
33. Petersen NH, Olsen OD, Groth-Pedersen L, Ellegaard AM, Bilgin M, Redmer S, Ostenfeld MS, Ulanet D, Dovmark TH, Lonborg A, Vindelov SD, Hanahan D, Arenz C, Ejsing CS, Kirkegaard T, Rohde M, Nylandsted J, Jaattela M (2013) Transformation-associated changes in sphingolipid metabolism sensitize cells to lysosomal cell death induced by inhibitors of acid sphingomyelinase. Cancer Cell 24 (3):379–393. https://doi.org/10.1016/j.ccr.2013.08.003
34. Chu YP, Hung YH, Chang HY, Yang WY (2017) Assays to monitor lysophagy. Methods Enzymol 588:231–244. https://doi.org/10.1016/bs.mie.2016.09.083
35. Ellegaard AM, Dehlendorff C, Vind AC, Anand A, Cederkvist L, Petersen NH, Nylandsted J, Stenvang J, Mellemgaard A, Osterlind K, Friis S, Jaattela M (2016) Repurposing cationic amphiphilic antihistamines for cancer treatment. EBioMedicine 9:130–139. https://doi.org/10.1016/j.ebiom.2016.06.013
36. Klutzny S, Lesche R, Keck M, Kaulfuss S, Schlicker A, Christian S, Sperl C, Neuhaus R, Mowat J, Steckel M, Riefke B, Prechtl S, Parczyk K, Steigemann P (2017) Functional inhibition of acid sphingomyelinase by Fluphenazine triggers hypoxia-specific tumor cell death. Cell Death Dis 8(3):e2709. https://doi.org/10.1038/cddis.2017.130
37. Pascua-Maestro R, Diez-Hermano S, Lillo C, Ganfornina MD, Sanchez D (2017) Protecting cells by protecting their vulnerable lysosomes: identification of a new mechanism for preserving lysosomal functional integrity upon oxidative stress. PLoS Genet 13(2):e1006603. https://doi.org/10.1371/journal.pgen.1006603
38. Sun L, Hu L, Cogdell D, Lu L, Gao C, Tian W, Zhang Z, Kang Y, Fleming JB, Zhang W (2017) MIR506 induces autophagy-related cell death in pancreatic cancer cells by targeting the STAT3 pathway. Autophagy 13 (4):703–714. https://doi.org/10.1080/15548627.2017.1280217
39. Villanueva-Paz M, Cordero MD, Pavon AD, Vega BC, Cotan D, De la Mata M, Oropesa-Avila M, Alcocer-Gomez E, de Lavera I, Garrido-Maraver J, Carrascosa J, Zaderenko AP, Muntane J, de Miguel M, Sanchez-Alcazar JA (2016) Amitriptyline induces mitophagy that precedes apoptosis in human HepG2 cells. Genes Cancer 7(7-8):260–277. https://doi.org/10.18632/genesandcancer.114
40. Stegmayr J, Lepur A, Kahl-Knutson B, Aguilar-Moncayo M, Klyosov AA, Field RA, Oredsson S, Nilsson UJ, Leffler H (2016) Low or no inhibitory potency of the canonical galectin carbohydrate-binding site by pectins and galactomannans. J Biol Chem 291 (25):13318–13334. https://doi.org/10.1074/jbc.M116.721464
41. Bidere N, Lorenzo HK, Carmona S, Laforge M, Harper F, Dumont C, Senik A (2003) Cathepsin D triggers Bax activation, resulting in selective apoptosis-inducing factor (AIF) relocation in T lymphocytes entering the early commitment phase to apoptosis. J Biol Chem 278(33):31401–31411. https://doi.org/10.1074/jbc.M301911200
42. Purschke M, Rubio N, Held KD, Redmond RW (2010) Phototoxicity of Hoechst 33342 in time-lapse fluorescence microscopy. Photochem Photobiol Sci 9(12):1634–1639. https://doi.org/10.1039/c0pp00234h

Chapter 22

Monitoring of Autophagy and Cell Volume Regulation in Kidney Epithelial Cells in Response to Fluid Shear Stress

Maria M. Lazari, Idil Orhon, Patrice Codogno, and Nicolas Dupont

Abstract

Fluidic shear stress applied to epithelial cells inside the kidney tubules affects cell size in an autophagy-related manner. Here, we describe the technical equipment that we routinely use to apply shear stress on cells, as well as immunoblotting, immunofluorescence, and three-dimensional cell volume reconstruction techniques used in analysis of the influence of this stress on cells and cellular components. By pointing out details of experimental techniques and potential pitfalls, this review will serve as a guide for those interested in study of how shear stress influences cells.

Key words Autophagy, Primary cilia, Shear stress, Fluid flow, Microslides, Cell volume, Immunoblot, Immunofluorescence, Microscopy

1 Introduction

Autophagy maintains homeostasis of numerous tissues and helps organisms cope with stress situations including mechanical stress [1–3]. Mechanical forces generated from flow of fluids such as blood or urine can be divided into three categories: (a) cycling stretch, (b) hydrostatic pressure, and (c) fluid shear stress. Shear stress is determined by flow rate, vessel geometry, and fluid viscosity and is known to regulate tissue morphogenesis [4] and, in particular, lumen diameter in the vasculature, and it also contributes to the progression of various human diseases such as atherosclerosis [5]. In the kidney, epithelial cell size is regulated in response to fluid flow through a primary cilia-dependent autophagy mechanism [6]. Primary cilia (PC) are microtubule-based structures present at the surface of almost all cell types that serve as mechanical sensors to detect the movement of body fluids, like urine. In the kidney, PC integrate variations in flow rates of the glomerular ultrafiltrate into cellular responses [7, 8].

In our laboratory we routinely use an in vitro model to apply shear stress on kidney epithelial cells under various conditions with

Nicholas Ktistakis and Oliver Florey (eds.), *Autophagy: Methods and Protocols*, Methods in Molecular Biology, vol. 1880, https://doi.org/10.1007/978-1-4939-8873-0_22, © Springer Science+Business Media, LLC, part of Springer Nature 2019

the goal of understanding how autophagy ensures cellular homeostasis in response to this stress. Fluid flow of 1 dyn cm^{-2} is applied on different kidney epithelial cell lines competent or incompetent for autophagy (genetically and pharmacologically). After shear stress, cells are fixed inside the special microslides of the flow chamber. Cells are then stained for autophagy markers for immunofluorescence microscopy or are harvested for immunoblotting. In order to measure the cell volume regulation of autophagy upon shear stress, cells are stained for specific actin markers, and, after cell imaging acquisition, the 3D images are reconstructed using Imaris software.

Inducing shear stress on kidney epithelial cells using a flow chamber mimics physiological conditions in the nephron, but application of the technique has potential pitfalls. In this review we describe the steps of the experimental procedures that we follow and highlight important details of the procedures used for monitoring of autophagy and cell volume regulation in kidney epithelial cells upon shear stress.

2 Materials

2.1 Cell Lines

Prepare all solutions required for the following cell culture protocols, and let them reach room temperature before use. Diligently follow all sterile handling guidelines as contamination can be easily spread through the flow chamber, especially during long shear stress experiments. Observe the cells microscopically during culture and throughout the experimental procedure specifically for differences in control cell survival and morphology.

1. Kidney mouse epithelial cells (KECs): Primary cilia-deficient KECs (obtained from mice with a hypomorphic deletion of the *Ift88* allele) were generated by G. J. Pazour (University of Massachusetts, Worcester, Massachusetts, USA). Wild-type (WT) KEC cells (Tg737$^{+/+}$) were transduced with *Lentivirus* (Sigma Mission Library, pLKO-puromycin-IPTG-3xLacO) in order to obtain stable cell lines expressing shRNA against the following mRNAs: *ATG5*, *ATG16L1*, *BECN1*, and *Kif3a*. KECs were cultured in DMEM glutamine that includes glucose and pyruvate (Thermo Fisher, Cat. No. 41966 with 10% fetal bovine serum (FBS) and 5 mM isopropyl β-D-1-thiogalactopyranoside (IPTG, Sigma-Aldrich) for 72 h.
2. Madin-Darby canine kidney (MDCK) epithelial cells: WT MDCK cells (obtained from American Type Culture Collection (ATCC)), MDCK GFP–LC3, and MDCK RFP–GFP–LC3 cells (stably transfected with expression vectors encoding GFP–LC3 and RFP–GFP–LC3, respectively, that also confer resistance to the antibiotic G418; kindly

provided by T. Yoshimori, Osaka University, Japan) were cultured in DMEM glutamine (Thermo Fisher, Cat. No. 41966-029/41966-052) supplemented with 10% FBS.

3. Murine inner medullary collecting duct (IMCD3) cells: IMCD3 cells were obtained from the ATCC. Medium used for culture of IMCD3 cells is DMEM/F-12 (Thermo Fisher, Cat. No. 31330) supplemented with 10% FBS.
4. Human renal proximal tubular cells (HK-2): HK-2 cells were obtained from the ATCC and cultured in DMEM glutamine (Thermo Fisher, Cat. No. 41966) with 10% FBS.

2.2 Microslides and Flow Chambers

1. Microslides (μ-slides): $I^{0.6}$ Luer μ-slides were purchased from Ibidi (Cat.No. 80186). The channel dimensions are 50 × 5 × 0.4 mm, and slides are coated with ibiTreat.
2. Fluidic units: The 10 mL fluidic units were purchased from Ibidi (Cat.No. 10903).
3. Perfusion Set BLUE: The system from Ibibi (Cat. No. 10961) is equipped with sterile air filters (0.2 μm, Teflon), 10 mL syringe reservoirs, silicon tubing, and female Luer coupler for setup without slide. The μ-slides should be connected to the flow chamber according to the manufacturer's instructions, and shear stress should be initiated at a rate of 1 dyn cm^{-2} (*see* **Note 1**) (Fig. 1).

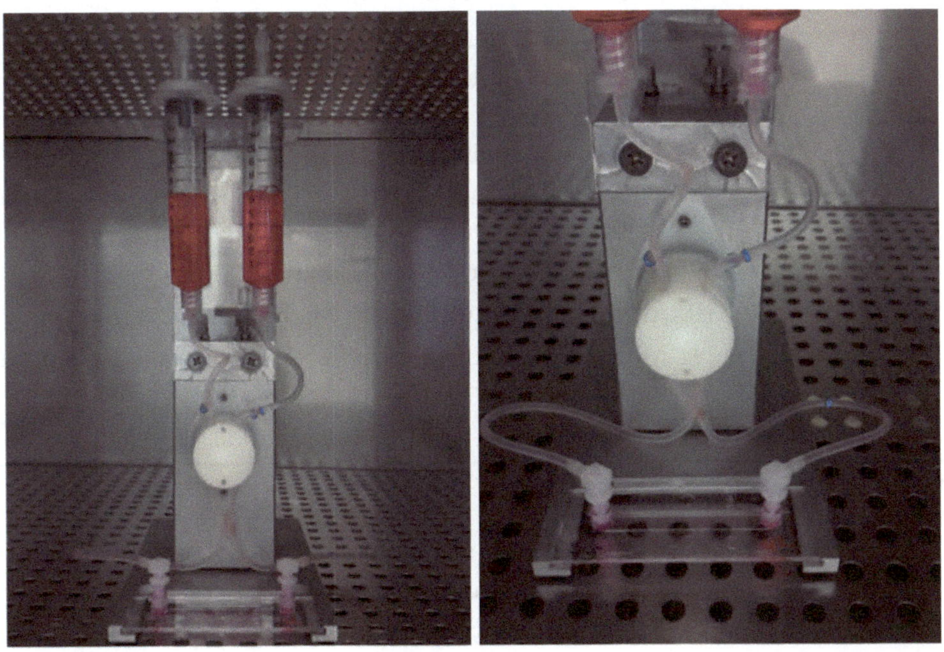

Fig. 1 Microslides and flow chambers. A μ-slide filled with KEC medium connected to the fluidic pump

2.3 Immunoblotting

1. DMEM (Invitrogen, Life Technologies, Cat. No. 31966) supplemented with 10% FBS (Invitrogen, Life Technologies).
2. Hanks' Balanced Salt Solution (HBSS) without sodium bicarbonate (Invitrogen, Life Technologies, Cat. No. 14025) or Earle's Balanced Salt Solution (EBSS) (Invitrogen, Life Technologies, Cat. No. 24010-043).
3. Phosphate-buffered saline (PBS): 137 mM NaCl, 2.7 mM KCl, 4.3 mM Na_2HPO_4, 1.4 mM KH_2PO_4, pH 7.3.
4. Laemmli (2×) sample buffer: 3.75% SDS, 25% glycerol, 25% β-mercaptoethanol, 0.125% bromophenol blue, and 31.5% 1.5 M buffer Tris-HCl, pH 8.8 (containing 0.4% SDS).
5. PBS-Tween 20 (0.1%) (Sigma-Aldrich Cat. No. P7949).
6. Blocking solution (5% milk PBS-Tween 20).
7. Antibody solution (2% milk PBS-Tween 20).
8. LC3B antibody (Sigma-Aldrich, Cat. No. L7543).
9. Drugs: Bafilomycin A1 (Sigma, Cat. No. B1793, final concentration 100 μM), chloroquine (Sigma, Cat. No. C6628, final concentration 50 μM), 3-methyladenine autophagy inhibitor (3MA) (Sigma, Cat. No. M9281, final concentration 10 mM).

2.4 Immunofluorescence

1. Paraformaldehyde (PFA, 4%).
2. PBS (1×)/FBS solution (10%).
3. Triton (0.1%) in PBS.
4. LC3 antibody (MBL, Cat. No. PM036).
5. E-cadherin antibody (BD Biosciences, Cat. No. 610182).
6. γ-Tubulin antibody (Santa Cruz Biotechnology, Cat. No. sc-10732).
7. Acetylated tubulin antibody (Sigma-Aldrich, Cat. No. T7451).
8. Fluorophore-conjugated secondary antibodies (Molecular Probes).
9. CellMask (Invitrogen, Cat. No. C10046).
10. Alexa Fluor 647 phalloidin (Molecular Probes).
11. Mounting medium containing 4,6-diamidino-2-phenylindole (DAPI) (SouthernBiotech Cat. No. 0100-20).

3 Methods

Approximately 700,000 cells (KEC, MDCK, IMCD3, or HK-2) are seeded into μ-slides and are cultured at 37 ° C, 5% CO_2 for 2–3 days before flow induction. This time period allows cells to become confluent and allows maximum ciliogenesis. The medium inside the μ-slides (150 μL) and the approximately 150 μL filling the two adapters must be changed twice a day using a 200 μL volume pipette (*see* **Note 2**).

3.1 Immunoblotting

1. During shear stress, cells treated with bafilomycin A1, chloroquine, or 3MA can be used as a negative control (*see* **Note 3**). The protease inhibitors are added to the medium inside the syringes and the tubing at concentrations and time points indicated in Subheading 2.3 and **Note 3**.
2. After cells have been subjected to shear stress, the μ-slides are removed from the flow chamber and washed twice with ice-cold PBS while on ice (*see* **Note 4**). After this step, dried cells can be kept at −80 °C until immunoblotting.
3. On the day of the Western blot experiment, thaw the stored cells on ice (this takes approximately 15 min). Add 150 μL of Laemmli 2× buffer per μ-slide, and incubate for 30 min at room temperature. It is important to note that cell lysis must be carried out on the same day as the Western blotting in order to avoid degradation of LC3 (*see* **Note 5**).
4. Place Laemmli buffer from each μ-slide into a 1.5 mL Eppendorf tube, and boil at 100 °C for 5 min.
5. Spin down the samples and load them onto a 12.5% SDS-PAGE gel. When autophagy is of interest, it is important to note that a gel less than 12.5% acrylamide is not be sufficient to separate LC3-I (16 kDa) and LC3-II (14 kDa).
6. Carry out a wet transfer onto membrane using a transfer buffer consisting of 24 mM Tris and 190 mM glycine for 1 h at 120 volts.

3.2 Immunofluorescence

Carry out all procedures at room temperature. At each step avoid creating air bubbles inside the μ-slides. Use a 1 mL pipette to remove the liquid all at once (*see* **Note 4**).

1. After shear stress remove the μ-slides from the fluidic unit and wash twice with PBS. Remove PBS, add 4% PFA, and incubate at room temperature for 15 min.
2. Remove PFA and wash twice with PBS. At this step you can leave cells in PBS at 4 °C for up to 1 month prior to immunofluorescence analyses. When MDCK GFP-LC3 and MDCK RFP-GFP-LC3 cells are used, proceed directly to step 10 of this method.
3. Remove PBS and add 0.1% Triton in PBS for 5 min.
4. Wash once with PBS.
5. Remove PBS, add PBS/FBS solution, and incubate for 30 min (or more) to block nonspecific sites in cells.

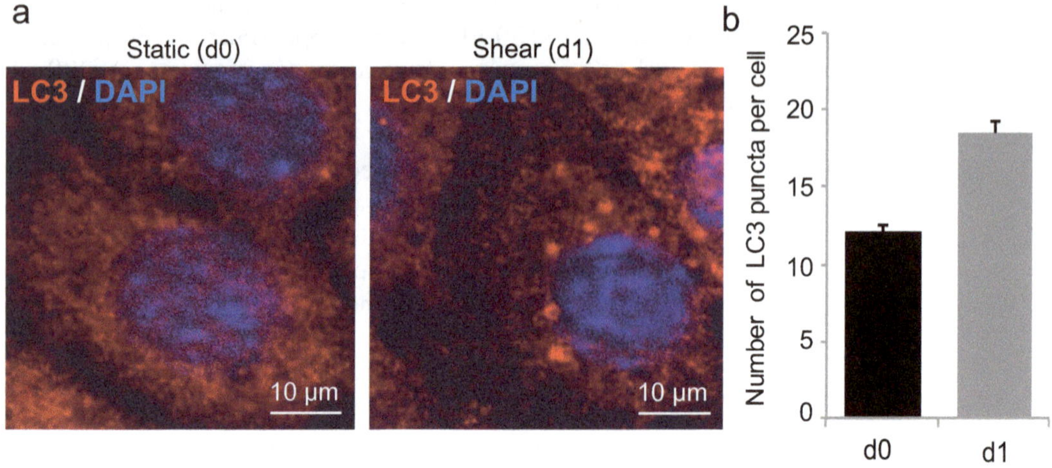

Fig. 2 Representative picture showing the effect of shear stress on the LC3 puncta formation in KEC cells. KEC cells were subjected to flow for 1 day or not (static) and then fixed and stained with LC3 antibody. LC3 puncta were analyzed by fluorescence microscopy as a measure of autophagic induction. (**a**) Representative images of cells subjected to flow. Scale bar indicates 10 μm. (**b**) Number of LC3 puncta per cell

6. Dilute primary antibodies in PBS/FBS (anti-LC3 at 1:150; acetylated tubulin, E-cadherin, and γ-tubulin at 1:100). Remove PBS/FBS; add antibodies, in master mix or individually, to μ-slides; and incubate for 1 h (*see* **Note 6**).
7. Wash once with PBS.
8. Dilute appropriate fluorophore-conjugated secondary antibody 1:250 in PBS/FBS, add solution to the μ-slides, and incubate for 30 min.
9. Wash once with PBS.
10. Remove PBS and add the mounting medium (*see* **Note 7**). At this step the μ-slides can be kept at 4 °C, protected from light, for up to 2 weeks. For LC3/p62 puncta identification and quantification, ImageJ or a similar software can be used (*see* **Note 8**) (Fig. 2).

3.3 Cell Volume Measurement

1. After shear stress, fix cells with 4% PFA as above, and use CellMask to stain the plasma membrane of the cells or Alexa Fluor 647 phalloidin to stain the cortical actin cytoskeleton (*see* **Note 9**).
2. Use an ApoTome or other confocal microscopes to acquire the *z*-stack images of the cells in different visual fields of the μ-slide.
3. Create 3D reconstruction images as mixed renderings using Imaris software (×64; version 8.4.1).
4. Calculate the mean cell volume using Imaris software (×64, version 8.4.1) (Fig. 3).

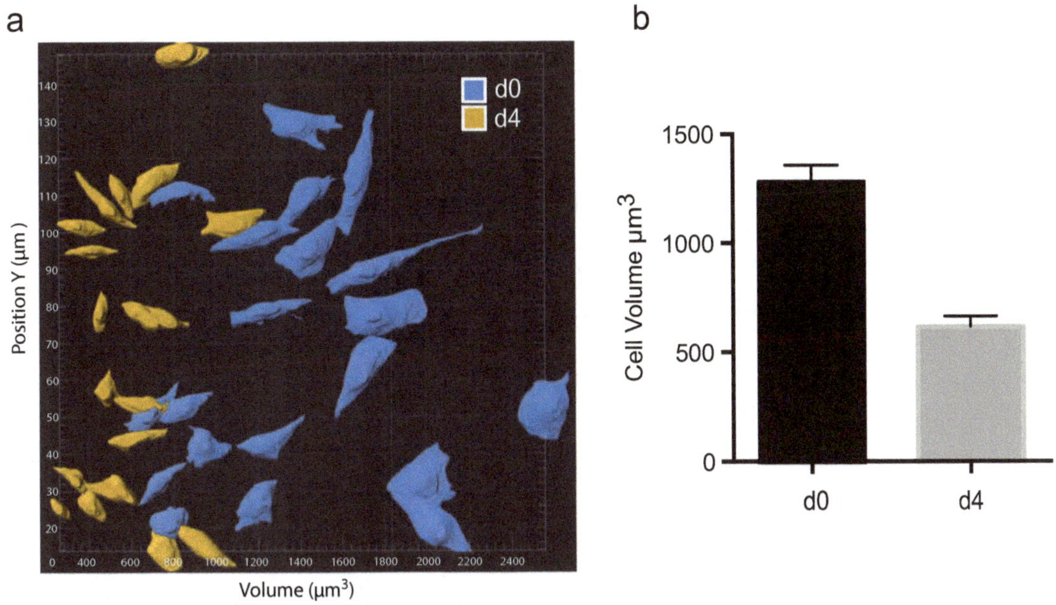

Fig. 3 Representative picture showing the 3D cell volume reconstruction and the effect of shear stress on the regulation of KEC cell size. KEC cells were subjected to flow for 4 days or not (static) and then fixed and analyzed by fluorescence microscopy to quantify cell volume. (**a**) Representative picture showing the 3D cell volume reconstruction. (**b**) Mean cell volumes of KEC cells underflow compared with cells not subjected to flow

4 Notes

1. If force higher than 1 dyn cm^{-2} is needed or for live cell microscopy, different μ-slides and a different perfusion set must be used. For more information visit the Ibidi website.
2. In order to change the medium inside the μ-slide, gently remove 150 μL from the right adapter, and add 150 μL of fresh medium to the left. Repeat two times so that all old medium (300 μL in total) is completely removed. Using this method, cells are not stressed or detached. Cells in this small volume of medium will rapidly consume the nutrients. The medium inside the μ-slide must be changed early in the morning and late in the evening to ensure that cells stay healthy.
3. Bafilomycin A1, a specific inhibitor of vacuolar type H^+-ATPase (V-ATPase), inhibits the acidification of lysosomes and therefore blocks the last step of autophagy (i.e., the fusion of the autophagosome with a lysosome), leading to an accumulation of early autophagosome vacuoles [9]. The time window for the incubation of cells with lysosomal inhibitors is critical and depends on the cell type involved. It is advisable to use a

saturating concentration of bafilomycin A1 [10]. The inhibitor treatment time is important for monitoring the blockade of the autophagic flux, because at early time points (2–4 h), the drug seems to have a greater blocking effect. However, for treatments lasting more than 6 h, impairment of the fusion of autophagosomes with late endosomes and lysosomes is observed [10]. It is advisable to use two different time points for the inhibitor treatment in order to observe the change in autophagic flux. Chloroquine can be used for longer incubations if needed. 3-MA inhibits autophagy by blocking autophagosome formation via inhibition of type III phosphatidylinositol 3-kinases (PI-3 K).

4. To wash cells inside the μ-slide, use a 1 mL pipette, and remove all medium from the one adapter at once. View a sample under a microscope to ensure that cells were not detached when the medium was removed.

5. LC3-II is sensitive to repeated freeze-thaw cycles of the samples, so it is advisable to use fresh samples in order to avoid any degradation of the protein. A third band due to LC3 may be observed in some experiments after the repeated use of lysates [11]. It is therefore advisable to use fresh cell lysates to analyze LC3 turnover.

6. The mounting medium is of high viscosity, and air bubbles may form inside the μ-slide when the mounting medium is introduced. Add the mounting medium slowly with a 200 μL pipette, controlling liquid movement by bending the μ-slide with your other hand. Handle μ-slides one by one so that cells do not dry out.

7. As mentioned above, cells must form a confluent monolayer inside the μ-slide in order to achieve maximum ciliogenesis. A way to verify this effect is by staining the cells with acetylated tubulin antibody to label the axoneme of the PC or with γ-tubulin antibody to label the PC basal body. Finally, E-cadherin staining can be used to identify the polarization levels of the ciliated monolayer in μ-slides.

8. In order to quantify changes in autophagy initiation, the levels of LC3-II are calculated relative to expression of actin (or another housekeeping protein). Autophagic flux is quantified from the ratio of LC3-II levels in the presence relative to the absence of lysosomal inhibitors. Another method of monitoring autophagic activity is to analyze LC3-II protein as a function of time. This method is useful to determine the rate of autophagosome biosynthesis overtime [12]. LC3 turnover is cell-type and stress dependent. It should be noted that the turnover of LC3 protein from LC3-I to LC3-II can also be used as a marker for autophagosome formation by determining

the LC3-II/LC3-I ratio; however, it is preferable to use a different method. Depending on the tissue or cell lines used, the abundance of LC3-I may change. The choice of LC3 antibody is also important, as the detection of the two forms of the LC3 protein depends on the affinity of the antibody. This means that it is preferable to quantify the turnover of LC3 based on the ratio of the intensity of LC3-II to that of a housekeeping gene rather LC3-II to LC3-I [11]. Readers can consult reviews for more information about the various aspects of the autophagic flux and the advantages and limitations of all autophagic read outs [13, 14].

9. When cell size measurement is required, ensure that the cells are confluent inside the μ-slides and no longer proliferating. Unless confluence is reached, cell volume changes can occur due to cell division.

Acknowledgments

Work in the Codogno laboratory is supported by funding from INSERM, CNRS, and Université Paris-Descartes-Sorbonne Paris Cité and by grants from INCa and ANR. M.L. and I.O. are supported by an ERASMUS internship and a FEBS postdoctoral fellowship, respectively.

References

1. Boya P, Reggiori F, Codogno P (2013) Emerging regulation and functions of autophagy. Nat Cell Biol 15(7):713–720
2. Mizushima N, Komatsu M (2011) Autophagy: renovation of cells and tissues. Cell 147 (4):728–741
3. Morel E, Mehrpour M, Botti J, Dupont N, Hamai A, Nascimbeni AC, Codogno P (2017) Autophagy: a druggable process. Annu Rev Pharmacol Toxicol 57:375–398
4. Mammoto T, Mammoto A, Ingber DE (2013) Mechanobiology and developmental control. Annu Rev Cell Dev Biol 29:27–61
5. Baeyens N, Bandyopadhyay C, Coon BG, Yun S, Schwartz MA (2016) Endothelial fluid shear stress sensing in vascular health and disease. J Clin Invest 126(3):821–828
6. Orhon I, Dupont N, Zaidan M, Boitez V, Burtin M, Schmitt A, Capiod T, Viau A, Beau I, Wolfgang Kuehn E, Friedlander G, Terzi F, Codogno P (2016) Primary-cilium-dependent autophagy controls epithelial cell volume in response to fluid flow. Nat Cell Biol 18(6):657–667
7. Praetorius HA (2015) The primary cilium as sensor of fluid flow: new building blocks to the model. A review in the theme: cell signaling: proteins, pathways and mechanisms. Am J Physiol Cell Physiol 308(3):C198–C208
8. Raghavan V, Weisz OA (2016) Discerning the role of mechanosensors in regulating proximal tubule function. Am J Physiol Renal Physiol 310(1):F1–F5
9. Yamamoto A, Tagawa Y, Yoshimori T, Moriyama Y, Masaki R, Tashiro Y (1998) Bafilomycin A1 prevents maturation of autophagic vacuoles by inhibiting fusion between autophagosomes and lysosomes in rat hepatoma cell line, H-4-II-E cells. Cell Struct Funct 23 (1):33–42
10. Klionsky DJ, Elazar Z, Seglen PO, Rubinsztein DC (2008) Does bafilomycin A1 block the

fusion of autophagosomes with lysosomes? Autophagy 4(7):849–950

11. Klionsky DJ, Abdalla FC, Abeliovich H, Abraham RT, Acevedo-Arozena A, Adeli K, Agholme L, Agnello M, Agostinis P, Aguirre-Ghiso JA, Ahn HJ, Ait-Mohamed O, Ait-Si-Ali S, Akematsu T, Akira S et al (2012) Guidelines for the use and interpretation of assays for monitoring autophagy. Autophagy 8 (4):445–544
12. Rubinsztein DC, Cuervo AM, Ravikumar B, Sarkar S, Korolchuk V, Kaushik S, Klionsky DJ (2009) In search of an "autophagometer". Autophagy 5(5):585–589
13. Dupont N, Leroy C, Hamai A, Codogno P (2017) Long-lived protein degradation during autophagy. Methods Enzymol 588:31–40
14. Klionsky DJ, Abdelmohsen K, Abe A, Abedin MJ, Abeliovich H, Acevedo Arozena A, Adachi H, Adams CM, Adams PD, Adeli K, Adhihetty PJ, Adler SG, Agam G, Agarwal R, Aghi MK et al (2016) Guidelines for the use and interpretation of assays for monitoring autophagy (3rd edition). Autophagy 12(1):1–222

Chapter 23

Identification and Regulation of Multimeric Protein Complexes in Autophagy via SILAC-Based Mass Spectrometry Approaches

Stéphanie Kaeser-Pebernard, Britta Diedrich, and Jörn Dengjel

Abstract

Mass spectrometry (MS)-based identification and characterization of protein complexes is becoming a prerequisite for in-depth biochemical analyses of intracellular processes. Here, we describe two state-of-the-art MS-based approaches to characterize protein-protein interactions and multi-protein complexes involved in autophagy in mammalian cells. The combination of affinity purification (AP)-MS, which identifies binary protein-protein interactions, with size-exclusion chromatography (SEC)-protein correlation profiling (PCP), which helps monitor protein complex assemblies, is a powerful tool to acquire a full overview of the interlinkage and regulation of novel multi-protein complexes that might play a role in autophagy.

Key words Macroautophagy, Autophagy, Starvation, SILAC, AP-MS, SEC-PCP-SILAC, Proteomics, Mass spectrometry

1 Introduction

Autophagy is a cellular "waste management" system, responsible for degrading intracellular compounds such as out-of-date organelles to regenerate basic nutrients and maintain cellular homeostasis (for review, *see* [1]). In eukaryotic cells, autophagy is constitutively active at basal levels [2] but is activated in response to metabolic or cellular stresses, including starvation, glucose deprivation, or ammonia accumulation [3]. Autophagy is regarded as a pro-survival response and directly linked to diseases, including cancer and neurodegeneration, but also to aging [3, 4], making it a potential target of choice for combination therapies. Although autophagy is extensively studied, there are still many open questions regarding its regulation. As high-throughput analytical techniques are being constantly developed and improved, increasing numbers of putative autophagy regulators are being identified and a need for powerful methods to characterize molecular interactions

Nicholas Ktistakis and Oliver Florey (eds.), *Autophagy: Methods and Protocols*, Methods in Molecular Biology, vol. 1880, https://doi.org/10.1007/978-1-4939-8873-0_23,

and functional networks addressing these novel regulators emerges. Here, we explain a two-step, mass spectrometry (MS)-based workflow to identify interactomes of putative novel autophagy regulators and to characterize their dynamic composition upon autophagy induction. These assays are amenable to endogenous proteins or to tagged versions of recombinant proteins of interest (baits) expressed in transgenic mammalian cell lines.

In a first set of experiments, binary protein-protein interactions are characterized by affinity purification-mass spectrometry (AP-MS) [5]. In a second set of experiments, the interlinkage and regulation of all the complex members identified by AP-MS are monitored by size-exclusion chromatography (SEC)-protein correlation profiling (PCP) [6]. Both experiments rely on metabolically labeled cells using stable isotope labeling by amino acids in cell culture (SILAC) to generate accurately quantifiable MS data [7]. In addition, the differentially labeled cells are exposed to stress-inducing conditions to trigger autophagy. By AP-MS, stable and labile interactors of the bait are identified from autophagy-induced SILAC-labeled cells [8]. APs of the bait are performed in two ways: (1) pooling the labeled samples before AP will identify the most stable interactors, whereas (2) pooling samples after AP will help identify more labile interactors. The purified proteins are fractionated by SDS-PAGE, digested to peptides, and analyzed by LC-MS/MS. By SEC-PCP-SILAC, native protein complexes are separated according to their size by SEC. Differentially SILAC-labeled fractions are mixed, proteins are therein digested in-solution, and again resulting peptides are analyzed by LC-MS/MS [9]. SEC elution profiles are recorded, and co-eluting proteins are assumed to form multimeric protein complexes. Altogether, this combination of state-of-the-art methods is ideal for a comprehensive characterization of protein complexes and their regulation in autophagy. The use of SILAC to metabolically label cells enables multiplexing, providing an additional layer of information via the identification of stable/labile and condition-dependent protein-protein interactions. In addition, AP-MS and SEC-PCP-SILAC are complementary. While AP-MS primarily identifies direct, binary protein-protein interactions, SEC-PCP-SILAC characterizes multimeric protein assemblies via co-elution.

2 Materials

2.1 Cell Culture and Harvest

1. These assays are established for mammalian cell lines, especially transgenic cell lines designed to ectopically express a tagged version of the bait protein (*see* **Note 1**). Ideally, a control cell line of the same origin, but not expressing any tagged bait, is used as negative control to monitor background binding to

magnetic beads in AP-MS experiments. We routinely use the MCF7 breast cancer cell line (ATCC #HTB-22).

2. SILAC Dulbecco's Modified Eagle Medium (SILAC-DMEM) deficient in arginine and lysine (Gibco).
3. Stable isotope-labeled amino acids: "medium-heavy" L-arginine-$^{13}C_6$ hydrochloride (Arg6) and L-lysine-4,4,5,5-*d*4 hydrochloride (Lys4); "heavy" L-arginine-$^{13}C_6$, $^{15}N_4$ hydrochloride (Arg10) and L-lysine-$^{13}C_6$, $^{15}N_2$ hydrochloride (Lys8) (all from Euriso-top GmbH, Saarbrücken, Germany). Stock solutions prepared in 1 × PBS: arginine isotopes, 0.4 M (2000×); lysine isotopes, 0.8 M (2000×).
4. Unlabeled amino acids: "light" L-arginine (Sigma-Aldrich), L-lysine (Sigma-Aldrich), and L-proline (Fluka). Stock solutions: L-arginine 0.4 M (2000×), L-lysine 0.8 M (2000×), and L-proline 0.17 M (750×), all in 1 × PBS.
5. Dialyzed Fetal Bovine Serum (dFBS, Gibco).
6. 200 mM L-glutamine (100× stock solution, PAN).
7. Penicillin/streptomycin 100× stock solution (10,000 U/mL and 10 mg/mL, respectively, PAN).
8. Sterile 1× Dulbecco's phosphate buffered saline (1 × DPBS, PAN).
9. Trypsin-EDTA solution (200 mg/L trypsin and 500 mg/L EDTA, PAN).
10. 10 and 15 cm tissue culture plates.
11. Cell scraper (Sarstedt).
12. Filter units, e.g., Filtropur series (Sarstedt).
13. Benchtop refrigerated centrifuge for 15 mL conical tubes, e.g., Eppendorf Centrifuge 5810R.

2.2 Autophagy Induction

1. Control, "untreated" SILAC-DMEM medium prepared as above.
2. Starvation medium: Hank's Balanced Salt Solution (HBSS, PAN) or other autophagy induction reagent (*see* **Note 2**).
3. Sterile 1× Dulbecco's phosphate buffered saline (1 × DPBS, PAN).
4. Concanamycin A stock solution: 20 μM in DMSO (10,000× stock solution) (Sigma-Aldrich)

2.3 AP-MS: Cell Lysis, Affinity Purification, Protein Fractionation, and In-Gel Digestion

1. Modified RIPA buffer: 50 mM Tris-HCl pH 7.5, 150 mM NaCl, 1% NP-40, 0.1% sodium deoxycholate, 1 mM EDTA. Store at 4 °C.
2. cOmplete Protease Inhibitors (Roche), aliquoted and stored at −20 °C.

3. PhosSTOP Phosphatase Inhibitors (Roche), aliquoted and stored at −20 °C.
4. Benchtop refrigerated centrifuge for 1.5 mL reaction tubes, e.g., Eppendorf Centrifuge 5415R.
5. BCA protein quantification kit (Thermo Fisher).
6. Plate reader spectrophotometer (absorbance 562 nm, e.g., BioTek KC4)
7. Magnetic beads for IPs:
   (a) Not coupled to antibodies: protein G- or protein A-coupled Dynabeads (Thermo Fisher).
   (b) Tag-specific beads, pre-coupled to antibodies, e.g., Pierce Anti-HA Magnetic Beads (Thermo Fisher); Anti-FLAG M2 Magnetic Beads (Sigma-Aldrich).
8. *Optional, for coupling to protein G/A beads only*: tag-specific antibody, 4–8 μg per AP.
9. Magnet for 2 or 15 mL conical tubes (Thermo Fisher). Rotating wheel for incubation.
10. 6× sample buffer: 0.3 M Tris-HCl pH 6.8, 12% SDS, 40% glycerol, 0.05% bromophenol blue.
11. Dithiothreitol (DTT): 100 mM stock solution in water, stored at −20 °C.
12. Iodoacetamide: 550 mM stock solution in water, stored at −20 °C (Sigma-Aldrich).
13. NuPAGE® Novex 4–12% Bis-Tris gradient gels (Thermo Fisher).
14. MOPS running buffer 20× solution (Thermo Fisher): 50 mM MOPS, 50 mM Tris Base, 0.1% SDS, 1 mM EDTA, pH 7.7.
15. Antioxidants (Thermo Fisher).
16. Colloidal Blue Stain (Thermo Fisher).
17. Scalpel (Braun).
18. Parafilm (Pechiney Plastic Packing).
19. ABC buffer: 100 mM ammonium bicarbonate, pH 7.5 (Sigma-Aldrich).
20. Ethanol (HPLC grade, Wako).
21. Trifluoroacetic acid (TFA), 2% (Sigma-Aldrich, MS grade).
22. Buffer A: 0.5% acetic acid in water.
23. Buffer A*: 3% acetonitrile (ACN) and 0.3% TFA in water.
24. Buffer B: 0.5% acetic acid and 80% ACN (Wako) in water.
25. StageTip material (Empore discs, C18 material from 3M).

26. Modified Sequencing-Grade Trypsin (Promega), 12.5 ng/μL in ABC buffer.
27. SpeedVac Concentrator, e.g., Savant SPD121P (Thermo Fisher).

2.4 SEC-PCP-SILAC: Cell Lysis, Ultracentrifugation, SEC, and In-solution Digestion

1. 1× SEC buffer: 50 mM KCl, 50 mM sodium acetate, pH 7.2.
2. SEC lysis buffer: 1× SEC buffer; 1× protease inhibitors (cOmplete Protease Inhibitors, Roche), 1× phosphatase inhibitors (PhosSTOP Inhibitors, Roche).
3. Rotor for ultracentrifugation (e.g., Rotor S100AT6; Thermo Fisher Scientific).
4. Ultracentrifugation tubes (e.g., 4PC tubes; Hitachi Koki).
5. Ultracentrifuge (e.g., Sorvall Discovery M150 SE; Thermo Fisher Scientific).
6. Molecular weight cutoff filter (100 kDa, Vivaspin 2, Sartorius AG).
7. Iodexanol OptiPrep (Sigma-Aldrich).
8. FPLC (e.g., ÄKTA Purifier 10, GE Healthcare).
9. SEC column 600 × 7.8 mm; resolving power 25,000 plates (BioSep SEC s4000, Phenomenex).
10. Gel Filtration Marker Kit for Protein Molecular Weights 29,000–700,000 Da (Sigma-Aldrich).
11. Sodium deoxycholate (DOC): 10% stock solution in water.
12. DTT: 100 mM stock solution in water, stored at −20 °C.
13. Iodoacetamide: 550 mM stock solution, stored at −20 °C (Sigma-Aldrich).
14. Modified Sequencing-Grade Trypsin (Promega), 12.5 ng/μL in ABC buffer.
15. 50% TFA (Sigma-Aldrich, MS grade).
16. Buffer A: 0.5% acetic acid (Sigma-Aldrich) in water.
17. Buffer B: 0.5% acetic acid and 80% acetonitrile (ACN) (Wako) in water.
18. StageTip material (Empore discs, C18 material from 3M).

2.5 LC-MS/MS Analysis

1. LC-MS/MS system (e.g., Q Exactive Plus Hybrid Quadrupole-Orbitrap, Thermo Fisher).
2. ReproSil-Pur C18 AQ 1.9 μm beads (Dr. Maisch GmbH).
3. SilicaTip emitters (New Objective).

3 Methods

The following methods imply the prior identification of a protein with putative functions in autophagy and the stable ectopic expression of a tagged version of this bait protein in a cell line of choice. Ideally, an isogenic cell line not expressing the tagged bait is used as control. In a first set of experiments, stable and labile interactors of the bait are identified by AP-MS. For this, separate APs versus combined APs are performed in SILAC-labeled cells under autophagy-inducing and control conditions (Fig. 1). Once binary protein-protein interactions of the bait are identified, SEC-PCP-SILAC is performed to characterize multimeric protein complexes and to study protein networking (Fig. 2). The combination of both approaches allows to comprehensively characterize interactomes and to define dynamics and regulation upon autophagy induction.

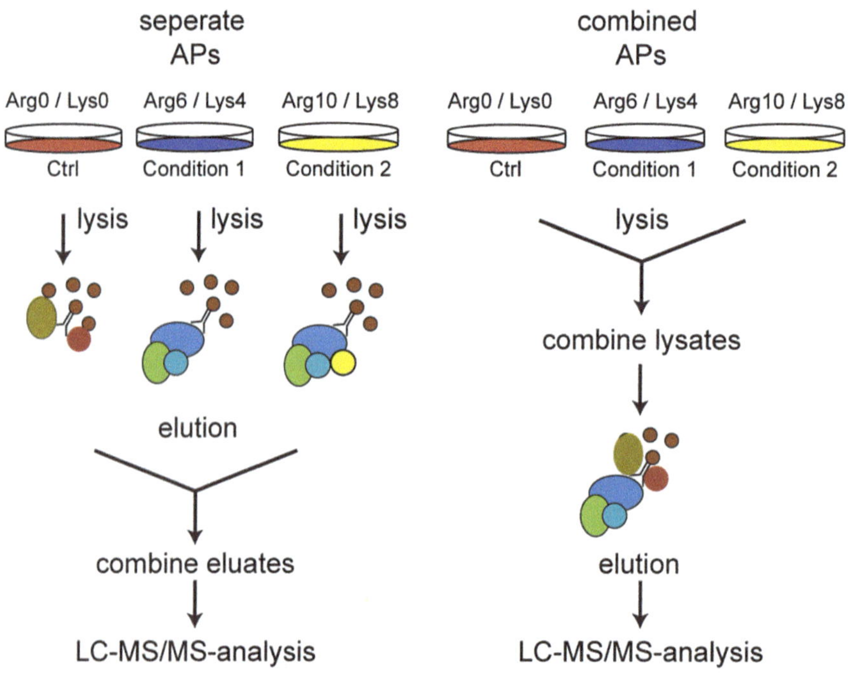

Fig. 1 Affinity purification (AP)-MS for the elucidation of protein-protein interactions. Shown are two SILAC-based approaches that allow the discrimination of (i) stimulus-specific and nonspecific interactions and (ii) of stable and transient interaction partners. Depending on the point of mixing, stable and transient (left) or only stable (right) interactions can be detected. Control samples using cells not expressing the tagged versions of proteins of interest (Ctrl in "light" label) are used to identify unspecific binding proteins. Cells expressing the tagged versions of proteins of interest are treated differently to identify stimulus-specific binding partners (Condition 1 in "medium-heavy" and Condition 2 in "heavy" label). Of note: by AP-MS only binary protein-protein interactions can be studied

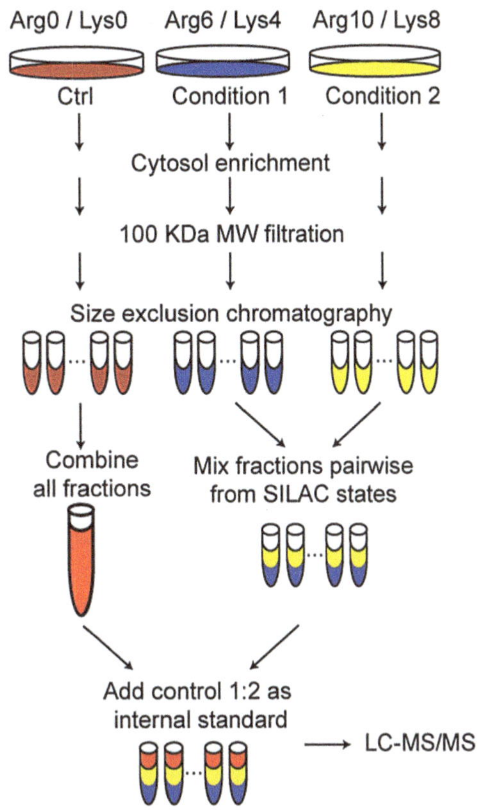

Fig. 2 Size-exclusion chromatography (SEC)-protein correlation profiling (PCP)-SILAC for the elucidation of multimeric protein complexes. Differently SILAC-labeled cells are treated and lysed. The cytosolic fraction is enriched, and multimeric protein complexes are separated by SEC. "Medium-heavy" (Arg6/Lys4) and "heavy" (Arg10/Lys8) labeled fractions are mixed and combined with a "light" labeled internal standard (Arg0/Lys0), which is spiked in a 1:2 ratio into the combined fractions. LC-MS/MS is used to generate elution profiles comparing medium-heavy and heavy to light SILAC ratios, respectively. Co-elution is used as a determinant to define multimeric protein complexes

3.1 Stable Isotope Labeling by Amino Acids in Cell Culture (SILAC)

Our experimental setup requires triple SILAC labeling. Commonly a biological replicate with swapped labels is performed in parallel to eliminate isotope-dependent bias. This SILAC labeling method is standard for both the AP-MS and SEC-PCP procedures; however the amounts of cells needed are different. The concentration of SILAC amino acids and proline in the medium is cell line-dependent and needs to be adjusted. This is an exemplified recipe for MCF7 cells.

1. Cells need to be cultured in SILAC medium containing "light" (Lys0 and Arg0), "medium-heavy" (Lys4 and Arg6), or "heavy" (Lys8 and Arg10) labeled amino acids. Prepare these

three SILAC-DMEM media in sufficient volumes for the whole duration of the experiment by dissolving lysine and arginine stocks to 1× concentration, in approximately 10 mL of SILAC medium. Add proline to 1× concentration (*see* **Note 3**), sterile filter the solution, and add it to the final volume of SILAC medium.

2. Complement the filtered SILAC-DMEM with 10% (v/v) dFBS (*see* **Note 4**), 1%(v/v) L-glutamine, and 1% (v/v) Pen/Strep.
3. To gain full incorporation of labeled amino acids, cells should be cultured for at least six cell doublings in SILAC medium [10] (*see* **Note 5**). Incomplete incorporation could lead to quantification inaccuracies. We recommend to start by maintaining each labeling condition in one 10 cm dish and to progressively amplify each population as needed, in order to reach a minimum of two 15 cm dishes per condition per AP, or eight 15 cm dishes per SEC, at the time of autophagy induction.
4. Labeling efficiency and arginine to proline conversion should be analyzed by MS prior to large-scale experiments.

3.2 Autophagy Induction

1. Experimental setup (Fig. 1):
   (a) Control: cell line not expressing the tagged bait, in "light" SILAC-DMEM.
   (b) Condition 1: cell line expressing the tagged bait, in "medium-heavy" SILAC-DMEM.
   (c) Condition 2: cell line expressing the tagged bait, expanded in "heavy" SILAC-DMEM, followed by induction of autophagy, e.g., by full starvation in HBSS for 3 h (*see* **Note 6**).

   To eliminate any putative bias caused by isotope labeling, a biological replicate with swapped labels is also prepared:

   (d) Control: cell line not expressing the tagged bait, in "light" SILAC-DMEM.
   (e) Condition 1: cell line expressing the tagged bait, expanded in "medium-heavy" SILAC-DMEM, followed by induction of autophagy.
   (f) Condition 2: cell line expressing the tagged bait, in "heavy" SILAC-DMEM.
2. Twenty-four hours before treatment, seed the appropriate amount of SILAC-labeled cells to reach 50–70% confluence the next day.
3. On induction day, prepare and warm up the required media. For DMEM-containing media, use the same SILAC-DMEM

(with light, medium-heavy, or heavy amino acids) that each line was maintained in.

4. At T = 0, treat cells with the chosen autophagy induction medium. For starvation conditions, wash cells twice with warm DPBS to remove respective nutrients. Incubate cells in a tissue culture incubator at 37 °C for 1–4 h for early autophagy induction (we routinely use 3 h) or for 7 h for long-term autophagosome accumulation (in the presence of 2 nM concanamycin A).

3.3 Cell Harvest

1. After induction, wash cells twice with ice-cold 1 × PBS in an ice bucket (*see* **Note 7**).
2. Scrape cells in 5 mL 1 × PBS on ice.
3. Pool harvest from all dishes of the same condition in one conical tube, and centrifuge to pellet cells at 1000 × *g* for 5 min at 4 °C.
4. Remove supernatant and immediately freeze pellets in liquid nitrogen (*see* **Note 8**).
5. Store at −80 °C until further use.

3.4 AP-MS

In *separate* APs, each sample is processed independently, and AP eluates of the three differentially labeled conditions are pooled for in-gel MS sample preparation, allowing the identification of both stable and transient interactions. In *combined* APs, the three differentially labeled conditions are mixed prior to performing one single AP, resulting in identifying the most abundant, hence the most stable, interactions formed with the bait (Fig. 1). Provided enough cells were seeded, the same lysates can be used for both separate and combined APs.

3.4.1 Cell Lysis and Affinity Purification

1. Prepare enough lysis buffer for the whole experiment (lysis, APs, and washes) by adding cOmplete protease inhibitors and PhosSTOP Phosphatase Inhibitors to an appropriate volume of modified RIPA buffer, mix, and keep on ice. This reconstituted lysis buffer can be kept at 4 °C for 24 h.
2. Lyse cells in at least three pellet volumes of lysis buffer. Vortex to resuspend, and incubate on ice for 15–45 min, vortexing every 5 min.
3. Centrifuge at 16,200 × *g* for 10 min at 4 °C in 1.5 mL reactions tubes or at 3,320 × *g* for 10 min at 4 °C in 15 mL conical tubes.
4. Transfer supernatants to fresh and labeled tubes, keep on ice.
5. Proceed to protein quantification using a BCA or another detergent-compatible detection kit in parallel with a BSA standard curve. Use the same amount of protein lysate for each AP.

6. Pre-wash magnetic beads: mix beads and pipet 100 μL slurry per AP into a new 1.5 mL reaction tube (prepare a master mix corresponding to $((N + 0.5)*100)$ μL, where $N =$ total number of APs). For volumes higher than 500 μL bead slurry, split in 2 or more tubes. Wash twice with 1 mL lysis buffer per tube, mix by inverting 5 times, and place tube on magnet for 2 min. Resuspend the pre-washed beads in $((N + 0.5)*100)$ μL lysis buffer, and store on ice until needed.
7. *Optional for Protein-G/A Dynabeads: pre-adsorption of the antibody.* (Do not do if using pre-coupled bead/antibody slurry and proceed directly to **step 8**). Mix $((N + 0.5)*100)$ μL (where $N =$ total number of APs) pre-washed beads with $((N + 0.5)*q)/[c])$ μL (where $q =$ amount of antibody per AP, usually 4–8 μg, and $[c] =$ antibody concentration in μg/μ L) of specific antibody, and complete to 500 μL with lysis buffer. Incubate 1–2 h at 4 °C with rotation. Place tube on magnet for 2 min, and then eliminate supernatant. Resuspend the bead/antibody mix in $((N + 0.5)*100)$ μL lysis buffer.
8. *Separate APs:* after quantification, prepare the AP tubes (1.5 mL reaction tubes or 15 mL conical tubes) by pipetting 2–5 mg lysate per AP tube (load the same protein amount in all AP samples) and adjusting all reactions to a standard volume (e.g., 1 or 2 mL or up to 10 mL) with lysis buffer. Add 100 μL of pre-washed and pre-coupled bead/antibody solution to each AP, taking care to carefully resuspend the beads in order to load equal bead amounts in all APs. Incubate all APs for up to 6 h at room temperature on a rotating wheel.
9. *Combined APs:* after quantification, pool equal protein amounts of the three SILAC lysates in one AP tube to reach 6–15 mg total protein content. Adjust volume of replicates with lysis buffer when necessary. Add 300 μL of pre-washed and pre-coupled bead/antibody solution to each combined AP. Incubate AP for up to 6 h at room temperature on a rotating wheel.
10. AP washes and elution. Wash APs three times with 1 mL lysis buffer. Mix by inverting 5 times at each wash. After the third wash, elute APs in 2 pellet volumes of 1× sample buffer containing 10 mM DTT by heating at 75 °C for 10 min on a shaker. Spin, place tubes on magnet for 2 min, and recover the eluates in fresh reaction tubes.
11. *Separate IPs only: pool the three eluates in one tube for each experiment.*
12. Store all eluates at −20 °C or proceed directly to in-gel digest.

3.4.2 AP-MS Sample Preparation: In-Gel Digest

1. Alkylate thiols by incubating all eluates with 5.5 mM iodoacetamide for 30 min at RT in darkness.
2. Separate each protein mixture on a NuPAGE® gel, fix the gel by incubation in 50% methanol/10% acetic acid in deionized water for 10 min, and stain/destain it using Colloidal Blue to assess the separation quality.
3. Cut each gel lane into slices of approximately equal sizes, taking care of cutting major bands resulting from the IPs in single slices. Cut each slice into 1 mm^3 cubes, transfer cubes into a reaction tube, and wash out remaining Colloidal Blue by three alternating 10 min incubations in 100 mM ammonium bicarbonate (ABC) buffer and 100% ethanol.
4. Let cubes swell in 80 μL trypsin solution (12.5 μg/mL 100 mM ABC) per slice for 10–15 min at room temperature, then add a minimal volume of ABC buffer until all gel is covered, and incubate overnight at 37 °C.
5. Stop trypsin activity by adding 50 μL of 2% TFA (or until all gel is covered). Shake gel cubes for 10 min at room temperature. Recover the supernatant and transfer solution to new tube. Wash cubes two times with 100–150 μL ethanol to extract all peptides. Combine supernatants of respective slices after each step.
6. Concentrate the collected peptide solution to less than 50 μL in a vacuum concentrator to remove ethanol, and add 200 μL buffer A.
7. Prepare StageTips [11] for desalting of peptide solutions, and proceed by centrifuging solutions through the discs in the following order: 50 μL buffer B to remove impurities and two times 50 μL buffer A to equilibrate and remove buffer B. Now load the sample, wash one time with 100 μL buffer A, and elute the sample with 50 μL buffer B into a new reaction tube.
8. Concentrate eluates to less than 5 μL to remove acetonitrile and add 10 μL buffer A/A* (75/25). The samples are now ready to be analyzed by LC-MS/MS (*see* **Note 9**).

3.5 SEC-PCP-SILAC

SEC-PCP-SILAC allows the accurate recording of SEC protein elution profiles (Fig. 2). Co-elution of proteins is regarded as a determinant to define multimeric protein complexes. In combination with AP-MS approaches, potentially interacting proteins are defined by AP-MS, whereas their spatial network is analyzed by SEC-PCP-SILAC.

3.5.1 Cell Lysis and SEC

Every step should be performed on ice and/or cold conditions to maintain assembled protein complexes.

1. Cool down all centrifuges to 4 °C.
2. Rinse SEC column with water for 1 min with 1 mL/min with valve in load position.
3. Rinse column with SEC buffer for 1 h with 1 mL/min with valve in inject position.
4. Thaw cell pellet on ice with 1.5 mL SEC lysis buffer added; pipet up/down and/or vortex shortly to resuspend cells.
5. Leave resuspended cells on ice for 10 min to swell.
6. Transfer mixture to a clean dounce homogenizer with a "tight" pestle (always keep the homogenizer on ice).
7. Dounce 250–300 times (about 5 min) to lyse the cells.
8. Transfer lysate to a fresh 15 mL conical tube; keep on ice until further usage.
9. Clear lysate from large debris by centrifugation at 4 °C, 3,220 × *g*, 10 min.
10. Add 1 mL of iodexanol to an ultracentrifugation tube.
11. Gently transfer the supernatant after centrifugation on top of the iodexanol, trying to avoid mixing.
12. Equilibrate weights before ultracentrifugation.
13. Ultracentrifuge at 100,000 × *g* 4 °C 20 min (including time to accelerate/decelerate; actual centrifugation time is 15 min). This step ensures that the cytosol is separated from the remaining organelles and membranes.
14. Transfer the upper phase to a 100 kDa MWCO, trying to avoid withdrawing any iodexanol.
15. Decrease sample volume (about 1.2 mL) by centrifugation at 3220 × *g* at 4 °C, until a final volume of 100–200 μL (depending on the protein concentration, this step may vary in duration).
16. Remove sample from the filter using a gel-loading tip, and transfer into a fresh reaction tube. Store on ice until further use.
17. Load sample to the HPLC and separate by SEC (4°-15 °C, flow rate of 0.5 mL/min).
18. Combine the corresponding "medium-heavy" and "heavy" labeled fractions pairwise (1:1).
19. Combine all "light" fractions, mix, and spike in "medium-heavy/heavy" pairs at a 1:2 ratio (1 volume "light", 2 volumes "medium-heavy/heavy"; *see* **Note 10**).

3.5.2 In-Solution Digestion of SEC Fractions

1. Add to each fraction 1% DOC (v/v) and 10 mM DTT.
2. Incubate 30 min at 75 °C.
3. Cool down samples to room temperature, and then add 5.5 mM iodoacetamide.
4. Incubate for 20 min at room temperature in the dark (*see* **Note 11**).
5. Add 2 μg trypsin (1:50 protein to trypsin), and vortex to mix.
6. Digest overnight at 37 °C.
7. Acidify with 15–30 μL of 50% TFA to precipitate DOC and to inactivate trypsin. Vortex thoroughly (*see* **Note 12**).
8. Centrifuge at maximum speed for 20 min at room temperature.
9. Transfer supernatants to fresh 1.5 mL reaction tubes.
10. Prepare StageTips with four disks of C18 materials [11].
11. Wash StageTips with 100 μL buffer B.
12. Wash with 150 μL buffer A.
13. Load the total volume of one fraction successively on one tip.
14. Wash with 150 μL buffer A.
15. Elute with 50 μL buffer B.
16. Concentrate eluates to less than 5 μL to remove acetonitrile and add 10 μL buffer A/A* (75/25). The samples are now ready to be analyzed by LC-MS/MS (*see* **Note 9**).

3.6 LC-MS/MS Analysis

Fractionate the peptide mixture by a reversed-phase chromatography column (SilicaTip emitters, New Objective; ID 75 μm) filled with C18 material (ReproSil-Pur C18 AQ 1.9 μm, Dr. Maisch GmbH), eluting the peptides directly into a mass spectrometer.

3.7 Data Analysis

For efficient analysis of large amounts of MS raw data with peptide identification and protein quantification, we recommend the freely available program MaxQuant, an integrated suite of algorithms specifically developed for high-resolution, high-accuracy quantitative MS data [12]. To assess the quality of a proteomic experiment and perform in-depth analysis of the data, we recommend to use the freely available software Perseus [13] or Graphical Proteomics Data Explorer (GProX) [14] platforms for comprehensive and integrated analysis and visualization of large proteomics datasets. Reproducibility of the experiment can be visualized by transforming the ratios by logarithm to the basis two and plotting two biological replicates in a scatterplot.

AP-MS data yields binary protein-protein interactions: (1) Comparing conditions 1 and 2 with the bead- control discriminates stimulus/autophagy-specific from stimulus-nonspecific interactions (Fig. 1). (2) Plotting separate versus combined APs (e.g., Log2 ratios of autophagy-induced versus control APs)

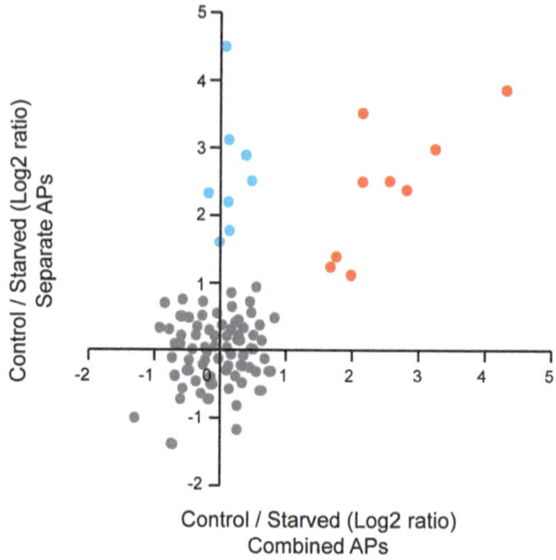

Fig. 3 AP-MS identification of stable and transient interactions (exemplified data). Log2 ratios of starved (Condition 1) versus bead-control protein ratios in combined versus separate APs are plotted. Gray dots: nonspecific interactions when compared to control AP. Red dots: interactors significantly enriched in both combined and separate APs, representing putative stable interactions with the bait. Blue dots: interactors only significantly enriched in separate APs, representing putative labile or transient interactions with the bait. A respective analysis comparing non-starved (Condition 2) versus bead-control protein ratios will identify stimulus-specific interactions

visualizes proteins that are stably enriched with the tagged bait (enriched in both separate and combined APs) and labile interactors (only enriched in separate APs). A theoretical example of respective AP-MS data is shown in Fig. 3. SEC-PCP-SILAC protein ratios are used to visualize co-eluting proteins and to characterize multimeric protein complexes. Figure 4 shows as an example elution profiles of the 14 proteasome core subunits. The 20S and 26S proteasome can be discriminated by overlaying elution profiles of core and accessory subunits (data not shown).

4 Notes

1. Analysis of cell lines expressing tagged versus non-tagged versions of proteins of interest allows the use of the same antibody for purification. This is critical as unspecific binding of background proteins differs between different antibody species. If endogenous proteins shall be analyzed, we recommend to use respective knockdown/knockout cells as control, again using the same antibody for AP [15].

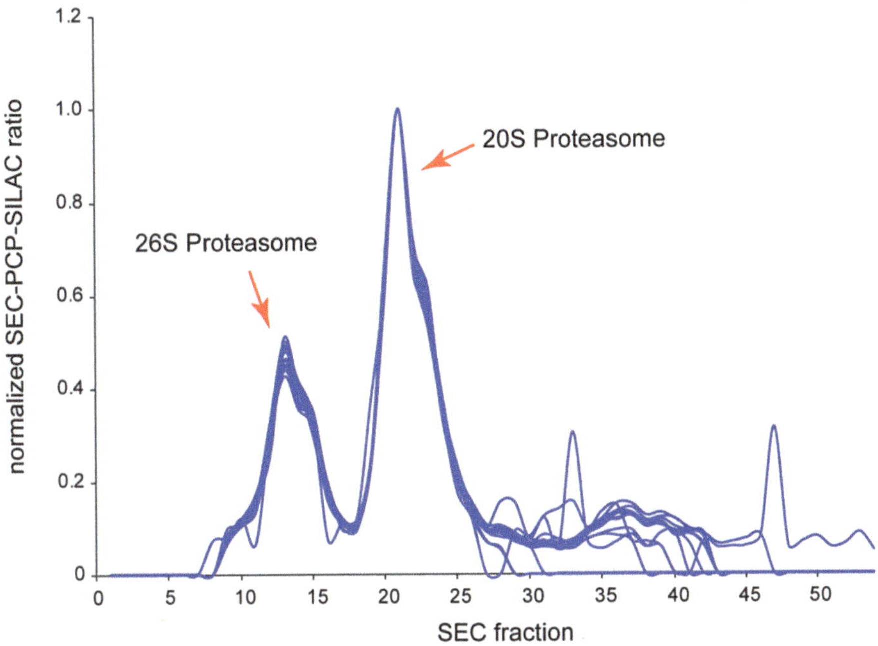

Fig. 4 SEC-PCP-SILAC data characterizing multimeric protein complexes. Shown are normalized SEC-PCP-SILAC ratios of the 14 proteasome core subunits (PSMA1–PSMA7, PSMB1–PSMB7). Two peaks are highlighted characterizing two distinct multimeric protein complexes, the 20S and 26S proteasome. The peak corresponding to the 26S proteasome is shared between core and accessory 19S regulatory subunits (data not shown)

2. Alternatively to full starvation in HBSS, autophagy can be successfully induced with the following reagents:
   (a) HBSS plus dFBS for amino acid starvation.
   (b) Rapamycin treatment: stock solution 100 μM in DMSO (1000×) (Sigma-Aldrich), aliquot, and store at −20 °C.
   (c) SILAC-DMEM without glucose.
3. Unlabeled proline can be added to the medium to avoid metabolic conversion of heavy arginine to heavy proline, which leads to quantification inaccuracies. Arginine and proline concentrations have to be adjusted for each cell line to ensure optimal labeling efficiency.
4. Dialyzed serum has to be used to ensure that only labeled variants of arginine and lysine are present.
5. In saturation-labeling experiments as described here, too short labeling periods result in incomplete label incorporation, which leads to quantification inaccuracies. Next to saturation labeling, dynamic pulse SILAC approaches exist, which allow the determination of protein synthesis, degradation, and turnover [16].

6. If protein complexes are studied that are potentially degraded by autophagy, we recommend to block lysosomal degradation by the addition of 2 nM concanamycin A [2].
7. Complete starvation in HBSS: cells have already been washed and are in serum-free medium. Hence, they do not have to be washed again. This might increase cell yield.
8. If possible, we recommend not freezing cells and directly continuing with AP or SEC as freezing might lead to loss of protein-protein interactions. If samples have to be frozen, this should be done at the stage of cell pellets. Cell lysates should not be stored and should be processed as quickly as possible.
9. If sample dries out while vacuum centrifuging, dissolve peptides with 20 μL buffer B and repeat steps: concentrate eluates to less than 5 μL to remove acetonitrile, and add 10 μL buffer A/A* (75/25).
10. To account for pipetting inaccuracies, increase the volume of the spike in standard by 2 mL of SEC buffer.

    This will ensure that the same volume can be added to all SEC fractions.
11. After IAA incubation, the solution becomes very viscous; this is not a problem for the subsequent steps.
12. Viscosity disappears as soon as TFA is added. Prior to acidification, it might help to freeze samples at −80 °C to help precipitate DOC.

Acknowledgments

This work was generously supported by the Swiss National Science Foundation (grants 31003A-166482/1, 316030-177088), by the Novartis Foundation for Medical-Biological Research (grant 16C221), by TRANSAUTOPHAGY (COST Action CA15138), and by the Alexander von Humboldt Foundation.

References

1. Ktistakis NT, Tooze SA (2016) Digesting the Expanding Mechanisms of Autophagy. Trends Cell Biol 26(8):624–635
2. Klionsky DJ, Abdelmohsen K, Abe A, Abedin MJ, Abeliovich H, Acevedo Arozena A, Adachi H, Adams CM, Adams PD, Adeli K et al (2016) Guidelines for the use and interpretation of assays for monitoring autophagy (3rd edition). Autophagy 12(1):1–222
3. Ryter SW, Cloonan SM, Choi AM (2013) Autophagy: a critical regulator of cellular metabolism and homeostasis. Mol Cells 36 (1):7–16
4. Choi AM, Ryter SW, Levine B (2013) Autophagy in human health and disease. N Engl J Med 368(7):651–662
5. Blagoev B, Kratchmarova I, Ong SE, Nielsen M, Foster LJ, Mann M (2003) A proteomics strategy to elucidate functional

protein-protein interactions applied to EGF signaling. Nat Biotechnol 21(3):315–318

6. Kristensen AR, Foster LJ (2014) Protein correlation profiling-SILAC to study protein-protein interactions. Methods Mol Biol 1188:263–270
7. Ong SE, Blagoev B, Kratchmarova I, Kristensen DB, Steen H, Pandey A, Mann M (2002) Stable isotope labeling by amino acids in cell culture, SILAC, as a simple and accurate approach to expression proteomics. Mol Cell Proteomics 1(5):376–386
8. Kuttner V, Mack C, Gretzmeier C, Bruckner-Tuderman L, Dengjel J (2014) Loss of collagen VII is associated with reduced transglutaminase 2 abundance and activity. J Invest Dermatol 134(9):2381–2389
9. Diedrich B, Rigbolt KT, Roring M, Herr R, Kaeser-Pebernard S, Gretzmeier C, Murphy RF, Brummer T, Dengjel J (2017) Discrete cytosolic macromolecular BRAF complexes exhibit distinct activities and composition. EMBO J 36(5):646–663
10. Becker AC, Dengjel J (2014) Autophagosomal proteome analysis by protein correlation profiling-SILAC. Methods Mol Biol 1188:271–279
11. Rappsilber J, Mann M, Ishihama Y (2007) Protocol for micro-purification, enrichment, pre-fractionation and storage of peptides for proteomics using StageTips. Nat Protoc 2 (8):1896–1906
12. Cox J, Mann M (2008) MaxQuant enables high peptide identification rates, individualized p.p.b.-range mass accuracies and proteome-wide protein quantification. Nat Biotechnol 26(12):1367–1372
13. Tyanova S, Temu T, Sinitcyn P, Carlson A, Hein MY, Geiger T, Mann M, Cox J (2016) The Perseus computational platform for comprehensive analysis of (prote)omics data. Nat Methods 13(9):731–740
14. Rigbolt KT, Vanselow JT, Blagoev B (2011) GProX, a user-friendly platform for bioinformatics analysis and visualization of quantitative proteomics data. Mol Cell Proteomics 10(8): O110 007450
15. Selbach M, Mann M (2006) Protein interaction screening by quantitative immunoprecipitation combined with knockdown (QUICK). Nat Methods 3(12):981–983
16. Gretzmeier C, Eiselein S, Johnson GR, Engelke R, Nowag H, Zarei M, Kuttner V, Becker AC, Rigbolt KTG, Hoyer-Hansen M et al (2017) Degradation of protein translation machinery by amino acid starvation- induced macroautophagy. Autophagy 13 (6):1064–1075

Chapter 24

Identification and Validation of Novel Autophagy Regulators Using an Endogenous Readout siGENOME Screen

Maria New, Tim Van Acker, Ming Jiang, Rebecca Saunders, Jaclyn S. Long, Jun-Ichi Sakamaki, Kevin M. Ryan, Michael Howell, and Sharon A. Tooze

Abstract

Autophagy is a highly regulated process, and its deregulation can contribute to various diseases, including cancer, immune diseases, and neurodegenerative disorders. Here we describe the design, protocol, and analysis of an imaging-based high-throughput screen with an endogenous autophagy readout. The screen uses a genome-wide siRNA library to identify autophagy regulators in mammalian cells.

Key words siGENOME screen, Autophagy, Autophagy regulation, Endogenous readout, LC3 puncta, Screen analysis, Large data set analysis, Hit identification

1 Introduction

Macroautophagy (hereafter referred to as autophagy) is an evolutionarily conserved homeostatic mechanism essential for degradation of damaged proteins and organelles that may otherwise diminish cellular fitness and integrity. Autophagy is a membrane-mediated process that delivers cytoplasmic contents to lysosomes for degradation and component recycling.

Upon autophagy induction, a double-membraned invagination or phagophore forms with the endoplasmic reticulum membrane acting as a membrane source [1]. The phagophore expands and eventually closes to form an autophagosome which engulfs cytoplasmic proteins and damaged organelles, either through a nonselective process or selective receptor-mediated autophagy, such as mitophagy [1]. The autophagosome fuses with lysosomes, and its contents are degraded.

Maria New and Tim Van Acker contributed equally to this work.

Nicholas Ktistakis and Oliver Florey (eds.), *Autophagy: Methods and Protocols*, Methods in Molecular Biology, vol. 1880, https://doi.org/10.1007/978-1-4939-8873-0_24, © Springer Science+Business Media, LLC, part of Springer Nature 2019

The autophagic process in mammals is mediated by at least 18 core autophagy genes (*ATG* genes) which act in a concerted hierarchy [2]. The ATG8 family of proteins, which include LC3B in mammals, can be used to monitor autophagosome formation and autophagic flux [2].

Although the importance of autophagy in cellular homeostasis and its role in disease is well known, the process is still not fully understood [3]. There is therefore a need for identification and study of novel autophagy regulators, particularly in mammalian cells, as the enhanced understanding of autophagy may yield new therapeutic targets.

A number of genome-wide siRNA screens for novel autophagy regulators have been performed prior to the genome-wide siRNA screen described in this chapter. Such genome-wide screens have been performed both under basal conditions [4], after autophagy induction with starvation [5] and under conditions of mitophagy induction to identify regulators for selective autophagy [6].

Fluorescently tagged GFP-LC3 puncta were used as a readout in all three screens mentioned above, as this is a convenient and easily quantifiable readout to assess autophagosome number, which in turn gives an indication of autophagy rate. However, there is concern that overexpression of tagged versions of LC3 to facilitate imaging may interfere with the autophagic process [7], which means that the use of endogenous LC3 staining is of particular interest, especially as not all cell types may be amenable to transfection of an exogenously tagged autophagic marker.

The protocol for a genome-wide siRNA screen where endogenous LC3 puncta are quantified as a readout is presented in this article. Data analysis from a HTS genome-wide screen is also discussed, in particular the selection strategy for positive autophagy regulators. Finally, assay development and troubleshooting are covered to enable researchers to adapt the protocol presented to their system of interest.

2 Materials

2.1 Genome-Wide Screen Transfection and Fixation

1. Phosphate-buffered saline (PBS): 137 mM NaCl, 3.4 mM KCl, 10 mM Na_2HPO_4, 1.8 mM KH_2PO_4, pH 7.2.
2. Distilled water.
3. Paraformaldehyde (PFA) 16% solution, methanol-free.
4. Bafilomycin A1 (Calbiochem).
5. Cell culture black 384-well microplate, clear F-bottom (Greiner).
6. Foil seal for 384-well plates (Pre-Cut Plate-Sized Microplate Sealing Tape).

7. Transfection reagent: INTERFERin-HTS.
8. siGENOME siRNA library (Dharmacon).
9. Control siRNA SMARTpools (Dharmacon): Human ULK1 siRNA, Human ULK2 siRNA, Human Ubb (ubiquitin B), Human ATG5, Human ATG7, RISC-Free Control siRNA, OTNT (ON-TARGETplus Non-targeting) pool siRNA.
10. Hank's Balanced Salt Solution.
11. Opti-MEM I Reduced Serum Medium.

2.2 Plate Staining (Reagent Quantity Given for a Batch of 24 Plates)

1. Methanol (230 mL).
2. 5% BSA (lyophilized powder) in PBS—380 mL of which 230 mL is used for blocking, the remainder for diluting primary and secondary antibodies.
3. Primary antibody solution in 1% BSA—230 mL 1% BSA + 2.3 mL mouse anti-LC3 5F10 antibody (nanoTools Antikörpertechnik) (used at 1:100 dilution) + 38.3 μL rabbit anti-p62 antibody ab91526 Abcam (used at 1:6000 dilution).
4. Secondary antibody solution in 1% BSA—230 mL 1% BSA + 230 μL donkey anti-mouse 488 (Life Technologies), (1:1000) + 230 μL donkey anti-rabbit 647 (Life Technologies), (1:1000) + 23 μL DAPI (1:10,000) (first resuspend DAPI in 1 mL to make sure it is not precipitated, and then add the 1 mL to the rest of the secondary antibody solution).

3 Methods

The protocol below is based on a successful 384-well format siGENOME screen performed under Bafilomycin A1 conditions to accumulate LC3 puncta and using endogenous LC3 immunofluorescence as an autophagy readout. An overview of the procedure is shown in Fig. 1 and assay development strategy in Fig. 2 (discussed in detail in Subheading 4 **Notes**). Note that the protocol for deconvolution (*see* Subheading 3.10) is identical to the protocol for the siGENOME screen (*see* Subheadings 3.1–3.6), with the exception of single siRNA duplexes being used rather than SMARTpools (mixtures of four siRNA duplexes).

The siGENOME screen protocol is divided into five sections: 384-well plate preparation (*see* Subheading 3.1), culturing sufficient cells (*see* Subheading 3.2), cell transfection (*see* Subheading 3.3), induction and fixation (*see* Subheading 3.4), and 384-well plate staining (*see* Subheading 3.5). While Subheadings 3.1–3.4 are performed with all 200 plates, the final staining step (*see* Subheading 3.5) is performed with batches of 24 plates. The protocol is followed by a description of siGENOME screen data analysis (*see* Subheadings 3.6–3.10).

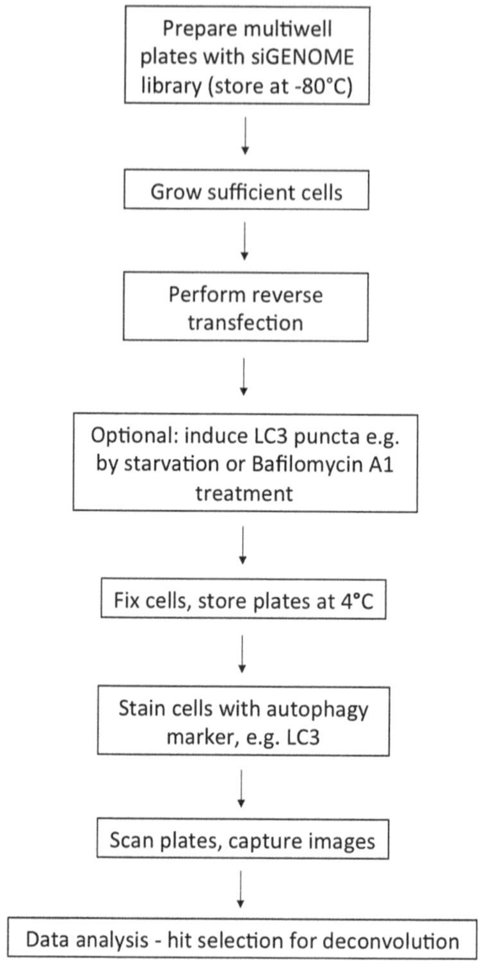

Fig. 1 siGENOME screen protocol overview

3.1 384-Well siGENOME Library Plate Preparation

After purchase of the siRNA genome library, reconstitute and aliquot into 384-well plates. Sealed with foil seals to prevent contamination and store at −80°C.

The screen should be done in triplicate—for a siGENOME screen with a human genome siRNA library this means 67 unique plates in triplicate, so 201 384-well plates in total. For other plate formats, please *see* Table 1. The control siRNAs can be added at this stage or on the day of transfection.

An example 384-well plate layout is presented in Fig. 3. Controls include known autophagy regulators, negative controls such as RISC-free and OTNT siRNA, and a siRNA which affects cell viability such as UBB.

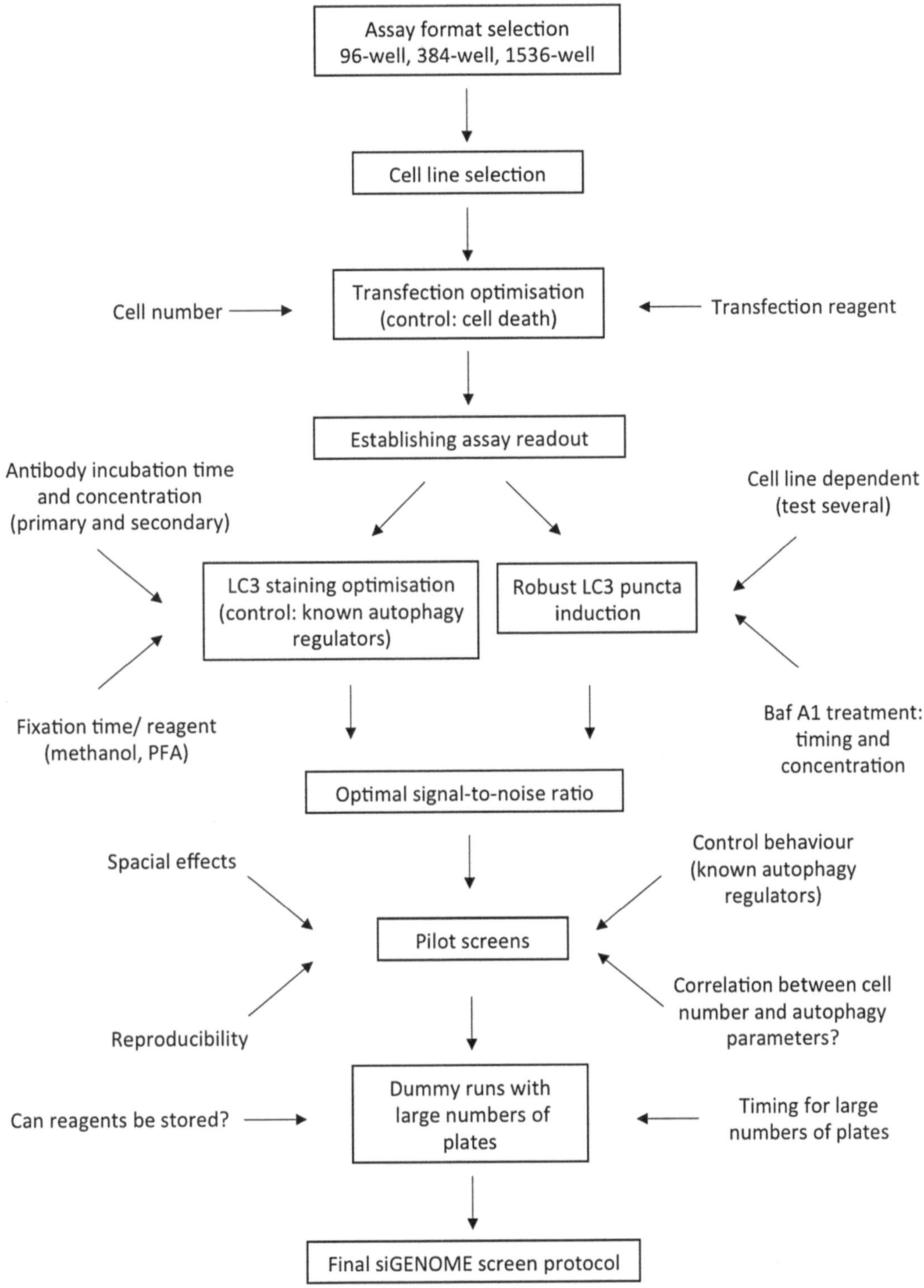

Fig. 2 siGENOME screen protocol development flow chart. Unboxed steps indicate issues researchers should consider when optimizing each step. The final result of this process is a complete optimized siGENOME screen protocol

Table 1
Approximate number of multi-well plates of various formats required for a genome-wide screen

Multi-well plate type	Number of plates for a genome-wide screen (approximate as number of control wells/plate may vary)
96-well	800
384-well	200
1536-well	50

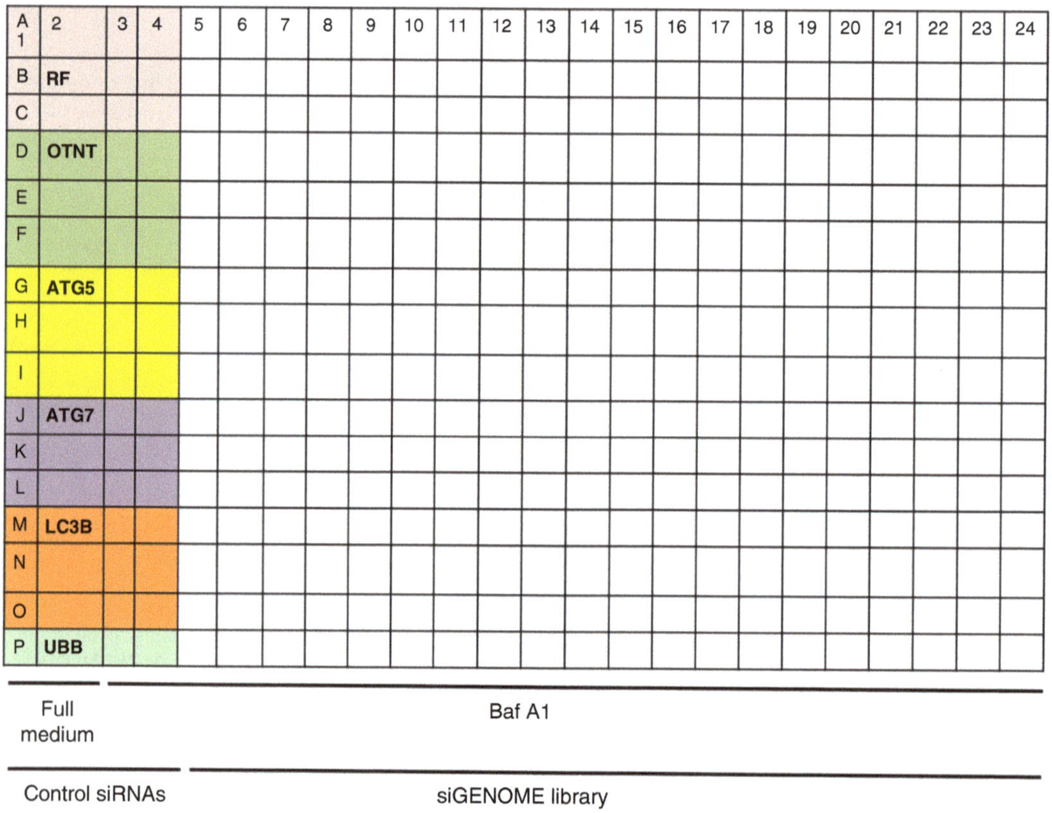

Fig. 3 Layout of the 384-well plates. Control siRNAs are in columns 1–4, and the siGENOME library SMART-pools are in remainder of the plate. Columns 1–2 were treated with full medium; columns 3–24 were treated with a final concentration of 50 nM Baf A1 for 6 h

3.2 Culture Cells (3 Days Prior to Transfection on Day 1)

Prepare sufficient cells for the screen. In our screen, we seed 1×10^6 cells/flask in 10 T175 flasks and grow for 3 days (72 h). Cells become approximately 90% confluent after 72 h, and one flask provides approximately 18×10^6 cells, giving sufficient cells (including excess) for the screen during which 600 cells/well are seeded into 201 384-well plates.

3.3 Transfection (Day 1)

1. Check cells to ensure expected confluency. Thaw the 201 plates with the siGENOME library at room temperature for 15 min, spin plates (800 rpm, 1 min), and remove the seal in hood. Spinning plates down and unsealing all 201 plates will take approximately 45 min if done by 4 people.
2. We used the FluidX Xrd-384 plate dispenser for liquid handling, hereafter described as "plate dispenser". Prime plate dispenser tubing with distilled water and PBS, and check nozzles are dispensing liquid correctly. Prime with control siRNAs and add 2.5 μL (375 nM in HBSS) control siRNAs in the control wells (column 1–4) using separate tubes and plate dispenser at high speed. Control siRNA addition takes 30 min with one plate dispenser.
3. Trypsinize cells, resuspend in full growth medium, and spin down (3 min, 900 rpm). After spinning, resuspend pellet thoroughly to ensure no cell clumps form and count cells. Dilute cell suspension into 4.2 L of pre-warmed full growth medium to give 600 cells in 45 μL.
4. Dilute the chosen transfection reagent, in our case INTERFERin-HTS (Polyplus-transfection; 0.0075 μL/well) into Opti-MEM I (705 μL of INTERFERin-HTS into 235 mL Opti-MEM I). Prime plate dispenser tubing with this mix. Add 2.5 μL/well sequentially into 201 plates containing the controls and siGENOME library using plate dispenser at high speed. Tap plates to ensure proper mixing of siRNA and transfection reagent.
5. Incubate plates with the mix of transfection reagent and siRNA at room temperature for 20 min prior to adding cell suspension—in our case 45 μL cell suspension/well (600 cells/well) using 2 plate dispensers at high speed (note it takes longer to add 45 μL cells per plate than to add 2.5 μL reagent per plate). Plate order at this stage must be the same as the plate order in Subheading 3.3, **step 4**.
6. After cell suspension addition, replace plate lids and leave plates at room temperature for 15 min, and then spin for 1 min at 800 rpm. Place in incubator for 72 h. Minimize incubator opening during this time.

3.4 Induction and Fixation (Day 4)

1. Prepare media: 2.2 L of 150 nM Baf A1 full growth media and 200 mL full growth media.
2. Treat cells in 384-well plates with 150 nM Baf A1 in batches of 30 plates at 30 min intervals, with each batch of cells being finally treated for 6 h prior to fixation with paraformaldehyde (PFA).
3. For the Baf A1 treatment: use 2 plate dispensers, prime both plate dispenser tubings with distilled water and PBS, followed by 3× Baf A1 media (plate dispenser 1) or full media (plate dispenser 2). Add 25 μL Baf A1 media (150 nM) to column 3–24; then full media to column 1–2 at high speed. Total volume per well is now 75 μL, and the final Baf A1 concentration is 50 nM. Replace plate lids after Baf A1 and full media addition, spin plates for 1 min, 800 rpm to mix and place each batch of plates in incubator for 6 h.
4. For each batch of plates: prime plate dispenser tubing with PBS, water, and 16% PFA. Add 25 μL of 16% PFA per well at high speed (final PFA concentration is 4%) to each plate at 2 min intervals. Spin plates for 1 min at 800 rpm.
5. After each plate has been fixed with PFA for 30 min, wash plate 3 times in PBS. We used a BioTek microplate washer for this step (subsequently referred to as plate washer), which was set to leave 100 μL PBS per well at the end of each washing step. Seal all plates with foil seal, close plate lids, and store at 4 °C.

3.5 Immunofluorescence Staining (Batches of 24 Plates)

All reagent addition should be done at medium speed and followed by a spin for 1 min at 800 rpm. The plate washers are set in such a way that a few μL of PBS is left in the well after the washes, so the well is not left completely dry during reagent addition. Each washing step starts with complete removal of well contents by the plate washer.

1. Remove the seals from 24 384-well plates and wash twice with PBS, using 2 plate washers, and then add 20 μL methanol per well. Each plate should be incubated with methanol for 30 min. This step takes approximately 25 min. Wash the tubing with distilled water after methanol addition, and wash the tubing with distilled water and PBS after all other liquid handling steps during the staining.
2. Wash 384-well plates twice in PBS and add 20 μL 5% BSA per well, in the same order as methanol addition. Leave for 1 h at room temperate.
3. Wash plates once in PBS and use plate dispenser to add primary antibody mix in 1% BSA (20 μL per well). Visually check wells to ensure primary antibody mix has been added to all wells and there are no air bubbles. Leave for 2 h at room temperature.

4. Wash plates 3 times in PBS using plate washers, and then add 20 μL secondary antibody/DAPI mix per well. Incubate for 1 h at room temperature.
5. Wash plates 3 times in PBS using plate washers, seal plates, and store in PBS (100 μL per well) at 4 °C until HTS data acquisition (*see* Subheading 3.6).

3.6 Genome-Wide Screen Data Acquisition

To collect antibody labelling data from the 201 384-well plates, a Thermo Scientific Cellomics Arrayscan Vti (20× magnification, 0.4 NA, BGRFR filter set, ×1 camera, LED illumination) and accompanying image analysis software were used (8 fields scanned per well). In our screen, the five autophagy-related parameters describing the properties of LC3 puncta per cell are total spot count, total spot area, average spot area, total spot intensity, and average spot intensity. Generic (non-autophagy-related) parameters measured are nuclear morphology parameters and total cell number per well (acquired by measuring nuclear DAPI staining with TTP Labtech's Acumen Scanning Cytometer running the Cellista analysis package). The total cell number is the parameter used as a viability indicator.

3.7 Data Analysis: Parameters Measured and Z-Score

There is no gold standard when it comes to analyzing high-throughput screening data. It is important to note that the method of analysis, the defined cutoffs, etc. are an interpretation by the scientists of how the "numbers" reflect the underlying biology. Therefore, it is crucial that one always views the original images collected during the data acquisition process to validate the analysis.

For our screen, analysis was based on normalized Z-scores calculated for each parameter. The Z-score allows to address the question of whether outliers (hits) are significantly different from an average siRNA (for more details *see* [8, 9]). Data from triplicate plates were combined by taking the median Z-score for each siRNA SMARTpool. For each autophagy-related parameter, such as LC3 intensity, three subparameters (which should be interpreted as "well summarizing" values) are defined: mean, %low, and %high. When defining the %low subparameter, e.g., total spot count, we record the % of cells in each well with a total spot count below a defined threshold. Positive autophagy regulators will score as wells with a high, positive Z-score in the %low subparameter and a low, negative Z-score in the mean subparameter.

3.8 Quality Analysis of HTS Data

Before starting the hit selection process, it is important to perform a quality analysis of the acquired HTS data. For our screen, quality analysis assessed the reproducibility between replicate plates, the distribution of normalized Z-scores, and the correlation between cell number and LC3/p62 puncta.

The reproducibility of the data from replicate plates was measured using the Spearman correlation coefficient for the various parameters, e.g., LC3 intensity (Fig. 4a).

The distribution of normalized Z-scores was plotted for each parameter, both for sample wells and for controls, and the data was found to follow a normal distribution (Fig. 4b).

The effect of cell number on LC3 and p62 puncta was examined using scatter plots. In our genome-wide siRNA screen, we found a negative correlation between cell number and spot parameters for all LC3 and p62-related parameters measured, although this varies in magnitude (Fig. 4c). p62 spot parameters show an increased negative correlation with cell number compared to LC3 spots parameters (Fig. 4c), so only LC3 parameters were analyzed during hit selection for our screen.

3.9 Hit Selection Process

Only wells with a Z-score between −1.5 and 1.5 for total cell number per well were considered as containing potential hits, as a total cell number per well exceeding this threshold causes difficulties during the validation process.

A schematic overview of the hit selection process is depicted in Fig. 5a. To select hits based on their ability to decrease LC3 puncta, a weighted binary scoring system was created. The data input for this scoring system was the ten subparameters that can be used to indicate a decrease in LC3 puncta: the mean and the %low Z-scores

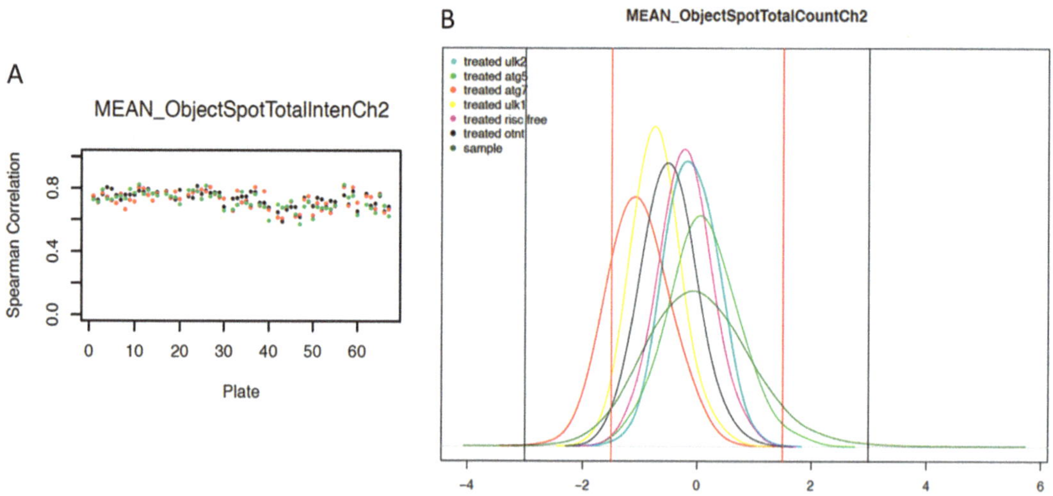

Fig. 4 Quality analysis of HTS data. (**a**) The reproducibility of replicate plates was measured using the Spearman correlation coefficient for the various subparameters scoring LC3 and p62 spots. In this example, Spearman correlation coefficients for LC3 total spot intensity are depicted for the three replicate plates (green, red, black). (**b**) Distribution of normalized Z-scores for LC3 total spot count for the different controls and samples. (**c**) Scatter plots representing the correlation between total LC3 spot intensity (Ch2 upper left), total p62 spot intensity (Ch3 lower left), average LC3 spot intensity (Ch2 upper right), average p62 spot intensity (Ch3 lower left), and cell number

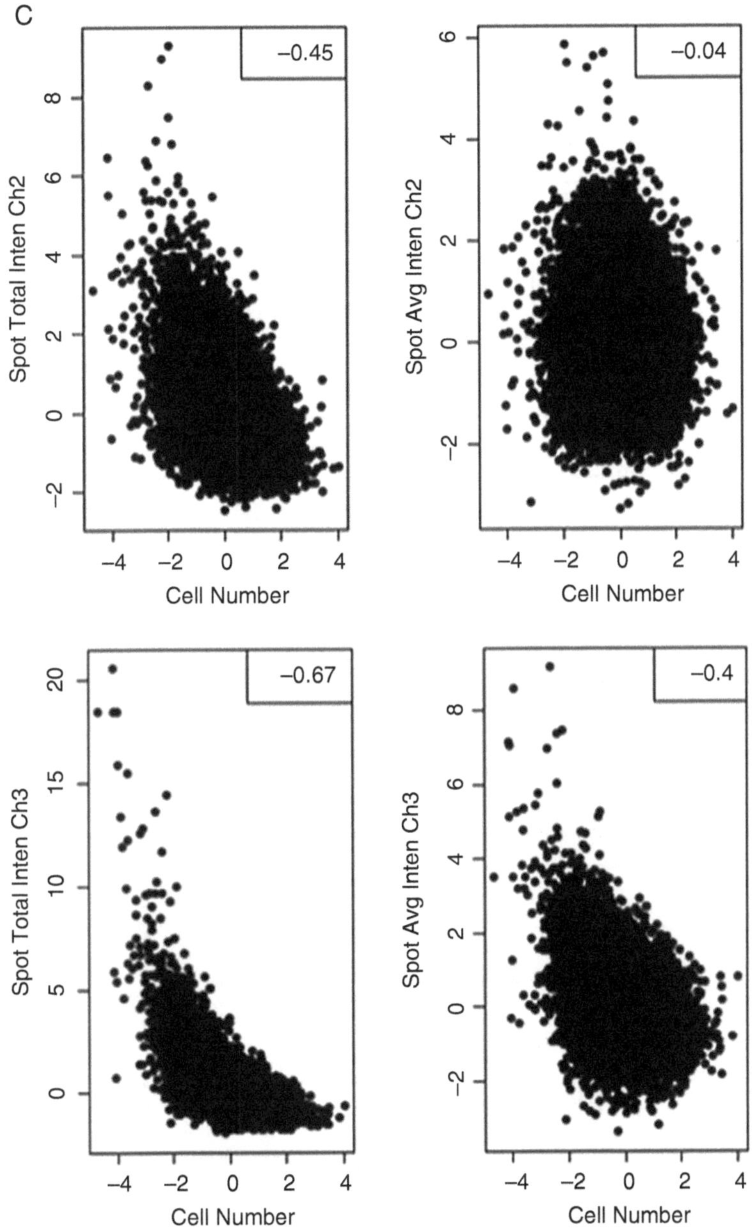

Fig. 4 (continued)

of the autophagy-related parameters total spot count, total spot area, average spot area, total spot intensity, and average spot intensity.

Based on visual image analysis, it was decided that hits with a mean Z-score ≤ -2 for a specific autophagy-related parameter were awarded a binary score of 2. Hits with a %low Z-score ≥ 2 were given a binary score of 1. The sum of these scores over all

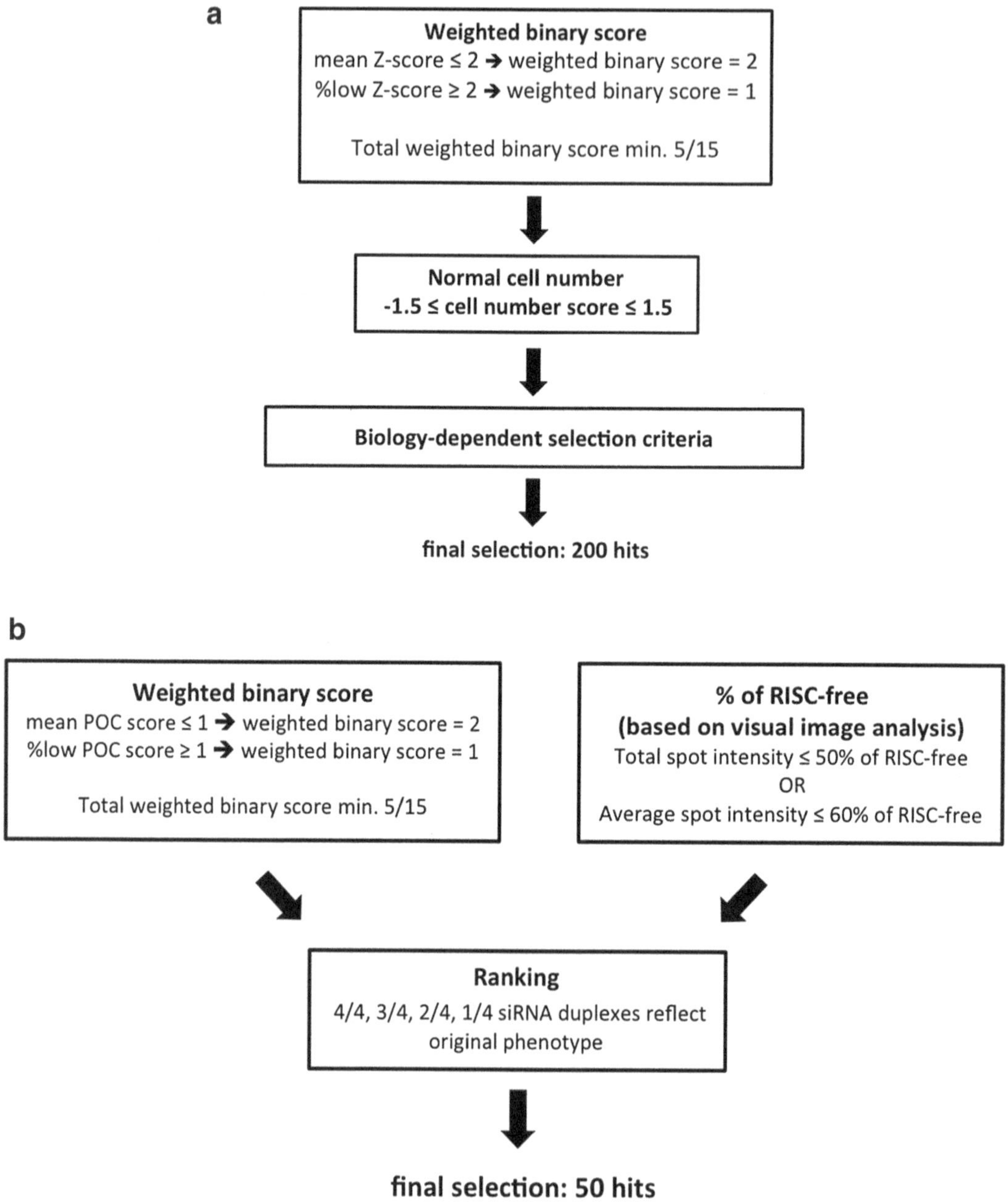

Fig. 5 Flow chart for hit selection after the siGENOME screen (**a**) and the deconvolution screen (**b**)

autophagy-related parameters produces a total weighted binary score with a maximum of 15. Two hundred hits with the highest weighted binary scores were taken forward into a deconvolution screen. Some hits were excluded based on biology-dependent criteria, e.g., low expression levels in a relevant cell line/tissue or competitive scientific landscape.

3.10 Deconvolution Screen Data Acquisition and Analysis

The goal of the deconvolution screen was to test whether the individual component oligonucleotide duplexes that make up the siRNA SMARTpools have the same phenotype as the siRNA SMARTpools.

Data acquisition for the deconvolution screen was performed identically to the data analysis for the genome-wide siRNA screen. However, the deconvolution screen was performed using a much smaller number of wells containing RISC-free control siRNAs compared to the genome-wide screen, which meant a Z-score could no longer be used to identify outliers. Instead, a POC (percentage of control) normalization was performed whereby every well is divided by the mean of the RISC-free wells on the plate and then multiplied by 100. Afterward, the median score of the three replicates for each parameter was calculated.

A flow chart summarizing hit selection in the deconvolution stage is presented in Fig. 5b. Two ranking methods were combined to decide whether to take a hit forward into the validation stage. The first method was based on a similar weighted binary scoring system as used for the genome-wide siRNA screen, albeit using POC values instead of Z-scores. The second method used to select siRNA duplexes based solely on total spot count and average spot intensity, as these two parameters reflected most accurately the biology observed in the original data images of the deconvolution screen. The cutoffs selected were chosen based on interrogation of the images acquired during the deconvolution screen. Any siRNA duplex with a mean subparameter score below the selected threshold for either the total spot intensity or the average spot intensity parameter was selected. Subsequently, we combined the single siRNA duplexes selected using the weighted binary scoring system and the ones selected based on the visual image analysis, allowing us to create a score out of 4 per hit.

Eventually, 50 hits were selected for further validation. Priority was given to hits score 4/4 and 3/4, followed by the remaining hits ranked on how they reflect the original SMARTpool phenotype in both analyses.

4 Notes: Assay Development and Where Things Go Wrong

There are a number of factors to consider during the development of a genome-wide screen protocol, which are summarized in Table 2. A general protocol development flow chart can be found in Fig. 2, although the order in which the various parameters to be considered are presented should not be regarded prescriptive, as optimization of several parameters at a time will likely be necessary. Parameter optimization can be considered an iterative procedure whereby when once each condition, such as cell number seeded, is established, the researcher then needs to go back and retest a range

Table 2
Factors to consider during genome-wide screen assay development

Assay development feature	Factors to consider
Assay format: 96-well, 384-well, or 1536-well	Larger well size—larger volume and easier liquid handling Smaller wells—reduction in cost due to lower reagent volume Smaller wells—lower number of plates in the genome-wide screen, less likelihood of needing to do transfection in batches
Cell line selection	Transfection efficiency Suitability for imaging Adherent and grows well in miniaturized format Cells should be frozen in large quantity, so the same batch and passage number are used for pilot experiments, screen, and deconvolution
Cell number selection	Lower cell number may improve transfection efficiency Cells which are too sparse take longer to image Cell number may correlate with LC3 puncta number
Reagent selection	Reagents to be used during screen, e.g., Bafilomycin A1, should be tested and acquired in a large batch sufficient for the entire screen and the deconvolution to increase reproducibility
Speed of reagent addition	High speed of addition may cause cell loss at the center of the well Slow addition speed may make timing of handling large numbers of plates impractical, as the protocol involves time-sensitive steps
Optimization of assay readout	To establish optimum signal-to-noise ratio, a range of conditions should be tested: Primary and secondary antibody concentrations Bafilomycin A1 concentrations and treatment time Methanol permeabilization times Paraformaldehyde fixation time (longer fixation may reduce LC3 intensity)
Control selection	siRNA treatment that results in cell death such as UBB (ubiquitin B), particularly useful during initial transfection efficiency optimization Known autography regulators—wells containing these need to be easily distinguishable
Pilot screens	Needed to test for: Spatial effects within the multi-well plate (such as edge effects) Reproducibility between the same conditions on different multi-well plates Reproducibility between pilot screens performed on different dates Reliable LC3 puncta induction Robust control behavior
Dummy runs	Involve using water/PBS and are required for: Testing transfection/staining timing To ensure a smooth transition from single plate experiments to dealing with hundreds of plates, e.g., addition of transfection reagent/Opti-MEM I mix to 200 plates may take 40 min, while the recommended incubation time of transfection reagent/Opti-MEM I mix is only 20 min Dummy runs identify the need to batch treat plates/test longer incubation periods

of options for the conditions previously selected, such as Bafilomycin A1 concentration, to give the final screen protocol.

During the months preceding the employment of a large-scale expensive siGENOME screen, it is important to foresee potential adverse eventualities and prepare where possible. For example, scanning and imaging the plates after staining may take many weeks even if the scanning goes to plan, but in the event of a power cut or the imager developing a fault, large delays may occur and plates may need to be stored for long periods of time. It is therefore prudent to test how long the antibody signal remains stable post-staining and how long the fixed plates can be stored prior to staining. Risk of contamination and infection such as yeast is also increased as a result of large-scale assays, so best practice careful cell culture technique is advised. Anti-fungal/anti-bacterial reagents can be tested but typically affect cell growth. Contamination may arise during lengthy storage of multi-well plates periods postfixation, which could be avoided through use of commercial sterilized PBS.

Spatial effects such as "edge effects" or "patches" on multi-well plates should be carefully controlled for. Possible measures to reduce such undesirable spatial effects include the use of a large properly functioning Cytomat incubator, in which plates are not stacked without spaces and which are maintained at the correct humidity. Ideally, the incubator should not be opened during the 72 h transfection period. Increasing the transfection volume per well can also reduce any undesirable special effects.

Other important issues that may affect the final results and should be tested for during the optimization procedure include the potential biological effect of cell number on the autophagic response. This can be determined by testing the correlation between cell number and LC3 puncta and avoided by selecting a cell number where this correlation is minimal (*see* Subheading 3.8). Another potential pitfall is the cross-reaction or "bleed-through" of the signal between various primary/secondary antibodies, if more than one primary antibody is used for staining. This should be tested for during the optimization procedure, by staining with each primary and secondary antibody independently and combination of the different antibodies with/without one primary/secondary antibody.

The importance of prescreen dummy runs to establish how experimental timing of a large-scale experiment with hundreds of plates differs from the test experiment that has already been addressed in Table 2. The optimal timing determined during the dummy run may lead the researcher to decide that plate transfection or Bafilomycin A1 treatment should be carried out in batches on the day of the screen. This raises the question of whether one large amount of reagent, such as Bafilomycin A1 media, should be made to treat all batches of plates on the day of the screen, or whether the reagent should be made fresh for each batch. If the former option is selected, it is important to ensure the reagent activity is not impacted adversely through storage for several hours on the day of the screen.

Despite all the preparation and assay development, unexpected issues could still arise on the day of the screen or become apparent when looking at the data obtained. In some cases it is possible to account for these at the analysis stage. For example, a slight spatial "edge effect" which was not apparent at the optimization stages but is present in the final screen could be improved by a normalization algorithm.

References

1. Gallagher LE, Williamson LE, Chan EYW (2016) Advances in autophagy regulatory mechanisms. Cell 5:24. https://doi.org/10.3390/cells5020024
2. Slobodkin MR, Elazar Z (2013) The Atg8 family: multifunctional ubiquitin-like key regulators of autophagy. Essays Biochem 55:51–64. https://doi.org/10.1042/bse0550051
3. Choi AMK, Ryter SW, Levine B (2013) Autophagy in human health and disease. N Engl J Med 368:651–662. https://doi.org/10.1056/NEJMra1205406
4. Lipinski MM, Zheng B, Lu T et al (2010) Genome-wide analysis reveals mechanisms modulating autophagy in normal brain aging and in Alzheimer's disease. Proc Natl Acad Sci U S A 107:14164–14169. https://doi.org/10.1073/pnas.1009485107
5. McKnight NC, Jefferies HBJ, Alemu EA et al (2012) Genome-wide siRNA screen reveals amino acid starvation-induced autophagy requires SCOC and WAC. EMBO J 31:1931–1946. https://doi.org/10.1038/emboj.2012.36
6. Orvedahl A, Sumpter R Jr, Xiao G et al (2011) Image-based genome-wide siRNA screen identifies selective autophagy factors. Nature 480:113–117. https://doi.org/10.1038/nature10546
7. Barth S, Glick D, Macleod KF (2010) Autophagy: assays and artifacts. J Pathol 221:117–124. https://doi.org/10.1002/path.2694
8. Joachim J, Jiang M, McKnight NC et al (2015) High-throughput screening approaches to identify regulators of mammalian autophagy. Methods 75:96–104. https://doi.org/10.1016/j.ymeth.2015.02.002
9. Malo N, Hanley JA, Cerquozzi S et al (2006) Statistical practice in high-throughput screening data analysis. Nat Biotechnol 24:167–175. https://doi.org/10.1038/nbt1186

Chapter 25

Autophagy Pathway Mapping to Elucidate the Function of Novel Autophagy Regulators Identified by High-Throughput Screening

Martina Wirth and Sharon A. Tooze

Abstract

Autophagy is an evolutionarily conserved degradative pathway, and the core autophagy machinery acts in a highly regulated, hierarchical manner to engulf cytoplasmic material in a double-membrane-bound organelle and deliver it to the lysosome. High-throughput screening approaches lead to the identification of novel autophagy regulators, and we describe an autophagy pathway mapping strategy to determine the stage of the autophagy pathway at which these novel candidate proteins function.

Key words Autophagy pathway mapping, mTORC1, ULK complex, WIPI2, ATG9, LC3, Autophagosome maturation, Autophagic flux

1 Introduction

Autophagy is a highly regulated process during which the sequential, hierarchical activation and recruitment of different autophagy-related (ATG) proteins ensure successful formation of autophagosomes and content degradation after fusion with lysosomes [1, 2]. In this chapter, we describe an autophagy pathway mapping strategy that allows researchers to efficiently establish whether novel autophagy regulators (e.g., identified by a genome-wide siRNA screen) function during early or late stages of the autophagy pathway.

In Fig. 1 we outline a decision tree to explain the pathway mapping process. We describe immunofluorescence (IF) and Western blot (WB) assays which can be routinely used to analyze key marker proteins required for the different stages of autophagosome formation, maturation, and autophagic flux.

In high-throughput screens LC3 puncta formation and lipidation are widely used as readouts to identify and validate novel autophagy regulators [3, 4]. LC3 labels both early-stage

Nicholas Ktistakis and Oliver Florey (eds.), *Autophagy: Methods and Protocols*, Methods in Molecular Biology, vol. 1880, https://doi.org/10.1007/978-1-4939-8873-0_25,

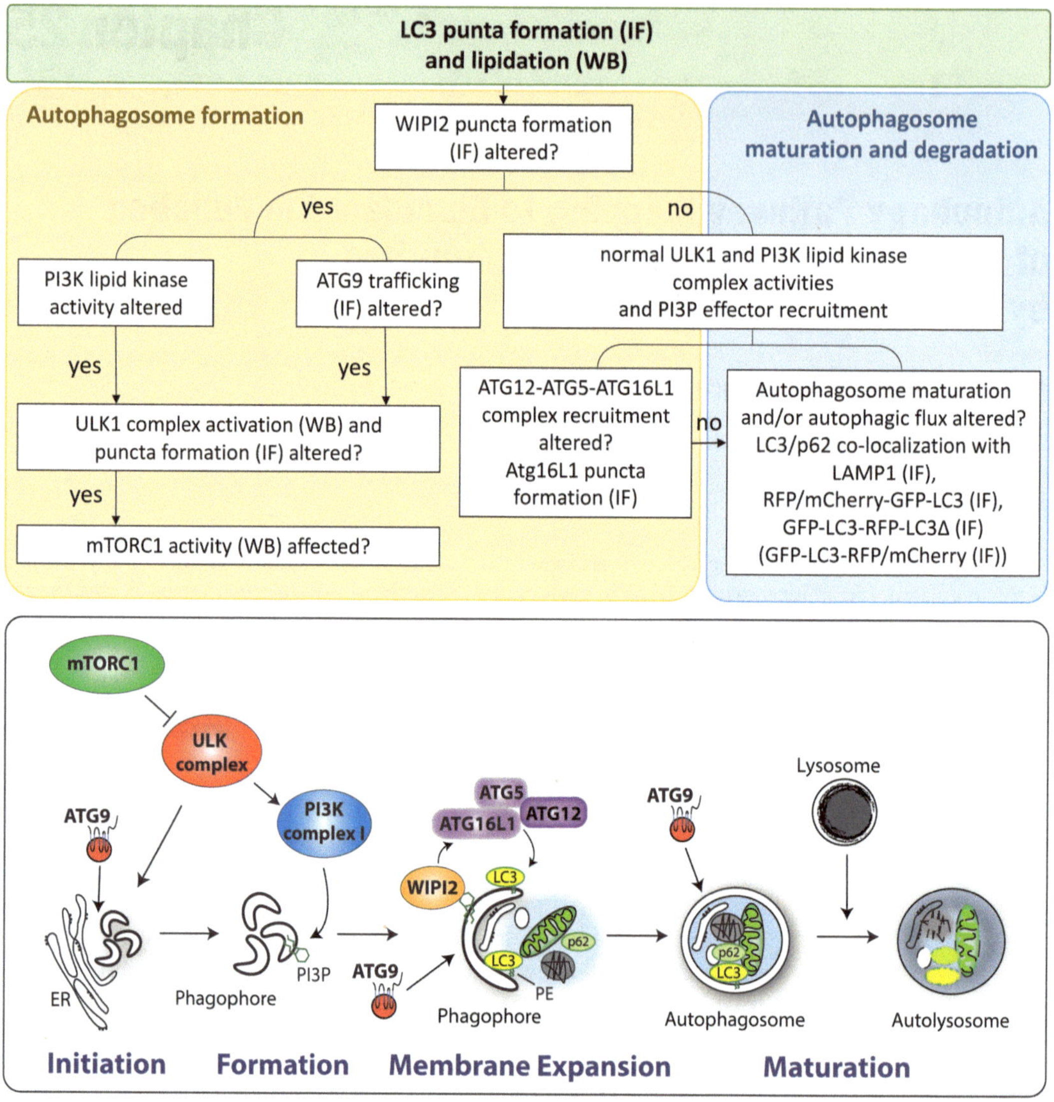

Fig. 1 Autophagic pathway mapping. Outline of the autophagic pathway mapping strategy showing the individual steps and assays involved to determine the stage at which novel autophagy regulatory proteins may function. (IF) immunofluorescence assay; (WB) Western blot assay. Below is a scheme of the current model of the autophagic pathway in mammalian cells and the key marker proteins mediating initiation and formation of the early autophagosome structure (called phagophore) as well as elongation and closure of the autophagosome. The ULK complex consists of ULK1/2, ATG13, FIP200, and ATG101. The PI3K lipid kinase complex I comprises VPS34, p150 (VPS15), BECLIN1, and ATG14. Lipidated LC3 and p62 are retained inside the autophagosome and degraded after fusion with late endosomes and lysosomes. (ER) endoplasmic reticulum, (PI3P) phosphatidylinositol-3-phosphate, (PE) phosphatidylethanolamine

(phagophore) and late-stage (closed and mature autophagosomes, autolysosomes) autophagosomal membranes [5] and is turned over by degradation upon fusion with the lysosome. Thus, changes in LC3 lipidation (measured by WB as an increase or decrease of

LC3-II) and LC3 puncta formation upon depletion of a candidate protein can be caused either by changes in autophagy initiation and autophagosome formation or by alterations in autophagosome maturation and autophagic flux (degradation of content). In contrast to LC3, most ATG proteins (including, e.g., WIPI2, ATG16L1, PI3K lipid kinase complex I, ULK kinase complex) transiently associate with the forming autophagosomal membrane, detaching prior (or immediately after) closure of the autophagosome [6].

The phosphatidylinositol-3-phosphate (PI3P) effector protein WIPI2 is recruited early to the forming autophagosome by PI3P, which is generated by the BECLIN 1 and ATG14 containing class III PI3K lipid kinase complex I [7]. WIPI2b directly binds to ATG16L1 mediating recruitment of the ATG12-ATG5-ATG16L1 complex to the membrane, which is essential for subsequent LC3 lipidation [8]. Quantification of WIPI2 puncta number by IF establishes whether early or later stages in the autophagic pathway are impaired.

For example, if loss of a candidate increases LC3 puncta number by IF and LC3-II levels by WB compared to controls, this could be due to either an increase in autophagosome formation or inhibition of turnover causing an accumulation of LC3-II. No changes in WIPI2 puncta number compared to controls would suggest that the autophagy-specific pool of PI3P has been generated and the activity of the BECLIN 1 and ATG14 containing class III PI3K lipid kinase complex I has not been affected. Moreover, normal recruitment of the ATG12-ATG5-ATG16L1 complex (assessed by quantification of ATG16L1 puncta by IF) supports the conclusion that the candidate protein may play a role in maturation of autophagosomes and/or autophagic flux. The increase in LC3-II observed would be likely due to inhibition of turnover in the lysosome as a result of failure to deliver the content to the lysosome or the lysosome being somehow unable to degrade the autophagic content. On the other hand, elevated or decreased WIPI2 puncta number (and ATG16L1 puncta number) compared to controls indicates altered PI3P levels and PI3K lipid kinase activity.

Trafficking of the only transmembrane autophagy protein ATG9 is important for formation of early-stage autophagosomes [9, 10]. Both ATG9 [10] and the ULK complex [10, 11] regulate PI3P production and WIPI2 recruitment during autophagy initiation, and their functions are intimately linked. ATG9 is required for formation of WIPI2 and ULK complex puncta [10, 12], and in turn ULK complex signaling regulates ATG9 trafficking [9]. During basal conditions ATG9 largely localizes to a juxtanuclear region and partially overlaps with Golgi (e.g., GM130) and TGN (trans-Golgi network, e.g., TGN46) markers. Both amino acid starvation and mTOR inhibition induce a redistribution of ATG9 which results in a marked decrease in signal intensity of juxtanuclear

ATG9 and a moderate increase of peripheral ATG9. Hence, ATG9 trafficking in response to autophagy induction can be measured by quantifying the juxtanuclear pool of ATG9 by IF ideally in combination with a Golgi marker. A candidate protein may affect ATG9 redistribution indirectly because it also impairs ULK complex activation (compared to controls) upon autophagy induction or its general trafficking through the secretory compartments [13].

The ULK complex has been shown to function downstream of the negative regulator mTORC1 and upstream of the BECLIN1-PI3K lipid kinase complex I. Inactivation of mTORC1 and activation of the ULK complex is one of the earliest events in initiation of starvation-induced autophagosome formation, and ATG9 vesicles, and assemblies of the ULK complex in association with the ER represent the earliest autophagy-specific structure forming de novo [12]. Altered formation of ULK1 puncta (IF) and phosphorylation of ATG13 at Ser318 by ULK1 (WB) suggest that the candidate protein affects ULK complex activity. Further upstream, Western blot analysis measuring the loss of phosphorylation of ULK1 at Ser757 (by mTORC1) and S6 ribosomal protein at Ser240/244 (by p70 S6 kinase) will reveal whether mTORC1 deactivation upon autophagy induction is changed compared to controls.

Several assays can be employed to assess whether novel candidate proteins function in autophagosome maturation (fusion with late endosomes and lysosomes) and autophagic flux (degradation by the lysosome). Examination of LC3 (and/or p62) colocalization (IF) with the lysosome-associated protein 1 (LAMP1) in the presence or absence of lysosomal inhibitors (e.g., Bafilomycin A1, chloroquine, or protease inhibitors) will determine whether autophagosomes have undergone fusion with late endosomes or lysosomes. However, a drawback of using lysosomal inhibitors is that they may cause undesirable mTORC1 inhibition and therefore autophagy [14, 15] or autophagy-independent LC3 lipidation (chloroquine) [16]. Alternatively, autophagosome maturation and autophagic flux can be monitored by expression of a RFP or mCherry-GFP tandem fluorescent-tagged LC3 (RFP/mCherry-GFP-LC3, tf-LC3). Whereas yellow puncta (colocalized GFP and RFP/mCherry) correspond to LC3 on autophagosomes, the acidic environment of the lysosome quenches the GFP signal faster than the red signal (RFP/mCherry), and autolysosomal LC3 only fluoresces red [17]. This probe is well suited to monitor maturation of autophagic structures within individual cells, but the eventual degradation of RFP/mCherry by the autolysosome means the precise amount of tandem fluorescent-tagged LC3 turned over by autophagy cannot be measured.

A recently developed probe by Mizushima's group overcomes this problem and also does not require treatment with lysosomal inhibitors [18]. When expressed in cells, the GFP-LC3-RFP-LC3ΔG probe is processed by endogenous ATG4 proteases into equimolar amounts of GFP-LC3 and RFP-LC3ΔG. The GFP-LC3 fraction undergoes lipidation as usual and will eventually be degraded by autophagy. However, the lipidation-deficient RFP-LC3ΔG fraction remains stable in the cytosol and serves as an internal control, reflecting the amount of GFP-LC3 before degradation by autophagy. Autophagic flux is calculated from the GFP:RFP signal ratio, e.g., a higher GFP:RFP ratio compared to controls indicates reduced autophagic flux. Since RFP or mCherry alone is stable in the cytosol the same way as RFP/mCherry-LC3ΔG, LC3ΔG is not essential, and autophagic flux can be quantified by a further modification of this probe (GFP-LC3-RFP/mCherry).

2 Materials

2.1 For Immunofluorescence Assays

1. Sterile, microscopy quality 13 mm round glass coverslips (pre-cleaned).
2. Frosted-edge microscope slides (pre-cleaned).
3. Poly-d-lysine—1 mg/mL in dH_2O.
4. Phosphate-buffered saline (PBS)—137 mM NaCl, 3.4 mM KCl, 10 mM Na_2HPO_4, 1.8 mM KH_2PO_4, pH 7.2.
5. Paraformaldehyde (PFA)—3% paraformaldehyde in phosphate-buffered saline (PBS), 0.9 mM $CaCl_2$, 0.5 mM $MgCl_2$.
6. Methanol.
7. 0.2% Triton X-100 (w/v) in PBS.
8. Digitonin—50 μg/mL in PBS.
9. BSA blocking buffer: 5% BSA fraction V (w/v) in PBS.
10. 50 mM NH_4Cl in PBS.
11. Hoechst dye.
12. Mowiol 4–88 (Calbiochem) in PBS.

2.2 For Western Blot Analysis

1. SDS-PAGE gels, ideally Bis-Tris 4–12% gradient gels run in MES buffer.
2. Lysis buffer: 20 mM Tris–HCl pH 7.2–7.5, 150 mM NaCl, 5 mM EDTA, 1% Triton X-100, 1× complete protease inhibitor cocktail (Roche), 1× PhosStop (Roche).
3. SDS-PAGE reagents and equipment.
4. 5× Laemmli sample buffer: 15% SDS (w/v), 312.5 mM Tris–HCl, pH 6.8, 50% glycerol (w/v), 16% β-mercaptoethanol, a sprinkle of bromophenol blue.
5. Western blot reagents and transfer equipment.

6. PVDF or nitrocellulose membrane.
7. Methanol.
8. Phosphate-buffered saline (PBS): 137 mM NaCl, 3.4 mM KCl, 10 mM Na_2HPO_4, 1.8 mM KH_2PO_4, pH 7.2.
9. Tris-buffered saline (TBS): 50 mM Tris, 150 mM NaCl, pH 7.5.
10. Blocking buffer for LC3 lipidation: 5% skimmed milk, 0.1% Tween-20, PBS.
11. Blocking buffer for phospho-specific antibodies: 5% BSA, 0.1% Tween-20, TBS.
12. Washing buffer for LC3 lipidation: PBST (PBS, 0.1% Tween-20).
13. Washing buffer for phospho-specific antibodies: TBST (TBS, 0.1% Tween-20).

2.3 Autophagy Inducers or Inhibitors

1. Earle's balanced salt solution (EBSS)—5.56 mM D-glucose, 123.08 mM NaCl, 5.37 mM KCl, 1.82 mM $CaCl_2$, 0.81 mM $MgSO_4$, 0.99 mM Na_2HPO_4, 13.10 mM $NaHCO_3$.
2. Rapamycin—11 μM in complete cell culture medium.
3. Torin 1–100 nM final in complete cell culture medium.
4. Bafilomycin A1 (Calbiochem, cat no. 196000)—100 nM final in complete cell culture medium or EBSS.
5. Wortmannin (Calbiochem)—100 nM final in complete cell culture medium or EBSS.

3 Methods

3.1 Immunofluorescence Microscopy to Detect and Quantify Autophagic Structures

All steps are performed at room temperature, and solutions are kept at room temperature unless otherwise indicated.

1. Prepare a 12-well tissue culture plate by adding glass coverslips to each well. For cells that are not very adherent, we recommend coating with poly-D-lysine as some cell lines tend to detach during treatments such as 2 h EBSS starvation. Add 1 mL poly-D-lysine solution to each well (enough to cover the bottom of well or coverslip). After incubation for 10 min, wash twice with 1 mL dH_2O and once with 1 mL PBS.
2. Seed cells into 12-well plate at a density that results in 60–70% confluency at the point of fixation (*see* **Note 1**). Return seeded plate to the incubator.
3. Perform experimental manipulations or starvation experiment as appropriate. All media should be pre-warmed to 37 °C. For

amino acid starvation, wash fed samples twice with 1 mL complete cell culture medium and starved samples twice with EBSS.

To wells washed with complete cell culture medium, add 1 mL growth medium (fed) or 1 mL growth medium containing 100 nM Bafilomycin A1 (fed plus BafA). To wells washed with EBSS, add 1 mL EBSS (starved) and 1 mL EBSS with 100 nM BafA (starved plus BafA). Incubate for appropriate time in incubator, e.g., for HEK293A cells 90–120 min (*see* **Note 2**).

4. In 12-well plate wash cells on glass coverslips once with 2 mL PBS and then fix in 3% PFA solution for 20 min.
5. Wash cells twice in PBS and incubate with 50 mM NH_4Cl in PBS for 10 min to quench aldehydes.
6. Permeabilize with room temperature methanol for 5 min (*see* **Note 3**).
7. Wash three times with PBS and block with BSA blocking buffer for at least 20 min.
8. Dilute primary antibodies in 1% BSA fraction V in PBS and set up a parafilm lined humidified chamber. Prepare 60–100 μL drops of antibody dilution (one drop per coverslip) on the parafilm.
9. Remove coverslip from 12-well plate using watchmaker's forceps, carefully drain excess liquid on a tissue, and invert coverslip (cell-side down) onto a drop of primary antibody solution. Incubate for 1 h at room temperature or overnight at 4 °C in the humidified chamber (*see* **Note 4**).
10. Place coverslips back into 12-well plate and wash three times with 1% BSA fraction V in PBS.
11. Dilute secondary antibodies in 1% BSA fraction V in PBS and add Hoechst dye at 1 μg/mL final concentration (for staining of nuclear DNA). Incubate coverslips with secondary antibody solution as described in **steps 8** and **9**.
12. Wash twice with 1% BSA fraction V in PBS and twice with PBS.
13. Pick up coverslips from 12-well plate using watchmaker's forceps and wash off PBS by dipping them into a beaker of dH_2O three to four times. Remove excess liquid and then mount the coverslips (cell-side down) onto a drop (10 μL) of Mowiol on a glass slide.
14. After Mowiol has hardened, store at 4 °C. Keep slides away from light.
15. Visualize using a confocal fluorescence microscope equipped with a 63×, 1.4 NA, oil lens. We typically acquire images at an optical slice thickness of 0.7–1 μm for spot counting and 2 μm for ATG9 trafficking. To ensure robust quantification, image

about ten random fields of view (from ~ 100 different cells) for each condition and keep laser settings constant for comparison of different conditions in an experiment. Use available analysis software such as ImageJ or Imaris for image analysis (*see* **Notes 5** and **6**).

3.2 Western Blot Analysis to Assay LC3 Lipidation and the Activity of mTORC1 and ULK Complexes

All steps are performed at room temperature, and working solutions are kept at room temperature unless otherwise indicated.

1. Perform experimental manipulation, e.g., knockdown or overexpression, as appropriate. To ensure robust protein detection for quantification, we recommend to plate cells in poly-D-lysine-coated 12-well plates (e.g., 1×10^5 HEK293A cells for DNA transfection the next day) and to prepare a minimum of duplicate samples for each condition (ideally triplicates).
2. On the day of inducing/inhibiting autophagy, cells should be at an appropriate density for experimental analysis, e.g., 80–100% confluent for amino acid starvation. Pre-warm all media to 37 °C.

   For a LC3-lipidation assay under amino acid starvation, wash fed samples twice with 1 mL complete cell culture medium and starved samples twice with EBSS. To wells washed with complete cell culture medium, add 1 mL growth medium (fed) or 1 mL growth medium containing 100 nM Bafilomycin A1 (fed plus BafA). To wells washed with EBSS, add 1 mL EBSS (starved) and 1 mL EBSS with 100 nM BafA (starved plus BafA). Incubate for appropriate time in incubator, e.g., for HEK293A cells 90–120 min.

   For analysis of ULK complex activation and mTOR inactivation, BafA treatment is not necessary, and we recommend to perform a time course of starvation or mTOR inhibition (treatment with 11 μM rapamycin or 100 nM Torin 1 in complete cell culture medium) for 0 min, 15 min, 30 min, and 60 min in HEK293A cells. This might need to be optimized when using other cell lines (*see* **Note 7**).
3. Place plate on ice and wash cells in each well three times with ice-cold PBS (*see* **Note 8**).
4. Remove excess liquid and lyse cells in 100 μL/well ice-cold lysis buffer containing 1× protease and phosphatase inhibitor cocktail on ice.
5. Transfer lysate to clean 1.5 mL tube and spin at maximum speed (~16.000 × *g*) for 15 min at 4 °C.
6. Mix 80 μL cleared lysate with 20 μL 5× SDS buffer and boil samples for 10 min at 100 °C (for ATG9 boil SDS samples at 65 °C) (*see* **Note 9**).

7. Load as much sample volume as possible, e.g., use 25 μL for a 20-well Bis-Tris 4–12% gradient gel (Invitrogen) (*see* **Note 10**).
8. Proteins are transferred to PVDF (polyvinylidene fluoride) membrane using wet transfer.
9. Stain membrane with Ponceau S to check equal loading. In order to detect proteins of different molecular weight, cut membrane horizontally, e.g., for LC3-lipidation analysis cut the membrane at the 30 kDa molecular weight marker into two halves. Use bottom half for LC3 and top half for actin/tubulin.
10. Block membrane with blocking buffer for 1 h at room temperature (*see* **Note 11**).
11. Incubate membrane with primary antibody (dilutions as indicated in Table 1) overnight at 4 °C.
12. Wash membrane at least three times with PBST for 10 min each before incubating with HRP-conjugated secondary antibodies for 1–2 h at room temperature (*see* **Note 12**).
13. Wash membrane three to four times, 10 min each with PBST. Detect protein using ECL reagent and X-ray film. Ensure that exposures are in the linear range.
14. Quantify bands using ImageJ or MetaMorph software.

4 Notes

1. For a 12-well plate, e.g., plate 5–7 × 10^4 HEK293A cells per well the day before fixation. Due to toxicity of transfection reagents such as Lipofectamine 2000, plate 1 × 10^5 HEK293A cells onto coverslips for DNA transfection the next day.
2. For LC3 spot formation in response to amino acid starvation or mTOR inhibition (using rapamycin or Torin 1), we recommend to include treatment with BafA to assess autophagic flux. For WIPI2, ULK1, and ATG16 spot formation assays, BafA treatment is not necessary as they dissociate from autophagosomes before they become LAMP1 positive (fusion with the lysosome) and do not mark autolysosomes. For analysis of ATG9 trafficking, BafA treatment is also not necessary.
3. The majority of antibodies listed in Table 1 work best using PFA fixation and methanol permeabilization. For ULK1 puncta staining, cells are permeabilized using 0.2% Triton X-100 (w/v) in PBS and for ATG9 immunostaining use 50 μg/mL digitonin in PBS for 5 min.
4. ULK1 antibody staining requires overnight incubation at 4 °C.

Table 1
List of antibodies to study autophagy using WB and IF

Antigen	Supplier	Species	Dilution IF	Dilution WB	Assay	Notes
Actin	Abcam (ab8227)	Rabbit		1:1000	LC3 lipidation	
ATG13	Tooze lab	Rabbit		1:500	ULK1 activity	Chan et al. (2009)
ATG13	Sigma (SAB4200100)	Rabbit		1:1000–1:500	ULK1 activity	
ATG13 pSer318	Rockland (600-401-C49)	Rabbit		1:500	ULK1 activity	MES buffer for SDS-PAGE, dilute in 5% skimmed milk in TBST
ATG9	Thermo Fisher (14F2 8B1)	Hamster (monoclonal IgM)	1:50–1:100	1:500	ATG9 trafficking	Young et al. (2006) WB: heat samples to 65 °C. Do not boil IF: digitonin permeabilization, use Cy™3-conjugated AffiniPure Goat Anti-Armenian Hamster IgG (Jackson Immuno Research, 127–165-160) as secondary antibody
ATG16L	Cosmo Bio (TMD-PH-AT16)	Rabbit	1:400		ATG16 puncta	Antibody discontinued
ATG16L	MBL (PM040)	Rabbit	1:400		ATG16 puncta	IF: Triton permeabilization
LC3B	Abcam (ab48394)	Rabbit	1:2000	1:1000	LC3 puncta/ lipidation	IF: methanol permeabilization
LC3B	Nanotools (LC3-5F10)	Mouse	1:100	1:200	LC3 puncta/ lipidation	IF: methanol permeabilization
GABARAP	Abgent (AP1821a)	Rabbit	1:250	1:250	Wipi2 puncta	IF: methanol permeabilization
GM130	BD (610822)	Mouse	1:250		ATG9 trafficking	
LAMP1	BD (CD107A)	Mouse	1:250–1:500		Autophagosome maturation	IF: methanol permeabilization
p62	BD (610832)	Mouse		1:500	p62 degradation	WB: incubate overnight
p62	Abnova (H00008878-M01)	Mouse		1:250	p62 degradation	WB: incubate overnight; detects murine p62

p62	Progen (GP62-N)	Guinea pig	1:300		Autophagosome maturation	IF: methanol permeabilization
p62	Abcam (ab91526)	Rabbit	1:6000			IF: methanol permeabilization
S6	Cell signaling (2217)	Rabbit		1:1000	mTOR activity	WB: according to datasheet instructions
S6 pSer240/244	Cell signaling (2215)	Rabbit		1:4000	mTOR activity	WB: according to datasheet instructions
ULK1	Cell signaling (D8H5)	Rabbit		1:500	mTOR activity	WB: incubate overnight
ULK1 pSer757	Cell signaling (6888)	Rabbit		1:1000	mTOR activity	WB: according to datasheet instructions (TBS not PBS)
ULK1	Cell signaling (D8H5)	Rabbit	1:500		ULK1 puncta	IF: Triton permeabilization, incubate overnight at 4 °C
TGN46	Serotec (AHP500G)	Sheep	1:500		ATG9 trafficking	
WIPI2	2A2 Abcam (ab105459)	Mouse	1:4000	1:1000	WIPI2 puncta	IF: methanol permeabilization

5. For a detailed description of quantifying autophagic structures using confocal microscopy and the Imaris 7.5 software package (Bitplane), we recommend reading [19].
6. We recommend to also determine whether knockdown/over-expression of candidate proteins affects total protein levels of LC3, WIPI2, ATG9, ATG16L1, and ULK1 (by Western Blot analysis), which can result in altered spot numbers or trafficking. *See* Subheading 3.2.
7. Run Bis-Tris 4–12% gradient gels in MES buffer for a good separation of lipidated and non-lipidated LC3 as well as ATG13 pS318 from a nonspecific band.
8. If cells tend to detach after EBSS treatment, wash starved cells only once with ice-cold PBS or directly lyse in ice-cold lysis buffer to reduce cell loss.
9. Ideally process samples as soon as possible. However, cells can be left in lysis buffer in the dish at −80 °C (put plate directly onto metal shelf in freezer). Boiled SDS samples should be stored at −80 °C as well.
10. For analysis of LC3 lipidation and ATG13 Ser318 phosphorylation (measuring ULK1 activity), use MES running buffer to obtain a good separation between LC3-I and LC3-II and to improve the separation of p-ATG13 Ser318 from a nonspecific band, respectively.
11. Phosphorylation-specific antibodies need to be diluted in 5% BSA (fraction V) in TBST as PBST and milk can cause a high background and obscure detection. However, ATG13 pSer318 antibody works best in 5% skimmed milk in TBST.
12. For detection of phosphorylated proteins, wash with TBST.

References

1. Gallagher LE, Williamson LE, Chan EYW (2016) Advances in autophagy regulatory mechanisms. Cell 5:24. https://doi.org/10.3390/cells5020024
2. Lamb CA, Yoshimori T, Tooze SA (2013) The autophagosome: origins unknown, biogenesis complex. Nat Rev Mol Cell Biol 14:759–774. https://doi.org/10.1038/nrm3696
3. McKnight NC, Jefferies HBJ, Alemu EA, Saunders RE, Howell M, Johansen T, Tooze SA (2012) Genome-wide siRNA screen reveals amino acid starvation-induced autophagy requires SCOC and WAC. EMBO J 31:1931–1946. https://doi.org/10.1038/emboj.2012.36
4. Lipinski MM, Zheng B, Lu T, Yan Z, Py BF, Ng A, Xavier RJ, Li C, Yankner BA, Scherzer CR et al (2010) Genome-wide analysis reveals mechanisms modulating autophagy in normal brain aging and in Alzheimer's disease. Proc Natl Acad Sci U S A 107:14164–14169. https://doi.org/10.1073/pnas.1009485107
5. Kabeya Y (2000) LC3, a mammalian homologue of yeast Apg8p, is localized in autophagosome membranes after processing. EMBO J 19:5720–5728. https://doi.org/10.1093/emboj/19.21.5720
6. Yoshii SR, Mizushima N (2017) Monitoring and measuring autophagy. Int J Mol Sci 18:1–13. https://doi.org/10.3390/ijms18091865
7. Polson HEJ, de Lartigue J, Rigden DJ, Reedijk M, Urbé S, Clague MJ, Tooze SA (2010) Mammalian Atg18 (WIPI2) localizes

to omegasome-anchored phagophores and positively regulates LC3 lipidation. Autophagy 6:506–522. https://doi.org/10.4161/auto.6.4.11863

8. Dooley HC, Razi M, Polson HEJ, Girardin SE, Wilson MI, Tooze SA (2014) WIPI2 links LC3 conjugation with PI3P, autophagosome formation, and pathogen clearance by recruiting Atg12–5-16L1. Mol Cell 55:238. https://doi.org/10.1016/j.molcel.2014.05.021
9. Young ARJ, Chan EYW, Hu XW, Köchl R, Crawshaw SG, High S, Hailey DW, Lippincott-Schwartz J, Tooze SA (2006) Starvation and ULK1-dependent cycling of mammalian Atg9 between the TGN and endosomes. J Cell Sci 119:3888–3900. https://doi.org/10.1242/jcs.03172
10. Orsi A, Razi M, Dooley HC, Robinson D, Weston AE, Collinson LM, Tooze SA (2012) Dynamic and transient interactions of Atg9 with autophagosomes, but not membrane integration, are required for autophagy. Mol Biol Cell 23:1860–1873. https://doi.org/10.1091/mbc.E11-09-0746
11. McAlpine F, Williamson LE, Tooze SA, Chan EYW (2013) Regulation of nutrient-sensitive autophagy by uncoordinated 51-like kinases 1 and 2. Autophagy 9:361–373. https://doi.org/10.4161/auto.23066
12. Karanasios E, Walker SA, Okkenhaug H, Manifava M, Hummel E, Zimmermann H, Ahmed Q, Domart M-C, Collinson L, Ktistakis NT (2016) Autophagy initiation by ULK complex assembly on ER tubulovesicular regions marked by ATG9 vesicles. Nat Commun 7:12420. https://doi.org/10.1038/ncomms12420
13. Lamb CA, Nühlen S, Judith D, Frith D, Snijders AP, Behrends C, Tooze SA (2016) TBC1D14 regulates autophagy via the TRAPP complex and ATG9 traffic. EMBO J 35:281–301. https://doi.org/10.15252/embj.201592695
14. Juhász G (2012) Interpretation of bafilomycin, pH neutralizing or protease inhibitor treatments in autophagic flux experiments: novel considerations. Autophagy 8:1875–1876. https://doi.org/10.4161/auto.21544
15. Zoncu R, Bar-Peled L, Efeyan A, Wang S, Sancak Y, Sabatini DM (2011) mTORC1 senses lysosomal amino acids through an inside-out mechanism that requires the vacuolar H+-ATPase. Science 334:678–683. https://doi.org/10.1126/science.1207056
16. Florey O, Gammoh N, Kim SE, Jiang X, Overholtzer M (2015) V-ATPase and osmotic imbalances activate endolysosomal LC3 lipidation. Autophagy 11:88–99. https://doi.org/10.4161/15548627.2014.984277
17. Kimura S, Noda T, Yoshimori T (2007) Dissection of the autophagosome maturation process by a novel reporter protein, tandem fluorescent-tagged LC3. Autophagy 3:452–460
18. Kaizuka T, Morishita H, Hama Y, Tsukamoto S, Matsui T, Toyota Y, Kodama A, Ishihara T, Mizushima T, Mizushima N (2016) An autophagic flux probe that releases an internal control. Mol Cell 64:835–849. https://doi.org/10.1016/j.molcel.2016.09.037
19. Lamb CA, Joachim J, Tooze SA (2017) Quantifying autophagic structures in mammalian cells using confocal microscopy. Methods Enzymol 587:21–42. https://doi.org/10.1016/bs.mie.2016.09.051

Chapter 26

In Vitro Screening Platforms for Identifying Autophagy Modulators in Mammalian Cells

Elena Seranova, Carl Ward, Miruna Chipara, Tatiana R. Rosenstock, and Sovan Sarkar

Abstract

Autophagy is a vital homeostatic pathway essential for cellular survival and human health. It primarily functions as an intracellular degradation process for the turnover of aggregation-prone proteins and unwanted organelles. Dysregulation of autophagy underlying diverse human diseases reduces cell viability, whereas stimulation of autophagy is cytoprotective in a number of transgenic disease models including neurodegenerative disorders. Thus, therapeutic exploitation of autophagy is considered a potential treatment strategy in certain human diseases, and therefore, chemical inducers of autophagy have tremendous biomedical relevance. In this review, we describe the in vitro screening platforms to identify autophagy modulators in mammalian cells using various methodologies including fluorescence and high-content imaging, flow cytometry, fluorescence and luminescence detection by microplate reader, immunoblotting, and immunofluorescence. The commonly used autophagy reporters in these screening platforms are either based on autophagy marker like LC3 or autophagy substrate such as aggregation-prone proteins or p62/SQSTM1. The reporters and assays for monitoring autophagy are evolving over time to become more sensitive in measuring autophagic flux with the capability of high-throughput applications for drug discovery. Here we highlight these developments and also describe the stringent secondary autophagy assays for characterizing the autophagy modulators arising from the primary screen. Since autophagy is implicated in myriad human physiological and pathological conditions, these technologies will enable identifying novel chemical modulators or genetic regulators of autophagy that will be of biomedical and fundamental importance to human health.

Key words Autophagy, Autophagy substrates, Autophagy modulators, Autophagy drug discovery, LC3, p62, Aggregation-prone protein, Chemical screen, Screening platform

1 Introduction

Macroautophagy (herein referred to as autophagy) is the main disposal route for aggregation-prone protein and damaged organelles in the cells. This evolutionarily conserved intracellular proteolytic pathway is essential for cellular homeostasis and organismal survival. Multiple lines of evidence have designated autophagy primarily as a cell survival mechanism [1]. Owing to its vital

Nicholas Ktistakis and Oliver Florey (eds.), *Autophagy: Methods and Protocols*, Methods in Molecular Biology, vol. 1880, https://doi.org/10.1007/978-1-4939-8873-0_26,

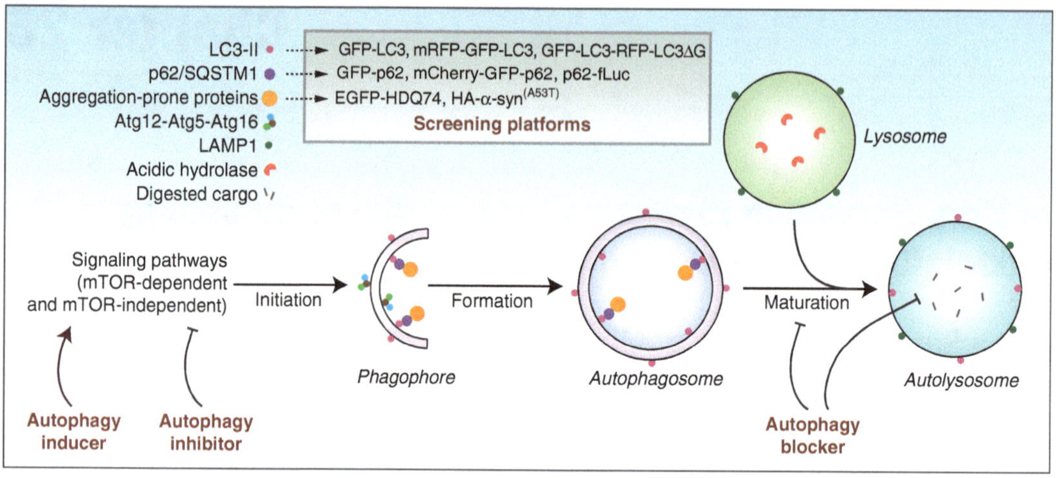

Fig. 1 Schematic representation of the autophagy pathway highlighting the screening platforms and the point of impact of the autophagy modulators. Autophagy is initiated by the formation of phagophores, which expand and engulf cytoplasmic materials to form autophagosomes that then fuse with the lysosomes to form autolysosomes where the autophagic cargo is degraded. Upstream signaling pathways via mTOR-dependent and mTOR-independent routes regulate autophagy. This process is amenable to chemical perturbations, and the points of action of autophagy inducers, inhibitors, and blockers are indicated. Some of the commonly used in vitro screening platforms to identify autophagy modulators in mammalian cells, based on LC3, p62/SQSTM1, and aggregation-prone proteins, are highlighted. LC3 (LC3-II form associated with autophagosomes throughout their life span) is used as a marker of autophagy, while p62/SQSTM1 and aggregation-prone proteins are autophagy substrates

housekeeping functions, autophagy has growing implications in myriad human physiology and diseases, such as but not limited to development, immunity, aging, cancer, and neurodegeneration [2–5]. Of biomedical interest, therapeutic modulation of autophagy is considered as a potential treatment strategy for diverse diseases including neurodegenerative disorders [6–11].

Autophagy is a dynamic process encompassing several intracellular vesicular compartments and vesicle fusion events (Fig. 1). It is initiated by the formation of double-membrane structures in the cytoplasm called phagophores, which expand and engulf cytoplasmic autophagic cargo to form double-membrane-bound vesicles called autophagosomes. Autophagosomes then fuse with the late endosomes to form single-membrane-bound vesicles called amphisomes, which then fuse with the acidic lysosomes to form the degradative autolysosomes where the autophagic cargo is degraded by the lysosomal hydrolytic enzymes [12–14]. Autophagosome maturation is normally mediated via this multistep vesicle fusion, but direct autophagosome-lysosome fusion also occurs [15, 16]. The turnover of autophagosomes and autophagic cargo is collectively termed as autophagic flux.

Several autophagy-related genes (*Atg*) orchestrate the process of autophagy [17]. Two ubiquitin-like conjugation systems are required for the initiation step of autophagy and phagophore formation: the Atg5-Atg12-Atg16 complex and the phosphatidylethanolamine-conjugated microtubule-associated protein 1 light chain 3 (LC3-II) [18, 19]. LC3-II remains associated with autophagosome membrane throughout their life span, and thus LC3-based reporters are widely used as a quantitative marker to study autophagy [20–23] (Fig. 1). Conventional or conditional autophagy-deficient mouse models have also been generated by knockout of essential autophagy genes such as *Atg5* or *Atg7* that influence the initiation stage of autophagy [24–28]. In addition, the clearance of autophagic cargo such as aggregation-prone proteins (a process termed aggrephagy), organelles like mitochondria and peroxisomes (processes termed mitophagy and pexophagy, respectively), and specific substrates like p62 can be used to monitor the flux through the autophagy pathway [29–33].

Several signaling pathways have been described to regulate autophagy. The mechanistic target of rapamycin complex 1 (mTORC1) pathway is the classical route that negatively regulates autophagy via the Atg1-ULK1-FIP200 complex [34–36]. In addition, a number of mTOR-independent pathways also negatively govern autophagy, such as elevated intracellular inositol, IP_3, cAMP, and Ca^{2+} levels [8, 37, 38]. Modulation of autophagy can be achieved by targeting these autophagy-regulating pathways with small molecules (Fig. 1). Few examples of autophagy inducers include the mTOR inhibitors, rapamycin and torin1 [35, 39], as well as mTOR-independent compounds such as carbamazepine, trehalose, and SMERs [37, 40, 41]. Since mTOR controls critical cellular functions like growth, proliferation, and translation and stimulation of autophagy by mTOR inhibitors could be associated with potential side effects, modulation of autophagy by mTOR-independent means is more desirable for clinical applications [6–8, 34]. Many of the autophagy inducers identified have demonstrated beneficial effects in transgenic models of various human diseases including a number of neurodegenerative diseases and certain liver diseases and myopathies [6–8, 10].

Autophagy modulators have tremendous biomedical relevance for therapeutic exploitation [6–10]. Over the last decade, several chemical screening methods have been utilized to identify autophagy modulators [42–45]. These include analyzing the perturbations in autophagic compartments via LC3-based reporters [20–22] and monitoring the levels of autophagic cargo via the clearance of autophagy substrates such as aggregation-prone proteins and p62 [29, 30] or organelle turnover such as mitochondria and peroxisomes [46, 47]. The screening assays for autophagy are evolving over time to be more sensitive with possible high-throughput applications for drug discovery. In this chapter, we describe the commonly used in vitro screening platforms for identifying autophagy modulators in mammalian cells using various methodologies including fluorescence and high-content image analysis, immunoblotting, flow cytometry, and fluorescence and luminescence detection by microplate reader (Table 1). We also highlight some secondary assays for characterizing the chemical modulators of autophagy as inducers, inhibitors, or blockers (Table 1).

2 Materials

2.1 Cell Culture

MEF medium—HeLa cells, mouse embryonic fibroblasts (MEFs), and neuroblastoma cells are cultured in MEF medium (composition for 500 mL):

1. Dulbecco's Modified Eagle Medium (DMEM) (435 mL, or the desired volume if selection antibiotics are added).
2. 10% fetal bovine serum (FBS) (50 mL).
3. 1% penicillin/streptomycin (100 U mL^{-1} final concentration) (5 mL of 10,000 U mL^{-1} solution).
4. 1% L-glutamine (2 mM final concentration) (5 mL of 200 mM solution).
5. 1% nonessential amino acids (1× or 100 μM final concentration) (5 mL of 100× or 10 mM solution).
6. Selection antibiotics for stable cell lines (lentiviral overexpression) expressing GFP-LC3, mRFP-GFP-LC3, GFP-LC3-RFP-LC3ΔG, inducible GFP-p62, or inducible p62-fLuc: 1 μg mL^{-1} puromycin (50 μL of 10 mg mL^{-1} stock, which is made in sterile distilled water and stored at −20 °C freezer as frozen aliquots).

PC12 medium—PC12 cells are cultured in PC12 medium (composition for 500 mL):

1. DMEM (415 mL or 413 mL if selection antibiotics are added).
2. 10% horse serum (50 mL).
3. 5% FBS (25 mL).
4. 1% penicillin/streptomycin (5 mL of 10,000 U mL^{-1} solution).
5. 1% L-glutamine (5 mL of 200 mM solution).
6. Selection antibiotics for stable inducible PC12 cell lines expressing EGFP-HDQ74 or HA-α-$syn^{(A53T)}$: 100 μg mL^{-1} G418/geneticin (500 μL of 100 mg mL^{-1} stock, which is made in sterile water and stored at −20 °C freezer as frozen aliquots) and 75 μg mL^{-1} hygromycin (1.44 mL of 26 mg mL^{-1} stock, which is made in DMEM and stored at −20 °C freezer as frozen aliquots).

Table 1
In vitro screening and assay platforms for autophagy commonly used to identify and characterize autophagy modulators in mammalian cells

Autophagy reporter screening and assay platforms	Expected readout for autophagy inducer	Expected readout for autophagy blocker	Expected readout for autophagy inhibitor	Approaches for analysis	Possible high-throughput applications
Autophagy reporter screening platforms					
GFP-LC3	Increase in GFP^+ vesicles	Increase in GFP^+ vesicles	Decrease in GFP^+ vesicles	Fluorescence microscopy	Yes
mRFP-EGFP-LC3	Increase in $mRFP^+/GFP^+$ and $mRFP^+/GFP^-$ vesicles	Increase in $mRFP^+/GFP^+$ and decrease in $mRFP^+/GFP^-$ vesicles	Decrease in $mRFP^+/GFP^+$ and $mRFP^+/GFP^-$ vesicles	Fluorescence microscopy	Yes
GFP-LC3-RFP-LC3ΔG	Decrease in GFP/RFP ratio	Increase in GFP/RFP ratio	Increase in GFP/RFP ratio	Fluorescence microscopy, Flow cytometry, Microplate reader	Yes
Inducible GFP-p62	Decrease in GFP-p62 levels/intensity	Increase in GFP-p62 levels/intensity	Increase in GFP-p62 levels/intensity	Flow cytometry, Microplate reader	Yes
Inducible p62-fLuc	Decrease in p62-fLuc levels/intensity	Increase in p62-fLuc levels/intensity	Increase in p62-fLuc levels/intensity	Microplate reader	Yes
Inducible EGFP-HDQ74	Decrease in EGFP-HDQ74 levels	Increase in EGFP-HDQ74 levels	Increase in EGFP-HDQ74 levels	Immunoblotting	No
Inducible HA-α-syn$^{(A53T)}$	Decrease in HA-α-syn$^{(A53T)}$ levels	Increase in HA-α-syn$^{(A53T)}$ levels	Increase in HA-α-syn$^{(A53T)}$ levels	Immunoblotting	No
Other autophagy assay platforms					
LC3-II	Increase in $LC3^+$ vesicles and LC3-II levels	Increase in $LC3^+$ vesicles and LC3-II levels	Decrease in $LC3^+$ vesicles and LC3-II levels	Immunofluorescence, Immunoblotting	No
LC3-II with bafilomycin A_1 treatment	Increase in LC3-II levels compared to bafilomycin A_1 alone	No change in LC3-II levels compared to bafilomycin A_1 alone	Decrease in LC3-II levels compared to bafilomycin A_1 alone	Immunoblotting	No

(continued)

Table 1
(continued)

Autophagy reporter screening and assay platforms	Expected readout for autophagy inducer	Expected readout for autophagy blocker	Expected readout for autophagy inhibitor	Approaches for analysis	Possible high-throughput applications
LC3 or p62 colocalization with LAMP1	Increase in colocalization	Decrease in colocalization	Decrease in colocalization	Immunofluorescence	No
GFP-p62	Decrease in GFP^+ puncta	Increase in GFP^+ puncta	Increase in GFP^+ puncta	Fluorescence microscopy	Yes
mCherry-GFP-p62	Decrease in $mCherry^+/GFP^+$ and $mCherry^+/GFP^-$ puncta	Increase in $mCherry^+/GFP^+$ and decrease in $mCherry^+/GFP^-$ puncta	Increase in $mCherry^+/GFP^+$ and decrease in $mCherry^+/GFP^-$ puncta	Fluorescence microscopy	Yes
EGFP-HDQ74	Decrease in $EGFP^+$ aggregates	Increase in $EGFP^+$ aggregates	Increase in $EGFP^+$ aggregates	Fluorescence microscopy	No
Atg5 or Atg16	Increase in $Atg5^+$ or $Atg16^+$ vesicles	No change in $Atg5^+$ or $Atg16^+$ vesicles	Decrease in $Atg5^+$ or $Atg16^+$ vesicles	Immunofluorescence	No
GFP-DFCP1	Increase in GFP^+ puncta	No change in GFP^+ puncta	Decrease in GFP^+ puncta	Fluorescence microscopy	Yes
GFP-WIPI-1	Increase in GFP^+ puncta	Increase in GFP^+ puncta	Decrease in GFP^+ puncta	Fluorescence microscopy	Yes

Expected readouts for autophagy modulators are indicated as a general guidance only; however, the actual readout may vary in some instances depending on the dose of the compound, the duration of the treatment, and the cell type used.

hESC medium (on feeders)—Human embryonic stem cells (hESCs) are maintained on mitomycin C-inactivated MEF feeder layer in hESC medium (composition for 500 mL):

1. DMEM/F12 (385 mL).
2. 15% hESC-qualified (HyClone) FBS (75 mL).
3. 5% knockout serum replacement (KSR) (25 mL).
4. 1% penicillin/streptomycin (5 mL of 10,000 U mL^{-1} solution).
5. 1% L-glutamine (5 mL of 200 mM solution).
6. 1% nonessential amino acids (5 mL of 100× solution).
7. 4 ng mL^{-1} human recombinant basic fibroblast growth factor (bFGF) (80 μL of 25 μg mL^{-1} stock, which is made in 0.1% BSA (bovine serum albumin) solution and stored at −20 °C freezer as frozen aliquots).
8. 0.1 mM 2-mercaptoethanol (3.5 μL of 14.3 M stock).
9. Selection antibiotics for stable hESC lines expressing mRFP-GFP-LC3: 1 μg mL^{-1} puromycin (50 μL of 10 mg mL^{-1} stock).

hESC feeder-free medium—During experimentation, hESCs are dissociated with 1 mg mL^{-1} collagenase from the MEF feeder layer and are cultured on basement membrane matrix (Matrigel) in hESC feeder-free culture medium (mTeSR1):

1. mTeSR1 basal medium (445 mL).
2. mTeSR1 5× supplement (50 mL).
3. 1% penicillin/streptomycin (5 mL of 10,000 U mL^{-1} solution).
4. Selection antibiotics for stable hESC lines expressing mRFP-GFP-LC3: 1 μg mL^{-1} puromycin (50 μL of 10 mg mL^{-1} stock).

Cell dissociation reagents and basement membrane matrix:

1. 0.25% trypsin-EDTA solution (for mammalian cells apart from hESCs).
2. Collagenase type IV (for hESCs cultured on mitomycin C-inactivated MEF feeders).
3. 500 μM EDTA solution: Prepared by diluting in PBS from 500 mM stock (for hESCs cultured feeder-free on Matrigel).
4. Basement membrane matrix (Matrigel; for culturing hESCs feeder-free).

Cell culture plasticware and reagents:

1. 75 cm^2 flasks (for mammalian cells except hESCs).
2. 15 cm dishes (for mammalian cells except hESCs).

3. 6-well plates (for hESCs).
4. 1.8 cm cell scrapers.
5. Phosphate buffered saline (PBS).
6. 0.2% gelatin in PBS.
7. Mitomycin C (dissolved in DMEM at 5 μg mL^{-1}; for MEF inactivation).
8. PBS+ (with $CaCl_2$ and $MgCl_2$; for MEF inactivation).
9. 1 μg mL^{-1} doxycycline (for inducing transgene synthesis in inducible cell lines): Dissolved in DMSO at 1 mg mL^{-1} and stored at −20 °C freezer as frozen aliquots. This compound is light-sensitive and is stored in dark Eppendorf tubes.
10. Dimethyl sulfoxide (DMSO; solvent for dissolving compounds).

2.2 Autophagy Modulators

The following autophagy modulators have been used as examples to describe the methodology. It is important to note that the effects of autophagy modulators might be dependent on the dose, the duration of treatment, and the cell type used (*see* **Note 1**).

1. Rapamycin: Autophagy inducer; mTOR inhibitor [35]; generally used at 200 nM. Rapamycin is light-sensitive and is stored in dark Eppendorf tubes. It is dissolved in DMSO and stored at −20 °C as 200 μM stock for short-term storage or as 200 mM stock for long-term storage.
2. Carbamazepine: mTOR-independent autophagy inducer; inositol-lowering agent [37]; used at 50 μM. It is dissolved in DMSO and stored at −20 °C as 50 mM stock.
3. Trehalose: mTOR-independent autophagy inducer [40]; inhibits SLC2A glucose transporters [48] and activates TFEB by Akt inhibition [49]; used at 100 mM. Although we previously demonstrated the autophagy-enhancing properties of trehalose that have been utilized in several studies both in vitro and in vivo [40, 48–58], few recent reports suggest that it may retard autophagic flux by impairing the lysosomal function [22, 59].
4. Bafilomycin A_1: Autophagy blocker; inhibits lysosomal vacuolar H^+-ATPase (v-ATPase), thereby preventing lysosomal cargo degradation and autophagosome-lysosome fusion [60–62]; used at 400 nM for 4 h (saturating concentration in most mammalian cell lines for the autophagosome formation assay) [40, 41]. This compound is light-sensitive and is stored in dark Eppendorf tubes. It is dissolved in DMSO and stored at −20 °C as 100 μM stock.

2.3 Transient Transfection of Autophagy Reporter Constructs

1. 6-well, 12-well, or 24-well plates.
2. Transfection reagent (Lipofectamine 2000).
3. Reduced serum medium (Opti-MEM).
4. Plasmid DNA: GFP-LC3 (from Noboru Mizushima, Japan), mRFP-GFP-LC3 (from Tamotsu Yoshimori, Japan), GFP-LC3-RFP-LC3ΔG (from Noboru Mizushima, Japan), and EGFP-HDQ74 (from David Rubinsztein, UK) constructs.
5. Vortex mixer.

2.4 Lentiviral Overexpression of Autophagy Reporter Constructs

1. HEK-293T packaging cells.
2. Target cell line.
3. 0.2% gelatin.
4. 10 cm dishes.
5. 1.5 mL Eppendorf tubes and 15 mL and 50 mL Falcon tubes.
6. FuGENE HD transfection reagent (Promega).
7. Serum-free DMEM.
8. Lentiviral plasmid DNA: LC3 reporter constructs or inducible p62 constructs.
9. Viral packaging and envelope plasmid DNA: psPAX2 (Gag, Pol, Rev, and Tat expressing packaging plasmid) and pMD2. G (VSV-G expressing envelope plasmid).
10. BSL2+ room for lentiviral cell culture facility.
11. 0.45 μm syringe filter.
12. 20 mL syringe.
13. SW-28 rotors with buckets and caps (Beckman).
14. SW-28 thin-walled ultra-clear centrifuge tubes (Beckman).
15. Benchtop centrifuge.
16. Opti-MEM medium.
17. 8 μg mL^{-1} Polybrene: Prepare 4 μg mL^{-1} stock in sterile water, filter-sterilize, and store at −20 °C freezer as frozen aliquots.
18. 0.25% trypsin-EDTA solution.
19. 1 μg mL^{-1} puromycin.

2.5 Fluorescence Microscopy and Immunofluorescence Analysis

1. 6-well, 12-well, or 24-well plates.
2. Glass coverslips of appropriate size.
3. Phosphate buffered saline (PBS).
4. 4% formaldehyde (methanol-free): Prepared by diluting 16% formaldehyde solution in PBS.
5. Anti-fade mountant with DAPI (4′,6-diamidino-2-phenylindole).

6. 0.5% Triton X-100 in PBS.
7. Chilled methanol at −20 °C (for LC3 immunostaining only).
8. 5% Goat Serum in PBS.
9. Primary antibodies: Anti-LC3 antibody (catalog number: 0231–100/LC3-5F10; NanoTools), anti-p62 antibody (catalog number: 610833; BD Biosciences; for human cells only); anti-p62 antibody (catalog number: GP62-C; Progen Biotechnik; for mouse cells only).
10. Secondary antibodies: Alexa Fluor 488 goat anti-mouse IgG H + L (catalog number: A-11001; Thermo Fisher Scientific) or Alexa Fluor 594 goat anti-mouse IgG H + L (catalog number: A-11005; Thermo Fisher Scientific) or Alexa Fluor 647 goat anti-mouse IgG H + L (catalog number: A32728; Thermo Fisher Scientific), Alexa Fluor 594 goat anti-guinea pig IgG H + L (catalog number: A-11076; Thermo Fisher Scientific).
11. Forceps.
12. Microscopic glass slides.
13. Nail varnish.
14. Software for image analysis such as ImageJ or CellProfiler.
15. Software for statistical analysis such as Microsoft Excel or GraphPad Prism 6.

2.6 High-Content Fluorescent Microscopy and Image Analysis

1. 96-well microplate with optical bottom.
2. Multi-channel pipette.
3. PBS.
4. 4% formaldehyde (methanol-free) in PBS.
5. DAPI staining agent (400 ng mL^{-1} in PBS).
6. Gelvatol (for 96-well plates): Prepare 50 mL Gelvatol by adding 4.2 g of polyvinyl alcohol (PVA) to 10.5 mL of glycerol, then add 30 mL of distilled water and a few crystals of sodium azide, and finally add 21 mL of 0.2 M Tris-HCl pH 8.5. Stir in low heat (45 °C) for a few hours or until the reagents dissolve, and then centrifuge at 5000 × *g* for 15 min. Aliquot and keep at 4 °C for short-term storage.
7. Magnetic stirrer with heat block and magnetic stir bars.
8. Thermo Scientific™ CellInsight CX5 or CellInsight CX7 High-Content Screening (HCS) Platform.
9. Thermo Scientific™ Orbitor RS Microplate Mover.
10. Thermo Scientific™ HCS Studio Cell Analysis Software.
11. Software for statistical analysis such as Microsoft Excel, GraphPad Prism 6, or appropriate analysis package for high-content screening data.

2.7 Fluorescence Analysis by Plate Reader

1. White flat bottom 96-well plates.
2. PBS.
3. 4% formaldehyde (methanol-free) in PBS.
4. A plate reader equipped with 488 nm_{ex}/509 nm_{em} and 584 nm_{ex}/607 nm_{em} lasers.
5. Software for statistical analysis such as Microsoft Excel or GraphPad Prism.

2.8 Flow Cytometry

1. 6-well or 12-well plates.
2. 0.25% trypsin-EDTA solution (for mammalian cells except hESCs).
3. 500 μM EDTA solution in PBS (for hESCs only).
4. Flow cytometry buffer: 2% FBS in PBS.
5. Benchtop centrifuge with temperature control.
6. Bottom round 5 mL flow cytometry tubes with filter caps.
7. Flow cytometer.
8. Flow cytometry analysis software such as Summit 4.3 or FlowJo software.
9. Software for statistical analysis such as Microsoft Excel or GraphPad Prism 6.

2.9 SDS-Polyacrylamide Gel Electrophoresis (SDS-PAGE) and Immunoblotting

1. Lysis buffer (2×): 20 mM Tris–HCl pH 6.8, 137 mM NaCl, 1 mM EGTA, 1% Triton X-100, 10% glycerol; stored at −20 °C as frozen aliquots. 1× lysis buffer is prepared fresh with distilled water where protease inhibitor is added just before harvesting the cell pellets for lysis.
2. Complete protease inhibitor cocktail tablet: 1 tablet is dissolved in 2 mL distilled water and stored at −20 °C. While making 1× lysis buffer, the complete protease inhibitor cocktail solution is added at 1:25 dilution.
3. Bradford Protein Assay Kit (Bio-Rad).
4. 1.8 cm cell scrapers.
5. Benchtop centrifuge with temperature control.
6. Glass plates and combs for casting the polyacrylamide gels (Bio-Rad).
7. Resolving gel (for making two 12% resolving gels; total volume 16 mL): 30% acrylamide and bis-acrylamide solution (19:1) (6.4 mL), 1.5 M Tris–HCl pH 8.8 (4 mL), 10% sodium dodecyl sulfate (SDS) (160 μL), 10% ammonium persulfate (APS) (160 μL; made fresh every time), *N*,*N*,*N*,*N*-′-tetramethylethylenediamine (TEMED) (16 μL), distilled water (5.26 mL).

8. Stacking gel buffer (for making two 5% stacking gels; total volume 5 mL): 30% acrylamide and bis-acrylamide solution (19:1) (1 mL), 0.5 M Tris–HCl pH 6.8 (1.25 mL), 10% SDS (200 μL), 10% APS (200 μL; made fresh every time), TEMED (10 μL), distilled water (2.34 mL).
9. Water-saturated isobutanol: Take equal volumes of distilled water and isobutanol, mix overnight by stirring, allow separating, and store at room temperature. Use the top isobutanol layer while casting the resolving gel.
10. SDS-PAGE apparatus (Bio-Rad).
11. Power pack (Bio-Rad).
12. Running buffer (10×): 250 mM Tris and 1.92 M glycine in distilled water; store at room temperature. While preparing 1× running buffer with distilled water, add 10% w/v SDS at 1:100 dilution.
13. 3× Laemmli buffer (for 10 mL): 187.5 mM Tris–HCl pH 6.8 (1.875 mL of 1 M Tris–HCl pH 6.8 stock), 6% w/v SDS (3 mL of 20% SDS solution), 30% glycerol (3 mL), 150 mM dithiothreitol (1.5 mL of 1 M DTT stock), 0.03% w/v bromophenol blue (2.07 mg), distilled water (625 μL). Store at −20 °C freezer as frozen aliquots.
14. Heat block.
15. Pre-stained protein molecular weight marker.
16. Wet blot transfer tank (Mini Trans-Blot Cell; Bio-Rad).
17. Transfer buffer (10×): 250 mM Tris and 1.92 M glycine in distilled water. Make 1× transfer buffer with distilled water.
18. Polyvinylidene fluoride (PVDF) microporous membrane (Amersham Hybond PVDF; GE Healthcare).
19. Methanol.
20. Extra thick filter paper.
21. PBS-T: PBS with 0.1% (v/v) Tween-20.
22. Blocking buffer: 6% (w/v) nonfat dry milk, 0.1% (v/v) Tween-20 in PBS.
23. Orbital shaker.
24. Primary antibodies: Anti-EGFP antibody (catalog number: MA1–952; Thermo Fisher Scientific), anti-HA antibody (catalog number: 26183; Thermo Fisher Scientific), anti-LC3B antibody (catalog number: NB100–2220; Novus Biologicals), anti-p62 antibody (catalog number: 610833; BD Biosciences; for human cells only); anti-p62 antibody (catalog number: GP62-C; Progen Biotechnik; for mouse cells only), anti-actin antibody (catalog number: A2066; Sigma-Aldrich/Merck),

anti-GAPDH antibody (catalog number: G8795; Sigma-Aldrich/Merck).

25. Secondary antibodies: Goat anti-rabbit IgG H&L chain-specific peroxidase conjugate (catalog number: 401393; Merck), goat anti-mouse IgG H&L chain-specific peroxidase conjugate (catalog number: 401253; Merck), goat anti-guinea pig IgG antibody HRP conjugate (catalog number: AP108P; Merck).
26. Cling film.
27. Western Blotting Detection Reagent (Amersham ECL; GE Healthcare).
28. Autoradiography films: For detecting chemiluminescent signal from protein blots.
29. Autoradiography cassettes.
30. Dark room and developer.
31. Software for statistical analysis such as Microsoft Excel or GraphPad Prism 6.

2.10 Luminescence Analysis by Plate Reader

1. White flat bottom 96-well plates.
2. PBS.
3. 4% formaldehyde (methanol-free) in PBS.
4. Aluminum foil.
5. ONE-Glo™ + Tox Luciferase Reporter and Cell Viability Assay Kit (Promega) comprising of CellTiter-Fluor Cell Viability Assay Reagent and ONE-Glo Reagent.
6. A luminometer or a plate reader equipped to read luminescence as well as with 380–400 nm_{ex}/505nm_{em} laser.
7. Software for statistical analysis such as Microsoft Excel or GraphPad Prism 6.

3 Methods

3.1 Cell Culture

3.1.1 Culture of Mammalian Cell Lines

1. HeLa cells, mouse embryonic fibroblasts (MEFs), and neuroblastoma cell lines expressing the autophagy reporters described below are maintained in full growth medium (MEF medium with appropriate selection antibiotics; described in Subheading 2.1) at 37 °C and 5% carbon dioxide (CO_2) in a humidified incubator. PC12 cell lines expressing the autophagy substrates described below are maintained in full growth medium (PC12 medium with appropriate selection antibiotics; described in Subheading 2.1) at 37 °C and 10% CO_2 in a humidified incubator.

2. The cells are passaged when they reach 90% confluency in 1:5–1:10 ratios depending on confluency. The cell culture medium from the flask is removed, and the cells are rinsed once with DMEM (without serum) or sterile PBS.
3. For cell dissociation, 0.25% trypsin-EDTA solution (2.5 mL for 75 cm^2 flasks) is added to cover the cell layer, and the flask is incubated at 37 °C for 5 min or until the cells are dissociated (but no longer than 15 min).
4. The enzyme activity is quenched by adding 7.5 mL of fresh MEF medium (for 2.5 mL trypsin-EDTA used in 75 cm^2 flasks), and the cells are dissociated by pipetting up and down few times. Any formation of cell clumps is dissociated by rigorous pipetting to a fine suspension before re-seeding them.
5. The cells are maintained at 37 °C in a 5% or 10% CO_2 humidified incubator for 3–4 days until next passaging.

3.1.2 Culture of Human Embryonic Stem Cells

1. Human embryonic stem cells (hESCs) are cultured as pluripotent colonies (without being differentiated) on MEF feeder layer in 6-well plates. The cells are maintained in hESC medium (described in Subheading 2.1) at 37 °C in a humidified incubator with 5% CO_2 and 5% O_2.
2. MEF feeder plate is prepared by seeding 3.5×10^6 inactivated MEFs in 0.2% gelatin-coated 6-well plate. For gelatin coating, add 500 μL of 0.2% gelatin per well for 30 min in the tissue culture hood; then air-dry in the hood for 15–30 min. For preparing inactivated MEFs, passage 2 or 3 MEFs are treated with 5 μg mL^{-1} mitomycin C for 2 h (to prevent proliferation), followed by washing the cells three times with sterile PBS+ and once with PBS. Generally, MEFs are grown in 15 cm tissue culture dishes for expansion and inactivation and then stored in liquid nitrogen as frozen aliquots. At least 1–4 days prior to passaging the hESCs, inactivated MEFs are thawed on new 6-well plates.
3. The hESCs are passaged every 5–7 days with 1 mg mL^{-1} collagenase, typically at 1:3–1:6 ratios depending on confluency.
4. Add 1 mL collagenase per well of a 6-well plate. Incubate at 37 °C in a humidified incubator with 5% CO_2 and 5% O_2 for 45–60 min.
5. When the hESC colonies start to lift from the MEF feeder layer, gently tap the plate to facilitate this process without dissociating the MEF feeder layer. Add 1 mL of hESC medium per well, and gently pipette the floating hESC colonies in a sterile 15 mL Falcon tube.

6. To remove any residual MEFs during dissociation, allow the hESC colonies to settle in the bottom for 3–5 min. Gently aspirate the medium and wash the hESC colonies with 5 mL of fresh hESC medium and repeat this step. To passage the hESC colonies on a new 6-well MEF feeder plate, hESC colonies are collected above from 1 to 3 wells (depending on confluency).
7. Aspirate the MEF medium in the new 6-well MEF feeder plate, and add 2 mL of hESC medium in each well.
8. Add 6 mL of hESC medium in the hESC colonies that have settled in the 15 mL Falcon tube after the washing step. Pipette gently few times to break the hESC colonies into smaller colonies. Add 1 mL of the resuspended hESCs per well (containing 2 mL of hESC medium) into the new 6-well MEF feeder plate.
9. Incubate at 37 °C in a humidified incubator with 5% CO_2 and 5% O_2 for 5–7 days. After placing the plate in incubator, gently move the plate back and forth and left to right a couple of times to evenly distribute the colonies across the well. If this step is omitted, colonies will settle in the center of the well. Change the cell culture medium every day with 3–4 mL of fresh hESC medium until the next passaging.
10. For experimentation, hESCs are cultured feeder-free in mTeSR1 medium (described in Subheading 2.1) on basement membrane matrix (Matrigel). Matrigel-coated 6-well plates are prepared fresh or 1–2 days prior to seeding. Thaw 1 mL of Matrigel on ice; then add to 19 mL of cold DMEM/F12 in chilled 15 mL Falcon tube. Mix well and immediately, add 1 mL of the mixture to each well of chilled 6-well plates. Incubate in the tissue culture hood at room temperature for 45–60 min. For fluorescence imaging, insert sterile glass coverslips in the wells prior to adding the Matrigel mixture. Use the Matrigel-coated plates in the same day, or store in the cold room for up to 7 days. Warm the plates to room temperature before using.
11. Dissociate the hESCs on MEF feeders with collagenase type IV as per the **steps 3–6** above.
12. To further remove any residual feeder MEFs, resuspend the hESC colonies in hESC medium, plate the hESCs on 0.2% gelatin-coated 10 cm tissue culture dishes, and incubate at 37 °C in a humidified incubator with 5% CO_2 and 5% O_2 for 30–40 min.
13. Collect the floating hESC colonies (gently tap the dish if necessary) into a 15 mL Falcon tube, and allow to settle down for 5 min. Wash the cells once with mTeSR1 medium, and allow to settle down for 5–10 min. For seeding in a 6-well Matrigel-coated plate, add 6 mL of mTeSR1 medium to the

cells, and pipette gently to break the hESC colonies into smaller colonies (*see* **Note 2**).

14. Aspirate the DMEM/F12 medium in the Matrigel-coated plates, and add 1 mL of mTeSR1 medium per well. Add 1 mL of the resuspended hESCs per well (containing 1 mL of mTeSR1 medium) into the Matrigel-coated 6-well plate.
15. Incubate at 37 °C in a humidified incubator with 5% CO_2 and 5% O_2 for 2–4 days. Change the cell culture medium every day with 2–3 mL (depending on confluency) of fresh mTeSR1 medium until the hESCs are 60–80% confluent for experiments.

3.2 Generation of Autophagy Reporter Cell Lines

Autophagy reporter cell lines can be generated either by transient transfection or stable lentiviral overexpression of the reporter constructs. The most commonly used are the LC3-based reporters [63], such as GFP-LC3 [20], mRFP-GFP-LC3 [21], and the newly-described GFP-LC3-RFP-LC3ΔG [22] constructs, or reporter systems based on autophagic substrate clearance, such as GFP-p62 [64], p62-fLuc [65], EGFP-HDQ74 (EGFP-tagged mutant huntingtin exon 1 with 74 polyglutamine repeats) [66], and HA-α-syn$^{(A53T)}$ (HA-tagged A53T α-synuclein mutant) [67]. Various mammalian cells lines can be subjected to transient transfection or stable lentiviral overexpression, but the latter method is to be used for hESCs.

3.2.1 Transient Transfection with Autophagy Reporter Constructs

1. Seed cells in 6-well, 12-well, or 24-well plates containing glass coverslips (for analyzing few compounds) in full growth medium (MEF medium; described in Subheading 2.1), 24 h prior to transfection (*see* **Note 3**). For mammalian cells to attain 70%–80% confluency, typical seeding density is 5×10^5 cells per well for 6-well plate, 2.5×10^5 cells per well for 12-well plate, or 1.25×10^5 cells per well for 24-well plate. For high-throughput chemical screening, stable cell lines with lentiviral overexpression of autophagy reporters in 96-well plates are ideal (described in Subheading 3.2.2).
2. For transient transfection with Lipofectamine 2000, use 1–2.5 μg of autophagy reporter plasmid DNA per well of a 6-well plate. The amount of DNA depends on how easy (1–1.5 μg for HeLa cells) or difficult (2–2.5 μg for neuroblastoma cells) to transfect the cell lines.
3. For each well of a 6-well plate, dilute the desired amount of plasmid DNA in reduced serum medium (Opti-MEM) to make the final volume of 200 μL.
4. For each well of a 6-well plate, dilute 10 μL Lipofectamine 2000 in 190 μL Opti-MEM.

5. Gently vortex both mixtures, and allow to stand for 5 min but no more than 10 min.
6. Mix the DNA mixture with the Lipofectamine 2000 mixture (in 1:1 ratio) while gently shaking the tube manually during the whole process, and then incubate for 20 min at room temperature.
7. Change medium in the 6-well plate to be transfected with 2 mL of fresh medium (MEF medium).
8. Add 400 μL of the DNA-Lipofectamine 2000 mixture dropwise to each well of the 6-well plate, and gently swirl to mix.
9. Incubate for 4–6 h (easy to transfect cells like HeLa cells) or overnight (difficult to transfect cells like neuroblastoma cells) at 37 °C in a humidified incubator with 5% CO_2.
10. After the incubation period, remove the transfection medium, and add fresh medium (MEF medium).
11. Post-transfection, treat cells with the compounds for identifying autophagy modulators. Depending on the screening method and analysis, the treatment period can vary between 4–6 h and 12–24 h (ideally for LC3-based reporters) or 48 h (ideally for autophagy substrate-based reporters).

3.2.2 Lentiviral Overexpression of Autophagy Reporter Constructs

1. Seed 2.5×10^6 HEK-293 T cells on a 0.2% gelatin-coated 10 cm dish in 12 mL of MEF medium, and incubate for 24 h in a 5% CO_2 humidified incubator. Cells should be about 80–85% confluent the next day at the time of transfection.
2. Add 40 μL of FuGENE HD transfection reagent (Promega) to 500 μL of serum-free DMEM, and then mix and incubate at room temperature for 5 min.
3. Prepare the plasmid DNA to be transfected to a total of 20 μg. Add 10 μg of the autophagy reporter plasmid DNA, 7.5 μg of psPAX2 DNA (viral packaging plasmid), and 2.5 μg of pMD2. G DNA (viral envelope plasmid) in 100 μL of serum-free DMEM.
4. Add the FuGENE/DMEM mix (from **step 2**) to the DNA/D-MEM mix (from **step 3**), mix well by tapping the tube, and incubate at room temperature for 15 min.
5. Replace the medium in the HEK-293 T cells (from **step 1**) with 10 mL of fresh MEF medium, then add the FuGENE/DNA mix dropwise to the cells, gently swirl the dish, and incubate for 16 h in a 5% CO_2 humidified incubator in a BSL2+ room for lentiviral cell culture facility. Next day, replace the transfection medium with 10 mL of fresh MEF medium in HEK-293 T cells, and incubate for 24 h.

6. After ~40 h from the start of transfection, collect the first viral harvest in a 15 cm Falcon tube, and store at 4 °C until use. Gently add 10 mL of fresh MEF medium to the HEK-293 T cells, and incubate for an additional 24 h for more virus production.
7. Collect the second viral harvest in a 15 mL Falcon tube, and store at 4 °C until use. For more virus production if required, repeat **steps 6** and 7, and collect the third viral harvest. Discard the HEK-293 T cells.
8. Filter all the lentiviral supernatant with 0.45 μm syringe filter using a 20 mL syringe to remove any HEK-293 T packaging cells.
9. Concentrate the lentivirus by spinning at 46,000 × *g* for 90 min at 4 °C using SW-28 rotors with buckets (Beckman).
10. Aspirate the supernatant and keep the pellet on ice for 2 h. Resuspend the pellet in 200 μL of Opti-MEM medium, and store at −80 °C freezer as frozen aliquots.
11. If required, titer the virus to calculate the amount of virus needed to infect the target cells for determining the multiplicity of infection (MOI).
12. A day prior to infection, seed 2.5×10^6 of the target cell line to be infected on a 10 cm dish in 12 mL of appropriate medium, and incubate for 24 h in a 5% CO_2 humidified incubator. Cells should be about 80–85% confluent the next day at the time of transduction.
13. Thaw the lentivirus on ice, and transfer the desired amount into a fresh Eppendorf tube. Add Opti-MEM medium to bring up the volume to 1 mL in which add 8 μg mL^{-1} polybrene (2 μL of 4 μg mL^{-1} stock) to increase the efficiency of transduction.
14. Replace the cell culture medium in the target cell line (from **step 12**) with 10 mL of fresh medium. Add 1 mL of the lentiviral medium (from **step 13**) dropwise to the target cell line, gently swirl the dish, and then incubate overnight in a 5% CO_2 humidified incubator.
15. Alternatively, the target cells can be transduced after every viral harvest in **steps 6** and 7. In this context, add equal volume of cell culture medium (8–10 mL) to the viral harvest along with 8 μg mL^{-1} polybrene, filter using 0.45 μm syringe filter and 20 mL syringe, and then add dropwise onto the target cell line where the medium has been completely removed prior to transduction.
16. Remove the lentiviral medium the next day after transduction, dissociate the transduced cells with trypsin-EDTA solution (described in Subheading 3.1.1), and spin down at 100 × *g*

for 5 min at room temperature. Discard the supernatant, and transfer the cell pellets to the BSL2 room for regular cell culture facility.

17. Resuspend the cell pellet in 12–15 mL of fresh cell culture medium in a 10 cm dish, and incubate for 48–72 h in a 5% CO_2 humidified incubator.
18. Replace the medium of the transduced cell line with fresh medium containing the selection antibiotic of the autophagy reporter plasmid DNA (such as puromycin; usually used at 1 μg mL^{-1}, but a kill curve can be performed to determine its lowest concentration that will completely kill the non-transduced cells). Culture and passage the transduced cells on the selection media for at least 2 weeks.

3.3 Screening Platforms for Identifying Autophagy Modulators

The autophagy reporter cell lines generated by transient transfection or lentiviral overexpression (described in Subheading 3.2) are treated with compounds for identifying autophagy modulators. Some of the screening platforms based on LC3 reporter or autophagy substrates generally used are described below:

3.3.1 Screening Platforms with GFP-LC3 and mRFP-GFP-LC3 Reporters

LC3 is the most commonly used autophagy marker [63, 68]. The phosphatidylethanolamine-conjugated LC3-II form associates with the autophagosome membrane from the formation of autophagosomes until their maturation with the lysosomes and degradation in the autolysosomes [20, 69]. GFP-tagged LC3 (GFP-LC3) reporter is the most straightforward assay platform to determine perturbations in autophagic compartments by assessing GFP^+ puncta by fluorescence microscopy [20]. In this screening platform, an autophagy inhibitor will decrease GFP^+ puncta, whereas both autophagy inducer and blocker will increase GFP^+ puncta (Fig. 2). Thus, this method will not allow distinguishing between an autophagy inducer and an autophagy blocker because both the scenarios will increase autophagosome numbers (*see* **Note 4**). Nonetheless, GFP-LC3 reporter can be used as a simple primary screen, and the screen hits can be subjected to secondary autophagy assays (described in Subheading 3.4) to characterize the autophagy modulators (*see* **Note 5**). A large number of image-based chemical and genetic screens including high-throughput approaches have been undertaken in mammalian cells using the GFP-LC3 screening platform to identify the modulators/regulators of autophagy [70–83].

A more specific readout for determining changes in autophagosome formation versus maturation is the tandem fluorescent-tagged LC3 (mRFP-GFP-LC3) reporter (Fig. 2). This reporter is pH-sensitive and can quantify autophagosomes and autolysosomes by fluorescence microscopy: the autophagosomes emit both $mRFP^+$ and GFP^+ signals ($mRFP^+/GFP^+$), whereas the acidic autolysosomes emit only the acid-stable $mRFP^+$ signal ($mRFP^+/GFP^-$)

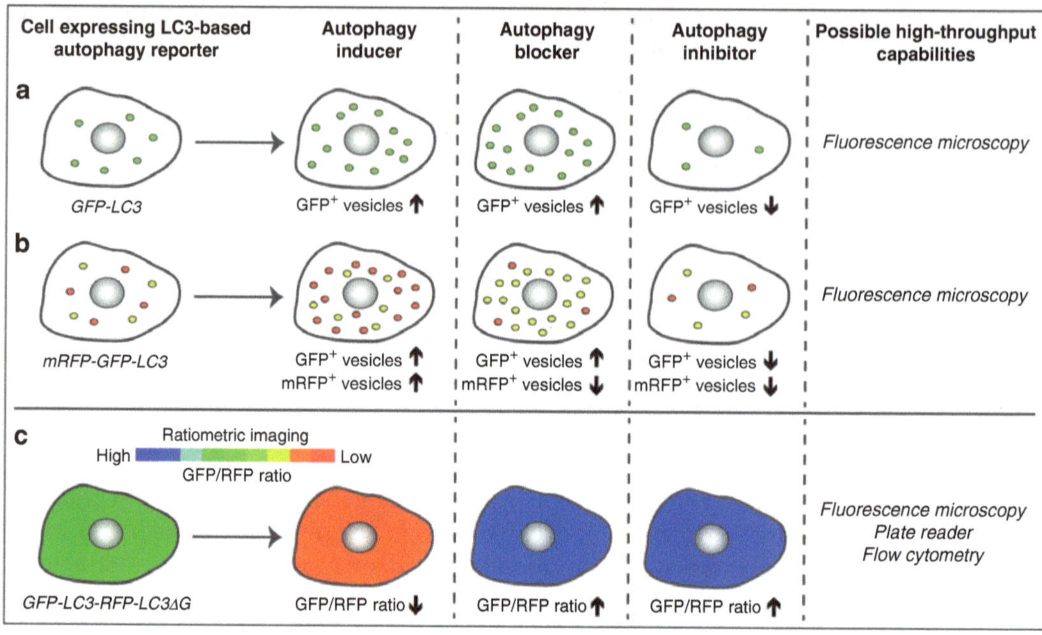

Fig. 2 LC3 reporter-based in vitro screening platforms. The transiently transfected or stable cell lines expressing fluorescent-tagged LC3 reporters as indicated can be utilized as a screening platform to assess the effects of autophagy modulators on the perturbations in autophagosome number (GFP-LC3), autophagosome maturation (mRFP-GFP-LC3), or autophagosome/autophagic flux (GFP-LC3-RFP-LC3ΔG). Possible high-throughput screening applications are indicated for each of the reporters. For GFP-LC3 reporter (**a**), an autophagy inducer will increase GFP^+ vesicles (by stimulating autophagosome formation), and an autophagy blocker will also increase GFP^+ vesicles (by preventing autophagosome maturation or lysosomal hydrolytic function), whereas an autophagy inhibitor will decrease GFP^+ vesicles (by inhibiting autophagosome formation). For mRFP-GFP-LC3 reporter (**b**) (where $mRFP^+/GFP^+$ vesicles are autophagosomes and $mRFP^+/GFP^-$ vesicles are autolysosomes), an autophagy inducer will increase $mRFP^+/GFP^+$ and $mRFP^+/GFP^-$ vesicles, and an autophagy blocker will increase $mRFP^+/GFP^+$ but decrease $mRFP^+/GFP^-$ vesicles, whereas an autophagy inhibitor will decrease $mRFP^+/GFP^+$ and $mRFP^+/GFP^-$ vesicles. For GFP-LC3-RFP-LC3ΔG reporter (**c**) (where GFP-LC3 undergoes degradation or recycling but RFP-LC3 remains in the cytosol as an internal control), an autophagy inducer will decrease GFP/RFP ratio (by enhancing autophagic flux), whereas an autophagy blocker or inhibitor will increase GFP/RFP ratio (by retarding autophagic flux)

because the GFP signal is acid-sensitive and gets quenched [21]. In this assay platform, an autophagy inhibitor (preventing autophagosome formation) will decrease both autophagosomes and autolysosomes, and an autophagy inducer (stimulating autophagosome formation) will increase autolysosomes but may or may not increase autophagosomes depending on the speed of autophagosome flux at the time point assessed, whereas an autophagy blocker (retarding autophagosome maturation) will increase autophagosomes but decrease autolysosomes [21, 84] (Fig. 2). Similarly to the GFP-LC3 reporter, it is pertinent to characterize the autophagy modulators by secondary assays (described in Subheading 3.4) after the primary screen with mRFP-GFP-LC3 reporter. Image-based chemical screens have also been undertaken in mammalian cells employing the mRFP-GFP-LC3 screening platform [83, 85, 86].

Both the LC3-based reporters described above can be utilized for manual, fluorescence microscopy-based screen as well as automated, high-throughput image-based screen. There are several variations of the tandem fluorescent-tagged LC3 reporter. For example in the mRFP-GFP-LC3 construct, mRFP could be replaced by mCherry owing to its superior photostability over mRFP [87–89]. Likewise, GFP can be substituted by mWasabi, which is a monomeric green fluorescent protein that is more acid-sensitive than GFP, and, therefore, might provide more accurate measurement of autophagosome flux by precisely distinguishing autophagosome and autolysosome compartments [90].

Fluorescence and Immunofluorescence Analysis

The GFP-LC3 and mRFP-GFP-LC3 reporters (described in Subheading 3.3.1) can be assessed by fluorescence analysis (immunofluorescence analysis is for secondary assay, described later in Subheading 3.4.2).

1. The volumes of buffers and reagents are indicated per well of a 6-well plate. Gently wash the cells on glass coverslips 2–3 times with 1 mL PBS. PBS can be stored at room temperature as 10× solution and then diluted to 1× solution with distilled water before use.
2. Incubate cells in 1 mL of 4% formaldehyde solution (diluted in PBS) for 15–20 min at room temperature.
3. Wash cells three times with 1 mL of PBS. Gently aspirate all liquid from the well.
4. For fluorescence analysis, mount the glass coverslips (cell surface facing down) using forceps onto 10 μL of anti-fade mountant with DAPI (4′,6-diamidino-2-phenylindole), to counterstain the nucleus) on microscopic slides. After 10–15 min, seal the coverslips with clear nail varnish, and store the slides at 4 °C or −20 °C.
5. For immunofluorescence analysis, continue after **step 3**.
6. Permeabilize cells with 1 mL of 0.5% Triton-X 100 in PBS for 10 min at room temperature. For immunostaining with LC3 antibody, permeabilize cells with 1 mL of chilled methanol (kept at −20 °C freezer) for 5 min.
7. Wash cells three times with 1 mL of PBS.
8. Block cells with blocking buffer (5% goat serum in PBS) for 30–45 min at room temperature on a rocker platform.
9. Add primary antibody (dilute in blocking buffer) and incubate overnight at 4 °C. For anti-LC3 and anti-p62 antibodies, use at 1:250 dilutions for both human and mouse cell lines.

10. Wash cells three times with 1 mL of PBS on a rocker platform at room temperature.
11. Add secondary antibody (dilute in blocking buffer at 1:1000 dilution), and incubate at room temperature in the dark (cover the plate in foil) for 1 h.
12. Wash cells three times with 1 mL of PBS on a rocker platform at room temperature in the dark (cover the plate in foil).
13. Mount the glass coverslips on anti-fade mountant with DAPI as described in **step 4**.
14. For analysis of fluorescent images taken by fluorescence or confocal microscopy, analyze images by ImageJ or CellProfiler software.
15. For ImageJ analysis, adjust the threshold intensity of the images (to be consistent for all images of the experiment), select the area for analysis, apply segmentation tool for individual cells to be distinguished by nucleus, set parameters for particle (puncta) count, set measurements, and analyze particles.
16. For GFP-LC3 reporter, quantify GFP^+ puncta by particle count in the green channel. Here, the GFP^+ puncta are autophagic vesicles that are primarily autophagosomes.
17. For mRFP-GFP-LC3 reporter, quantify GFP^+ and $mRFP^+$ puncta in the green and red channels, respectively; then use image calculator to subtract GFP^+ puncta from $mRFP^+$ puncta. Here, all the GFP^+ puncta are autophagosomes (that are also $mRFP^+$), whereas the $mRFP^+$ only puncta (obtained by subtracting $mRFP^+ - GFP^+$ puncta) are autolysosomes.
18. For statistical analysis, perform the experiment in triplicate. Analyze multiple images from the control and treatment samples. Calculate *p* value by unpaired Student's *t*-test.

High-Content Microscopy and Image Analysis

For high-content microscopy using automated image-based analysis, Thermo Fisher Scientific CellInsight CX5 or CellInsight CX7 High-Content Screening (HCS) Platform and HCS Studio Cell Analysis Software can be used for the GFP-LC3 and mRFP-GFP-LC3 reporters. Parameters to consider for undertaking high-throughput chemical or genetic screens to identify autophagy modulators are reviewed elsewhere [43].

1. The volumes of buffers and reagents are indicated per well of a 96-well microplate with optical bottom. However, screens can also be performed in 384-well plates with automated liquid handling workstations.
2. After treatment with the compounds from the chemical library, fix cells as described in Subheading 3.3.1.1 using 100 μL of

PBS for washing and 100 μL of 4% formaldehyde solution for fixing. Use multi-channel pipette for experimentation in 96-well plates.

3. Stain nuclei with DAPI solution (400 ng mL^{-1} in PBS) for 10 min at room temperature. Wash cells with 100 μL of PBS.
4. Mount cells with 100 μL of Gelvatol (a stable, self-mounting medium) for 30 min at room temperature or overnight at 4 °C to allow Gelvatol to harden. If required, the 96-well plates can be transferred to a different location for high-content image analysis or can be stored at 4 °C for future analysis.
5. Perform automated image analysis using the Thermo Fisher Scientific HCS Studio Cell Analysis Software. Select the "autophagy module" that performs the intensity thresholding, spot (puncta) count measurement parameters, as well as cell segmentation based on DAPI nuclear staining detected via the Hoechst-associated channel. The GFP^+ and $mRFP^+$ puncta are measured by spot count via the FITC- and Texas Red-associated channels, respectively, dependent upon size, shape, and intensity thresholding. The mean number of spot count (puncta) per object (cell) is calculated by the software as the total number of puncta per field divided by the total number of objects per field (*see* **Note 6**).
6. The number of autophagosomes per cell is quantified by the GFP^+ puncta (for both GFP-LC3 and mRFP-GFP-LC3 reporters), whereas the number of autolysosomes per cell is quantified by subtracting the GFP^+ puncta from the $mRFP^+$ puncta (for mRFP-GFP-LC3 reporter).
7. Apart from spot count per object (SCPO), measure additional parameters such as spot total intensity per object (STIPO) and spot total area per object (STAPO). These can be used to determine the size of the autophagic compartments, for example, by measuring the STAPO in FITC- or Texas Red-associated channels and then dividing the value by the SCPO in the respective channels.
8. Each compound treatment should be at least in triplicate or quadruplicate samples to obtain the mean ± s.d. Perform statistical analysis (factorial ANOVA, Student's *t*-test, or via appropriate analysis package for high-content screening data) to determine the high-confidence hits that score significantly either compared to the untreated or vehicle-treated condition or away from the screen median.
9. For high-throughput screens, calculate the Z-factor that provides an indicator of the quality of the high-throughput screening system [91, 92].

3.3.2 Screening Platform with GFP-LC3-RFP-LC3ΔG Reporter

Although the GFP-LC3 and the mRFP-GFP-LC3 reporters can be used for low-, medium-, or high-throughput screens, the readouts can only be quantified by fluorescence image-based analysis. Recently, a new autophagic flux probe, GFP-LC3-RFP-LC3ΔG, has been designed that can have versatile high-throughput applications where the readouts can be quantified by fluorescence microscopy, microplate reader, or flow cytometry [22] (Fig. 2). This construct has GFP-LC3 and RFP-LC3ΔG fused in tandem that can be cleaved by endogenous ATG4 family proteases into equimolar amounts. During autophagy, GFP-LC3 on the inner or outer autophagosomal membrane undergoes degradation or recycling, respectively, after fusion with the lysosomes. However, RFP-LC3ΔG has a deletion in the C-terminal glycine and cannot be lipidated; therefore, it remains in the cytosol and serves as an internal control [22]. Thus, the readout of GFP/RFP ratio inversely correlates with the extent of autophagic flux. In this screening platform, an autophagy inhibitor or blocker will reduce autophagic flux and increase the GFP/RFP ratio, whereas an autophagy inducer will increase autophagic flux and reduce the GFP/RFP ratio (Fig. 2). These perturbations can be quantified by microplate reader, by flow cytometry, or by ratiometric imaging with fluorescence microscopy. A chemical screen has been undertaken in mammalian cells utilizing the GFP-LC3-RFP-LC3ΔG reporter via a microplate reader [22].

Fluorescence Analysis by Microplate Reader

1. Seed 3×10^5 cells in 96-well white flat bottom plates, and culture for 24 h prior to drug treatment.
2. Treat the cells for 6 h or 24 h (or for the desired time point) with compounds in 100 μL of cell culture medium. Use known autophagy modulators, such as rapamycin and bafilomycin A_1, as positive and negative controls, respectively. Keep untreated and DMSO (vehicle)-treated cells in every plate.
3. Wash cells with 100 μL of PBS per well of a of 96-well plate, and then add 100 μL of PBS to each of the wells before measurement.
4. Adjust the plate reader parameters for excitation and emission using 488 nm_{ex}/509 nm_{em} for GFP measurement and 584 nm_{ex}/607 nm_{em} for RFP measurement.
5. Remove PBS from the wells, and then add 100 μL of culture medium to the cells with or without the drug treatment to perform the measurements over time.
6. Calculate the GFP/RFP ratio to measure the effects of compounds on autophagic flux.
7. Perform statistical analysis by Student's *t*-test, and represent the data as mean ± s.e.m.

Flow Cytometry Analysis

1. The method described here is based on 12-well plate, and the cell seeding density and the volume of buffers will differ for other plate formats.
2. Seed 2.5×10^5 cells per well for 12-well plate, and culture for 24 h prior to drug treatment.
3. Treat the cells for 6 or 24 h (or for the desired time point) with compounds in 1 mL of cell culture medium. Use known autophagy modulators, such as rapamycin and bafilomycin A_1, as positive and negative controls, respectively. Keep untreated and DMSO (vehicle)-treated cells in every plate.
4. Wash the cells with 1 mL of PBS per well.
5. Dissociate with 100 μL of 0.25% trypsin-EDTA solution in room temperature for 2 min.
6. Quench trypsin with 1 mL of ice-cold flow cytometry buffer, which consists of 2% FBS in PBS.
7. Transfer the cells in each well onto individual round bottom flow cytometry tubes, and keep on ice for analysis by flow cytometry.
8. The reading is performed using a flow cytometer equipped with a 488_{nm} and 561_{nm} lasers, such as the Beckman Coulter CyAn ADP. The signal collection is done on the FITC and PE channels in the flow cytometer for the GFP^+ and RFP^+ signals, respectively.
9. Apply gating to discriminate the cells according to their size and granularity and for excluding the cell debris. After gating, measure the fluorescence signals.
10. Acquire at least 10,000 events per sample.
11. Both cell gating and the analysis are performed using the Summit 4.3 or FlowJo software (or other software of the flow cytometer model to be used).
12. Calculate the GFP/RFP ratio which is indicative of autophagic flux.
13. Perform statistical analysis by Student's *t*-test, and represent the data as mean ± s.e.m.

3.3.3 Screening Platforms Based on Autophagic Clearance of Aggregation-Prone Proteins

As an alternative to the LC3-based reporters, screening platforms pertaining to autophagic substrate degradation can also be used where the clearance of aggregation-prone proteins is measured as a readout [29, 93] (Fig. 3). This assay depicts autophagic substrate flux and is informative on the overall autophagic flux. The aggregation-prone proteins, such as mutant huntingtin (with expanded polyglutamine repeats) or mutant α-synuclein (with A53T or A30P mutation) that are associated with neurodegenerative diseases like Huntington's and Parkinson's disease,

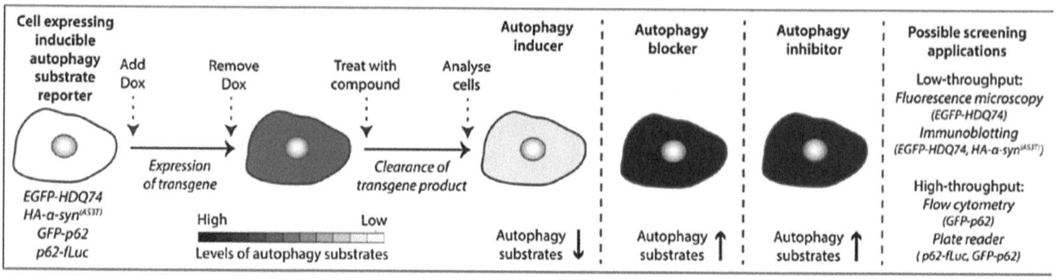

Fig. 3 Autophagy substrate-based in vitro screening platforms. The stable inducible cell lines expressing autophagy substrates as indicated can be utilized as a screening platform to assess the effects of autophagy modulators on autophagic cargo flux. In this context, the cells are initially induced with doxycycline (Dox) for transgene expression, after which the cells are treated with compounds to specifically analyze their effects on the degradation of the transgene product. The method of analysis and the possible screening applications vary with the platform to be used, as indicated. An autophagy inducer will decrease the protein levels of the autophagy substrates by enhancing their clearance, whereas an autophagy blocker or inhibitor will increase the protein levels of the autophagy substrates by retarding their clearance

respectively, are predominantly degraded by autophagy, whereas their wild-type counterparts are poor autophagy substrates [8, 10, 94]. In cell-based in vitro contexts, mutant huntingtin forms intracellular aggregates, and its clearance can be quantified by fluorescence microscopy (aggregated form) or immunoblotting (monomeric form) [29, 66], whereas mutant α-synuclein can be quantified by immunoblotting (monomeric form) [29, 67]. For immunoblotting assays, inducible overexpression systems are used where the transgene expression is "switched on" with doxycycline and then "switched off" by removing the doxycycline, followed by treatment with autophagy modulators to influence the substrate clearance that can be monitored over time [29, 93] (Fig. 3). An autophagy inducer will decrease the aggregates and increase the clearance of these aggregation-prone proteins, whereas an autophagy inhibitor or blocker will increase the aggregates and decrease their clearance. Previously, these screening platforms based on autophagic substrate clearance have been employed to identify a number of autophagy modulators [37, 38, 40, 41, 95].

Fluorescence Analysis

This method is described by using EGFP-tagged huntingtin exon 1 with 74 polyglutamine repeats (EGFP-HDQ74) construct that is overexpressed in cells by transient transfection.

1. Transfect cells (such as HeLa cells) using Lipofectamine 2000 (as described in Subheading 3.2.1) with the EGFP-HDQ74 construct. 4–6 h post-transfection, treat cells with or without compounds for 48 h. Post-treatment, fix the cells with 4% formaldehyde solution, and mount the glass coverslips on DAPI (as described in Subheading 3.3.1.1).

2. Assess the proportion of $EGFP^+$ cells with EGFP-HDQ74 aggregates (green fluorescent aggregates) by fluorescence microscopy (since not all the transiently transfected cells will exhibit mutant huntingtin aggregation). Count at least 200 $EGFP^+$ cells, in which give a score of "zero" if a cell has no aggregates, whereas give a score of "one" if a cell has one or more aggregates.
3. Perform the experiment in triplicate, and do statistical analysis by odds ratio with 95% confidence interval.

Immunoblotting Analysis

This method is described by using doxycycline-inducible EGFP-HDQ74 or HA-tagged A53T α-synuclein (HA-α-$syn^{(A53T)}$) that is stably overexpressed in cells, such as in stable PC12 cell lines as reported previously [66, 67].

1. Seed stable inducible PC12 cell lines ($1–2 \times 10^5$ cells per well of 6-well plate or $0.5–1 \times 10^5$ cells per well of 12-well plate) expressing EGFP-HDQ74 or HA-α-$syn^{(A53T)}$, and culture in PC12 medium for 24 h prior to doxycycline treatment.
2. Induce transgene expression with 1 μg mL^{-1} doxycycline treatment for 8 h (for EGFP-HDQ74) or 48 h (for HA-α-$syn^{(A53T)}$).
3. Switch off transgene expression by removing the medium containing doxycycline followed by three washes with fresh doxycycline-free PC12 medium.
4. Treat the cells with the desired concentration of compounds for 72–120 h (for EGFP–HDQ74) or 24 h (for HA-α-$syn^{(A53T)}$) in 2 mL of PC12 medium. Use known autophagy modulators, such as rapamycin and bafilomycin A_1, as positive and negative controls, respectively, as well as keep untreated and DMSO (vehicle)-treated cells. Replenish cells with the compounds every 24 h (for EGFP–HDQ74).
5. After the compound treatment, wash the cells once with PBS, supplement each well with 1 mL of fresh PBS, and gently remove the cells from the wells using a cell scraper.
6. After collecting the cells into Eppendorf tubes, centrifuge at 2400 × *g* at 4 °C for 5 min. Remove the supernatant completely, and either store the cell pellets at −80 °C freezer or proceed to the lysis step.
7. Resuspend the cell pellets in Eppendorf tubes on ice in appropriate volumes of lysis buffer. Add complete protease inhibitor cocktail in the final 1× lysis buffer before using. Vortex each sample after adding the lysis buffer for 5 s, and incubate on ice for at least 30 min (but no more that 45–60 min).

8. Centrifuge at 9600 × *g* at 4 °C for 5 min. Transfer the supernatant (cell lysate) into new Eppendorf tubes, and discard the pellets (cell debris).
9. Perform Bradford protein assay (Bio-Rad) for detecting the amount of protein in each sample. Store the lysates at −80 °C freezer, or proceed to SDS-PAGE electrophoresis of the immunoblotting step.
10. Add 3× Laemmli buffer to the cell lysates in Eppendorf tubes on ice to a final concentration of 1×. Vortex each sample after adding the Laemmli buffer for 5 s. Boil the samples for 4–8 min at 95 °C on a hot block, and briefly centrifuge at 10,000 rpm for 30 s. Load the samples immediately onto the wells of the polyacrylamide gel in **step 14**.
11. Perform SDS-PAGE (sodium dodecyl sulfate polyacrylamide gel electrophoresis) with the samples for separating the proteins. Cast 12% polyacrylamide gel using a glass plate sandwich with an integrated spacer of 1–1.5 mm to pour the resolving gel (12%) up to approximately 75%. Slowly add 1 mL of water-saturated isobutanol on top of the resolving gel after pouring, and wash it off with distilled water after 20–30 min when the resolving gel has polymerized.
12. Pour the stacking gel (3.5%) on top of the resolving gel, and insert 10-well or 15-well combs immediately. Wait for 20–30 min for the stacking gel to polymerize. Remove the combs carefully, and wash the wells with distilled water.
13. Fit the gels casted into the SDS-PAGE apparatus (Bio-Rad), and add 1× running buffer till the top covering all the wells.
14. Add 8–10 μL of protein ladder in the first well of the gel, and then add the desired volume (corresponding to 15–30 μg of protein) of each of the boiled samples (cell lysates) from **step 10** into individual wells.
15. Run the protein gel at 100 V constant voltage until the blue dye (bromophenol blue in Laemmli buffer) reaches the bottom of the gel (to separate the proteins by size). This usually takes ~2 h.
16. After running the SDS-PAGE, transfer the proteins electrophoretically onto a polyvinylidene fluoride (PVDF) microporous membrane (Amersham Hybond PVDF; GE Healthcare) using a wet transfer apparatus (Mini Trans-Blot Cell; Bio-Rad). Disassemble the glass gel plates to obtain the SDS-PAGE gel.
17. Cut the PVDF membrane to match the size of the gel, soak in methanol for 30–60 s, and rinse twice with distilled water. Equilibrate the PVDF membrane along with the SDS-PAGE gel, filter papers (four pieces cut to size), and foam pads (two pieces) in 1× transfer buffer for ~5 min.

18. Assemble the wet transfer sandwich in the following order (from bottom to top) placing the negative electrode in the bottom and then on top one foam pad, two filter papers, SDS-PAGE gel, PVDF membrane, two filter papers, one foam pad, and finally the positive electrode. The surface of the transfer sandwich is rolled gently with a plastic pipette to remove any air bubbles.
19. Put the transfer sandwich into the transfer tank, and run at 100 V constant voltage for 90–120 min at 4 °C in the cold room. After the transfer, disassemble the transfer sandwich to obtain the PVDF membrane.
20. Block the PVDF membrane in blocking buffer (5% milk in PBS-T) for 1 h at room temperature on a shaker.
21. Prepare the primary antibodies by diluting in the blocking buffer (1:2000 dilution for anti-EGFP antibody for detecting EGFP-HDQ74 and 1:1000 dilution for anti-HA antibody for detecting HA-α-syn$^{(A53T)}$; for detecting LC3 and p62 described in Subheading 3.4, use 1:4000 dilution for anti-LC3 antibody and 1:1000 dilution for anti-p62 antibody, respectively; for loading control like actin and GAPDH, use 1:5000 dilution for anti-actin and anti-GAPDH antibodies, respectively). Incubate the PVDF membrane with the primary antibodies overnight on a roller or shaker at 4 °C in the cold room.
22. Wash the PVDF membrane (immunoblot) in PBS-T for 1 h (six times wash for 10 min each) on a shaker in room temperature.
23. Prepare the secondary antibodies (as appropriate) by diluting at 1:10000 in blocking buffer. Incubate the PVDF membrane with the respective secondary antibodies for 1 h on a shaker at room temperature.
24. Wash the PVDF membrane in PBS-T for 1 h (six times wash for 10 min each) on a shaker in room temperature.
25. Develop the PVDF membrane (immunoblot) in a dark room under safe light conditions. For one immunoblot, mix 1.5 mL each of the Amersham ECL Western Blotting Detection Reagents 1 and 2, and add to the immunoblot for 1 min while ensuring coverage of the entire surface. Discard the detection reagents, wrap the immunoblot in cling film, place in a hypercassette along with a hyperfilm with appropriate exposure times, and develop in an automated Kodak developer.
26. Scan the protein bands in the hyperfilm on a scanner, and perform densitometric analysis from three independent experiments using the ImageJ software.

27. Calculate the ratio of band intensities of the proteins of interest (EGFP-HDQ74 or HA-α-syn$^{(A53T)}$) versus a loading control (such as actin or GAPDH). Set the control condition value to 100% for comparing with the experimental samples. Represent the data as mean s.e.m., and calculate the *p* values by two-way ANOVA analysis (GraphPad Prism 6 software).

3.3.4 Screening Platforms Based on the Autophagic Clearance of p62

One of the very few known autophagy substrate is p62/SQSTM1 (Sequestosome-1), which is also an adaptor protein for recognizing specific autophagic cargo during selective autophagy [30, 87, 96, 97]. It has been reported to be specifically degraded by autophagy in most mammalian cells. The levels of p62 inversely correlate with the extent of autophagic flux; for example, high flux reduces p62, whereas low flux accumulates p62 (Fig. 3). The inducible p62 reporter constructs that have been established constitute attractive screening platforms for assessing autophagy modulators via flow cytometry, microplate reader, or immunoblotting. These include the inducible GFP-p62 and p62-fLuc (Firefly luciferase) constructs [64, 65]. Chemical and genetic screens have been undertaken in mammalian cells using the p62/SQSTM1-based reporter platforms [80, 98–101]. The clearance of GFP-p62 can be monitored by immunoblotting (described in Subheading 3.3.3.2), fluorescence imaging (described in Subheadings 3.3.1.1 and 3.3.1.2), or flow cytometry (described in Subheading 3.3.2.2) methods as described above (Fig. 3). The clearance of p62-fLuc can be measured by luminescence analysis as described below. Luminescence-based chemical screening approaches have also been described for LC3 tagged to *Renilla Reniformis* luciferase (rLuc-LC3) or Firefly luciferase (fLuc-LC3) [102, 103].

However, it is possible that certain known autophagy modulators can influence the mRNA levels of p62 by affecting transcription (thereby complicating the readout at the protein level), and this effect can depend on the dose, time point, and cell type used [63]. This could potentially complicate the autophagy analysis when p62 protein levels are measured as a readout. Therefore, it is important to assess the autophagy-specific degradation of p62 in the cell type to be used, as well as characterize the positive and negative controls of the assay (with known autophagy inducers and inhibitors/blockers, respectively) that only affect p62 levels by regulating its autophagic degradation.

Luminescence Analysis by Luminometer

1. Seed 5000–10,000 cells per well (p62-fLuc stable inducible cell line) in white, polystyrene, 96-well flat bottom plates in 100 μL MEF medium with 1 μg mL^{-1} puromycin (selection marker).
2. After 24 h, treat the cells with 1 μg mL^{-1} doxycycline for 24 h to induce p62-fLuc expression.

3. Wash cells three times with PBS after 24 h of doxycycline treatment to stop p62-fLuc expression.
4. Treat cells with compounds (diluted in 100 μL medium) for the desired time period (24–72 h). Autophagic substrate clearance generally takes longer time than alterations seen with LC3-based reporters, and hence the assay should be initially titrated for the time point to be assessed.
5. For luminescence measurement, this assay is described with the ONE-Glo™ + Tox Luciferase Reporter and Cell Viability Assay (Promega). It is important to simultaneously measure cell viability as well as luciferase activity in all the wells to quantify the readout based on the number of live cells present (*see* **Note 7**).
6. Add 20 μL of 5× CellTiter-Fluor Cell Viability Assay Reagent to all the wells, and incubate for 30 min at 37 °C in an incubator with 5% CO_2. Wrap the plate in aluminum foil to protect from light. To prepare the 5× solution, add 10 μL of the GF-AFC substrate into 2 mL of assay buffer (for one 96-well plate).
7. For cell viability, measure fluorescence at 380–400 nm_{ex}/505 nm_{em} in the microplate reader.
8. Add 100 μL of ONE-Glo Reagent to all the wells, and incubate for 3 min at room temperature.
9. Measure luminescence using the luminometer.
10. Plot the luminescence and the cell viability values to compare between treated and untreated samples. Evaluate statistical significance using one-way ANOVA with Tukey's multiple comparison test (GraphPad Prism 6 software).

3.4 Secondary Assays for Characterizing Autophagy Modulators

3.4.1 Autophagosome Formation Assay Using Bafilomycin A_1

After identifying the potential autophagy modulators through the screening platforms described above, it is pertinent to characterize the compounds as autophagy inducer, blocker, or inhibitor (*see* **Notes 8** and **9**). Since the steady-state levels of autophagosomes (LC3-II levels) are affected both by their synthesis and degradation, one of the stringent secondary assays to be performed after the primary screen is the analysis for autophagosome formation using a saturating concentration of bafilomycin A_1 [29, 63, 104] (Fig. 4). Bafilomycin A_1 inhibits lysosomal v-ATPase and clamps autophagic flux at the late stage by preventing lysosomal cargo degradation and autophagosome-lysosome fusion (and thus leads to accumulation of LC3-II levels) [60–62]. In this context, an autophagy enhancer (which increases LC3-II levels by stimulating autophagosome formation at the early stage) will cause a further elevation in LC3-II levels in the presence of bafilomycin A_1 compared to bafilomycin A_1 alone. In contrast, an autophagy blocker (which also increases LC3-II levels by retarding autophagosome-

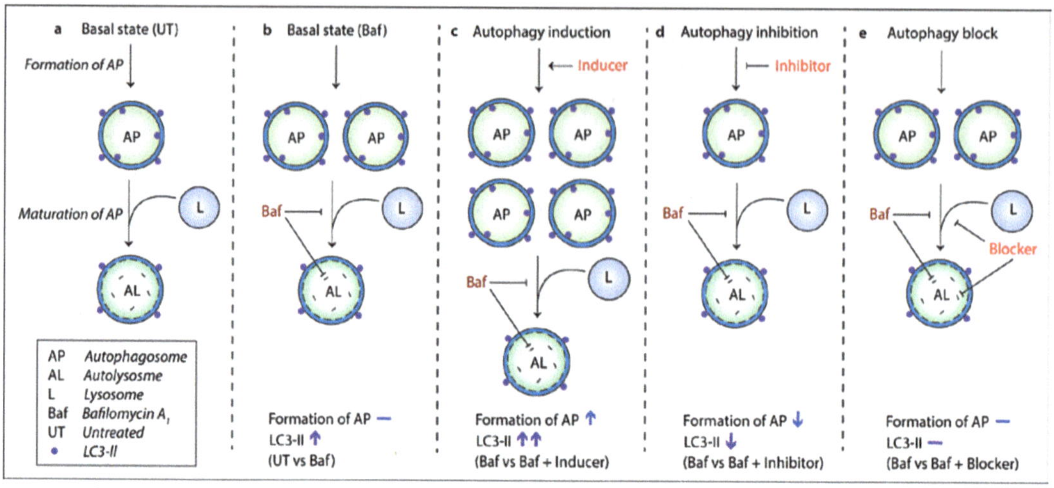

Fig. 4 Autophagosome formation analysis with bafilomycin A_1 for characterizing autophagy modulators. In basal state (untreated condition), autophagosome formation and maturation occur under basal autophagy (**a**). To determine the effects of autophagy modulators on autophagosome formation, the autophagy pathway is first clamped with a saturating concentration of bafilomycin A_1 (Baf), which accumulates autophagosomes (LC3-II levels) by preventing lysosomal cargo degradation and autophagosome maturation (**b**). An autophagy inducer (which normally increases autophagosomes on its own) will cause greater accumulation of autophagosomes in Baf-treated cells compared to Baf treatment alone due to increased autophagosome formation (**c**). An autophagy inhibitor (which normally decreases autophagosomes on its own) will cause lesser accumulation of autophagosomes in Baf-treated cells compared to Baf treatment alone due to decreased autophagosome formation (**d**). An autophagy blocker (which normally increases autophagosomes on its own) will have no change in the number of autophagosomes in Baf-treated cells compared to Baf treatment alone because Baf causes maximal block in autophagy (**e**)

lysosome fusion or lysosomal cargo degradation at the late stage) will have no change in LC3-II levels in the presence of bafilomycin A_1 compared to bafilomycin A_1 alone because the saturating dose of bafilomycin A_1 used causes maximal block in autophagy. Moreover, an autophagy inhibitor (which decreases LC3-II levels by reducing autophagosome formation at the early stage) will cause a reduction in LC3-II levels in the presence of bafilomycin A_1 compared to bafilomycin A_1 alone (Fig. 4).

We have previously characterized a number of autophagy modulators using this assay [38, 40, 41, 84, 105], which is being routinely used in the field of autophagy [63]. This method involves treating the cells with a saturating concentration of bafilomycin A_1 (usually 400 nM in most mammalian cell lines but could be also determined in the cell type to be used) for the last 4 h of the compound treatment period and then subjected to immunoblotting analysis with anti-LC3 antibody (described in Subheading 3.3.3.2).

3.4.2 Other Secondary Assays for Measuring Autophagic Flux

Autophagic flux encompasses the turnover of autophagosomes (autophagosome flux) and autophagic cargo (autophagic cargo flux). Therefore, both parameters should be evaluated for the characterization of the autophagy modulators. Depending on the nature of the primary screening platform, various secondary assays can be undertaken (Table 1). These include the analysis of autophagosome maturation and turnover with the pH-sensitive tandem fluorescent-tagged LC3 reporters, such as mRFP-GFP-LC3 [21] and GFP-LC3-RFP-LC3ΔG [22], respectively (described in Subheadings 3.3.1 and 3.3.2), and autophagic cargo flux via the clearance of autophagy substrates such as p62 or aggregation-prone proteins (described in Subheadings 3.3.3 and 3.3.4) or via organelle turnover such as mitophagy and pexophagy (*see* **Notes 10** and **11**). Similar to the pH-sensitive LC3-based reporters, those for autophagic substrate/cargo like p62 (mCherry-GFP-p62) [87] and mitochondria (mCherry-GFP-FIS1$_{101-152}$ or mt-Keima) [46, 106] can also be employed for monitoring their clearance through the autophagy pathway. Immunofluorescence analysis (described in Subheading 3.3.1.1) for the colocalization of autophagosomes (staining with anti-LC3 antibody) or autophagy substrates (staining with anti-p62 antibody) with the lysosomes (staining with anti-LAMP1 antibody) can be used to confirm the data arising with the pH-sensitive LC3 or p62 reporters (indicating delivery to the lysosomes). In addition, early markers of autophagy can be assessed to measure perturbations in the autophagy initiation process and phagophore formation [107]. These include Atg5 [108], Atg16 [109], DFCP1 (double FYVE domain-containing protein 1) [110], and WIPI-1 (WD-repeat protein interacting with phosphoinositides) [111]. Various other methods are outlined in the Autophagy Guidelines [63].

4 Notes

1. Autophagy modulators may or may not influence autophagy in all cell types, and it is also possible that they can have opposite effects that are cell type-, dose-, and time-dependent. Therefore, pilot experiments should be done with established autophagy modulators in the desired cell line to be utilized for screening to determine the positive and negative controls.
2. For transitioning hESCs from feeder layer to feeder-free conditions, avoid breaking down the hESC colonies into single cell suspension but rather seed them as smaller colonies by pipetting slowly and gently.
3. For mammalian cell lines, coating the experimental plates with 0.2% gelatin will ensure that the cells remain adherent to the

plates during the compound treatment and the washing steps in the screening assays.

4. The GFP-LC3 screening platform will not distinguish between an autophagy inducer and an autophagy blocker since both the conditions will increase GFP^+ puncta. Therefore, further assays pertaining to autophagosome formation and maturation as well as autophagic cargo clearance must be performed for further characterization.
5. To achieve high-confidence hits, it is rational to consider a two-tier primary and secondary screening approaches. All the significant hits from the primary screen could be subjected to a smaller secondary screen to obtain high-confidence hits that can be further characterized by stringent autophagy assays.
6. For high-content image analysis, seeding the optimal density of cells and appropriately using the segmentation tool in the image analysis software are important to ensure that only the individual cells are assessed (and not a group of cells clustered together is identified as one cell).
7. For the method related to luminescence analysis, starvation of cells is not compatible with the ONE-Glo™ + Tox Luciferase Reporter and Cell Viability Assay Kit due to increased metabolism affecting the fluorescence readout.
8. The potential hits arising from the primary screen must be subjected to stringent secondary autophagy assays for characterizing them as autophagy inducer, blocker, or inhibitor.
9. The time resolution could vary for the effects of autophagy modulators on different LC3-based reporters, such as between GFP^+ puncta formation using the GFP-LC3 platform and GFP/RFP ratio changes using the GFP-LC3-RFP-LC3ΔG platform. Likewise, for autophagy substrate-based reporters, the half-life of these substrates could vary. These parameters should be taken into account while designing the screen for the duration of compound treatment, as well as for the secondary assays afterward.
10. Certain autophagy modulators can potentially modulate the protein levels of p62 by affecting its transcription. This effect can be cell type-, dose-, and time-dependent and could complicate the readout when p62-based reporters are used as a screening platform. Thus, autophagy-specific degradation of p62 should be initially assessed with the positive and negative controls of the screen in the cell type to be used, and afterward, the effects of the screen hits on p62 mRNA levels must be measured by qPCR.
11. For the screen hits exhibiting increased autophagosomes or autophagy substrates and appearing as an autophagy blocker,

it is important to assess the hydrolytic function of the lysosomes because lysosomotropic agents containing a lipophilic group and a basic amine can interfere with lysosomal function and autophagic flux. Thus, this secondary assay can potentially determine whether the compound is directly affecting autophagy maturation in the late stage of autophagy or indirectly preventing autophagic flux by impairing lysosomal function.

Acknowledgments

We thank the funding agencies for supporting our research. SS is funded by Wellcome Trust Seed Award (109626/Z/15/Z), UKIERI (UK-India Education and Research Initiative) DST Thematic Partnership Award (2016-17-0087), and Birmingham Fellowship. TRR is funded by Fundação de Amparo à Pesquisa do Estado de São Paulo (FAPESP; 2015/02041-1), Fundação de Apoio à Pesquisa da Faculdade de Ciências Médicas da Santa Casa de São Paulo (FAP FCMSCSP), and Conselho Nacional de Desenvolvimento Científico e Tecnológico (CNPq)/Coordenação de Aperfeiçoamento de Pessoal de Nível Superior (CAPES). SS is also a Former Fellow for life at Hughes Hall, University of Cambridge, UK.

References

1. Mizushima N, Levine B, Cuervo AM, Klionsky DJ (2008) Autophagy fights disease through cellular self-digestion. Nature 451:1069–1075
2. Ravikumar B, Sarkar S, Davies JE, Futter M, Garcia-Arencibia M, Green-Thompson ZW, Jimenez-Sanchez M, Korolchuk VI, Lichtenberg M, Luo S, Massey DC, Menzies FM, Moreau K, Narayanan U, Renna M, Siddiqi FH, Underwood BR, Winslow AR, Rubinsztein DC (2010) Regulation of mammalian autophagy in physiology and pathophysiology. Physiol Rev 90:1383–1435
3. Levine B, Kroemer G (2008) Autophagy in the pathogenesis of disease. Cell 132:27–42
4. Boya P, Reggiori F, Codogno P (2013) Emerging regulation and functions of autophagy. Nat Cell Biol 15:713–720
5. Jiang P, Mizushima N (2014) Autophagy and human diseases. Cell Res 24:69–79
6. Rubinsztein DC, Codogno P, Levine B (2012) Autophagy modulation as a potential therapeutic target for diverse diseases. Nat Rev Drug Discov 11:709–730
7. Levine B, Packer M, Codogno P (2015) Development of autophagy inducers in clinical medicine. J Clin Investig 125:14–24
8. Sarkar S (2013) Regulation of autophagy by mTOR-dependent and mTOR-independent pathways: autophagy dysfunction in neurodegenerative diseases and therapeutic application of autophagy enhancers. Biochem Soc Trans 41:1103–1130
9. Nixon RA (2013) The role of autophagy in neurodegenerative disease. Nat Med 19:983–997
10. Sarkar S, Ravikumar B, Floto RA, Rubinsztein DC (2009) Rapamycin and mTOR-independent autophagy inducers ameliorate toxicity of polyglutamine-expanded huntingtin and related proteinopathies. Cell Death Differ 16:46–56
11. Renna M, Jimenez-Sanchez M, Sarkar S, Rubinsztein DC (2010) Chemical inducers of autophagy that enhance the clearance of

mutant proteins in neurodegenerative diseases. J Biol Chem 285:11061–11067

12. Yang Z, Klionsky DJ (2010) Mammalian autophagy: core molecular machinery and signaling regulation. Curr Opin Cell Biol 22:124–131
13. Mizushima N (2007) Autophagy: process and function. Genes Dev 21:2861–2873
14. Ktistakis NT, Tooze SA (2016) Digesting the expanding mechanisms of autophagy. Trends Cell Biol 26:624–635
15. Ganley IG (2013) Autophagosome maturation and lysosomal fusion. Essays Biochem 55:65–78
16. Reggiori F, Ungermann C (2017) Autophagosome maturation and fusion. J Mol Biol 429:486–496
17. Mizushima N, Yoshimori T, Ohsumi Y (2011) The role of Atg proteins in autophagosome formation. Annu Rev Cell Dev Biol 27:107–132
18. Mizushima N, Noda T, Yoshimori T, Tanaka Y, Ishii T, George MD, Klionsky DJ, Ohsumi M, Ohsumi Y (1998) A protein conjugation system essential for autophagy. Nature 395:395–398
19. Klionsky DJ, Schulman BA (2014) Dynamic regulation of macroautophagy by distinctive ubiquitin-like proteins. Nat Struct Mol Biol 21:336–345
20. Kabeya Y, Mizushima N, Ueno T, Yamamoto A, Kirisako T, Noda T, Kominami E, Ohsumi Y, Yoshimori T (2000) LC3, a mammalian homologue of yeast Apg8p, is localized in autophagosome membranes after processing. EMBO J 19:5720–5728
21. Kimura S, Noda T, Yoshimori T (2007) Dissection of the autophagosome maturation process by a novel reporter protein, tandem fluorescent-tagged LC3. Autophagy 3:452–460
22. Kaizuka T, Morishita H, Hama Y, Tsukamoto S, Matsui T, Toyota Y, Kodama A, Ishihara T, Mizushima T, Mizushima N (2016) An autophagic flux probe that releases an internal control. Mol Cell 64:835–849
23. Mizushima N, Yoshimori T, Levine B (2010) Methods in mammalian autophagy research. Cell 140:313–326
24. Mizushima N, Levine B (2010) Autophagy in mammalian development and differentiation. Nat Cell Biol 12:823–830
25. Kuma A, Hatano M, Matsui M, Yamamoto A, Nakaya H, Yoshimori T, Ohsumi Y, Tokuhisa T, Mizushima N (2004) The role of autophagy during the early neonatal starvation period. Nature 432:1032–1036
26. Komatsu M, Waguri S, Ueno T, Iwata J, Murata S, Tanida I, Ezaki J, Mizushima N, Ohsumi Y, Uchiyama Y, Kominami E, Tanaka K, Chiba T (2005) Impairment of starvation-induced and constitutive autophagy in Atg7-deficient mice. J Cell Biol 169:425–434
27. Hara T, Nakamura K, Matsui M, Yamamoto A, Nakahara Y, Suzuki-Migishima R, Yokoyama M, Mishima K, Saito I, Okano H, Mizushima N (2006) Suppression of basal autophagy in neural cells causes neurodegenerative disease in mice. Nature 441:885–889
28. Komatsu M, Waguri S, Chiba T, Murata S, Iwata J, Tanida I, Ueno T, Koike M, Uchiyama Y, Kominami E, Tanaka K (2006) Loss of autophagy in the central nervous system causes neurodegeneration in mice. Nature 441:880–884
29. Sarkar S, Ravikumar B, Rubinsztein DC (2009) Autophagic clearance of aggregate-prone proteins associated with neurodegeneration. Methods Enzymol 453:83–110
30. Bjorkoy G, Lamark T, Pankiv S, Overvatn A, Brech A, Johansen T (2009) Monitoring autophagic degradation of p62/SQSTM1. Methods Enzymol 452:181–197
31. Zhu Y, Chen G, Chen L, Zhang W, Feng D, Liu L, Chen Q (2014) Monitoring mitophagy in mammalian cells. Methods Enzymol 547:39–55
32. Ezaki J, Komatsu M, Yokota S, Ueno T, Kominami E (2009) Method for monitoring pexophagy in mammalian cells. Methods Enzymol 452:215–226
33. Lystad AH, Simonsen A (2015) Assays to monitor aggrephagy. Methods 75:112–119
34. Kim YC, Guan KL (2015) mTOR: a pharmacologic target for autophagy regulation. J Clin Investig 125:25–32
35. Blommaart EF, Luiken JJ, Blommaart PJ, van Woerkom GM, Meijer AJ (1995) Phosphorylation of ribosomal protein S6 is inhibitory for autophagy in isolated rat hepatocytes. J Biol Chem 270:2320–2326
36. Mizushima N (2010) The role of the Atg1/ULK1 complex in autophagy regulation. Curr Opin Cell Biol 22:132–139
37. Sarkar S, Floto RA, Berger Z, Imarisio S, Cordenier A, Pasco M, Cook LJ, Rubinsztein DC (2005) Lithium induces autophagy by inhibiting inositol monophosphatase. J Cell Biol 170:1101–1111

38. Williams A, Sarkar S, Cuddon P, Ttofi EK, Saiki S, Siddiqi FH, Jahreiss L, Fleming A, Pask D, Goldsmith P, O'Kane CJ, Floto RA, Rubinsztein DC (2008) Novel targets for Huntington's disease in an mTOR-independent autophagy pathway. Nat Chem Biol 4:295–305
39. Thoreen CC, Kang SA, Chang JW, Liu Q, Zhang J, Gao Y, Reichling LJ, Sim T, Sabatini DM, Gray NS (2009) An ATP-competitive mammalian target of rapamycin inhibitor reveals rapamycin-resistant functions of mTORC1. J Biol Chem 284:8023–8032
40. Sarkar S, Davies JE, Huang Z, Tunnacliffe A, Rubinsztein DC (2007) Trehalose, a novel mTOR-independent autophagy enhancer, accelerates the clearance of mutant huntingtin and alpha-synuclein. J Biol Chem 282:5641–5652
41. Sarkar S, Perlstein EO, Imarisio S, Pineau S, Cordenier A, Maglathlin RL, Webster JA, Lewis TA, O'Kane CJ, Schreiber SL, Rubinsztein DC (2007) Small molecules enhance autophagy and reduce toxicity in Huntington's disease models. Nat Chem Biol 3:331–338
42. Sarkar S (2013) Chemical screening platforms for autophagy drug discovery to identify therapeutic candidates for Huntington's disease and other neurodegenerative disorders. Drug Discov Today Technol 10:e137–e144
43. Joachim J, Jiang M, McKnight NC, Howell M, Tooze SA (2015) High-throughput screening approaches to identify regulators of mammalian autophagy. Methods 75:96–104
44. Fleming A, Noda T, Yoshimori T, Rubinsztein DC (2011) Chemical modulators of autophagy as biological probes and potential therapeutics. Nat Chem Biol 7:9–17
45. Shu CW, Liu PF, Huang CM (2012) High throughput screening for drug discovery of autophagy modulators. Comb Chem High Throughput Screen 15:721–729
46. Allen GF, Toth R, James J, Ganley IG (2013) Loss of iron triggers PINK1/Parkin-independent mitophagy. EMBO Rep 14:1127–1135
47. Mishra P, Dauphinee AN, Ward C, Sarkar S, Gunawardena A, Manjithaya R (2017) Discovery of pan autophagy inhibitors through a high-throughput screen highlights macroautophagy as an evolutionarily conserved process across 3 eukaryotic kingdoms. Autophagy 13:1556–1572
48. DeBosch BJ, Heitmeier MR, Mayer AL, Higgins CB, Crowley JR, Kraft TE, Chi M, Newberry EP, Chen Z, Finck BN, Davidson NO, Yarasheski KE, Hruz PW, Moley KH (2016) Trehalose inhibits solute carrier 2A (SLC2A) proteins to induce autophagy and prevent hepatic steatosis. Sci Signal 9:ra21
49. Palmieri M, Pal R, Nelvagal HR, Lotfi P, Stinnett GR, Seymour ML, Chaudhury A, Bajaj L, Bondar VV, Bremner L, Saleem U, Tse DY, Sanagasetti D, Wu SM, Neilson JR, Pereira FA, Pautler RG, Rodney GG, Cooper JD, Sardiello M (2017) mTORC1-independent TFEB activation via Akt inhibition promotes cellular clearance in neurodegenerative storage diseases. Nat Commun 8:14338
50. Maetzel D, Sarkar S, Wang H, Abi-Mosleh L, Xu P, Cheng AW, Gao Q, Mitalipova M, Jaenisch R (2014) Genetic and chemical correction of cholesterol accumulation and impaired autophagy in hepatic and neural cells derived from Niemann-Pick Type C patient-specific iPS cells. Stem Cell Reports 2:866–880
51. Perucho J, Casarejos MJ, Gomez A, Solano RM, de Yebenes JG, Mena MA (2012) Trehalose protects from aggravation of amyloid pathology induced by isoflurane anesthesia in APP(swe) mutant mice. Curr Alzheimer Res 9:334–343
52. Schaeffer V, Lavenir I, Ozcelik S, Tolnay M, Winkler DT, Goedert M (2012) Stimulation of autophagy reduces neurodegeneration in a mouse model of human tauopathy. Brain 135:2169–2177
53. Rodriguez-Navarro JA, Rodriguez L, Casarejos MJ, Solano RM, Gomez A, Perucho J, Cuervo AM, Garcia de Yebenes J, Mena MA (2010) Trehalose ameliorates dopaminergic and tau pathology in parkin deleted/tau overexpressing mice through autophagy activation. Neurobiol Dis 39:423–438
54. Castillo K, Nassif M, Valenzuela V, Rojas F, Matus S, Mercado G, Court FA, van Zundert B, Hetz C (2013) Trehalose delays the progression of amyotrophic lateral sclerosis by enhancing autophagy in motoneurons. Autophagy 9:1308–1320
55. Zhang X, Chen S, Song L, Tang Y, Shen Y, Jia L, Le W (2014) MTOR-independent, autophagic enhancer trehalose prolongs motor neuron survival and ameliorates the autophagic flux defect in a mouse model of amyotrophic lateral sclerosis. Autophagy 10:588–602
56. Belzile JP, Sabalza M, Craig M, Clark E, Morello CS, Spector DH (2015) Trehalose, an mTOR-independent inducer of autophagy, inhibits human cytomegalovirus infection in multiple cell types. J Virol 90:1259–1277

57. Kruger U, Wang Y, Kumar S, Mandelkow EM (2012) Autophagic degradation of tau in primary neurons and its enhancement by trehalose. Neurobiol Aging 33:2291–2305
58. Aguib Y, Heiseke A, Gilch S, Riemer C, Baier M, Schatzl HM, Ertmer A (2009) Autophagy induction by trehalose counteracts cellular prion infection. Autophagy 5:361–369
59. Tien NT, Karaca I, Tamboli IY, Walter J (2016) Trehalose alters subcellular trafficking and the metabolism of the Alzheimer-associated amyloid precursor protein. J Biol Chem 291:10528–10540
60. Yamamoto A, Tagawa Y, Yoshimori T, Moriyama Y, Masaki R, Tashiro Y (1998) Bafilomycin A1 prevents maturation of autophagic vacuoles by inhibiting fusion between autophagosomes and lysosomes in rat hepatoma cell line, H-4-II-E cells. Cell Struct Funct 23:33–42
61. Fass E, Shvets E, Degani I, Hirschberg K, Elazar Z (2006) Microtubules support production of starvation-induced autophagosomes but not their targeting and fusion with lysosomes. J Biol Chem 281:36303–36316
62. Klionsky DJ, Elazar Z, Seglen PO, Rubinsztein DC (2008) Does bafilomycin A1 block the fusion of autophagosomes with lysosomes? Autophagy 4:849–850
63. Klionsky DJ et al (2016) Guidelines for the use and interpretation of assays for monitoring autophagy (3rd edition). Autophagy 12:1–222
64. Larsen KB, Lamark T, Overvatn A, Harneshaug I, Johansen T, Bjorkoy G (2010) A reporter cell system to monitor autophagy based on p62/SQSTM1. Autophagy 6:784–793
65. Brown A, Patel S, Ward C, Lorenz A, Ortiz M, DuRoss A, Wieghardt F, Esch A, Otten EG, Heiser LM, Korolchuk VI, Sun C, Sarkar S, Sahay G (2016) PEG-lipid micelles enable cholesterol efflux in Niemann-Pick Type C1 disease-based lysosomal storage disorder. Sci Rep 6:31750
66. Wyttenbach A, Swartz J, Kita H, Thykjaer T, Carmichael J, Bradley J, Brown R, Maxwell M, Schapira A, Orntoft TF, Kato K, Rubinsztein DC (2001) Polyglutamine expansions cause decreased CRE-mediated transcription and early gene expression changes prior to cell death in an inducible cell model of Huntington's disease. Hum Mol Genet 10:1829–1845
67. Webb JL, Ravikumar B, Atkins J, Skepper JN, Rubinsztein DC (2003) Alpha-Synuclein is degraded by both autophagy and the proteasome. J Biol Chem 278:25009–25013
68. Yoshii SR, Mizushima N (2017) Monitoring and measuring autophagy. Int J Mol Sci 18: E1865
69. Kabeya Y, Mizushima N, Yamamoto A, Oshitani-Okamoto S, Ohsumi Y, Yoshimori T (2004) LC3, GABARAP and GATE16 localize to autophagosomal membrane depending on form-II formation. J Cell Sci 117:2805–2812
70. Zhang L, Yu J, Pan H, Hu P, Hao Y, Cai W, Zhu H, Yu AD, Xie X, Ma D, Yuan J (2007) Small molecule regulators of autophagy identified by an image-based high-throughput screen. Proc Natl Acad Sci U S A 104:19023–19028
71. Balgi AD, Fonseca BD, Donohue E, Tsang TC, Lajoie P, Proud CG, Nabi IR, Roberge M (2009) Screen for chemical modulators of autophagy reveals novel therapeutic inhibitors of mTORC1 signaling. PLoS One 4:e7124
72. Li Y, McGreal S, Zhao J, Huang R, Zhou Y, Zhong H, Xia M, Ding WX (2016) A cell-based quantitative high-throughput image screening identified novel autophagy modulators. Pharmacol Res 110:35–49
73. McKnight NC, Jefferies HB, Alemu EA, Saunders RE, Howell M, Johansen T, Tooze SA (2012) Genome-wide siRNA screen reveals amino acid starvation-induced autophagy requires SCOC and WAC. EMBO J 31:1931–1946
74. Lipinski MM, Hoffman G, Ng A, Zhou W, Py BF, Hsu E, Liu X, Eisenberg J, Liu J, Blenis J, Xavier RJ, Yuan J (2010) A genome-wide siRNA screen reveals multiple mTORC1 independent signaling pathways regulating autophagy under normal nutritional conditions. Dev Cell 18:1041–1052
75. Chan EY, Kir S, Tooze SA (2007) siRNA screening of the kinome identifies ULK1 as a multidomain modulator of autophagy. J Biol Chem 282:25464–25474
76. He P, Peng Z, Luo Y, Wang L, Yu P, Deng W, An Y, Shi T, Ma D (2009) High-throughput functional screening for autophagy-related genes and identification of TM9SF1 as an autophagosome-inducing gene. Autophagy 5:52–60
77. Orvedahl A, Sumpter R Jr, Xiao G, Ng A, Zou Z, Tang Y, Narimatsu M, Gilpin C, Sun Q, Roth M, Forst CV, Wrana JL, Zhang YE, Luby-Phelps K, Xavier RJ, Xie Y, Levine B (2011) Image-based genome-wide siRNA

screen identifies selective autophagy factors. Nature 480:113–117

78. Li M, Yang Z, Vollmer LL, Gao Y, Fu Y, Liu C, Chen X, Liu P, Vogt A, Yin XM (2015) AMDE-1 is a dual function chemical for autophagy activation and inhibition. PLoS One 10:e0122083
79. Szyniarowski P, Corcelle-Termeau E, Farkas T, Hoyer-Hansen M, Nylandsted J, Kallunki T, Jaattela M (2011) A comprehensive siRNA screen for kinases that suppress macroautophagy in optimal growth conditions. Autophagy 7:892–903
80. Pietrocola F, Lachkar S, Enot DP, Niso-Santano M, Bravo-San Pedro JM, Sica V, Izzo V, Maiuri MC, Madeo F, Marino G, Kroemer G (2015) Spermidine induces autophagy by inhibiting the acetyltransferase EP300. Cell Death Differ 22:509–516
81. Lao Y, Wan G, Liu Z, Wang X, Ruan P, Xu W, Xu D, Xie W, Zhang Y, Xu H, Xu N (2014) The natural compound oblongifolin C inhibits autophagic flux and enhances antitumor efficacy of nutrient deprivation. Autophagy 10:736–749
82. Peppard JV, Rugg C, Smicker M, Dureuil C, Ronan B, Flamand O, Durand L, Pasquier B (2014) Identifying small molecules which inhibit autophagy: a phenotypic screen using image-based high-content cell analysis. Curr Chem Genom Transl Med 8:3–15
83. Kuo SY, Castoreno AB, Aldrich LN, Lassen KG, Goel G, Dancik V, Kuballa P, Latorre I, Conway KL, Sarkar S, Maetzel D, Jaenisch R, Clemons PA, Schreiber SL, Shamji AF, Xavier RJ (2015) Small-molecule enhancers of autophagy modulate cellular disease phenotypes suggested by human genetics. Proc Natl Acad Sci U S A 112:E4281–E4287
84. Sarkar S, Korolchuk V, Renna M, Winslow A, Rubinsztein DC (2009) Methodological considerations for assessing autophagy modulators: a study with calcium phosphate precipitates. Autophagy 5:307–313
85. Chauhan S, Ahmed Z, Bradfute SB, Arko-Mensah J, Mandell MA, Won Choi S, Kimura T, Blanchet F, Waller A, Mudd MH, Jiang S, Sklar L, Timmins GS, Maphis N, Bhaskar K, Piguet V, Deretic V (2015) Pharmaceutical screen identifies novel target processes for activation of autophagy with a broad translational potential. Nat Commun 6:8620
86. Pampaloni F, Mayer B, Kabat Vel-Job K, Ansari N, Hotte K, Kogel D, Stelzer EHK (2017) A novel cellular spheroid-based autophagy screen applying live fluorescence microscopy identifies nonactin as a strong inducer of autophagosomal turnover. SLAS Discov 22:558–570
87. Pankiv S, Clausen TH, Lamark T, Brech A, Bruun JA, Outzen H, Overvatn A, Bjorkoy G, Johansen T (2007) p62/SQSTM1 binds directly to Atg8/LC3 to facilitate degradation of ubiquitinated protein aggregates by autophagy. J Biol Chem 282:24131–24145
88. N'Diaye EN, Kajihara KK, Hsieh I, Morisaki H, Debnath J, Brown EJ (2009) PLIC proteins or ubiquilins regulate autophagy-dependent cell survival during nutrient starvation. EMBO Rep 10:173–179
89. Shaner NC, Steinbach PA, Tsien RY (2005) A guide to choosing fluorescent proteins. Nat Methods 2:905–909
90. Zhou C, Zhong W, Zhou J, Sheng F, Fang Z, Wei Y, Chen Y, Deng X, Xia B, Lin J (2012) Monitoring autophagic flux by an improved tandem fluorescent-tagged LC3 (mTagRFP-mWasabi-LC3) reveals that high-dose rapamycin impairs autophagic flux in cancer cells. Autophagy 8:1215–1226
91. Zhang JH, Chung TD, Oldenburg KR (1999) A simple statistical parameter for use in evaluation and validation of high throughput screening assays. J Biomol Screen 4:67–73
92. Malo N, Hanley JA, Cerquozzi S, Pelletier J, Nadon R (2006) Statistical practice in high-throughput screening data analysis. Nat Biotechnol 24:167–175
93. Ravikumar B, Sarkar S, Rubinsztein DC (2008) Clearance of mutant aggregate-prone proteins by autophagy. Methods Mol Biol 445:195–211
94. Rubinsztein DC (2006) The roles of intracellular protein-degradation pathways in neurodegeneration. Nature 443:780–786
95. Ravikumar B, Duden R, Rubinsztein DC (2002) Aggregate-prone proteins with polyglutamine and polyalanine expansions are degraded by autophagy. Hum Mol Genet 11:1107–1117
96. Bjorkoy G, Lamark T, Brech A, Outzen H, Perander M, Overvatn A, Stenmark H, Johansen T (2005) p62/SQSTM1 forms protein aggregates degraded by autophagy and has a protective effect on huntingtin-induced cell death. J Cell Biol 171:603–614
97. Stolz A, Ernst A, Dikic I (2014) Cargo recognition and trafficking in selective autophagy. Nat Cell Biol 16:495–501
98. Hale CM, Cheng Q, Ortuno D, Huang M, Nojima D, Kassner PD, Wang S, Ollmann MM, Carlisle HJ (2016) Identification of modulators of autophagic flux in an image-

based high content siRNA screen. Autophagy 12:713–726

99. DeJesus R, Moretti F, McAllister G, Wang Z, Bergman P, Liu S, Frias E, Alford J, Reece-Hoyes JS, Lindeman A, Kelliher J, Russ C, Knehr J, Carbone W, Beibel M, Roma G, Ng A, Tallarico JA, Porter JA, Xavier RJ, Mickanin C, Murphy LO, Hoffman GR, Nyfeler B (2016) Functional CRISPR screening identifies the ufmylation pathway as a regulator of SQSTM1/p62. eLife 5:e17290
100. Strohecker AM, Joshi S, Possemato R, Abraham RT, Sabatini DM, White E (2015) Identification of 6-phosphofructo-2-kinase/fructose-2,6-bisphosphatase as a novel autophagy regulator by high content shRNA screening. Oncogene 34:5662–5676
101. Tsuganezawa K, Shinohara Y, Ogawa N, Tsuboi S, Okada N, Mori M, Yokoyama S, Noda NN, Inagaki F, Ohsumi Y, Tanaka A (2013) Two-colored fluorescence correlation spectroscopy screening for LC3-P62 interaction inhibitors. J Biomol Screen 18:1103–1109
102. Farkas T, Hoyer-Hansen M, Jaattela M (2009) Identification of novel autophagy regulators by a luciferase-based assay for the kinetics of autophagic flux. Autophagy 5:1018–1025
103. Farkas T, Daugaard M, Jaattela M (2011) Identification of small molecule inhibitors of phosphatidylinositol 3-kinase and autophagy. J Biol Chem 286:38904–38912
104. Rubinsztein DC, Cuervo AM, Ravikumar B, Sarkar S, Korolchuk V, Kaushik S, Klionsky DJ (2009) In search of an "autophagomometer". Autophagy 5:585–589
105. Sarkar S, Korolchuk VI, Renna M, Imarisio S, Fleming A, Williams A, Garcia-Arencibia M, Rose C, Luo S, Underwood BR, Kroemer G, O'Kane CJ, Rubinsztein DC (2011) Complex inhibitory effects of nitric oxide on autophagy. Mol Cell 43:19–32
106. Sun N, Malide D, Liu J, Rovira II, Combs CA, Finkel T (2017) A fluorescence-based imaging method to measure in vitro and in vivo mitophagy using mt-Keima. Nat Protoc 12:1576–1587
107. Shibutani ST, Yoshimori T (2014) A current perspective of autophagosome biogenesis. Cell Res 24:58–68
108. Mizushima N, Yamamoto A, Hatano M, Kobayashi Y, Kabeya Y, Suzuki K, Tokuhisa T, Ohsumi Y, Yoshimori T (2001) Dissection of autophagosome formation using Apg5-deficient mouse embryonic stem cells. J Cell Biol 152:657–668
109. Fujita N, Itoh T, Omori H, Fukuda M, Noda T, Yoshimori T (2008) The Atg16L complex specifies the site of LC3 lipidation for membrane biogenesis in autophagy. Mol Biol Cell 19:2092–2100
110. Axe EL, Walker SA, Manifava M, Chandra P, Roderick HL, Habermann A, Griffiths G, Ktistakis NT (2008) Autophagosome formation from membrane compartments enriched in phosphatidylinositol 3-phosphate and dynamically connected to the endoplasmic reticulum. J Cell Biol 182:685–701
111. Proikas-Cezanne T, Pfisterer SG (2009) Assessing mammalian autophagy by WIPI-1/Atg18 puncta formation. Methods Enzymol 452:247–260

Chapter 27

Automated Detection of Autophagy Response Using Single Cell-Based Microscopy Assays

Amelie J. Mueller and Tassula Proikas-Cezanne

Abstract

The fluorescence microscopy-based detection of intracellular LC3, p62, and/or WIPI punctate structures is a robust tool to monitor and assess macroautophagy/autophagy in single cells. This method was established for automated high-throughput/content analysis to reliably detect narrow differences in autophagy activity/capacity and to provide screening opportunities for biological and chemical libraries.

Key words Autophagy, Automated high-throughput analysis, LC3, p62, WIPI

1 Introduction

The process of macroautophagy (referred to as autophagy hereafter), a lysosomal degradation pathway for a variety of cytoplasmic material, including proteins, lipids, and organelles, is characterized by the formation of multi-membrane vesicles, called autophagosomes [1]. An autophagosome sequesters cytoplasmic cargo upon evolving from a precursor membrane, called phagophore that forms around the cargo [1–4]. Subsequently, autophagosomes acquire acidic hydrolases required for cargo degradation by fusing with lysosomes to vesicles termed autolysosomes or autophagolysosomes [1, 5, 6].

The process of autophagy can be assessed using a variety of different techniques [7]. Currently, standard measures for autophagy employ the fluorescence-based detection of specific autophagy-related (ATG) proteins in single cells [8–14]. Employed autophagy marker proteins either become membrane proteins of nascent autophagosomes, e.g., LC3 [8, 12] and WIPI β-propellers [15], or localize to autophagosomal membranes through specific protein-protein interaction, e.g., p62/SQSTM1 that directly binds to LC3 [16] and the ATG16L complex that directly binds to WIPI2 [17]. Moreover, autophagosomes can also be visualized by

Nicholas Ktistakis and Oliver Florey (eds.), *Autophagy: Methods and Protocols*, Methods in Molecular Biology, vol. 1880, https://doi.org/10.1007/978-1-4939-8873-0_27,

fluorescence microscopy using probes or proteins that detect phospholipids specific for autophagosomal membranes, e.g., DFCP1 [18] and WIPI β-propellers that specifically bind to phosphatidylinositol 3-phosphate (PtdIns3P) [19]. The detection of autophagosomal membranes by direct (e.g., GFP-based) or indirect (e.g., using ATG antibodies) fluorescence microscopy is based on the appearance of specific punctate structures, also simply referred to as puncta or dots, during autophagy initiation [7]. Although it is not possible to distinguish between phagophores, autophagosomes, and autolysosomes at this resolution, the employment of ATG proteins characterized as early or late autophagy markers can distinguish the different steps in the process of autophagy.

Traditionally, LC3 is the most commonly used biological marker to detect autophagosomal membranes by fluorescence microscopy [8, 20]. Upon autophagy initiation, either during basal autophagy or upon the induction of autophagy during periods of starvation, cytoplasmic LC3 becomes conjugated to phosphatidylethanolamine (PE) at the nascent autophagosome [8, 20]. Hence, cytoplasmic LC3 (also referred to as LC3-I) becomes membrane-bound LC3 (also referred to as LC3-II) at the onset of autophagy [8, 20]. By visualizing LC3 through fluorescence microscopy, one can distinguish between the overall distributed LC3 fluorescence in the cytoplasm, indicating the absence of autophagy, and autophagosomal membrane-bound LC3, appearing as LC3 puncta and indicating the formation of autophagosomal membranes [7]. Therefore, defining the number of LC3 puncta per single cell by appropriate quantifications provides a strong and robust measurement for autophagy. Of note, as LC3 is able to localize to non-autophagic membranes, it is advisable to use additional ATG marker proteins for parallel, quantitative autophagy assessments [7].

Alternative markers for phagophores and autophagosomes are the WIPI (WD-repeat protein interacting with phosphoinositides) β-propellers, the human/mouse group of ATG proteins belonging to the PROPPIN family that evolved from the yeast ATG18 ancestor [21]. Similar to yeast ATG18, the four WIPI β-propellers (WIPI 1–4) function as PtdIns3P-binding and effector proteins at the nascent autophagosome [19]. Functionally, it was recently shown that all four WIPI members scaffold AMPK- and TORC1-driven signal control of autophagy with the formation of functional autophagosomes [19]. Important for their employment as specific autophagosomal marker proteins, it was demonstrated that all four WIPI β-propellers colocalize at the nascent autophagosome [19].

The employment of WIPI protein detection for fluorescence-based autophagy assessments was first demonstrated by quantitative comparisons of WIPI1 (endogenous as well as overexpressed GFP- and myc-tagged) and LC3 puncta formation at the onset of autophagy [22]. Initially, it was shown that (i) the formation of

WIPI1 puncta in dividing human cells reflects basal autophagy, (ii) treatments known to induce autophagy above basal level (e.g., starvation, rapamycin administration) provoke a significant increase in the number of cells displaying WIPI1 puncta, and (iii) treatments known to inhibit autophagy (e.g., wortmannin or LY249002 administration) nullified the presence of WIPI1 puncta in the majority of cells [22]. Moreover, it was observed that both WIPI1 and WIPI2 become membrane proteins of phagophores and autophagosomes [22, 23]. Based on these observations and quantifications, both WIPI1 and WIPI2 protein detection for fluorescence-based autophagy assessments in high-throughput/content fashion was employed for screening purposes, e.g., lentiviral shRNA library targeting the human kinome [19]. Hence, equivalent to LC3 puncta assessment, assessing the number of WIPI1 and/or WIPI2 puncta per cell provides a robust and reliable evaluation opportunity of autophagy. An additional benefit of using WIPI1 and/or WIPI2 as autophagy marker proteins is that simply counting the number of WIPI puncta-positive cells provides also a reliable measure important for low-throughput microscopy applications [24]. In order to assess the autophagic flux, treatments that inhibit lysosomal function (e.g., bafilomycin A1 [25] administration) may be employed when detecting WIPI1 and/or WIPI2 puncta, equivalent to standard LC3 puncta assessments [7, 24], but it should be noted that proteins that do not accumulate on autophagosomes also do not build up during autophagy induction.

It is also possible to reliably monitor the autophagic flux by the detection of endogenous p62/SQSTM1 for both low- and high-throughput autophagy assessments [7]. p62/SQSTM1 connects ubiquitin-associated cargo with the autophagosomal membrane by direct protein-protein interaction with LC3 [16]. Thereby p62 becomes part of the inner autophagosome and becomes degraded in the autolysosome. An increase in p62 puncta therefore argues for an inhibition of the autophagic flux, while a decrease argues for an induction of autophagy [7].

Automated, high-throughput/content image acquisition and analysis of ATG proteins (e.g., LC3, p62/SQSTM1, WIPI proteins) using cell-based fluorescence microscopy enable unbiased assessments important for the detection of narrow differences often appearing in screening procedures using, e.g., RNAi technology or compound libraries. Several platforms for automated imaging have been employed for this purpose in the past (Table 1). In the following examples, experimental procedures for such high-throughput/content autophagy assessments are summarized for automated imaging using 96-well plates and the In Cell Analyzer (GE Healthcare). In addition, experimental procedures for low-throughput autophagy assessments are summarized for using 6-well plates and laser-scanning confocal microscopy (Carl Zeiss).

Table 1
Devices employed for automated, fluorescence-based imaging of autophagy

Device	References
In Cell Analyzer (GE Healthcare)	[19, 26–33]
Opera Phenix High Content Screening Systems (PerkinElmer)	[34–36]
Cellomics Array Scan (Thermo Scientific)	[36–43]
BD Pathway 855 (BD Imaging Systems)	[44–46]

2 Materials

2.1 Cell Culture

1. Cells: G-361, human skin, malignant melanoma (ATCC, CRL-1424); U-2 OS, human bone, osteosarcoma (ATCC, HTB-96).
2. Cell culture medium: Dulbecco's Modified Eagles Medium (DMEM) with 4500 mg/L glucose, GlutaMAX I, pyruvate (Life Technologies/Thermo Fisher Scientific), 10% fetal calf serum (FCS) (PAA Laboratories), 100 U/mL penicillin, and 100 μg/mL streptomycin solution (Life Technologies/Thermo Fisher Scientific), optional: 5 μg/mL plasmocin ant-mpp (InvivoGen).

2.2 Autophagy Assays

1. Cell wash solution: Dulbecco's Phosphate-Buffered Saline (PBS) without calcium or magnesium (Life Technologies/Thermo Fisher Scientific).
2. Nutrient-rich media: Dulbecco's Modified Eagles Medium (DMEM) with 4500 mg/L glucose, GlutaMAX I, pyruvate (Life Technologies/Thermo Fisher Scientific; 31966047) with or without 10% FCS; Dulbecco's Modified Eagle Medium (DMEM) (Life Technologies/Thermo Fisher Scientific; 11966025) with or without 10% FCS.
3. Starvation media: Dulbecco's Modified Eagle Medium (DMEM) without glucose (Life Technologies/Thermo Fisher Scientific; A1443001); Dulbecco's Modified Eagle Medium (DMEM) without glucose and glutamine (Life Technologies/Thermo Fisher Scientific; 41965039); Earl's Balanced Salt Solution (EBSS) (Life Technologies/Thermo Fisher Scientific; 24010043).
4. Compounds: bafilomycin A1 (AppliChem; 24010-043); wortmannin (Sigma-Aldrich; W1628); LY294002 (Sigma-Aldrich; L9908); AICAR (Selleckchem; S1802).

2.3 Immunostaining of Endogenous LC3, p62/SQSTM1, and WIPI1/WIPI2

1. Cell fixation: 3.7% paraformaldehyde dissolved in PBS (pH 7.4).
2. Blocking solution (PBS/B/T): PBS (AppliChem; A0965,9010, stock solution 10×, filter-sterilized), 1% BSA (stock solution 10%, filter-sterilized), 0.1% Tween-20 (stock solution 10%).
3. Wash solution (PBS/T): PBS (stock solution 10×, filter-sterilized), 1% BSA (stock solution 10%, filter-sterilized), 0.1% Tween-20 (stock solution 10%).
4. Primary antibodies: LC3 (Nanotools, 0231-100/LC3-5F10), dilution 1:5 in PBS/T); p62 antibody (Santa Cruz; sc-28359), dilution 1:50 in PBS/T; WIPI1 (Abnova, H00055062-M02), dilution 1:25 in PBS/T; WIPI2 (Abgent, AP9559b), dilution 1:25 in PBS/T.
5. Secondary antibodies: Alexa Fluor 488 goat anti-mouse IgG (Life Technologies/Thermo Fisher Scientific, A-11008), dilution 1:200 in PBS/T; Alexa Fluor 488 goat anti-rabbit IgG (Life Technologies/Thermo Fisher Scientific, A-11010), dilution 1:200 in PBS/T.
6. Nuclear staining: DAPI (AppliChem, 28718-90-3), 5 μg/mL in PBS.
7. Optional: Image-iT FX signal enhancer (Life Technologies/Thermo Fisher Scientific, I36933).
8. Mounting medium ProLong Gold Antifade Mountant (Life Technologies/Thermo Fisher Scientific, P36930).

2.4 Laser-Scanning Confocal Microscopy

1. Axiovert 200 M microscope (Carl Zeiss GmbH).
2. Confocal laser-scanning microscope (Carl Zeiss GmbH, LSM 510).

2.5 Automated High-Throughput/Content Image Acquisition and Analysis

1. IN Cell Analyzer 1000 (GE Healthcare).
2. IN Cell Analyzer Workstation 3.4 software (GE Healthcare).

3 Methods

3.1 Cell Culture

1. Culture the cell lines at 37 °C and 5% CO2 in Dulbecco's Modified Eagles Medium with 4500 mg/L glucose, GlutaMAX I, pyruvate (DMEM), supplemented with 10% FCS,

100 U/mL penicillin, and 100 mg/mL streptomycin solution, 5 mg/mL plasmocin ant-mpp, using, e.g., 10 cm cell culture dishes.

2. For long-term storage freeze cells (1×10^7 cells in 1 mL of DMEM, 10% FCS, 10% DMSO) in liquid nitrogen.
3. For low-throughput autophagy assessments, seed the cells on coverslips in 6-well tissue culture plates; for high-throughput autophagy assessments, seed the cells in 96-well plates. Incubate the cells for 16 h at 37 °C, 5% CO2.
4. At this time point, cells should have reached a confluency of 70–85%; hence, depending on the cell type, different numbers of cells are seeded, e.g., 10,000 U2-OS cells per well/96-well plate.

3.2 Autophagy Assays

1. Pre-warm all media and solutions to 37 °C.
2. Mock treatment control: aspirate off the culture medium, wash the cells three times with PBS, and add fresh nutrient-rich medium (DMEM/FCS).
3. Amino acid/serum deprivation: aspirate off the culture medium, wash the cells three times with PBS, and add EBSS.
4. Serum deprivation: aspirate off the culture medium, wash the cells three times with PBS, and add nutrient-rich medium lacking FCS (DMEM).
5. Glucose deprivation: aspirate off the culture medium, wash the cells three times with PBS, and add DMEM lacking glucose.
6. Glucose and glutamine deprivation: aspirate off the culture medium, wash the cells three times with PBS, and add DMEM lacking glucose and glutamine.
7. Bafilomycin A1 treatment: aspirate off the culture medium, wash the cells three times with PBS, and add the appropriate medium containing 200 nM bafilomycin A1.
8. Wortmannin treatment: aspirate off the culture medium, wash cells three times with PBS, and add the appropriate medium containing 200 nM wortmannin.
9. LY294002 treatment: aspirate off the culture medium, wash cells three times with PBS, and add the appropriate medium containing 50–100 μM LY294002.
10. AICAR treatment: aspirate off the culture medium, wash cells three times with PBS, and add the appropriate medium containing 1 mM AICAR (freshly made).
11. Routinely, assays are conducted for 2–3 h at 37 °C, 5% CO2.

3.3 Immunostaining of Endogenous LC3, p62/SQSTM1, and WIPI1/WIPI2 for Indirect Immunofluorescence Microscopy

1. Take the cell culture plate out of the incubator and aspirate of the medium.
2. Wash the cells twice with pre-warmed (37 °C) PBS.
3. To fix the cells, add pre-warmed (37 °C) 3.7% PFA to each well and incubate for 20 min at room temperature.
4. Aspirate of the fixative and wash the cells twice with PBS (room temperature).
5. Block the cells by adding PBS/B/T and incubate for 1–16 h at 4 °C.
6. Wash the cells twice with PBS (4 ° C).
7. Incubate the cells with primary antibodies, diluted in PBS/T (4 °C), and incubate the cells for 1–16 h at 4 °C.
8. Wash the cells twice with PBS (4 °C).
9. Incubate the cells with secondary antibodies, diluted in PBS/T (4 °C), and incubate the cells for 1–16 h at 4 °C in the dark.
10. Wash the cells twice with PBS/T and once with PBS (both at 4 °C).
11. To visualize the nuclei, add DAPI to the cells and incubate for 10 min in the dark.
12. Finally, wash the cells once with PBS.
13. Store 96-well plates at 4 ° C in the dark (add 200 μL PBS per well, seal plates with parafilm), or mount the cells grown on cover slips using ProLong and store the samples at 4 °C in the dark.

3.4 Visualizing GFP-Tagged LC3 or WIPI Proteins for Direct Fluorescence Microscopy

1. Fix the cells with pre-warmed 3.7% PFA (37 °C), incubate for 20 min at room temperature.
2. Aspirate of the PFA and wash the cells twice with PBS (room temperature).
3. Stain the cell nuclei with DAPI for 10 min in the dark and wash again twice with PBS.
4. Store 96-well plates at 4 ° C in the dark (add 200 μL PBS per well, seal plates with parafilm), or mount the cells grown on cover slips using ProLong and store the samples at 4 ° C in the dark.

3.5 Laser-Scanning Confocal Microscopy

1. For confocal laser microscopy, use, e.g., a 63x 1.4 DIC Plan-Apochromat oil immersion objective and a helium neon laser for excitations at 543 or 633 nm and an argon laser for excitations at 488 nm.
2. Take a series of 20–30 image sections with 0.2- to 0.5-μm spacing along the z-axis.

3. Document at least 50–100 cells for each assay condition in each independent experiment ($n \geq 3$).
4. Regarding laser intensity, sections should be analyzed by using the integrated range indicator tool to avoid oversaturation of images.

3.6 Automated High-Throughput/Content Image Acquisition and Analysis

1. For automated image acquisition, bring the plates to room temperature, aspirate off the PBS, and add 100 μL fresh PBS (room temperature) per well. Wipe the bottom of the plate with 70% EtOH to remove, e.g., fingerprints that may disturb the acquisition procedure.
2. Load the plate on to the plate holder of the IN Cell Analyzer 1000 (e.g., equipped with a Nikon Plan Fluor ELWD 40×0.6 objective).
3. Define the wells for image acquisition, the excitation/emission filter, exposure time, and the focus for optimal imaging.
4. Automatically acquire 20–25 random image fields per well.
5. Analyze ATG punctate structures using the IN Cell Analyzer Workstation 3.4 software as follows:
   (a) Automated cell recognition: set up the decision tree to recognize individual cells by both nuclei (DAPI) detection (top-hat segmentation algorithm, minimum size of 100 μm^2 per nucleus) and overall fluorescence (e.g., GFP) intensity (e.g., cell area detected with a characteristic area of 1500 μm^2 per U-2 OS cell).
   (b) Automated puncta recognition: ATG puncta are recognized by inclusion size (set between 0.5 and 5 μm) and inclusion fluorescence intensity (>200) versus overall cell intensity (>1.15).
6. Extract at least the following data (for each image field and each well): number of cells counted, number of puncta-positive and puncta-negative cells, number of puncta per cell.

3.7 Quantifications

1. For low-throughput approaches, count 100 cells per assay condition and repeat this series of experiments 3–5 times. Count cells positive for harboring ATG puncta when they display a minimum of 2–3 puncta. Present results as percentage of cells displaying ATG puncta. If desired, express results in addition as ratio of ATG puncta/non-puncta cells. In general, the complete experiment should be conducted by at least two independent scientists.
2. For high-throughput approaches, count at least 300 cells per assay condition (in triplicate sets), and repeat this series of experiments 3–5 times for a final cell count number of at least

1000 (if possible). Present results as percentage of cells displaying ATG puncta, as well as the number of ATG puncta per cell.

4 Additional Notes

1. If using overexpressed ATG proteins for high-throughput approaches, monoclonal cell lines should be generated and low-expressing clones only employed. Monoclonal cell lines should be appropriately characterized, including assessing whether the overexpression interferes with the process of autophagy.
2. Due to their low content of endomembranes, U-2 OS cells exhibit a higher signal to noise ratio when it comes to imaging autophagy-related structures. As such, they are more suited for monitoring autophagy using automated high-throughput imaging; automated analysis can be conducted more precisely when compared to cells with high endomembrane content (e.g., G361 cells).
3. Neither WIPI1 nor WIPI2 respond to glucose starvation and are therefore not recommended as markers for studies including this treatment. Instead, WIPI4 can be employed to study glucose starvation, as it was recently shown that AMPK and AMPK-related kinases directly signal toward WIPI4 which in turn translocates to the nascent autophagosome [19].

Acknowledgments

We apologize to researchers whose work we were unable to cite due to length constraints. Amelie J. Mueller received a predoctoral stipend from the International Max Planck Research School "From Molecules to Organisms." Tassula Proikas-Cezanne receives grant support from the Deutsche Forschungsgemeinschaft (DFG), SFB/TRR 209 (project B02), and FOR 2625 (project 1).

References

1. Ohsumi Y (2014) Historical landmarks of autophagy research. Cell Res 24(1):9–23. https://doi.org/10.1038/cr.2013.169
2. Roberts R, Ktistakis NT (2013) Omegasomes: PI3P platforms that manufacture autophagosomes. Essays Biochem 55:17–27. https://doi.org/10.1042/bse0550017
3. Rubinsztein DC, Marino G, Kroemer G (2011) Autophagy and aging. Cell 146 (5):682–695. https://doi.org/10.1016/j.cell.2011.07.030
4. Tooze SA, Jefferies HB, Kalie E, Longatti A, McAlpine FE, McKnight NC, Orsi A, Polson HE, Razi M, Robinson DJ, Webber JL (2010) Trafficking and signaling in mammalian autophagy. IUBMB Life 62(7):503–508. https://doi.org/10.1002/iub.334
5. Boya P, Reggiori F, Codogno P (2013) Emerging regulation and functions of autophagy. Nat Cell Biol 15(7):713–720. https://doi.org/10.1038/ncb2788

6. Carlsson SR, Simonsen A (2015) Membrane dynamics in autophagosome biogenesis. J Cell Sci 128(2):193–205. https://doi.org/10.1242/jcs.141036
7. Klionsky DJ, Abdelmohsen K, Abe A, Abedin MJ, Abeliovich H, Acevedo Arozena A, Adachi H, Adams CM, Adams PD, Adeli K, Adhihetty PJ, Adler SG, Agam G, Agarwal R, Aghi MK, Agnello M, Agostinis P, Aguilar PV, Aguirre-Ghiso J, Airoldi EM, Ait-Si-Ali S, Akematsu T, Akporiaye ET, Al-Rubeai M, Albaiceta GM, Albanese C, Albani D, Albert ML, Aldudo J, Algul H, Alirezaei M, Alloza I, Almasan A, Almonte-Beceril M, Alnemri ES, Alonso C, Altan-Bonnet N, Altieri DC, Alvarez S, Alvarez-Erviti L, Alves S, Amadoro G, Amano A, Amantini C, Ambrosio S, Amelio I, Amer AO, Amessou M, Amon A, An Z, Anania FA, Andersen SU, Andley UP, Andreadi CK, Andrieu-Abadie N, Anel A, Ann DK, Anoopkumar-Dukie S, Antonioli M, Aoki H, Apostolova N, Aquila S, Aquilano K, Araki K, Arama E, Aranda A, Araya J, Arcaro A, Arias E, Arimoto H, Ariosa AR, Armstrong JL, Arnould T, Arsov I, Asanuma K, Askanas V, Asselin E, Atarashi R, Atherton SS, Atkin JD, Attardi LD, Auberger P, Auburger G, Aurelian L, Autelli R, Avagliano L, Avantaggiati ML, Avrahami L, Awale S, Azad N, Bachetti T, Backer JM, Bae DH, Bae JS, Bae ON, Bae SH, Baehrecke EH, Baek SH, Baghdiguian S, Bagniewska-Zadworna A, Bai H, Bai J, Bai XY, Bailly Y, Balaji KN, Balduini W, Ballabio A, Balzan R, Banerjee R, Banhegyi G, Bao H, Barbeau B, Barrachina MD, Barreiro E, Bartel B, Bartolome A, Bassham DC, Bassi MT, Bast RC Jr, Basu A, Batista MT, Batoko H, Battino M, Bauckman K, Baumgarner BL, Bayer KU, Beale R, Beaulieu JF, Beck GR Jr, Becker C, Beckham JD, Bedard PA, Bednarski PJ, Begley TJ, Behl C, Behrends C, Behrens GM, Behrns KE, Bejarano E, Belaid A, Belleudi F, Benard G, Berchem G, Bergamaschi D, Bergami M, Berkhout B, Berliocchi L, Bernard A, Bernard M, Bernassola F, Bertolotti A, Bess AS, Besteiro S, Bettuzzi S, Bhalla S, Bhattacharyya S, Bhutia SK, Biagosch C, Bianchi MW, Biard-Piechaczyk M, Billes V, Bincoletto C, Bingol B, Bird SW, Bitoun M, Bjedov I, Blackstone C, Blanc L, Blanco GA, Blomhoff HK, Boada-Romero E, Bockler S, Boes M, Boesze-Battaglia K, Boise LH, Bolino A, Boman A, Bonaldo P, Bordi M, Bosch J, Botana LM, Botti J, Bou G, Bouche M, Bouchecareilh M, Boucher MJ, Boulton ME, Bouret SG, Boya P, Boyer-Guittaut M, Bozhkov PV, Brady N, Braga VM, Brancolini C, Braus GH, Bravo-San Pedro JM, Brennan LA, Bresnick EH, Brest P, Bridges D, Bringer MA, Brini M, Brito GC, Brodin B, Brookes PS, Brown EJ, Brown K, Broxmeyer HE, Bruhat A, Brum PC, Brumell JH, Brunetti-Pierri N, Bryson-Richardson RJ, Buch S, Buchan AM, Budak H, Bulavin DV, Bultman SJ, Bultynck G, Bumbasirevic V, Burelle Y, Burke RE, Burmeister M, Butikofer P, Caberlotto L, Cadwell K, Cahova M, Cai D, Cai J, Cai Q, Calatayud S, Camougrand N, Campanella M, Campbell GR, Campbell M, Campello S, Candau R, Caniggia I, Cantoni L, Cao L, Caplan AB, Caraglia M, Cardinali C, Cardoso SM, Carew JS, Carleton LA, Carlin CR, Carloni S, Carlsson SR, Carmona-Gutierrez D, Carneiro LA, Carnevali O, Carra S, Carrier A, Carroll B, Casas C, Casas J, Cassinelli G, Castets P, Castro-Obregon S, Cavallini G, Ceccherini I, Cecconi F, Cederbaum AI, Cena V, Cenci S, Cerella C, Cervia D, Cetrullo S, Chaachouay H, Chae HJ, Chagin AS, Chai CY, Chakrabarti G, Chamilos G, Chan EY, Chan MT, Chandra D, Chandra P, Chang CP, Chang RC, Chang TY, Chatham JC, Chatterjee S, Chauhan S, Che Y, Cheetham ME, Cheluvappa R, Chen CJ, Chen G, Chen GC, Chen G, Chen H, Chen JW, Chen JK, Chen M, Chen M, Chen P, Chen Q, Chen Q, Chen SD, Chen S, Chen SS, Chen W, Chen WJ, Chen WQ, Chen W, Chen X, Chen YH, Chen YG, Chen Y, Chen Y, Chen Y, Chen YJ, Chen YQ, Chen Y, Chen Z, Chen Z, Cheng A, Cheng CH, Cheng H, Cheong H, Cherry S, Chesney J, Cheung CH, Chevet E, Chi HC, Chi SG, Chiacchiera F, Chiang HL, Chiarelli R, Chiariello M, Chieppa M, Chin LS, Chiong M, Chiu GN, Cho DH, Cho SG, Cho WC, Cho YY, Cho YS, Choi AM, Choi EJ, Choi EK, Choi J, Choi ME, Choi SI, Chou TF, Chouaib S, Choubey D, Choubey V, Chow KC, Chowdhury K, Chu CT, Chuang TH, Chun T, Chung H, Chung T, Chung YL, Chwae YJ, Cianfanelli V, Ciarcia R, Ciechomska IA, Ciriolo MR, Cirone M, Claerhout S, Clague MJ, Claria J, Clarke PG, Clarke R, Clementi E, Cleyrat C, Cnop M, Coccia EM, Cocco T, Codogno P, Coers J, Cohen EE, Colecchia D, Coletto L, Coll NS, Colucci-Guyon E, Comincini S, Condello M, Cook KL, Coombs GH, Cooper CD, Cooper JM, Coppens I, Corasaniti MT, Corazzari M, Corbalan R, Corcelle-Termeau E, Cordero MD, Corral-Ramos C, Corti O, Cossarizza A, Costelli P, Costes S, Cotman SL, Coto-Montes A, Cottet S, Couve E, Covey LR, Cowart LA, Cox JS, Coxon FP, Coyne CB, Cragg MS, Craven RJ, Crepaldi T, Crespo JL,

Criollo A, Crippa V, Cruz MT, Cuervo AM, Cuezva JM, Cui T, Cutillas PR, Czaja MJ, Czyzyk-Krzeska MF, Dagda RK, Dahmen U, Dai C, Dai W, Dai Y, Dalby KN, Dalla Valle L, Dalmasso G, D'Amelio M, Damme M, Darfeuille-Michaud A, Dargemont C, Darley-Usmar VM, Dasarathy S, Dasgupta B, Dash S, Dass CR, Davey HM, Davids LM, Davila D, Davis RJ, Dawson TM, Dawson VL, Daza P, de Belleroche J, de Figueiredo P, de Figueiredo RC, de la Fuente J, De Martino L, De Matteis A, De Meyer GR, De Milito A, De Santi M, de Souza W, De Tata V, De Zio D, Debnath J, Dechant R, Decuypere JP, Deegan S, Dehay B, Del Bello B, Del Re DP, Delage-Mourroux R, Delbridge LM, Deldicque L, Delorme-Axford E, Deng Y, Dengjel J, Denizot M, Dent P, Der CJ, Deretic V, Derrien B, Deutsch E, Devarenne TP, Devenish RJ, Di Bartolomeo S, Di Daniele N, Di Domenico F, Di Nardo A, Di Paola S, Di Pietro A, Di Renzo L, DiAntonio A, Diaz-Araya G, Diaz-Laviada I, Diaz-Meco MT, Diaz-Nido J, Dickey CA, Dickson RC, Diederich M, Digard P, Dikic I, Dinesh-Kumar SP, Ding C, Ding WX, Ding Z, Dini L, Distler JH, Diwan A, Djavaheri-Mergny M, Dmytruk K, Dobson RC, Doetsch V, Dokladny K, Dokudovskaya S, Donadelli M, Dong XC, Dong X, Dong Z, Donohue TM Jr, Doran KS, D'Orazi G, Dorn GW 2nd, Dosenko V, Dridi S, Drucker L, Du J, Du LL, Du L, du Toit A, Dua P, Duan L, Duann P, Dubey VK, Duchen MR, Duchosal MA, Duez H, Dugail I, Dumit VI, Duncan MC, Dunlop EA, Dunn WA Jr, Dupont N, Dupuis L, Duran RV, Durcan TM, Duvezin-Caubet S, Duvvuri U, Eapen V, Ebrahimi-Fakhari D, Echard A, Eckhart L, Edelstein CL, Edinger AL, Eichinger L, Eisenberg T, Eisenberg-Lerner A, Eissa NT, El-Deiry WS, El-Khoury V, Elazar Z, Eldar-Finkelman H, Elliott CJ, Emanuele E, Emmenegger U, Engedal N, Engelbrecht AM, Engelender S, Enserink JM, Erdmann R, Erenpreisa J, Eri R, Eriksen JL, Erman A, Escalante R, Eskelinen EL, Espert L, Esteban-Martinez L, Evans TJ, Fabri M, Fabrias G, Fabrizi C, Facchiano A, Faergeman NJ, Faggioni A, Fairlie WD, Fan C, Fan D, Fan J, Fang S, Fanto M, Fanzani A, Farkas T, Faure M, Favier FB, Fearnhead H, Federici M, Fei E, Felizardo TC, Feng H, Feng Y, Feng Y, Ferguson TA, Fernandez AF, Fernandez-Barrena MG, Fernandez-Checa JC, Fernandez-Lopez A, Fernandez-Zapico ME, Feron O, Ferraro E, Ferreira-Halder CV, Fesus L, Feuer R, Fiesel FC, Filippi-Chiela EC, Filomeni G, Fimia GM, Fingert JH, Finkbeiner S, Finkel T, Fiorito F, Fisher PB, Flajolet M, Flamigni F, Florey O, Florio S, Floto RA, Folini M, Follo C, Fon EA, Fornai F, Fortunato F, Fraldi A, Franco R, Francois A, Francois A, Frankel LB, Fraser ID, Frey N, Freyssenet DG, Frezza C, Friedman SL, Frigo DE, Fu D, Fuentes JM, Fueyo J, Fujitani Y, Fujiwara Y, Fujiya M, Fukuda M, Fulda S, Fusco C, Gabryel B, Gaestel M, Gailly P, Gajewska M, Galadari S, Galili G, Galindo I, Galindo MF, Galliciotti G, Galluzzi L, Galluzzi L, Galy V, Gammoh N, Gandy S, Ganesan AK, Ganesan S, Ganley IG, Gannage M, Gao FB, Gao F, Gao JX, Garcia Nannig L, Garcia Vescovi E, Garcia-Macia M, Garcia-Ruiz C, Garg AD, Garg PK, Gargini R, Gassen NC, Gatica D, Gatti E, Gavard J, Gavathiotis E, Ge L, Ge P, Ge S, Gean PW, Gelmetti V, Genazzani AA, Geng J, Genschik P, Gerner L, Gestwicki JE, Gewirtz DA, Ghavami S, Ghigo E, Ghosh D, Giammarioli AM, Giampieri F, Giampietri C, Giatromanolaki A, Gibbings DJ, Gibellini L, Gibson SB, Ginet V, Giordano A, Giorgini F, Giovannetti E, Girardin SE, Gispert S, Giuliano S, Gladson CL, Glavic A, Gleave M, Godefroy N, Gogal RM Jr, Gokulan K, Goldman GH, Goletti D, Goligorsky MS, Gomes AV, Gomes LC, Gomez H, Gomez-Manzano-C, Gomez-Sanchez R, Goncalves DA, Goncu E, Gong Q, Gongora C, Gonzalez CB, Gonzalez-Alegre P, Gonzalez-Cabo P, Gonzalez-Polo RA, Goping IS, Gorbea C, Gorbunov NV, Goring DR, Gorman AM, Gorski SM, Goruppi S, Goto-Yamada S, Gotor C, Gottlieb RA, Gozes I, Gozuacik D, Graba Y, Graef M, Granato GE, Grant GD, Grant S, Gravina GL, Green DR, Greenhough A, Greenwood MT, Grimaldi B, Gros F, Grose C, Groulx JF, Gruber F, Grumati P, Grune T, Guan JL, Guan KL, Guerra B, Guillen C, Gulshan K, Gunst J, Guo C, Guo L, Guo M, Guo W, Guo XG, Gust AA, Gustafsson AB, Gutierrez E, Gutierrez MG, Gwak HS, Haas A, Haber JE, Hadano S, Hagedorn M, Hahn DR, Halayko AJ, Hamacher-Brady A, Hamada K, Hamai A, Hamann A, Hamasaki M, Hamer I, Hamid Q, Hammond EM, Han F, Han W, Handa JT, Hanover JA, Hansen M, Harada M, Harhaji-Trajkovic L, Harper JW, Harrath AH, Harris AL, Harris J, Hasler U, Hasselblatt P, Hasui K, Hawley RG, Hawley TS, He C, He CY, He F, He G, He RR, He XH, He YW, He YY, Heath JK, Hebert MJ, Heinzen RA, Helgason GV, Hensel M, Henske EP, Her C, Herman PK, Hernandez A, Hernandez C, Hernandez-Tiedra S, Hetz C, Hiesinger PR, Higaki K, Hilfiker S, Hill BG, Hill JA, Hill WD,

Hino K, Hofius D, Hofman P, Hoglinger GU, Hohfeld J, Holz MK, Hong Y, Hood DA, Hoozemans JJ, Hoppe T, Hsu C, Hsu CY, Hsu LC, Hu D, Hu G, Hu HM, Hu H, Hu MC, Hu YC, Hu ZW, Hua F, Hua Y, Huang C, Huang HL, Huang KH, Huang KY, Huang S, Huang S, Huang WP, Huang YR, Huang Y, Huang Y, Huber TB, Huebbe P, Huh WK, Hulmi JJ, Hur GM, Hurley JH, Husak Z, Hussain SN, Hussain S, Hwang JJ, Hwang S, Hwang TI, Ichihara A, Imai Y, Imbriano C, Inomata M, Into T, Iovane V, Iovanna JL, Iozzo RV, Ip NY, Irazoqui JE, Iribarren P, Isaka Y, Isakovic AJ, Ischiropoulos H, Isenberg JS, Ishaq M, Ishida H, Ishii I, Ishmael JE, Isidoro C, Isobe KI, Isono E, Issazadeh-Navikas S, Itahana K, Itakura E, Ivanov AI, Iyer AK, Izquierdo JM, Izumi Y, Izzo V, Jaattela M, Jaber N, Jackson DJ, Jackson WT, Jacob TG, Jacques TS, Jagannath C, Jain A, Jana NR, Jang BK, Jani A, Janji B, Jannig PR, Jansson PJ, Jean S, Jendrach M, Jeon JH, Jessen N, Jeung EB, Jia K, Jia L, Jiang H, Jiang H, Jiang L, Jiang T, Jiang X, Jiang X, Jiang X, Jiang Y, Jiang Y, Jimenez A, Jin C, Jin H, Jin L, Jin M, Jin S, Jinwal UK, Jo EK, Johansen T, Johnson DE, Johnson GV, Johnson JD, Jonasch E, Jones C, Joosten LA, Jordan J, Joseph AM, Joseph B, Joubert AM, Ju D, Ju J, Juan HF, Juenemann K, Juhasz G, Jung HS, Jung JU, Jung YK, Jungbluth H, Justice MJ, Jutten B, Kaakoush NO, Kaarniranta K, Kaasik A, Kabuta T, Kaeffer B, Kagedal K, Kahana A, Kajimura S, Kakhlon O, Kalia M, Kalvakolanu DV, Kamada Y, Kambas K, Kaminskyy VO, Kampinga HH, Kandouz M, Kang C, Kang R, Kang TC, Kanki T, Kanneganti TD, Kanno H, Kanthasamy AG, Kantorow M, Kaparakis-Liaskos M, Kapuy O, Karantza V, Karim MR, Karmakar P, Kaser A, Kaushik S, Kawula T, Kaynar AM, Ke PY, Ke ZJ, Kehrl JH, Keller KE, Kemper JK, Kenworthy AK, Kepp O, Kern A, Kesari S, Kessel D, Ketteler R, Kettelhut ID, Khambu B, Khan MM, Khandelwal VK, Khare S, Kiang JG, Kiger AA, Kihara A, Kim AL, Kim CH, Kim DR, Kim DH, Kim EK, Kim HY, Kim HR, Kim JS, Kim JH, Kim JC, Kim JH, Kim KW, Kim MD, Kim MM, Kim PK, Kim SW, Kim SY, Kim YS, Kim Y, Kimchi A, Kimmelman AC, Kimura T, King JS, Kirkegaard K, Kirkin V, Kirshenbaum LA, Kishi S, Kitajima Y, Kitamoto K, Kitaoka Y, Kitazato K, Kley RA, Klimecki WT, Klinkenberg M, Klucken J, Knaevelsrud H, Knecht E, Knuppertz L, Ko JL, Kobayashi S, Koch JC, Koechlin-Ramonatxo C, Koenig U, Koh YH, Kohler K, Kohlwein SD, Koike M, Komatsu M, Kominami E, Kong D, Kong HJ, Konstantakou EG, Kopp BT, Korcsmaros T, Korhonen L, Korolchuk VI, Koshkina NV, Kou Y, Koukourakis MI, Koumenis C, Kovacs AL, Kovacs T, Kovacs WJ, Koya D, Kraft C, Krainc D, Kramer H, Kravic-Stevovic T, Krek W, Kretz-Remy C, Krick R, Krishnamurthy M, Kriston-Vizi J, Kroemer G, Kruer MC, Kruger R, Ktistakis NT, Kuchitsu K, Kuhn C, Kumar AP, Kumar A, Kumar A, Kumar D, Kumar D, Kumar R, Kumar S, Kundu M, Kung HJ, Kuno A, Kuo SH, Kuret J, Kurz T, Kwok T, Kwon TK, Kwon YT, Kyrmizi I, La Spada AR, Lafont F, Lahm T, Lakkaraju A, Lam T, Lamark T, Lancel S, Landowski TH, Lane DJ, Lane JD, Lanzi C, Lapaquette P, Lapierre LR, Laporte J, Laukkarinen J, Laurie GW, Lavandero S, Lavie L, MJ LV, Law BY, Law HK, Law KB, Layfield R, Lazo PA, Le Cam L, Le Roch KG, Le Stunff H, Leardkamolkarn V, Lecuit M, Lee BH, Lee CH, Lee EF, Lee GM, Lee HJ, Lee H, Lee JK, Lee J, Lee JH, Lee JH, Lee M, Lee MS, Lee PJ, Lee SW, Lee SJ, Lee SJ, Lee SY, Lee SH, Lee SS, Lee SJ, Lee S, Lee YR, Lee YJ, Lee YH, Leeuwenburgh C, Lefort S, Legouis R, Lei J, Lei QY, Leib DA, Leibowitz G, Lekli I, Lemaire SD, Lemasters JJ, Lemberg MK, Lemoine A, Leng S, Lenz G, Lenzi P, Lerman LO, Lettieri Barbato D, Leu JI, Leung HY, Levine B, Lewis PA, Lezoualc'h F, Li C, Li F, Li FJ, Li J, Li K, Li L, Li M, Li M, Li Q, Li R, Li S, Li W, Li W, Li X, Li Y, Lian J, Liang C, Liang Q, Liao Y, Liberal J, Liberski PP, Lie P, Lieberman AP, Lim HJ, Lim KL, Lim K, Lima RT, Lin CS, Lin CF, Lin F, Lin F, Lin FC, Lin K, Lin KH, Lin PH, Lin T, Lin WW, Lin YS, Lin Y, Linden R, Lindholm D, Lindqvist LM, Lingor P, Linkermann A, Liotta LA, Lipinski MM, Lira VA, Lisanti MP, Liton PB, Liu B, Liu C, Liu CF, Liu F, Liu HJ, Liu J, Liu JJ, Liu JL, Liu K, Liu L, Liu L, Liu Q, Liu RY, Liu S, Liu S, Liu W, Liu XD, Liu X, Liu XH, Liu X, Liu X, Liu X, Liu Y, Liu Y, Liu Z, Liu Z, Liuzzi JP, Lizard G, Ljujic M, Lodhi IJ, Logue SE, Lokeshwar BL, Long YC, Lonial S, Loos B, Lopez-Otin C, Lopez-Vicario C, Lorente M, Lorenzi PL, Lorincz P, Los M, Lotze MT, Lovat PE, Lu B, Lu B, Lu J, Lu Q, Lu SM, Lu S, Lu Y, Luciano F, Luckhart S, Lucocq JM, Ludovico P, Lugea A, Lukacs NW, Lum JJ, Lund AH, Luo H, Luo J, Luo S, Luparello C, Lyons T, Ma J, Ma Y, Ma Y, Ma Z, Machado J, Machado-Santelli GM, Macian F, MacIntosh GC, MacKeigan JP, Macleod KF, MacMicking JD, MacMillan-Crow LA, Madeo F, Madesh M, Madrigal-Matute J, Maeda A, Maeda T, Maegawa G, Maellaro E, Maes H, Magarinos M, Maiese K, Maiti TK, Maiuri L, Maiuri MC, Maki CG, Malli R, Malorni W,

Maloyan A, Mami-Chouaib F, Man N, Mancias JD, Mandelkow EM, Mandell MA, Manfredi AA, Manie SN, Manzoni C, Mao K, Mao Z, Mao ZW, Marambaud P, Marconi AM, Marelja Z, Marfe G, Margeta M, Margittai E, Mari M, Mariani FV, Marin C, Marinelli S, Marino G, Markovic I, Marquez R, Martelli AM, Martens S, Martin KR, Martin SJ, Martin S, Martin-Acebes MA, Martin-Sanz P, Martinand-Mari C, Martinet W, Martinez J, Martinez-Lopez N, Martinez-Outschoorn U, Martinez-Velazquez M, Martinez-Vicente M, Martins WK, Mashima H, Mastrianni JA, Matarese G, Matarrese P, Mateo R, Matoba S, Matsumoto N, Matsushita T, Matsuura A, Matsuzawa T, Mattson MP, Matus S, Maugeri N, Mauvezin C, Mayer A, Maysinger D, Mazzolini GD, McBrayer MK, McCall K, McCormick C, McInerney GM, McIver SC, McKenna S, McMahon JJ, McNeish IA, Mechta-Grigoriou F, Medema JP, Medina DL, Megyeri K, Mehrpour M, Mehta JL, Mei Y, Meier UC, Meijer AJ, Melendez A, Melino G, Melino S, de Melo EJ, Mena MA, Meneghini MD, Menendez JA, Menezes R, Meng L, Meng LH, Meng S, Menghini R, Menko AS, Menna-Barreto RF, Menon MB, Meraz-Rios MA, Merla G, Merlini L, Merlot AM, Meryk A, Meschini S, Meyer JN, Mi MT, Miao CY, Micale L, Michaeli S, Michiels C, Migliaccio AR, Mihailidou AS, Mijaljica D, Mikoshiba K, Milan E, Miller-Fleming L, Mills GB, Mills IG, Minakaki G, Minassian BA, Ming XF, Minibayeva F, Minina EA, Mintern JD, Minucci S, Miranda-Vizuete A, Mitchell CH, Miyamoto S, Miyazawa K, Mizushima N, Mnich K, Mograbi B, Mohseni S, Moita LF, Molinari M, Molinari M, Moller AB, Mollereau B, Mollinedo F, Mongillo M, Monick MM, Montagnaro S, Montell C, Moore DJ, Moore MN, Mora-Rodriguez R, Moreira PI, Morel E, Morelli MB, Moreno S, Morgan MJ, Moris A, Moriyasu Y, Morrison JL, Morrison LA, Morselli E, Moscat J, Moseley PL, Mostowy S, Motori E, Mottet D, Mottram JC, Moussa CE, Mpakou VE, Mukhtar H, Mulcahy Levy JM, Muller S, Munoz-Moreno-R, Munoz-Pinedo C, Munz C, Murphy ME, Murray JT, Murthy A, Mysorekar IU, Nabi IR, Nabissi M, Nader GA, Nagahara Y, Nagai Y, Nagata K, Nagelkerke A, Nagy P, Naidu SR, Nair S, Nakano H, Nakatogawa H, Nanjundan M, Napolitano G, Naqvi NI, Nardacci R, Narendra DP, Narita M, Nascimbeni AC, Natarajan R, Navegantes LC, Nawrocki ST, Nazarko TY, Nazarko VY, Neill T, Neri LM, Netea MG, Netea-Maier RT, Neves BM, Ney PA, Nezis IP, Nguyen HT, Nguyen HP, Nicot AS, Nilsen H, Nilsson P, Nishimura M, Nishino I, Niso-Santano M, Niu H, Nixon RA, Njar VC, Noda T, Noegel AA, Nolte EM, Norberg E, Norga KK, Noureini SK, Notomi S, Notterpek L, Nowikovsky K, Nukina N, Nurnberger T, O'Donnell VB, O'Donovan T, O'Dwyer PJ, Oehme I, Oeste CL, Ogawa M, Ogretmen B, Ogura Y, Oh YJ, Ohmuraya M, Ohshima T, Ojha R, Okamoto K, Okazaki T, Oliver FJ, Ollinger K, Olsson S, Orban DP, Ordonez P, Orhon I, Orosz L, O'Rourke EJ, Orozco H, Ortega AL, Ortona E, Osellame LD, Oshima J, Oshima S, Osiewacz HD, Otomo T, Otsu K, Ou JJ, Outeiro TF, Ouyang DY, Ouyang H, Overholtzer M, Ozbun MA, Ozdinler PH, Ozpolat B, Pacelli C, Paganetti P, Page G, Pages G, Pagnini U, Pajak B, Pak SC, Pakos-Zebrucka K, Pakpour N, Palkova Z, Palladino F, Pallauf K, Pallet N, Palmieri M, Paludan SR, Palumbo C, Palumbo S, Pampliega O, Pan H, Pan W, Panaretakis T, Pandey A, Pantazopoulou A, Papackova Z, Papademetrio DL, Papassideri I, Papini A, Parajuli N, Pardo J, Parekh VV, Parenti G, Park JI, Park J, Park OK, Parker R, Parlato R, Parys JB, Parzych KR, Pasquet JM, Pasquier B, Pasumarthi KB, Patschan D, Patterson C, Pattingre S, Pattison S, Pause A, Pavenstadt H, Pavone F, Pedrozo Z, Pena FJ, Penalva MA, Pende M, Peng J, Penna F, Penninger JM, Pensalfini A, Pepe S, Pereira GJ, Pereira PC, Perez-de la Cruz V, Perez-Perez ME, Perez-Rodriguez D, Perez-Sala D, Perier C, Perl A, Perlmutter DH, Perrotta I, Pervaiz S, Pesonen M, Pessin JE, Peters GJ, Petersen M, Petrache I, Petrof BJ, Petrovski G, Phang JM, Piacentini M, Pierdominici M, Pierre P, Pierrefite-Carle V, Pietrocola F, Pimentel-Muinos FX, Pinar M, Pineda B, Pinkas-Kramarski R, Pinti M, Pinton P, Piperdi B, Piret JM, Platanias LC, Platta HW, Plowey ED, Poggeler S, Poirot M, Polcic P, Poletti A, Poon AH, Popelka H, Popova B, Poprawa I, Poulose SM, Poulton J, Powers SK, Powers T, Pozuelo-Rubio M, Prak K, Prange R, Prescott M, Priault M, Prince S, Proia RL, Proikas-Cezanne T, Prokisch H, Promponas VJ, Przyklenk K, Puertollano R, Pugazhenthi S, Puglielli L, Pujol A, Puyal J, Pyeon D, Qi X, Qian WB, Qin ZH, Qiu Y, Qu Z, Quadrilatero J, Quinn F, Raben N, Rabinowich H, Radogna F, Ragusa MJ, Rahmani M, Raina K, Ramanadham S, Ramesh R, Rami A, Randall-Demllo S, Randow F, Rao H, Rao VA, Rasmussen BB, Rasse TM, Ratovitski EA, Rautou PE, Ray SK, Razani B, Reed BH, Reggiori F, Rehm M, Reichert AS, Rein T, Reiner DJ,

Reits E, Ren J, Ren X, Renna M, Reusch JE, Revuelta JL, Reyes L, Rezaie AR, Richards RI, Richardson DR, Richetta C, Riehle MA, Rihn BH, Rikihisa Y, Riley BE, Rimbach G, Rippo MR, Ritis K, Rizzi F, Rizzo E, Roach PJ, Robbins J, Roberge M, Roca G, Roccheri MC, Rocha S, Rodrigues CM, Rodriguez CI, de Cordoba SR, Rodriguez-Muela N, Roelofs J, Rogov VV, Rohn TT, Rohrer B, Romanelli D, Romani L, Romano PS, Roncero MI, Rosa JL, Rosello A, Rosen KV, Rosenstiel P, Rost-Roszkowska M, Roth KA, Roue G, Rouis M, Rouschop KM, Ruan DT, Ruano D, Rubinsztein DC, Rucker EB 3rd, Rudich A, Rudolf E, Rudolf R, Ruegg MA, Ruiz-Roldan C, Ruparelia AA, Rusmini P, Russ DW, Russo GL, Russo G, Russo R, Rusten TE, Ryabovol V, Ryan KM, Ryter SW, Sabatini DM, Sacher M, Sachse C, Sack MN, Sadoshima J, Saftig P, Sagi-Eisenberg R, Sahni S, Saikumar P, Saito T, Saitoh T, Sakakura K, Sakoh-Nakatogawa M, Sakuraba Y, Salazar-Roa M, Salomoni P, Saluja AK, Salvaterra PM, Salvioli R, Samali A, Sanchez AM, Sanchez-Alcazar JA, Sanchez-Prieto-R, Sandri M, Sanjuan MA, Santaguida S, Santambrogio L, Santoni G, Dos Santos CN, Saran S, Sardiello M, Sargent G, Sarkar P, Sarkar S, Sarrias MR, Sarwal MM, Sasakawa C, Sasaki M, Sass M, Sato K, Sato M, Satriano J, Savaraj N, Saveljeva S, Schaefer L, Schaible UE, Scharl M, Schatzl HM, Schekman R, Scheper W, Schiavi A, Schipper HM, Schmeisser H, Schmidt J, Schmitz I, Schneider BE, Schneider EM, Schneider JL, Schon EA, Schonenberger MJ, Schonthal AH, Schorderet DF, Schroder B, Schuck S, Schulze RJ, Schwarten M, Schwarz TL, Sciarretta S, Scotto K, Scovassi AI, Screaton RA, Screen M, Seca H, Sedej S, Segatori L, Segev N, Seglen PO, Segui-Simarro JM, Segura-Aguilar J, Seki E, Seiliez I, Sell C, Semenkovich CF, Semenza GL, Sen U, Serra AL, Serrano-Puebla A, Sesaki H, Setoguchi T, Settembre C, Shacka JJ, Shajahan-Haq AN, Shapiro IM, Sharma S, She H, Shen CJ, Shen CC, Shen HM, Shen S, Shen W, Sheng R, Sheng X, Sheng ZH, Shepherd TG, Shi J, Shi Q, Shi Q, Shi Y, Shibutani S, Shibuya K, Shidoji Y, Shieh JJ, Shih CM, Shimada Y, Shimizu S, Shin DW, Shinohara ML, Shintani M, Shintani T, Shioi T, Shirabe K, Shiri-Sverdlov R, Shirihai O, Shore GC, Shu CW, Shukla D, Sibirny AA, Sica V, Sigurdson CJ, Sigurdsson EM, Sijwali PS, Sikorska B, Silveira WA, Silvente-Poirot S, Silverman GA, Simak J, Simmet T, Simon AK, Simon HU, Simone C, Simons M, Simonsen A, Singh R, Singh SV, Singh SK, Sinha D, Sinha S, Sinicrope FA, Sirko A, Sirohi K, Sishi BJ, Sittler A, Siu PM, Sivridis E, Skwarska A, Slack R, Slaninova I, Slavov N, Smaili SS, Smalley KS, Smith DR, Soenen SJ, Soleimanpour SA, Solhaug A, Somasundaram K, Son JH, Sonawane A, Song C, Song F, Song HK, Song JX, Song W, Soo KY, Sood AK, Soong TW, Soontornniyomkij V, Sorice M, Sotgia F, Soto-Pantoja DR, Sotthibundhu A, Sousa MJ, Spaink HP, Span PN, Spang A, Sparks JD, Speck PG, Spector SA, Spies CD, Springer W, Clair DS, Stacchiotti A, Staels B, Stang MT, Starczynowski DT, Starokadomskyy P, Steegborn C, Steele JW, Stefanis L, Steffan J, Stellrecht CM, Stenmark H, Stepkowski TM, Stern ST, Stevens C, Stockwell BR, Stoka V, Storchova Z, Stork B, Stratoulias V, Stravopodis DJ, Strnad P, Strohecker AM, Strom AL, Stromhaug P, Stulik J, Su YX, Su Z, Subauste CS, Subramaniam S, Sue CM, Suh SW, Sui X, Sukseree S, Sulzer D, Sun FL, Sun J, Sun J, Sun SY, Sun Y, Sun Y, Sun Y, Sundaramoorthy V, Sung J, Suzuki H, Suzuki K, Suzuki N, Suzuki T, Suzuki YJ, Swanson MS, Swanton C, Sward K, Swarup G, Sweeney ST, Sylvester PW, Szatmari Z, Szegezdi E, Szlosarek PW, Taegtmeyer H, Tafani M, Taillebourg E, Tait SW, Takacs-Vellai K, Takahashi Y, Takats S, Takemura G, Takigawa N, Talbot NJ, Tamagno E, Tamburini J, Tan CP, Tan L, Tan ML, Tan M, Tan YJ, Tanaka K, Tanaka M, Tang D, Tang D, Tang G, Tanida I, Tanji K, Tannous BA, Tapia JA, Tasset-Cuevas I, Tatar M, Tavassoly I, Tavernarakis N, Taylor A, Taylor GS, Taylor GA, Taylor JP, Taylor MJ, Tchetina EV, Tee AR, Teixeira-Clerc F, Telang S, Tencomnao T, Teng BB, Teng RJ, Terro F, Tettamanti G, Theiss AL, Theron AE, Thomas KJ, Thome MP, Thomes PG, Thorburn A, Thorner J, Thum T, Thumm M, Thurston TL, Tian L, Till A, Ting JP, Titorenko VI, Toker L, Toldo S, Tooze SA, Topisirovic I, Torgersen ML, Torosantucci L, Torriglia A, Torrisi MR, Tournier C, Towns R, Trajkovic V, Travassos LH, Triola G, Tripathi DN, Trisciuoglio D, Troncoso R, Trougakos IP, Truttmann AC, Tsai KJ, Tschan MP, Tseng YH, Tsukuba T, Tsung A, Tsvetkov AS, Tu S, Tuan HY, Tucci M, Tumbarello DA, Turk B, Turk V, Turner RF, Tveita AA, Tyagi SC, Ubukata M, Uchiyama Y, Udelnow A, Ueno T, Umekawa M, Umemiya-Shirafuji R, Underwood BR, Ungermann C, Ureshino RP, Ushioda R, Uversky VN, Uzcategui NL, Vaccari T, Vaccaro MI, Vachova L, Vakifahmetoglu-Norberg H, Valdor R, Valente EM, Vallette F, Valverde AM, Van den Berghe G, Van Den Bosch L, van den Brink

GR, van der Goot FG, van der Klei IJ, van der Laan LJ, van Doorn WG, van Egmond M, van Golen KL, Van Kaer L, van Lookeren Campagne M, Vandenabeele P, Vandenberghe W, Vanhorebeek I, Varela-Nieto I, Vasconcelos MH, Vasko R, Vavvas DG, Vega-Naredo I, Velasco G, Velentzas AD, Velentzas PD, Vellai T, Vellenga E, Vendelbo MH, Venkatachalam K, Ventura N, Ventura S, Veras PS, Verdier M, Vertessy BG, Viale A, Vidal M, Vieira H, Vierstra RD, Vigneswaran N, Vij N, Vila M, Villar M, Villar VH, Villarroya J, Vindis C, Viola G, Viscomi MT, Vitale G, Vogl DT, Voitsekhovskaja OV, von Haefen C, von Schwarzenberg K, Voth DE, Vouret-Craviari V, Vuori K, Vyas JM, Waeber C, Walker CL, Walker MJ, Walter J, Wan L, Wan X, Wang B, Wang C, Wang CY, Wang C, Wang C, Wang C, Wang D, Wang F, Wang F, Wang G, Wang HJ, Wang H, Wang HG, Wang H, Wang HD, Wang J, Wang J, Wang M, Wang MQ, Wang PY, Wang P, Wang RC, Wang S, Wang TF, Wang X, Wang XJ, Wang XW, Wang X, Wang X, Wang Y, Wang Y, Wang Y, Wang YJ, Wang Y, Wang Y, Wang YT, Wang Y, Wang ZN, Wappner P, Ward C, Ward DM, Warnes G, Watada H, Watanabe Y, Watase K, Weaver TE, Weekes CD, Wei J, Weide T, Weihl CC, Weindl G, Weis SN, Wen L, Wen X, Wen Y, Westermann B, Weyand CM, White AR, White E, Whitton JL, Whitworth AJ, Wiels J, Wild F, Wildenberg ME, Wileman T, Wilkinson DS, Wilkinson S, Willbold D, Williams C, Williams K, Williamson PR, Winklhofer KF, Witkin SS, Wohlgemuth SE, Wollert T, Wolvetang EJ, Wong E, Wong GW, Wong RW, Wong VK, Woodcock EA, Wright KL, Wu C, Wu D, Wu GS, Wu J, Wu J, Wu M, Wu M, Wu S, Wu WK, Wu Y, Wu Z, Xavier CP, Xavier RJ, Xia GX, Xia T, Xia W, Xia Y, Xiao H, Xiao J, Xiao S, Xiao W, Xie CM, Xie Z, Xie Z, Xilouri M, Xiong Y, Xu C, Xu C, Xu F, Xu H, Xu H, Xu J, Xu J, Xu J, Xu L, Xu X, Xu Y, Xu Y, Xu ZX, Xu Z, Xue Y, Yamada T, Yamamoto A, Yamanaka K, Yamashina S, Yamashiro S, Yan B, Yan B, Yan X, Yan Z, Yanagi Y, Yang DS, Yang JM, Yang L, Yang M, Yang PM, Yang P, Yang Q, Yang W, Yang WY, Yang X, Yang Y, Yang Y, Yang Z, Yang Z, Yao MC, Yao PJ, Yao X, Yao Z, Yao Z, Yasui LS, Ye M, Yedvobnick B, Yeganeh B, Yeh ES, Yeyati PL, Yi F, Yi L, Yin XM, Yip CK, Yoo YM, Yoo YH, Yoon SY, Yoshida KI, Yoshimori T, Young KH, Yu H, Yu JJ, Yu JT, Yu J, Yu L, Yu WH, Yu XF, Yu Z, Yuan J, Yuan ZM, Yue BY, Yue J, Yue Z, Zacks DN, Zacksenhaus E, Zaffaroni N, Zaglia T, Zakeri Z, Zecchini V, Zeng J, Zeng M, Zeng Q, Zervos AS, Zhang DD, Zhang F, Zhang G, Zhang GC, Zhang H, Zhang H, Zhang H, Zhang H, Zhang J, Zhang J, Zhang J, Zhang J, Zhang JP, Zhang L, Zhang L, Zhang L, Zhang L, Zhang MY, Zhang X, Zhang XD, Zhang Y, Zhang Y, Zhang Y, Zhang Y, Zhang Y, Zhao M, Zhao WL, Zhao X, Zhao YG, Zhao Y, Zhao Y, Zhao YX, Zhao Z, Zhao ZJ, Zheng D, Zheng XL, Zheng X, Zhivotovsky B, Zhong Q, Zhou GZ, Zhou G, Zhou H, Zhou SF, Zhou XJ, Zhu H, Zhu H, Zhu WG, Zhu W, Zhu XF, Zhu Y, Zhuang SM, Zhuang X, Ziparo E, Zois CE, Zoladek T, Zong WX, Zorzano A, Zughaier SM (2016) Guidelines for the use and interpretation of assays for monitoring autophagy (3rd edition). Autophagy 12(1):1–222. https://doi.org/10.1080/15548627.2015.1100356

8. Kabeya Y, Mizushima N, Ueno T, Yamamoto A, Kirisako T, Noda T, Kominami E, Ohsumi Y, Yoshimori T (2000) LC3, a mammalian homologue of yeast Apg8p, is localized in autophagosome membranes after processing. EMBO J 19(21):5720–5728. https://doi.org/10.1093/emboj/19.21.5720
9. Mizushima N, Sugita H, Yoshimori T, Ohsumi Y (1998) A new protein conjugation system in human. The counterpart of the yeast Apg12p conjugation system essential for autophagy. J Biol Chem 273(51):33889–33892
10. Mizushima N, Yoshimori T, Ohsumi Y (2011) The role of Atg proteins in autophagosome formation. Annu Rev Cell Dev Biol 27:107–132. https://doi.org/10.1146/annurev-cellbio-092910-154005
11. Itakura E, Mizushima N (2010) Characterization of autophagosome formation site by a hierarchical analysis of mammalian Atg proteins. Autophagy 6(6):764–776
12. Slobodkin MR, Elazar Z (2013) The Atg8 family: multifunctional ubiquitin-like key regulators of autophagy. Essays Biochem 55:51–64. https://doi.org/10.1042/bse0550051
13. Behrends C, Sowa ME, Gygi SP, Harper JW (2010) Network organization of the human autophagy system. Nature 466(7302):68–76. https://doi.org/10.1038/nature09204
14. Wild P, McEwan DG, Dikic I (2014) The LC3 interactome at a glance. J Cell Sci 127 (Pt 1):3–9. https://doi.org/10.1242/jcs.140426
15. Proikas-Cezanne T, Takacs Z, Donnes P, Kohlbacher O (2015) WIPI proteins: essential PtdIns3P effectors at the nascent autophagosome. J Cell Sci 128(2):207–217. https://doi.org/10.1242/jcs.146258

16. Stolz A, Ernst A, Dikic I (2014) Cargo recognition and trafficking in selective autophagy. Nat Cell Biol 16(6):495–501. https://doi.org/10.1038/ncb2979
17. Dooley HC, Razi M, Polson HE, Girardin SE, Wilson MI, Tooze SA (2014) WIPI2 links LC3 conjugation with PI3P, autophagosome formation, and pathogen clearance by recruiting Atg12-5-16L1. Mol Cell 55(2):238–252. https://doi.org/10.1016/j.molcel.2014.05.021
18. Axe EL, Walker SA, Manifava M, Chandra P, Roderick HL, Habermann A, Griffiths G, Ktistakis NT (2008) Autophagosome formation from membrane compartments enriched in phosphatidylinositol 3-phosphate and dynamically connected to the endoplasmic reticulum. J Cell Biol 182(4):685–701. https://doi.org/10.1083/jcb.200803137
19. Bakula D, Muller AJ, Zuleger T, Takacs Z, Franz-Wachtel M, Thost AK, Brigger D, Tschan MP, Frickey T, Robenek H, Macek B, Proikas-Cezanne T (2017) WIPI3 and WIPI4 beta-propellers are scaffolds for LKB1-AMPK-TSC signalling circuits in the control of autophagy. Nat Commun 8:15637. https://doi.org/10.1038/ncomms15637
20. Kabeya Y, Mizushima N, Yamamoto A, Oshitani-Okamoto S, Ohsumi Y, Yoshimori T (2004) LC3, GABARAP and GATE16 localize to autophagosomal membrane depending on form-II formation. J Cell Sci 117 (Pt 13):2805–2812. https://doi.org/10.1242/jcs.01131
21. Proikas-Cezanne T, Waddell S, Gaugel A, Frickey T, Lupas A, Nordheim A (2004) WIPI-1alpha (WIPI49), a member of the novel 7-bladed WIPI protein family, is aberrantly expressed in human cancer and is linked to starvation-induced autophagy. Oncogene 23(58):9314–9325. https://doi.org/10.1038/sj.onc.1208331
22. Proikas-Cezanne T, Ruckerbauer S, Stierhof YD, Berg C, Nordheim A (2007) Human WIPI-1 puncta-formation: a novel assay to assess mammalian autophagy. FEBS Lett 581 (18):3396–3404. https://doi.org/10.1016/j.febslet.2007.06.040
23. Proikas-Cezanne T, Robenek H (2011) Freeze-fracture replica immunolabelling reveals human WIPI-1 and WIPI-2 as membrane proteins of autophagosomes. J Cell Mol Med 15(9):2007–2010. https://doi.org/10.1111/j.1582-4934.2011.01339.x
24. Thost AK, Donnes P, Kohlbacher O, Proikas-Cezanne T (2015) Fluorescence-based imaging of autophagy progression by human WIPI protein detection. Methods 75:69–78. https://doi.org/10.1016/j.ymeth.2014.11.011
25. Yamamoto A, Tagawa Y, Yoshimori T, Moriyama Y, Masaki R, Tashiro Y (1998) Bafilomycin A1 prevents maturation of autophagic vacuoles by inhibiting fusion between autophagosomes and lysosomes in rat hepatoma cell line, H-4-II-E cells. Cell Struct Funct 23 (1):33–42
26. Grotemeier A, Alers S, Pfisterer SG, Paasch F, Daubrawa M, Dieterle A, Viollet B, Wesselborg S, Proikas-Cezanne T, Stork B (2010) AMPK-independent induction of autophagy by cytosolic Ca2+ increase. Cell Signal 22(6):914–925. https://doi.org/10.1016/j.cellsig.2010.01.015
27. Pfisterer SG, Mauthe M, Codogno P, Proikas-Cezanne T (2011) Ca2+/calmodulin- dependent kinase (CaMK) signaling via CaMKI and AMP-activated protein kinase contributes to the regulation of WIPI-1 at the onset of autophagy. Mol Pharmacol 80(6):1066–1075. https://doi.org/10.1124/mol.111.071761
28. Mauthe M, Jacob A, Freiberger S, Hentschel K, Stierhof YD, Codogno P, Proikas-Cezanne T (2011) Resveratrol-mediated autophagy requires WIPI-1-regulated LC3 lipidation in the absence of induced phagophore formation. Autophagy 7 (12):1448–1461
29. Mauthe M, Yu W, Krut O, Kronke M, Gotz F, Robenek H, Proikas-Cezanne T (2012) WIPI-1 positive autophagosome-like vesicles entrap pathogenic Staphylococcus aureus for lysosomal degradation. Int J Cell Biol 2012:179207. https://doi.org/10.1155/2012/179207
30. Gaugel A, Bakula D, Hoffmann A, Proikas-Cezanne T (2012) Defining regulatory and phosphoinositide-binding sites in the human WIPI-1 beta-propeller responsible for autophagosomal membrane localization downstream of mTORC1 inhibition. J Mol Signal 7(1):16. https://doi.org/10.1186/1750-2187-7-16
31. Peppard JV, Rugg C, Smicker M, Dureuil C, Ronan B, Flamand O, Durand L, Pasquier B (2014) Identifying small molecules which inhibit autophagy: a phenotypic screen using image-based high-content cell analysis. Curr Chem Genom Transl Med 8(Suppl 1):3–15. https://doi.org/10.2174/2213988501408010003
32. Burbulla LF, Fitzgerald JC, Stegen K, Westermeier J, Thost AK, Kato H, Mokranjac D, Sauerwald J, Martins LM, Woitalla D, Rapaport D, Riess O, Proikas-Cezanne T, Rasse TM, Kruger R (2014) Mitochondrial proteolytic stress induced by loss of

mortalin function is rescued by Parkin and PINK1. Cell Death Dis 5:e1180. https://doi.org/10.1038/cddis.2014.103

33. Walter C, Clemens LE, Muller AJ, Fallier-Becker P, Proikas-Cezanne T, Riess O, Metzger S, Nguyen HP (2016) Activation of AMPK-induced autophagy ameliorates Huntington disease pathology in vitro. Neuropharmacology 108:24–38. https://doi.org/10.1016/j.neuropharm.2016.04.041
34. Hale CM, Cheng Q, Ortuno D, Huang M, Nojima D, Kassner PD, Wang S, Ollmann MM, Carlisle HJ (2016) Identification of modulators of autophagic flux in an image-based high content siRNA screen. Autophagy 12 (4):713–726. https://doi.org/10.1080/15548627.2016.1147669
35. Jung J, Nayak A, Schaeffer V, Starzetz T, Kirsch AK, Muller S, Dikic I, Mittelbronn M, Behrends C (2017) Multiplex image-based autophagy RNAi screening identifies SMCR8 as ULK1 kinase activity and gene expression regulator. eLife 6:e23063. https://doi.org/10.7554/eLife.23063
36. Dupont N, Chauhan S, Arko-Mensah J, Castillo EF, Masedunskas A, Weigert R, Robenek H, Proikas-Cezanne T, Deretic V (2014) Neutral lipid stores and lipase PNPLA5 contribute to autophagosome biogenesis. Curr Biol 24(6):609–620. https://doi.org/10.1016/j.cub.2014.02.008
37. Zhang L, Yu J, Pan H, Hu P, Hao Y, Cai W, Zhu H, Yu AD, Xie X, Ma D, Yuan J (2007) Small molecule regulators of autophagy identified by an image-based high- throughput screen. Proc Natl Acad Sci U S A 104 (48):19023–19028. https://doi.org/10.1073/pnas.0709695104
38. Balgi AD, Fonseca BD, Donohue E, Tsang TC, Lajoie P, Proud CG, Nabi IR, Roberge M (2009) Screen for chemical modulators of autophagy reveals novel therapeutic inhibitors of mTORC1 signaling. PLoS One 4(9):e7124. https://doi.org/10.1371/journal.pone.0007124
39. Chauhan S, Ahmed Z, Bradfute SB, Arko-Mensah J, Mandell MA, Won Choi S, Kimura T, Blanchet F, Waller A, Mudd MH, Jiang S, Sklar L, Timmins GS, Maphis N, Bhaskar K, Piguet V, Deretic V (2015) Pharmaceutical screen identifies novel target processes for activation of autophagy with a broad translational potential. Nat Commun 6:8620. https://doi.org/10.1038/ncomms9620
40. Li Y, McGreal S, Zhao J, Huang R, Zhou Y, Zhong H, Xia M, Ding WX (2016) A cell-based quantitative high-throughput image screening identified novel autophagy modulators. Pharmacol Res 110:35–49. https://doi.org/10.1016/j.phrs.2016.05.004
41. Mauthe M, Langereis M, Jung J, Zhou X, Jones A, Omta W, Tooze SA, Stork B, Paludan SR, Ahola T, Egan D, Behrends C, Mokry M, de Haan C, van Kuppeveld F, Reggiori F (2016) An siRNA screen for ATG protein depletion reveals the extent of the unconventional functions of the autophagy proteome in virus replication. J Cell Biol 214(5):619–635. https://doi.org/10.1083/jcb.201602046
42. McKnight NC, Jefferies HB, Alemu EA, Saunders RE, Howell M, Johansen T, Tooze SA (2012) Genome-wide siRNA screen reveals amino acid starvation-induced autophagy requires SCOC and WAC. EMBO J 31 (8):1931–1946. https://doi.org/10.1038/emboj.2012.36
43. Strohecker AM, Joshi S, Possemato R, Abraham RT, Sabatini DM, White E (2015) Identification of 6-phosphofructo-2-kinase/fructose-2,6-bisphosphatase as a novel autophagy regulator by high content shRNA screening. Oncogene 34(45):5662–5676. https://doi.org/10.1038/onc.2015.23
44. Niso-Santano M, Malik SA, Pietrocola F, Bravo-San Pedro JM, Marino G, Cianfanelli V, Ben-Younes A, Troncoso R, Markaki M, Sica V, Izzo V, Chaba K, Bauvy C, Dupont N, Kepp O, Rockenfeller P, Wolinski H, Madeo F, Lavandero S, Codogno P, Harper F, Pierron G, Tavernarakis N, Cecconi F, Maiuri MC, Galluzzi L, Kroemer G (2015) Unsaturated fatty acids induce non-canonical autophagy. EMBO J 34(8):1025–1041. https://doi.org/10.15252/embj.201489363
45. Orvedahl A, Sumpter R Jr, Xiao G, Ng A, Zou Z, Tang Y, Narimatsu M, Gilpin C, Sun Q, Roth M, Forst CV, Wrana JL, Zhang YE, Luby-Phelps K, Xavier RJ, Xie Y, Levine B (2011) Image-based genome-wide siRNA screen identifies selective autophagy factors. Nature 480(7375):113–117. https://doi.org/10.1038/nature10546
46. Shen S, Niso-Santano M, Adjemian S, Takehara T, Malik SA, Minoux H, Souquere S, Marino G, Lachkar S, Senovilla L, Galluzzi L, Kepp O, Pierron G, Maiuri MC, Hikita H, Kroemer R, Kroemer G (2012) Cytoplasmic STAT3 represses autophagy by inhibiting PKR activity. Mol Cell 48 (5):667–680. https://doi.org/10.1016/j.molcel.201.09.013

Chapter 28

Methods for the Study of Entotic Cell Death

Jens C. Hamann, Sung Eun Kim, and Michael Overholtzer

Abstract

Entosis is a mechanism of cell competition occurring in cancers that involves the engulfment and killing of neighboring cells. The death of ingested cells, called entotic cell death, usually occurs in a non-apoptotic, autophagy protein-dependent manner, where microtubule-associated protein light chain 3 (LC3) is lipidated onto entotic vacuoles. Here we present methods to quantify entotic cell death and its associated LC3 lipidation.

Key words Entosis, Cell-in-cell, Cannibalism, Engulfment, Time-lapse, Soft agar, Anoikis

1 Introduction

Entosis was first discovered in epithelial and cancer cell populations cultured in the absence of matrix adhesion [1]. Loss of matrix attachment is known to induce an apoptotic form of cell death called anoikis, but non-apoptotic forms of cell death such as entotic cell death can also be induced [2, 3]. In suspension conditions, cells that adhere to each other through E- or P-cadherin-dependent junctions engage in entosis when one cell invades into another, an activity that results from increased RhoA-GTPase and Rho-kinase (ROCK)-dependent actomyosin contraction within the ingested cell [1, 4] (*see* Fig. 1a). Stiffer cells are therefore ingested by softer cells through this mechanism, a relationship that can promote competition in heterogeneous cell populations [5]. While matrix detachment was the first reported inducer of this process, new triggers have recently been found that can induce entosis even under adherent conditions, including long-term starvation for glucose [6], and also mitosis [7], where cell rounding and Rho activity, particularly during aberrant mitoses, can lead to the death of dividing cells through this mechanism (*see* Fig. 1b, c).

During entosis induced by matrix detachment, glucose starvation, or mitosis, entotic cells become engulfed while they are alive and remain viable inside of their neighbors, until lysosomes fuse to

Nicholas Ktistakis and Oliver Florey (eds.), *Autophagy: Methods and Protocols*, Methods in Molecular Biology, vol. 1880, https://doi.org/10.1007/978-1-4939-8873-0_28,

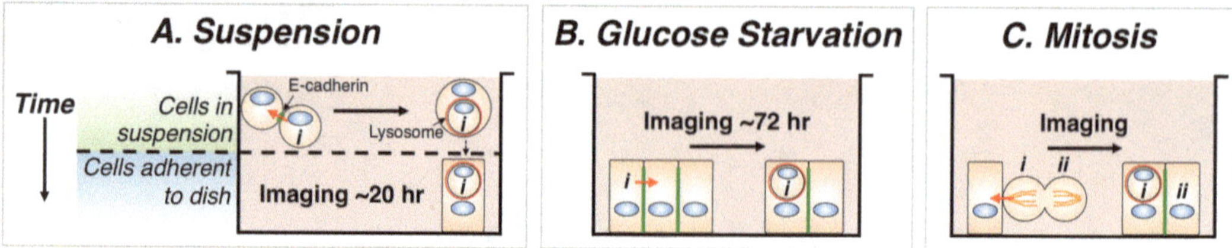

Fig. 1 Entosis can be induced by multiple mechanisms. (**a**) Matrix detachment of cells can lead to entosis. Culturing cells in matrix-detached conditions, or even trypsinizing cells and readhering them to matrix, can result in entotic cell internalization. Tension differences, driven by Rho-mediated actomyosin contractility between neighboring cells that have formed adherens junctions (green line), lead to the invasion of one cell ("i") into another (top). Entotic structures can be analyzed for cell fate, including cell death and lysosome fusion (red), by imaging after cells have become attached to the plate (bottom). Red arrow depicts loser cell invading into winner cell. (**b**) Prolonged glucose starvation induces entosis. Under adherent conditions, cells deprived of glucose for several days undergo high levels of entosis. (**c**) Entosis can result from mitosis. Cell rounding and increased Rho activity during mitoses can result in the engulfment of live dividing cells into adherent neighbors

the membrane that surrounds them, called the entotic vacuole. Lysosome fusion and entotic cell death are promoted by autophagy protein-dependent lipidation of the microtubule-associated protein light chain 3 (LC3) onto entotic vacuoles, a process resembling LC3 lipidation onto phagosomes, called LC3-associated phagocytosis, or LAP [8, 9]. Entotic cells can also undergo apoptosis, resulting from prolonged entrapment inside of engulfing cells, or by inhibition of key survival pathways such as canonical autophagy, which internalized cells use to survive inside of the nutrient-limited entotic vacuole [8]. Here we present methods to quantify entotic cell death that involves endocytic LC3 lipidation.

2 Materials

1. Glass-bottom culture dishes (e.g., P06G-1.5-20-F; MatTek).
2. Growth medium. Use media formulations as optimized for each cell line. For MCF-7 cells, use DMEM high glucose plus 10% heat-inactivated fetal bovine serum (FBS) and 100 I.U./mL penicillin and 100 μg/mL streptomycin. MCF-10A cells are cultured as described [10].
3. 0.25% trypsin/1 mM EDTA solution.
4. 1× phosphate-buffered saline (PBS) solution, pH = 7.4, Ca^{2+}-free.
5. Mineral oil.
6. Confocal or widefield microscope with live-cell imaging capabilities and a motorized stage (e.g., Ultraview Vox spinning-disk confocal system equipped with a Yokogawa CSU-X1

spinning-disk head and an electron-multiplying charge-coupled device camera (Hamamatsu C9100-13) coupled to a Nikon Ti-E microscope) equipped with a 37 °C and 5% CO_2 incubation chamber.

7. Software for imaging and analysis (e.g., Volocity; PerkinElmer).

3 Methods

3.1 Time-Lapse Imaging

1. Pre-warm cell culture medium, PBS, and trypsin in 37 °C water bath.
2. Trypsinization of cells: Wash cells once with pre-warmed PBS, and incubate with pre-warmed trypsin at 37 °C and 5% CO_2 for 5–15 minutes (depending on cell line) to dissociate cells.
3. Pipette cells using a P1000 pipetman to achieve a single-cell suspension, add 5 mL growth medium to quench trypsin, pellet cells, and resuspend cells in fresh growth medium.
4. Count cells using a hemacytometer and plate cells at a density of ~$3 \times 10^4/\text{cm}^2$ (e.g., 250,000 cells per 35 mm diameter dish). Incubate cells overnight at 37 °C and 5% CO_2 to allow cells to adhere to the glass dish. Entotic cell-in-cell structures will form during this time, while cells are in suspension (*see* **Note 1** and Fig. 1).
5. Wash cells once with 2 mL pre-warmed PBS to remove any non-adherent cells, and add 2 mL fresh, pre-warmed growth medium.
6. Spread 1 mL mineral oil on top of media to prevent evaporation, and mount sample on confocal or widefield microscope in cell incubation chamber for live-cell imaging (*see* **Note 2**). Typically widefield imaging is used to quantify cell death frequencies in cell populations, while confocal imaging is used to quantify LC3 lipidation onto entotic vacuoles.
7. Identify microscopic fields of interest containing entotic cell structures for time-lapse analysis. Multiple points containing entotic engulfments throughout the sample should be picked to maximize the number of cell events that can be quantified (*see* **Notes 3** and **4**).
8. For imaging and cell death analysis, expression of a fluorescent nuclear marker (e.g., H2B-mCherry), as well as GFP-tagged light chain 3 (GFP-LC3) to monitor LC3 lipidation onto entotic vacuoles, is optimal. Both fluorescence and differential interference contrast (DIC) images should be acquired at intervals of 5–10 min for a total duration of up to 24 h (or longer, if needed), preferably using a 20× objective for widefield

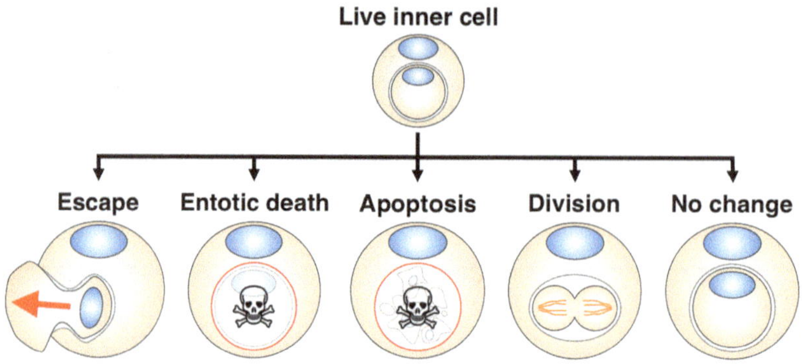

Fig. 2 Fates of live internalized cells. Cells internalized by entosis can have different fates. Upon completion of engulfment, cells can either escape from their hosts and remain viable, or they can undergo entotic or apoptotic death after internalization (*see* Subheading 3.2 above). Alternatively, inner cells can also divide, after which they can further undergo the abovementioned fates. Depending on the length of time-lapse analysis, some cells can remain unchanged. Entotic cell death is typically the most common fate that is observed

microscopy-based quantification of cell death frequencies in cell populations (*see* Subheading 3.2) or a 63× objective for confocal analysis of LC3 lipidation (*see* Subheading 3.3).

3.2 Analysis of Entotic Cell Death

1. Internalized entotic cells can have one of several different fates: they can undergo cell death, they can escape from their hosts, or they can remain viable inside of their hosts for long periods of time, sometimes even for longer than 24 h (*see* Fig. 2). In some cases entotic cells can also divide inside of their hosts, and host cells can therefore contain more than one internalized cell. Typically, the different cell fates for entotic cells are recorded for populations examined by time-lapse microscopy as relative percentages of "death," "escape," and "no change," which refers to cells that remain inside of their hosts and do not die. The category of cell death can be further subcategorized as entotic cell death or apoptotic cell death, as discussed below. Cell fates should be recorded for control and experimental conditions to determine relative rates of each cell fate. Typically, an imaging duration of at least 10–20 h is used to determine cell fate frequencies. Cell fate quantifications are performed from a minimum of three independent experiments and are based on analyses of large populations of entotic cells identified through many different microscopic fields of view (*see* **Notes 5** and **6**).
2. To distinguish between apoptotic and entotic cell death of internalized cells, time-lapse movies can be analyzed for changes in cell and nuclear morphology in the DIC and fluorescent channels, respectively (*see* Fig. 3).
3. For entotic cell deaths, LC3 lipidation typically occurs prior to the internalized cell undergoing death, and lipidation is

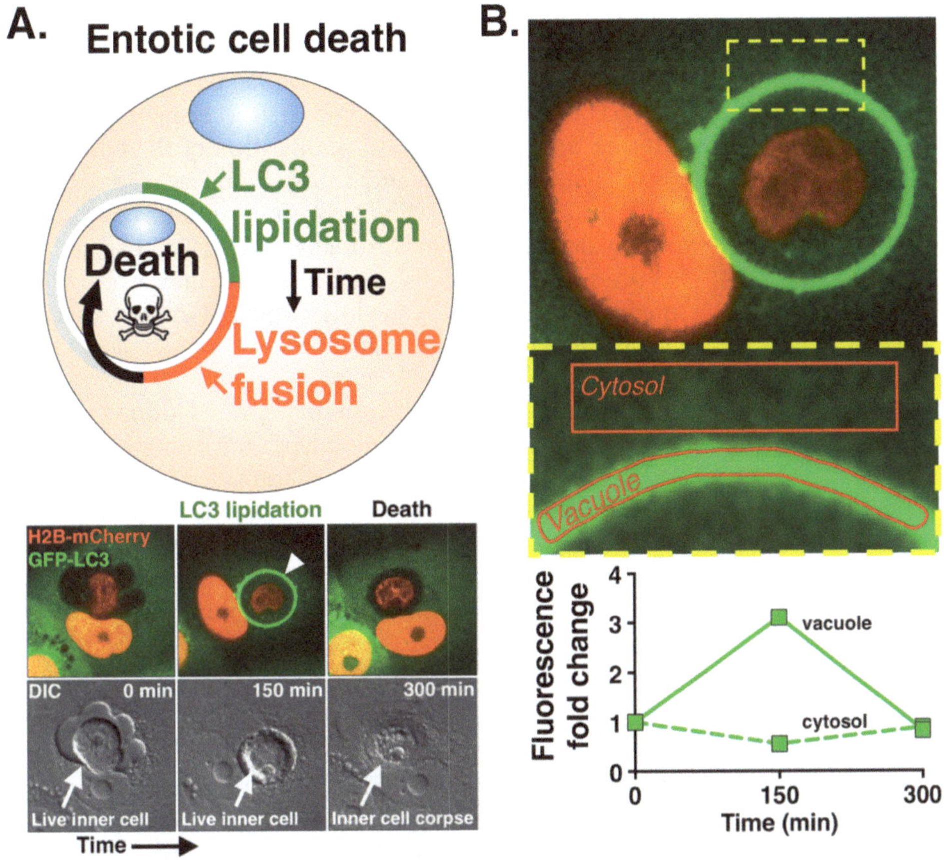

Fig. 3 Cell death following entosis. (**a**) Maturation of the entotic vacuole involves transient LC3 lipidation prior to entotic cell death. After engulfment, host cells mature the single-membrane vacuole, which involves LC3 lipidation and subsequent lysosome fusion that results in acidification and non-apoptotic cell death. Representative image sequence depicts maturation steps beginning after completion of engulfment. Note the clear recruitment of GFP-LC3 (white arrowhead) onto the vacuole and the subsequent cell death. The inner cell undergoes non-apoptotic cell death, as evidenced by the absence of nuclear fragmentation and condensation. The corpse will continue to degrade and shrink after this time. (**b**) Quantification of LC3 recruitment to entotic vacuoles. Top image shows GFP-LC3 lipidation onto entotic vacuole to be analyzed. Inset shows higher magnification of vacuole membrane area, including ROIs of both cytosol and vacuole, of which mean fluorescence intensity is quantified (*see* Subheading 3.3 above). Graph shows quantification of image sequence shown in (**a**)

followed by fusion of lysosomes that acidify the entotic vacuole, leading to the non-apoptotic cell death. In some cases, the inner cell may undergo apoptosis, as evidenced by the traditional criteria of this type of death, including cell blebbing and nuclear condensation or fragmentation [1]. Apoptotic cell death in these cases typically precedes LC3 recruitment to the vacuoles (*see* **Note 7**). By time-lapse analysis, the degradation and shrinkage of cells that is observable by DIC microscopy can also be used as an indicator of cell death.

4. As an alternative approach, monitoring acidification of the vacuoles that contain internalized cells can be used as a marker for entotic cell death. Acidification can be monitored by adding LysoTracker fluorescent dyes (Thermo Fisher) to the medium during time-lapse analysis; acidification of the entotic vacuole results in an increased and diffuse LysoTracker staining within internalized cells [1]. Similarly, time-lapse imaging of fluorescently tagged lysosomal proteins (e.g., LAMP1 or cathepsin B) expressed in host cells can be used as indicator of lysosome fusion to the entotic vacuole, which is known to cause cell death [8].

3.3 Analysis of LC3 Lipidation

1. To quantify LC3 lipidation that occurs during entotic cell death, typically confocal time-lapse microscopy is used with a 63× objective. Entotic cell structures for time-lapse imaging should be identified as described in Subheading 3.2. Following time-lapse imaging, identify cells that have undergone cell death, as determined by DIC and fluorescence imaging, to be used for LC3 quantification (*see* Subheading 3.2 above and Fig. 3).
2. To quantify relative GFP-LC3 lipidation at entotic vacuoles, use a single confocal imaging plane from a midplane region of the vacuole (if z-sections have been acquired in the time-lapse analysis), and draw three free-form regions of interest (ROI) along vacuole border, as well as adjacent cytoplasmic regions as controls (*see* Fig. 3b and **Notes 8** and **9**). Also identify an area of the image that is free of cells to use as the background ROI. We typically use a minimum size of 8 μm^2 for all regions.
3. Measure and calculate the mean fluorescence intensity in the GFP channel of all of the ROI and record them. Calculate the average GFP intensity of the three ROI along the entotic vacuole, as well as of the cytoplasmic controls, subtract the background fluorescence intensity from all values, and plot the data normalized to the first timepoint (*see* Fig. 3 and **Note 10**).

4 Notes

1. For maximum entosis induction, typically cells are plated on untreated glass; pre-coating with serum or matrix proteins to enhance cell adherence to the glass may reduce the frequency of entotic events, as these occur while cells are in suspension, prior to adhering. For cells that will not adhere to untreated glass, pre-coating may be required.
2. The exact volume of mineral oil needed to completely cover the culture may vary; spread until entire surface of medium is covered.

3. This method describes analysis of internalized cell fates for entotic structures that form while cells are in suspension and then adhere to the glass surface. To selectively monitor entotic structures that form while cells are adherent, during glucose starvation, for example (*see* Fig. 1), we suggest initially plating cells in the presence of 10 μM Y-27632 (Y0503; Sigma-Aldrich) (a Rho-kinase inhibitor that blocks entosis) to prevent entosis from occurring as cells adhere to the glass surface. To then monitor entotic structures that form as a result of glucose starvation, remove Y-27632 by washing cells several times with PBS and replacing with starvation medium, and proceed to time-lapse analysis [6].
4. When selecting fields of view with entotic cells to image through time, we suggest focusing on cell-in-cell structures where the vacuole of the host cell is clearly visible, which is indicative of completed engulfment (*see* Fig. 3 as an example). If further verification of the completion of engulfment is needed, cell membranes can be stained with the FM4-64 fluorescent dye (*N*-(3-triethylammoniumpropyl)-4-(6-(4-(diethylamino)phenyl)hexatrienyl)pyridinium dibromide; T11320, Thermo Fisher), which will be excluded from entotic vacuoles when cell-in-cell formation is completed, as reported [8].
5. To quantify the "no change" category, typically only those cells that remain within the observable field of view for the duration of the experiment are counted.
6. For quantification of internalized cell fates in a population, the deaths of two internalized cells within the same entotic vacuole (typically resulting from cell division of an internalized cell) will often occur at the same time and are counted as one event. For host cells with multiple internalized cells that are contained within separate entotic vacuoles, the fates of these cells are typically counted as individual events.
7. For MCF-7 and MCF-10A cells, entotic cell death is typically the most common type of cell death that occurs, while apoptotic death of internalized cells occurs at low frequency. For other cell lines, the ratio between entotic cell death and apoptotic cell death events may be different, and more apoptotic events may be observed.
8. When quantifying fluorescence intensity in cytosolic regions, it is important to select areas that do not contain any GFP-LC3 puncta. When quantifying regions of the vacuole, it is important to carefully select regions that do not overlap with cytosol.
9. When quantifying fluorescence intensity, make sure that the signal in the respective channel is not saturated to obtain values within the appropriate range.

10. As an alternative to the method described in Subheading 3.3 to quantify LC3 lipidation to entotic vacuoles, a line scan measuring fluorescence intensity from the cytosol to across the entotic vacuole membrane can be used [6].

References

1. Overholtzer M, Mailleux AA, Mouneimne G, Normand G, Schnitt SJ, King RW, Cibas ES, Brugge JS (2007) A nonapoptotic cell death process, entosis, that occurs by cell-in-cell invasion. Cell 131(5):966–979. https://doi.org/10.1016/j.cell.2007.10.040
2. Buchheit CL, Weigel KJ, Schafer ZT (2014) Cancer cell survival during detachment from the ECM: multiple barriers to tumour progression. Nat Rev Cancer 14(9):632–641. https://doi.org/10.1038/nrc3789
3. Martins I, Raza SQ, Voisin L, Dakhli H, Law F, De Jong D, Allouch A, Thoreau M, Brenner C, Deutsch E, Perfettini JL (2017) Entosis: the emerging face of non-cell-autonomous type IV programmed death. Biom J 40(3):133–140. https://doi.org/10.1016/j.bj.2017.05.001
4. Sun Q, Cibas ES, Huang H, Hodgson L, Overholtzer M (2014) Induction of entosis by epithelial cadherin expression. Cell Res 24 (11):1288–1298. https://doi.org/10.1038/cr.2014.137
5. Sun Q, Luo T, Ren Y, Florey O, Shirasawa S, Sasazuki T, Robinson DN, Overholtzer M (2014) Competition between human cells by entosis. Cell Res 24(11):1299–1310. https://doi.org/10.1038/cr.2014.138
6. Hamann JC, Surcel A, Chen R, Teragawa C, Albeck JG, Robinson DN, Overholtzer M (2017) Entosis is induced by glucose starvation. Cell Rep 20(1):201–210. https://doi.org/10.1016/j.celrep.2017.06.037
7. Durgan J, Tseng YY, Hamann JC, Domart MC, Collinson L, Hall A, Overholtzer M, Florey O (2017) Mitosis can drive cell cannibalism through entosis. Elife 6:e27134. https://doi.org/10.7554/eLife.27134
8. Florey O, Kim SE, Sandoval CP, Haynes CM, Overholtzer M (2011) Autophagy machinery mediates macroendocytic processing and entotic cell death by targeting single membranes. Nat Cell Biol 13(11):1335–1343. https://doi.org/10.1038/ncb2363
9. Florey O, Overholtzer M (2012) Autophagy proteins in macroendocytic engulfment. Trends Cell Biol 22(7):374–380. https://doi.org/10.1016/j.tcb.2012.04.005
10. Debnath J, Muthuswamy SK, Brugge JS (2003) Morphogenesis and oncogenesis of MCF-10A mammary epithelial acini grown in three-dimensional basement membrane cultures. Methods 30(3):256–268

Chapter 29

MHC Class I Internalization via Autophagy Proteins

Monica Loi, Laure-Anne Ligeon, and Christian Münz

Abstract

Macroautophagy is a ubiquitous degradative pathway involved in innate and adaptive immunity. Its molecular machinery has been described to deliver intracellular and extracellular antigens to MHC class II loading compartment by regulating autophagosome and phagosome maturation. We recently found that the respective Atg proteins can contribute to MHC class I-restricted antigen presentation to $CD8^+$ T cells by regulating MHC class I surface levels in mouse dendritic cell. Indeed, we determined that MHC class I molecules are stabilized on the cell surface of murine antigen presenting cells deficient for core components of the macroautophagy machinery such as Atg5 and Atg7. This stabilization seems to result from defective internalization of MHC class I molecules dependent on adaptor protein kinase 1 (AAK1), involved in clathrin-mediated endocytosis. Moreover, macroautophagy-dependent stabilization of MHC class I molecules leads to enhanced $CD8^+$ T cell priming during influenza A virus infection in vivo, resulting in decreased pathology. In this chapter, we describe four experiments to monitor, characterize, and quantify the effect of macroautophagy deficiency on MHC class I molecule trafficking and the subsequent $CD8^+$ T cell priming. First, we will show how to monitor MHC class I internalization in lung $CD11c^+$ cells from mice lacking key components of the macroautophagy machinery. Then, we will propose a method to characterize the interaction between either MHC class I or Atg8/LC3 with AAK1. Finally, we will describe how to evaluate the influenza A-specific $CD8^+$ T cell response in mice conditionally depleted for *Atg5* in their DC compartment. This set of experiments allows to characterize MHC class I internalization with the help of the molecular machinery of macroautophagy.

Key words MHC class I, Autophagy, Atg5, Atg7, Dendritic cells

1 Introduction

Macroautophagy is a highly conserved degradative pathway among all eukaryotic cells, which mediates delivery of nonfunctional cytosolic proteins or organelles to the lysosome for degradation. The molecular mechanisms involved during macroautophagy are fairly well known, and more than 30 autophagy-related proteins (Atg) have been described to play a role to generate a double-membrane vesicle, called autophagosome [1]. The microtubule-associated protein light chain LC3 (LC3, mammalian orthologue of yeast Atg8) is by far the best characterized of the Atg proteins and

Nicholas Ktistakis and Oliver Florey (eds.), *Autophagy: Methods and Protocols*, Methods in Molecular Biology, vol. 1880, https://doi.org/10.1007/978-1-4939-8873-0_29, © Springer Science+Business Media, LLC, part of Springer Nature 2019

generally considered as the hallmark of the autophagosome. Indeed, LC3 exists in two forms, a cytosolic form (LC3-I) and a lipidated form associated with autophagosomal membranes (LC3-II). LC3 lipidation is under the control of a molecular machinery including the two proteins Atg5 and Atg7 [2]. These proteins are frequently targeted to compromise macroautophagy by blocking LC3 lipidation. Macroautophagy has first emerged as an important catabolic process involved during innate immunity, and more recently it was described to play a role also during adaptive immunity. The fact that approximately 20% of the natural major histocompatibility complex (MHC) class II ligands originate from cytosolic and nuclear proteins, including the macroautophagy marker LC3, suggests a possible link between macroautophagy and antigen processing for MHC presentation [3]. It has been shown that the macroautophagy machinery assists and facilitates MHC presentation to T cells through different pathways. Its molecular machinery was described to deliver intracellular antigens to MHC class II loading compartments [4], to enhance exogenous antigen delivery for MHC class II presentation [5], but also described to facilitate efficient antigen cross-presentation on MHC class I molecules [6]. Macroautophagy plays also an important role in endogenous self-protein processing involved in the negative and positive selection of $CD4^+$ T cell in the thymus [7, 8]. In addition to the involvement of classical macroautophagy in MHC class II antigen processing, we have shown that exogenous antigen processing for MHC class II presentation can also be regulated via a non-canonical macroautophagy pathway, called LC3-associated phagocytosis or LAP [9]. Briefly, in this pathway, LC3 is directly coupled to the phagosomal membrane and requires reactive oxygen species production by NADPH oxidase 2 (NOX2), which, depending on the cell type, seems to accelerate or attenuate the fusion of phagosomes with the lysosomes [9–11]. In case of human macrophages and conventional and plasmacytoid dendritic cells (DCs), the fusion between the phagosome and lysosome is attenuated, resulting in prolonged MHC class II presentation or cargo delivery to Toll-like receptor (TLR) containing endosomes [9]. Compared to the role of macroautophagy during antigen processing for MHC class II presentation, little is known of how this pathway can influence antigen processing for MHC class I presentation. Recently, we demonstrated that the macroautophagy machinery contributes to MHC class I-restricted antigen presentation to $CD8^+$ T cells by regulating MHC class I surface levels in mouse cells leading to enhanced $CD8^+$ T cell priming during influenza A virus infection in vivo [12]. In fact, mouse dendritic cells with a compromised macroautophagy machinery showed an elevated MHC class I surface expression, which seems to result from a defective internalization of MHC class I molecule dependent on adaptor protein kinase 1 (AAK1). Indeed, AAK1 can interact with the cytosolic form of

LC3, but it fails to be recruited to the MHC class I internalization machinery when LC3 lipidation is blocked. Furthermore, we showed that the impairment of LC3 lipidation by *atg5* or *atg7* gene know-out in murine lung DCs and macrophages decreases the intracellular pool of internalized MHC class I molecules. Interestingly, in the absence of Atg-dependent MHC class I internalization, dendritic cells stimulate influenza-specific $CD8^+$ T cell responses more efficiently. This was also associated with better immune control of influenza infection in vivo, showing that the macroautophagy machinery orchestrates T cell immunity by attenuating MHC class I surface expression levels [12].

In order to monitor how MHC class I internalization is affected by the macroautophagy machinery, we will describe four methods. First, we will propose two independent but complementary experiments to assess the level of MHC class I internalization in Atg-deficient cells. The first method is a flow cytometry-based assay allowing to follow the internalization of antibody-labelled MHC class I molecules over time and compare the rate of internalization between Atg-sufficient and Atg-deficient cells. The obtained data can be validated by visualizing the internalized MHC class I molecules with a confocal microscopy approach. These two methods allow to assess the impact of a compromised macroautophagy machinery on the internalization of MHC class I molecules, but do not give information about the molecular mechanisms involved. To address this point, we investigated the interaction between either MHC class I or LC3 with the adaptor molecule AAK1. Finally, we describe an assay to show the effect of the impairment of Atg-dependent MHC class I internalization on $CD8^+$ T cell response during influenza A virus infection. In this assay, we take advantage of an influenza A virus infection model in mice in order to study if the macroautophagy machinery can shape the antiviral $CD8^+$ T cell response by regulating MHC class I surface levels on antigen-presenting cells. These basic experiments allow monitoring of the effects of an impaired macroautophagy machinery on MHC class I internalization and to characterize the implications of such regulation in vivo.

2 Materials

2.1 MHC Class I Internalization by FACS

2.1.1 Mice

C57BL/6 (Janvier), *atg5*$^{flox/flox}$ (kindly provided by Dr. Mizushima, Tokyo Metropolitan Institute of Medical Science, Tokyo 113-8613, Japan) [13], and CD11c-Cre Tg (CD11c$^{cre+/-}$, Jackson).

Conditional knockout mice for disruption of autophagy in the $CD11c^+$ cellular compartment, designated *atg5*$^{-/-}$ DC mice, are generated by crossing *atg5*$^{fl/fl}$ with CD11c$^{cre+/-}$ mice. As littermate control mice, *atg5*$^{fl/fl}$ × CD11c$^{cre-/-}$ (*atg5*$^{+/+}$ DC) mice are used.

2.1.2 Cell Culture

Cell culture medium: RPMI-1640 supplemented with heat-inactivated (*see* **Note 1**) 10% fetal calf serum (FCS) (R10).

2.1.3 Reagents

1. DNase I grade II.
2. Hepes 1 M.
3. Collagenase A.
4. CD11c beads.
5. Percoll.

2.1.4 Probes

1. Biotin anti-mouse H2-K^b (clone AF6-88.5, BioLegend 116504).
2. Biotin anti-mouse H2-D^b (clone KH95 BioLegend 111504 diluted).
3. PE-streptavidin (BioLegend 405203).
4. Pacific Blue anti-I-A/I- E (clone M5/114.15.2, BioLegend 107620).
5. PE-Cy7 anti-mouse CD11c (clone N418, BioLegend 117318).
6. Live/Dead Fixable Aqua Dead Cell Stain Kit (Invitrogen L34957).

2.1.5 Buffers

1. MACS buffer: 1% bovine serum albumin (BSA), 2 mM EDTA in PBS, to be filtered 0.22 μm.
2. FACS buffer: PBS supplemented with 2% FCS and 0.01% sodium azide.

2.2 MHC Class I Internalization by Immunofluorescence

2.2.1 Mice

C57BL/6 (Janvier), *atg5*$^{flox/flox}$ (kindly provided by Dr. Mizushima, Tokyo Metropolitan Institute of Medical Science, Tokyo 113-8613, Japan) [13], *atg7*$^{fl/fl}$ (kindly provided by Dr. Komatsu, Tokyo Metropolitan Institute of Medical Science, Tokyo 113-8613, Japan) [14], and CD11c-Cre Tg (CD11c$^{cre+/-}$, Jackson).

Conditional ***knockout*** mice for disruption of autophagy in CD11c$^+$ compartment, designated *atg5*$^{-/-}$ or *atg7*$^{-/-}$ DC mice, are generated by crossing *atg5*$^{fl/fl}$ or *atg7*$^{flox/flox}$ with CD11c$^{cre+/-}$ mice. As littermate control mice, *atg5*$^{fl/fl}$ × CD11c$^{cre-/-}$ (*atg5*$^{+/+}$ DC) and *atg7*$^{fl/fl}$-CD11c$^{cre-/-}$ (*atg7*$^{+/+}$ DC) mice are used.

2.2.2 Cell Culture

1. Cell culture medium: RPMI-1640 supplemented with heat-inactivated (*see* **Note 1**) 10% FCS (R10).
2. Multi-chamber 8-wells (IBIDI).

2.2.3 Reagents

1. DNase I grade II (Roche 7600105).
2. Hepes 1M (Invitrogen 15630-056).
3. Collagenase A (Sigma C9891-500MG).
4. CD11c beads (Miltenyi Biotec 130-052-001).
5. Percoll (GE Healthcare 17-0891-01).
6. Poly-L-lysine.
7. 4% paraformaldehyde solution in PBS.
8. Dako fluorescence mounting medium.

2.2.4 Probes

1. Purified rat anti-mouse CD16/32 (mouse BD Fc block) (BD 553141).
2. Biotin anti-mouse H2-K^b (clone AF6-88.5, BioLegend 116504).
3. PE anti-mouse H2 (clone M1/42, BioLegend 125506).
4. Alexa Fluor® 555-conjugated goat anti-rat IgG H&L (Invitrogen A21434).
5. Alexa Fluor® 555-conjugated goat anti-mouse IgG H&L (Invitrogen A21422).
6. 4,6-diamidino-2-phenylindole (DAPI) nucleic acid stain, stock solution 5 mg/mL.

2.2.5 Buffers

1. MACS buffer: 1% BSA, 2 mM EDTA in PBS, filtered 0.22 μm.
2. FACS buffer: PBS supplemented with 2% FCS and 0.01% sodium azide.
3. Coated solution: Poly-L-lysine (Sigma) diluted at 1:10 in sterile water.
4. Permeabilization solution: 0.1% Triton X-100 in PBS.
5. Blocking buffer: PBS supplemented with 1% of bovine serum albumin (BSA).

2.3 Immunoprecipitation for MHC Class I and LC3, Followed by Western Blotting for AAK1

2.3.1 Mice

C57BL/6 (Janvier), *atg5*$^{flox/flox}$ (kindly provided by Dr. Mizushima, Tokyo Metropolitan Institute of Medical Science, Tokyo 113-8613, Japan) [13], *atg7*$^{fl/fl}$ (kindly provided by Dr. Komatsu, Tokyo Metropolitan Institute of Medical Science, Tokyo 113-8613, Japan) [14], and CD11c-Cre Tg (CD11c$^{cre+/-}$, Jackson).

Conditional knockout mice for disruption of autophagy in CD11c^{+} compartment, designated *atg5*$^{-/-}$ or *atg7*$^{-/-}$ DC mice, are generated by crossing *atg5*$^{fl/fl}$ or *atg7*$^{flox/flox}$ with CD11c$^{cre+/-}$ mice. As littermate control mice, *atg5*$^{fl/fl}$ × CD11c$^{cre-/-}$ (*atg5*$^{+/+}$ DC) and *atg7*$^{fl/fl}$-CD11c$^{cre-/-}$ (*atg7*$^{+/+}$ DC) mice are used.

2.3.2 Cell Culture

1. BM-DC medium filtered sterile RPMI-1640 supplemented with penicillin/streptomycin 50 μM β-mercaptoethanol (*see* **Note 2**), 20 ng/mL GM-CSF and 20% heat-inactivated FCS (*see* **Note 3**).

2.3.3 Reagents

1. ACK lysis buffer: 0.15 M NH4 Cl, 1 mM KHCO3, and 0.1 mM EDTA, pH 7.2.
2. Protein A beads.
3. BCA protein assay kit.
4. Complete protease inhibitor cocktail tablets.
5. ECL detection reagents.
6. Acrylamide/Bis-acrylamide: 30%, ratio 29:1.
7. TEMED (*N,N,N,N*-Tetramethylethylenediamine).
8. Protein marker, dual color.
9. 0.45 μm PVDF membrane.
10. 100MM blotting paper.

2.3.4 Probes and Cytokine

1. Normal rabbit antiserum (NRS) diluted 1:20 in cold IP lysis buffer.
2. Rabbit anti-mouse H2-K^b (Exon-8) (a gift from Dr. Jack Bennink, Bethesda, MD) used 5 μL.
3. Biotin anti-mouse H2-K^b (clone AF6-88.5, BioLegend 116504) diluted 1:1000 in in PBS-T-1% BSA.
4. Rabbit anti-LC3 (MBL PM036) used 5 μL for IP and diluted 1:1000 in PBS-T-1% skimmed milk for WB.
5. Rabbit anti-AAK1 (Abcam ab77082) diluted 1:1000 in in PBS-T-1% BSA.
6. Peroxidase-AffiniPure Goat Anti-Rabbit IgG (H + L) (HRP) (Jackson) diluted 1:50000 in in PBS-T-1% BSA.
7. Streptavidin HRP diluted 1:1000 in in PBS/T-1% BSA.
8. Human granulocyte-macrophage colony-stimulating factor (GM-CSF) (BioLegend 576,302), stock solution 0.20 mg/mL, used at 20 ng/mL.
9. Pacific Blue anti-I-A/I-E (clone M5/114.15.2,BioLegend 107,620) diluted 1:400 in FACS buffer.
10. PE-Cy7 anti-mouse CD11c (clone N418, BioLegend 117,318) diluted 1:400 in FACS buffer.
11. Live/Dead Fixable Aqua Dead Cell Stain Kit (Invitrogen L34957) diluted 1:500 in FACS buffer.

2.3.5 Buffers

1. IP lysis buffer: 50 mM Tris–HCl pH 7.5, 150 mM NaCl, 1% NP-40.
2. SDS-PAGE loading buffer: 50 mL dH_2O, 16 mL 1.5 M Tris–HCl pH 6.8, 15 mL glycerol, 5 g SDS, few flakes of bromophenol blue; store aliquots at −20 °C, and add 1% of β-mercaptoethanol to the sample buffer before use.
3. Resolving gel buffer: 1.5 M Tris–HCl pH 8.8; store at room temperature.
4. Stacking gel buffer: 0.5 M Tris–HCl pH 6.8; store at room temperature.
5. Initiator (APS): 10% ammonium persulfate solution in dH_2O; prepare just prior to use, or store aliquots at −20 °C.
6. SDS-PAGE Running buffer: dilute 10× Tris/glycine/SDS (TGS) with dH_2O to obtain 1× TGS final - 25 mM Tris, 192 mM glycine, 0.1% SDS, pH 8.3.
7. Transfer buffer: dilute 20× NuPAGE® transfer buffer (20×) (Life Technologies NP0006-1) with dH_2O to obtain 1× TG final (25 mM Tris, 92 mM glycine, pH 8.3), and add 20% of methanol.
8. PBS/T: 0.1% Tween 20 in PBS.
9. Blocking buffer: 5% skimmed milk powder in PBS/T.

2.4 Influenza Infection of $CD11c^{cre+/-}$ × $Atg5^{fl/fl}$ Mice and Readout for $CD8^+$ T Cell Responses in the Lung

2.4.1 Mice

C57BL/6 (Janvier), *atg5*$^{flox/flox}$ (kindly provided by Dr. Mizushima, Tokyo Metropolitan Institute of Medical Science, Tokyo 113-8613, Japan) [13], and CD11c-Cre Tg ($CD11c^{cre+/-}$, Jackson).

Conditional knockout mice for disruption of autophagy in the $CD11c^+$ cellular compartment, designated *atg5*$^{-/-}$ DC mice, are generated by crossing *atg5*$^{fl/fl}$ × $CD11c^{cre+/-}$ mice. As littermate control mice, *atg5*$^{fl/fl}$ × $CD11c^{cre-/-}$ (*atg5*$^{+/+}$ DC) mice are used.

2.4.2 Cell Culture

1. Cell culture medium: RPMI-1640 supplemented with heat-inactivated (*see* **Note 1**) 10% FCS (R10).

2.4.3 Reagents

1. DNase I grade II.
2. Hepes 1 M.
3. Collagenase A.
4. CD11c beads.
5. Percoll.
6. Brefeldin A.
7. BD Cytofix/Cytoperm Fixation/Permeabilization Kit (554714).

2.4.4 Probes

1. APC-Cy7 anti-mouse CD45.
2. Pacific Blue anti-mouse CD4.
3. APC anti-mouse CD8.
4. Anti-mouse CD28.
5. Anti-mouse CD3.
6. PE-anti-IFN-γ, XMG1.2 (BD 554412) diluted 1:50 in 1× Perm/Wash solution (BD kit).
7. Eight Live/Dead Fixable Aqua Dead Cell Stain Kit (Invitrogen L34957).

2.4.5 Virus and Peptide

1. Influenza A/PR8 virus (H1N1) Charles River.
2. Peptides $NP1_{366–374}$, $HA_{211–225}$, $NP2_{311–325}$, and NY-ESO-$1_{157-170}$ were synthesized by GL Biochem (Shanghai) at a purity >90%, used at 20 μg/mL.

2.4.6 Buffers

1. FACS buffer PBS supplemented with 2% FCS and 0.01% sodium azide.

3 Methods

3.1 MHC Class I Internalization by FACS

3.1.1 Lung Digestion

1. Euthanize *atg5*$^{-/-}$ or *atg7*$^{-/-}$ DC mice and their littermates with CO_2 inhalation (*see* **Note 4**).
2. Perfuse immediately the lungs with 10 mL PBS from the left ventricle of the heart with 18G needle.
3. Remove the lung in 5 mL in R10 and keep on ice.
4. Transfer the lungs in a 1.5 mL collection tube for mechanical predigestion by dissecting it with scissors.
5. Transfer the obtained lung preparation in a 6-well plate (one lung/well). To easier collect the lung suspension, cut the tip of the P1000.
6. Wash the collection tube with 1 mL of R10, and transfer the remaining 4 mL of R10 into the well.
7. Add the following digestion cocktail: 400 μg/mL collagenase A, 50 μg/mL DNase I, and 25 mM Hepes, and incubate for 45 min at 37 °C on shaker.
8. Incubate for additional 5 min at 37 °C with 10 mM EDTA to stop the reaction.
9. Put a 40 μm strainer on top of a 50 mL Falcon tube, and equilibrate the strainer with 10 mL cold PBS. Keep samples on ice. Press the tissue through the strainer using the piston of the 2.5/5 mL syringe.

10. Wash the cell strainer with 10 mL cold PBS, centrifuge at 500 × *g* for 10 min, and aspirate the supernatant.
11. Prepare 15 mL Falcon tube with 70% Percoll, and overlay it with the cells previously resuspended in 4 mL of 40% Percoll (*see* **Note 5**).
12. Add 1 mL of RPMI-1640 on top. Spin 1500 × *g* with an acceleration of 4 and deceleration of 1 during 30 min at 4 °C.
13. Carefully collect the tubes out from the centrifuge to not disrupt the gradient. Remove the top ring (CD45-negative fraction, or collect it if needed), and collect the intermediate ring (CD45-positive fraction) in 50 mL Falcon tube.
14. Wash with 40 mL cold PBS, 500 × *g* for 10 min (*see* **Note 6**).
15. Resuspend the pellet in 5 mL PBS per 1 mL of R10 based on the further use and count the cells. Average yield from one mouse: 1–4 × 10^7 CD45-positive cells total.

3.1.2 CD11c+ MACS Separation

1. Resuspend cells isolated from lung (*see* Subheading 3.1.1) at a concentration of 2.5 × 10^8 cells/mL with cold MACS buffer, and add 100 μL CD11c beads per 10^8 cells at 4 °C.
2. After 20 min, wash in cold MACS buffer, 500 × *g* for 5 min, and then resuspend 2 × 10^8 cells/mL in cold MACS buffer.
3. Proceed with positive magnetic cell separation in the autoMACS cell separator (Miltenyi Biotec). Choose positive selection "posseld" program (*see* **Note 7**).
4. Take the positive fraction; centrifuge at 400 × *g* at 4 °C for 10 min.
5. Carefully remove the supernatant, and resuspended cells in 100 μL of PBS, and then count cells. These cells are enriched for CD11c+ cells with purity of 72–80%.

3.1.3 Internalization Assay

1. For each time point, transfer 2 × 10^5 CD11c sorted dendritic cells (*see* Subheading 3.1.1) in a FACS tube and wash with 2 mL PBS once (500 × *g* for 5 min).
2. Cells are stained in FACS tube with either anti-H2-D^b or anti-H2-K^b biotinylated antibodies, or their respective isotypes, diluted at 1:50 in FACS buffer for 30 min at 4 °C (*see* **Note 8**).
3. Wash cells two times in 2 mL PBS, 500 × *g* for 5 min, and resuspend in 200 μL R10. Keep cells at 4 °C, or incubate at 37 °C, for 10, 30, 60, and 90 min (*see* **Note 9**).
4. After the incubation time, wash cells with 2 mL PBS, 500 × *g* for 5 min, and perform surface staining at 4 °C for 30 min with FACS buffer with following antibody mix:
   (a) 1:400 PE-coupled streptavidin
   (b) 1:400 PE-Cy7-anti-CD11c

   (c) 1:400 PB- anti-I-A/I-E
   (d) 1:500 Aqua
5. Wash cells with 2 mL PBS 500 × *g* for 5 min, resuspend in 100 μL FACS buffer, and acquire samples at the BD FACS-Canto-II.
6. Analyze the data in two steps:
   (a) Subtract the mean fluorescence intensity (MFI) of the PE fluorescence of the isotype from the MFI of the MHC class I.
   (b) Evaluate the decrease of MFI intensity compared to the MFI of control DCs incubated at 4 °C set as a reference at 100%.

3.2 MHC Class I Internalization by Immune Fluorescence Microscopy

3.2.1 Staining of MHC Class I

1. 8-well chamber slides are coated by adding 50 μL of poly-L-lysine and incubated for 10 min at room temperature (RT) (*see* **Note 10**). Then, they are gently washed three times with sterile PBS. Prepare some extra wells, which will be used for control conditions and stainings (*see* **Note 11**).
2. Wild-type or *atg7*-deficient lung CD11c-positive cells are selected as described in the Subheading 3.1.1 and are plated into pre-coated wells at the density of 2 × 10^5 cells/well.
3. After overnight cultured at 37 °C and 5% CO_2, the cells are spun down at 500 rpm during 3 min and then gently and carefully washed with sterile PBS 1 time (*see* **Notes 12** and **13**).
4. Directly add Fc blocker 2.4G2Fcc III/II diluted at 1:250 in PBS and incubates cells during 30 min at 4 °C.
5. The cells are directly stained with 20 μL of the primary antibody against MHC class I for 30 min at 4 °C (dilution 1:50) and then transferred to 37 °C for 1 h. MHC class I can be stained by using an anti-H2-K^b or anti-H2 antibody.
6. Labelled cells are spun down at 500 rpm for 3 min and gently washed 1 time with PBS.
7. After a centrifugation step at 500 rpm for 3 min the cells are fixed with 4% paraformaldehyde during 20 min at room temperature.
8. Fixed cells are carefully washed three times with PBS and then permeabilized with 0.1% Triton X-100 for 5 min at RT; afterward the cells are washed with PBS (*see* **Note 14**).
9. Add the blocking buffer (PBS-BSA 1%) and incubate during 1 h at RT.
10. Alexa Fluor 555 is used as a secondary antibody (*see* **Note 15**). The secondary antibody (1:500) and DAPI nucleic acid stain (1:10 0000) are diluted into the blocking buffer, and then 20 μL is added to each well.

11. After 1 h of incubation at RT in the dark, the cells are washed three times with PBS.
12. Remove any trace of PBS, and then carefully detach the plastic chamber from the slide (*see* **Note 16**).
13. Add a drop of the mounting medium Dako and then on top a coverslip of 1.5 mm thickness. Each coverslip is carefully pressed down and dried at room temperature in the dark (*see* **Note 17**). Slides could be stored for several months at 4 °C in the dark.

3.2.2 Immune Fluorescence Analysis

1. Cells were analyzed with an upright confocal laser-scanning microscope (SP8, Leica), using a 63×, NA 1.4 oil immersion objective (*see* **Note 18**). The excitation is performed at 405 nm to elicit the DAPI fluorescence (emission max 470 nm, blue fluorescence) and an excitation at 543 nm elicits the Alexa Fluor® 555 (emission max 580 nm, red fluorescence). Images are acquired using Leica software (Leica) and then analyzed and assembled using ImageJ software.
2. To appreciate and quantify the internalization of MHC class I, the intensity across trajectories through vesicular MHC class I is recorded by using the ImageJ software. The obtained data are represented by profile plot of MHC class I fluorescence intensity. The number of MHC class I vesicle can be quantified by hand or using an available plugging from ImageJ.
3. Figure 1 shows the internalization of MHC class I assessed by immunofluorescence. $CD11c^+$ cells from lungs were treated and analyzed following the procedure described above. Figure 1a shows the localization of MHC class I in Atg-sufficient $CD11c^+$ cells. Figure 1b shows the localization of MHC class I in *atg7*$^{-/-}$ $CD11c^+$ cells. Figure 1c shows the profile plot of MHC class I intensity in the cytoplasm.

3.3 Immunoprecipitation for MHC Class I and LC3, Followed by Western Blotting for AAK1

3.3.1 Isolation of DCs from the Bone Marrow

1. Euthanize *atg5*$^{-/-}$ and their littermates with CO_2 inhalation, place the mouse in dorsal recumbency on a clean dissection board, and spray legs with 70% ethanol.
2. With help of scissors and forceps, remove skin and muscle overlaying femur and tibia.
3. Remove the bones by cutting between femur and hip joint, and remove the remaining muscle with the help of a paper tissue. Then, transfer into Falcon tubes with phosphate-buffered saline (PBS) on ice (*see* **Note 19**).
4. Cut the femur from tibia, cut open distally, and introduce a syringe with a 30G × 12 mm needle into the bone marrow channel to flush out the bone marrow with sterile PBS.
5. Wash with PBS 500 × *g* for 10 min.

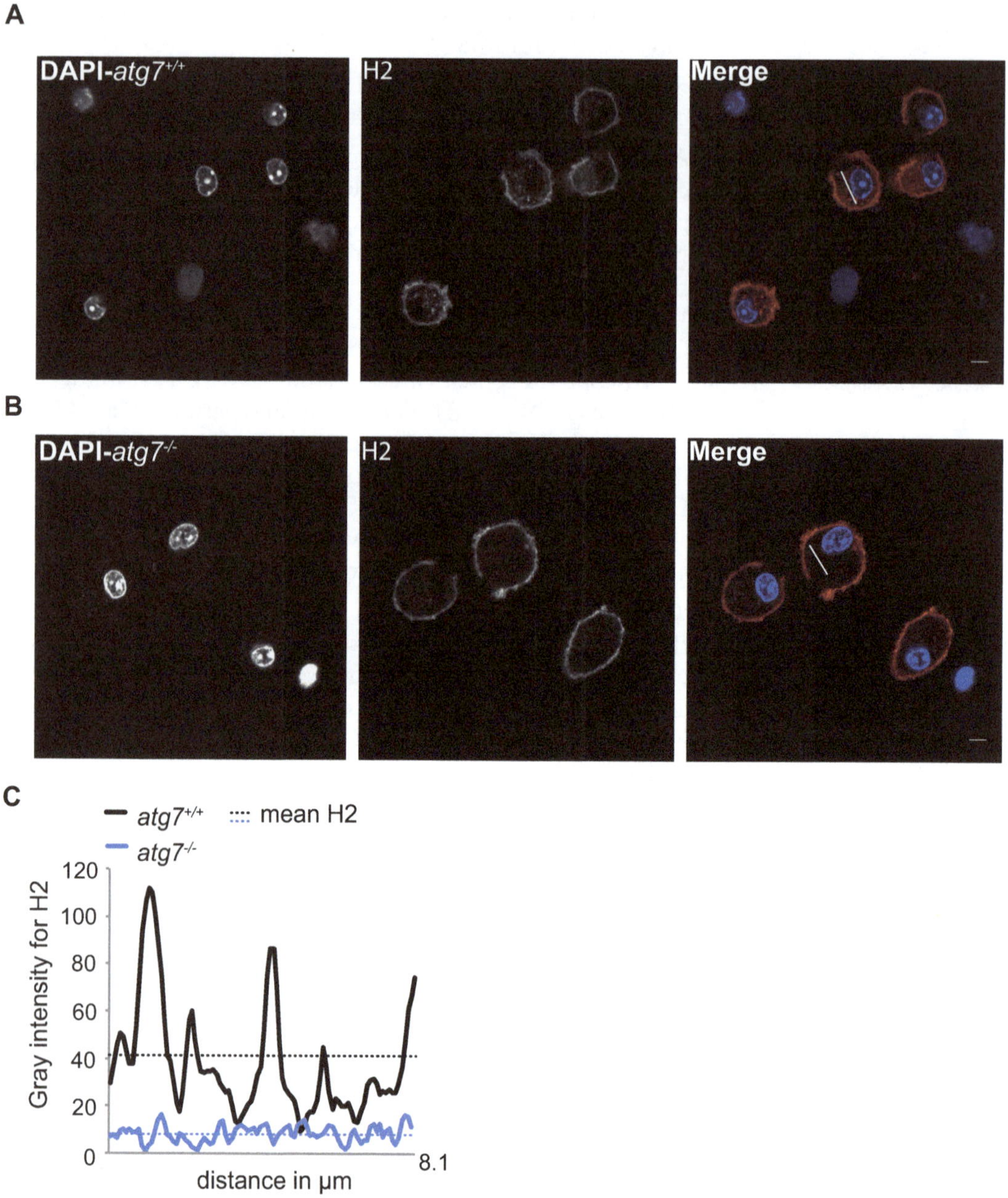

Fig. 1 Influence of the macroautophagy pathway on the internalization of MHC class I molecules in murine lung CD11c$^+$ cells, analyzed by confocal microscopy. Lung CD11c$^+$ cells from wild-type mice (*atg7*$^{+/+}$) (**a**) and mice with a deficiency in the macroautophagy machinery (*atg7*$^{-/-}$) (**b**) were stained for MHC class I with anti-H2 antibody (red channel). Nuclear DNA was counterstained with DAPI (blue channel). Merged images, corresponding to the compilation of the two individual staining, clearly show how the cytoplasmic pool of MHC class I molecules after internalization is compromised in *atg7*-deficient CD11c-positive cells. Scale bar indicates 5 μm. (**c**) The fluorescence intensity of vesicular MHC class I was recorded along the trajectories, as depicted in the merged images of (**b**)

6. Force cell suspension through a 70 μm strainer, and centrifuge 500 × *g* for 5 min.
7. Lyse erythrocytes by adding 1 mL ACK lysis buffer (*see* Subheading 2.3.5), and incubate 3 min at RT, and then wash with PBS at 500 × *g* for 5 min (*see* **Note 20**).
8. Plate cells at 3 × 10^6/mL BM-DC medium (*see* Subheading 2.3.2) in 100 × 15 mm non-tissue-coated petri dishes (*see* **Note 21**).
9. Perform a full medium change every second day by collecting the cells, spinning them down, and resuspending them in fresh BM-DC medium (*see* **Note 22**). Place them back into their original dish.
10. On day 8–10 for mature DCs (*see* **Note 23**), remove all the culture medium from each plate (*see* **Note 24**).
11. Wash the plate with 2–3 mL PBS, trypsinize cells with 2 mL trypsin for 5 min at 37 °C, inactivate the trypsin with 8 mL medium, and collect it.
12. Add 5 mL of ice-cold PBS and mechanically detach the cells from the plate by cell scraper.
13. Collect, centrifuge (500 × *g* for 5 min), and count them. From one mouse, 5–8 × 10^7 total cells can be isolated (*see* **Note 25**).

3.3.2 Preparation Stock IP Beads

1. For a final volume of 2 mL beads, add 40 mL of distilled water (dH_2O) to 400 mg of lyophilized protein A (*see* **Note 26**).
2. Incubate 1 h at RT under rotation.
3. Spin down at 300 × *g* for 3 min or allow beads to settle. Decant off supernatant. Wash three times with 50 mM Tris buffer pH 7.0.
4. Resuspend in equal medium volume of 50 mM Tris buffer pH 7.0 + 20% ethanol (EtOH) for long-term storage at 4 °C (if we have 2 mL of beads, add 2 mL of EtOH/Tris buffer).

3.3.3 IP Beads Washing

1. For one IP experiment, transfer 300 μL of beads to an Eppendorf tube.
2. Wash three times with 500 μL PBS, and spin at 1500 × *g* for 1 min.
3. Resuspend them in 300 μL cold lysis IP buffer (*see* Subheading 2.3.5).

3.3.4 Immunoprecipitation for MHC I and LC3

1. The whole cell lysis and IP procedures has to be done at 4 °C.
2. Grow BM-DCs from *atg5*$^{-/-}$ and *atg5*$^{+/+}$ as described in (*see* Subheading 3.1.1). After 10 days post differentiation, collect and count BM-DCs. For MHC class I and LC3 pulldown, the use of at least 2 × 10^7 cells is recommended (*see* **Note 27**).

3. Wash the cells three times with PBS, and transfer the cells to a 1.5 mL collection tube.
4. Resuspend with cold IP lysis buffer (1 × 10^6 cells/50 μL) containing appropriate protease inhibitors (*see* Subheading 2.3.5). Incubate at 4 °C for 30 min, then vortex briefly (up to 10 s).
5. Centrifuge tubes at 13000 × *g* for 10 min at 4 °C to remove nuclei and transfer the supernatant to another tube.
6. Measure protein concentration by BCA to be sure to have the same amount of proteins between *atg5*$^{-/-}$ *and atg5*$^{+/+}$ lysates.
7. To preclear the cell lysate from protein that non-specifically bind the beads, add 50 μL of washed beads (*see* Subheading 3.3.3) (*see* **Note 28**) to the cell lysate and incubate them for 1 h at 4 °C with rotation.
8. Spin at 1500 × *g* for 10 min to pull down beads and harvest supernatant. Be careful to keep sample free of beads. Keep 50 μL of precleared lysate as control for the WB.
9. Divide your cell lysate in two collection tubes.
10. Add primary antibody or its isotype/normal serum. For instance, 5 μL anti-EXON 8 or 2 μL anti-LC3 and the corresponding amount of NRS diluted in IP lysis buffer (*see* Subheading 2.3.4) are added to cell lysate and mix well.
11. Immunoprecipitate your sample overnight at 4 °C under rotation (*see* **Note 29**).
12. The next day, add 50 μL of beads resuspended in the cold lysis buffer. Mix well and incubate with gentle rotation at 4 °C for 1 h.
13. Spin beads down at 2500 × *g* for 1 min, remove the supernatant, and add 1:20 cold lysis buffer (*see* **Note 30**). Repeat this **steps 3–5** times.
14. After washes, denature the beads in 20 μL SDS-PAGE loading buffer (*see* Subheading 2.3.5), 3 min at 95 °C.
15. Centrifuge at 12000 × *g* for 10 min at 4 °C and collect the supernatant (*see* **Note 31**).
16. Load 20 μL of the sample per lane in a 1.5 mm thick SDS-polyacrylamide gel for electrophoresis.

3.3.5 Western Blotting for AAK1

1. All procedures have to be carried out at RT unless otherwise specified.
2. Prepare 7% and 12.5% gel mixture for analyzing association between MHC class I and AAK1 and LC3 and AAK1, respectively. See recipe in Table 1.

   Cast 7.5 mL of gel mixture within a 7.25 cm × 10 cm × 1.5 mm gel cassette (Bio-Rad MINIPROTEAN III system). Gently

Table 1
Recipe for resolving gel

	7%	12.5%
Resolving buffer	2.5 mL	2.5 mL
Acrylamide mixture	2.33 mL	4.2 mL
dH_2O	5 mL	3.15 mL
10% SDS	100 μL	100 μL
TEMED	15 μL	15 μL
APS	75 μL	75 μL

overlay with water, and wait until the gel mixture is polymerized (*see* **Note 32**).

3. To prepare the stacking gel mix 750 μL of stacking gel buffer, 325 μL of acrylamide mixture, 1.5 mL of water, 50 μL of 10% SDS, 18.75 μL of APS, and 3.75 μL of TEMED. Insert a 10- or 15-well gel comb immediately without introducing air bubbles.
4. Remove the water from the gel cassette by simply flipping the system, and cast 2 mL of stacking gel as prepared above.
5. Perform electrophoresis at constant amperage with SDS-PAGE running buffer. Start with 60 mA until the sample has entered the gel, and then continue at 90 mA till the dye front has reached the bottom of the gel.
6. Cut PVDF membrane to the size of the gel, and immerse it in 100% methanol for 5 min. Rinse once with transfer buffer for 5 min.
7. Immediately after electrophoresis, separate the glass gel plates with the help of a spatula or similar tool.
8. Perform semidry transfer with Mini Trans-Blot Electrophoretic Transfer Cell, using standard procedure with transfer buffer at 15 mA for 1 h (*see* **Note 33**).
9. Block the membrane with blocking solution at least 45 min at RT or overnight at 4 °C under agitation.
10. Cut the membrane according to the molecular weight (MW) of the interested proteins, and incubate it with the corresponding primary antibody diluted 1:1000 in PBS/T-1% BSA for 3 h at RT or overnight at 4 °C under rotation. Rabbit anti-AAK1 (MW 94 kDa), biotin anti-mouse H2-K^b (MW 46 kDa), and rabbit anti-LC3 Ab (MW 18–15 kDa) (*see* Subheading 2.3.4).

11. Wash with PBS/T once for 15 min and three times for 5 min.
12. Add the secondary antibody, goat anti-rabbit IgG HRP-conjugate (dilution 1:50000 in PBS/T-1% BSA), or streptavidin HRP (diluted 1:1000 in PBS/T-1% BSA (*see* Subheading 2.3.4), and incubate for 1 h at RT.
13. Wash with PBS/T: once for 15 min and four times for 5 min.
14. Reveal using the ECL kit following the manufacturer's instructions.

3.4 Influenza Infection of CD11c-cre × atg5$^{fl/fl}$ Mice and Readout for CD8^{+} T Cell Responses in the Lung

3.4.1 Influenza Virus Infection In Vivo

1. Thaw virus on ice.
2. Prepare 10 HAU virus suspension and keep it on ice.
3. Anesthetized mice using isoflurane (*see* **Note 34**). After the mouse is deeply anesthetized, confirmed by the absence of reflex on the footpad, hold the mouse by its ears, and let it lean along your hand.
4. Gradually inject 25 μL (*see* **Note 35**) of virus suspension into the nostrils (divide the volume in the two nostrils) with the help of micropipette with filtered tip. Adjust the rate of release to allow the mouse to inhale without forming bubbles. Inoculate control mice with 25 μL of PBS.
5. Hold the mouse in the hanging position for another couple of minutes until its breathing gradually returns normal.
6. Weight the mouse. Weight loss is used to monitor the development of the infection.

3.4.2 Evaluation of CD8^{+} T Cell Response by IFN-γ Quantification by FACS

1. Label four sterile FACS tubes with cap the day before (*see* **Note 36**) for each condition.
2. Incubate lung single cell suspensions obtained after Percoll gradient centrifugation (*see* Subheading 3.1.1) in R10 at 37 °C for 5 h in the presence of 4 μg/mL anti-mouse CD28 (*see* Subheading 2.4.4) and 20 μg/mL influenza-specific peptides. Use approximately 1×10^6 cells in 250 μL per staining.
3. Add 250 μL CD28/influenza peptide solution in RPMI without serum, and incubate for 1 h at 37 °C. Setup three controls:
   (a) 10 μg/mL anti-mouse CD3 (*see* Subheading 2.4.4),
   (b) PMA 1 μg/mL and ionomycin 1 μM final concentration, and (c) no peptide/no anti-CD28.
4. Add 50 μL/tube of Brefeldin A solution (10 μg/mL in RPMI), and incubate for additional 4 h at 37 °C.
5. Wash cells once with at least 2 mL of cold PBS (*see* **Note 37**), spin at 500 × *g* for 5 min, and discard the supernatant.
6. Use 50 μL/tube of the following antibody mix diluted in FACS buffer:
   (a) 1:500 Pacific Blue anti-mouse CD4.

(b) 1:125 APC anti-mouse CD8.

(c) 1:50 APC-Cy7 anti-mouse CD45.

(d) 1:500 Aqua Live/Dead diluted in FACS buffer.

Incubate 20 min on ice.

7. Wash with 2 mL cold PBS, and spin at 500 × *g* for 5 min 4 °C.
8. Resuspend cell pellet in 250 μL Cytofix/Cytoperm (BD Kit), and incubate for 20 min on ice, and then wash with cold PBS.
9. Resuspend cell pellet in FACS buffer; store o/n in the fridge.
10. Wash cells with 1 mL of 1× Perm/Wash buffer (BD kit, 1:10 dilution in dH_2O). Spin at 400 × *g* for 5 min, and discard supernatant.
11. Add 50 μL of PE-anti-mouse IFN-γ antibody (*see* Subheading 2.4.4) diluted 1:50 in 1× Perm/Wash solution (BD kit). Incubate for 25–30 min.
12. Wash with 1× Perm/Wash.
13. Resuspend in FACS buffer and acquire the sample the same day using a cytometer.

4 Notes

1. FCS batches need to be heat inactivated before use in culture to denature complement that may be present in serum. Place thawed bottle of FCS in 56 °C water bath for 40 min. Be sure to mix contents manually by swirling regularly to ensure even temperature distribution and reduce precipitation of the FCS. FCS being added to RPMI should be passed through a 0.2 μm filter to ensure sterility and reduce precipitates within Culture Media.
2. β-Mercaptoethanol is used as reducing agent because cells at 20% O_2 produce ROS that leads to oxidation of antioxidant pools. To avoid oxidative stress, most cultured cells amplify antioxidant defense; however, certain cell types are unable to do so, and thus β-mercaptoethanol is added to aid in maintaining a reducing environment. It is recommended to add β-mercaptoethanol freshly.
3. FCS is one component that profoundly influences BM-DC generation. Indeed, it has been shown that even different FCS lots or batches of the same product could drastically affect BM-DC generation. Therefore, it is highly recommended to make test cultures for a side-by-side comparison with different FCS [15].
4. Avoid cervical dislocation because it results in rapid death but is traumatic. This technique damages the trachea and cervical

region and confounds the results by causing hemorrhage into various tissues and contaminating the lung tissue.

5. Carefully first prepare 90% Percoll with 10× PBS in order to have further dilutions that are isotonic.
6. After centrifugation, make sure Percoll has been completely washed out because it might be toxic for the cells.
7. Filter the cells before going to the autoMACS because cells could aggregate. Moreover, to keep the cells at 4 °C put the collecting tubes in a refrigerator rack.
8. To be more consistent and be sure that all cells get exactly the same staining, you could also stain them in one tube, and divide them after washing for the different time points.
9. The incubation time has to finish at the same moment to achieve better consistency. Start with the 90 min incubation, keeping the other tubes at 4 °C. Put the next tube after 30 min (time point 60 min) and so on.
10. Poly-L-lysine is a charged enhancer used for coating the wells to promote better cells adhesion. The poly-L-lysine is diluted at 1:10 in water and has to be freshly made every time. Some cell types are able to digest the poly-L-lysine; in this case, it is preferable to use the poly-D-lysine.
11. For correct interpretation of the result, the following controls should be included:
    (a) One well should be incubated without any primary or secondary antibody but only with the blocking buffer to determine the cells auto fluorescence.
    (b) One slide should be incubated only with the secondary antibody to assess the background due to the secondary antibody. Background of the secondary antibody should be low.

    In case of multiple stainings, include slides to perform single labelling of cells, and check the leak-through fluorescent in the adjacent channel. For instance, Alexa Fluor® 555 will be analyzed with the green channel, and no significant signal should be detected.
12. The overnight culture is an important step to give time to the cells to nicely adhere and spread the cytoplasm on the well surface.
13. CD11c-positive cells do not adhere very strongly to the bottom of the well; in order to limit the loss of cells, every single step before the fixation must be preceded by a centrifugation at 500 rpm for 3 min. The liquid (medium, PBS, etc.) has to be gently and carefully removed with a vacuum suction flask.

14. From the fixation step forward, cells could be handled outside a sterile biosafety hood on a laboratory bench. Use a vacuum suction flask to change solutions, and exchange the plastic tip of the suction devise between different solutions and conditions.
15. Secondary antibody coupled with red fluorochromes is preferred to green ones due to the high level of cellular autofluorescence in the green channel.
16. The plastic chambers are carefully detached from the slide, and with the help of a scalpel, remove any traces of glue before adding the mounting medium.
17. Dako mounting medium contains an anti-fading agent, which retards the fading of fluorescence and allows the analysis of the slides long after mounting. During the overnight incubation time in the dark, the mounting medium will polymerize, and it will not be necessary to seal the coverslips with nail polish.
18. The slide could also be analyzed with an inverted confocal laser-scanning microscope having the right lens and laser.
19. It is possible to keep bones at 4 °C for 1–2 days in PBS. Cells are not affected, but it might be a bit harder to flush the bones out, since they become softer over time in solution.
20. Do not incubate cells for long periods of time with ACK lysis buffer because it is harmful to the cells.
21. Be careful with respect to the concentrations of your cells. DCs love cell-to-cell contact. Therefore, they do not mind dense conditions. Also, after 3 days of culture, most nonmyeloid cells die off leaving extra space for the DCs. On the other hand, too dense conditions, like 5×10^6/mL and above, would lead to spontaneous maturation.
22. A complete change to fresh medium is not necessary and a half-old-half-new protocol works well too.
23. After 3 days of culture, it is already possible to detect around 20% of $CD11c^+$ MHC II^{high} cells. The percentage increases over time up to 80% at days 8–10. Before using the cells, always confirm purity and maturation status of DC cultures by FACS staining: 1:400 PE-Cy7-anti-CD11c, 1:400 PB-anti-I-A/I-E, and 1:500 Aqua (*see* Subheading 2.3.4) in FACS buffer for 30 min. 80% of $CD11c^+$ MHC II^{high} DCs should be achieved.
24. For a correct interpretation of the data, be aware that BM-DC populations are heterogeneous. Indeed, it has been shown that the $CD11c^+MHCII^+$ fraction of GM-CSF cultures comprises at least two cell types that by ontogenetic and gene expression criteria correspond to monocyte-derived macrophages and CDP-derived DCs [16].

Table 2
Protein A/G affinities for polyclonal sera

Ab Isotype	Affinity
Human	Protein A or G
Horse	Protein G
Cow	Protein G
Pig	Protein A or G
Sheep	Protein G (weakly)
Goat	Protein G (weakly)
Rabbit	Protein A or G
Chicken	Protein G (weakly)
Hamster	Protein G (weakly)
Guinea Pig	Protein A
Rat	Protein G (weakly)
Mouse	Protein A or G (both weakly)

25. It is possible to freeze the cells at any time during their maturation process. We recommend to freeze the cells down right after the isolation. Resuspend them in freezing medium at a concentration of 2×10^6 per mL and freeze the cells in cryotubes (1 mL per tube).
26. Protein G is often considered the more universal IgG binding protein compared to protein A, but different species, and isotypes of species, do vary in their binding to these proteins; *see* Table 2. Therefore, choose wisely with respect to protein A versus G usage.
27. The minimum input for the co-immunoprecipitation is 1 g of protein per condition. Usually, for 3×10^6 cells, 250–300 μg of total protein is expected.
28. The preclearing step is optional and depends on the nonspecific binding due to charges and/or contact of hydrophobic surfaces.
29. In case of high levels of unspecific binding, shorten the incubation step of the primary antibody to 45 min.
30. Other possible changes to reduce the unspecific binding are to adjust washing stringency and washing steps. Move toward a higher stringency buffer (i.e., use up to 1% Tween 20, a nonionic detergent, up to 0.2% SDS, an anionic charged detergent, or up to 1M NaCl), and increase the number of washing steps.

31. Be careful not to take beads in order to avoid carry-over of contamination.
32. Assemble the system, and before pouring the resolving buffer, fill the system with dH_2O, and leave it for 5 min. Sometimes the system is not perfectly sealed, and it could leak out. If there is no leakage, remove the water by just flicking the system; otherwise reassemble the system.
33. If you are not confident in your transfer, it is possible to incubate the PVDF membrane with Ponceau S (0.1% (w/v) in 5% acetic acid) staining, a reversible staining to check protein bands. After less than 5 min of incubation, it will be possible to visualize the bands as a red staining that can be removed simply by washing with dH_2O.
34. The type of anesthesia during intranasal instillation is an important variable that could have a significant impact on its efficiency for delivery of inocula to the lungs. In fact, mice can swallow, and the gastrointestinal tract is heavily influenced by level of anesthesia and the volume of the inoculum. Several studies have shown that the infection is significantly more efficient when the injection is performed under isoflurane-inhaled anesthesia in comparison to the parenteral-administered ketamine/xylazine. It has been speculated that this could be because the irregular Cheyne-Stokes type respiratory pattern that is typically observed following inhalation of anesthesia causes transient hypoxia which results in deeper inhalation of larger volumes of inoculum per breath, facilitating more efficient delivery of the material to the lower respiratory tract [17]. Accordingly, increasing the depth of isoflurane anesthesia (230 $\mu L/dm^3$) improved the infectivity of the large-volume inoculum, probably because of suppression of swallow and sneeze reflexes [18]. In contrast, mice that receive injectable anesthesia breathe in a more regular and more shallow pattern, resulting in a more coating of the upper respiratory tract surface with the inoculum that leads to an inefficiently delivery of the inoculum to the lower tract.
35. Another important variable for the intranasal inoculation of anesthesia is the volume. It is relatively well established that intranasal instillation for delivery to the upper respiratory tract requires a low administration volume ($\leq$10 μL). However, no real consensus has been reached with respect to dose volume for delivery to the lower respiratory tract. It has been shown that a brief respiratory distress occurs in mice that have received intranasal instillation volumes $\geq$50 μL [19].

36. For easier and faster handling, it is possible to perform the assay in U-bottom 96-well plate. Perform washing steps with 200 μL and centrifugation step at 1000 × *g* for 2 min. Volumes: stimulation in 100 μL, add 20 μL Brefeldin A, and extracellular and IFN-γ staining in 20 μL.

37. Make sure the medium is washed out. RPMI is known to inhibit Aqua staining.

References

1. Mizushima N, Yoshimori T, Ohsumi Y (2011) The role of Atg proteins in autophagosome formation. Annu Rev Cell Dev Biol 27:107–132
2. Ohsumi Y (2001) Molecular dissection of autophagy: two ubiquitin-like systems. Nat Rev Mol Cell Biol 2:211–216
3. Dengjel J, Schoor O, Fischer R, Reich M, Kraus M, Müller M, Kreymborg K, Altenberend F, Brandenburg J, Kalbacher H, Brock R, Driessen C, Rammensee RG, Stevanovic S (2005) Autophagy promotes MHC class II presentation of peptides from intracellular source proteins. PNAS 102 (22):7922–7927
4. Schmid D, Pypaert M, Münz C (2007) Antigen-loading compartments for major histocompatibility complex class II molecules continuously receive input from autophagosomes. Immunity 26:79–92
5. Lee HK, Mattei LM, Steinberg BE, Alberts P, Lee YH, Chervonsky A, Mizushima N, Grinstein S, Iwasaki A (2010) In vivo requirement for Atg5 in antigen presentation by dendritic cells. Immunity 32:227–239
6. Li Y, Wang LX, Yang G, Hoa F, Ubra WJ, Hu HM (2008) Efficient cross-presentation depends on autophagy in tumor cells. Cancer Res 68:6889–6895
7. Aichinger M, Wu C, Nedjic J, Klein L (2013) Macroautophagy substrates are loaded onto MHC class II of medullary thymic epithelial cells for central tolerance. J Exp Med 210 (2):287–300
8. Nedjic J, Aichinger M, Emmerich J, Mizushima N, Klein L (2008) Autophagy in thymic epithelium shapes the T-cell repertoire and is essential for tolerance. Nature 455:396–400
9. Romao S, Gasser N, Becker AC, Guhl B, Bajagic M, Vanoaica D, Ziegler U, Roesler J, Dengjel J, Reichenbach J, Münz C (2013) Autophagy proteins stabilize pathogen-containing phagosomes for prolonged MHC II antigen processing. J Cell Biol 203:757–766
10. Sanjuan MA, Dillon CP, Tait SW, Moshiach S, Dorsey F, Connell S, Komatsu M, Tanaka K, Cleveland JL, Withoff S, Green DR (2007) Toll-like receptor signalling in macrophages links the autophagy pathway to phagocytosis. Nature 450:1253–1257
11. Martinez J, Malireddi RK, Lu Q, Cunha LD, Pelletier S, Gingras S, Orchard R, Guan JL, Tan H, Peng J, Kanneganti TD, Virgin HW, Green DR (2015) Molecular characterization of LC3-associated phagocytosis reveals distinct roles for Rubicon, NOX2 and autophagy proteins. Nat Cell Biol 17:893–906
12. Loi M, Müller A, Steinbach K, Barreira da Silva R, Paul P, Ligeon L-A, Caruso A, Albercht RA, Becker AC, Annaheim N, Nowag H, Dengjel J, Garcia-Saster A, Merkler D, Münz C, Gannagé M (2016) Macroautophagy proteins control MHC class I levels on dendritic cells and shape anti-viral $CD8^+$ T cells responses. Cell Rep 15 (5):1076–1087
13. Hara T, Nakamura K, Matsui M, Yamamoto A, Nakahara Y, Suzuki-Migishima R, Yokoyama M, Mishima K, Saito I, Okano H, Mizushima N (2006) Suppression of basal autophagy in neural cells causes neurodegenerative disease in mice. Nature 441 (7095):885–889
14. Komatsu M, Waguri S, Chiba T, Murata S, Iwata J, Tanida I et al (2006) Loss of autophagy in the central nervous system causes neurodegeneration in mice. Nature 441 (7095):880–884
15. Lutz MB, Rössner S (2007) Factors influencing the generation of murine dendritic cells from bone marrow: the special role of fetal calf serum. Immunobiology 212 (9–10):855–862
16. Helft J, Böttcher J, Chakravarty P, Zelenay S, Huotari J, Schraml BU, Goubau D, Reis e Sousa C (2015) GM-CSF mouse bone marrow cultures comprise a heterogeneous population of $CD11c^+MHCII^+$ macrophages and dendritic cells. Immunity 42(6):1197–1211

17. Miller MA, Stabenow JM, Parvathareddy J, Wodowski AJ, Fabrizio TP, Bina XR, Zalduondo L, Bina JE (2012) Visualization of murine intranasal dosing efficiency using luminescent Francisella tularensis: effect of instillation volume and form of anesthesia. PLoS One 7(2):e31359
18. Rosseels V, Nazé F, De Craeye S, Francart A, Kalai M, Van Gucht S (2011) A non-invasive intranasal inoculation technique using isoflurane anesthesia to infect the brain of mice with rabies virus. J Virol Methods 173(1):127–136
19. Southam DS, Dolovich M, O'Byrne PM, Inman MD (2002) Distribution of intranasal instillations in mice: effects of volume, time, body position, and anesthesia. Am J Physiol Lung Cell Mol Physiol 282(4):L833–L839

Part IV

Measuring and Imaging Autophagy In Vitro

Chapter 30

Analysis of Autophagy for Liver Pathogenesis

Nazmul Huda, Hui Zou, Shengmin Yan, Bilon Khambu, and Xiao-Ming Yin

Abstract

The autophagy pathway in hepatocytes is well characterized. Autophagy plays a critical role in the normal function of the liver. A growing number of studies suggest that there is a mechanistic relationship between autophagy and the pathogenesis of human diseases including liver diseases. Here we focus on the methods assessing the level of lipids, lipid peroxidation, and lipophagy in the liver, which would be particularly relevant to the study of fatty liver diseases.

Key words Autophagy, Fatty liver diseases, Hepatocytes, Immunofluorescence, Lipid droplets, Lipophagy, Reactive oxygen species

1 Introduction

Autophagy (*self-eating* in Greek) is a finely regulated degradative mechanism, by which cells recycle macromolecules for nutrients or to remove dysfunctional organelles and aggregated proteins. There are three forms of autophagy: macroautophagy, microautophagy, and chaperone-mediated autophagy [1, 2]. One of the main inducers of autophagy is food restriction or starvation, which activates autophagy to provide essential nutrients through the degradation of intracellular components [1]. Research in hepatocellular autophagy has been very active from the beginning of this field, primarily due to the fact that the liver is the most important metabolic organ for the homeostasis of the body [1, 3]. Hepatic autophagy is also critical for maintaining physiological turnover of subcellular organelles like lipid droplets, endoplasmic reticulum, and mitochondria.

Hepatic steatosis develops when an excessive amount of fat accumulates in hepatocytes in the form of lipid droplets, which can be caused by uptake of high carbohydrate- or high fat-containing diet. A fatty liver condition develops when the fat content accounts for more than 5% of the liver [4]. The liver can become vulnerable to injury signals once there is an excessive storage of fat, which may promote inflammation, steatohepatitis,

Nicholas Ktistakis and Oliver Florey (eds.), *Autophagy: Methods and Protocols*, Methods in Molecular Biology, vol. 1880, https://doi.org/10.1007/978-1-4939-8873-0_30,

and fibrosis. The so-called fatty liver diseases can be caused by alcohol or other nonalcoholic conditions [5]. Alcoholic fatty liver disease (AFLD) and nonalcoholic fatty liver disease (NAFLD) share common pathological features including steatosis, inflammation, fibrosis, and cirrhosis. Our group [6] and others [7] have reported that hepatic autophagy can be activated by acute alcohol treatment both in vivo and in vitro. On the other hand, autophagy may be suppressed in chronic alcohol treatment. In this chapter, we discuss the basic methods adapted in our laboratory to study steatosis and the function of autophagy in regulating steatosis in the context of fatty liver diseases.

2 Materials

2.1 Analysis of the Level of Lipids and Lipophagy

2.1.1 Analysis of Steatosis and Lipophagy Based on Immunofluorescence Staining in the Liver

1. 1× phosphate buffered saline (PBS).
2. BODIPY 581/591 C11 (1 μmol/L).
3. Hoechst dye 33258.
4. Super PAP pen.
5. Mounting medium.
6. Nail polish (colorless).
7. Cover glass (22 × 22 #1).
8. Humidified chamber.

2.1.2 Analysis of Steatosis and Lipophagy Based on Immunofluorescence Staining in Hepatic Cells

1. AML12 cells, immortalized mouse hepatocytes (ATCC CRL-2254).
2. AML12 cell culture medium: Dulbecco's phosphate-buffered saline (DBPS) with 5% fetal bovine serum (FBS), insulin, and dexamethasone.
3. Cover slip: Sterile 25 mm round glass coverslips.
4. Culture plates.
5. 4% paraformaldehyde.
6. Anti-LC3 antibody and Alexa 488-conjugated secondary antibody.
7. BODIPY 581/591 C11 (1.0 μmol/L).

2.1.3 Measurement of Triglycerides

1. Triglycerides (Cat# T7532) and cholesterol (Cat# C7510) kits are available at Pointe Scientific, Canton, MI, USA. All other reagents are of analytical grade.
2. RIPA buffer (100 mL): 10× PBS 10 mL, 1% Igepal CA 630 1 mL, 0.5% deoxycholate acid 0.5 g, 0.1% SDS (10%) 1 mL, and deionized H2O to 100 mL stored at 4 °C (stable for 4 weeks). To 1 mL of NP-40 lysis buffer, add the following

chemicals: 2.5 μL PMSF (20 mg/mL in ethanol), 1.0 μL aprotinin (1.7 mg/mL), 1.0 μL leupeptin (2.5 mg/mL in methanol), 2.0 μL EDTA (0.5 M, pH 8.0).

2.2 Assessing Oxidative Stress in the Liver

2.2.1 Evaluation of Lipid Peroxidation

1. 96-well plate with clear flat bottom.
2. Dounce homogenizer.
3. Glacial acetic acid.
4. Malondialdehyde (MDA) lysis buffer, phosphotungstic acid solution, butylated hydroxytoluene (BHT), thiobarbituric acid (TBA), and standard MDA were purchased from BioVision, CA, USA.

2.2.2 Evaluation of the Damage to the DNA by Assessing the Presence of 8-OxoG

All solutions are freshly prepared in ultrapure water (sensitivity, 18 MΩ-cm at 25 °C) at room temperature.

1. Phosphate buffer saline (1 × PBS), pH 7.2.
2. Sodium citrate buffer (10 mM, pH 6.0): Dissolve 2.94 g trisodium citrate (dihydrate) in water, adjust pH to 6.0 with 1 N HCl, and make 1000 mL solution.
3. Fresh frozen sections of liver tissues.
4. Blocking buffer: Mix 1× PBS (947 μL), goat serum (50 μL), and Triton X-100 (3 μL), to make 1 mL solution.
5. RNase A (DNase free) solution: Prepare 100 μg/mL RNase A solution in 10 mM Tris–HCl (pH 7.5), 15 mM NaCl.
6. 2 N HCl.
7. Neutralization solution: 50 mM Tris-base in water.
8. Antibody diluent: Mix 1× PBS (997 μL), BSA 10 mg, and Triton X-100 (3 μL) to make a 1 mL solution.
9. Primary antibody: Mouse monoclonal anti-8-oxoG (clone 15A3, Santa Cruz, Cat# 66036), dilution 1:250.
10. Secondary antibody: Goat anti-mouse conjugated with Cy3 (Ex 550; Em 570 nm) (Jackson ImmunoResearch Laboratories, Inc., Cat# 115-165-146), dilution 1:500.
11. Hoechst 33342.
12. Mounting solution: 70% glycerol in PBS.
13. Nail polish (colorless).

3 Methods

3.1 Analysis of Steatosis and Lipophagy in the Liver and in Hepatocytes

Autophagy can degrade lipid droplets through a process termed as lipophagy [8]. Lipid droplets (in separate parts or as a whole) can be engulfed by autophagosomes and degraded by lysosomal acid lipases after lysosomal fusion. The level of lipids and the extent of lipophagy in the liver can be studied using microscopic colocalization analysis of the lipid droplets and autophagosomes. In addition, the influence of autophagy on the lipid content can be assessed using a biochemical approach.

3.1.1 Analysis of Steatosis and Lipophagy Based on Immunofluorescence Staining in the Liver

Lipid droplets in steatotic liver can be observed by staining the hepatic cryosections with lipophilic dyes such as BODIPY 581/591-C11, BODIPY 493/503, or Oil Red O [6, 8]. Lipophagy can be observed by co-staining with the BODIPY dye together with antibodies against autophagosome markers like LC3 or p62. Lipid droplet markers like perilipin 1 can also be used to examine the colocalization with autophagosome markers.

1. Cryosections (5.0 μM thickness) of GFP-LC3 transgenic livers can be used for colocalization analysis of GFP-LC3 and lipid droplets. Cryosections can be stored temporarily at −80 °C (*see* **Note 1**).
2. Soak the cryosection slides in 1× PBS for 5 min at room temperature.
3. Use PAP pen to encircle the tissue section.
4. Stain the section with 100 μL of BODIPY 581/591 C11 (1 μmol/L) for 15 min at room temperature.
5. Add Hoechst 33258 to the sections for visualization of individual hepatocyte nuclei for 5 min at room temperature.
6. Wash the slides three times, 5 min each, in 1× PBS.
7. Tap off excess PBS onto paper towel. Remove any remaining PBS and PAP barriers with a folded Kimwipe.
8. Mount the section by adding few drops of mounting medium to the slide, and cover with cover slip, and seal the edges of cover slip with a nail polish.
9. Store the slides at 4 °C in the dark.
10. Fluorescence images (60×) are digitally acquired with a confocal fluorescence microscope (*see* **Note 2**).

3.1.2 Analysis of Steatosis and Lipophagy Based on Immunofluorescence Staining in Hepatic Cells

Here the method described is based on the study on the lipid elevation by ethanol treatment in AML12 cells [9]. AML12 cells are immortalized mouse hepatocytes, and they have retained the ability to metabolize ethanol. These cells are therefore suitable to study the metabolic effects of ethanol.

1. AML12 are cultured in 12-well plate on cover slips (2 × 10^5/well).
2. Treat the cells with 50–80 mM of ethanol for 24 h.
3. Retrieve the coverslips with cells. Fix cells with 4% paraformaldehyde for 10 min in the dark at room temperature.

4. Incubate the coverslips with an anti-LC3 antibody, followed by Alexa 488-conjugated secondary antibody. After washing, the coverslips are further stained with 100 μL of BODIPY 581/591 C11 (1 μmol/L) for 15 min at RT (*see* **Notes 3** and **4**).
5. Add Hoechst 33258 for 5 min followed by washing with 1× PBS 3 times for 5 min each.
6. Mount the coverslips onto glass slides and allow drying for 1 h at room temperature.
7. Examine the slides under a fluorescence microscope.

3.1.3 Measurement of Triglycerides

The major lipids in the cells are in the form of triglycerides and stored in the lipid droplets. The level of triglycerides can be determined using biochemical methods. The following protocol is based on a published method [10] with some modifications.

Measurement of Triglycerides in the Liver

1. 20–80 mg (*see* **Note 5**) of frozen liver is weighed (*see* **Note 6**) and incubated with 1 mL of chloroform-methanol mix (2:1) for 1 h at room temperature with shaking to extract the lipids.
2. After addition of 200 μL of H_2O, samples are vortexed and centrifuged for 5 min at 3000 × *g*.
3. The lower lipid phase is collected and dried at room temperature in a chemical hood (*see* **Note 7**).
4. The lipid pellet is re-suspended in 60 μL of tert-butanol and 40 μL of a Triton X-114-methanol (2:1) mix (*see* **Note 7**).
5. 2 μL lipid solution is used to determine triglyceride and cholesterol content using triglyceride reagent set or cholesterol reagent set, respectively. Lipid contents were normalized with the respective tissue weight.

Measurement of Triglycerides in Cultured Cells

1. Cells are collected using 1 mL of PBS.
2. 200 μL of the cell suspension is used for protein concentration measurement after the cells are collected and lysed in the RIPA buffer.
3. Total protein content is measured using bicinchoninic acid (BCA) assay method. Centrifuge cell suspension for 12 min at 13,800 × *g*, 4 °C. Collect the supernatant and estimate protein content using the BCA method. The other 800 μL of cell suspension are used for lipid content determination after the cells are spun down. Triglyceride and cholesterol contents are measured using triglyceride reagent set or cholesterol reagent set, respectively, following manufacturers' protocols. At the last step, lipid contents are normalized with the protein level.

3.2 Oxidative Stress in the Liver

Reactive oxygen species (ROS) at a basal (or physiological) level are necessary for cellular homeostasis, but an excessive level of ROS is associated with pathogenesis including carcinogenesis [11]. Mitochondria are the main source of intracellular ROS. If autophagy mechanism is defective, or dysfunctional, mitochondria can accumulate which may lead to an altered level of ROS in cells [11]. In addition, when lipids accumulate in cells, they are prone to oxidation. Peroxidized lipids can become a source of free radicals as well. As the half-life of ROS in biological systems is very short (ranging from nanoseconds to seconds) [12], it is not easy to directly determine ROS in tissue samples. A variety of indirect methods have been used for assessing oxidative stress in cells and tissues. One is to measure the level of peroxidated lipids, and the other is to measure the level of oxidative DNA.

3.2.1 Colorimetric Measurement of Peroxidized Lipids

Malondialdehyde (MDA) is a natural bi-product of lipid peroxidation, a useful indicator of oxidative stress. The MDA generated in the tissue samples due to lipid peroxidation is allowed to react with thiobarbituric acid (TBA) to generate MDA-TBA complex. This complex can be quantified spectrophotometrically by taking absorbance at 532 nm.

1. Prepare standard MDA solutions for standard curve: Take 10 μL of standard MDA solution (4.17 M) in an Eppendorf tube, and add 407 μL water to make 0.1 M MDA solution. Use 0.1 M MDA solution to make 200 μL of 0, 4, 8, 12, 16, and 20 nmol standard solutions.
2. Add 7.5 mL of glacial acetic acid to 1 bottle of TBA and mix well. Transfer the semiliquid to a 50 mL tube and add water to adjust the volume to 25 mL. Mix properly to dissolve completely.
3. Mix 300 μL of MDA lysis buffer with 3 μL of BHT (100×). Take approximately 10.0 mg liver tissue in a douncer, and add the MDA lysis buffer containing BHT, and homogenize.
4. Centrifuge the lysate at 21,100 × *g* for 10 min. Transfer 200 μL of the clear lysate to a fresh tube for MDA analysis.
5. Add 600 μL of TBA reagent to both standards and samples, mix, and incubate at 95 °C for 1 h in a water bath.
6. Bring back the tubes to room temperature by incubating on ice for 10 min.
7. Transfer 200 μL of both standards and samples to a microplate and take absorbance at 532 nm.

3.2.2 Determination of Oxidative DNA Damage

An important target of ROS in cells is the DNA. Oxidative DNA damage triggered by ROS can produce multiple base modifications, which have been associated with mutagenesis, cancer, and aging.

ROS-mediated DNA alterations can happen in the form of base modifications. A common oxidative change of DNA is the hydroxylation of guanine at C-8, leading to the formation of 8-oxo-7, 8-dihydro-2′-deoxyguanosine (8-OxoG) [13–15]. Guanine has the lowest redox potential of the four DNA bases, which is thus more susceptible to oxidative damage by ROS [15]. An immunofluorescence technique has been developed to detect nucleus positive for 8-OxoG containing DNA.

1. Rinse slides once with 1× PBS for 5 min at room temperature.
2. Place slides on a slide holder, submerge in citrate buffer, and perform antigen retrieval in a microwave oven at high-power setting for 20 min (*see* **Note 8**).
3. Allow the slides to cool down in the buffer to room temperature (about 1 h).
4. Rinse the slides first with water and then with 1× PBS for 5 min each.
5. Mark the area of liver tissues with a liquid blocker super PAP pen.
6. Add approximately 90 μL of RNase solution to the section and incubate for 1 h at 37 °C in a humidified chamber.
7. Wash twice with 1× PBS 5 min each.
8. Denature tissues DNA with 2 N HCl for 5 min at room temperature (*see* **Note 9**).
9. Neutralize with the neutralization buffer for 10 min at room temperature.
10. Wash thrice with 1× PBS 5 min each.
11. Apply the blocking solution for 1 h at room temperature.
12. Dilute the anti-8-OxoG antibody with the antibody diluent (1:250) and add approx. 90 μL to the section. Incubate in a humidified chamber at 4 °C for overnight.
13. Wash thrice with 1× PBS containing 0.1% Triton X-100 for 20 min each.
14. Dilute the secondary antibody (Cy3-Goat anti-mouse IgG) with the antibody diluent (1:500) and add approx. 90 μL to the section. Incubate in a humidified chamber in dark for 1 h at room temperature (*see* **Note 10**).
15. Wash thrice with 1× PBS containing 0.1% Triton X-100 for 20 min each.
16. Counterstain the sections with Hoechst 33342 (10 μg/mL in 1× PBS) for 10 min at room temperature.
17. Wash with 1× PBS for 5 min at room temperature.

18. Apply the mounting solution, place coverslips, and seal the coverslips with nail polish.
19. Examine under a fluorescence microscope to check the presence of 8-OxoG positive DNA containing nuclei.

4 Notes

1. Cryosections are preferable for preserving the fluorescence signals.
2. Fluorescence images at 60× magnification are preferred to visualize the lipid droplets. Wider microscopic field covering a large number of hepatocytes could be observed by taking image at 20×, but the image resolution would be lower.
3. A humidified chamber should be used while working with cover slips. The slides should not be allowed to dry out at any point of the experiment.
4. Perform staining in a dark place and protect stained slides from direct light source.
5. The amount of sample required for this assay depends on the level of lipids in it. It is recommended to use approximately 20 mg of liver tissue for the trial experiment. Adjustment of the quantity of liver tissue is necessary depending on the lipid content in the respective tissue.
6. The tissue samples may be cut and grinded into smaller pieces to facilitate a better extraction of lipids.
7. Working under a chemical hood is recommended.
8. Triton X-100 causes leakages in the border line drawn around the tissue section. Mark multiple border lines around the section and check frequently if there is leakage.
9. During each experiment, prepare at least one slide for a negative control. Treat the tissue section with DNase I solution at Subheading 3.2.2, **step 6**. This will remove the DNA from the nucleus, and it will appear as an empty space with no fluorescence signal.
10. After addition of secondary antibody, all the steps onward must be performed in dark to avoid the bleach of the fluorescence signals of the Cy3-conjugated secondary antibody.

References

1. Yin XM, Ding WX, Gao W (2008) Autophagy in the liver. Hepatology 47(5):1773–1785. https://doi.org/10.1002/hep.22146
2. Jiang P, Mizushima N (2014) Autophagy and human diseases. Cell Res 24(1):69–79. https://doi.org/10.1038/cr.2013.161

3. De Duve C (1967) Lysosomes and phagosomes. The vacuolar apparatus. Protoplasma 63(1):95–98
4. Lebovics E, Rubin J (2011) Non-alcoholic fatty liver disease (NAFLD): why you should care, when you should worry, what you should do. Diabetes Metab Res Rev 27(5):419–424. https://doi.org/10.1002/dmrr.1198
5. Duan XY, Zhang L, Fan JG, Qiao L (2014) NAFLD leads to liver cancer: do we have sufficient evidence? Cancer Lett 345(2):230–234. https://doi.org/10.1016/j.canlet.2013.07.033
6. Ding WX, Li M, Chen X, Ni HM, Lin CW, Gao W, Lu B, Stolz DB, Clemens DL, Yin XM (2010) Autophagy reduces acute ethanol-induced hepatotoxicity and steatosis in mice. Gastroenterology 139(5):1740–1752. https://doi.org/10.1053/j.gastro.2010.07.041
7. Dolganiuc A, Thomes PG, Ding WX, Lemasters JJ, Donohue TM Jr (2012) Autophagy in alcohol-induced liver diseases. Alcohol Clin Exp Res 36(8):1301–1308. https://doi.org/10.1111/j.1530-0277.2012.01742.x
8. Singh R, Kaushik S, Wang Y, Xiang Y, Novak I, Komatsu M, Tanaka K, Cuervo AM, Czaja MJ (2009) Autophagy regulates lipid metabolism. Nature 458(7242):1131–1135. https://doi.org/10.1038/nature07976
9. Wang L, Zhou J, Yan SM, Lei GS, Lee CH, Yin XM (2017) Ethanol-triggered lipophagy requires SQSTM1 in AML12 hepatic cells. Sci Rep 7:12307. https://doi.org/10.1038/s41598-017-12485-2
10. Buettner R, Newgard CB, Rhodes CJ, O'Doherty RM (2000) Correction of diet-induced hyperglycemia, hyperinsulinemia, and skeletal muscle insulin resistance by moderate hyperleptinemia. Am J Physiol Endocrinol Metabol 278(3):E563–E569
11. Kongara S, Karantza V (2012) The interplay between autophagy and ROS in tumorigenesis. Front Oncol 2:17. https://doi.org/10.3389/fonc.2012.00171
12. Dikalov SI, Harrison DG (2014) Methods for detection of mitochondrial and cellular reactive oxygen species. Antioxid Redox Signal 20(2):372–382. https://doi.org/10.1089/ars.2012.4886
13. Achanta G, Huang P (2004) Role of p53 in sensing oxidative DNA damage in response to reactive oxygen species-generating agents. Cancer Res 64(17):6233–6239. https://doi.org/10.1158/0008-5472.can-04-0494
14. Wiseman H, Halliwell B (1996) Damage to DNA by reactive oxygen and nitrogen species: role in inflammatory disease and progression to cancer. Biochem J 313(Pt 1):17–29
15. Breen AP, Murphy JA (1995) Reactions of oxyl radicals with DNA. Free Radic Biol Med 18(6):1033–1077

Chapter 31

Autophagy in 3D In Vitro and Ex Vivo Cancer Models

Carlo Follo, Dario Barbone, William G. Richards, Raphael Bueno, and V. Courtney Broaddus

Abstract

Three-dimensional (3D) models are acquiring importance in cancer research due to their ability to mimic multiple features of the tumor microenvironment more accurately than standard monolayer two-dimensional (2D) cultures. Several groups, including our laboratory, are now accumulating evidence that autophagy in solid tumors is also better represented in 3D than in 2D. Here we detail how we generate 3D models, both in vitro multicellular spheroids generated from cell lines and ex vivo tumor fragment spheroids generated from tumor samples, and how autophagy can be measured in 3D cultures.

Key words ATG13, Autophagy initiation, Autophagic flux, Three-dimensional, Ex vivo, Mesothelioma, Spheroids

1 Introduction

Three-dimensional (3D) models are gaining momentum in cancer research because they may recapitulate the behavior of the in situ tumor better than traditional two-dimensional (2D) cultures, with less complexity and cost than with xenografts in animal models [1–4]. When compared to cells grown in standard 2D monolayer cultures, cells allowed to grow in 3D experience a spatiotemporal gradient of nutrients and chemicals [5], are engaged in cell-cell, or cell-extracellular matrix interactions [6], and assume a cell shape more similar to that of the solid tumor in vivo. For these reasons, 3D models have shown a different biology at baseline and more realistic responses to stress such as chemotherapy than 2D cultures [4, 7, 8]. In particular, ex vivo culture provides a realistic model for studying cancer. In contrast to established cell lines representing selective clonal expansion, ex vivo culture includes unselected malignant and nonmalignant cells and components of the extracellular matrix derived from the in situ tumor [9]. Thus, in the case of ex vivo tumor cultures, cancer cells are subjected to heterotypic cross talk not present in 2D [10, 11].

Nicholas Ktistakis and Oliver Florey (eds.), *Autophagy: Methods and Protocols*, Methods in Molecular Biology, vol. 1880, https://doi.org/10.1007/978-1-4939-8873-0_31,

Autophagy is one area of study that may especially benefit from studies in 3D [12–16]. Autophagy may be better represented in 3D than in 2D, due to the fact that, similar to the avascular unit of tumor, 3D cultures rely on the diffusion of nutrients and oxygen, which are known to regulate autophagy [17]. In addition, the canonical autophagy signaling pathways in 3D are likely different from those documented in 2D; for example, the mammalian target of rapamycin (mTOR), a known important regulator of autophagy, has been reported to have a much lower activity in spheroids than in monolayers [18]. Moreover in 3D ex vivo cultures, autophagy may be modulated by input from the multiple cell types present in the tumor. Importantly, studying autophagy in an ex vivo tumor provides a window on this process in the original tumor, something currently hampered by the inability of static markers to measure such a dynamic process.

In our hands, 3D models of mesothelioma have proven to be more representative of the autophagy of the actual tumor than 2D cultures [19]. As measured by LC3 accumulation [20], the autophagic flux differs significantly in the same cell line whether it is grown as 3D *multicellular spheroids* (MCS) or as 2D monolayers. As measured by ATG13 puncta [21], autophagy initiation status correlates with the autophagic flux in both MCS and ex vivo *tumor fragment spheroids* (TFS) 3D models, but not in 2D. ATG13 positivity measured in TFS matched that detected in fixed surgical specimens from the original tumors employed to generate the TFS, indicating that ATG13 is a static marker of the autophagic flux of the actual tumor. Analysis of ATG13 puncta in tumors represented on tissue microarray showed that autophagy initiation, and therefore presumably the level of autophagy, is remarkably different among the mesothelioma tumors. ATG13 positivity identified two groups of mesotheliomas with either low or high level of autophagy, with higher ATG13 positivity correlating with longer overall patient survival [19]. Our work supports the value of using 3D models to study autophagy.

Here we describe in detail the procedures to generate both MCS and TFS and to measure autophagy in 3D models.

2 Materials

All the solutions should be prepared using ultrapure water (resistivity of 18.2 MΩ/cm at 25 °C). Unless specified otherwise, all solutions should be kept at room temperature.

2.1 Non-adsorbent Poly-HEMA Plates for Generation of Multicellular Spheroids

1. Poly-HEMA coating stock solution: in a 50 mL tube, dissolve 120 mg/mL solution of poly-HEMA (Sigma-Aldrich Corp, P3932) in 95% ethanol (99% diluted with water to 95% to aid dissolution of poly-HEMA; Thermo Fisher Scientific, A433F). Rotate on a wheel until all material dissolves (approx. 24 h). Store the solution for up to 1 month.

2. Poly-HEMA coating working solution: dilute stock solution to 5 mg/mL of poly-HEMA in 95% ethanol. Prepare the working solution the same day as plate coating.
3. Plates: U-bottomed 96-well plates (e.g., Greiner, 650,180) and 12-well plates (e.g., Corning, 3512).

2.2 Agar-Coated Plates for Generation of Tumor Fragment Spheroids

1. Noble agar (Sigma-Aldrich Corp, A5431).
2. Dulbecco's Modified Eagle's medium (DMEM) without serum, glutamine, and antibiotics.
3. 10 cm^2 Petri dishes.

2.3 Culture Media

Spheroid and monolayer cultures: high glucose DMEM (GE Healthcare, SH30243) supplemented with 10% fetal bovine serum (FBS), 4 mM glutamine, and 1% antibiotics (penicillin and streptomycin).

2.4 Autophagy Studies

1. Lysosomal inhibitor stock solutions: dissolve ammonium chloride (NH_4^+; Sigma-Aldrich Corp, A9434) or hydroxychloroquine (HCQ; Sigma-Aldrich Corp, H0915) in water to a concentration of 1 M and 50 mM, respectively, aliquot and store the inhibitors at −20 °C.
2. LC3 immunoblotting reagents:
   (a) Harvesting: cold phosphate-buffered saline (DPBS; HyClone, SH30028), RIPA buffer (1% Nonidet P-40 [Sigma-Aldrich Corp, 4385], 0.5% sodium deoxycholate [Sigma-Aldrich Corp, 30,970], 1% sodium dodecyl sulfate [SDS; Sigma-Aldrich Corp, L3771] in DPBS). Store RIPA buffer at 4 °C and supplement it with protease and phosphatase inhibitor cocktail (Thermo Scientific, 78,442) immediately before use.
   (b) Protein concentration method compatible with RIPA buffer composition (e.g., DC Protein Assay [Bio-Rad, 500-0111]).
   (c) Laemmli buffer 5×: 62.5 mM Tris-HCl pH 6.8 (Sigma-Aldrich Corp., T1503), 25% glycerol (Sigma-Aldrich Corp., G-6279), 2% SDS, 0.01% bromophenol blue (Sigma-Aldrich Corp., B8026). Store aliquoted solution at −20 °C and supplement with 350 mM 1,4-dithioerythritol (Sigma-Aldrich Corp, D8255) before use.
   (d) PVDF membrane (Bio-Rad, 162-0177).
   (e) Blocking solution: 5% nonfat dry milk (Santa Cruz Biotechnology, sc-2324) in DPBS.

(f) Washing solution: TBS (Amresco, 0788)-0.1% Tween 20 (Fisher Scientific, BP337).

(g) Stripping solution (e.g., EMD Millipore Corporation, 2502).

(h) Antibody diluent solutions: TBS-0.1% Tween/5% BSA (for primary antibodies), TBS-0.1% Tween/3% BSA (for secondary antibodies).

(i) Primary antibodies: rabbit monoclonal anti-LC3B (Cell Signaling Technology, 3868), mouse monoclonal anti-- α-tubulin (Sigma-Aldrich Corp, T6074).

(j) Secondary antibodies: horseradish peroxidase-conjugated goat anti-mouse IgG (Bio-Rad, 170-6516), horseradish peroxidase-conjugated goat anti-rabbit IgG (Bio-Rad, 170-6515).

(k) Enhanced chemiluminescence substrate (Thermo Scientific, 34,080).

3. Immunofluorescence reagents for MCS:

(a) Harvesting: cold DPBS, Accumax (Innovative Cell Technologies Inc., AM-105).

(b) Fixing solution: 100% cold methanol to fix cells for LC3 immunofluorescence, 4% cold paraformaldehyde in DPBS for ATG13 immunofluorescence.

(c) Antibody diluent solution (e.g., EMD Millipore, 21,544).

(d) Washing solution: TBS (Amresco, 0788)-0.1% Tween 20 (Fisher Scientific, BP337).

(e) Blocking solution: 1% BSA (HyClone, SH30574) in TBS-0.1% Tween 20.

(f) Primary antibodies: rabbit monoclonal anti-ATG13 (Cell Signaling Technology, 13468), rabbit monoclonal anti-LC3B (Cell Signaling Technology, 3868).

(g) Secondary antibody or fluorescent dye: Alexa Fluor 546 goat anti-rabbit IgG (Life Technologies, A11010), TO-PRO-3 iodide (Life Technologies, T3605).

(h) ProLong Gold antifade (Life Technologies, P36930).

NOTE: cell lines are plated on 12 mm cover slips (Fisher Scientific, 12-545-80) for monolayer immunofluorescence studies.

4. Immunofluorescence reagents for TFS:

(a) Harvesting: cold DPBS.

(b) Fixing solution: 10% cold buffered formalin phosphate.

(c) Embedding: 70% cold ethanol, 3% agarose solution in DPBS, paraffin.

(d) TFS sections processing: xylene; ethanol gradient solutions (100%, 80%, 70%, 50% in water); citrate buffer (Sigma-Aldrich Corp, C999).

After antigen retrieval, all the reagents used for TFS immunofluorescence are the same as for MCS immunofluorescence, with the addition of the following antibodies and fluorescent dye to identify the mesothelioma cells: primary antibody mouse monoclonal anti-cytokeratin clones AE1/AE3 (Dako North America, M3515), secondary antibody biotinylated sheep anti-mouse IgG (GE Healthcare, RPN1001V), and NeutrAvidin Oregon Green 488 conjugate (Life Technologies, A6374). For other tumors, a cytokeratin marker may be appropriate to identify the tumor cells; if not, another marker can be substituted.

3 Methods

3.1 Preparation of Non-adsorbent Poly-HEMA Plates for Generation of Multicellular Spheroids

In our laboratory, after testing other methods, we generate MCS from cell lines in 96-well plates coated with poly-HEMA [7, 18, 19, 22–25]. Poly-HEMA reduces the adhesivity of tissue culture plastic allowing the cells to adhere preferentially to each other and thus to form aggregates [26]. We have not needed to use matrix proteins for mesothelioma cultures, which generate their own matrix; however, if spheroids do not form readily for other solid tumors, adding matrix proteins known to be present in the tumor may be useful.

1. Plate coating (96 well): these plates will be used to generate the MCS (*see* Subheading 3.2). Add the poly-HEMA working solution to each well of U-bottomed 96-well plates. The optimal volume of poly-HEMA should be assessed for each cell type; for mesothelioma cell lines, add 50–60 μL per well. Let the plates dry with lids on at 37 °C in a dry oven for at least 48 h. The ethanol in the poly-HEMA working solution will dry and leave a thin uniform coating of poly-HEMA in the wells. Avoid tilting or vibrating the plate during the drying steps. Vibrations can disturb the drying poly-HEMA layer, yielding a rough surface that will prevent optimal spheroid formation. Vibrations can be minimized by using shock-absorbing material (e.g., foam or folded tissue) to support the incubator or the wire shelf on which the plates are drying. When the plates are dry (48 h), seal the plates with their lids on with parafilm and store them at 4 °C.
2. Plate coating (12 well): these plates will be used for autophagy studies using fully formed MCS (24 h old MCS; *see* Subheadings 3.3 and 3.4) and TFS (2–3 weeks old TFS, *see* Subheadings 3.7, 3.8 and 4). For coating 12-well plates, add 250–300 μL of

poly-HEMA working solution to each well of the 12-well plates and proceed as for the 96-well plates.

3. Before using in experiments: remove the Parafilm and sterilize the plates and the inverted lids under a biosafety cabinet under UV light for 30 min.

NOTE: alternatively, coated plates are available for purchase (e.g., Costar, Ultra-low attachment 96-plates, 7007). However, in our experience, these give inconsistent results and are expensive. Use them only if spheroids fail to form in the homemade plates.

3.2 Generation of Multicellular Spheroids

1. Detach monolayer culture of mesothelioma cells with trypsin, resuspend the cells in MCS medium, and determine and adjust cell concentration to a range of 5×10^4 to 5×10^5 cells/mL (*see* Subheading 4). Add 200 μL of cell suspension (corresponding to 10^4 or 10^5 cells/well) to each well of poly-HEMA-coated 96-well plates.
2. Centrifuge the plates after plating and again after 4–5 h at 800 rpm for 5 min to pull the cells to the bottom of the well so that they are in contact with each other. Within 6 h, mesothelioma cells will begin to attach to each other as aggregates; at this early stage, these aggregates will fall apart if pipetted.
3. By 24 h, mesothelioma cells will have formed into a disk-like shape called a multicellular spheroid (MCS). At this point, MCS will hold their shape and can be transferred or collected using a pipette with a P1000 tip that has been cut approximately 2–3 mm from the tip to enlarge the opening.
4. After 24 h, the MCS are transferred to poly-HEMA-coated 12-well plates for experimentation.

3.3 Measurement of the Autophagic Flux in MCS

We measure autophagic flux in MCS by blocking lysosomal proteases and measuring the accumulation of LC3B-II. In the presence of the lysosomal inhibitor, the autophagy proteins LC3 will accumulate in the cells in proportion to the autophagic flux [20]. Lysosomal inhibitors should be employed accordingly to the recent autophagy guidelines ([20], *see* **Note 1**). Determine the autophagic flux as described next.

1. MCS incubation: transfer MCS ($n = 32$) from the 96-well plate to each well of a poly-HEMA-coated 12-well plate and maintain them for the next 24 h in fresh medium. To block the lysosomal proteases and thus measure autophagy, add the lysosomal inhibitor NH_4^+ (10 mM) or HCQ (20 μM) to the medium for the last 4–8 h before harvesting the spheroids; be sure to have wells without lysosomal inhibitors as controls.
2. LC3 immunoblotting: transfer the MCS ($n = 32$) from each well of the poly-HEMA-coated 12-well plate to a 1.5 mL tube,

wash them (3 times) with cold DPBS and resuspend them in cold RIPA buffer (150 μL/32 MCS). Homogenize the spheroids using an ultrasonic cell disruptor (e.g., Fisher Scientific, Sonic Dismembrator model 100). Measure the protein concentration and aliquot an equal amount of total lysate (from 30 to 50 μg) of each sample into 1.5 mL tubes. Depending on the cell line, we expect an average total lysate concentration of approximately 1 μg/μL (obtained from 32 MCS collected in 150 μL of RIPA buffer). Keep the samples on ice during the processing. Samples can be directly loaded on SDS-PAGE (after adding Laemmli buffer and boiling for 5 min at 95 °C) or stored at −80 °C. After separation on SDS-PAGE, transfer the proteins onto a PVDF membrane. After transfer, block the filter (1 h), and incubate it with the anti-LC3B antibody (1:1000, overnight, 4 °C) and then the secondary antibody (1:10,000, 1 h). After each incubation, wash the filter three times. After completion of LC3 immunoblotting, strip the membrane, block it, and incubate it with the anti-α-tubulin antibody (1:1000, overnight, 4 °C) and then the secondary antibody (1:10,000, 1 h), with appropriate washing steps. Develop the filter with an enhanced chemiluminescence substrate. Image and perform densitometry analysis of the bands using an imaging apparatus equipped with an image analysis software (e.g., UVP LLC, BioSpectrum imaging system apparatus, equipped with the Vision-WorksLS software). The autophagic flux is expressed as a ratio of normalized LC3B-II band intensities with NH_4^+ to without NH_4^+ (Fig. 1a).

NOTE: In mesothelioma, we found that α-tubulin is a suitable loading control in spheroids, with consistent expression among the cell lines and experimental conditions.

3. LC3 immunofluorescence: transfer the MCS ($n = 32$) from each well of the poly-HEMA-coated 12-well plate to a 1.5 mL tube and wash (3 times) with cold DPBS. Disaggregate the spheroids, count the cells and adjust the cell concentration with media to 5×10^5 cells/mL. Cytospin 2×10^4 cells (200 μL of cell suspension) onto glass slides. Fix the cells with cold methanol for 20 min, and wash them once with DPBS for 10 min. Block nonspecific binding (1 h). Perform washes (×3) after each antibody incubation. Incubate with the anti-LC3B antibody (1:100, overnight at 4°C) and then with the secondary antibody (1:200) together with TO-PRO-3 (1:1000) for 2 h. After mounting, capture images at 63× magnification with a confocal microscope (e.g., Nikon C1, Nikon Instruments, Inc.) for later identification of LC3 puncta. A punctum is defined as a LC3-positive mainly circular cytoplasmic structure of approximately 1 μm in diameter; this is thought to

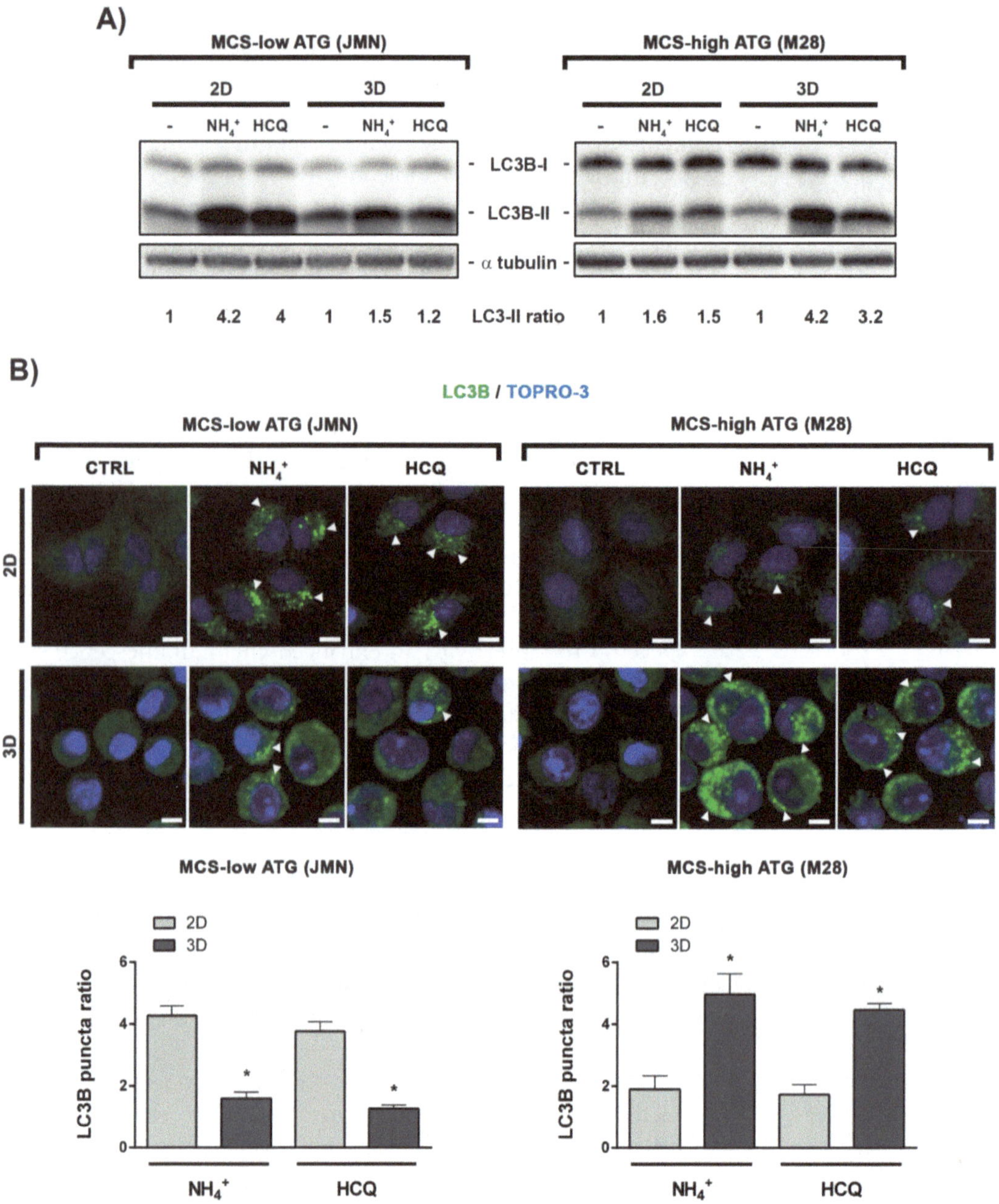

Fig. 1 Autophagic flux significantly differs when cells are grown in 2D versus 3D. (**a**) Mesothelioma cells (JMN and M28) were grown as monolayers (2D) or MCS (3D). Where indicated, the cells were exposed to 10 mM ammonium chloride (NH_4^+) or 20 μM hydroxychloroquine (HCQ) for 4 h. LC3B expression was assessed by immunoblotting. As a loading control, filters were probed with anti-α-tubulin antibody. Band intensities were determined by densitometric analysis. The autophagic flux is expressed as a ratio of normalized LC3B-II band intensities with NH_4^+ to without NH_4^+ (LC3-II ratio). A representative immunoblot of three independent experiments is shown, with ratios shown below. LC3B-II ratios show that the autophagic flux differs between 2D and 3D cultures. (**b**) Mesothelioma cells (JMN and M28) were grown as monolayers (2D) on cover slips or as MCS (3D). Where indicated, the cells were exposed to the lysosomal inhibitor as in (**a**). Spheroid cells were

correspond to an autophagic vesicle (autophagosome or autolysosome). Only the lipidated LC3, or LC3-II, is associated with the autophagic vesicles, thus only LC3-II is present on the puncta [27]. Count LC3 puncta/cell for at least a total of 100 cells for each condition in three different experiments and calculate the average value. The autophagic flux is expressed as a ratio of LC3B-II puncta/cell with NH_4^+ to without NH_4^+ (Fig. 1b).

NOTE: If the experiment will compare 3D to 2D, treat monolayer cultures in the same manner as the MCS. For LC3 immunofluorescence, one can plate the cells on cover slips and process them as the MCS. We found no significant difference in the number of puncta detected in cells that are cytospun compared to the cells that are plated on cover slips and not cytospun.

3.4 Assessment of Autophagy Initiation in MCS

Autophagy initiation in MCS is assessed by analyzing the presence of ATG13 puncta at baseline, using immunofluorescence. Lysosomal inhibitors are not required. The presence of puncta of the subunits of the ULKl complex, including ATG13, represents the relocation of the complex to the autophagosome and the earliest event in autophagy initiation ([21] *see* **Note 2**).

ATG13 immunofluorescence: Transfer the MCS ($n = 32$) into each well of a poly-HEMA-coated 12-well as described in Subheading 3.3. MCS are collected and processed for immunofluorescence as described for LC3 (*see* Subheading 3.3, **step 2**) aside from the cell fixation (4% paraformaldehyde for 10 min at room temperature, or overnight at 4 °C) and the primary antibody (anti-ATG13 antibody, 1:50, overnight at 4 °C). A punctum is defined as an ATG13-positive mainly circular cytoplasmic structure of approximately 1 μm in diameter; this is thought to correspond to an early autophagy structure (omegasome or phagophore). In mesothelioma cells (either MCS or monolayer cultures), we did not need to count the ATG13 puncta/cell because the puncta were either detectable in almost all the cells (MCS or monolayer with high autophagy initiation) or absent (MCS or monolayers with low autophagy initiation) (Fig. 2).

Fig. 1 (continued) disaggregated and cytospun on glass slides. Cells adherent on cover slips or glass slides were then fixed, stained for LC3B (green) and nuclei (blue), and imaged by confocal microscopy. Arrowheads indicate representative LC3 puncta. Scale bars: 10 μm. Bars show the ratio between LC3 puncta counted in cells grown in the presence or absence of the indicated lysosomal inhibitor (NH_4^+, HCQ). Asterisks indicate significantly different LC3 puncta ratios between 2D and 3D ($P < 0.05$). Error bars, S.D. LC3 puncta ratios confirm that the autophagic flux differs between 2D and 3D as shown in (**a**)

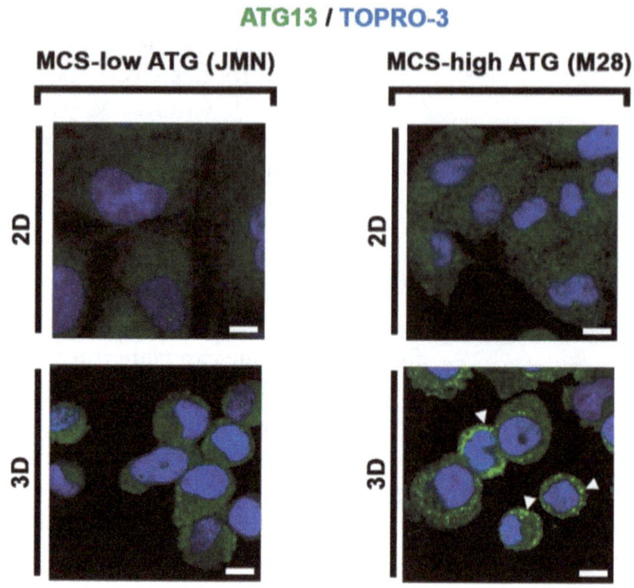

Fig. 2 In multicellular spheroids, autophagy initiation can be assessed by ATG13 immunofluorescence. Mesothelioma cells (JMN and M28) were grown as monolayers (2D) on cover slips or as MCS (3D). Spheroid cells were disaggregated and cytospun on glass slides. Cells adherent on cover slips or glass slides were then fixed, stained for ATG13 (green) and nuclei (blue), and imaged by confocal microscopy. Representative ATG13 puncta are indicated by arrowheads. Scale bars: 10 μm. ATG13 puncta reflect the autophagic flux in 3D, but not in 2D

3.5 Preparation of Agar-Coated Plates for Generation of Tumor Fragment Spheroids

We use these plates to generate and maintain ex vivo tumors. Agar proved to be an inexpensive coating for the larger plates needed and suitable for the tumor tissue fragments which did not require smooth layers to form into spheroids. For experiments, however, TFS are transferred into the poly-HEMA-coated 12-well plates.

1. Agar working solution (0.77%): add 1.54 g of Noble agar to 200 mL of DMEM. Dissolve the agar in a bottle with the lid removed or loosened using a microwave at the defrost setting for approximately 20 min. Work with the agar quickly in the next steps before it cools and starts to solidify. Working under a biosafety cabinet, restore the volume to 200 mL by adding DMEM to compensate for the evaporation in the microwave and mix the solution by gently swirling the bottle (avoid creating bubbles).
2. Plate coating: working under a biosafety cabinet, gently pour (do not pipet) the DMEM/agar solution while still warm into 10 cm^2 petri dishes until there is a 0.3/0.4 cm thickness of agar in each plate (200 mL of agar solution will coat approximately 7–8 plates). Let the agar solidify in the plates for 30–40 min while under UV light to sterilize the plates and inverted lids.

Agar will become opaque when fully solidified. Put the lids on the plates, seal the lidded plates with Parafilm, and store them inverted (bottom up) at 4 °C.

NOTE: before adding the TFS, the plates should be sterilized again by removing the lids and exposing the plates and inverted lids to UV light under a biosafety cabinet for 20 min. Equilibrate the sterilized agar-coated plates with TFS culture medium by adding 10 mL of medium to each plate, replacing the lids and keeping them in the incubator (37°C, 5% CO_2) for 15 min.

3.6 Generation of Tumor Fragment Spheroids

We receive fresh mesothelioma resection samples from our collaborators at the Brigham and Women's Hospital. Samples are sent by overnight courier in media at ambient temperature. We have shown these to be viable on arrival when sent in 50 mL falcon tubes with 30 mL of media. On arrival, the tumor is diced as described below and then maintained on agar-coated plates.

1. Transfer a small piece of tumor to a 15 cm^2 petri dish. Cut the tumor using two #10 scalpels into pieces of approximately 1 cm^3. Save 1–2 pieces for dicing. Store the other pieces in a 50 mL tube with 20–30 mL of TFS medium until dicing of the previous piece has been completed.
2. Keep the tumor moist by the frequent addition of small quantities of media (250–500 μL) on the top of the tumor piece placed on the 15 cm^2 petri dish.
3. Finely dice the tumor piece with two scalpels by using a "scissor-like" technique (holding the scalpels with opposite hands and slicing the tissue by moving the hands apart while the blades touch each other). You can alternate the "scissor-like" technique by holding the two scalpels together with one hand and cut down upon the tumor with chopping movements. At the end of the dicing process, tumor fragments should be around 1 mm in diameter or less. At this point, they appear to be grainy paste which can be resuspended in medium with a 25 mL pipette.
4. Aspirate the resuspended TFS with the 25 mL pipette and transfer them to an agar-coated plate. TFS generated from 1 to 2 cm^3 of original tumor will generally fit in one plate, which allows the fragments to disperse in the media without touching each other. Add media to fill the plate to approximately 25–30 mL/plate and place the plate in the incubator (37 °C, 5% CO_2). Change the medium the next day and every 3 days routinely. Medium is changed by gently tilting the agar-coated plate, aspirating ~80% of the medium in the plate and replacing with new culture medium. We use a glass pipette with a 10 μL pipette tip placed over the tip to minimize the chance of suctioning up tumor fragments.

NOTE: the day after TFS generation, some fragments may float on top of the medium. These TFS normally have fat tissue in them and we recommend discarding them. Do not allow fragments to protrude up into the air; these fragments may develop mold. It is normal if the viscosity of the medium increases during the weeks of incubation due to the hyaluronic acid that may be produced by mesothelioma.

3.7 Measurement of the Level of Autophagy in TFS

We measure autophagy in the TFS in a similar fashion as for the MCS. As for the MCS, when the autophagic flux is blocked by addition of the lysosomal inhibitor, the autophagy proteins LC3 will accumulate in the cells in proportion to the autophagic flux [20]. However, unlike in MCS, immunoblotting for LC3 in TFS cannot distinguish the mesothelioma-specific LC3 from that of other cell types. Therefore, we measure the level of autophagy in mesothelioma cells within the tumor by performing a dual immunofluorescence staining of LC3 and cytokeratin (to identify mesothelioma cells), as described below.

1. TFS incubation: let the TFS grow until they become rounded (from 3 to 4 weeks; *see* Subheading 4). At this point, the TFS can be transferred or collected using a pipette with a P1000 tip that has been cut to enlarge the orifice, as described for the MCS. In preparation for experiments using TFS, transfer the TFS (n = 20–30) into each well of a poly-HEMA-coated 12-well plate and maintain them for the next 24 h in fresh medium. To measure autophagy, add the lysosomal inhibitor NH_4^+ (10 mM) or HCQ (20 μM) to the medium for the last 12 h before harvesting the TFS, remembering to have wells without lysosomal inhibitors as a control.
2. TFS sections: at the end of the 24 h exposure, transfer the TFS from each well to a 10 mL tube. Remove the media, wash the TFS with DPBS, and fix them in 10% formalin (overnight, 4 °C). Remove the formalin and wash the TFS with 70% ethanol solution in water. Using a glass pipette with a 10 μL pipette tip placed over the tip to minimize the chance of suctioning up tumor fragments, remove the 70% ethanol solution, and embed the TFS in a 3% agarose solution in DPBS. For 20 to 30 fragments, pour approximately 200 μl of warm agarose solution in each 10 mL tube and quickly flick the tube to allow the agarose to distribute evenly among the fragments and then tap the tube to collect the fragments at the bottom of the tube and eliminate air bubbles. Place the tubes for 10 min in ice to solidify the agarose. Carefully remove the solidified agar pellets using a spatula (e.g., Grainger, CG-1983-12) and place each pellet sideways into a tissue cassette. Embed the agarose pellets in paraffin, and cut the TFS into 5 μm sections.

NOTE: the agarose pellet facilitates the positioning of the 20–30 fragments in the tissue cassette. A 3% agarose concentration produces a pellet of hardness comparable to that of the fragment tissue, helping the later cutting of homogeneous sections.

3. TFS sections processing for immunofluorescence: remove the paraffin from TFS sections with xylene (5 min incubation for 3 times) and rehydrate the sections using an ethanol gradient (100-80-70-50% ethanol in water) and a final rehydration step in water (3 min ×2). Unmask the antigen by incubating the TFS sections in antigen retrieval buffer and by boiling them for 10 min in a pressure cooker. After cooling, draw a water-repellent circle around the tissue using a hydrophobic barrier pen (PAP pen) and wash with DPBS for 10 min. Block non-specific binding (1 h) and proceed to the immunofluorescence.
4. LC3/cytokeratin dual staining: perform washes (×3) after each antibody incubation. Incubate with the mouse anti-cytokeratin antibody (1:200) for 1 h, followed with the biotinylated secondary anti-mouse antibody (1:200) for 1 h, and then with the rabbit anti-LC3B antibody (1:50) overnight at 4°C. Incubate the sections with the secondary anti-rabbit antibody (1:200) for 1 h, together with NeutrAvidin Oregon Green (1:200) and TO-PRO-3 (1:1000). After mounting, capture images at 63× magnification with a confocal microscope (e.g., Nikon C1, Nikon Instruments, Inc.). Because the LC3 puncta overlap each other in the tissue sections and cannot be individually counted, we have measured the accumulation of LC3 by quantifying the percentage of mesothelioma cells positive for LC3 puncta (LC3-positive mesothelioma cells). Count the LC3-positive cells in at least at total of 100 cytokeratin-positive, mesothelioma cells for each condition in three different TFS. The level of autophagy is expressed as the percentage of cytokeratin-positive cells containing LC3 puncta (LC3B-positive mesothelioma cells) (Fig. 3). To display the level of autophagy, because we are dealing with percentages, we prefer to show both values (with and without NH_4^+) instead of calculating a ratio.

3.8 Assessment of Autophagy Initiation in TFS

We measure autophagy initiation in the TFS by the assessment of ATG13 puncta specifically in the mesothelioma cells. Lysosomal inhibitors are not required to determine the level of autophagy initiation. We measure autophagy initiation in TFS by performing a dual immunofluorescence staining of ATG13 and cytokeratin (to identify mesothelioma cells), as described below.

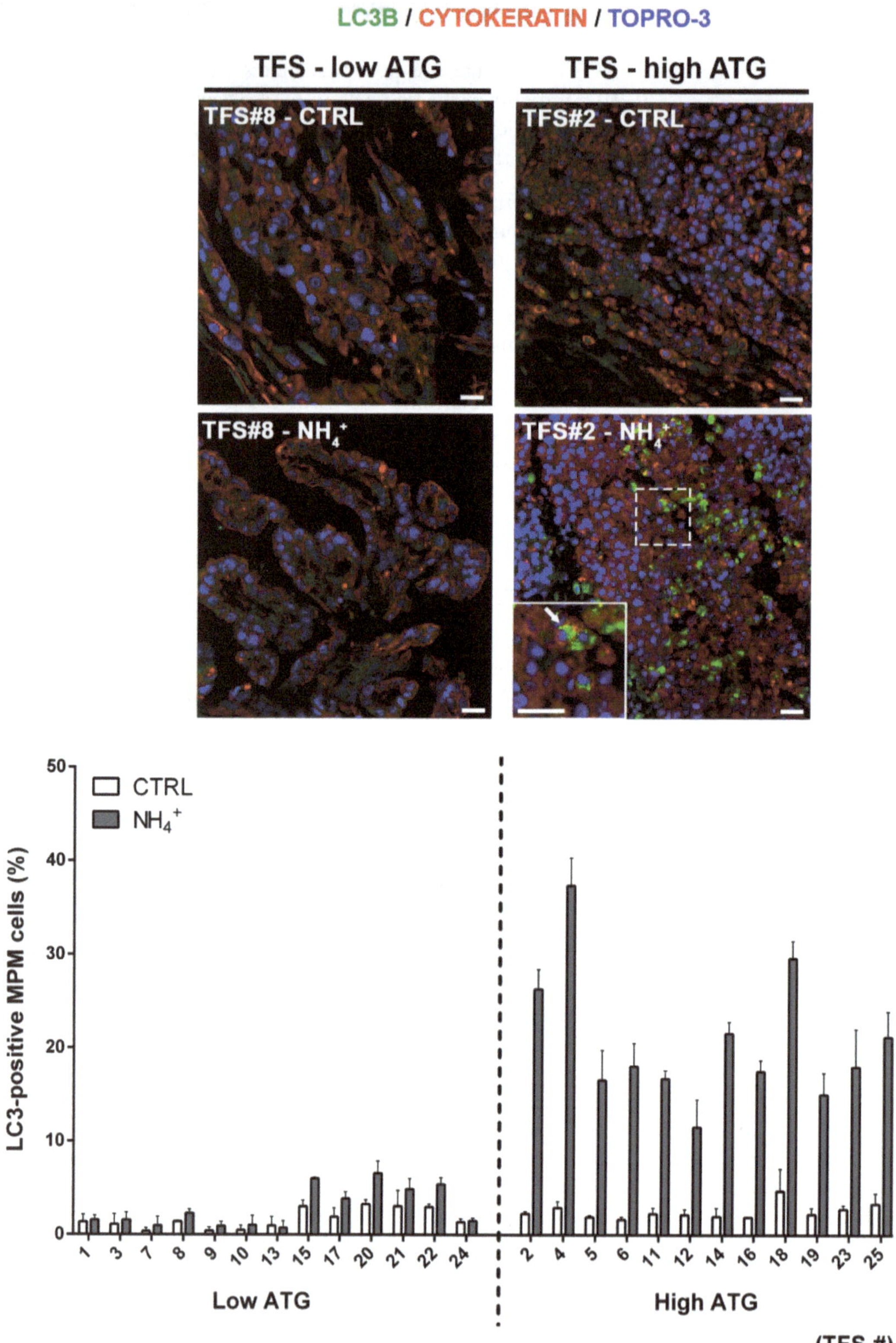

Fig. 3 In tumor fragment spheroids, the level of autophagy can be determined by dual immunostaining for LC3 and cytokeratin. TFS were generated from tumor biopsies obtained from 25 chemonaive mesothelioma patients and grown in the presence or absence of 10 mM ammonium chloride (NH_4^+) for 12 h. TFS were then fixed, embedded in paraffin, stained for LC3B (green), cytokeratin to identify mesothelioma cells (red) and

1. Transfer the TFS ($n = 20$–30) as described in Subheading 3.7, **step 1** and maintain them for the next 24 h in fresh medium.
2. Prepare 5 μm sections of TFS as described in Subheading 3.7, **step 2**.
3. Process TFS sections for immunofluorescence as described in Subheading 3.7, **step 3**.
4. ATG13/cytokeratin dual staining: TFS sections are processed for immunofluorescence as described for LC3 (*see* Subheading 3.7, **step 3**) aside from using a different primary antibody (anti-ATG13 antibody, 1:50, overnight at 4 °C). Because the ATG13 puncta overlap each other in the tissue sections and cannot be individually counted, we measure the percentage of mesothelioma cells positive for ATG13 puncta. Count the ATG13-positive cells in at least at total of 100 mesothelioma cells for each condition in 3 different TFS. The level of autophagy initiation is expressed as the percentage of cytokeratin-positive cells containing ATG13 puncta (ATG13-positive mesothelioma cells). For mesothelioma, we determined an optimal cut-point of 6% ATG13 positivity to identify TFS with either low autophagy initiation (<6% of ATG13-positive mesothelioma cells) or high autophagy initiation (≥6% of ATG13-positive mesothelioma cells). The cutoff point was selected from our previous results showing a strong correlation between ATG13 positivity ≥6% and a better patient outcome in mesothelioma [19] (Fig. 4).

4 Notes

1. Lysosomal inhibitors: be aware that prolonged exposure to lysosomal inhibitors (>6 h for in vitro cultures; >12 h for ex vivo cultures) may trigger a secondary autophagic response and result in an incorrect measurement of the autophagic flux [20]. Also, the autophagic flux should be confirmed using different lysosomal inhibitors. We have confirmed the measurement of the autophagic flux in MCS and monolayers by 4 h and 8 h inhibition experiments and by using two different lysosomal inhibitors [19]. TFS were exposed to the lysosomal

Fig. 3 (continued) TO-PRO-3 to detect nuclei (blue), and imaged by confocal microscopy. Representative images of TFS with either low (TFS #8) or high (TFS #2) autophagy levels are shown. Zoom-in view of the region in the dashed box shows representative cells with LC3 puncta (arrow). Scale bars: 10 μm. Bars represent the mean percentage of LC3-positive mesothelioma cells measured in TFS grown in the presence (gray bars) or absence (white bars) of NH_4^+. Error bars, S.E.M. LC3 immunofluorescence identifies tumors with either low or high levels of autophagy

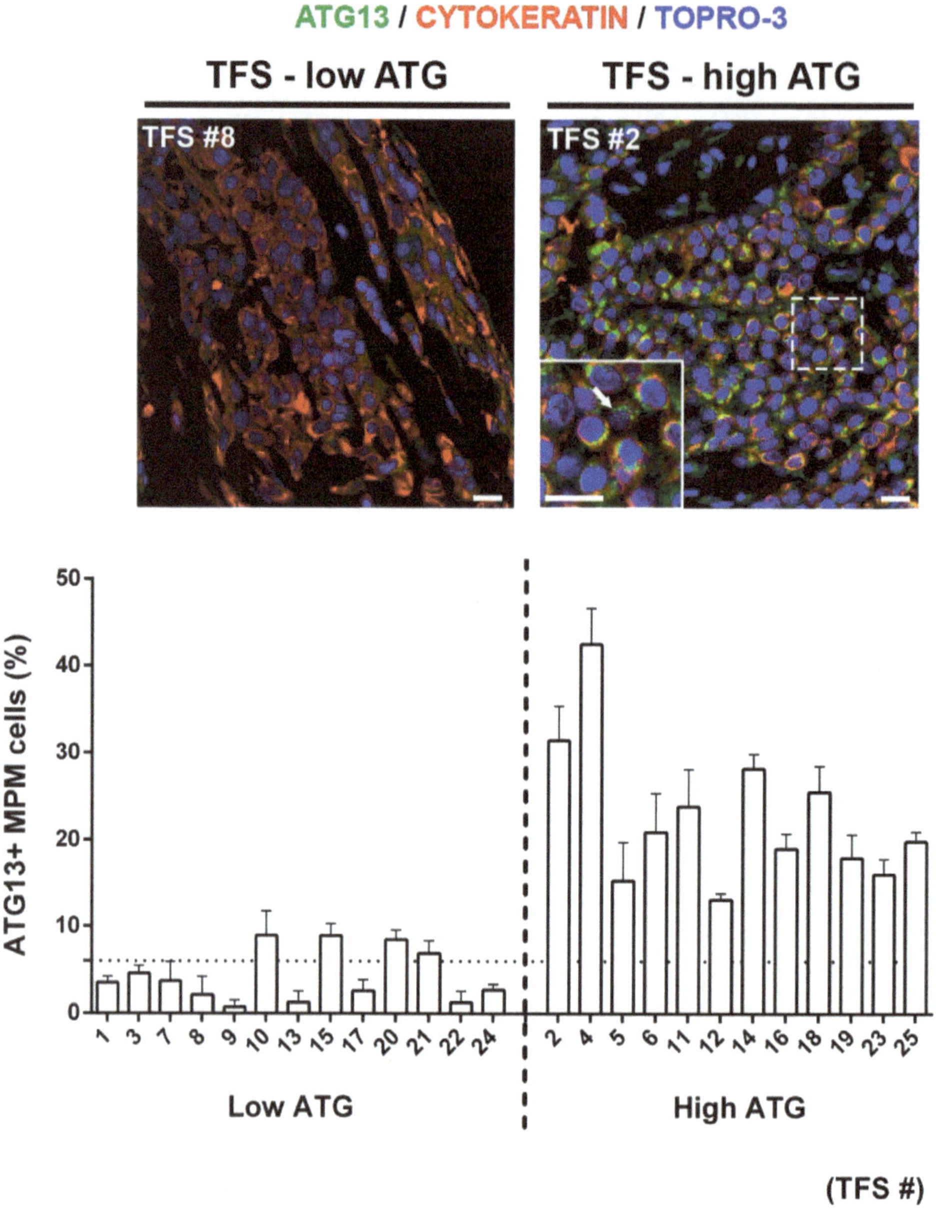

Fig. 4 In tumor fragment spheroids, the level of autophagy initiation can be determined by dual immunostaining for ATG13 and cytokeratin. The TFS shown in Fig. 3 and not exposed to ammonium chloride were fixed, embedded in paraffin, stained for ATG13 (green), cytokeratin to identify mesothelioma cells (red) and TO-PRO-3 to detect nuclei (blue), and imaged by confocal microscopy. Representative TFS with either low autophagy (TFS #8) or high autophagy (TFS #2) levels are shown. Zoom-in view of the region in the dashed box shows representative cells with ATG13 puncta (arrowhead). Scale bars: 10 μm. Bars represent the mean percentage of ATG13-positive mesothelioma cells for the TFS determined in Fig. 3 to be low or high ATG. Error bars, S.E.M. dotted line represents the ATG13-positivity cutoff (6%) and shows that the level of autophagy initiation (determined by ATG13 immunofluorescence) correlates with the level of autophagy (determined by LC3 immunofluorescence) in TFS

inhibitor for a longer time (12 h), as suggested in the autophagy guidelines [20].

2. Cautionary notes about measurement of autophagy in 3D: as in traditional monolayer cultures, multiple markers of autophagy must also be investigated in 3D. In mesothelioma 3D models, we have measured autophagy by determining the autophagic flux and by assessing the presence of ATG13 puncta at steady state. We also checked other known autophagy markers such as p62 accumulation following the inhibition of lysosomal proteases or Beclin 1 levels at steady state [20]. However, neither of these markers correlated with the level of autophagy as measured by both LC3 and ATG13 analysis in 3D models of mesothelioma [19], suggesting that canonical autophagy signaling pathways in 3D may differ from those known to be active in 2D. Since autophagy clearly differs from 2D to 3D, it is reasonable to think that also other cellular processes could differ between the two settings. For example, the activity or substrate specificity of the proteasome may change between 2D and 3D, and this, in turn, may influence the level of autophagy proteins such as p62. Thus, extra caution should be taken when studying autophagy in 3D models.

3D culture notes: size and time of culture should be taken in consideration for both in vitro and ex vivo 3D models. In the case of our in vitro model MCS, we have determined that from 10^4 to 10^5 cells per each spheroid can be employed to generate spheroids for autophagy studies, with minimal baseline apoptosis for up to 48 h from the plating in the poly-HEMA-coated 96-well plates (not shown). In our hands, we observed that MCS cultured for more than 48 h may develop a dark core in the center, which may correspond to a necrotic core [28–30]. In the case of our ex vivo model TFS, we choose to study them after they become rounded, as evidence of remodeling by living cells, at a time around 3–4 weeks. However, TFS can be cultured for up to 3 months [9]. We also try to dice the TFS always at a comparable size of about 1 mm, which is similar to that of the MCS generated from 10^5 cells per each spheroid (Fig. 5).

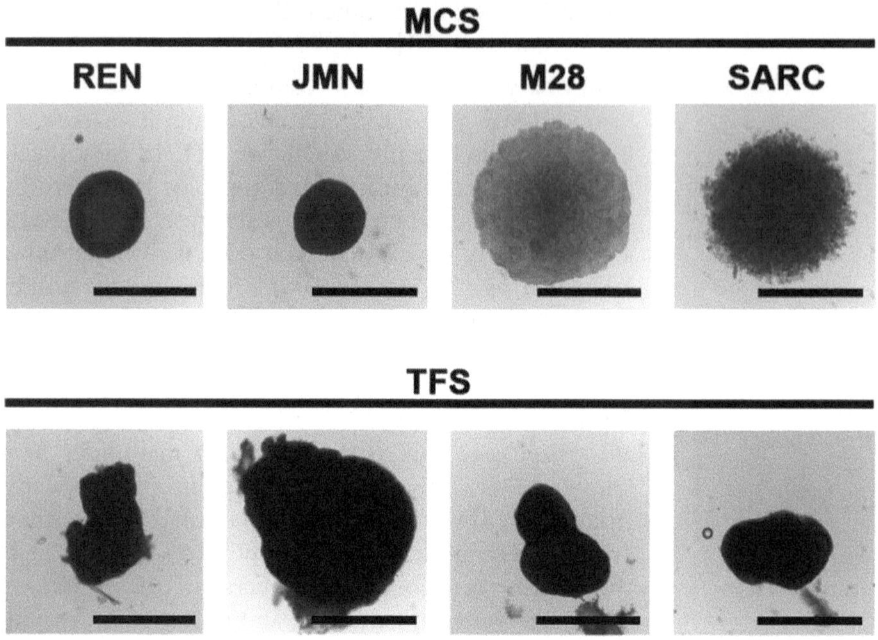

Fig. 5 Multicellular spheroids generated from 10^5 cells, and tumor fragment spheroids have comparable size. Bright-field images of representative MCS generated from 10^5 mesothelioma cells show a comparable size to that of representative TFS. Scale bars: 1 mm

Acknowledgments

This work was supported by the Simmons Mesothelioma Foundation; CF was supported also by the Meso Foundation under grant 383573.

References

1. Pampaloni F, Reynaud EG, Stelzer EH (2007) The third dimension bridges the gap between cell culture and live tissue. Nat Rev Mol Cell Biol 8(10):839–845. https://doi.org/10.1038/nrm2236
2. Nyga A, Cheema U, Loizidou M (2011) 3D tumour models: novel in vitro approaches to cancer studies. J Cell Commun Signal 5(3):239–248. https://doi.org/10.1007/s12079-011-0132-4
3. Hoffmann OI, Ilmberger C, Magosch S, Joka M, Jauch KW, Mayer B (2015) Impact of the spheroid model complexity on drug response. J Biotechnol 205:14–23. https://doi.org/10.1016/j.jbiotec.2015.02.029
4. Imamura Y, Mukohara T, Shimono Y, Funakoshi Y, Chayahara N, Toyoda M, Kiyota N, Takao S, Kono S, Nakatsura T et al (2015) Comparison of 2D- and 3D-culture models as drug-testing platforms in breast cancer. Oncol Rep 33(4):1837–1843. https://doi.org/10.3892/or.2015.3767
5. Baker BM, Chen CS (2012) Deconstructing the third dimension: how 3D culture microenvironments alter cellular cues. J Cell Sci 125(Pt 13):3015–3024. https://doi.org/10.1242/jcs.079509
6. Antoni D, Burckel H, Josset E, Noel G (2015) Three-dimensional cell culture: a breakthrough in vivo. Int J Mol Sci 16(3):5517–5527. https://doi.org/10.3390/ijms16035517

7. Barbone D, Cheung P, Battula S, Busacca S, Gray SG, Longley DB, Bueno R, Sugarbaker DJ, Fennell DA, Broaddus VC (2012) Vorinostat eliminates multicellular resistance of mesothelioma 3D spheroids via restoration of Noxa expression. PLoS One 7(12):e52753. https://doi.org/10.1371/journal.pone.0052753
8. Edmondson R, Broglie JJ, Adcock AF, Yang L (2014) Three-dimensional cell culture systems and their applications in drug discovery and cell-based biosensors. Assay Drug Dev Technol 12(4):207–218. https://doi.org/10.1089/adt.2014.573
9. Kim KU, Wilson SM, Abayasiriwardana KS, Collins R, Fjellbirkeland L, Xu Z, Jablons DM, Nishimura SL, Broaddus VC (2005) A novel in vitro model of human mesothelioma for studying tumor biology and apoptotic resistance. Am J Respir Cell Mol Biol 33 (6):541–548. https://doi.org/10.1165/rcmb.2004-0355OC
10. Thoma CR, Zimmermann M, Agarkova I, Kelm JM, Krek W (2014) 3D cell culture systems modeling tumor growth determinants in cancer target discovery. Adv Drug Deliv Rev 69–70:29–41. https://doi.org/10.1016/j.addr.2014.03.001
11. Tanner K, Gottesman MM (2015) Beyond 3D culture models of cancer. Sci Transl Med 7 (283):283–289. https://doi.org/10.1126/scitranslmed.3009367
12. do Amaral JB, Rezende-Teixeira P, Freitas VM, Machado-Santelli GM (2011) MCF-7 cells as a three-dimensional model for the study of human breast cancer. Tissue Eng Part C Methods 17(11):1097–1107. https://doi.org/10.1089/ten.tec.2011.0260
13. Ma XH, Piao S, Wang D, McAfee QW, Nathanson KL, Lum JJ, Li LZ, Amaravadi RK (2011) Measurements of tumor cell autophagy predict invasiveness, resistance to chemotherapy, and survival in melanoma. Clin Cancer Res 17(10):3478–3489. https://doi.org/10.1158/1078-0432.CCR-10-2372
14. Gomes LR, Vessoni AT, Menck CF (2015) Three-dimensional microenvironment confers enhanced sensitivity to doxorubicin by reducing p53-dependent induction of autophagy. Oncogene 34(42):5329–5340. https://doi.org/10.1038/onc.2014.461
15. Koehler BC, Jassowicz A, Scherr AL, Lorenz S, Radhakrishnan P, Kautz N, Elssner C, Weiss J, Jaeger D, Schneider M et al (2015) Pan-Bcl-2 inhibitor Obatoclax is a potent late stage autophagy inhibitor in colorectal cancer cells independent of canonical autophagy signaling. BMC Cancer 15:919. https://doi.org/10.1186/s12885-015-1929-y
16. Bingel C, Koeneke E, Ridinger J, Bittmann A, Sill M, Peterziel H, Wrobel JK, Rettig I, Milde T, Fernekorn U et al (2017) Three-dimensional tumor cell growth stimulates autophagic flux and recapitulates chemotherapy resistance. Cell Death Dis 8(8):e3013. https://doi.org/10.1038/cddis.2017.398
17. Russell RC, Yuan HX, Guan KL (2014) Autophagy regulation by nutrient signaling. Cell Res 24(1):42–57. https://doi.org/10.1038/cr.2013.166
18. Barbone D, Yang TM, Morgan JR, Gaudino G, Broaddus VC (2008) Mammalian target of rapamycin contributes to the acquired apoptotic resistance of human mesothelioma multicellular spheroids. J Biol Chem 283 (19):13021–13030. https://doi.org/10.1074/jbc.M709698200
19. Follo C, Barbone D, Richards WG, Bueno R, Broaddus VC (2016) Autophagy initiation correlates with the autophagic flux in 3D models of mesothelioma and with patient outcome. Autophagy 12(7):1180–1194. https://doi.org/10.1080/15548627.2016.1173799
20. Klionsky DJ, Abdelmohsen K, Abe A, Abedin MJ, Abeliovich H, Acevedo Arozena A, Adachi H, Adams CM, Adams PD, Adeli K et al (2016) Guidelines for the use and interpretation of assays for monitoring autophagy (3rd edition). Autophagy 12(1):1–222. https://doi.org/10.1080/15548627.2015.1100356
21. Karanasios E, Stapleton E, Manifava M, Kaizuka T, Mizushima N, Walker SA, Ktistakis NT (2013) Dynamic association of the ULK1 complex with omegasomes during autophagy induction. J Cell Sci 126(Pt 22):5224–5238. https://doi.org/10.1242/jcs.132415
22. Barbone D, Ryan JA, Kolhatkar N, Chacko AD, Jablons DM, Sugarbaker DJ, Bueno R, Letai AG, Coussens LM, Fennell DA et al (2011) The Bcl-2 repertoire of mesothelioma spheroids underlies acquired apoptotic multicellular resistance. Cell Death Dis 2:e174. https://doi.org/10.1038/cddis.2011.58
23. Xiang X, Phung Y, Feng M, Nagashima K, Zhang J, Broaddus VC, Hassan R, Fitzgerald D, Ho M (2011) The development and characterization of a human mesothelioma in vitro 3D model to investigate immunotoxin therapy. PLoS One 6(1):e14640. https://doi.org/10.1371/journal.pone.0014640
24. Barbone D, Follo C, Echeverry N, Gerbaudo VH, Klabatsa A, Bueno R, Felley-Bosco E, Broaddus VC (2015) Autophagy correlates with the therapeutic responsiveness of malignant pleural mesothelioma in 3D models. PLoS

One 10(8):e0134825. https://doi.org/10.1371/journal.pone.0134825

25. Broaddus VC, Follo C, Barbone D. 3D models of mesothelioma in the study of mechanisms of cell survival. In Asbestos and mesothelioma. Testa JR,. Springer International Publishing, New York 2017:237–257
26. Folkman J, Moscona A (1978) Role of cell shape in growth control. Nature 273 (5661):345–349
27. Barth S, Glick D, Macleod KF (2010) Autophagy: assays and artifacts. J Pathol 221 (2):117–124. https://doi.org/10.1002/path.2694
28. Kunz-Schughart LA, Kreutz M, Knuechel R (1998) Multicellular spheroids: a three-dimensional in vitro culture system to study tumour biology. Int J Exp Pathol 79(1):1–23
29. Kunz-Schughart LA, Freyer JP, Hofstaedter F, Ebner R (2004) The use of 3-D cultures for high-throughput screening: the multicellular spheroid model. J Biomol Screen 9 (4):273–285. https://doi.org/10.1177/1087057104265040
30. Hirschhaeuser F, Menne H, Dittfeld C, West J, Mueller-Klieser W, Kunz-Schughart LA (2010) Multicellular tumor spheroids: an underestimated tool is catching up again. J Biotechnol 148(1):3–15. https://doi.org/10.1016/j.jbiotec.2010.01.012

Chapter 32

Autophagy in Platelets

Meenakshi Banerjee, Yunjie Huang, Madhu M. Ouseph, Smita Joshi, Irina Pokrovskaya, Brian Storrie, Jinchao Zhang, Sidney W. Whiteheart, and Qing Jun Wang

Abstract

Anucleate platelets are produced by fragmentation of megakaryocytes. Platelets circulate in the bloodstream for a finite period: upon vessel injury, they are activated to participate in hemostasis; upon senescence, unused platelets are cleared. Platelet hypofunction leads to bleeding. Conversely, pathogenic platelet activation leads to occlusive events that precipitate strokes and heart attacks. Recently, we and others have shown that autophagy occurs in platelets and is important for platelet production and normal functions including hemostasis and thrombosis. Due to the unique properties of platelets, such as their lack of nuclei and their propensity for activation, methods for studying platelet autophagy must be specifically tailored. Here, we describe useful methods for examining autophagy in both human and mouse platelets.

Key words Platelets, Autophagy, Hemostasis, Live imaging, Electron microscopy

1 Introduction

1.1 Discovery of Autophagy Machinery and Autophagic Structures in Platelets

Platelets are circulating first responders which act upon blood vessel damage and facilitate hemostasis. Compromised platelet functions can cause bleeding diatheses. Conversely, inappropriate platelet activation can result in spurious thrombosis that causes acute vascular obstruction, precipitating strokes, heart attacks, and other ischemic pathologies. Platelets are released under shear conditions from megakaryocytes as anucleate cellular fragments [1–3] and remain in circulation for about 4–5 (for mouse) or 7–10 days (for human) [3–5]. As their functions decay over time [6], the aged platelets are cleared by the liver and spleen (reviewed in Ref. [7]).

We and others [8, 9] have reported that as detected by immunoblotting, resting mouse and human platelets express numerous

Meenakshi Banerjee, Yunjie Huang, and Madhu M. Ouseph contributed equally to this work.

Nicholas Ktistakis and Oliver Florey (eds.), *Autophagy: Methods and Protocols*, Methods in Molecular Biology, vol. 1880, https://doi.org/10.1007/978-1-4939-8873-0_32, © Springer Science+Business Media, LLC, part of Springer Nature 2019

components of the major autophagy protein complexes. These include ULKl, FIP200, Beclin 1, VPS34, VPS15, ATG14, NRBF2, UVRAG, ATG7, the ATG12-ATG5 conjugate, ATG3, and LC3II (summarized in the **Supplemental Table 1** in [9]). In addition to the protein data, microscopy clearly shows the presence of autophagy-related structures in platelets. Resting platelets, isolated from *GFP-LC3/+* [10, 11], *Becn1-EGFP/+* [12], and *EGFP-Atg5/+* [9] transgenic mice, display distinct GFP-positive puncta by confocal fluorescence microscopy [9] or GFP-positive structures resembling incomplete rings by super-resolution microscopy (Fig. 3), demonstrating the presence of phagophores and autophagosomes. Endogenous LC3-positive puncta were also detected in human platelets by immunocytochemistry/immunofluorescence (ICC/IF) [8]. Further, we visualized phosphatidylinositol 3-phosphate (PI3P), the product of the class III phosphatidylinositol 3-kinase (PI3K) VPS34, as puncta in mouse platelets, using GST-2×FYVE-mediated ICC/IF (Fig. 1b). Autophagic structures, e.g., double-membraned phagophore-like structures wrapping around portions of cytosol, granules, and mitochondria, are also clearly visible by electron microscopy ([9] and unpublished data; representative micrographs shown in Fig. 2). Mitochondria-containing autophagosomes were observed in mouse platelets subjected to hypoxic conditions either ex vivo or in vivo [13]. Besides human and mouse, autophagosome-like structures were also seen in platelets from dogs with severe non-regenerative anemia [14].

Taken as a whole, the above descriptive data clearly show that both the machinery and cellular structures associated with autophagy are readily detectible in resting platelets. These data were initially surprising since it was unclear why platelets, with their short life-span, would retain such an energy-requiring, degradative system.

1.2 Functional Significance of Autophagy in Platelets

In nucleated eukaryotic cells, basal autophagy is constitutively active to maintain cellular homeostasis. Under stress (e.g., nutrient deprivation), autophagy can be further induced to meet increasing cellular needs for amino acids, nucleotides, sugars, and fatty acids as metabolic fuels or anabolic building blocks. In addition to the presence of both the machinery and cellular structures associated with autophagy in resting platelets, we also demonstrated the occurrence of basal autophagy in resting platelets by monitoring autophagic flux, using both LC3II immunoblotting and an imaging assay that counts GFP-LC3 puncta ([9], reproduced in Fig. 1d). In resting platelets, autophagy can be induced by starvation or by treatment with the mTOR complex 1 (mTORC1) inhibitor, rapamycin [8]. Autophagy, particularly mitophagy (i.e., autophagy of mitochondria), can also be induced by hypoxia [13]. We further showed that autophagy can be induced during

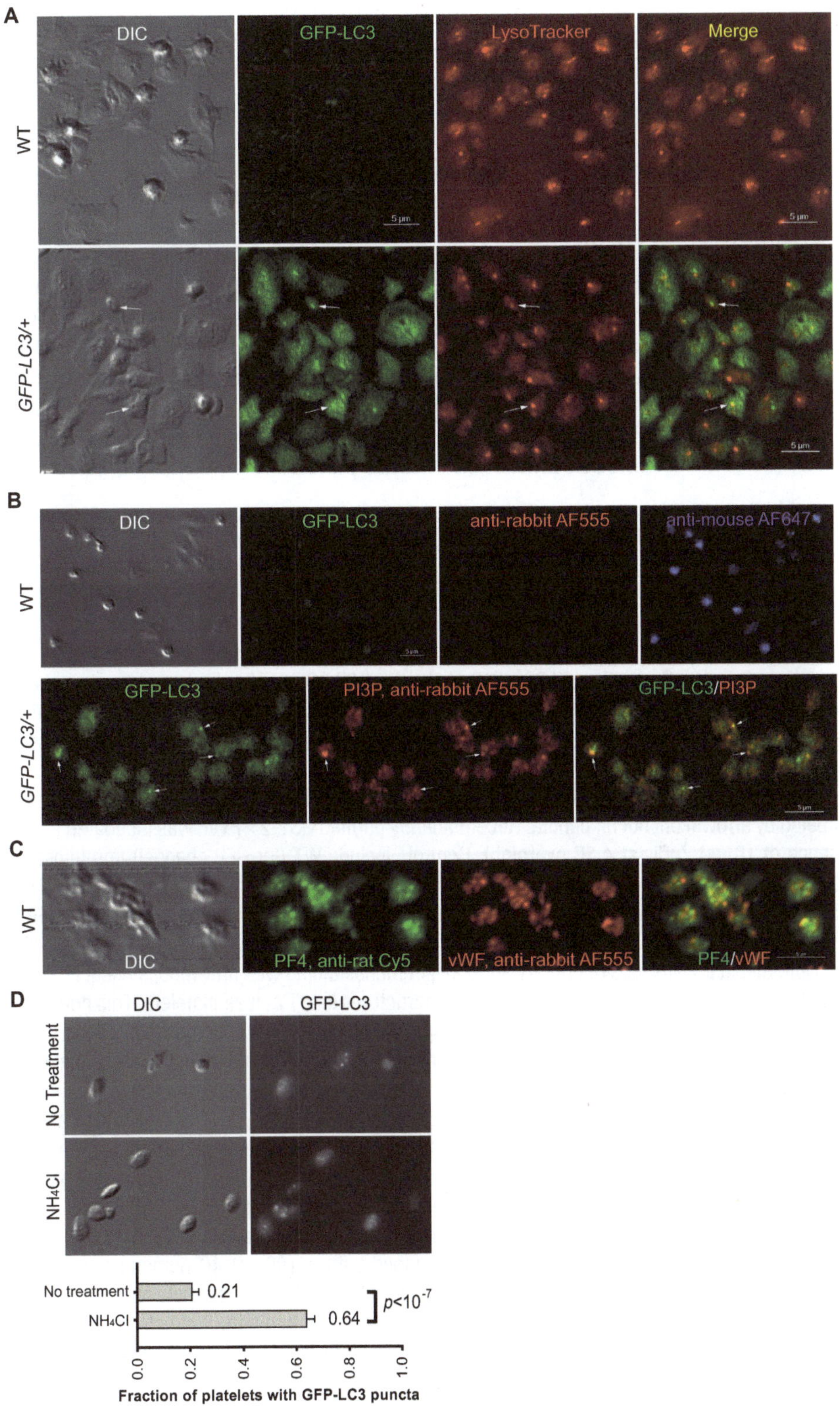

Fig. 1 Imaging platelet autophagy using light microscopy. (**a**) Confocal and DIC images of GFP-LC3 (GFP channel) in WT and *GFP-LC3/+* mouse platelets stained live with 50 nM LysoTracker® Red DND-99 (Thermo

platelet activation by hemostatic agonists [9], suggesting an involvement of autophagy in platelet functions. We and others further showed that genetic impairment of autophagy, including mitophagy, leads to defective platelet functions ex vivo and in vivo in *Atg7*- [9], *Atg5*- [13], *Becn1*- (Ouseph and Huang et al., manuscript in preparation), *Pik3c3/VPS34*- ([15, 16], and Ouseph and Huang et al., manuscript in preparation), and *FUNDC1*- [13] deficient mice.

Together, both general and selective autophagy (e.g., mitophagy) occur in platelets and are essential for cellular functions of platelets as well as hemostasis and thrombosis. The presence and importance of autophagy in platelets warrants further investigation since autophagy could be a viable target for therapeutic intervention of occlusive thrombotic diseases. Here, we describe some useful methods that are specifically tailored for studying autophagy in both human and mouse platelets.

Fig. 1 (continued) Fisher Scientific, L7528) (Cy3 channel) at 37 °C for 30 min. Images were collected on a Nikon A1R inverted confocal microscope. Arrows point to two puncta positive for both GFP-LC3 and LysoTracker. Scale bar: 5 μm. (**b**) Confocal and DIC images of fixed WT and *GFP-LC3/+* mouse platelets incubated with PI3P-binding GST-2xFYVE and then immunostained with GST antibody (Santa Cruz, sc-459). These images show punctate structures for both GFP-LC3 and PI3P and colocalization of GFP-LC3 and PI3P on some (labeled by arrows) but not all puncta. Recombinantly purified GST-2×FYVE was incubated at 1 μg/mL in the presence of 10 μM $ZnCl_2$ at 4 °C overnight. Controls include WT (for GFP channel) immunostained with anti-rabbit Alexa Fluor® 555 (Thermo Fisher Scientific, A31572) or anti-mouse Alexa Fluor® 647 (Thermo Fisher Scientific, A21235), but without primary antibody. Images were collected on a Nikon A1R inverted confocal microscope. Images were acquired and processed using the same settings for GFP-LC3 and WT samples. Note that in the absence of primary antibody, immunostaining with anti-mouse Alexa Fluor® 647, but not with anti-rabbit Alexa Fluor® 555, shows punctate structures in WT mouse platelets. This nonspecific anti-mouse Alexa Fluor® 647-staining is likely due to nonspecific binding of the anti-mouse secondary antibody to endogenous mouse immunoglobulins that are naturally packaged in the platelet α-granules. Scale bar: 5 μm. (**c**) Confocal and DIC images of WT and *GFP-LC3/+* mouse platelets immunostained with rabbit anti-vWF (Dako, A0082) and rat anti-PF4 (R&D Systems, DY595). Note that PF4 and vWF, both of which are α-granule cargos, appear not perfectly colocalized, suggesting presence of sub-domains within the α-granules. Images were collected on a Nikon A1R inverted confocal microscope. Scale bar: 5 μm. (**d**) Autophagic flux in mouse platelets as monitored by GFP-LC3 puncta in the presence and absence of NH_4Cl (20 mM, 2 h) using an upright wide-field microscope. Samples were visualized for DIC and GFP fluorescence (GFP-LC3). Images were acquired using a Nikon Eclipse E600 microscope equipped with a 100×/1.40 numeric aperture DIC H oil objective lens and a Zeiss AxioCam MR camera. Ten images (17 and 74 platelets/field for no treatment versus NH_4Cl treatment) were obtained at random and quantified for each condition. Platelets containing GFP-LC3-positive puncta are 21 ± 8% and 64 ± 10% without and with NH_4Cl treatment, respectively. Statistical significance was evaluated with the Student's *t* test. This panel is reproduced from Fig. 2c in [9] with publisher's permission

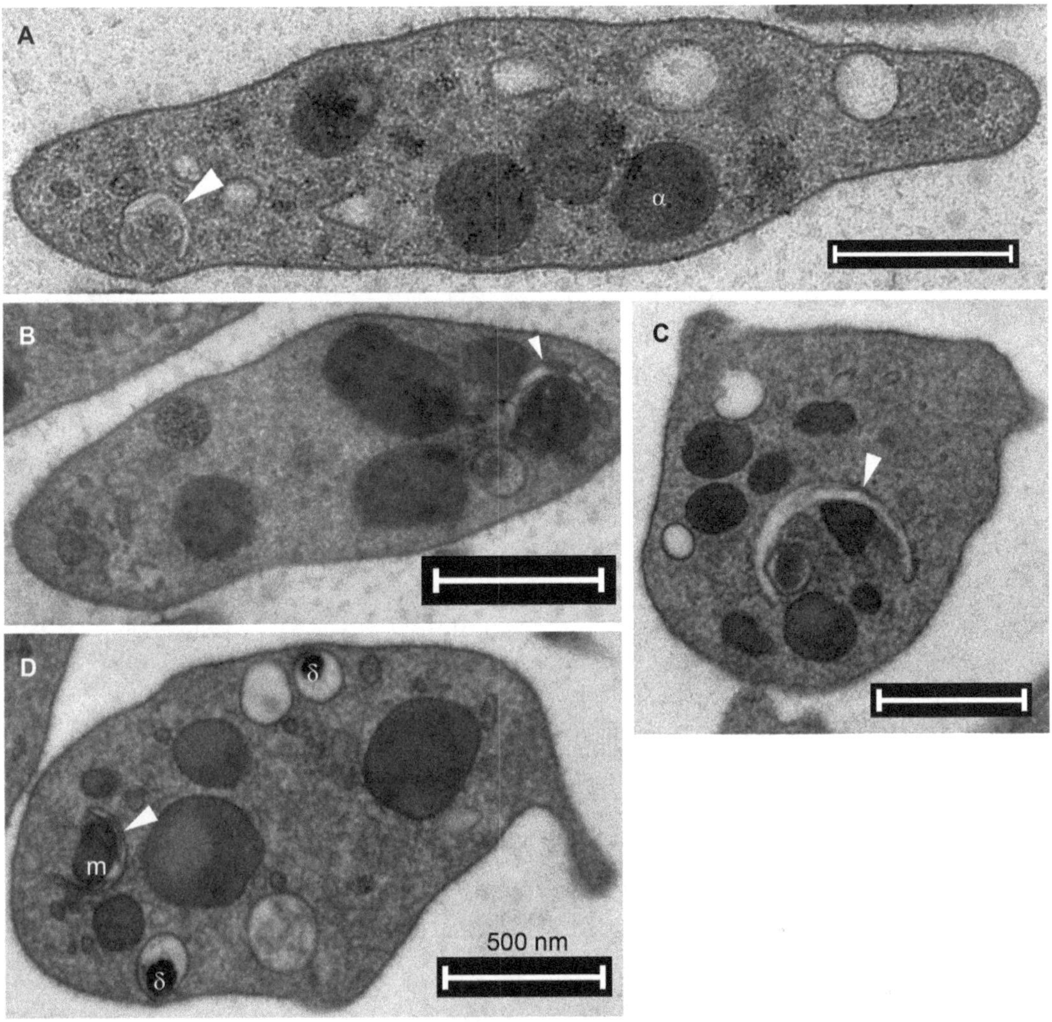

Fig. 2 Electron micrographs of autophagosome-related structures in mouse platelets. Double-membraned phagophore-like structures (arrow heads) wrapping bulk of cytosol and/or granules in (**a**, **b**) resting and (**c**, **d**) thrombin-stimulated (0.1 U/mL, 10–30 s) mouse platelets. Labels: m, mitochondria; α, α granules; δ, dense granules. Scale bars: 500 nm. (**a**, **b**) are reproduced from Ref. [9] with publisher's permission

2 Materials

2.1 Mouse and Human Platelet Preparation

1. Anticoagulant master mix: For each mouse, prepare 100 μL 3.8% (w/v) sodium citrate; add 1 μL apyrase (an ATP/ADP diphosphohydrolase) stock; add 1 μL prostacyclin (PGI_2) stock.
2. Apyrase stock: 200 U/mL in HEPES-Tyrode buffer (pH 7.4), stored at −20 °C until use.
3. Protacyclin PGI_2 stock: 10 μg/mL in 50 mM Tris/HCl pH 9.5, stored at −20 °C until use.

4. Acid citrate dextrose (ACD) collection tubes.
5. HEPES-Tyrode buffers: 20 mM HEPES, 128 mM NaCl, 2.8 mM KCl, 1 mM $MgCl_2$, 5 mM D-glucose, 12 mM $NaHCO_3$ and 0.4 mM NaH_2PO_4, titrated to pH 7.4 and pH 6.5 using KOH and HCl, respectively. Store HEPES-Tyrode buffers at 4 °C and pre-warm them to room temperature before use.
6. 1× phosphate buffer saline (PBS): containing 8 mM Na_2HPO_4, 2 mM KH_2PO_4, 137 mM NaCl, and 2.7 mM KCl, with pH adjusted to 7.4 by HCl.
7. 0.5 M EGTA.
8. 500 mM $CaCl_2$.
9. 1 mL Tuberculin slip tip syringe.
10. 26G 3 1/8 needle.
11. Euthanization chamber.
12. Compressed CO_2 gas tank.
13. 15 and 50 mL conical polypropylene or polyethylene tubes. Polystyrene or glass tubes will activate platelets and should be avoided.
14. Pasteur pipettes.
15. Particle Count and Size Analyzer, for example, Beckman Z2 Coulter Counter.

2.2 Visualizing Autophagic Structures in Platelets by Light Microscopy Using Autophagy Reporter Mice

1. Glass-bottom 35 mm culture dishes.
2. Glass microscope slide.
3. 1.5-mm-thick cover slips.
4. 0.1 mg/mL poly-D-Lysine.
5. 1× PBS.
6. 4% paraformaldehyde in 1× PBS. Make fresh.
7. Permeabilization buffer: 1× PBS containing 50 μg/mL digitonin, 2% BSA, and 2% goat serum.
8. Blocking buffer: 1× PBS containing 2% bovine serum albumin (BSA) and 2% goat serum.
9. HEPES-Tyrode buffer, pH 7.4.
10. Lysotracker® Red DND-99.
11. Primary antibodies.
12. Fluorescently labeled secondary antibodies.
13. Anti-fade mounting medium.
14. Clear nail polish.
15. Inverted confocal microscope and upright bright-field microscope.

2.3 Fixing Platelets for Electron Microscopy

1. 2× fixative stock A: 6% paraformaldehyde and 0.2% glutaraldehyde in 1× PBS. Make fresh.
2. ACD: 85 mM trisodium citrate, 83 mM dextrose, and 21 mM citric acid.
3. 1 mL Tuberculin slip tip syringe.
4. 26G 3 1/8 needle.
5. 5% glutaraldehyde in 1× PBS.
6. HEPES-Tyrode buffer, pH 7.4.
7. 2× fixative stock B: 2% paraformaldehyde and 5% glutaraldehyde in 1× PBS.
8. $CaCl_2$.
9. Liquid nitrogen.
10. Cryovials.
11. 2% agarose with ultralow gelling temperature in 1× PBS.
12. Acetone.
13. Acetone solution A: 2% OsO_4, 0.1% glutaraldehyde, and 1% H_2O, in acetone.
14. Acetone solution B: 1% OsO_4/1% H_2O, in acetone.
15. 1% tannic acid/1% H_2O in acetone.
16. EMBED-resin 812.
17. Uranyl acetate.
18. Reynold's lead citrate.
19. Ultramicrotome.
20. Electron microscope.
21. High-pressure freezer with rapid transfer system.
22. Freeze substitution and low-temperature embedding system.

3 Methods

Anucleate platelets do not transcribe nuclear DNA, despite being capable of some level of cell division (i.e., increased cell number without DNA replication) [17]. Thus, for studying platelets, freshly isolated primary cells are generally used for in vitro experiments. Since washed platelets lose optimal functionality over time, one should strive to use them within 2–3 h post-isolation. Moreover, as platelets can be easily activated and desensitized, caution is necessary to make good preparations for assays (*see* **Note 1**).

3.1 Preparing Washed Platelets from Fresh Mouse Blood

From one adult mouse, ideally 0.6–1 mL of whole blood can be drawn, and $0.6–1 \times 10^9$ washed platelets can be obtained. Factors affecting yield include mouse age and size, and importantly, whether platelets are activated during the preparation. All steps are performed at room temperature.

3.1.1 Blood Harvesting from Mice

1. Draw 100 μL of the anticoagulant master mix into a 1 mL Tuberculin slip tip syringe with a 26G 3 1/8 needle. In 1 mL of citrated blood, the final concentrations should be 0.38% sodium citrate, 0.2 U/mL apyrase, and 10 ng/mL PGI_2.
2. Fill the euthanization chamber with CO_2 from a compressed gas tank with continuous influx of CO_2 at a rate of 20% chamber volumes per min. Transfer one mouse at a time into the euthanization chamber. Mouse will stop breathing within ~60 s. Ensure that the mouse is not responsive by toe or tail pinching. Immediately harvest blood by cardiac puncture as described below. Minimal delay between losing toe/tail pinching response and blood harvesting is important for getting good quality platelets.
3. Pin the limbs loosely on a foam board. Cut through abdominal skin, muscle and diaphragm sequentially to expose the heart. Pin the xiphoid process loosely on the foam board.
4. Stick the 26G 3 1/8 needle into the exposed ventricle in the direction along the longitudinal axis of the heart and slowly draw blood. Blood drawing should be with little resistance while avoiding bubbles. To retain platelets in their resting state, complete the blood harvesting in one draw and do not re-stick the heart.
5. Take off the needle, gently expel the harvested blood into a 1.5 mL Eppendorf tube and mix.

3.1.2 Preparing Platelet-Rich Plasma (PRP) and Washed Platelets from Mouse Blood

1. (Optional) Pool blood from more than one mouse with the same genotype into 15 or 50 mL conical polypropylene or polyethylene tubes.
2. Dilute the blood 1:1 (v/v) with 1× PBS or HEPES-Tyrode buffer (pH 6.5). Supplement the diluted blood with 0.2 U/mL apyrase and 10 ng/mL PGI_2 and incubate for 5 min to allow the platelet antagonists to take effect.
3. Balance the tubes and centrifuge at $215 \times g$ in a swinging-bucket rotor (e.g., Thermo Sorvall LEGEND RT Centrifuge, with acceleration and deceleration set as 5 and 2, respectively) for 5–10 min (depending on the volumes of the samples, e.g., 6 min for 3 mL diluted blood). With an optimal centrifugation time, red blood cells and leukocytes will be in the bottom layers, and platelet-rich plasma (PRP) will be slightly yellowish platelet suspension in the top layer.

4. Without agitating the "buffy coat" interface layer of leukocytes, transfer the top PRP layer to new 15 or 50 mL tubes, starting from the top using large disposable transfer pipets followed by 200 μL pipet tips with tip cut.
5. (Optional) After adding 0.2 U/mL apyrase and 10 ng/mL PGI_2 to the PRP layer collected in the new tubes, balance the new tubes and centrifuge again at 215 × *g* for 5–10 min to remove residual red blood cells and leukocytes. Transfer the supernatants to new 15 or 50 mL tubes using large disposable transfer pipets.
6. After adding 0.2 U/mL apyrase and 10 ng/mL PGI_2 to the collected PRP, balance the new tubes and centrifuge at 675 × *g* for 7–10 min to pellet the platelets.
7. Gently resuspend the platelet pellets in 1–2 mL HEPES-Tyrode buffer (pH 6.5) supplemented with 1 mM EGTA, 0.2 U/mL apyrase, and 10 ng/mL PGI_2, using large disposable transfer pipets.
8. Balance the tubes and centrifuge again at 675 × *g* for 7 min to wash the platelets.
9. Gently resuspend the platelet pellets in 1–2 mL HEPES-Tyrode buffer (pH 7.4), using large disposable transfer pipets.
10. Measure platelet concentrations using an appropriate Particle Count and Size Analyzer, following manufacturer's protocols.
11. Adjust platelet concentration as needed using HEPES-Tyrode buffer (pH 7.4).
12. Incubate washed platelets at room temperature or 37 °C for 30 min to dissipate the residual effects of PGI_2 before assaying platelet functions. PGI_2 is quickly degraded in aqueous buffers, with a $t_{1/2}$ of ~3 min at 37 °C/pH 7.4 [18].
13. At the end of incubation and right before experiments, add $CaCl_2$ to a final concentration of 1 mM.
14. (Optional) A quick lumi-aggregometry measurement can be run to monitor the quality of the platelet preparations. This method can be found in [19].

3.2 Preparing Washed Platelets from Banked Human PRP and Fresh Human Blood

Human platelets can be prepared from 1 unit of human PRP (40–50 mL) purchased from a local blood bank. Usually 2–6 × 10^9 platelets can be isolated from 100 mL of banked human PRP. Alternatively, fresh whole blood can be drawn from healthy volunteers, with appropriate local approvals. All steps are performed at room temperature.

3.2.1 Preparing Washed Platelets from Banked Human PRP

1. Divide banked human PRP into several 50 mL conical polypropylene or polyethylene centrifuge tubes (15–20 mL per tube). Supplement each tube with apyrase and PGI_2 at the

final concentrations of 0.2 U/mL and 10 ng/mL, respectively. Incubate samples for 5 min to allow the platelet antagonists to take effect.

2. Balance the tubes and centrifuge at 155 × *g* in a swinging-bucket rotor for 13 min to remove residual red blood cells and leukocytes. Transfer the PRP to new 50 ml tubes using large disposable transfer pipets.
3. After adding 0.2 U/mL apyrase and 10 ng/mL PGI_2 to the PRP, balance the new tubes and centrifuge at 750 × *g* for 13 min to pellet the platelets.
4. Gently resuspend the platelet pellets in 1–2 mL HEPES-Tyrode buffer (pH 6.5) containing 1 mM EGTA, 0.2 U/mL apyrase, and 10 ng/mL PGI_2.
5. Balance the tubes and centrifuge again at 750 × *g* for 13 min to wash the platelets.
6. Gently resuspend the platelet pellets in 1–2 mL HEPES-Tyrode buffer (pH 7.4).
7. Measure platelet concentrations using a Particle Count and Size Analyzer.
8. Adjust platelet concentration as needed using HEPES-Tyrode buffer (pH 7.4).
9. Incubate washed platelets at room temperature or 37 °C for 30 min before use to ensure that residual effects of PGI_2 are dissipated.
10. At the end of incubation and right before experiments, add $CaCl_2$ to a final concentration of 1 mM.
11. (Optional) A quick lumi-aggregometry measurement can be run to monitor the quality of the platelet preparations. This method can be found in [19].

3.2.2 Preparing Washed Platelets from Fresh Human Blood

1. Draw fresh whole blood from healthy volunteers into ACD collection tubes.
2. Divide whole blood into several 50 mL conical polypropylene or polyethylene tubes (15–20 mL per tube). Dilute each tube in 1:1 (v/v) ratio with 1× PBS or HEPES-Tyrode buffer (pH 6.5). Supplement diluted blood with 0.2 U/mL apyrase and 10 ng/mL PGI_2. Incubate the samples at room temperature for 5 min to allow the platelet antagonists to take effect.
3. Balance the tubes and centrifuge at 215 × *g* in a swinging-bucket rotor for 20 min.
4. Transfer the top layers (i.e., PRP) to new 50 mL tubes using large disposable transfer pipets.
5. (Optional) After adding 0.2 U/mL apyrase and 10 ng/mL PGI_2 to the PRP, balance the new tubes and centrifuge again

at 155 × *g* for 10 min to remove residual red blood cells and leukocytes. Transfer the new PRP fractions to new 50 mL tubes using large disposable transfer pipets.

6. After adding 0.2 U/mL apyrase and 10 ng/mL PGI_2 to the new PRP fractions, balance the new tubes and centrifuge at 750 × *g* for 13 min to pellet the platelets.
7. Gently resuspend the platelet pellets in 1–2 mL HEPES-Tyrode buffer (pH 6.5) supplemented with 1 mM EGTA, 0.2 U/mL apyrase, and 10 ng/mL PGI_2.
8. Balance the tubes and centrifuge again at 750 × *g* for 13 min to wash the platelets.
9. Gently resuspend the platelet pellets in 1–2 mL HEPES-Tyrode buffer (pH 7.4).
10. Measure platelet concentrations using a Particle Count and Size Analyzer.
11. Adjust platelet concentration as needed using HEPES-Tyrode buffer (pH 7.4).
12. Incubate washed platelets at room temperature or 37 °C for 30 min before use to ensure that residual effects of PGI_2 are dissipated.
13. At the end of incubation and right before experiments, add $CaCl_2$ to a final concentration of 1 mM.
14. (Optional) A quick lumi-aggregometry measurement can be run to monitor the quality of the platelet preparations. This method can be found in [19].

3.3 Visualizing Autophagic Structures in Platelets by Light Microscopy Using Autophagy Reporter Mice

GFP-LC3/+ transgenic mice (made by Dr. Noboru Mizushima, University of Tokyo, Japan) express *GFP-LC3* transgene under an actin promoter [10, 11]. This mouse strain is currently available through Riken BioResource Center. Both *Becn1-EGFP/+* [12] and *EGFP-Atg5/+* [9] strains were made using bacterial artificial chromosome (BAC) transgenics [20] to ensure that tagged genes are expressed under the endogenous *Becn1* and *Atg5* promoters and regulatory elements (*see* **Note 2**).

3.3.1 Visualizing Live Platelets Using Autophagy Reporter Mice

Imaging Live Platelets Stained with LysoTracker Using an Inverted Confocal Microscope

1. Prepare washed platelets from the autophagy reporter mice (e.g., *GFP-LC3/+*, *Becn1-EGFP/+*, *EGFP-Atg5/+*) and their wild type (WT) littermates.
2. Coat glass-bottom 35 mm culture dishes with 0.1 mg/mL poly-D-lysine at 4 °C overnight (*see* **Note 3**); rinse dishes with 1 mL HEPES-Tyrode buffer (pH 7.4) twice.
3. Add 100 μL of 1 × 10^9/mL platelet suspension to the center of the poly-D-lysine-coated glass-bottom dishes; wait for at least 15 min for platelets to attach; aspirate unattached cells.

4. Add 1 mL HEPES-Tyrode buffer containing 50 nM LysoTracker® Red DND-99 and incubate at 37 °C for 30 min.
5. Aspirate LysoTracker-containing buffer, and wash with 1 mL HEPES-Tyrode buffer (pH 7.4) two times; after the second wash, cover the platelets with 1 mL HEPES-Tyrode buffer for imaging.
6. Immediately image fluorescently labeled autophagy protein (e.g., GFP-LC3), LysoTracker® Red DND-99, and differential interference contrast (DIC) using an inverted confocal microscope (*see* Fig. 1a for representative images).

Imaging Live Platelets Stained with LysoTracker Using an Upright Bright-Field Microscope

1. Prepare washed platelets from either WT or the autophagy reporter mice.
2. For LysoTracker staining, incubate resting washed platelets (100 μL of 1×10^9/mL) with LysoTracker® Red DND-99 (final concentration 50 nM) in a 1.5 mL Eppendorf tube at 37 °C for 30 min.
3. Place 6 μL of platelets onto a glass microscope slide.
4. Overlay with a 1.5-mm-thick cover slip.
5. Allow platelets to settle for 5–15 min.
6. Seal the cover slip to the slide with clear nail polish.
7. Using an upright microscope, for example, a Nikon Eclipse E600 microscope equipped with a 100×/1.40 numeric aperture DIC H oil objective lens and a Zeiss AxioCam MR camera, image DIC, fluorescently labeled autophagy protein (e.g., GFP-LC3) (GFP channel), and LysoTracker® Red DND-99 (Cy3 channel) sequentially with appropriate exposure times. Representative images for live imaging of *GFP-LC3/+* platelets using an upright bright-field microscope are shown in [9] (reproduced in Fig. 1d).

Monitoring Autophagy Flux in Platelets Using GFP-LC3 Imaging

1. Prepare washed platelets from the *GFP-LC3/+* autophagy reporter mice and their WT littermates.
2. Pretreat unstimulated GFP-LC3 platelets either with or without the lysosome inhibitor, NH_4Cl (20 mM, 2 h) (*see* **Note 4**).
3. Acquire DIC and GFP fluorescence images of live platelets following either of the protocols described in Subheadings 3.3.1.1 or 3.3.1.2.
4. Count the number of GFP-LC3 puncta in each platelet in each field.
5. Quantify fractions of platelets with different numbers of GFP-LC3 puncta to create a distribution histogram.
6. Average the results from multiple fields for each condition. Representative images and quantification are shown in [9] (reproduced in Fig. 1d).

3.3.2 Visualizing Fixed Platelets by ICC/IF and Confocal Microscopy

1. Prepare washed platelets from either WT or the autophagy reporter mice.
2. Coat glass-bottom 35 mm culture dishes with 0.1 mg/mL poly-D-lysine at 4 °C overnight (*see* **Note 3**); rinse dishes with 1 mL HEPES-Tyrode buffer (pH 7.4) twice.
3. Add 100 μL of 1×10^9/mL platelet suspension to the center of the glass-bottom dishes; incubate at 37 °C for 30 min; aspirate unattached cells.
4. Fix with 500 μL 4% paraformaldehyde for 15 min at room temperature.
5. Wash with 200 μL 1× PBS three times.
6. Incubate cells in 200 μL permeabilization buffer at room temperature for 10–15 min.
7. Block with 200 μL blocking buffer at room temperature for 30 min.
8. Incubate cells with 150 μL primary antibodies in blocking buffer at 4 °C overnight (*see* **Note 5**).
9. Wash with 500 μL 1× PBS, three times.
10. Incubate with 150 μL secondary antibodies in blocking buffer at room temperature for 40 min to 1 h (*see* **Note 5**).
11. Wash four times, each time with 500 μL 1× PBS. During the last wash, wait for 10 min.
12. After the last wash, aspirate liquid, add one drop of anti-fade mounting medium (warmed to room temperature before use) to the center of each glass-bottom dish, wait for 1–2 min until the mounting medium has spread throughout the glass bottom, and slowly place a 1.5 mm cover glass on top of the mounting medium.
13. Cure for 24 h and seal all edges with a clear nail polish.
14. Image using an inverted confocal microscope (*see* Fig. 1b, c for representative images).

3.4 Visualizing Autophagic Structures in Platelets by Electron Microscopy

3.4.1 Fixation of Fresh Mouse Resting Platelets

1. For each mouse, prefill a 1 mL Tuberculin slip tip syringe with 85 μL of ACD supplemented with 0.2 U/mL apyrase and 10 ng/mL PGI_2 and 500 μL 2× fixative stock A.
2. Draw blood by cardiac puncture as described above in Subheading 3.1.1 into the fixative-prefilled syringe. Take off the needle, gently expel the harvested blood into a 1.5 mL Eppendorf tube and mix.
3. Fix for 20 min at room temperature.
4. Pool the fixed blood from the animals with the same genotype; and prepare washed platelets as described in Subheading 3.1.2.

5. Resuspend the platelet pellet in HEPES-Tyrode buffer (pH 7.4), and fix a second time by adding equal volume of 5% glutaraldehyde (final 2.5% glutaraldehyde) at room temperature for 20 min.

3.4.2 Fixation of Washed Mouse or Human Platelets

1. For analysis of agonist/inhibitor effects on platelets by electron microscopy, prepare washed platelets (2.5×10^8/mL) as described in Subheadings 3.1 and 3.2.
2. Supplement platelet suspension with 0.7 mM $CaCl_2$ (final concentration) and incubate at room temperature for 5–30 min before stimulation with desired agonist/inhibitor.
3. Fix the platelet suspension with equal volume of 2× fixative stock B (final concentration 1% paraformaldehyde and 2.5% glutaraldehyde) at room temperature for 30 min.
4. Centrifuge at 1000 × *g* and resuspend platelets in 1× PBS and wash three times to remove fixative.

3.4.3 High-Pressure Freezing, Freeze Substitution Dehydration, and Low-Temperature Embedding

This protocol is based on our published work [21] (*see* **Note 6**).

1. Resuspend platelets in 2% agarose with ultralow gelling temperature in 1× PBS.
2. Freeze cryoprotected platelets under high pressure (2100 bar), for instance, in a Leica Microsystems EM PACT2 high-pressure freezer with rapid transfer system.
3. Transfer the frozen samples under liquid nitrogen to cryovials containing acetone solution A (2% OsO_4, 0.1% glutaraldehyde and 1% H_2O in acetone).
4. Carry out freeze substitution dehydration, for instance, in the Leica Microsystems EM AFS2 freeze substitution and low-temperature embedding system, following the schedule below: −90 °C for 22 h, −60 °C for 8 h, −30 °C for 8 h, and 0 °C, with transitions of 3 °C/h between each temperature change.
5. Rinse samples with acetone three times.
6. Incubate samples in 1% tannic acid/1% H_2O in acetone at 4 °C for 1 h.
7. Rinse samples with acetone three times.
8. Incubate samples in acetone solution B (1% OsO_4/1% H_2O in acetone) at 4 °C for 1 h.
9. Rinse samples with acetone three times.
10. Embed samples in EMBED-812 resin starting at a 25% concentration and proceeding through 50%, 76%, and 100% resin steps.

3.4.4 Transmission Electron Microscopy (TEM)

1. Prepare ultrathin (50 nm) sections using an ultramicrotome.
2. Post-stain sections with uranyl acetate and Reynold's lead citrate.
3. Acquire micrographs using, for instance, a FEI Tecnai TF20 FEG transmission electron microscope at 80 keV.

4 Notes

1. Platelets can be easily activated during isolation. Therefore, caution must be taken to minimize unintended activation while handling platelets, particularly during blood harvesting and platelet preparation. For instance, the one-stick policy during cardiac puncture should be emphasized to avoid platelet activation by repeated piercing-induced vessel injury; polypropylene or polyethylene tubes, but not polystyrene or glass tubes, should be used to avoid contact activation. Whenever possible, centrifugation steps should be performed at low centrifugal forces to minimize shearing which activates platelets; centrifugation tubes containing platelets should be balanced within 0.1 g, and acceleration and deceleration should be set at low settings to minimize vibrations that may activate platelets during acceleration and deceleration. Pipetting should be gentle and minimal; either pipet tips (1 mL or 200 μL) with the tips cut or large disposable transfer pipets should be used to minimize shearing during pipetting. Additionally, shaking of the sample tubes should be avoided at any time for handling blood and live platelets.
2. Mice either expressing autophagy reporter (*GFP-LC3/+*, *Becn1-EGFP/+*, *EGFP-Atg5/+*) or with genetically deficient autophagy genes have been utilized to study autophagy in platelets [8, 9, 13, 22]. *GFP-LC3/+* mice have been extremely useful in imaging autophagy in platelets using both traditional wide-field/confocal microscopes [9] and super-resolution microscopes (Fig. 3, and Banerjee et al., manuscript in preparation). In *Becn1-EGFP/+* and *EGFP-Atg5/+* BAC transgenic mice, Beclin 1-EGFP and EGFP-ATG5 are expressed at low levels (thus giving out low GFP fluorescent signals), therefore limiting the usage of these autophagy reporter mice in live imaging of platelet autophagy (Ouseph and Huang et al., unpublished results). However, *Becn1-EGFP/+* and *EGFP-Atg5/+* mice are still valuable for immunofluorescence imaging or immuno-electron microscopic studies with anti-GFP antibodies. Of note, endogenous *Becn1* and *Atg5* genes are still expressed in *Becn1-EGFP/+* and *EGFP-Atg5/+* BAC transgenic mice.

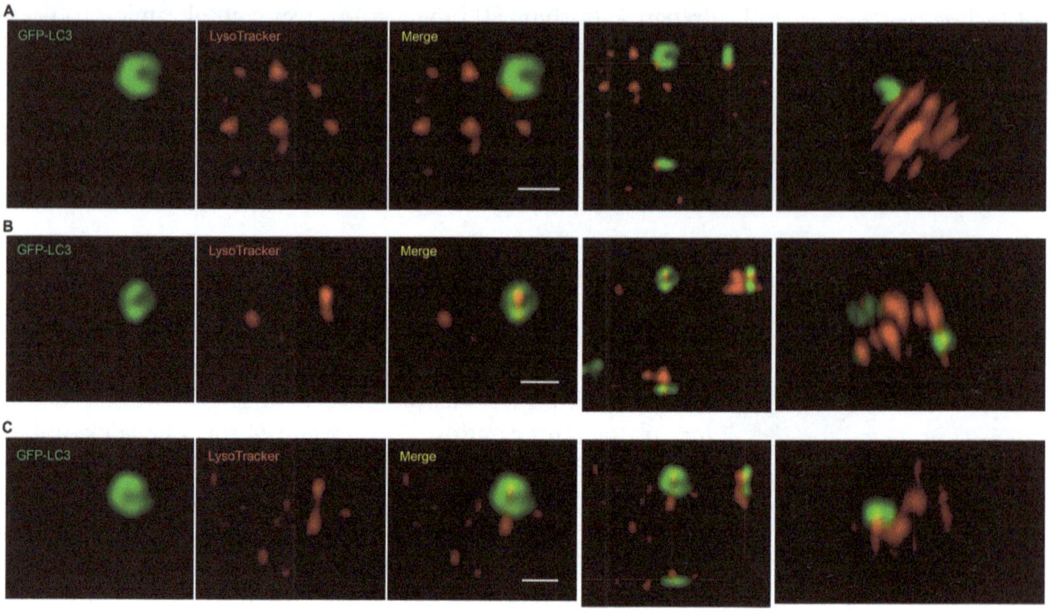

Fig. 3 Super-resolution microscopy of platelet autophagy. 3D-SIM images of GFP-LC3 (GFP channel) and live-stained LysoTracker Blue (405 channel, pseudo-color in red) in *GFP-LC3/+* mouse platelets show incomplete ring-like GFP-LC3 structures resembling isolation membranes/phagophores as observed by electron microscopy in Fig. 2. These 3D-SIM images show an interesting spatial relationship between the GFP-LC3-positive and LysoTracker-positive structures in platelets: The LysoTracker-positive structures appear to be associated with the incomplete, ring-shaped GFP-LC3-positive structures, i.e., the LysoTracker-stained, tubular structures are either on the outside (**a**) or inside (**b**, **c**) of the GFP-LC3-positive rings. Images were collected on a Nikon Ti-E N-STORM/N-SIM super-resolution microscope that was equipped with an Apo SR 100X/1.49 NA oil objective and an A1R camera. Images were processed using Nikon NIS-Elements v3.2 N-SIM/STORM software and Adobe Photoshop. Scale bars: 1 μm

Mice with their autophagy genes deleted should be used cautiously. First, as normal hemostasis relies on coordination of several cell types, e.g., platelets and endothelial cells, bleeding diatheses observed in whole body autophagy-deficient animals (e.g., *Becn1*$^{+/-}$ [8]) may result, in part, from deficiency in the endothelium. Second, platelets are originated from megakaryocytes, thus autophagy deficiency in megakaryocytes may lead to defects in megakaryopoiesis and thrombopoiesis, in addition to platelet functions.

3. Alternative to poly-D-lysine, fibrinogen coating (incubating with 50 μg/mL fibrinogen at room temperature for 2 h) can be used to facilitate platelet adhesion. However, to avoid any form of platelet activation, coating with poly-D-lysine is recommended.

4. Caution is needed when using common autophagy inducers and inhibitors to study platelets, as these autophagy inducers and inhibitors may have additional targets in platelets besides the autophagy pathway. For example, as platelet secretion from

α-granules, dense granules, and lysosomes is essential for platelet activation and hemostasis, it is possible that lysosomal inhibitors, such as NH_4Cl, chloroquine, and bafilomycin A1, may alter lysosomal exocytosis in addition to autophagy.

5. Nonspecific binding of primary antibodies on platelets usually has rarely been determined for commercial antibodies against autophagy proteins and must be tested before use. Moreover, platelets stained with secondary antibodies but without primary antibodies are needed as controls to monitor nonspecific binding of secondary antibodies. In particular, anti-mouse secondary antibodies often nonspecifically stain α-granules in mouse platelets (Fig. 2b), as immunoglobulins are α-granule cargos. In contrast, anti-rabbit, anti-rat, or anti-goat secondary antibodies have often been cross-adsorbed against human and mouse sera and thus do not stain human or mouse platelets nonspecifically.
6. The high-pressure freezing followed by freeze substitution is used to give a very gentle dehydration procedure that limits the formation of electron-dense, α-granule nucleoids, an artifact of dehydration [23]. High-pressure freezing is less practical for complex mixtures of cells as in freshly drawn blood or PRP. High-pressure freezing could also be used as a rapid "fixation" technique to terminate in vitro experiments with highly purified platelets.

Acknowledgments

The authors thank the laboratory personnel and collaborators who conducted the research on platelet autophagy over the years. The authors thank Dr. Zhenyu Li for helpful discussion. The authors also thank Dr. Harry Chanzu and Laura Tichachek for their careful perusal of this manuscript. This work was supported by a New Scholar in Aging award from the Ellison Medical Foundation (to Q.J.W.), Grant-in-Aid awards from the American Heart Association (AHA16GRNT31310020 to Q.J.W. and AHA16GRNT27620001 to S.W.W.), Predoctoral Fellowships from the American Heart Association (AHA 15PRE25550020 to S.J. and AHA 11PRE7500051 to Y.H.), National Institutes of Health (HL56652 and HL138179 to S.W.W., HL119393 to B. S.), and a Veterans Affairs Merit Award (to S.W.W.).

References

1. Pease DC (1956) An electron microscopic study of red bone marrow. Blood 11 (6):501–526
2. Junt T et al (2007) Dynamic visualization of thrombopoiesis within bone marrow. Science 317(5845):1767–1770
3. Machlus KR, Italiano JE Jr (2013) The incredible journey: from megakaryocyte development to platelet formation. J Cell Biol 201 (6):785–796
4. Harker LA (1977) The kinetics of platelet production and destruction in man. Clin Haematol 6(3):671–693
5. Ault KA, Knowles C (1995) In vivo biotinylation demonstrates that reticulated platelets are the youngest platelets in circulation. Exp Hematol 23(9):996–1001
6. Thompson CB et al (1984) Platelet size and age determine platelet function independently. Blood 63(6):1372–1375
7. Grozovsky R, Hoffmeister KM, Falet H (2010) Novel clearance mechanisms of platelets. Curr Opin Hematol 17(6):585–589
8. Feng W et al (2014) Dissection of autophagy in human platelets. Autophagy 10(4):642–651
9. Ouseph MM et al (2015) Autophagy is induced upon platelet activation and is essential for hemostasis and thrombosis. Blood 126 (10):1224–1233
10. Kuma A, Mizushima N (2008) Chromosomal mapping of the GFP-LC3 transgene in GFP-LC3 mice. Autophagy 4(1):61–62
11. Mizushima N et al (2004) In vivo analysis of autophagy in response to nutrient starvation using transgenic mice expressing a fluorescent autophagosome marker. Mol Biol Cell 15 (3):1101–1111
12. Arsov I et al (2008) BAC-mediated transgenic expression of fluorescent autophagic protein Beclin 1 reveals a role for Beclin 1 in lymphocyte development. Cell Death Differ 15(9):1385–1395
13. Zhang W et al (2016) Hypoxic mitophagy regulates mitochondrial quality and platelet activation and determines severity of I/R heart injury. eLife 5:e21407
14. Pieczarka EM et al (2014) Platelet vacuoles in a dog with severe nonregenerative anemia: evidence of platelet autophagy. Vet Clin Pathol 43 (3):326–329
15. Valet C et al (2017) A dual role for the class III PI3K, Vps34, in platelet production and thrombus growth. Blood 130:2032
16. Liu Y et al (2017) Class III PI3K positively regulates platelet activation and thrombosis via PI(3)P-directed function of NADPH oxidase. Arterioscler Thromb Vasc Biol 37:2075
17. Schwertz H et al (2010) Anucleate platelets generate progeny. Blood 115(18):3801–3809
18. Moore PK (1982) Prostaglandins, prostacyclin and thromboxanes. Biochem Educ 10 (3):82–87
19. Ren QS et al (2007) Endobrevin/VAMP-8 is the primary v-SNARE for the platelet release reaction. Mol Biol Cell 18(1):24–33
20. Heintz N (2001) BAC to the future: the use of bac transgenic mice for neuroscience research. Nat Rev Neurosci 2(12):861–870
21. Pokrovskaya ID et al (2016) STEM tomography reveals that the canalicular system and alpha-granules remain separate compartments during early secretion stages in blood platelets. J Thromb Haemost 14(3):572–584
22. Cao Y et al (2015) Autophagy regulates the cell cycle of murine HSPCs in a nutrient-dependent manner. Exp Hematol 43 (3):229–242
23. Yadav S, Storrie B (2017) The cellular basis of platelet secretion: emerging structure/function relationships. Platelets 28(2):108–118

Chapter 33

Methods to Image Macroautophagy in the Brain In Vivo

Xigui Chen, Kanoh Kondo, and Hitoshi Okazawa

Abstract

Macroautophagy is the process to remove intracellular organelles or proteins by using autophagosome that is composed of autophagy proteins such as atg3, atg7, and atg8/LC3 (Mizushima, et al. Annu Rev Cell Dev Biol. 27:107–132, 2011). Here, we develop a useful method for in vivo imaging of autophagosome under the two-photon microscopy. Time-lapse imaging of LC3-ECFP enables us to quantify the dynamics of number, size, and signal intensity of autophagosomes in neurons or in other types of cells in the brain.

Key words Macroautophagy, Autophagosome, Two-photon microscopy

1 Introduction

Autophagy, especially macroautophagy, has been implicated in the clearance of abnormal proteins or abnormal organelles in neurons or glia under neurodegenerative diseases, while the exact role of macroautophagy in each disease remains largely obscure. LC3-EGFP has been used for imaging or pathological examination of autophagosomes in culture cells or transgenic mice [1]. To observe macroautophagy directly in the brain of living animals, we developed an in vivo imaging technique to visualize autophagosomes by expressing LC3-EGFP in neurons with lentiviral vector. This method would be useful for observation of various physiological and pathological conditions including Alzheimer's disease pathology [2].

Nicholas Ktistakis and Oliver Florey (eds.), *Autophagy: Methods and Protocols*, Methods in Molecular Biology, vol. 1880, https://doi.org/10.1007/978-1-4939-8873-0_33,

2 Materials

2.1 Lentivirus Preparation

1. For construction of pEGFP-LC3/pLVSIN-CAG, NheI-BamHI fragment was digested from pEGFP-LC325 and subcloned into pLVSIN-CMV. CMV promoter was replaced by CAG promoter with ClaI-XhoI.
2. Lipofectamine 2000 transfection reagent.
3. HEK293T cells.

2.2 Lentivirus Injection

1. Make capillary injection through capillary puller under the following conditions: heat at 741, pull power at 150, velocity at 75, delay at 1000, and pressure at 400.
2. Sharpen the capillary using a micromotor grinder.
3. Micromotor drill and bar.
4. Mice stereotaxic instruments.
5. Stereotaxic manipulator.
6. Stereo microscopy.
7. Povidone-iodine.
8. Microinjector.
9. Safety cabinet.
10. Razor blades.
11. Antibiotic.

2.3 Observation

1. Super glue.
2. Stereo microscopy.
3. Mice stereotaxic instruments.
4. Headplate (order-made, Fig. 1).
5. Observation of mice stage.

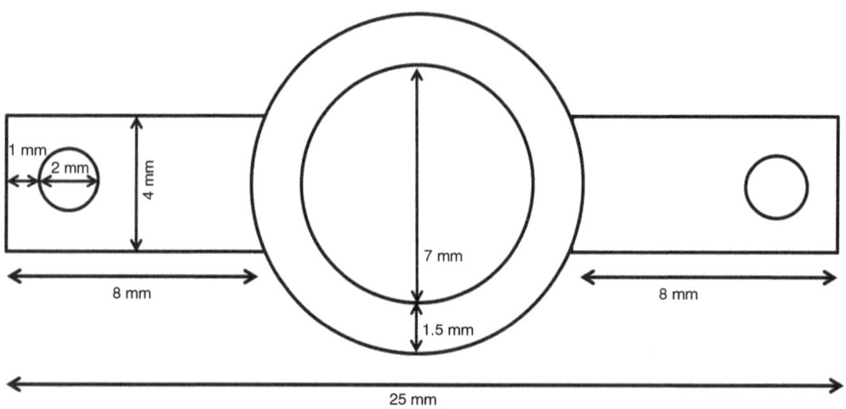

Fig. 1 The map of mouse headplate for in vivo observation

6. Shaving blades.
7. Isoflurane and anesthesia machine.
8. Antibiotic.
9. Eye ointment.
10. Two-photon microscopy.
11. Animal temperature controller.
12. Headplate.
13. Stage of observation.

3 Methods

3.1 Lentiviral Preparation

1. Using Lenti-XTM Expression Systems (Clontech Laboratories, Inc.), 6 μg of pEGFP-LC3/pLVSIN-CAG in Lenti-X HTX Packaging Mix (components of Lenti-X HTX Packaging Mix) was transfected into 293T cells (plated at 4×10^6 cells 16 h before) using Lipofectamine® 2000 following the manufacturer's instruction. 48 h after transfection, the medium of the transfected cells was collected and centrifugated at 8000 G for 5 min at 4 °C.
2. The supernatant was ultracentrifuged at $48000 \times g$ for 2 h at 4 °C. The pellet was suspended in PBS and centrifuged overnight at 8000 G and 4 °C. The pellet was suspended in 160 μL of PBS by pipetting 100 times, kept at 4 °C for 1 h, pipetted 100 times, kept at 4 °C for 1 h, and centrifuged at 8000 G for 5 min at 4 °C, and the final supernatant was kept at −80 °C until the experiment.
3. Viral genomes were quantified using a qPCR cycler (ABI 7500HT, Applied Biosystems) and Lenti-X™ qRT-PCR Titration Kit (Clontech Laboratories). Samples were diluted to an appropriate concentration. For the preparation of PCR sample, lentivirus was incubated with DNase I (Lenti-X™ qRT-PCR Titration Kit component). Lentiviral genomic RNA was purified and reverse-transcribed. Viral DNA was diluted to an appropriate concentration.

3.2 Injection of Lentivirus into the Mouse Brain

1. Prepare the mouse at suitable age. Firstly, anesthetize the mouse using 5% isoflurane, and then inject intraperitoneally 10–25 mg/kg body weight of pentobarbital and antibiotic (the volume should follow the antibiotic information), place the ear bar, and set the mice on stereotaxic instrument supplied with 0.5–1% isoflurane (*see* **Note 1**).
2. Cover the mouse eye with eye ointment, shave the hair of the parietal head using a sharp blade, and clean the skin with 70% ethanol and 1% povidone-iodine. Make a sagittal cut in the

parietal skin, expose the skull, and remove the connective tissue and the periosteum of the skull gently with cotton swabs. Finally, clean the skull with 70% ethanol again.

3. Adjust the stereotaxic instrument to keep bregma and lambda at the same horizontal level.
4. Following the mouse atlas, measure and mark the position for injection. Then, open a small hole using micromotor drill (*see* **Note 2**).
5. In a safety cabinet, fill the capillary with LC3-EGFP lentivirus (titer: higher than 5.0×10^6 vector genomes/mL, 0.5–5 μL). Inject the lentivirus at the selected position and suitable depth, at a speed of 1 μL/5 min. Hold the capillary for 5 min after injecting.
6. After the injection, apply instant adhesive block in the injected hole, and then sew the skin using surgical thread. After operation, release the mice from stereotaxic instruments into the home cage.

3.3 Observation of Autophagy In Vivo with Two-Photon Microscopy

1. Three weeks after virus injection, anesthetize the mice again, the same procedure described at Subheading 3.1, **step 1**.
2. Cover the mice eye with ointment, and shave the hair of parietal head again, the same procedure described at Subheading 3.1, **step 2**.
3. Confirm the position of the headplate in order to keep the injection point in the center of the plate hole, and then apply super glue on the skull to fix the headplate solidly [3]. Then, wait for the super glue to dry completely. It could take more than 1 h (*see* **Note 3**).
4. Set the slim tip drill bar into the micromotor and adjust the motor speed. Supply the skull with PBS, thin the skull gently, and then dry the skull to check the thinned skull condition; repeat these three steps until the skull thin is 20–50 μm (*see* **Note 4**).
5. Turn on the two-photon microscopy, and open the two-photon microscopy software. Place the mouse on the observation stage by fixing headplate, and anesthetize the mice with 0.5–1% isoflurane. Set the thermometers' sensor into the rectum of the mouse, and keep the body temperature of the mouse using heat-sheet.
6. Set the parameter following the user guide; select the suitable laser wavelength for observation (e.g., 890 nm for observation of LC3-EGFP). Here, a laser-scanning microscopy system FV1000MPE2 equipped with an upright microscopy, a water-immersion objective lens (XLPlanN25xW; numerical aperture, 1.05), was used. The Ex/Em of EGFP was 890 nm and

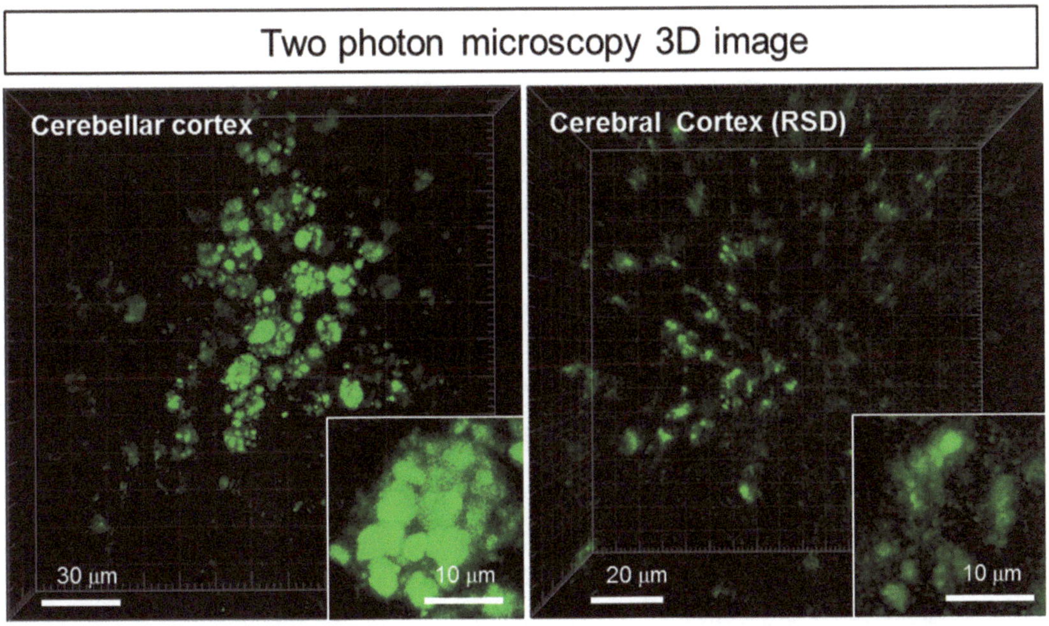

Fig. 2 EGFP-LC3 lentivirus was injected into the cerebellar cortex (left panel) or retrosplenial dysgranular cortex (RSD, right panel) of wild-type mouse (C57BL/6 × SJL) at 3 months old. Twenty days later, LC3-EGFP signals were observed by two-photon microscopy. The EGFP-positive grouped vesicles indicate autophagosomes in a cell

500–550 nm. Select an area for three-dimensional imaging (100 × 100 µm × 10–50 µm depth, 1 µm Z steps, 1024 × 1024 pixels, and digital zoom ×3). The LC3-labeled autophagosome in the neuron of the mouse was detected at the cortical layer I or layer II (*see* **Note 5**) (Fig. 2).

3.4 Analysis

Images of LC3-EGFP-positive vesicles were analyzed according to the number, size, and signal intensity of autophagosome using IMARIS 7.2.2 (Bitplane, Switzerland). In order to locate the autophagosome in the same cell using time-lapse imaging, the trans-image of blood vessel can serve as coordinate.

4 Notes

All experimental procedures should be performed in accordance with the protocols approved by The Committees on Human Ethics and Animal Experiments of the university or institution.

1. Pentobarbital volume should be decreased for aged mice. For example, mouse older than 1 year old, start with 5% isoflurane, and keep at 1% during operation; aged mice should be handled carefully.

2. Adjust the drill speed. If it is too fast, it would increase the risk of bleeding.
3. If super glue cannot dry completely, it will form a membrane on the lens at observation.
4. The PBS supply will reduce the heat of the drill and keep the skull under normal temperature, and smooth the thinned-skull surface.
5. During the observation, keep checking the buffer between the lens and thinned-skull cranial window. Supply buffer when it is insufficient, and keep monitoring the body temperature carefully.

Acknowledgments

This work was supported by "Brain Mapping by Integrated Neurotechnologies for Disease Studies (Brain/MINDS)," "Strategic Research Program for Brain Sciences (SRPBS)," and a Grant-in-Aid for Scientific Research on Innovative Areas (Foundation of Synapse and Neurocircuit Pathology) from the Ministry of Education, Culture, Sports, Science and Technology of Japan. We also thank Drs. Kazuhiko Tagawa and Kyota Fujita for supporting ethics approval, technical development, and discussion.

References

1. Mizushima N, Yoshimori T, Ohsumi Y (2011) The role of Atg proteins in autophagosome formation. Annu Rev Cell Dev Biol 27:107–132
2. Chen X, Kondo K, Motoki K, Homma H, Okazawa H (2015) Fasting activates macroautophagy in neurons of Alzheimer's disease mouse model but is insufficient to degrade amyloid-beta. Sci Rep 5:12115
3. Sato M, Kawano M, Yanagawa Y, Hayashi Y (2016) *In vivo* two-photon imaging of striatal neuronal circuits in mice. Neurobiol Learn Mem 135:146–151

Chapter 34

Measuring Nonselective and Selective Autophagy in the Liver

Takashi Ueno and Masaaki Komatsu

Abstract

Administration of leupeptin, a specific inhibitor of lysosomal cysteine proteinases, to starved rats or mice inhibits autolysosomal protein degradation and results in accumulation of autolysosomes in their livers. Immunoblotting of liver homogenates to examine autophagic flux in vivo reveals elevated levels of the selective autophagy substrate p62 and the autophagosomal membrane protein LC3-II in the livers of leupeptin-treated animals. Percoll density gradient centrifugation can be used to isolate autolysosomes from the livers of untreated and leupeptin-treated animals. Moreover, autolysosomes can be examined for the presence of sequestered cytoplasmic proteins as well as degradation intermediates.

Key words Liver, Leupeptin, Autolysosome, Percoll, Selective autophagy, Nonselective autophagy

1 Introduction

Researchers have used lysosomal proteolysis inhibitors to examine autophagy flux in vivo and in vitro [1]. Leupeptin, a specific inhibitor of lysosomal cysteine proteinases, including cathepsin L, cathepsin B, cathepsin H, and legumain, penetrates through hepatocyte plasma membranes and incorporates into lysosomes [2–5]. Inhibition of autophagic proteolysis results in accumulation of autolysosomes with sequestered cytoplasmic proteins in the lumen [6–8]. Therefore, we hypothesized that liver homogenate obtained from leupeptin-administered rats or mice should have elevated levels of the selective autophagy substrate p62 and the autophagosomal membrane marker LC3-II due to accumulated autolysosomes. Post-nuclear supernatants (PNS) were prepared by low-speed centrifugation of liver homogenate to remove nuclei. Immunoblotting analysis of PNS samples prepared from mouse and rat liver with or without leupeptin-treatment is a convenient technique to estimate in vivo autophagy flux.

Given that leupeptin has no effect on lysosomal lipolysis, the density of accumulated autolysosomes is greater than those of any

Nicholas Ktistakis and Oliver Florey (eds.), *Autophagy: Methods and Protocols*, Methods in Molecular Biology, vol. 1880, https://doi.org/10.1007/978-1-4939-8873-0_34, © Springer Science+Business Media, LLC, part of Springer Nature 2019

other cytoplasmic organelles. Therefore, leupeptin-induced autolysosomes can be separated from other cell organelles by centrifugation in Percoll gradients [6, 7, 9–11]. Interestingly, the amount of leupeptin-induced autolysosomes is increased by starvation [11]. The number of isolated autolysosomes, therefore, becomes a useful indicator of starvation-induced autophagy in the liver. Moreover, these denser autolysosomes are useful for the examination of cytoplasmic substrates that are sequestered in autophagosomes via both selective and nonselective autophagy.

2 Materials

Prepare all solutions using ultrapure distilled and deionized water and analytical grade reagents. Unless otherwise stated, store all reagents and solutions at 4 °C.

2.1 Preparation of PNS from Starved Mouse Livers

1. Leupeptin solution, 3.5 mg/mL leupeptin in 0.9% NaCl, for rats. Dissolve 90 mg of NaCl to 10 mL of water. Add 35 mg of leupeptin to 10 mL of 0.9% NaCl to give a final concentration of 3.5 mg/mL.
2. Leupeptin solution, 0.8 mg/mL leupeptin in 0.9% NaCl, for mice: Add 8 mg of leupeptin to 10 mL of 0.9% NaCl.
3. 1 M NaOH: Slowly dissolve 40 g of NaOH in water with gentle stirring, and then adjust the volume to 1 L.
4. 0.2 M TES-NaOH (pH 7.4): Weigh out 4.59 g of N-tris (hydroxymethyl)methyl-2-aminoethanesulfonic acid (TES), and add it to a glass beaker. Add about 50 mL water to the beaker. Stir gently using a magnetic stirrer to dissolve TES. Titrate with 1 M NaOH to adjust pH of the solution to 7.4. Add water to adjust the final volume to 100 mL.
5. Extraction buffer (5 mM TES-NaOH, pH 7.4, 0.3 M sucrose): Weigh out 51.3 g of sucrose, and add it to a 500 mL glass beaker. Add 12.5 mL of 0.2 M TES-NaOH, pH 7.4, and about 300 mL of water, and dissolve sucrose using a magnetic stirrer. Add water to adjust the volume to 500 mL.

2.2 Separation of Autolysosomes by Centrifugation in Percoll Density Gradients

1. 55% Percoll in extraction buffer: Weigh out 51.3 g of sucrose and add it to 500 mL beaker. Add 12.5 mL of 0.2 M TES-NaOH, pH 7.4, 275 mL Percoll, and ~100 mL water. Dissolve sucrose with gentle stirring. Adjust the volume of the solution to 500 mL by adding water.

3 Methods

Use male Wistar rats weighing 200–300 g and 8- to 12-week-old male C57BL/6J mice. Weigh out animals on the morning of the intraperitoneal injections to determine the appropriate volume of leupeptin solution required for the dose of 0.02 mg/g body weight. Leupeptin solutions of 0.8 mg/mL and 3.5 mg/mL are used for the mice and rats, respectively. One hour after the administration of leupeptin, kill the animals by inhalant anesthesia with isoflurane (rats) or cervical dislocation (mice). Harvest their livers and immediately store them on ice (*see* **Note 1**).

3.1 Preparation of PNS

1. Cut livers from control and leupeptin-treated animals into small pieces with scissors while stored on ice.
2. Suspend the liver samples in five volumes of extraction buffer (5 mL/g of liver).
3. Homogenize the samples using a motor-driven loose-fitting glass/Teflon homogenizer with 4–5 up-down strokes (*see* **Note 2**).
4. Centrifuge the homogenates at 700 × *g*, 5 min at 4 °C.
5. Transfer the supernatants (PNS) to a beaker and store them on ice.
6. Resuspend the pellets in the same volume of extraction buffer as described in **step 2**.
7. Centrifuge the suspension at 700 × *g*, 5 min.
8. Combine the supernatants with the PNS in the beaker as noted at **step 5** (*see* **Note 3**).

3.2 Separation of Autolysosomes by Centrifugation in 50% Percoll Gradient

1. Centrifuge PNS (**step 8** of Subheading 3.1) prepared from leupeptin-treated livers at 9000 × *g*, 15 min.
2. Suspend the pellet (mitochondrial/lysosomal fraction) with a small volume (1–2 mL for mice and 3–5 mL for rats) of extraction buffer.
3. Gently mix the mitochondrial/lysosomal fraction with 10 volumes of 55% Percoll in extraction buffer. Alternatively overlay the mitochondrial/lysosomal suspension onto 10 volumes of 55% Percoll in extraction buffer.
4. Centrifuge at 50,000 × *g*, for 1 h.
5. The upper half of the mixture contains mitochondria, while the lower half of the mixture contains autolysosomes. Remove the upper half of the mixture using an aspirator. Pool autolysosome layer separated in the lower half portion.
6. Centrifuge pooled autolysosome layer at 100,000 × *g*, for 1 h.

7. Aspirate autolysosome layer concentrated in the middle of centrifugal tubes using a Pasteur pipette, and transfer it to a beaker. Note that Percoll sediments as transparent precipitate at the bottom of centrifugal tube (*see* **Note 4**).
8. Dilute autolysosomes with tenfold volumes of extraction buffer.
9. Centrifuge the autolysosomal suspension at 9000 × *g*, for 15 min.
10. Discard the supernatant. Suspend the pelleted autolysosomes with a minimal volume of extraction buffer.
11. Alternatively, if the pellet is too loose, suspend the precipitate with tenfold volumes of extraction buffer, and centrifuge at 9000 × *g* for 15 min.
12. Remove supernatant. Suspend the pellet with a minimal volume of extraction buffer.

4 Notes

1. Although the exact reason is not understood, the reproducibility of harvesting denser autolysosomes by Percoll gradient centrifugation is significantly better in rat livers than in mouse livers.
2. Recommended rotation is ~700 to ~800 rpm.
3. In Fig. 1, one typical example of immunoblotting analysis of PNS for LC3-II and p62 is presented.
4. Of note, contamination of Percoll in the final autolysosome fraction should be avoided as much as possible because residual Percoll forms insoluble precipitate when the sample is suspended in SDS-sample buffer containing dithiothreitol/β-mercaptoethanol and heated at 100 °C. These precipitates prevent clear separation of polypeptides by SDS-PAGE.

Acknowledgments

M.K. is supported by Grant-in-Aid for Scientific Research on Innovative Areas (25111006 and 25111001, to M.K) and Japan Society for the Promotion of Science (an A3 foresight program). T.U. is supported by High Technology Research Center Grant, Strategic Research Foundation at Private Universities. T.U. and M.K. are supported by Grant-in-Aid for Scientific Research on Priority Areas (18076005) and the Takeda Science Foundation.

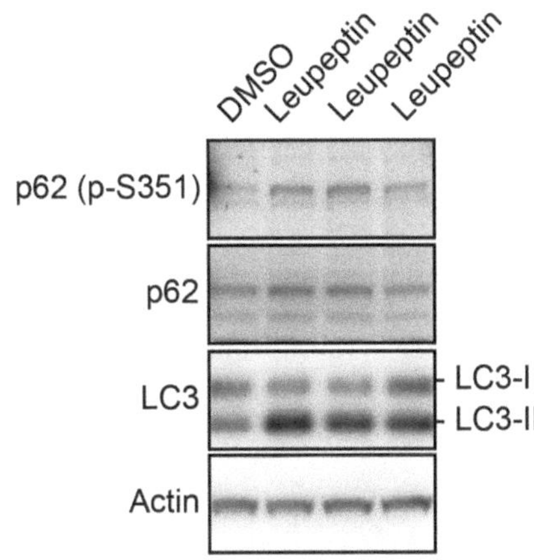

Fig. 1 Immunoblot analysis with homogenates prepared from leupeptin-treated mouse livers. PNS from control mice without leupeptin administration (left-most lane) and three PNS samples isolated from three mice after leupeptin administration. LC3-I is converted into LC3-II, transferring to inner and outer membrane of autophagosomes. After the fusion of autophagosomes with lysosomes, the LC3-II localized on the inner membrane is degraded by lysosomal hydrolases irrespective of selectivity [12]. Meanwhile, only when p62 is phosphorylated (e.g., Serine 349, corresponding to mouse Serine 351) [13–15], it is translocated to selective cargos and subsequently degraded by autophagy. On the basis of the comparison of the amount of LC3-II and phosphorylated p62 in leupeptin-treated and non-treated mouse livers, we are able to evaluate both selective and nonselective liver autophagy

References

1. Klionsky DJ, Abdelmohsen K, Abe A, Abedin MJ, Abeliovich H, Acevedo AA et al (2016) Guidelines for the use and interpretation of assays for monitoring autophagy (3rd edition). Autophagy 12:1–222
2. Kirschke H, Langner J, Wiederanders B, Ansorge S, Bohley P (1977) Cathepsin L. A new proteinase from rat-liver lysosomes. Eur J Biochem 74:293–301
3. Seglen PO, Grinde B, Solheim AE (1979) Inhibition of the lysosomal pathway of protein degradation in isolated rat hepatocytes by ammonia, methylamine, chloroquine and leupeptin. Eur J Biochem 95:215–225
4. Aronson NN Jr, Dennis PA, Dunn WA (1981) Metabolism of leupeptin and its effect on autophagy in the perfused rat liver. Acta Biol Med Ger 40:1531–1538
5. Tanaka K, Ikegaki N, Ichihara A (1981) Effects of leupeptin and pepstatin on protein turnover in adult rat hepatocytes in primary culture. Arch Biochem Biophys 208:296–304
6. Furuno K, Ishikawa T, Kato K (1982) Appearance of autolysosomes in rat liver after leupeptin treatment. J Biochem 91:1485–1494
7. Kominami E, Hashida S, Khairallah EA, Katunuma N (1983) Sequestration of cytoplasmic enzymes in an autophagic vacuole-lysosomal system induced by injection of leupeptin. J Biol Chem 258:6093–6100
8. Ishikawa T, Furuno K, Kato K (1983) Ultrastructural studies on autolysosomes in rat hepatocytes after leupeptin treatment. Exp Cell Res 144:15–24
9. Kopitz J, Kisen GO, Gordon PB, Bohley P, Seglen PO (1990) Nonselective autophagy of cytosolic enzymes by isolated rat hepatocytes. J Cell Biol 111:941–953
10. Ueno T, Ishidoh K, Mineki R, Tanida I, Murayama K, Kadowaki M et al (1999) Autolysosomal membrane-associated betaine homocysteine methyltransferase. Limited degradation fragment of a sequestered cytosolic enzyme monitoring autophagy. J Biol Chem 274:15222–15229

11. Ueno T, Muno D, Kominami E (1991) Membrane markers of endoplasmic reticulum preserved in autophagic vacuolar membranes isolated from leupeptin-administered rat liver. J Biol Chem 266:18995–18999
12. Tanida I, Minematsu-Ikeguchi N, Ueno T, Kominami E (2005) Lysosomal turnover, but not a cellular level, of endogenous LC3 is a marker for autophagy. Autophagy 1:84–91
13. Matsumoto G, Wada K, Okuno M, Kurosawa M, Nukina N (2011) Serine 403 phosphorylation of p62/SQSTM1 regulates selective autophagic clearance of ubiquitinated proteins. Mol Cell 44:279–289
14. Pilli M, Arko-Mensah J, Ponpuak M, Roberts E, Master S, Mandell MA et al (2012) TBK-1 promotes autophagy-mediated antimicrobial defense by controlling autophagosome maturation. Immunity 37:223–234
15. Ichimura Y, Waguri S, Sou YS, Kageyama S, Hasegawa J, Ishimura R et al (2013) Phosphorylation of p62 activates the Keap1-Nrf2 pathway during selective autophagy. Mol Cell 51:618–631

Chapter 35

Measuring Autophagy in Pancreatitis

Alejandro Ropolo, Daniel Grasso, and Maria I. Vaccaro

Abstract

Acute pancreatitis is one of the first pathological processes where autophagy has been described in a human tissue. Autophagy, autodigestion, and cell death are early cellular events in acute pancreatitis. Recent advances in understanding autophagy highlight its importance in pathological conditions. However, methods for monitoring autophagic activity during complex diseases, involving highly differentiated secretory cells, are complicated, and the results are sometimes misinterpreted. Here, we describe methods used to identify autophagic structures and to measure autophagic flux in cultured cell models and animal models of pancreatitis. We also briefly describe the pancreas specific autophagy mouse model that was useful to understand the actual role of autophagy in pancreatitis and to identify a novel selective autophagy pathway named zymophagy. Lastly, we describe the immunomagnetic isolation of autophagosomes and the detection of autophagy in pancreatic tissue samples derived from humans.

Key words Pancreatitis, Caerulein, Selective autophagy, Autophagosomes, Ela1-VMP1 transgenic mouse

1 Introduction

1.1 Acute Pancreatitis

Defined as the pancreas self-digestion, it is the most frequent disease of the pancreas. During pancreatitis, ultrastructural alterations of zymogen granules are produced in a yet undefined way. These alterations are characterized by premature activation of trypsinogen to trypsin within pancreatic acinar cells leading to the progression of the disease. The pathophysiology involves a complex cascade of events initializing in pancreatic acinar cells causing hemorrhage, necrosis, edema, and complete destruction of pancreatic parenchyma [1, 2]. The presence of autophagy has been described in dying acinar cells from human pancreatitis. Helin et al. in 1980 showed ultrastructural alterations in pancreatic acini from patients operated for acute necrotizing pancreatitis. Their findings in acinar cells included changes in zymogen granules and an increased autophagocytosis indicated by several autophagic vesicles [3].

Nicholas Ktistakis and Oliver Florey (eds.), *Autophagy: Methods and Protocols*, Methods in Molecular Biology, vol. 1880,
https://doi.org/10.1007/978-1-4939-8873-0_35,

1.2 Pathologically Activated Zymogen Granules

The pancreatic acinar cell is a highly polarized, differentiated cell whose primary function is the synthesis and secretion of digestive enzymes into the pancreatic juice. Pancreatic digestive enzymes are produced as inactive enzymes (zymogens) and stored in subcellular structures called zymogen granules, until exocytosis [4, 5]. Zymogen granules are secreted to the gut where zymogen is physiologically activated to digest food. In pancreatitis, an unknown trigger leads to the premature conversion of digestive proenzymes (zymogens) into their active form (digestive enzymes) within the acinar cell. Activated pancreatic digestive enzymes are able to hydrolyze tissue parenchyma and eventually trigger a severe disease [1, 2].

1.3 Zymophagy

It is a novel selective form of autophagy that works as a protective cell response to disease [6, 7]. This new selective autophagic pathway is triggered in pancreatic acinar cells during pancreatitis in order to sequester and degrade potentially deleterious activated zymogen granules.

Pancreatitis Model. Cholecystokinin (CCK) is a secretagogue secreted by duodenal endocrine cells that interacts with Gq-coupled receptors in the pancreatic acinar cell to induce pancreatic secretion in physiological conditions [8]. On the other hand, the hyperstimulation of CCK receptors (CCK-R) with its analogue caerulein leads to intracellular proteolytic enzyme activation, autophagic morphological features, and other cellular events characteristic of pancreatitis [9]. Further, the secretagogue-induced model is the most commonly employed and best characterized experimental model of acute pancreatitis [10]. Interestingly, early during this experimental model, CCK-R hyperstimulation triggers the selective autophagic pathway zymophagy [6, 11].

1.4 The Vacuole Membrane Protein 1

VMP1 is a transmembrane protein highly activated in acinar cells early during pancreatitis-induced autophagy. VMP1 expression is able to trigger autophagy in mammalian cells, even under nutrient-replete conditions, and it remains in autophagosomal membrane [12]. VMP1 expression is not detectable in healthy pancreas tissue, but it is rapidly induced in pancreas with pancreatitis and correlates with morphological features resembling autophagy [13, 14]. VMP1 does not have any known homologue in yeast, but its expression is required to start the autophagic process in mammalian cells [15]. VMP1 is required for autophagosome biogenesis by means of the interaction of its ATG domain with the BH3 domain of BECN1 [16]. Further, VMP1 is related to ER-plasma membrane contact site through the interaction with Esyt3 in the autophagosome formation site [17].

1.5 The Ela1-VMP1 Mouse

It is an animal model of autophagy in pancreas. The ElaI-VMP1 is a tissue-specific transgenic mouse in which acinar cell-specific constitutive expression of a VMP1-EGFP chimera induces the formation of autophagosomes within the acinar cell [12]. We use this unique tool to investigate the VMP1 pathway in autophagy during acute pancreatitis [6]. The fact that VMP1 is a transmembrane protein of the autophagosomal membrane allows the isolation of autophagosomes using magnetic beads fused to specific antibodies. Here, we describe methods used to measuring autophagy in pancreatitis in cell models, animal models, and human tissue specimens. We describe the immunomagnetic isolation of autophagosomes and the detection of autophagy in pancreatic tissue samples derived from humans. We also briefly describe the pancreas-specific autophagy mouse model that was useful to understand the actual role of autophagy in pancreatitis and to identify a novel selective autophagy pathway named zymophagy.

2 Materials

2.1 Measuring Autophagy in an Experimental Mouse Model of Pancreatitis

1. Caerulein stock (74 μM): Dissolve 1 mg caerulein in 10 mL 50% ethanol (5 mL absolute ethanol +5 mL milli-Q H_2O). Store at −20 °C.
2. Sterile non-pyrogenic saline solution (0.9% NaCl).
3. Phosphate-buffered saline (PBS): 5 M NaCl, 1 M KCl, 1 M Na_2HPO_4, 1 M KH_2PO_4, pH 7.4. Weight 8.16 g NaCl, 0.2 g KCl, 3.58 g Na_2HPO_4, and 0.25 g KH_2PO_4, and transfer to a beaker. Add distilled water to a volume of 900 mL. Mix and adjust to pH 7.4 with 10 N NaOH. Make up to 1 L with distilled water. Sterilize and store at 4 °C.
4. 4% p-formaldehyde in PBS: Weigh 4 g p-formaldehyde and transfer to a 250 mL glassware (*see* **Note 1**). Add 100 mL PBS and 40 μL 10 N NaOH. Heat at 55 °C until complete dissolution. Let cool at room temperature. Adjust to pH 7.4 with diluted HCl. Store at 4 °C. Do not sterilize.
5. 0.5 M EDTA (pH 8): Add 18.61 g EDTA.2H_2O to 80 mL H_2O milli-Q. Stir vigorously. Adjust the pH to 8.0 with NaOH pellets. Make up to 1 L with H_2O milli-Q. Dispense into aliquots of 5 mL and sterilize by autoclaving. Store at 4 °C.
6. 5 M NaCl: Dissolve 292 g NaCl in 1 L H_2O milli-Q. Dispense into aliquots of 50 mL and sterilize by autoclaving. Store at 4 °C.
7. HEPES lysis buffer: 50 mM HEPES, 100 mM NaCl, 1 mM EDTA, 20 mM NaF, 20 mM $Na_4P_2O_7$, 1 mM $NaVO_3$. Weight 2.98 g HEPES, 0.21 g NaF, 2.23 g $Na_4P_2O_7$.10H_2O, and 0.049 g $NaVO_3$.4H_2O, and transfer to a beaker. Add 200 mL H_2O milli-Q, 5 mL 5 M NaCl, 0.5 mL 0.5 M EDTA (pH 8). Mix and adjust pH to 7.4 with NaOH. Make up to 250 mL

with H_2O milli-Q. Sterilize by filtration. Store in aliquots of 50 mL at −20 °C.

8. Triton X-100 for molecular biology.
9. Phenylmethylsulfonyl fluoride (PMSF) stock (10 mM): Dissolve 7 mg PMSF in 4 mL isopropanol. Store in aliquots of 50 μL at −20 °C.
10. Protease inhibitor mix (1000X stock). Dissolve 10 mg pepstatin, 10 mg leupeptin, and 10 mg aprotinin in 1 mL DMSO. Store in aliquots 50 μL at −20 °C.

2.2 Measuring Autophagy in a Cell Model of Pancreatitis Using RFP-LC3 Transfection

1. Rat pancreatic acinar AR42J cells (pancreatoma, ATCC CRL 1492).
2. Dulbecco's modified Eagle's medium (DMEM).
3. Fetal bovine serum (FBS).
4. Antibiotics solution – penicillin-streptomycin (Pen/Strep) 100x.
5. Dexamethasone stock. To prepare 4 mg/mL stock solution, add 1 mL absolute ethanol per 4 mg product; gently swirl to dissolve. Store at −20 °C in aliquots of 500 μL.
6. Caerulein stock (74 μM): Dissolve 1 mg caerulein in 10 mL ethanol 50 (5 mL absolute ethanol +5 mL milli-Q H_2O). Store at −20 °C.
7. pRFP-C1 containing full-length rat LC3.
8. TransIT®-LT1 Transfection Reagent.
9. Opti-MEM I Reduced-Serum Medium.
10. PBS (Phosphate-Buffered Saline) for cell culture commercially available.
11. 4% p-formaldehyde in PBS.
12. Glutaraldehyde solution Grade I, 50% in water (GA): 1% glutaraldehyde, 1% p-formaldehyde in PBS. Mix 12.5 mL 4% p-formaldehyde in PBS, 1 mL 50% GA, and adjust to 50 mL with H_2O milli-Q.
13. DABCO mounting medium: Weigh 0.233 g 1,4-DiazaBicyclo [2.2.2] octane, and dissolve in 1 mL Tris pH 8 (200 μL Tris 1 M pH 8 + 800 μL de H_2O). Add 9 mL glycerol to this solution. Store in aliquots of 1 mL at −20 °C.
14. Fluorescence microscope.

2.3 Immunoisolation of Zymophagosomes from Ela1-VMP1 Mouse Pancreas

1. ElaI-VMP1 transgenic mice.
2. Animal care and experimentation facility.
3. 10 cm petri dish.
4. Surgical resection instruments.

5. Phosphate-buffered saline (PBS): 5 M NaCl, 1 M KCl, 1 M Na_2HPO_4, 1 M KH_2PO_4, pH 7.4. Weight 8.16 g NaCl, 0.2 g KCl, 3.58 g Na_2HPO_4, and 0.25 g KH_2PO_4, and transfer to a beaker. Add distilled water to a volume of 900 mL. Mix and adjust to pH 7.4 with 10 N NaOH. Make up to 1 L with distilled water. Sterilize and store at 4 °C.
6. 250-STM buffer: 250 mM sucrose, 50 mM Tris pH 7.4, 5 mM $MgCl_2$. First, prepare the following stock solutions, 1 M $MgCl_2 \cdot 6H_2O$ and 1 M Tris pH 7.4. For 1 M $MgCl_2.6H_2O$ stock solution, dissolve 203.3 g of $MgCl_2.6H_2O$ in 800 mL of distilled H_2O. Adjust the volume to 1 L with distilled H_2O, aliquot, and sterilize by autoclaving. For 1 M Tris pH 7.4 stock solution, dissolve 121.1 g of Tris base in 800 mL of distilled H_2O, adjust pH to 7.4 by adding concentrated HCl, aliquot, and sterilize by autoclaving. To prepare 250-STM buffer mix, dissolve 8.56 g sucrose in 80 mL distilled H_2O plus 5 mL 1 M Tris pH 7.4 and 0.5 mL 1 M $MgCl_2.6H_2O$ stock solutions. Adjust the volume to 100 mL and sterilize by passing it through a 0.22 μm filter, and store at 4 °C.
7. EDTA-free protease inhibitor cocktail commercially available.
8. Benzamidine hydrochloride hydrate.
9. Tosylated superparamagnetic beads.
10. Borate buffer: 10 mM H_3BO_3, pH 9.5. Dissolve 0.62 g of boric acid in 400 mL distilled H_2O, and adjust to pH 9.5 with 1 M NaOH. Make up 1 L with distilled H_2O, and sterilize by passing it through a 0.22 μm filter. Store at 4 °C.
11. Rabbit anti-mouse linker antibody (ZyMax, Zymed Laboratories).
12. Magnetic rack device.
13. PBS-BSA: PBS 0.1% BSA. Dissolve 0.1 g bovine serum albumin (BSA) in 100 mL PBS while stirring. Sterilize by passing it through a 0.45 μm filter. Store at −20 °C for long-term storage.
14. 0.2 M Tris pH 8.5 0.1% BSA: Dissolve 2.42 g of Tris base in 80 mL of distilled H_2O; adjust pH to 8.5 by adding concentrated HCl. Dissolve 0.1 g bovine serum albumin (BSA) in 100 mL of this buffer while stirring. Sterilize by passing it through a 0.45 μm filter. Store at −20 °C for long-term storage.
15. Mouse anti-GFP monoclonal antibody.
16. Buffer A: PBS 2 mM EDTA 5% FBS. Dissolve 0.74 g of disodium $EDTA.2H_2O$ in 800 mL PBS, add 5 mL fetal bovine serum (FBS), and adjust to 1 L with PBS. Sterilize by passing it through a 0.25 μm filter. Store at −20 °C for long-term storage.

2.4 Measuring Zymophagy in Human Pancreas with Pancreatitis

1. Human pancreas specimens (10 micron-thick cuts) obtained from surgical resection of patients with pancreatitis.
2. Polyclonal rabbit antiserum to VMP1 at 1:100 dilution.
3. The following commercial polyclonal antibodies suitable for human immunohistochemistry: goat anti-LC3, goat anti-p62, goat anti-USP9x, goat anti-Lamp2, rabbit anti-trypsinogen, goat anti-amylase.
4. Alexa Fluor 488 and 594 antibodies.

3 Methods

3.1 Measuring Autophagy in an Experimental Mouse Model of Pancreatitis

1. Use male mice weighing 20–25 g. Previous to the experiment, house animals with free access to food and water. Perform the experiments according to the standard ethical and legal guidelines of your region.
2. Prepare fresh caerulein working solution. Mix 0.5 mL caerulein stock and 4.5 mL sterile non-pyrogenic saline solution.
3. Inject intraperitoneally 0.1 mL caerulein working solution every 10 g mouse body weight (50 μg/Kg caerulein) for seven times giving a 1-h interval between each injection [9, 12]. Control mice were injected intraperitoneally with equal volume of sterile non-pyrogenic saline solution.
4. Proceed to euthanize mice at different times after first injection (*see* **Notes 2** and **3**).
5. Place animal in a surgical plate, and make an epigastric incision opening by planes. Perceive the stomach and observe the pancreas boundaries from the spleen through the stomach to the duodenum.
6. Carefully and quickly remove the pancreas (*see* **Note 4**). Place resected pancreas into a 6 mm petri dish filled previously with ice-cold PBS.
7. Place the pancreas into a 6 mm petri dish filled previously with ice-cold PBS. Separate the pancreatic tissue from remains of spleen and fat (*see* **Note 5**).
8. For tissue specimens, one piece from each pancreas is minced and fixed during 24 h with 4% p-formaldehyde. Then proceed to tissue inclusion into paraffin. With this fixed sample, you can perform an immunohistochemistry or immunofluorescence to determine the expression of the autophagy-related protein VMP1 (*see* **Note 6**) (Fig. 1).

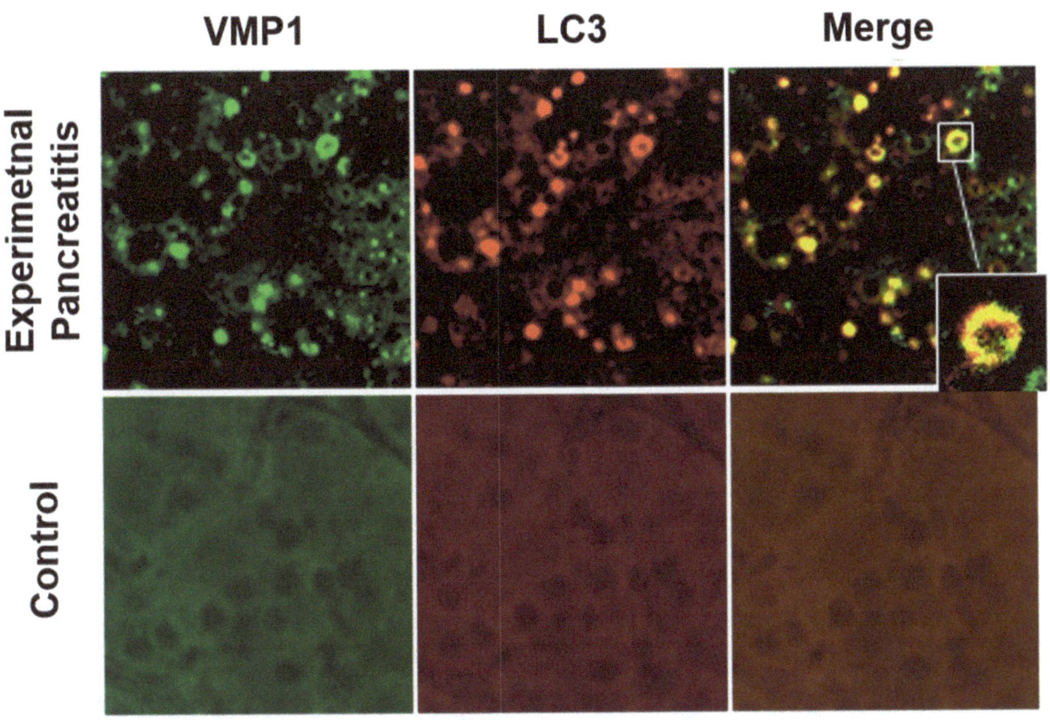

Fig. 1 Example of immunofluorescence of endogenous VMP1 and LC3 in pancreatic tissue from control and caerulein-treated animals. This research was originally published in Journal of Biological Chemistry. Ropolo A. et al. The Pancreatitis-induced Vacuole Membrane Protein 1 Triggers Autophagy in Mammalian Cells. *J. Biol. Chem.* 2007; 282:37124–37,133. © the American Society for Biochemistry and Molecular Biology

9. Fix another piece of tissue specimen in 1% glutaraldehyde, 1% p-formaldehyde in PBS. Keep on ice and process for transmission electron microscopy by standard procedures to determine the presence of autophagic structures.
10. Another piece is processed immediately for protein extraction. Add 10 μL Triton X-100 (1% final concentration) and 1 μl protease inhibitors mix per 1 mL HEPES buffer, pH 7.4. Mix and keep on ice.
11. Just before proceeding to homogenize pancreas tissue, add 100 μL PMSF stock to 1 mL HEPES buffer, pH 7.4, containing protease inhibitors (*see* **Note 7**) (final concentration: 1 mM PMSF, 10 μg/mL pepstatin, 10 μg/mL leupeptin, and 10 μg/mL aprotinin). Homogenize the piece of pancreas tissue with a motor-driven Teflon pestle homogenizer in 1 mL ice-cold HEPES buffer, pH 7.4, containing protease inhibitors during 1 min. Then, centrifuge at 16,873 × *g* by 3 min at 4 °C.
12. Place the supernatant in a new tube and process for Western blot analysis of the LC3 processing and other autophagy-related proteins such as VMP1 (*see* **Note 8**) to determine autophagy.

3.2 Measuring Autophagy in a Cell Model of Pancreatitis Using RFP-LC3 Transfection

1. Culture AR42J cells routinely at 37 °C in a 5% CO_2, 95% air atmosphere in DMEM supplemented with 10% FBS and antibiotics (100 units/mL penicillin and 100 g/mL streptomycin).
2. The day before transfection dissociate cells that are 60–70% confluent in T-75 flask.
3. Seed 175,000 cells per well on round coverslips in 24-well plate.
4. Six hours later, treat cells with dexamethasone at a final concentration of 100 nM for a total experiment time of 48 h to induce their differentiation [6]. Make a dilution 1/1000 of dexamethasone stock in sterile DMEM supplemented with 10% FBS. Add 9.9 μL dexamethasone dilution 1/1000 per 1 mL of DMEM supplemented with 10% FBS to achieve a final concentration of 100 nM.
5. Carefully pour of the media from the flask or plate into a waste pot containing some disinfectant. Immediately replace the media with the appropriate volume of fresh pre-warmed at 37 °C DMEM supplemented with 10% FBS and 100 nM dexamethasone. Return to CO_2 37 °C incubator.
6. After 18 h, transfect cells with pRFP-LC3 vector (*see* **Note 9**). Here, we use TransIT®-LT1 Transfection Reagent (Mirus) in a transfection reagent-DNA ratio 3:1 (*see* **Note 10**).
7. Warm TransIT-LT1 Reagent to room temperature, and vortex gently before using.
8. Place 50 μL of Opti-MEM I Reduced-Serum Medium in a sterile tube.
9. Add 0.5 μg pRFP-LC3 plasmid DNA per well. Pipet gently to mix completely.
10. Add 1.5 μL transfection reagent to the diluted DNA mixture. Avoid any contact of the transfection reagent with the sides of the plastic tube. Pipet gently to mix completely.
11. Incubate at room temperature for 20 min.
12. Add the transfection reagent-DNA complexes dropwise to different areas of the well.
13. Gently rock the plate back and forth and from side to side to evenly distribute the transfection reagent-DNA complexes.
14. Incubate for 24 h (*see* **Note 11**).
15. Treat cells with 3.5 μM Caerulein (Sigma) to induce CCK-R hyperstimulation.
16. Incubate between 15 min to 3 h depending on your experiment (*see* **Note 12**).
17. Wash three times with PBS.
18. Fix cells during 10 min with cold 4% p-formaldehyde in PBS.

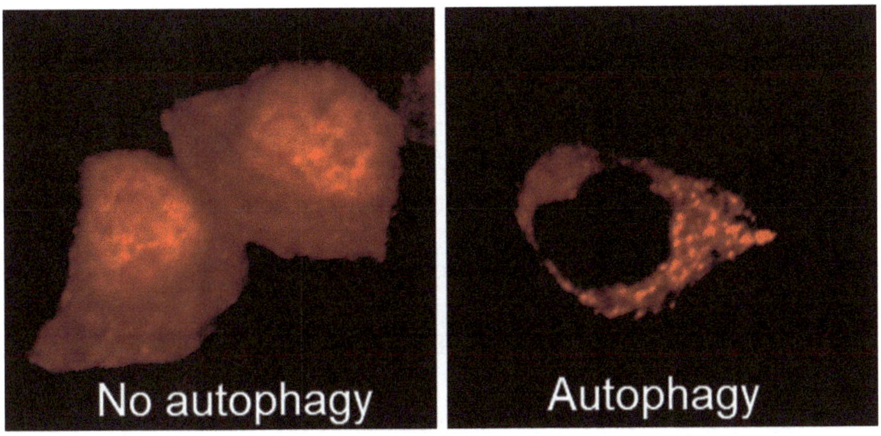

Fig. 2 AR42J pancreatic cells transfected with RFP-LC3 expression vector. Left, control cells not developing autophagy. Right, recruitment of RFP-LC3 in autophagy positive cells treated with caerulein

19. Wash three times with PBS.
20. Mount cells grown on round coverslips onto a slide using DABCO mounting medium (*see* **Note 13**).
21. Let dry for 10 min at room temperature.
22. Observe transfected cells under a fluorescence microscope. Caerulein treatment induces autophagy in pancreatic cells. Therefore, recruitment of RFP-LC3 will be observed. Consider RFP-LC3 recruitment staining when all the red fluorescence is present as punctate and no diffused protein remains (Fig. 2).
23. Determine the number of cells with punctate staining per 100 fluorescent RFP-LC3 transfected cells (*see* **Note 14**).

3.3 Measuring Zymophagy: The Selective Autophagy of Zymogen Granules During Pancreatitis

3.3.1 Mice Pancreas Resection

1. Use male mice weighing 20–25 g. Previous to the experiment, house animals with free access to food and water. Perform the experiments according to the standard ethical and legal guidelines of your region.
2. Proceed to euthanize mice (*see* **Note 2**).
3. Place animal in a surgical plate and make a large midline incision. Perceive the stomach, and observe the pancreas boundaries from the spleen through the stomach to the duodenum.
4. Quickly remove the pancreas and place it into a 6 mm petri dish filled previously with ice-cold PBS.
5. In petri dish, cut off all fatty tissue that frequently comes with the organ (*see* **Notes 5, 15** and **16**).

3.3.2 Postnuclear Supernatant

1. After tissue resection, immediately homogenize the organ for 1 min in 1 mL of ice-cold 250-STM buffer plus inhibitor cocktail with 1 mM benzamidine (*see* **Notes 17** and **18**).
2. Place homogenate into 1.5 mL microtube and centrifugate at 800 × *g* at 4 °C for 10 min.
3. Transfer supernatants to a new microtube and keep them in ice.

3.3.3 Magnetic Bead Preparation

1. Gently homogenize the tosylated superparamagnetic beads with pipette and vortex for 1–2 min.
2. Take 150 μL (4.5 mg) beads and deposit into a 0.6 mL microtube.
3. Thoroughly resuspend beads with 93.75 μL 0.1 M borate buffer pH 9.5.
4. Incorporate the rabbit anti-mouse linker antibody to the mix (10 μg Ab/mg beads).
5. Mix thoroughly for 1–2 min.
6. Incubate on rotator overnight at 4 °C.
7. Place microtube in a magnetic rack and discard supernatant.
8. Wash pellet twice with PBS-BSA.
9. Resuspend beads in 200 μL ice-cold 0.2 M Tris pH 8.5 0.1% BSA (*see* **Note 19**).
10. Discard Tris buffer and resuspend beads in 150 μL ice-cold PBS-BSA (*see* **Note 20**).
11. Split suspension in two fraction: 120 μL CBds (zymophagosomes isolation) y 30 μL Ctl (negative controls).
12. Add 120 μL of diluted anti-GFP antibody (75 μL antibody plus 45 μL PBS-BSA) to CBds fraction and 30 μl of PBS-BSA only to the Ctl fraction (*see* **Notes 21** and **22**).
13. Incubate both fractions on rotator overnight at 4 °C.
14. Place tubes in magnetic rack for 2–3 min and discard supernatant.
15. Wash three times with 200 μL PBS-BSA.
16. Keep the pellets as the "coated beads."

3.3.4 Immunomagnetic Isolation of Zymophagosomes from the ELA1-VMP1 Mouse Pancreas

1. In a fresh 1.5 mL microtube, mix 800 μL of postnuclear with 30 μL either CBds beads or Ctl beads.
2. Incubate on rotator overnight at 4 °C.
3. Wash three times with buffer A (15 min each time).
4. Resuspend beads in glutaraldehyde for 30 min.
5. Keep in ice-cold PBS, and follow the standard electronic microscopy techniques; *see* Fig. 3.

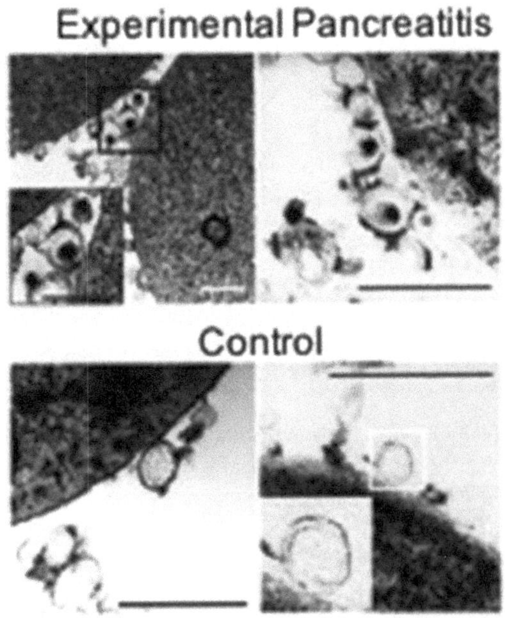

Fig. 3 Isolation of autophagosomes using anti-GFP antibody fused to magnetic beads. The same procedure is possible using anti-VMP1 antibody fused to magnetic beads. This figure was originally published in *The Journal of Biological Chemistry*. Grasso. et al. 2011. Zymophagy, a novel selective autophagy pathway mediated by VMP1-USP9x-p62, prevents pancreatic cell death. Journal of Biological Chemistry 286, 8308–8324. © the American Society for Biochemistry and Molecular Biology

3.4 Measuring Zymophagy in Human Pancreas with Pancreatitis

On tissue specimens obtained from pancreas surgically resected from patients with pancreatitis [6]:

1. Perform immunofluorescence assays of LC3 and VMP1 (*see* **Note 6**) to identify autophagosomes in pancreas tissue with pancreatitis.
2. Perform immunofluorescence assays of Trypsinogen and LC3 (Fig. 4); VMP1 and p62 to identify zymophagy in tissue suffering pancreatitis.
3. Perform immunofluorescence assays of Lamp2 and Trypsinogen to identify autolysosomes in tissue suffering pancreatitis.

4 Notes

1. P-formaldehyde is toxic. Please, gloves and safety glasses should be worn and solutions made inside a fume hood.
2. Euthanasia should perform by CO_2 inhalation.
3. It is important to bleed the animal because blood in the pancreas can mask some detections.

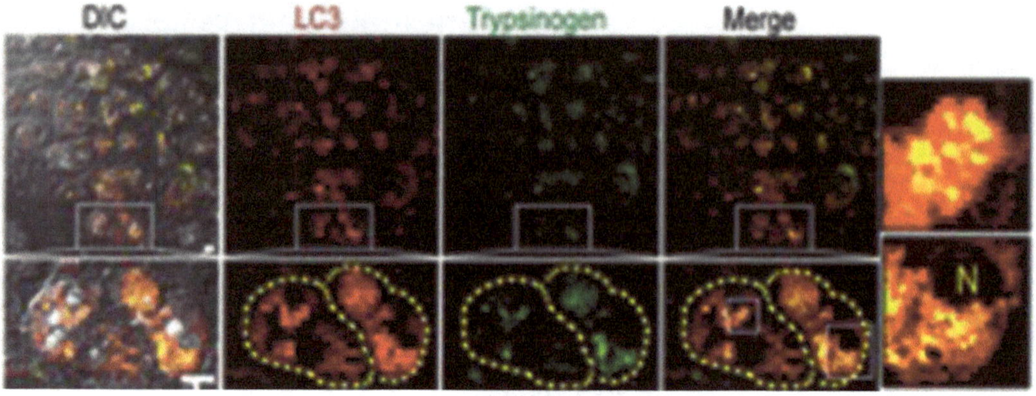

Fig. 4 Identification of zymophagy in human pancreas with pancreatitis. Immunofluorescence of trypsinogen as a marker of zymogen granules and LC3 as a marker of autophagosomes. This figure was originally published in *The Journal of Biological Chemistry*. Grasso. et al. 2011. Zymophagy, a novel selective autophagy pathway mediated by VMP1-USP9x-p62, prevents pancreatic cell death. Journal of Biological Chemistry 286, 8308–8324. © the American Society for Biochemistry and Molecular Biology

4. Remove the pancreas in less than 3 min after the animal was killed, in order to avoid the activation of proteases and nucleases present as zymogens in the pancreas.
5. Be careful because mouse pancreas is a limp whitish tissue, easy to confuse with fat tissue with a naked eye. But it is easy to differentiate fatty tissue from pancreas in aqueous medium. The fatty tissue floats and has a true white color in cold PBS.
6. VMP1 is an autophagy-related transmembrane protein which expression is highly induced during pancreatitis [12]. Polyclonal rabbit antiserum to VMP1 against the peptide MAQSYAKRIQQRLNSEEKTK (residues 385–406) was obtained in Vaccaro lab. This VMP1 antibody is suitable for immunofluorescence on human and rodent cells and tissue, and it is commercial as vacuole membrane protein 1 polyclonal antibody ALX-215-068-C200 (ENZO).
7. Add 50 μL PMSF stock by each 1 mL of homogenate buffer. Add PMSF just before homogenizing, because it is stable for only about 30 min in aqueous solution.
8. VMP1 is an autophagy-related transmembrane protein which expression is highly induced during pancreatitis [12]. Antibody for Western blot: TMEM49/VMP1 (D1Y3E) rabbit mAb #12929 (Cell Signaling).
9. Ideally, cells should be at 50–70% confluence.
10. AR42J cells are not easy to transfect. We use TransIT®-LT1 Transfection Reagent (Mirus) because we obtain a good percentage of transfection for these cells, around 30%. A similar

percentage of transfection is obtained using FuGENE® HD Transfection Reagent (Promega).

11. It is not necessary to replace the complete growth medium with fresh medium.
12. Note that 3.5 μM caerulein kills 50% of the cells in 3.5 h.
13. Put coverslip over a drop of about 2.5 μL of DABCO solution.
14. Stimulation of CCK-R with caerulein in AR42J cells significantly increase cells with punctate staining compared to control.
15. This is the moment, if it needed, to pick up a little piece of pancreas for protein extraction, histology, or whatever.
16. Because of its physiological role in food digestion, the pancreas is plenty of proteolytic enzymes. Therefore, all the procedure including animal euthanasia and pancreas sample obtention (Subheading 3.3.1, **steps 3–5**) must be perfomed in less than 3 minutes time on ice.
17. Add the proteases inhibitors just prior to homogenization. Alternatively to the commercial protease inhibitor cocktail, it is possible to add separately 1 mM PMSF, 10 g/mL leupeptin, 0.5 mg/mL soybean trypsin inhibitor, 20 nM aprotinin, 1 μg/mL pepstatin, and 1 mM benzamidine.
18. It is important to add benzamidine to the inhibitor cocktail in order to block trypsin activity in homogenate which is capable to activate the rest of pancreatic zymogens.
19. Do not avoid this step, because Tris/BSA buffer react with remaining tosyl groups and block the bead surface.
20. At this point, it is possible to stop and continue later or even the next day. The beads can be stored in this state for some months at 4 °C. Sodium azide to a final concentration of 0.02% can be added for avoid microorganisms colonization.
21. Diluted in buffer A.
22. Concentration final of anti-GFP antibody, 10 μg Ab/mg beads.

References

1. Sah RP, Saluja A (2011) Molecular mechanisms of pancreatic injury. Curr Opin Gastroenterol 27:444–451
2. Lankisch PG, Apte M, Banks PA (2015) Acute Pancreatitis. Lancet 386:85–96
3. Helin H, Mero M, Markkula H, Helin M (1980) Pancreatic acinar ultrastructure in human acute pancreatitis. Virchows Arch 387:259–270
4. Gorelick FS, Adler G, Kern HF et al (1993) Cerulein-induced pancreatitis. In: Go VLW, DiMagno EP, Gardner JD et al (eds) The pancreas: biology, pathophysiology and disease, 2nd edn. Raven Press, New York, NY, pp 501–526
5. Chandra R, Liddle RA (2015) Regulation of pancreatic secretion. Pancreapedia. https://doi.org/10.3998/panc.2015.38

6. Grasso D, Ropolo A, Ré L et al (2011) Zymophagy, a novel selective autophagy pathway mediated by VMP1-USP9x-p62, prevents pancreatic cell death. J Biol Chem 286:8308–8324
7. Vaccaro MI (2012) Zymophagy: selective autophagy of secretory granules. Int J Cell Biol 2012:396705. https://doi.org/10.1155/2012/396705
8. Williams JA (2008) Receptor-mediated signal transduction pathways and the regulation of pancreatic acinar cell function. Curr Opin Gastroenterol 24:573–579
9. Niederau C, Ferrell LD, Grendell JH (1985) Caerulein-induced acute necrotizing pancreatitis in mice: protective effects of proglumide, benzotript, and secretin. Gastroenterology 88:1192–1204
10. Gorelick FS, Lerch MM (2017) Do animal models of acute pancreatitis reproduce human disease? Cell Mol Gastroenterol Hepatol 4:251–262
11. Klionsky D et al (2016) Guidelines for the use and interpretation of assays for monitoring autophagy (3rd edition). Autophagy 12:1–222
12. Ropolo A, Grasso D, Pardo R et al (2007) The pancreatitis-induced vacuole membrane protein 1 triggers autophagy in mammalian cells. J Biol Chem 282:37124–37133
13. Vaccaro MI, Grasso D, Ropolo A, Iovanna JL, Cerquetti MC (2003) VMP1 expression correlates with acinar cell cytoplasmic vacuolization in arginine-induced acute pancreatitis. Pancreatology 3:69–74
14. Vaccaro MI (2008) Autophagy and pancreas disease. Pancreatology 8:425–429
15. Molejon MI, Ropolo A, Vaccaro MI (2013) VMP1 is a new player in the regulation of the autophagy-specific phosphatidylinositol 3-kinase complex activation. Autophagy 9:933–935
16. Molejon MI, Ropolo A, Re AL, Boggio V, Vaccaro MI (2013) The VMP1-Beclin 1 interaction regulates autophagy induction. Sci Rep 3:1055. https://doi.org/10.1038/srep01055
17. Nascimbeni AC, Giordano F, Dupont N, Grasso D, Vaccaro MI, Codogno P, Morel E (2017) ER-plasma membrane contact sites contribute to autophagosome biogenesis by regulation of local PI3P synthesis. EMBO J 36:2018–2033

Chapter 36

Characterization of the "Autophagic Flux" in Prostate Cancer Tissue Biopsies by LC3A/LAMP2a Immunofluorescence and Confocal Microscopy

D. Kalamida, A. Giatromanolaki, and M. I. Koukourakis

Abstract

A method is described for the immunostaining of human prostate cancer biopsies for the autophagic marker LC3 and the lysosomal protein LAMP2A. Using this combination we can provide some information on autophagic flux, and specific patterns of staining have been recognized for normal prostate tissue and prostatic carcinomas.

Key words LC3, Prostate cancer, Immunostaining, Autophagic flux

1 Introduction

Prostate cancer is the most common neoplasia of men, as the incidence ranges from 90 to 200 cases per 100,000 people per year [1]. Local tumor invasion, Gleason histopathological score, and plasma PSA levels are the parameters used for therapeutic decisions, providing also an estimate of the risk to develop metastasis. Radiotherapy is an overall equally effective alternative to radical prostatectomy [2]. The development of biomarkers that would predict increased or reduced radioresistance would help to identify subgroups of patients that would rather be treated with surgery instead of radiotherapy and vice versa.

Assessment of the autophagic activity of human carcinomas can be done by immunohistochemistry for major proteins of the autophagosomal membrane, like the autophagy-related protein 8 (Atg8)/MAP1LC3, which is a ubiquitin-like protein essential for the formation of autophagosomal membranes. Three different proteins are produced by three different genes (LC3A from 20q11.22; LC3B from 16q24.2; LC3C from 1q43). A pro-LC3 (~30 kDa) is produced as a precursor, which is subsequently cleaved by the protease Atg4 to produce the LC3-I soluble form (~18 kDa)

Nicholas Ktistakis and Oliver Florey (eds.), *Autophagy: Methods and Protocols*, Methods in Molecular Biology, vol. 1880,
https://doi.org/10.1007/978-1-4939-8873-0_36,

of LC3. The exposed C-terminus of LC3-I is conjugated to the head group amine of phosphatidylethanolamine (PE) through an amide bond by a sequence of ubiquitination-like reactions that involve an E1 (ATG7), an E2 (ATG3), and an E3 (a complex including ATG5, ATG12, and ATG16L). This generates the LC3-II membrane-bound form that resides to the outer and inner membranes of the autophagosomes [3].

Using specific antibodies recognizing the LC3 forms, immunohistochemistry can be applied in paraffin-embedded tissue sections from transrectal biopsies performed in patients with localized prostate carcinomas. It is important that researchers choose and validate their antibodies as for the isoform of LC3 that they recognize. The three LC3 proteins have different subcellular localization and, eventually, distinct biological functions [4].

Detection of autophagosomes using LC3 immunostaining gives an estimate of the cytoplasmic autophagosomal density, but this does not inform on the autophagy flux. High autophagosomal content may be a result of either increased formation or of blockage of the auto-lysosomal fusion or lysis. p62/sequestosome expression is an overall used marker of autophagy flux as it is degraded following auto-lysosomal fusion. An additional method developed in our lab is the assessment of auto-lysosomal fusion using double staining for the LC3A and the lysosomal membrane protein LAMP2a. Specific patterns of staining have been recognized for normal prostate tissue and prostatic carcinomas [5].

2 Materials

1. Prostate tissue paraffin biopsies mounted on poly-L-lysin-coated slides:
   (a) In a series of patients with localized prostate tumors treated with radical radiotherapy.
   (b) Control prostate tissues with prostatic hyperplasia.
2. Heat-induced epitope retrieval solution, citrate buffer (1:10 dilution, pH 7.2).
3. PBS pH 7.4.
4. 5% w/v BSA pH 7.4 in PBS.
5. Anti-LC3A rabbit polyclonal antibody (1:200, ab62720, Abcam).
6. Anti-LAMPA2A rabbit polyclonal antibody (1:100, ab18528, Abcam).
7. Hoechst 33342 (Sigma-Aldrich).
8. CF488A donkey anti-rabbit (Biotium).

9. CF568 goat anti-rabbit (Biotium).
10. Homemade Mowiol mounting medium (Mowiol 4-88, Sigma-Aldrich).

3 Methods

3.1 Mowiol Mounting Medium

2.4 gr Mowiol 4-88.
6gr glycerol.
6 mL ddH_2O.
12 mL 0.2 M Tris pH 8.5.

- Place glycerol in 50 mL Falcon tube.
- Add Mowiol and stir thoroughly.
- Add ddH_2O and leave 2 h at RT.
- Add Tris buffer, and incubate the tube at approximately 53°C until the Mowiol has dissolved; stir occasionally. (You can use a small glass beaker, in which the Falcon fits and stands, filled with water and heated on the stirrer; check temperature using a thermometer. Thus you can heat and stir at the same time.)
- Clarify by centrifugation at 4000 × *g*, 20 min, RT.
- Collect and aliquot the supernatant into 1 mL each.
- Store at −20 °C. Once defrosted store at 4 °C.

3.2 Tissue Permeabilization and Antigen Retrieval

Sections are cut at 3 μm from formalin-fixed paraffin-embedded tissues and dried for 30 min at 80 °C and mounted on poly-L-lysin-coated slides.

- Following deparaffinization and rehydration.
- For heat-induced epitope retrieval, the sections were placed in citrate buffer (1:10 dilution, pH 7.2) and heated at 120 °C for three times 5 min.

3.3 Antibody Staining (See Notes 1 and 2)

The samples can either be stained in *a single staining* in the case of two antibodies raised in different animals or, in the case here, with a *sequential immunofluorescence staining* protocol in two steps, first for LC3A (1:200, Abcam) and then LAMP2a (1:100, Abcam), given that both antibodies were raised in rabbit.

For *single staining* co-localization:

- Block the tissues slides in PBS/5% w/v BSA pH 7.4 for 30 min.
- Add primary antibodies for 1 h at room temperature.
- Wash samples in PBS pH 7.4 (three 5-min washes).
- Incubate with the appropriate secondary antibody (1:250; Biotium) for 30 min at room temperature.

- Wash in PBS pH 7.4 (three 5-min washes).
- Counterstain the DNA with Hoechst 33342 (1 μg/mL; Sigma-Aldrich) for 30 min at room temperature.
- Wash again with PBS,
- Mount coverslip samples using homemade Mowiol mounting medium.

For *sequential staining* for co-localization imaging:

- Block the tissues slides in PBS/5% w/v BSA pH 7.4 for 30 min.
- Add the *first primary antibody* for 1 h at room temperature.
- Wash the slides in PBS pH 7.4 (three 5-min washes).
- Incubate with the appropriate secondary antibody (1:250; Biotium) for 30 min at room temperature.
- Wash in PBS pH 7.4 (three 5-min washes).
- Block in PBS/5% w/v BSA pH 7.4 for 1 h.
- Perform the *second step of the sequential staining* using the second primary antibody for 1 h at room temperature, RT.
- Wash in PBS pH 7.4 (three 5-min washes).
- Incubate with the appropriate secondary antibody (1:250; Biotium).
- Counterstain DNA with Hoechst 33342 (1 μg/mL; Sigma-Aldrich) for 30 min at room temperature.
- Wash again with PBS.
- Mount coverslips using homemade Mowiol mounting medium.

3.4 Confocal Microscopy and Image Analysis

Imaging is performed on a customized Andor Revolution Spinning Disk Confocal System built around a stand (IX81; Olympus) with a 40× lens and a digital camera (Andor Ixon+885) (CIBIT Facility, MBG-DUTH). Image acquisition is performed using Andor IQ 2 software. Optical sections are recorded every 0.3 μm. Confocal microscopy images can be presented as 2D maximum intensity projections of z-stack images (ImageJ 1.47v National Institute of Health, USA).

The tissue immunofluorescence protocol is frequently used in our lab; an application in prostate tissue biopsies has been already published [4], and some representative images of the method are presented in Fig. 1.

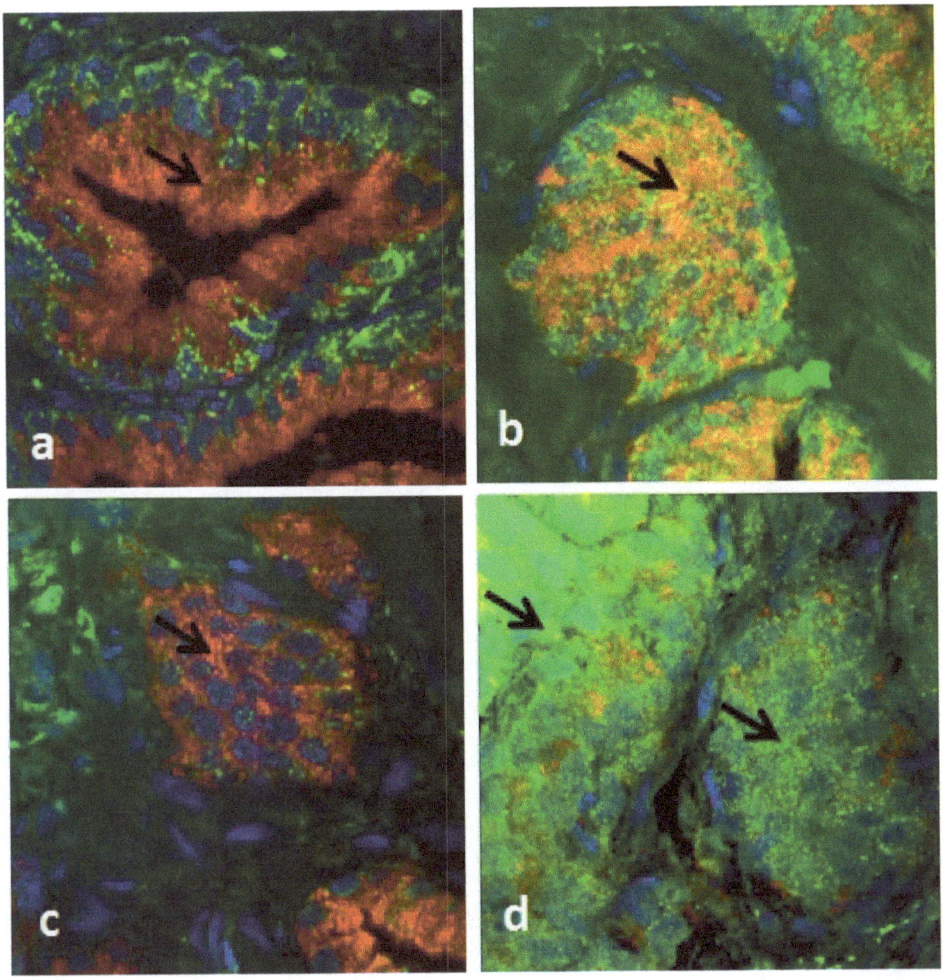

Fig. 1 (**a**) Normal prostate gland (arrow) with intense expression of LAMP2a (red) and less intense expression of LC3A (green) and poor co-localization. (**b**) Prostate cancer with intense expression of LC3A (green) and LAMP2a (red), with intense co-localization (arrow) of the proteins (orange), suggestive of an intense autophagy flux. (**c**) Prostate cancer with intense expression of LAMP2a (red), but poor expression of LC3A (green) (arrow), simulating a normal prostate-like staining pattern, reflecting a relatively benign prostate cancer phenotype. (**d**) Prostate cancer cells with intense expression of LC3A (green) but poor expression of LAMP2a (red) (arrow), reflecting either a blockage of the LC3A autophagy flux or exhaustive consumption of lysosomes [4]

4 Notes

1. Extra care should be taken, with the appropriate controls in single staining and sequential staining for each antibody to make sure that any potential co-localization is actually real and not a side effect of the staining process.
2. In the case of LC3A staining, extra care should be taken with the selected antibody. In our lab we have tested several

antibodies to make sure that we do not have unspecific binding with LC3B or LC3C, and this work has been published [5]. It is vital to select the correct antibody with high specificity especially in tissue studies.

References

1. http://www.wcrf.org/int/cancer-facts-figures/data-specific-cancers/prostate-cancer-statistics
2. Hamdy FC et al (2016) 10-year outcomes after monitoring, surgery, or radiotherapy for localized prostate Cancer. N Engl J Med 375 (15):1415–1424
3. Kabeya Y, Mizushima N, Ueno T, Yamamoto A et al (2000) LC3, a mammalian homologue of yeast Apg8p, is localized in autophagosome membranes after processing. EMBO J 19 (21):5720–5728
4. Koukourakis MI, Kalamida D, Mitrakas A et al (2015) Intensified autophagy compromises the efficacy of radiotherapy against prostate cancer. Biochem Biophys Res Commun 461 (2):268–274
5. Koukourakis MI, Kalamida D, Giatromanolaki A et al (2015) Autophagosome proteins LC3A, LC3B and LC3C have distinct subcellular distribution kinetics and expression in cancer cell lines. PLoS One 10(9):e0137675

Chapter 37

Methods to Determine the Role of Autophagy Proteins in *C. elegans* Aging

Sivan Henis-Korenblit and Alicia Meléndez

Abstract

This chapter describes methods for the analysis of autophagy proteins in *C. elegans* aging. We discuss the strains to be considered, the methods for the delivery of double-stranded RNA, and the methods to measure autophagy levels, autophagic flux, and degradation by autophagy.

Key words *Caenorhabditis elegans*, L4, Double-stranded RNA, dsRNA, RNA interference, HT115, Transgene, GFP::LGG-1, SQST-1::GFP, Autophagy, Aging, Longevity

1 Introduction

Aging is the process of physiological deterioration over time, starting at the cellular level, eventually leading to the death of the organism. Research from the last few decades proved that aging is not just a passive, entropic, and inevitable process of physiological deterioration. Specifically, groundbreaking work in genetic model organisms proved that the rate of aging can be regulated by simple genetic and environmental manipulations [1]. Importantly, these manipulations that alter the aging rate can, simultaneously, extend life span and reduce morbidity. Genetic studies have identified a variety of evolutionarily conserved signaling pathways, which can activate defense and stress resistance mechanisms to combat aging. These include regulatory pathways that decrease insulin/IGF-1 signaling [2, 3], reduce mitochondrial function [4–6], decrease protein synthesis [7–9], mediate nutrient sensing and energy homeostasis [10–14], and interpret signals from the reproductive system [15]. Many of these alterations were first discovered in model organisms such as worms, whose short life span makes them most suitable for these studies. Importantly, the majority of these aging regulatory pathways are evolutionary conserved and extend life span in mammals as well.

Nicholas Ktistakis and Oliver Florey (eds.), *Autophagy: Methods and Protocols*, Methods in Molecular Biology, vol. 1880,
https://doi.org/10.1007/978-1-4939-8873-0_37,

One of the hallmarks of the aging process is a progressive failure to maintain and repair cellular homeostasis [16]. The decline in cellular homeostasis is accompanied by a gradual accumulation of misfolded, or aggregated, macromolecules and defective organelles that can have deleterious effects on cellular homeostasis and tissue integrity and contributes to multiple age-related diseases such as Alzheimer's (AD), Parkinson's (PD), or Huntington's disease (HD) [17, 18]. One of the ways to counteract the accumulation of these misfolded proteins, and damaged organelles, is to remove them through lysosome-mediated autophagy [18, 19].

Autophagy is an evolutionarily conserved process of degradation, which can be induced in response to nutrient deprivation [20]. It involves the formation of a double-membrane structure, an autophagosome, which engulfs cytoplasmic components, such as organelles and long-lived proteins. The autophagosome fuses with the acidified lysosomes, containing hydrolases, and forms the autolysosome, where the sequestered cargo is degraded.

Degradation by autophagy enables the recycling of amino acids, carbohydrates, lipids, and nucleotides. Although autophagy has been considered to be nonselective in its degradation of bulk cytoplasm, it can also be selective and target cargo such as mitochondria, peroxisomes, and endoplasmic reticulum. In addition to nutrient deprivation, autophagy is induced in response to other stresses, such as organelle dysfunction, DNA damage, osmotic stress, oxidative stress, and resistance to pathogens [17]. More than 30 *ATG* (autophagy-related) genes have been identified in genetic screens in *Saccharomyces cerevisiae* to be required for the process of autophagy (Table 1), most of which are conserved from yeast to mammals. Autophagy proteins encoded by these genes form distinct complexes that act at different steps of the pathway to form and degrade the contents of the autophagosome [20, 71, 72].

Autophagy is induced by the activation of the serine/threonine protein kinase complex, which includes UNC-51 and ATG-13. Nucleation of the isolation membrane or phagophore requires the class III phosphatidylinositol 3-kinase VPS-34 complex, which includes BEC-1, VPS-34, VPS-15, and EPG-8. Expansion of the phagophore in yeast requires two ubiquitin-like conjugation systems resulting in the ubiquitin-like protein Atg8 to be conjugated to phosphatidylethanolamine (PE). In *C. elegans* two orthologs exist for yeast Atg8: LGG-1 and LGG-2 [67]. In the biogenesis of the autophagosome, the posttranslational modifications of LGG-1 or LGG-2 result in its insertion into the outer and inner membrane of the autophagosome, and thus GFP-tagged versions of LGG-1 or LGG-2 proteins (and their orthologs in other systems) have been used to assess steady-state levels of autophagy [67]. Completion of the autophagosomes results in LGG-1 or

Table 1
***A*utophagy genes in *C. elegans*, *S. cerevisiae*, and mammals**

C. elegans atg gene	Mutant allele	*S. cerevisiae* ortholog	Mammalian ortholog	References
atg-2	*bp576*	*ATG2*	*ATG2A, ATG2B*	[21, 22]
atg-3 [b]	*bp412*	*ATG3*	*ATG3*	[22, 23]
atg-4.1	*bp501*	*ATG4*	*ATG4A, ATG4B*	[22, 24]
atg-4.2	*tm3948*	*ATG4*	*ATG4C, ATG4D*	[22, 24]
atg-5 [b]	*bp484*	*ATG5*	*ATG5*	[22, 25]
atg-7 [b]	*bp422, bp290*	*ATG7*	*ATG7*	[21, 26, 27]
atg-9 [b]	*bp564, gk421128*	*ATG9*	*ATG9A/ATG9B*	[21, 22]
atg-10 [b]	*bp421, bp588*	*ATG10*		[23]
atg-16.1	*gk668615*	*ATG16*	*ATG16L1, ATG16L2*	[22, 25]
atg-16.2	*bp636, ok3224*	*ATG16*	*ATG16L1, ATG16L2*	[22, 25, 27]
atg-18 [ab]	*gk378, bp594*	*ATG18*	*WIPI1/WIPI2/ WIPI3/WIPI4*	[27–30]
epg-1	*bp414, bp417*	*KIAA0652*	*ATG13*	[22, 31, 32]
epg-2 [b]	*bp444*	–		[30]
epg-3 [b]	*bp405*	–	*VMP1*	[30, 33]
epg-4 [b]	*bp425*	–	*EI24*	[30]
epg-5 [b]	*bp450, tm3425*	–	*EPG5*	[22, 30, 35–36]
epg-6	*bp242*	–	*WIPI3,4*	[21, 22]
epg-7	*tm2508*	–	*FIP200*	[37]
epg-8 [ab]	*bp251, ok2561*	*ATG14*	*ATG14L/Barkor*	[22, 38]
epg-9	*bp320*	–	*ATG101*	[22, 32]
bec-1 [ab]	*ok691, ok700, bp613*	*ATG6/VPS30*	*BECN1*	[26–28, 39–41]
unc-51 [b]	*e369*	*ATG1*	*ULK1/ULK2*	[22, 29, 42, 43]
vps-15 [ab]	*ok3132*	*VPS15*	*PIK3R4*	[44–46]
vps-34 [ab]	*h741*	*VPS34*	*PIK3C3*	[26, 47]

(continued)

Table 1
(continued)

C. elegans atg gene	Mutant allele	*S. cerevisiae* ortholog	Mammalian ortholog	References
let-363 [ab]	*h98*	*TOR1,2*	*TOR*	[48–51]
lgg-1 [ab]	*bp500, tm3489*	*ATG8*	*LC3, GABARAP, GABARAPL1, GABARAPL2*	[22, 23, 29, 30, 42, 52–58]
lgg-2 [b]	*tm5755*	*ATG8*	*LC3, GABARAP, GABARAPL1, GABARAPL2*	[22, 55–59]
lgg-3 [ab]	*tm1462*	*ATG12*	*ATG12*	[22, 23, 29]
sepa-1	*bp456*	–	–	[30, 31]
lmp-1 [b]	*nr2045*	–	*LAMP1*	[60, 61]
rab-7 [b]	*ok511, tm3300*	*YPT7*	*RAB7*	[62–66]
sqst-1	*ok2892*	–	*SQSTM1/p62*	[28, 37, 66–70]

[a] Null mutants are lethal
[b] RNAi has a phenotype

LGG-2 being cleaved off by the cysteine protease ATG-4 on the outer membrane.

Autophagy plays a crucial role in the process of aging. Experiments in this small round nematode were the first to show a role for autophagy in the extension of life span [29]. Although the levels of autophagy normally drop with age [73], all long-lived *C. elegans* mutants examined to date display elevated numbers of autophagic vesicles at steady state in their hypodermis, during larval development, and/or in their intestine during adulthood [48–50, 74, 75]. More recently, unique age- and tissue-specific changes in autophagic activity were detected in distinct long-lived mutants [73]. Accordingly, many longevity mutants have increased transcript levels of several autophagy genes [48–50, 73, 74, 76]. Most importantly, all of these longevity mutants rely on their ability to execute autophagy for their longevity. Likewise, it has been shown that overexpression of critical autophagy components is sufficient for life span extension in model organisms, including *D. melanogaster* [77, 78] and mice [79]. Altogether, these findings suggest that autophagic activity is elevated in long-lived animals and that this increase is critical, and in some cases sufficient, for life span extension.

This chapter describes techniques to investigate the role of autophagy proteins in aging in *C. elegans*. We give a step-by-step description on how to measure life span, investigate the connection between autophagy genes and longevity mutants, and, finally, visually detect autophagy. We emphasize that given the complexity of autophagy, and given the inherent limitations of the different assays, a combination of assays is required for measuring autophagy activity with high confidence.

2 Materials

1. Petri dishes (60 mm × 15 mm).
2. 5 mg/mL cholesterol stock solution dissolved in ethanol. Cholesterol in ethanol. Store at 4 °C. Do not autoclave.
3. Phosphate buffer solution, pH 6.0: Dissolve 102.2 g KH_2PO_4 and 57.06 g K_2HPO_4 in 1 L of distilled water for 1 M. Autoclave to sterilize and store at room temperature.
4. Magnesium sulfate solution: Dissolve 30.09 g $MgSO_4$ in 250 mL for 1 M. Filter-sterilize and store at room temperature.
5. Calcium chloride stock solution: Dissolve 27.7 g $CaCl_2$ in 250 mL of distilled water for 1 M. Filter-sterilize and store at room temperature.
6. Nematode Growth Medium (NGM) agar plates: Mix 3 g NaCl, 17 g agar, 2.5 g Peptone, 1 mL cholesterol (5 mg/mL in EtOH), and 975 mL H_2O. Autoclave, and then add the following sterile solutions, mixing after each solution is added: 1 mL of 1 M $CaCl_2$, 1 mL of 1 M $MgSO_4$, and 25 mL of 1 M potassium phosphate (pH 6.0). Pour plates. Plates can be stored at room temperature to allow the plates to dry for several days before using or at 4 °C for several days to keep. If kept at 4 °C, equilibrate at room temperature for several hours before using.
7. NGM agar plates + carbenicillin: Prepare NGM plates and add 25–50 μg/mL carbenicillin. Pour plates 4–7 days before using, allowing to dry.
8. M9 buffer: 3 g KH_2PO_4, 5 g NaCl, add H_2O up to 1 L, and sterilize. Add 1 mM $MgSO_4$ after sterilization. pH near 7.0 should be attained.
9. Luria Broth (LB): 10 g Tryptone, 5 g of yeast extract, 5 g of NaCl, 10 mL 1 M Tris, and pH 8.0, per liter.
10. LB plates + tetracycline + ampicillin: LB plates contain 10 μg/mL tetracycline and 100 μg/mL ampicillin (ampicillin 100 mg/mL stock, sterile-filtered and stored at −20 °C) (*see* **Note 1**).
11. NGM plates with final concentration of 1–2 mM IPTG and 25–50 μg/mL carbenicillin (*see* **Note 2**).

12. PBST buffer: 137 mM NaCl, 2.7 mM KCl, 20 mM dibasic sodium phosphate, 2 mM monobasic potassium phosphate, and 0.1% Tween-20.
13. Bleach solution (10 mL solution: 5.5 mL distilled water, 3 mL sodium hypochlorite (bleach), 1.5 mL of 5 N KOH).
14. SDS sample buffer: To make 10 mL of 4× stock, add 2 mL Tris-HCl 1M pH6.8, 0.8 g SDS, 4 mL glycerol (100%), 0.4 mL 14.7 M β-mercaptoethanol, 1.0 mL 0.5 M EDTA, and 8 mg bromophenol blue (0.5%). Final concentration 1×: 50 mM Tris–HCl pH 6.8, 2% SDS, 10% glycerol, 1% β-mercaptoethanol, 12.5 mM EDTA, and 0.02% bromophenol blue.

3 Methods

3.1 Maintenance of *C. elegans* on NGM and RNAi Plates

3.1.1 Plate Preparation

1. For 1 L of NGM medium, add 20 g agar, 2.5 g Peptone, and 3 g NaCl, and autoclave.
2. When temperature drops to 55–60 °C, add 1 mL of 5 mg/mL cholesterol (dissolved in ethanol), 1 mL of 1 M $CaCl_2$, 1 mL of 1 M $MgSO_4$, and 25 mL of potassium phosphate buffer (0.69 M KH_2PO_4, 0.15 M K_2HPO_4, pH 6.0). All reagents added after the autoclave should be filter sterilized.
3. If plates are intended for RNAi, add 25–50 μg/mL carbenicillin (to select for plasmid expression) and 1–2 mM IPTG (2 mM commonly used to induce RNAi expression from the plasmid).
4. Using careful sterile technique, pour the medium into petri dishes of the required size (3 cm diameter plates are usually convenient).
5. Allow the plates to solidify. Unseeded plates can be stored at 4 °C for 2–3 weeks.
6. Seed ~100 μL of overnight cultures of OP50- or RNAi-expressing bacteria at the center of each plate, avoiding the spread of bacteria to the edges of the plates.
7. Incubate the seeded plates at room temperature (RT) for at least 24 h prior to experimentation.

3.1.2 Growing RNAi Bacteria

Gene-specific RNAi knockdown is achieved by feeding *C. elegans* with HT115 bacteria containing a plasmid encoding double-stranded RNA corresponding to the gene to be knocked down, expressed from flanking T7 promoters. Two commercially genome-wide RNAi bacterial libraries covering most of the *C. elegans* genome are available [6, 80, 81] (*see* **Note 3**).

1. Inoculate a fresh, single colony of RNAi bacteria in LB medium containing antibiotics (100 μg/mL carbenicillin +12.5 μg/mL tetracycline).

2. Incubate overnight at 37 °C with shaking (*see* **Note 4**).
3. Seed ~100 μL of the culture onto the center of the RNAi plates [NGM plates + 25–50 μg/mL carbenicillin (to select for plasmid expression) + 1–2 mM IPTG (to induce RNAi expression from the plasmid)].
4. Allow for overnight growth at room temperature.

3.2 Care of Animals

The N2 wild-type strain, as well as long-lived mutants and autophagy mutants, can be obtained from the *Caenorhabditis* Genetics Center (CGC, Minneapolis, MN; https://cgc.umn.edu/). Worms are maintained typically at 20 °C in petri dishes containing Nematode Growth Medium (NGM), seeded with *Escherichia coli* strain OP50 (also available from the CGC). A group of worms should be transferred to fresh plates every 2–3 days, prior to the depletion of the bacterial lawn.

3.3 Egg Collection by Hand

Life span studies follow the survival of an isogenic age-matched population of 100 individuals or more, from day 0 of adulthood. Isogenic age-matched populations are obtained by isolating a batch of eggs that will hatch into a synchronized population. These can be obtained in several ways:

3.3.1 Egg Collection by Hand

In this method, eggs that have already been laid, but are yet to hatch, are picked by hand and transferred to a plate for analysis (*see* **Note 5**).

1. Using a worm pick, pick eggs from a plate with growing worms of the desired genotype.
2. Place 100–200 eggs per plate on fresh plates seeded with bacteria. A minimum of three plates per strain should be prepared, to account for low hatching rates or contamination.
3. Allow the eggs to develop into L4s at 20 degrees for 2 days.
4. The age of the population should be resynchronized at the L4 stage.

3.3.2 Obtaining Eggs by Bleaching

In this method, many gravid adults of the desired genotype are dissolved in a bleaching solution. Eggs, which are resistant to the bleaching procedure, will develop into a synchronized population with an age range spanning a few hours. This approach yields a very large number of eggs (useful for setting up multiple life span experiments in parallel), but it also stresses the eggs (and hence it creates a stress history for the future progeny).

1. Wash worms off crowded but not starving plates (plates should contain many gravid adults) with 5–6 mL M9 buffer (3 g KH_2PO_4, 6 g Na_2HPO_4, 5 g NaCl, 1 mL 1 M $MgSO_4$, H_2O to 1 L) and pipette into a conical tube. To concentrate animals,

spin down for 30 s, at 2000 rpm =~700g. Aspirate liquid leaving 0.5 mL cushion.

2. Wash worms with another 10 mL of M9, and spin down again for 30 s at 2000 rpm =~700g. Aspirate liquid leaving 0.5 mL cushion.
3. Suspend worms in 5 mL bleach solution (10 mL solution: 5.5 mL distilled water, 3 mL sodium hypochlorite, 1.5 mL of 5 N KOH), and spin down immediately at 3000 rpm =~1500g. Aspirate liquid leaving 0.5 mL cushion.
4. Add another 10 mL bleach solution, and incubate worms, shaking or vortexing regularly for up to 4–5 min (longer incubations may stress and harm the eggs). Bleach incubation is stopped once most adult worms dissolve, leaving only eggs intact. This is determined by monitoring the worms under the microscope. Collect the eggs by centrifugation (30 s at 3000 rpm), and aspirate the bleach. From this point, sterile conditions should be employed.
5. Wash worms three times in 10 mL M9 solution. After the last wash, suspend the eggs in 1 mL M9.
6. Count the density of the eggs by dropping three drops of 1–5 μL of the solution onto an unseeded NGM plate (eggs should look smooth and oval-shaped and remain intact when liquid is absorbed).
7. Seed 100–150 eggs per plate. Allow the eggs to develop into adults. Three plates should be plated per strain to account for low hatching rate or contaminations. Allow the eggs to develop into adults. The age of the population should be resynchronized at the L4 stage.

3.3.3 Obtaining Eggs by a Timed Egg-Lay ("Hatch-Off")

In this method gravid adults of the desired genotype are transferred to fresh plates for a few hours and then removed. Eggs laid within this time course will develop into a synchronized population with an age range spanning just a few hours. This approach does not stress the eggs in any way; however the success of this approach relies on the ability of the adult hermaphrodites to lay a large number of eggs in a short time. It is hard to estimate how many eggs have actually been laid on the plates. This may vary depending on the genetic background of the animals used and their age.

1. Place 10–15 young adults of the desired genotype on fresh plates seeded with bacteria. 5–10 independent egg-laying plates should be prepared in parallel, to account for low laying and low hatching rates or contamination.
2. After a defined time period of 2–6 h, remove the adult animals from the plate (the smaller the time window, the tighter the synchronization).

3. Allow the eggs that have been laid on the plates during this time period to develop into L4s at 20 degrees.
4. The age of the population should be resynchronized at the L4 stage.

3.3.4 Synchronized L1 Larvae from "Hatch-Offs"

This protocol is adapted from a method described by Pepper et al. [82].

1. Place ~30 gravid adult worms (with eggs) on a plate for each strain that you wish to synchronize.
2. Allow worms to lay eggs overnight. Plates should have many eggs and some L1s by morning; otherwise do not proceed, and consider restarting with more adult worms per plate.
3. Pour 1–2 mL of M9 buffer onto the plates and swirl. The adults and young animals will come off the plate together with the liquid, while the eggs will stick to the bacteria. Pour off and discard the M9 with the worms. Inspect the plate under a microscope to ensure all worms have come off and only eggs remain (*see* **Note 6**).
4. Wait 1 h for new L1 animals to hatch before collecting animals. To collect "hatch-off" animals, pour 1–2 mL of M9 buffer onto plates, and swirl. Collect all L1 larvae into a 15 mL tube. "Hatch-offs" are all synchronized. The collection period may vary depending on how tightly synchronized you wish the animals to be.
5. Spin down in a clinical centrifuge at 3000 rpm for 5 min.
6. Ensure that the worms are at the bottom before discarding supernatant. Remove as much supernatant as possible without disturbing pellet with Pasteur pipette, and discard supernatant. Leaving roughly 500 mL of M9 with the worms is suggested.
7. Rinse the pellet with 1.5 mL M9 by spinning down and discarding supernatant again.
8. Aspirate the pelleted worms from the tube with a Pasteur pipette, and squirt worms onto fresh plate. Label the plate with the time, and consider that hour zero post-hatching.
9. **Steps 4–10** can be repeated every hour to collect additional batches of synchronized worms with roughly 1 h spacing between them. Do not exceed three collections from the same plate because the eggs will no longer stick to the thinning bacterial lawn.

3.4 Life Span Assay and Scoring

Life span experiments are usually performed at 20 °C, including all strains whose life spans are to be compared. Life span analyses should be repeated at least three times to show consistency in trends. Life span comparisons should be performed within a batch of life span experiments performed in parallel and not between experiments that have been performed independently.

1. Collect around 100 synchronized day 0 (L4 stage) animals. Place 10–15 animals on each of 6–10 seeded NG plates seeded with OP50 or on plates supplemented with IPTG and carbenicillin seeded with dsRNA-expressing HT115 bacteria for RNAi studies.
2. The life history of all animals on each plate is documented at least every 2 days by marking number of live/dead/censored animals on each plate at a given age. Dead and censored animals are documented and removed from the plate. Live animals are documented and left on the plate for further analysis.

An animal is defined as dead when it no longer responds to gentle prodding with a hot platinum pick.

Censored worms include animals with hatched live progeny within them ("bagging"), animals with eruption of intestines from their vulva ("explosion"), animals that cannot be found ("missing"), and those that have desiccated on the sides of the plates ("dried"). In addition, experiments with animals that contaminated by non-OP50/HT115 bacteria, by fungi or by yeast, must be censored as well ("contaminated").

3. During the reproductive period of the animals, adults must be gently transferred away from their progeny onto fresh plates at least every other day. One should completely avoid transferring progeny along with the adults (*see* **Note 7**).
4. Scoring should be performed at least every other day for all strains at once, until all worms are dead.

3.5 Analysis and Interpretation of Life Span Data

Life span assays are statistically treated as a survival analysis that compares the survival of two populations. Several statistical packages are available for survival analysis, including Prism 5, JMP8, and the STATA package. For each individual animal, the life span data is reported as its day of death or its day of censoring. Censored animals are considered alive until the day of censoring, in which they are excluded from the report. For reliable life span analysis, at least 40 non-censored death events per strain should be reported. Life span survival curves are typically presented in the form of a Kaplan-Meier curve, and mean life span values are reported along with statistical significance between groups examined.

To determine statistical significance for survival analysis, log-rank and Wilcoxon tests are commonly used. These tests account for censored animals and do not make assumptions about the distributions of survival measurements. The log-rank test is the

Table 2
Longevity mutant strains

Signaling pathway	Gene	Allele	Strain	Instructions on how to grow	References
Insulin IGF-like	*daf-2*	*e1370* *e1368*	CB1370 DR1572	Strain is temperature sensitive Grow at 20 °C or at 15 °C	[3, 29, 48, 73, 85]
	age-1	*hx546*	TJ1052	Not temperature sensitive. Normal fertility	[2]
Tor	*let363/ tor-1*	*h98* *h502* *ok3018*	KR344 KR827 VC2312	All *let-363/tor-1* mutants are lethal as L3 larvae. RNAi treatment has to be employed for longevity studies	[48, 49, 51]
Dietary restriction	*eat-2*	*ad1116*	CF1908	*eat-2(ad1116)* mutants are developmentally delayed. Alternatively, animals can be dietary restricted using different diet regimens. Note that life span determinants differ between the different regimens	[7, 13, 86]
Germlineless	*glp-1*	*e2141*	CB4037	Strain is temperature sensitive and sterile at 25 °C. Maintain strain at 15–20 °C. For sterility induction, place eggs at 25 °C, and shift animals to 20 °C after 48 h Alternatively, laser-mediated ablation of the Z2 and Z3 germline precursor cells can be performed in L1 animals	[49, 73, 87, 88]
Mitochondrial dysfunction	*isp-1*	*qm150*	MQ887	Very slow development, with 3-day delay at 20 °C	[4, 5, 50, 89, 90]
	clk-1	*e2519* *qm30*	CB4876 MQ130	Very slow development, with 1-day delay at 20 °C	[4, 91]
	atp-3			RNAi (during development, prior to L4 stage) targeting respiratory chain components interferes with mitochondrial function and extends life span	[4, 90]
	nuo-2				[4, 90, 92]
	cyc-1				[4]
	cco-1				[4, 90]
Protein translation	*ife-2*	*ok306*	KX15, RB579		[8, 93]
	rsks-1	*sv33* *ok1255* *tm1714*	RB1206	*rsks-1(sv33)* is developmentally delayed (approximately 16h delayed at 20 °C)	[93–96]

most commonly used test and assumes that the number of deaths per time is the same at all the measured time points. Wilcoxon on the other hand gives more importance to earlier fatalities [83, 84].

3.6 Autophagy Requirements for Life Span

The question of the importance of autophagy to the longevity of long-lived mutants can be addressed in several ways.

1. *Does autophagy contribute to the longevity of the animals?* To address this, the life spans of four strains should be analyzed in parallel: (a) wild-type animals, (b) an autophagy mutant of interest, (c) a long-lived mutant of interest (*see* Table 2 for commonly used longevity mutants), and (d) a double mutant carrying a mutation in the corresponding longevity and autophagy genes (*see* Table 1 for autophagy genes and mutations). The following autophagy genes have been tested with different longevity mutants: *bec-1*, *unc-51*, *lgg-1*, *atg-7*, *atg-18*, and *lgg-3/ATG12* [29, 48, 49, 85]. Upon completion of the life span experiment, the percentage of mean life span extension achieved in animals with intact autophagy can be compared to that achieved in animals with perturbed autophagy. If the percentage of life span extension in animals with intact autophagy consistently exceeds the percentage of life span extension of animals with perturbed autophagy, this argues that autophagy contributes to the examined longevity pathway.
2. *When does autophagy contribute to the longevity of the animals?* Once it is established that a longevity pathway requires autophagy, one can ask whether autophagy is required throughout the full life span of the animals or whether it is required during specific periods in the animal's lifetime.

   The simplest way to address this is to perform life span experiments using a genetic mutant with an extended life span and to inactivate autophagy, using RNAi against a specific autophagy gene. At each time point, one can decide whether the animals are grown on RNAi plates seeded with control bacteria or on RNAi plates seeded with bacteria-expressing dsRNA against an autophagy gene.

   To inactivate autophagy throughout life, eggs are placed on autophagy gene RNAi plates and maintained on these plates henceforth. To initiate RNAi treatment later in life, eggs are placed on control RNAi plates and transferred to autophagy gene RNAi plates at later time points.

   To halt an RNAi treatment that was initiated early in life, eggs are initially placed and maintained on autophagy gene RNAi plates. To inactivate the RNAi treatment, animals are transferred to RNAi plates seeded with bacteria-expressing dsRNA against *dicer*. The latter RNAi treatment curtails the exogenous RNAi pathway, which is essential for the silencing of autophagy genes by RNAi.

Table 3
RNAi-restricted strains

Tissue	Genotype	Strain	References
Germline-sensitive*	*rrf-1(pk1417)*	MAH23	[98]
Hypodermis-sensitive	*rde-1(ne219)V; kzIs9[(pKK1260) lin-26p::NLS::GFP + (pKK1253) lin-26p::rde-1 + rol-6(su1006)]*	NR222	[73, 99]
Hypodermis-sensitive	*rde-1(ne219)V; kzIs16* [pPD93.97(*myo-3p::nls::gfp*), pKK1253 (*lin-26p::rde-1*) + *rol-6(su1006)]*	NR229	[99]
Muscle-specific	*rde-1(ne219)V; kzIs20[hlh-1p::rde-1 + sur-5p::NLS::GFP]*	NR350	[73, 99]
Muscle-specific	*rde-1(ne300); neIs9[myo- 3::HA::RDE-1 + pRF4, rol-6 (su1006)]*	WM118	[100–102]
Muscle-specific	*sid-1(qt9) V; uthIs237[myo-3p::dtomato + myo-3p::sid-1]*	AGD855	[73, 86]
Muscle-specific	*eat-2(ad1116)III; sid-1(qt9)V; uthEx228[myo-3p::dtomato +myo-3p::sid-1]*	AGD620	[86]
Intestine-specific	*sid-1(qt9)V; alxIs6[vha-6::sid-1::SL2::GFP]*	MGH167	[73, 86]
Intestine-specific	*glp-1(e2141)III; sid-1(qt9)V; alxIs6[vha-6::sid-1::SL2::GFP]*	MAH330	[73]
Intestine-specific	*daf-2(e1370)III; sid-1(qt9)V; alxIs6[vha-6::tomato + myo-3p:: sid-1]*	MAH410	[73]
Intestine-specific	*sid-1(qt9)V; mgEx769[vha-6p::sid-1::SL2::GFP]*	GR1639	[73, 103]
Intestine-specific	*rde-1(ne219)V; kbIs7[nhx-2p::rde-1 + rol-6(su1006)]*	VP303	[86, 104]
Intestine-specific	*eat-2(ad1116)II; rde-1(ne219)V; kbIs7[nhx-2p::rde-1 + rol-6 (su1006)]*	MAH64	[86]

*= C. elegans rrf-1 OK mutants maintain some RNAi activity in some parts of the soma [98].

3. *Where does autophagy contribute to the longevity of the animals?* Once it is established that a longevity pathway requires autophagy, one can ask whether autophagy is required throughout the animal to promote longevity or whether it is specifically required in specific tissues. This question can be addressed by two complementary approaches:

   (a) Rescue of autophagy gene activity in specific tissues or cells

   This can be achieved by generating transgenic lines in double mutants, which carry a long-lived mutation and a mutation in an autophagy gene. By generating transgenic animals that express a rescuing transgene of the corresponding autophagy gene driven by tissue-/cell-specific promoters, one can restore autophagy gene activity in the corresponding tissues/cells. If the life span of the double-mutant animals is restored due to the expression of the transgene in a specific tissue/cell, this indicates that autophagy at that anatomical site is sufficient for that longevity of the animals.

(b) Depletion of autophagy gene activity in specific tissues or cells

This can be achieved by generating strains in which a longevity pathway has been activated by a longevity-promoting mutation, which responds to RNAi in specific tissues. The response to RNAi in specific tissues can be achieved by inactivating the RNAi machinery throughout the animal (i.e., by introducing a mutation in *rde-4* or in *sid-1*) and restoring the expression of the gene essential for RNAi in the tissue of interest, with a rescuing transgene driven by a cell-/tissue-specific promoter. Such animals can then be placed on plates seeded with bacteria that produce RNAi against an autophagy gene or on plates seeded with bacteria that produce control RNAi. If the life span of animals in which autophagy has been inactivated in specific tissues is curtailed to the same extent as animals that respond to RNAi in all tissues, this indicates that autophagy at that site is important for that longevity of the animals. A list of strain with spatially restricted RNAi can be found in Table 3.

3.6.1 Quantification of Autophagic Vesicles

To measure autophagy levels in *C. elegans*, researchers rely heavily on the expression pattern of reporters for the Atg8/LC3 orthologs LGG-1/LGG-2, both of which are involved in autophagy [20, 23, 29, 30]. Several reporter strains are available and listed in Table 4.

The phosphatidylethanolamine (PE)-conjugated form of Atg8/LC3 is localized to the inner and outer membranes of the phagophore, and thus Atg8/LC3 is associated with autophagosomal membranes upon induction [107–110]. However, this assay has limitations, as the reporter only shows autophagy levels at steady state. In addition, under some circumstances, lipidated LGG-1 can also accumulate on unknown structures [67, 111]. Hence, an increase in autophagosomes and/or LGG-1-labeled structures does not establish whether autophagy activity is elevated or impaired. If an increase in the number of structures labeled by GFP::LGG-1 or mCherry::GFP::LGG-1 reporters is observed, researchers are advised to also determine if autophagic flux is affected (*see* below).

Table 4
***A*utophagy reporters**

C. elegans strain name	Genotype	References
QU1	*dzEx1[lgg-1p::gfp::lgg-1, rol-6(su1006)]*	[29]
DA2123	*adIs2122[pLGG-1::GFP::LGG-1, rol-6]*	[52]
MAH14	*daf-2(e1370)III; adIs2122*	[105]
MAH44	*glp-1(e2141)III; adIs2122*	[49]
MAH167	*sqEx13[lgg-1p::mCherry::gfp::lgg-1, rol-6(su1006)]*	[73]
MAH215	*sqIs11[lgg-1p::mCherry::gfp::lgg-1, rol-6(su1006)]*	[73]
MAH241	*daf-2(e1370)III; sqIs11*	[73]
MAH 341	*glp-1(e2141)III; sqIs11*	[73]
DLM1	*ttTi5605 II; unc-119(ed3) III; uwaEx1[eft-3p::CERULEAN-VENUS:: lgg-1 + unc-119(+)]*	[106]
DLM3	*ttTi5605 II; unc-119(ed3) III; uwaEx2[vha-6p::CERULEAN-VENUS::lgg-1 + unc-119(+)]*	[106]
DLM5	*ttTi5605 II; unc-119(ed3) III; uwaEx3[dpy-7p::CERULEAN-VENUS::lgg-1 + unc-119(+)]*	[106]
DLM7	*ttTi5605 II; unc-119(ed3) III; uwaEx4[myo-2p::CERULEAN-VENUS::lgg-1 + unc-119(+)]*	[106]
DLM9	*ttTi5605 II; unc-119(ed3) III; uwaEx5[myo-4p::CERULEAN-VENUS::lgg-1 + unc-119(+)]*	[106]
DLM11	*ttTi5605 II; unc-119(ed3) III; uwaEx6[rab-3p::CERULEAN-VENUS::lgg-1 + unc-119(+)]*	[106]
DLM2	*uwaSi1[eft-3p::Cerulean-Venus::LGG-1, unc-119(+)] II; unc-119(ed3) III* *Note: MoSCI insertion into ttTi5605 II to monitor ubiquitous expression*	[106]
DLM4	*uwaSi2[vha-6p::Cerulean-Venus::LGG-1, unc-119(+)] II; unc-119 (ed3)III* *Note: MoSCI insertion into ttTi5605 II to monitor expression in intestine*	[106]
DLM6	*uwaSi3[dpy-7p::Cerulean-Venus::LGG-1, unc-119(+)] II; unc-119(ed3) III* *Note: MoSCI insertion into ttTi5605 II to monitor expression in hypodermis*	[106]
DLM8	*uwaSi4[myo-2p::Cerulean-Venus::LGG-1, unc-119(+)] II; unc-119 (ed3)III* *Note: MoSCI insertion into ttTi5605 II to monitor expression in pharyngeal muscle*	[106]
DLM10	*uwaSi5[myo-3p::Cerulean-Venus::LGG-1, unc-119(+)] II; unc-119 (ed3)III*	[106]

(continued)

Table 4 (continued)

C. elegans strain name	Genotype	References
	Note: MoSCI insertion into ttTi5605 II to monitor expression in body wall muscle	
DLM12	*uwaSi6[rab-3p::Cerulean-Venus::LGG-1, unc-119(+)] II; unc-119 (ed3)III* *Note: MoSCI insertion into ttTi5605 II to monitor expression in neurons*	[106]

Measuring Autophagosomes in Steady-State by Microscopy

1. Mount live *C. elegans* on a 2% agarose pad in M9 medium containing 0.1% NaN3. Animals can also be immobilized by mounting using 0.5 μl of 0.1 μm diameter polystyrene microspheres (e.g., Polysciences 00876-15, 2.5% w/v suspension).
2. Image using an LSM Zeiss 710 scanning confocal microscope. Z-stack images should be acquired at 0.5 microns slice intervals at 63×. GFP excitation/emission is set to 493/517 nm to eliminate background autofluorescence.
3. GFP::LGG-1 (GFP) [29, 52] or mCherry::GFP::LGG-1 [73] (mCherry/GFP or mCherry only)-positive punctate structures (or foci) are counted in the hypodermal seam cells, or in the intestine, as the number per cell. GFP fluorescence becomes quenched in lysosomes due to the lysosomal acidified compartment, whereas mCherry fluorescence will remain unaffected. Thus, the mCherry::GFP::LGG-1 will allow for the counting of autophagosomes (yellow) and autolysosomes (red) (*see* **Note 8**).
4. Z-stacks should be converted to maximum intensity projections using either the GFP or RFP/mCherry channel and adjusted using the subtract background option in FIJI. Foci/ punctate structures are isolated by thresholding-based segmentation.
5. Statistical analysis of positive labeled structures should be performed using Poisson regression calculated with the R Core Team (2015). The average and SEM are calculated using one-way analysis of variance (ANOVA) or two-way ANOVA as applicable (GraphPad Prism, La Jolla, CA).

Immunoblotting to Determine Steady-State Autophagy Levels

Autophagy levels can be tracked using a dual fluorescent protein (dFP) tag fused to LGG-1 (dFP::LGG-1) that releases monomeric protein (mFP) when the dFP is cleaved by proteases in the lysosome [106]. The different forms can be visualized by gel electrophoresis. The dFP::LGG-1 reporter was expressed under the endogenous promoter *lgg-1*p, a ubiquitously active promoter *eft-3p*, or under the control of promoters driving expression specifically in the hypodermis (*dpy-7*p), muscles (*myo-3*p), pharynx (*myo-2*p), intestine (*vha-6*p), or neurons (*rab-3*p) [106]. Single-copy chromosomal insertions or tandem, multicopy extrachromosomal arrays of the dFP::LGG-1 are available and can be visualized by immunoblot with an anti-FP antibody. The following protocol is adapted from [106].

1. Synchronize animals by "hatch-off" method to have enough animals and to avoid stressful conditions.
2. Animals to be analyzed should be harvested by centrifugation.
3. Lysates are prepared in SDS sample buffer by repeated cycles of freezing and boiling, pelleted by centrifugation at $8000 \times g$ for 5 min.
4. Collect the supernatant, and run on a 12% electrophoresis polyacrylamide gel containing sodium dodecyl sulfate (SDS-PAGE).
5. Transfer to a nitrocellulose membrane (Bio-Rad Laboratories).
6. Block the membrane with Odyssey blocking buffer (LI-COR Biosciences) for 1 h.
7. Incubate membrane with anti-GFP monoclonal antibody (Roche Applied Science) to detect FP or GFP and a polyclonal anti-actin antibody as a control overnight.
8. Wash membrane with PBST buffer 3×10 min with gentle agitation.
9. Incubate membrane with species-appropriate, Alexa-conjugated secondary antibody (IRDye 800 goat anti-mouse from Rockland Immunochemicals or Alexa Fluor 680 goat anti-rabbit from Invitrogen).
10. Wash membrane with PBST 4×10 min.
11. Visualize membrane using the Odyssey Imaging System (LI-COR Biosciences).
12. Use ImageJ software (National Institutes of Health) to calculate densitometry, and use a two-tailed t-test for statistical comparisons.

3.6.2 Measuring Autophagic Flux

Several methods have been described to measure autophagic flux in *C. elegans* [73, 106, 112]. These are based on exposing animals to a lysosomal inhibitor such as chloroquine or bafilomycin A and measuring the number of autophagosomes labeled with either GFP::LGG-1 or the tandem construct mCherry::GFP::LGG-1. Similarly, the dFP::LGG-1 and mFP can be visualized by immunoblotting after incubating animals with chloroquine [106].

Chloroquine Treatment to Measure Flux

The following protocol is adapted from [112].

1. Standard NGM plates should be seeded with OP50 and UV-treated at 500 J/cm^2 for 10 min using a UV crosslinker (Fisher Scientific).
2. UV-treated plates are then supplemented with either 30 mM chloroquine (Sigma) or dH_2O as a control.
3. Animals should be raised on live OP50 NGM plates at 20 °C and synchronized as L1 larvae by "hatch-offs (*see* above)." L1 larvae are transferred onto UV-treated plates with either chloroquine or control and allowed to grow at 20 °C until they reached the L3 larval stage.
4. To image, mount live worms on 3% agarose pads with 50 mM sodium azide (Sigma) in M9 buffer.

   Fluorescence images can be obtained using an Axio Imager M2 microscope equipped with a CCD camera (Carl Zeiss Microscopy), captured using the AxioVision release 4.8 software (Carl Zeiss Microscopy), and deconvolved using the AxioVision 3D deconvolution module (Carl Zeiss Microscopy). Confocal images can also be acquired as z-stacks at a thickness of 0.5 microns.
5. Quantification of images can be performed using the open-source software FIJI (FIJI is just ImageJ, National Institute of Health).

   For quantification of fluorescently GFP::LGG-1 or labeled punctate structures, in seam cells and/or intestinal cells, z-stacks should be converted to maximum intensity projections using either the GFP or RFP/mCherry channel and adjusted using the subtract background option in FIJI. Foci/punctate structures are isolated by thresholding-based segmentation (*see* **Note 9**).

Bafilomycin A Treatment to Measure Flux

The following protocol was adapted from [73]. Bafilomycin A (BafA; BioViotica, Dransfeld, Germany) can be injected into the body cavity of *C. elegans* to test for autophagic flux [73, 113].

1. Inject 50 mM of BafA in 0.2% DMSO (25 mM stock in DMSO) or 0.2% DMSO (control) together with 2.5 ug/mL Texas Red Dextran or Cascade Blue Dextran, 3000 Daltons (resuspended in water at a stock concentration of 25 ug/mL; Thermo Fisher Scientific, Waltham, MA), into the anterior body cavity, intestine, or intestinal lumen (*see* **Note 10**).
2. Animals are allowed to recover for 2 h before they can be imaged by confocal microscopy.

3.6.3 Monitoring the Clearance of Autophagy Substrates

A variety of protein substrates, including the germline P granule components PGL-1 and PGL-3 and the receptor protein SEPA-1 and SQST-1, are degraded by autophagy [23, 30, 67]. Thus, monitoring the levels of these autophagy substrates can also be used to determine autophagy levels. This can be done using strains expressing a fluorescent marker that is fused to the autophagy substrate. Under normal growth conditions, these markers are expressed at very low levels. However, a block in autophagy results in the accumulation of the labeled substrate, whereas enhancement of autophagy further decreases their basal levels. In general, the SQST-1 substrate is more relevant for aging studies, as it is expressed in the soma of adult animals (*see* **Note 11**).

1. Synchronize animals expressing the SQST-1::GFP transgene by "hatch-off" method to have enough animals and to avoid stressful conditions.
2. Wait until the eggs develop into adults, resynchronizing them at L4 stage. Commence imaging (as described below) as soon as the age of interest is reached.
3. Mount live *C. elegans* on a 2% agarose pad in M9 medium containing 0.1% NaN_3.
4. Use 10× or 20× magnifications to take images of the fluorescent protein.

4 Notes

1. Ampicillin can be replaced for carbenicillin. Carbenicillin is more stable. Carbenicillin 100 mg/mL stock should be sterile-filtered and stored at −20 °C.
2. IPTG 1 M stock should be sterile-filtered and stored at −20 °C. IPTG can be excluded and added 1 day before using plates.
3. To ensure accurate targeting, the RNAi clone should be sequenced using the M13fwd primer. As a negative control, HT115 bacteria expressing an empty vector or irrelevant RNAi construct should be used. RNAi efficacy should be verified by testing the expression of GFP fusion protein, if available, or by qRT-PCR.
4. For stronger expression of the dsRNA, grow the bacteria for 1–2 days at room temperature, and then add the IPTG to the bacterial lawn.
5. This is the simplest approach, but the age range of the resulting population may span 10 h. This approach does not stress the eggs in any way.

6. In all cases, eggs intended for life span experiments should come from clean non-contaminated plates, in which animals have been growing under contamination-free non-crowded conditions for at least two generations.
7. To avoid separating animals from their progeny along the life span experiment, sterile animals (devoid of progeny) can be used. For example, one can use temperature-sensitive sterile mutants (such as *glp-4* or *fer-15* mutants). Some sterile animals (such as *glp-1* mutants) are themselves longevity mutants; hence they cannot be used for this purpose. Alternatively, the cell division-arresting chemical FUDR (2′ fluro-5′ deoxyuridine, 100 μg/mL) can be added to the plates, and animals can be transferred to FUDR-containing plates at late L4 stage to arrest progeny production. In both cases, control experiments must be performed to verify that the sterility mutations and/or the FUDR treatment do not alter life span on their own. Hence, since these approaches introduce new variants into the experiment, they should be applied only when it is not feasible to carry out life span experiments in fertile animals (e.g., in strains with high censoring due to bagging or when dealing with life spans of more than ten strains in parallel). Furthermore, any genetic perturbation of life span using sterile animals must be tested in fertile animals as well.
8. If the intensity of red fluorescence is weaker than that of green fluorescence, the gain on the red channel can be set lower to observe mCherry-positive foci clearly. If the intensity of red fluorescence is greater so that the red punctate structures are overexposed, the cytoplasmic mCherry::GFP::LGG-1 can appear as yellow [73]. As a control, the GFP::LGG-1 (G116A) [73] transgene, which is expected to be defective in lipidation and autophagosome membrane targeting and thus should not form positive punctate, should also be analyzed.
9. For quantification of intestinal foci, a consistent region of interest should be selected from the anterior and/or posterior of the intestine. To analyze the mCherry::LGG-1 expression in the intestine, the number and size of mCherry;:LGG-1-positive foci in the posterior region of the intestine only are detected, since this region consistently expresses the mCherry::LGG-1 transgenic reporter. In the seam cells of dauer animals, the size of GFP::LGG-1 foci can be quantified by scanning through z-slices and adjusting the threshold to highlight the entire area of individual foci found. The "add to" manager tool in FIJI should be used to ensure a single foci was not counted more than once. Autofluorescence can be detected by the DAPI channel, which normally detects autofluorescence. Any DAPI-positive punctate structures should be subtracted from any counts detected with the green or red

channels, using the image calculator feature. Both channels should be adjusted using the subtract background option.

10. Injection of BafA can have variable results when counting LGG-1::GFP-positive punctate structures; however, the variability is likely due to variation per animal rather than variability introduced by the injection of BafA.
11. Accumulation of an autophagy substrate, such as SQST-1::GFP, may also reflect defects in a process unrelated to autophagy, such as protein aggregates or damaged organelles, and thus defects in autophagy must be verified further. The *rpl-43* gene encodes the 60S ribosomal protein L37, and the *bp399* mutation or RNAi against *rpl-43* results in the accumulation of SQST-1::GFP-positive aggregates, independently of autophagy [28]. In *rpl-43* mutants, a second mutation that elevates autophagy levels can suppress the accumulation of SQST-1::GFP aggregates. Thus, to investigate if a mutation induces autophagy, test that mutation or RNAi treatment in a strain that carries the SQST-1::GFP in the background of the *rpl-43* mutation. Clearance of the SQST-1::GFP may demonstrate an increase in autophagy function [28].

To further analyze autophagy flux, one can also measure the colocalization between an autophagosome GFP::LGG-1 and a lysosomal LMP-1::tagRFP reporters [112].

Acknowledgments

Work in the Henis-Korenblit laboratory is supported by grants: no. 2013188 from the Binational Science Foundation, no. I-1211-309.13/2012 from the German-Israeli Foundation for Scientific Research and Development, and no. 3-12066 from the Israeli Ministry of Science, Technology and Space, and work in the Meléndez Lab is supported by grants R15GM102846 from the National Institute of Health, a PSC-CUNY Research award, and A.M. was an Ellison Medical Foundation New Scholar in Aging (AG-NS-0521-0).

References

1. Kenyon CJ (2010) The genetics of ageing. Nature 464:504–512
2. Friedman DB, Johnson TE (1988) A mutation in the age-1 gene in Caenorhabditis elegans lengthens life and reduces hermaphrodite fertility. Genetics 118:75–86
3. Kenyon C, Chang J, Gensch E, Rudner A, Tabtiang R (1993) A C. elegans mutant that lives twice as long as wild type. Nature 366:461–464
4. Dillin A, Hsu AL, Arantes-Oliveira N, Lehrer-Graiwer J, Hsin H, Fraser AG, Kamath RS, Ahringer J, Kenyon C (2002) Rates of behavior and aging specified by mitochondrial function during development. Science 298:2398–2401

5. Feng J, Bussiere F, Hekimi S (2001) Mitochondrial electron transport is a key determinant of life span in Caenorhabditis elegans. Dev Cell 1:633–644
6. Lee SS, Lee RY, Fraser AG, Kamath RS, Ahringer J, Ruvkun G (2003) A systematic RNAi screen identifies a critical role for mitochondria in C. elegans longevity. Nat Genet 33:40–48
7. Hansen M, Taubert S, Crawford D, Libina N, Lee SJ, Kenyon C (2007) Lifespan extension by conditions that inhibit translation in Caenorhabditis elegans. Aging Cell 6:95–110
8. Syntichaki P, Troulinaki K, Tavernarakis N (2007) eIF4E function in somatic cells modulates ageing in Caenorhabditis elegans. Nature 445:922–926
9. Syntichaki P, Troulinaki K, Tavernarakis N (2007) Protein synthesis is a novel determinant of aging in Caenorhabditis elegans. Ann N Y Acad Sci 1119:289–295
10. Apfeld J, O'Connor G, McDonagh T, DiStefano PS, Curtis R (2004) The AMP-activated protein kinase AAK-2 links energy levels and insulin-like signals to lifespan in C. elegans. Genes Dev 18:3004–3009
11. Greer EL, Dowlatshahi D, Banko MR, Villen J, Hoang K, Blanchard D, Gygi SP, Brunet A (2007) An AMPK-FOXO pathway mediates longevity induced by a novel method of dietary restriction in C. elegans. Curr Biol 17:1646–1656
12. Klass MR (1977) Aging in the nematode Caenorhabditis elegans: major biological and environmental factors influencing life span. Mech Ageing Dev 6:413–429
13. Lakowski B, Hekimi S (1998) The genetics of caloric restriction in Caenorhabditis elegans. Proc Natl Acad Sci U S A 95:13091–13096
14. Mair W, Morantte I, Rodrigues AP, Manning G, Montminy M, Shaw RJ, Dillin A (2011) Lifespan extension induced by AMPK and calcineurin is mediated by CRTC-1 and CREB. Nature 470:404–408
15. Hsin H, Kenyon C (1999) Signals from the reproductive system regulate the lifespan of C. elegans. Nature 399:362–366
16. Lopez-Otin C, Blasco MA, Partridge L, Serrano M, Kroemer G (2013) The hallmarks of aging. Cell 153:1194–1217
17. Levine B, Kroemer G (2008) Autophagy in the pathogenesis of disease. Cell 132:27–42
18. Vilchez D, Saez I, Dillin A (2014) The role of protein clearance mechanisms in organismal ageing and age-related diseases. Nat Commun 5:5659
19. Nixon RA (2013) The role of autophagy in neurodegenerative disease. Nat Med 19:983–997
20. Klionsky DJ, Abdelmohsen K, Abe A, Abedin MJ, Abeliovich H, Acevedo Arozena A, Adachi H, Adams CM, Adams PD, Adeli K et al (2016) Guidelines for the use and interpretation of assays for monitoring autophagy (3rd edition). Autophagy 12:1–222
21. Lu Q et al (2011) The WD40 repeat PtdIns(3)P-binding protein EPG-6 regulates progression of omegasomes to autophagosomes. Dev Cell 21(2):343–357
22. Stavoe AK et al (2016) KIF1A/UNC-104 transports ATG-9 to regulate neurodevelopment and autophagy at synapses. Dev Cell 38(2):171–185
23. Zhang Y, Yan L, Zhou Z, Yang P, Tian E, Zhang K, Zhao Y, Li Z, Song B, Han J et al (2009) SEPA-1 mediates the specific recognition and degradation of P granule components by autophagy in C. elegans. Cell 136:308–321
24. Wu F et al (2012) Differential function of the two Atg4 homologues in the aggrephagy pathway in Caenorhabditis elegans. J Biol Chem 287(35):29457–29467
25. Zhang H et al (2013) The two C. elegans ATG-16 homologs have partially redundant functions in the basal autophagy pathway. Autophagy 9(12):1965–1974
26. Lapierre LR et al (2013) Autophagy genes are required for normal lipid levels in C. elegans. Autophagy 9(3):278–286
27. Ames K et al (2017) A non-cell-autonomous role of BEC-1/BECN1/Beclin1 in coordinating cell-cycle progression and stem cell proliferation during germline development. Curr Biol 27(6):905–913
28. Guo B, Huang X, Zhang P, Qi L, Liang Q, Zhang X, Huang J, Fang B, Hou W, Han J et al (2014) Genome-wide screen identifies signaling pathways that regulate autophagy during Caenorhabditis elegans development. EMBO Rep 15:705–713
29. Melendez A, Talloczy Z, Seaman M, Eskelinen EL, Hall DH, Levine B (2003) Autophagy genes are essential for dauer development and life-span extension in C. elegans. Science 301:1387–1391
30. Tian Y, Li Z, Hu W, Ren H, Tian E, Zhao Y, Lu Q, Huang X, Yang P, Li X et al (2010) C. elegans screen identifies autophagy genes specific to multicellular organisms. Cell 141:1042–1055
31. Tian E et al (2009) epg-1 functions in autophagy-regulated processes and may encode a

highly divergent Atg13 homolog in C. elegans. Autophagy 5(5):608–615
32. Liang Q et al (2012) The C. elegans ATG101 homolog EPG-9 directly interacts with EPG-1/Atg13 and is essential for autophagy. Autophagy 8(10):1426–1433
33. Zhao YG et al (2017) The ER-localized transmembrane protein EPG-3/VMP1 regulates SERCA activity to control ER-isolation membrane contacts for autophagosome formation. Mol Cell 67(6):974–989 e6
34. Wang Z et al (2016) The vici syndrome protein EPG5 Is a Rab7 effector that determines the fusion specificity of autophagosomes with late endosomes/lysosomes. Mol Cell 63 (5):781–795
35. Cullup T et al (2013) Recessive mutations in EPG5 cause Vici syndrome, a multisystem disorder with defective autophagy. Nat Genet 45(1):83–87
36. Herpin A et al (2015) Defective autophagy through epg5 mutation results in failure to reduce germ plasm and mitochondria. FASEB J 29(10):4145–4161
37. Lin L et al (2013) The scaffold protein EPG-7 links cargo-receptor complexes with the autophagic assembly machinery. J Cell Biol 201(1):113–129
38. Yang P, Zhang H (2011) The coiled-coil domain protein EPG-8 plays an essential role in the autophagy pathway in C. elegans. Autophagy 7(2):159–165
39. Ruck A et al (2011) The Atg6/Vps30/Beclin 1 ortholog BEC-1 mediates endocytic retrograde transport in addition to autophagy in C. elegans. Autophagy 7(4):386–400
40. Takacs-Vellai K et al (2005) Inactivation of the autophagy gene bec-1 triggers apoptotic cell death in C. elegans. Curr Biol 15 (16):1513–1517
41. Ames K, Melendez A (2017) Non-autonomous autophagy in germline stem cell proliferation. Cell Cycle 16(16):1481–1482
42. Toth ML et al (2007) Influence of autophagy genes on ion-channel-dependent neuronal degeneration in Caenorhabditis elegans. J Cell Sci 120(Pt 6):1134–1141
43. Rowland AM et al (2006) Presynaptic terminals independently regulate synaptic clustering and autophagy of GABAA receptors in Caenorhabditis elegans. J Neurosci 26 (6):1711–1720
44. Lee HC et al (2012) Depletion of mboa-7, an enzyme that incorporates polyunsaturated fatty acids into phosphatidylinositol (PI), impairs PI 3-phosphate signaling in Caenorhabditis elegans. Genes Cells 17(9):748–757
45. Lowry J et al (2015) High-throughput cloning of temperature-sensitive caenorhabditis elegans mutants with adult syncytial germline membrane architecture defects. G3 (Bethesda) 5(11):2241–2255
46. Silhankova M et al (2010) Wnt signalling requires MTM-6 and MTM-9 myotubularin lipid-phosphatase function in Wnt-producing cells. EMBO J 29(24):4094–4105
47. Roggo L et al (2002) Membrane transport in Caenorhabditis elegans: an essential role for VPS34 at the nuclear membrane. EMBO J 21 (7):1673–1683
48. Hansen M, Chandra A, Mitic LL, Onken B, Driscoll M, Kenyon C (2008) A role for autophagy in the extension of lifespan by dietary restriction in *C. elegans*. PLoS Genet 4: e24
49. Lapierre LR, Gelino S, Melendez A, Hansen M (2011) Autophagy and lipid metabolism coordinately modulate life span in germline-less C. elegans. Curr Biol 21:1507–1514
50. Toth ML, Sigmond T, Borsos E, Barna J, Erdelyi P, Takacs-Vellai K, Orosz L, Kovacs AL, Csikos G, Sass M et al (2008) Longevity pathways converge on autophagy genes to regulate life span in Caenorhabditis elegans. Autophagy 4:330–338
51. Vellai T, Takacs-Vellai K, Zhang Y, Kovacs AL, Orosz L, Muller F (2003) Genetics: influence of TOR kinase on lifespan in C. elegans. Nature 426:620
52. Kang C, You YJ, Avery L (2007) Dual roles of autophagy in the survival of Caenorhabditis elegans during starvation. Genes Dev 21:2161–2171
53. Sato M, Sato K (2011) Degradation of paternal mitochondria by fertilization-triggered autophagy in C. elegans embryos. Science 334(6059):1141–1144
54. Samokhvalov V, Scott BA, Crowder CM (2008) Autophagy protects against hypoxic injury in C. elegans. Autophagy 4 (8):1034–1041
55. Hashimoto Y, Ookuma S, Nishida E (2009) Lifespan extension by suppression of autophagy genes in Caenorhabditis elegans. Genes Cells 14(6):717–726
56. Al Rawi S et al (2011) Postfertilization autophagy of sperm organelles prevents paternal mitochondrial DNA transmission. Science 334(6059):1144–1147
57. Alberti A et al (2010) The autophagosomal protein LGG-2 acts synergistically with LGG-1 in dauer formation and longevity in C. elegans. Autophagy 6(5):622–633

58. Manil-Segalen M et al (2014) The C. elegans LC3 acts downstream of GABARAP to degrade autophagosomes by interacting with the HOPS subunit VPS39. Dev Cell 28 (1):43–55
59. Kozlowski L et al (2014) The Caenorhabditis elegans HP1 family protein HPL-2 maintains ER homeostasis through the UPR and hormesis. Proc Natl Acad Sci U S A 111 (16):5956–5961
60. Nilsson L, Jonsson E, Tuck S (2011) Caenorhabditis elegans numb inhibits endocytic recycling by binding TAT-1 aminophospholipid translocase. Traffic 12(12):1839–1849
61. Kostich M, Fire A, Fambrough DM (2000) Identification and molecular-genetic characterization of a LAMP/CD68-like protein from Caenorhabditis elegans. J Cell Sci 113 (Pt 14):2595–2606
62. Gutierrez MG et al (2004) Rab7 is required for the normal progression of the autophagic pathway in mammalian cells. J Cell Sci 117(Pt 13):2687–2697
63. Jager S et al (2004) Role for Rab7 in maturation of late autophagic vacuoles. J Cell Sci 117 (Pt 20):4837–4848
64. Li W et al (2012) Autophagy genes function sequentially to promote apoptotic cell corpse degradation in the engulfing cell. J Cell Biol 197(1):27–35
65. Grant B, Hirsh D (1999) Receptor-mediated endocytosis in the Caenorhabditis elegans oocyte. Mol Biol Cell 10(12):4311–4326
66. Schimmoller F, Riezman H (1993) Involvement of Ypt7p, a small GTPase, in traffic from late endosome to the vacuole in yeast. J Cell Sci 106(Pt 3):823–830
67. Zhang H, Chang JT, Guo B, Hansen M, Jia K, Kovacs AL, Kumsta C, Lapierre LR, Legouis R, Lin L et al (2015) Guidelines for monitoring autophagy in Caenorhabditis elegans. Autophagy 11:9–27
68. Palmisano NJ, Melendez A (2016) Detecting autophagy in Caenorhabditis elegans embryos using markers of P granule degradation. Cold Spring Harb Protoc 2016(1):pdb. prot086504
69. Zhang G et al (2017) The composition of a protein aggregate modulates the specificity and efficiency of its autophagic degradation. Autophagy 13(9):1487–1495
70. Zhang P, Zhang H (2013) Autophagy modulates miRNA-mediated gene silencing and selectively degrades AIN-1/GW182 in C. elegans. EMBO Rep 14(6):568–576
71. Feng Y, He D, Yao Z, Klionsky DJ (2014) The machinery of macroautophagy. Cell Res 24:24–41
72. Nakatogawa H, Suzuki K, Kamada Y, Ohsumi Y (2009) Dynamics and diversity in autophagy mechanisms: lessons from yeast. Nat Rev Mol Cell Biol 10:458–467
73. Chang JT, Kumsta C, Hellman AB, Adams LM, Hansen M (2017) Spatiotemporal regulation of autophagy during Caenorhabditis elegans aging. elife 6:e18459
74. Lapierre LR, Kumsta C, Sandri M, Ballabio A, Hansen M (2015) Transcriptional and epigenetic regulation of autophagy in aging. Autophagy 11:867–880
75. Melendez A, Hall DH, Hansen M (2008) Monitoring the role of autophagy in C. elegans aging. Methods Enzymol 451:493–520
76. Jia K, Hart AC, Levine B (2007) Autophagy genes protect against disease caused by polyglutamine expansion proteins in Caenorhabditis elegans. Autophagy 3:21–25
77. Bai H, Kang P, Hernandez AM, Tatar M (2013) Activin signaling targeted by insulin/dFOXO regulates aging and muscle proteostasis in Drosophila. PLoS Genet 9:e1003941
78. Simonsen A, Cumming RC, Brech A, Isakson P, Schubert DR, Finley KD (2008) Promoting basal levels of autophagy in the nervous system enhances longevity and oxidant resistance in adult drosophila. Autophagy 4:176–184
79. Pyo JO, Yoo SM, Jung YK (2013) The interplay between autophagy and aging. Diabetes Metab J 37:333–339
80. Kamath RS, Fraser AG, Dong Y, Poulin G, Durbin R, Gotta M, Kanapin A, Le Bot N, Moreno S, Sohrmann M et al (2003) Systematic functional analysis of the Caenorhabditis elegans genome using RNAi. Nature 421:231–237
81. Rual JF, Hill DE, Vidal M (2004) ORFeome projects: gateway between genomics and omics. Curr Opin Chem Biol 8:20–25
82. Pepper AS, Killian DJ, Hubbard EJ (2003) Genetic analysis of Caenorhabditis elegans glp-1 mutants suggests receptor interaction or competition. Genetics 163:115–132
83. Jiang H, Fine JP (2007) Survival analysis. Methods Mol Biol 404:303–318
84. Jung SH, Jeong JH (2003) Rank tests for clustered survival data. Lifetime Data Anal 9:21–33
85. Hars ES, Qi H, Ryazanov AG, Jin S, Cai L, Hu C, Liu LF (2007) Autophagy regulates ageing in C. elegans. Autophagy 3:93–95

86. Gelino S, Chang JT, Kumsta C, She X, Davis A, Nguyen C, Panowski S, Hansen M (2016) Intestinal Autophagy Improves Healthspan and Longevity in C. elegans during Dietary Restriction. PLoS Genet 12:e1006135
87. Lapierre LR, De Magalhaes Filho CD, McQuary PR, Chu CC, Visvikis O, Chang JT, Gelino S, Ong B, Davis AE, Irazoqui JE, Dillin A, Hansen M (2013) The TFEB orthologue HLH-30 regulates autophagy and modulates longevity in Caenorhabditis elegans. Nat Commun 4:2267
88. Wang MC, O'Rourke EJ, Ruvkun G (2008) Fat metabolism links germline stem cells and longevity in C. elegans. Science 322:957–960
89. Felkai S, Ewbank JJ, Lemieux J, Labbe JC, Brown GG, Hekimi S (1999) CLK-1 controls respiration, behavior and aging in the nematode Caenorhabditis elegans. EMBO J 18:1783–1792
90. Rea SL, Ventura N, Johnson TE (2007) Relationship between mitochondrial electron transport chain dysfunction, development, and life extension in Caenorhabditis elegans. PLoS Biol 5:e259
91. Lakowski B, Hekimi S (1996) Determination of life-span in Caenorhabditis elegans by four clock genes. Science 272:1010–1013
92. Tsang WY, Sayles LC, Grad LI, Pilgrim DB, Lemire BD (2001) Mitochondrial respiratory chain deficiency in Caenorhabditis elegans results in developmental arrest and increased life span. J Biol Chem 276:32240–32246
93. Sheaffer KL, Updike DL, Mango SE (2008) The Target of Rapamycin pathway antagonizes pha-4/FoxA to control development and aging. Curr Biol 18:1355–1364
94. Chen D, Li PW, Goldstein BA, Cai W, Thomas EL, Chen F, Hubbard AE, Melov S, Kapahi P (2013) Germline signaling mediates the synergistically prolonged longevity produced by double mutations in daf-2 and rsks-1 in C. elegans. Cell Rep 5:1600–1610
95. McQuary PR, Liao CY, Chang JT, Kumsta C, She X, Davis A, Chu CC, Gelino S, Gomez-Amaro RL, Petrascheck M, Brill LM, Ladiges WC, Kennedy BK, Hansen M (2016) C. elegans S6K Mutants Require a Creatine-Kinase-like Effector for Lifespan Extension. Cell Rep 14:2059–2067
96. Pan KZ, Palter JE, Rogers AN, Olsen A, Chen D, Lithgow GJ, Kapahi P (2007) Inhibition of mRNA translation extends lifespan in Caenorhabditis elegans. Aging Cell 6:111–119
97. Seo K, Choi E, Lee D, Jeong DE, Jang SK, Lee SJ (2013) Heat shock factor 1 mediates the longevity conferred by inhibition of TOR and insulin/IGF-1 signaling pathways in C. elegans. Aging Cell 12:1073–1081
98. Kumsta C, Hansen M (2012) C. elegans rrf-1 mutations maintain RNAi efficiency in the soma in addition to the germline. PLoS One 7:e35428
99. Qadota H, Inoue M, Hikita T, Koppen M, Hardin JD, Amano M, Moerman DG, Kaibuchi K (2007) Establishment of a tissue-specific RNAi system in C. elegans. Gene 400:166–173
100. Bolz DD, Tenor JL, Aballay A (2010) A conserved PMK-1/p38 MAPK is required in caenorhabditis elegans tissue-specific immune response to Yersinia pestis infection. J Biol Chem 285:10832–10840
101. Cai L, Phong BL, Fisher AL, Wang Z (2011) Regulation of fertility, survival, and cuticle collagen function by the Caenorhabditis elegans eaf-1 and ell-1 genes. J Biol Chem 286:35915–35921
102. Silva MC, Amaral MD, Morimoto RI (2013) Neuronal reprograming of protein homeostasis by calcium-dependent regulation of the heat shock response. PLoS Genet 9: e1003711
103. Melo JA, Ruvkun G (2012) Inactivation of conserved C. elegans genes engages pathogen- and xenobiotic-associated defenses. Cell 149:452–466
104. Espelt MV, Estevez AY, Yin X, Strange K (2005) Oscillatory Ca2+ signaling in the isolated Caenorhabditis elegans intestine: role of the inositol-1,4,5-trisphosphate receptor and phospholipases C beta and gamma. J Gen Physiol 126:379–392
105. Egan D, Kim J, Shaw RJ, Guan KL (2011) The autophagy initiating kinase ULK1 is regulated via opposing phosphorylation by AMPK and mTOR. Autophagy 7:643–644
106. Chapin HC, Okada M, Merz AJ, Miler DL (2015) Tissue-specific autophagy responses to aging and stress in *C. elegans*. Aging (Albany NY) 7(6):419–434
107. Kabeya Y, Mizushima N, Yamamoto A, Oshitani-Okamoto S, Ohsumi Y, Yoshimori T (2004) LC3, GABARAP and GATE16 localize to autophagosomal membrane depending on form-II formation. J Cell Sci 117:2805–2812
108. Kirisako T, Baba M, Ishihara N, Miyazawa K, Ohsumi M, Yoshimori T, Noda T, Ohsumi Y (1999) Formation process of autophagosome is traced with Apg8/Aut7p in yeast. J Cell Biol 147:435–446
109. Kirisako T, Ichimura Y, Okada H, Kabeya Y, Mizushima N, Yoshimori T, Ohsumi M,

Takao T, Noda T, Ohsumi Y (2000) The reversible modification regulates the membrane-binding state of Apg8/Aut7 essential for autophagy and the cytoplasm to vacuole targeting pathway. J Cell Biol 151:263–276

110. Mizushima N, Yamamoto A, Matsui M, Yoshimori T, Ohsumi Y (2004) In vivo analysis of autophagy in response to nutrient starvation using transgenic mice expressing a fluorescent autophagosome marker. Mol Biol Cell 15:1101–1111

111. Liang Q, Yang P, Tian E, Han J, Zhang H (2012) The C. elegans ATG101 homolog EPG-9 directly interacts with EPG-1/Atg13 and is essential for autophagy. Autophagy 8:1426–1433

112. Palmisano NJ, Rosario N, Wysocki M, Hong M, Grant B, Meléndez A (2017) The recycling endosome protein RAB-10 promotes autophagic flux and localization of the transmembrane protein ATG-9. Autophagy 13 (10):1742–1753

113. Wilkinson DS, Jariwala JS, Anderson E, Mitra K, Meisenhelder J, Chang JT, Ideker T, Hunter T, Nizet V, Dillin A et al (2015) Phosphorylation of LC3 by the hippo kinases STK3/STK4 is essential for autophagy. Mol Cell 57:55–68

Part V

Mitophagy and Other Selective Autophagy Pathways

Chapter 38

Investigating Non-selective Autophagy in *Drosophila*

Szabolcs Takáts, Sarolta Tóth, Győző Szenci, and Gábor Juhász

Abstract

Drosophila melanogaster is a popular model organism in molecular genetics and cell biology. Various *Drosophila* tissues have been successfully used for studying autophagy, and our knowledge about the genetic regulation of this process is constantly growing. It is important to use assays that distinguish between non-selective autophagy and the selective forms. Here we introduce a selection of proven methods, which, taking into account their limitations, are suitable to measure non-selective autophagy in *Drosophila* fat and other tissues.

Key words Drosophila, Non-selective autophagy, Fluorescent dyes

1 Introduction

During the main form of autophagy, a phagophore membrane cistern captures parts of the cytoplasm (including organelles). This process can be very selective during basal conditions, while starvation triggers a massive induction of autophagy in tissues such as the larval fat in *Drosophila* to ensure the more or less random turnover of bulk cytoplasm, ensuring this way survival via recycling stored nutrients and energy [1, 2]. Considering that *Drosophila* possesses an ortholog of the p62 receptor for selective autophagy of ubiquitinated proteins and organelles (termed Ref(2)P in flies) and a complete Atg8a conjugation system (Atg8a is the main LC3/Atg8 homolog in *Drosophila*), it is obvious that fruit fly cells can perform selective autophagy in addition to non-selective forms [3, 4]. As the activity of selective and non-selective pathways is not separated either spatially or temporally, it is still a challenge to clearly distinguish between them. Some selectivity of cargo degradation is likely always retained [5]. Using different genetic backgrounds is one possible strategy to deal with this problem. For example, *Atg13* and *Atg14* mutant flies that are defective for the upstream-acting Atg1 protein kinase and Vps34 lipid kinase complexes, respectively, show a complete autophagy block. In contrast,

Nicholas Ktistakis and Oliver Florey (eds.), *Autophagy: Methods and Protocols*, Methods in Molecular Biology, vol. 1880,
https://doi.org/10.1007/978-1-4939-8873-0_38,

mutations in *Atg8a* or genes encoding members of the Atg8a lipidation system permit some residual autophagic degradation [6]. Because selective autophagy is known to depend on Atg8a and its homologs, this residual autophagy must represent the non-selective process. Nonetheless, it is important to note that Atg8 family proteins are important for autophagosome biogenesis/closure [6], so using these genetic manipulations to follow non-selective autophagic activity has limitations.

The larval fat tissue is the best-characterized organ to investigate non-selective autophagy in *Drosophila* [7, 8]. Upon starvation, extremely high levels of autophagy are induced in the fat cells of early L3 stage larvae. In this chapter, we give a practical summary of the most frequently used methods for studying non-selective autophagy in starved larvae, focusing on their fat cells. We introduce the common autophagy reporters, fluorescent dyes, and imaging techniques that are suitable to establish the overall autophagic activity. Combining these methods with different *Atg* mutant genetic backgrounds may highlight the different contributions of Atg protein complexes to selective and non-selective autophagic degradation.

2 Materials

2.1 Starving Larvae

1. Starvation solution: 20% sucrose dissolved in distilled water.
2. Forceps.
3. 1.5 mL Eppendorf tube.
4. Petri dish.

2.2 Fluorescent Microscopy

1. LysoTracker Red (LTR) dye solution: 100 nM LTR dissolved in phosphate-buffered saline (PBS).
2. Mounting solution: 1:1 PBS-glycerine solution containing 0.2 μg/mL DAPI (4′,6-diamidino-2-phenylindole).
3. Slide and coverslip.

2.3 Transmission Electron Microscopy

1. PBS.
2. Glass vial.
3. Chemical fixative solution: 0.5% glutaraldehyde, 3.2% paraformaldehyde, 1% sucrose, 2 mM $CaCl_2$, 0.1 M sodium cacodylate, pH 7.4.
4. Sodium cacodylate buffer: sodium cacodylate dissolved in distilled water (0.2 M), pH of 7.4 with HCl.
5. Rotator.
6. Osmium tetroxide solution: 1 g osmium tetroxide dissolved in 200 mL 0.1 M sodium cacodylate, pH 7.4.

7. 10% acetone.
8. 2% uranyl acetate solution in water.
9. Ethanol in various concentrations: 50%, 70%, and 96% and absolute ethanol.
10. Propylene oxide.
11. Ultramicrotome.
12. Glass knife breaker.
13. Glass knife/diamond knife for ultrasectioning.
14. Toluidine blue dye: 1% toluidine blue, 1% Azure II, 1% Borax dissolved in distilled water.
15. Grids (e.g., 300 mesh copper grids).
16. Reynolds' lead citrate solution. Mix 1.33 g lead nitrate and 1.75 g sodium citrate in 30 mL distilled water, allowing time for lead citrate to form, and then add 8 mL sodium hydroxide and bring up to 50 mL.

2.4 Western Blot

1. 2× Laemmli buffer (Sigma).
2. Running buffer: 0.025 M Tris base, 0.2 M glycine, 0.5% SDS dissolved in distilled water.
3. Transfer buffer: 0.025 M Tris base, 0.2 M glycine, 10% methanol dissolved in distilled water.
4. TBST (pH 7.4): 0.025 M Tris-HCl, 0.9% NaCl, and 0.1% Tween 20 dissolved in distilled water.
5. Blocking solution (pH 7.5): 0.5% casein from bovine milk (Sigma) and 0.1 M Tris-HCl, pH 7.4.
6. First antibody solution: first antibody dissolved in 1:1 Blocking solution-TBST.
7. Secondary antibody solution: alkaline phosphatase (AP) conjugated secondary antibody dissolved in 1:1 blocking solution-TBST.
8. AP buffer: 100 mM Tris-HCl, pH 9.5, 100 mM NaCl, 5 Mm $MgCl_2$, 0.05% Tween 20.
9. Nitro blue tetrazolium chloride (NBT) solution: 5% NBT in 70% dimethylformamide.
10. 5-Bromo-4-chloro-3-indolyl-phosphate (BCIP) solution: 2.5% BCIP in 100% dimethylformamide.
11. Developing solution: 66 μL NBT and 66 μL BCIP solutions mixed in 10 mL AP buffer.
12. 10% polyacrylamide separating gel and a stacking gel, both containing SDS.
13. Polyvinylidene fluoride (PVDF) membrane.

14. Gel electrophoresis and Western blotting equipment.
15. Orbital shaker.
16. Plastic box.

3 Methods

3.1 Starving *Drosophila* Larvae

To prepare starved larvae for the methods described below, we suggest using the following starvation protocol:

1. Scoop feeding larvae into a Petri dish (*see* **Note 1**).
2. Add starvation solution to the dish, and crush the fly food into small pieces (*see* **Note 2**).
3. Pipette 1 mL starvation solution into an empty Eppendorf tube (*see* **Note 3**).
4. Transfer early L3 stage larvae from the dish to a fresh drop of starvation solution, and wash them (*see* **Note 3**).
5. Transfer five to ten washed larvae into the Eppendorf tube (do not close!), and incubate for 3–4 h (*see* **Note 4**).

3.2 Investigating Autophagy Using a Fluorescent Confocal Microscope

Fluorescent reporters and vital dyes are routinely used to visualize autophagic structures in larval fat tissue.

3.2.1 Labeling Autolysosomes with LysoTracker Red (LTR)

LTR staining is a common method for the detection of acidic lysosomes. Upon starvation, induction of autophagy results in the appearance of numerous LTR-positive granules in fat cells, which can be considered as autolysosomes in this tissue [9, 10] (*see* **Note 5**).

1. Dissect larvae under a stereomicroscope, and incubate the fat bodies for 5 min in LTR dye solution (*see* **Note 6**).
2. Aspirate the staining solution and wash the sample with PBS (*see* **Note 7**).
3. Mount the sample in 30 μL mounting solution (*see* **Note 8**).

Optional: LTR labeling combined with clonal analysis

By using a *Drosophila* stock carrying proper transgenes (such as *hs-FLP, Act-FRT-CD2-FRT-Gal4, UAS-GFP*), it is possible to generate GFP-positive Gal4 expressing clones of cells surrounded by GFP-negative control cells [9] (*see* **Note 9**). This clonal system can be combined not only with LTR staining (Fig. 1) but also with fluorescent reporters to visualize cell autonomous effects of gene overexpression or silencing.

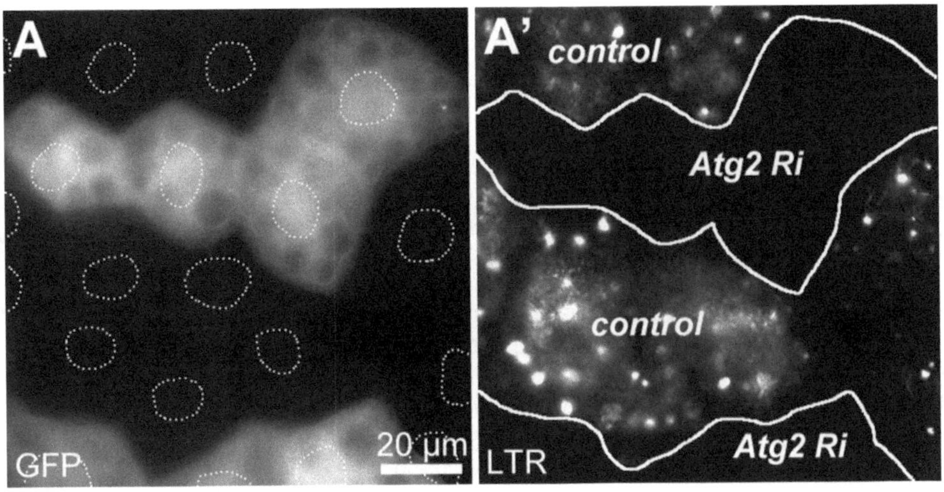

Fig. 1 Detection of acidic autolysosomes using LysoTracker Red (LTR) in the fat body of starved *Drosophila* larvae. (**a**) *Atg2* was silenced by expressing a transgenic RNAi construct in cells co-expressing GFP. Neighboring GFP-negative cells are controls. Dashed circles indicate the positions of nuclei. (**a′**) LTR-positive autolysosomes are seen in control cells, but not in *Atg2* knockdown (Ri) cells

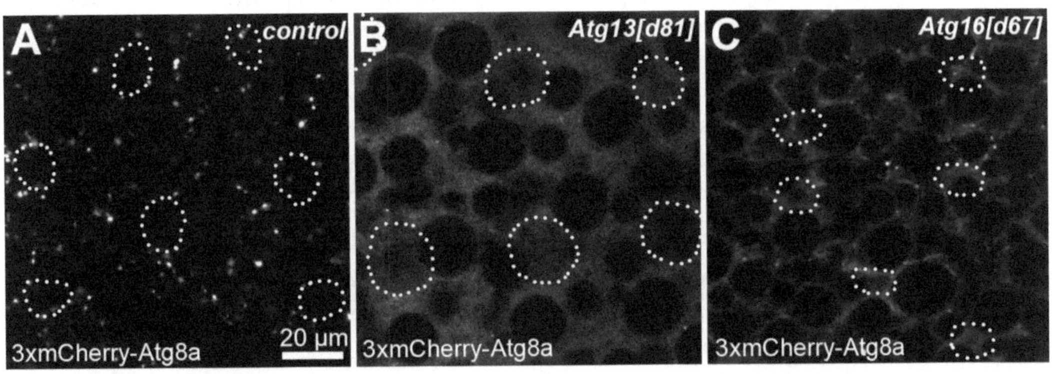

Fig. 2 Labeling autophagic structures with endogenous promoter-driven 3xmCherry-Atg8a. (**a**) Upon starvation, large 3xmCherry-Atg8a-positive structures appear in control cells (*see* **Note 10**). (**b**) Its punctate signal is almost completely lacking from fat cells of larvae lacking the Atg1 kinase complex subunit Atg13 [13]. A few small 3xmCherry-Atg8a-positive dots can be detected in an *Atg16* null mutant (Atg8a conjugation-deficient) tissue [20], which likely represent residual non-selective autophagy (**c**). Dashed circles mark the positions of nuclei

3.2.2 Labeling Autophagic Structures with Fluorescently Tagged Atg8a

The absence or accumulation of tagged Atg8a allows drawing conclusions about autophagic activity (Fig. 2), when used in combination with other assays [11–13].

1. After 4 h of starvation, dissect larvae in mounting solution and immediately examine by a fluorescent confocal microscope (*see* **Notes 8** and **11**).

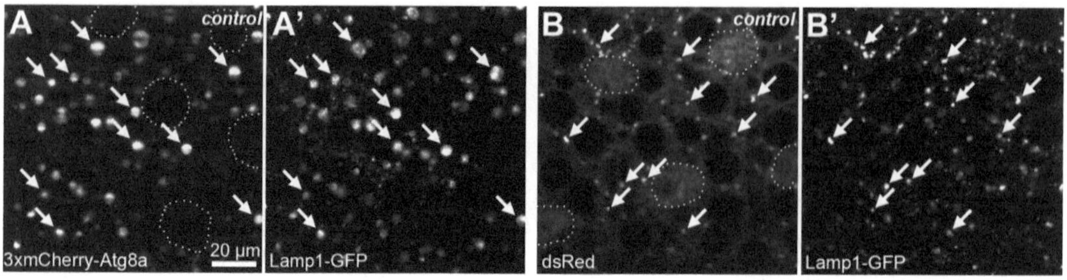

Fig. 3 Autophagic and lysosomal markers overlap in fat cells of starved larvae. Autophagic 3xmCherry-Atg8a (**a**) and the lysosomal Lamp1-GFP reporters (**a′**) show a high colocalization rate in fat cells of starved *Drosophila* larvae. (**b**) The cytosolic form of the RFP family dsRed fluorescent protein is captured into autophagic compartments in a non-selective manner, and it colocalizes with Lamp1-GFP (**b′**)-positive structures representing autolysosomes. Arrows label 3xmCherry-Atg8a and dsRed dots, respectively, which overlap with Lamp1-GFP. Dashed circles mark the positions of nuclei

3.2.3 Analysis of Autophagosome-Lysosome Fusion by Colocalization of Fluorescent Reporters

Examination of the colocalization rate between the autophagic and lysosomal reporters mCherry-Atg8a and Lamp1-GFP, respectively, is useful for the analysis of autophagosome-lysosome fusion and autolysosome formation [8, 14]. After 4 h of starvation, the majority of Atg8a-positive compartments overlap with Lamp1, indicating that autophagosomes fused with lysosomes, so autolysosome formation normally occurred (Fig. 3a).

Starvation-induced autophagy of soluble cytosolic-free dsRed results in a shift from diffuse cytosolic to punctate distribution in larval fat cells [11], and the overlap of dsRed dots with Lamp1-GFP is a clear indication for non-selective autophagy (Fig. 3b) (*see* **Note 12**).

1. Co-express mCherry-Atg8a or UAS-dsRed and UAS-Lamp1-GFP using an appropriate Gal4 driver (such as cg-Gal4 or fb-Gal4 for fat tissue).
2. Starve early L3 stage larvae as before.
3. Dissect the fat body in mounting solution, and immediately examine by a fluorescent confocal microscope (*see* **Note 8**).

3.3 Observing Autophagic Structures by Transmission Electron Microscopy

Ultrastructural analysis of L3 stage *Drosophila* larval fat tissue is a routine method for the evaluation of autophagy. This method gives a very detailed view of autophagic structures (Fig. 4), and the contents of autophagosomes (and to some extent even autolysosomes) can be recognized [15, 16].

3.3.1 Chemical Fixation of Samples for Electron Microscopy

1. Dissect starved larvae in a drop of fixative. Take care that fat body lobes are exposed to the solution. After a few seconds, you can transfer samples into glass vials for continued fixation overnight at 4 °C.

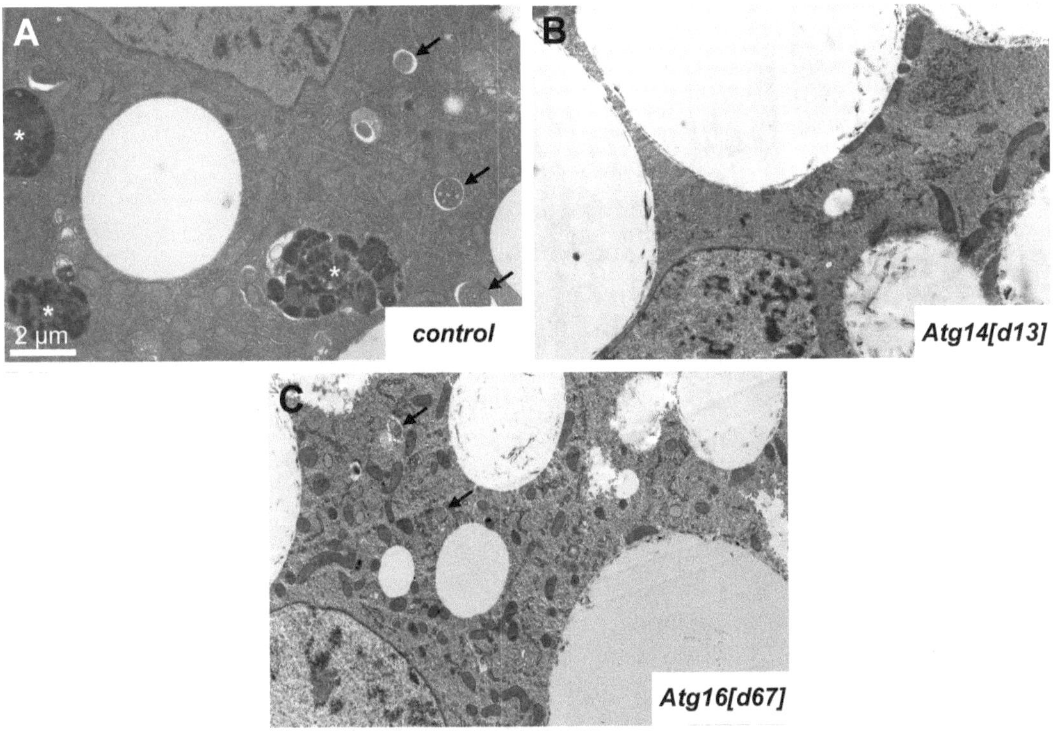

Fig. 4 Ultrastructural analysis of starvation-induced autophagic structures in fat cells. (**a**) Numerous autophagosomes (arrows) and autolysosomes (asterisks) are seen in control cells. (**b**) These organelles are lacking in the absence of the core autophagy gene *Atg14* [11], which encodes a subunit of the autophagic Vps34 lipid kinase complex. (**c**) In contrast, Atg16 null mutant larvae [20] show some residual autophagic activity, as a few autophagic structures can still be observed

2. On the next day, wash with 0.1 M sodium cacodylate buffer 3× at room temperature.
3. The samples can be stored in sodium cacodylate buffer up to a few weeks at 4 °C before embedding (*see* **Note 13**).

3.3.2 Embedding

1. Incubate the samples in 0.5% osmium tetroxide solution in the dark for 1 h at room temperature to postfix the tissue and increase the contrast of biological membranes. Use a rotator for all of the incubation steps during embedding.
2. Rinse 2× with 10% acetone.
3. Incubate in 2% sterile filtered uranyl acetate for 20–30 min in the dark (*see* **Note 14**).
4. Dehydration with increasing concentrations of ethanol in the following order:
   (a) 50% ethanol 15 min
   (b) 70% ethanol 15 min
   (c) 96% ethanol 2 × 15 min
   (d) Absolute ethanol 2 × 15 min

5. Infiltration with resin. First incubate the samples with propylene oxide, and then use different proportion mixtures of propylene oxide and araldite resin as follows:
   (a) Propylene oxide 15 min
   (b) 2 propylene oxide: 1 araldite 1 h
   (c) 1 propylene oxide: 1 araldite 1 h
   (d) 1 propylene oxide: 2 araldite overnight
6. Change the propylene oxide-araldite mixture to fresh araldite and incubate for 1 h.
7. Pour araldite resin into embedding molds and transfer the samples using a toothpick. Take care of proper orientation to make the trimming easier.
8. Polymerize the resin for 2 days at 60 °C.

3.3.3 Ultrasectioning

1. Trim away unnecessary empty resin using a sharp blade.
2. Cut semithin (0.5 μm) sections using an ultramicrotome equipped with a diamond knife or a glass knife that can be made using knife breaker equipment. Pick up the sections with a thin brush and transfer these into a water drop on a microscope slide. Dry it on a hot plate and stain with toluidine blue, by holding the slide above a flame for about 10 s (until a gold rim appears around the dye drop). Wash off the dye with water. Dry the slide and check the sections in a light microscope to check for section quality and identify the tissue(s) present in the block face. If you see that the sections contain fat body cells, proceed with cutting ultrathin sections.
3. Cut ultrathin sections (50–80 nm) and collect them on empty grids (*see* **Note 15**).

3.3.4 Staining Ultrathin Sections to Increase Contrast

1. Place the grids in a drop of Reynolds' lead citrate on a piece of parafilm. Incubate for 8 min and wash 3× with double distilled, autoclaved, sterile filtered water, followed by drying the grids on clean filter paper (*see* **Note 16**).

3.3.5 Analyze Samples in the Transmission Electron Microscope

Evaluate autophagic and other structures based on their morphology. This requires experience. As a ground rule, phagophores are usually curved membrane cisterns devoid of ribosomes, and an empty-looking space is often seen between the two membranes due to a chemical fixation/embedding artifact. This feature is also often seen for autophagosomes: a characteristic cleft appears between the two limiting membranes of these double-membrane vesicles. Autolysosomes contain cytoplasmic material in various stages of degradation (Fig. 4).

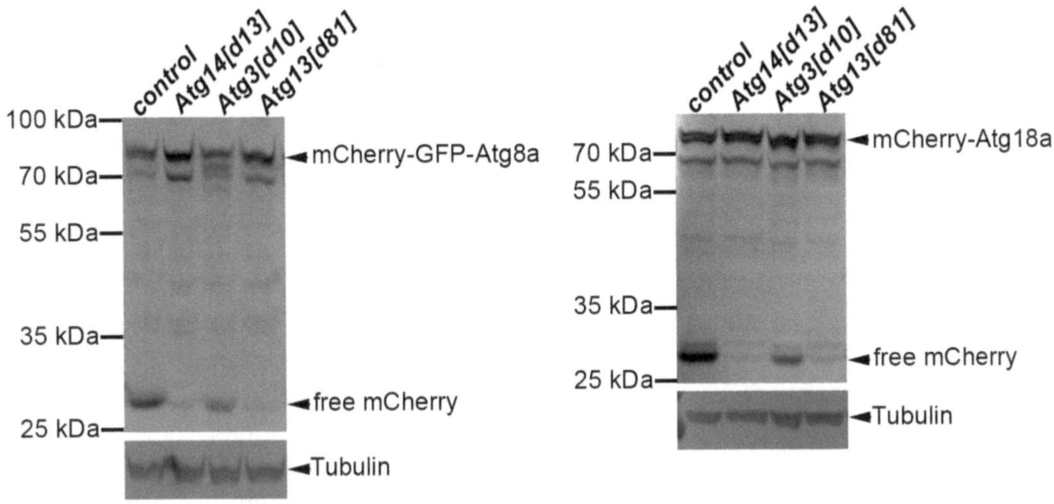

Fig. 5 Estimating autophagic flux by Western blotting. The levels of free mCherry in mCherry-GFP-Atg8a (left panel) or mCherry-Atg18a (right panel) expressing larvae clearly show that the importance of various Atg genes for the autolysosomal conversion of these reporters into free tag is different. Free mCherry is practically undetectable in *Atg14* and *Atg13* null mutants [11, 13], suggesting a complete block of autophagic flux. In contrast, larvae lacking *Atg3* encoding the E2-like enzyme for Atg8a lipidation [21], which are thus defective in Atg8a-dependent selective autophagy, show reduced but obvious levels of free mCherry. This suggests that these mutants can still perform non-selective autophagy, unlike the case of selective autophagy of p62/Ref(2)P where there is a similar defect in different *Atg* mutants [7]. Tubulin serves as loading control

3.4 Measuring Autophagic Flux by Western Blot

Western blotting can be used to measure free mCherry levels in starved larvae expressing, for example, a UAS-mCherry-GFP-Atg8a or UAS-mCherry-Atg18a reporter driven by Actin-Gal4 (Fig. 5) (*see* **Notes 17** and **18**).

1. Analyze samples by Western blot using standard techniques.

4 Notes

1. Larvae will compete with each other and/or exhaust the medium too soon and may begin to starve if the culture is even slightly crowded. To avoid that, transfer a few larvae to fresh medium at least 2 h before the beginning of the starvation protocol. To make sure that larvae are properly fed, pre-feed larvae by adding plenty of yeast paste to the top of the fly food in the culture vial, and let them eat for at least 2 h. This treatment ensures a maximal autophagic response to starvation.
2. The advantage of using 20% sucrose solution for starvation is that larvae float in it, but the fragments of the medium sink to the bottom.

3. Keeping larvae in 20% sucrose leads to amino acid/nitrogen starvation, which triggers high levels of autophagy (and at the same time also large-scale glycogen synthesis) in fat cells. If you want to perform full starvation, use PBS or water instead of sucrose. Note that larvae may sink and suffocate in these, so use a thin layer of water in a glass vial sealed with parafilm with a few airholes punched in it with a needle.
4. The number of autolysosomes peaks between 3 and 4 h of starvation. You can vary starvation length if shorter or longer treatment is more appropriate for your experiment.
5. LTR is a fluorescent vital dye, which preferentially accumulates in the lumen of acidic compartments. Thus, it could mark all acidic lysosomes including auto- and endolysosomes. Interestingly, fat cells of well-fed *Drosophila* larvae are almost completely negative for punctate LTR signal, indicating a low basal lysosomal activity.
6. Always use fresh LTR dye solution. For repeated experiments, a 5 μM LTR-PBS stock solution can be stored at 4 °C in the dark for up to a week.
7. Avoid sucking up fat body lobes with the pipette.
8. DAPI needs 5–10 min to penetrate and stain the nuclei.
9. FLP is a yeast-derived recombinase that recognizes FRT (FLP recombination target) sequences. The minimal *hsp70* promoter preceding FLP is leaky, so keeping flies at 25 °C leads to random loss of the neutral CD2 cassette from *Act-FRT-CD2-FRT-Gal4*, and the constitutive *Actin* promoter will drive expression of the yeast-derived Gal4 transcription factor in these cells. Upon crossing this stock with a fly line carrying a Gal4-responding UAS-regulated (upstream activating sequence, the binding site of Gal4) transgene for gene overexpression or RNAi, GFP-positive overexpression or knockdown clone cells will be generated in the progeny, respectively.
10. Fluorescently tagged Atg8a forms are the most common markers to investigate autophagy by fluorescent microscopy. As mCherry is relatively resistant to low pH, it remains fluorescent in the acidic lysosomal lumen. Thus, mCherry-Atg8a labels all autophagic structures from phagophores to autolysosomes, unlike GFP-Atg8a which is more selective for autophagosomes because it is quenched more rapidly in autolysosomes [17].
11. Alternatively, a tandem mCherry- and GFP-tagged Atg8a can be used to follow autophagic flux. With this reporter, autolysosomes are mostly positive for mCherry and show little to no GFP fluorescence, while autophagosomes and phagophores are positive for both mCherry and GFP.

12. During non-selective autophagy, any kind of cytoplasmic protein is randomly sequestered and delivered into the lysosomal lumen. Consequently, due to the low-pH-resistant nature of RFP family proteins (such as mCherry or dsRed), overexpressed reporters with an RFP-like tag will be delivered to and accumulate in autolysosomes, seen by their extensive overlap with lysosomal Lamp1-GFP.
13. The sample can also be stored in fixative for up to a week if necessary.
14. The uranyl acetate solution can be used several times, but make sure to filter the solution before each use.
15. If you plan to perform immunogold labeling, use a nickel grid because it is more inert and more suitable for immunoreactions or enzyme reactions compared to the copper grid. Take into account that it is magnetic, though.
16. During the staining of ultrathin sections, cover the grids while they are in the lead citrate drops (e.g., with a Petri dish) because CO_2 contamination from the air may result in the formation of precipitate. Do not exhale onto the samples at this stage.
17. Atg8a, and to a lesser extent other Atg factors such as Atg18/WIPI family proteins [18], remain associated to autophagic membranes after autophagosome closure; hence, significant amounts of these proteins are delivered for degradation within autolysosomes. If these proteins are tagged with mCherry/RFP or GFP, partial degradation releases the globular and thus protease-resistant fluorescent tag [19]. Thus, the amount of free mCherry or GFP serves as a good indicator of autophagic degradation/flux.
18. This protocol is for lysates of whole larvae and not for fat tissues.

References

1. Mulakkal NC et al (2014) Autophagy in Drosophila: from historical studies to current knowledge. Biomed Res Int 2014:273473
2. Lorincz P, Mauvezin C, Juhasz G (2017) Exploring autophagy in Drosophila. Cells 6 (3):E22
3. Nezis IP et al (2008) Ref(2)P, the Drosophila melanogaster homologue of mammalian p62, is required for the formation of protein aggregates in adult brain. J Cell Biol 180 (6):1065–1071
4. Juhász G et al (2007) Atg7-dependent autophagy promotes neuronal health, stress tolerance, and longevity but is dispensable for metamorphosis in Drosophila. Genes Dev 21 (23):3061–3066
5. Zaffagnini G, Martens S (2016) Mechanisms of selective autophagy. J Mol Biol 428(9 Pt A):1714–1724
6. Tsuboyama K et al (2016) The ATG conjugation systems are important for degradation of the inner autophagosomal membrane. Science 354(6315):1036–1041

7. Nagy P et al (2015) How and why to study autophagy in Drosophila: it's more than just a garbage chute. Methods 75:151–161
8. Mauvezin C et al (2014) Assays to monitor autophagy in Drosophila. Methods 68 (1):134–139
9. Scott RC, Schuldiner O, Neufeld TP (2004) Role and regulation of starvation-induced autophagy in the Drosophila fat body. Dev Cell 7(2):167–178
10. Rusten TE et al (2004) Programmed autophagy in the Drosophila fat body is induced by ecdysone through regulation of the PI3K pathway. Dev Cell 7(2):179–192
11. Hegedus K et al (2016) The Ccz1-Mon1-Rab7 module and Rab5 control distinct steps of autophagy. Mol Biol Cell 27(20):3132–3142
12. Takats S et al (2013) Autophagosomal Syntaxin17-dependent lysosomal degradation maintains neuronal function in Drosophila. J Cell Biol 201(4):531–539
13. Chang YY, Neufeld TP (2009) An Atg1/Atg13 complex with multiple roles in TOR-mediated autophagy regulation. Mol Biol Cell 20(7):2004–2014
14. Pulipparacharuvil S et al (2005) Drosophila Vps16A is required for trafficking to lysosomes and biogenesis of pigment granules. J Cell Sci 118(Pt 16):3663–3673
15. Takats S et al (2014) Interaction of the HOPS complex with Syntaxin 17 mediates autophagosome clearance in Drosophila. Mol Biol Cell 25(8):1338–1354
16. Rusten TE et al (2007) ESCRTs and Fab1 regulate distinct steps of autophagy. Curr Biol 17(20):1817–1825
17. Kimura S, Noda T, Yoshimori T (2007) Dissection of the autophagosome maturation process by a novel reporter protein, tandem fluorescent-tagged LC3. Autophagy 3 (5):452–460
18. Proikas-Cezanne T et al (2007) Human WIPI-1 puncta-formation: a novel assay to assess mammalian autophagy. FEBS Lett 581 (18):3396–3404
19. Klionsky DJ et al (2016) Guidelines for the use and interpretation of assays for monitoring autophagy (3rd edition). Autophagy 12 (1):1–222
20. Varga K et al (2016) Loss of Atg16 delays the alcohol-induced sedation response via regulation of Corazonin neuropeptide production in Drosophila. Sci Rep 6:34641
21. Scott RC, Juhász G, Neufeld TP (2007) Direct induction of autophagy by Atg1 inhibits cell growth and induces apoptotic cell death. Curr Biol 17(1):1–11

Chapter 39

Imaging the Dynamics of Mitophagy in Live Cells

Andrew S. Moore and Erika L. F. Holzbaur

Abstract

Investigating the precise spatiotemporal dynamics of mitophagy can provide insights into how mitochondrial quality control is regulated in different tissues and organisms. Here, we outline live imaging assays to quantitatively assess mitophagy dynamics in real time. This protocol describes both chemical and optogenetic techniques to induce mitochondrial damage with high spatial and temporal control. Using these assays, mitochondria can be tracked from before they sustain damage up to their engulfment by autophagosomes and acidification by lysosomes.

Key words Mitophagy, Parkin, LC3, OPTN, Autophagy, Mitochondria, Ubiquitin, Lysosome

1 Introduction

Mitochondrial autophagy, termed mitophagy, is a tightly regulated degradative process by which damaged mitochondria are identified, sequestered, and then degraded. Within minutes of an initial precipitating injury leading to mitochondrial depolarization, a feed-forward cascade is activated involving the ubiquitin kinase PINK1 and the E3 ubiquitin ligase Parkin [1]. PINK1 levels are normally low on healthy mitochondria, but mitochondrial depolarization leads to the accumulation of this kinase on the outer mitochondrial membrane and activation via autophosphorylation. The PINK1-dependent phosphorylation of ubiquitin, present at basal levels on the outer mitochondrial membrane, leads to recruitment of Parkin, which binds phospho-ubiquitin (p65-Ub) with high affinity [1]. Once recruited, Parkin rapidly ubiquitinates mitochondrial proteins. The resulting high local concentration of ubiquitinated proteins actively recruits ubiquitin-binding adaptor proteins, including optineurin (OPTN), NDP-52, TAX1BP1, and NBR1 [2–4]. The recruitment of these adaptor proteins has been visualized by live cell imaging [2, 4] and occurs within 45 min of damage to the mitochondria by either uncouplers such as CCCP or local generation of ROS by optogenetic activation of Mito-KillerRed

Nicholas Ktistakis and Oliver Florey (eds.), *Autophagy: Methods and Protocols*, Methods in Molecular Biology, vol. 1880, https://doi.org/10.1007/978-1-4939-8873-0_39,

[2, 5]. Multicolor live imaging has established the relative timing of receptor recruitment in HeLa cells, as well as the parallel recruitment of the kinase TBK1 [4].

Activation of the local autophagy machinery leads to the generation of LC3-positive isolation membranes. The mitophagy receptors OPTN and NDP52 have LC3-binding LIR motifs that mediate the association of these isolation membranes with damaged mitochondria. The engulfment of damaged mitochondria can be readily visualized by live imaging using a spinning disk confocal microscope [2, 4]. Of note, mutations in either OPTN or TBK1 are causative for rare forms of familial ALS as well as glaucoma. Live imaging reveals that ALS-linked mutations in OPTN or TBK1 significantly disrupt LC3 recruitment and thus the effective engulfment of damaged mitochondria [2, 4].

Once a damaged mitochondrion is engulfed, the fully formed autophagosome will fuse with lysosomes over a period of several hours, facilitating the degradation of the organelle. This fusion can be visualized using a tandem mCherry-EGFP-LC3 construct [6]. Initially, this construct will be visible in both the mCherry and GFP channels, but as the autophagosome becomes acidified by lysosomal fusion, the GFP fluorescence will be preferentially quenched, and only the mCherry fluorescence will remain. The acidification of the autophagosome can also be monitored by a pH-sensitive, cell-permeant reporter such as LysoTracker.

Here we detail three protocols using multicolor live cell imaging to monitor the dynamics of mitophagy. We describe protocols to induce mitophagy either by chemical or optogenetic strategies. Also, we outline the use of pH-sensitive reporters to monitor autophagosome formation and acidification by lysosomes.

2 Materials

2.1 HeLa Cell Culture

1. HeLa cells (CCL-2) from the American Type Culture Collection (ATCC) (*see* **Note 1**).
2. 1× Dulbecco's phosphate-buffered saline (dPBS).
3. 0.25% trypsin.
4. Dulbecco's Modified Eagle Medium (DMEM) with 4.5 g/L glucose and L-glutamine (Corning, 10-017-CM).
5. Fetal bovine serum (FBS).
6. Glutamax.
7. Growth medium: DMEM +10% (v/v) FBS + 2 mM Glutamax.
8. Imaging medium: DMEM +10% (v/v) FBS + 2 mM Glutamax +25 mM HEPES.

9. Plastic tissue culture dishes, 100 mm.
10. Glass-bottom microwell dishes, 35 mm, No. 1.5 cover glass (e.g., MatTek, P35G-1.5-20-C).

2.2 Transfection

1. FuGENE 6 Transfection Reagent (Promega, E2692).
2. Opti-MEM Reduced-Serum Media (Invitrogen, 31,985,070).
3. Mammalian expression constructs (*see* **Note 2**).
   (a) Mito-TagBFP (Addgene, 49,151).
   (b) EGFP-OPTN (Addgene, 27,052).
   (c) Mito-KillerRed (Evrogen, FP964).
   (d) HaloTag-LC3B.
   (e) mCherry-GFP-LC3B (Addgene, 22,418),
   (f) Untagged Parkin (*see* **Note 3**).
4. Far-red HaloTag ligand (*see* **Note 4**).

2.3 Live Imaging

1. Spinning disk confocal microscope (e.g., UltraVIEW VoX 3D live cell imaging system (PerkinElmer) on a Nikon Eclipse Ti microscope).
2. Perfect focus system.
3. EM-CCD camera (e.g., Hamamatsu, C9100-50).
4. 100×/1.49 NA apochromat objective lens.
5. 405 nm, 488 nm, 561 nm, 640 nm laser lines.
6. Photobleaching module (e.g., UltraVIEW PhotoKinesis accessory (PerkinElmer)).
7. Volocity acquisition software (PerkinElmer).

2.4 Chemicals and Dyes

1. CCCP (Sigma, C2759).
2. TMRE (Thermo Fisher, T669).
3. LysoTracker Deep Red (Thermo Fisher, L12492).

2.5 Image Analysis Programs

1. FIJI [7].

3 Methods

3.1 HeLa Cell Passaging Protocol

1. Warm complete DMEM, 0.25% trypsin, and 1× dPBS in 37 °C water bath for at least 15 min.
2. Remove 10 cm dish of HeLa cells from incubator and place in biosafety cabinet.
3. Aspirate media and wash with 10 mL of warm dPBS.
4. Aspirate dPBS and replace with 1 mL of 0.25% trypsin. Place dish in 37 °C incubator for 5–7 min.

5. While cells are trypsinizing, add 12 mL of warm, complete DMEM to a new 10 cm dish and 2 mL complete medium to a 35 mm glass-bottom dish.
6. Remove dish of trypsinized HeLa cells from the incubator and transfer to biosafety cabinet. Add 4 mL of warm, complete DMEM to the HeLa cells, and pipette up and down to dislodge remaining adherent cells. Transfer 5 mL of cell suspension to 15 mL conical tube and centrifuge for 2 min at 500 × *g*.
7. Aspirate supernatant and resuspend cell pellet in 1 mL complete DMEM.
8. Count cells using a hemocytometer. Seed 2,000,000 cells on the 10 cm dish. These will be ready to split again in 48 h. Seed 300,000 cells on the 35 mm glass-bottom dishes. These cells can be transfected in 24 h.

3.2 Chemical Depolarization Mitophagy Protocol

3.2.1 Transfection

1. Before transfection, inspect the plate to ensure cells are ~40–60% confluent.
2. In a biosafety cabinet, add 100 μL Opti-MEM to a 1.5 mL microcentrifuge tube.
3. Add 500 ng of each plasmid you plan to transfect. Tap to mix (*see* **Note 5**).
   (a) Mito-TagBFP.
   (b) EGFP-optineurin.
   (c) HaloTag-LC3B.
   (d) Untagged Parkin.
4. Add 6 μL FuGENE 6 Transfection Reagent for a 3:1 ratio to plasmid DNA. Tap to mix.
5. Incubate at room temperature for 10 min.
6. Add transfection mix to your cells in a dropwise manner and return plates to 37 °C incubator.

3.2.2 Imaging

1. Approximately 23 h after transfection, remove cells from incubator, and gently replace medium with 2 mL pre-warmed imaging medium containing 30 nM TMRE and 100 nM far-red HaloTag ligand (*see* **Note 6**).
2. Allow cells to equilibrate for 45 min at 37 °C (*see* **Note 7**).
3. Remove 500 μL of medium from the plate and transfer to a microcentrifuge tube. Add 2 μL of 20 mM CCCP and incubate at 37 °C.
4. Transfer cells to 37 °C heated microscope stage.
5. Manually adjust the objective height with the focus knob until reaching the desired focal plane.

6. Identify a cell expressing the constructs of interest with clear TMRE labeling of mitochondria (*see* **Note 8**).
7. Add back the 500 μL of medium containing CCCP for a final concentration of 20 μM. This concentration of CCCP should robustly depolarize mitochondria as indicated by loss of TMRE intensity (*see* **Note 9**).
8. Acquire a four-color time-lapse movie of the cell at one frame per minute for 90 min.

3.3 Optogenetic Mitophagy Protocol

3.3.1 Transfection

1. Before transfection, inspect the plate to ensure cells are ~40–60% confluent.
2. In a biosafety cabinet, add 100 μL Opti-MEM to a 1.5 mL microcentrifuge tube.
3. Add 500 ng of each plasmid you plan to transfect. Tap to mix.
   (a) Mito-TagBFP.
   (b) EGFP-optineurin.
   (c) Mito-KillerRed (Mito-KR).
   (d) HaloTag-LC3B.
   (e) Untagged Parkin.
4. Add 7.5 μL FuGENE 6 Transfection Reagent for a 3:1 ratio to plasmid DNA. Tap to mix.
5. Incubate at room temperature for 10 min.
6. Add transfection mix to your cells in a dropwise manner and return plates to 37 °C incubator.

3.3.2 Imaging

1. Approximately 23 h after transfection, remove cells from incubator, and gently replace medium with 2 mL pre-warmed imaging medium containing 100 nM far-red HaloTag ligand. Allow cells to equilibrate for 45 min at 37 °C.
2. Transfer cells to 37 °C heated microscope stage.
3. Manually adjust the objective height with the focus knob until reaching the desired focal plane.
4. Identify a cell expressing the constructs of interest. Take care to minimize green light exposure when scanning the plate as Mito-KR is a potent photosensitizer.
5. Using a 561 nm laser, bleach a ~225 μm^2 region of the mitochondrial network. Typically, we use 100% laser power and bleach for no longer than 1 min.
6. Acquire a four-color time-lapse movie of the cell at one frame per minute for 90 min. Within 45 min we generally begin to observe LC3 recruitment to mitochondria within the bleach window (Fig. 1).

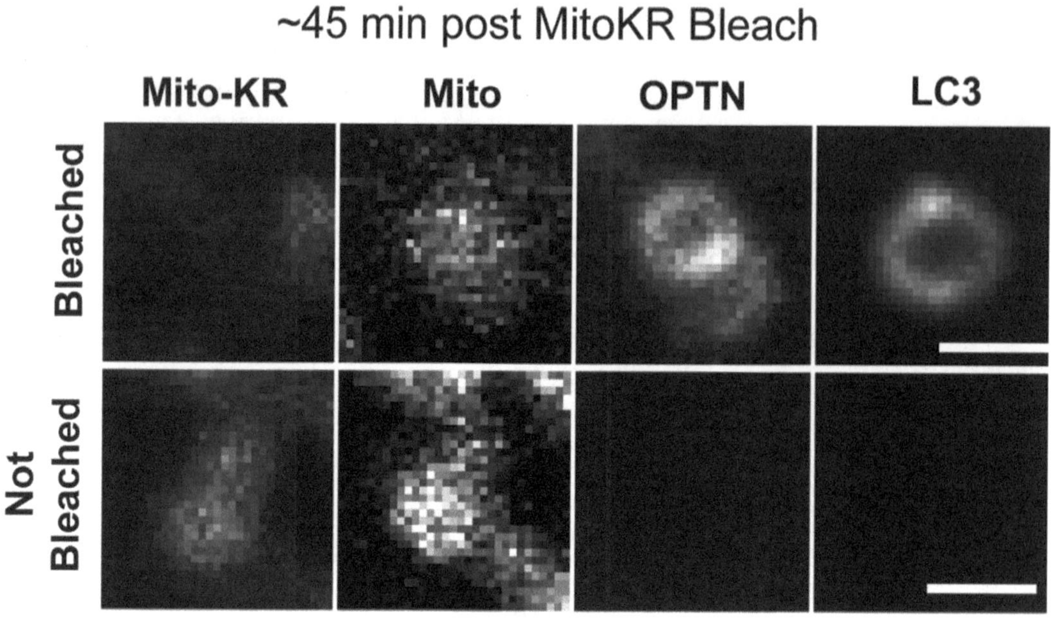

Fig. 1 45 min after bleach of Mito-KR, OPTN, and LC3 can be visualized around the stable Mito marker (top column). In the same cell, unbleached mitochondria show robust Mito-KR fluorescence and no OPTN or LC3 recruitment (right column). Scale = 1 μm

3.4 LysoTracker Staining of Acidified Autophagosomes

3.4.1 Transfection

1. Before transfection, inspect the plate to ensure cells are ~40–60% confluent.
2. In a biosafety cabinet, add 100 μL Opti-MEM to a 1.5mL microcentrifuge tube.
3. Add 500 ng of each plasmid you plan to transfect. Tap to mix.
   (a) Mito-TagBFP.
   (b) EGFP-mCherry-LC3B.
   (c) Untagged Parkin.
4. Add 4.5 μL FuGENE 6 Transfection Reagent for a 3:1 ratio to plasmid DNA. Tap to mix.
5. Incubate at room temperature for 10 min.
6. Add transfection mix to your cells in a dropwise manner and return plates to 37 °C incubator.

3.4.2 Imaging

1. Approximately 22 h after TFX, replace medium with pre-warmed complete DMEM containing 75 nM LysoTracker Deep Red. Allow cells to incubate for 1 h.
2. Replace staining solution with 2 mL pre-warmed imaging medium. Allow cells to equilibrate for 30 min at 37 °C.

3. Remove 500 μL of medium from the plate and transfer to a microcentrifuge tube. Add 2 μL of 20 mM CCCP and incubate at 37 °C.
4. Transfer cells to 37 °C heated microscope stage.
5. Manually adjust the objective height with the focus knob until reaching the desired focal plane.
6. Identify a cell expressing the constructs of interest.
7. Add back the 500 μL of medium containing CCCP for a final concentration of 20 μM.
8. Acquire a four-color time-lapse movie of the cell at one frame per 2 min for 3 h (*see* **Note 10**).

3.5 Assessing the Autophagic Engulfment of Depolarized Mitochondria with FIJI

1. Open time-lapse movie of CCCP-treated cell in FIJI.
2. At each time point, use the multipoint tool to count the total number of (a) mitochondria, (b) OPTN-positive mitochondria, and (c) LC3-positive mitochondria (*see* **Note 11**).
3. For each time point, divide the number of OPTN-positive mitochondria by the total number of mitochondria. Repeat for LC3-positive mitochondria.
4. Plot the ratio of OPTN-positive mitochondria and LC3-positive mitochondria over time.

3.6 Assessing the Autophagic Engulfment of Mito-KR-Bleached Mitochondria in FIJI

1. Open time-lapse movie in FIJI.
2. Identify a bleached mitochondrion labeled with Mito-TagBFP but not Mito-KR.
3. Using the line tool, draw a 2 μm line scan through the organelle. Open the ROI manager (Analyze > tools > ROI manager), and select “Add” to save the position on the line scan.
4. Measure the mean intensity (Analyze > measure) along the line scan for each channel.
5. Move to the next time point and adjust the location of the line scan to correct for movement of the mitochondria. Add the position of the new line scan to the ROI manager and measure the intensity of all channels. Repeat for all time points.
6. Next, repeat this protocol for a Mito-KR-labeled mitochondrion that was not bleached.
7. Plot the fluorescence intensity over time for OPTN and LC3. Compare between bleach and unbleached conditions.

3.7 Assessing the Lysosomal Acidification of Mitochondria

1. Open time-lapse movie in FIJI.
2. Identify a mitochondrion and use the line tool to draw a 2 μm line scan through the organelle. Add the line scan to the ROI manager as previously described.

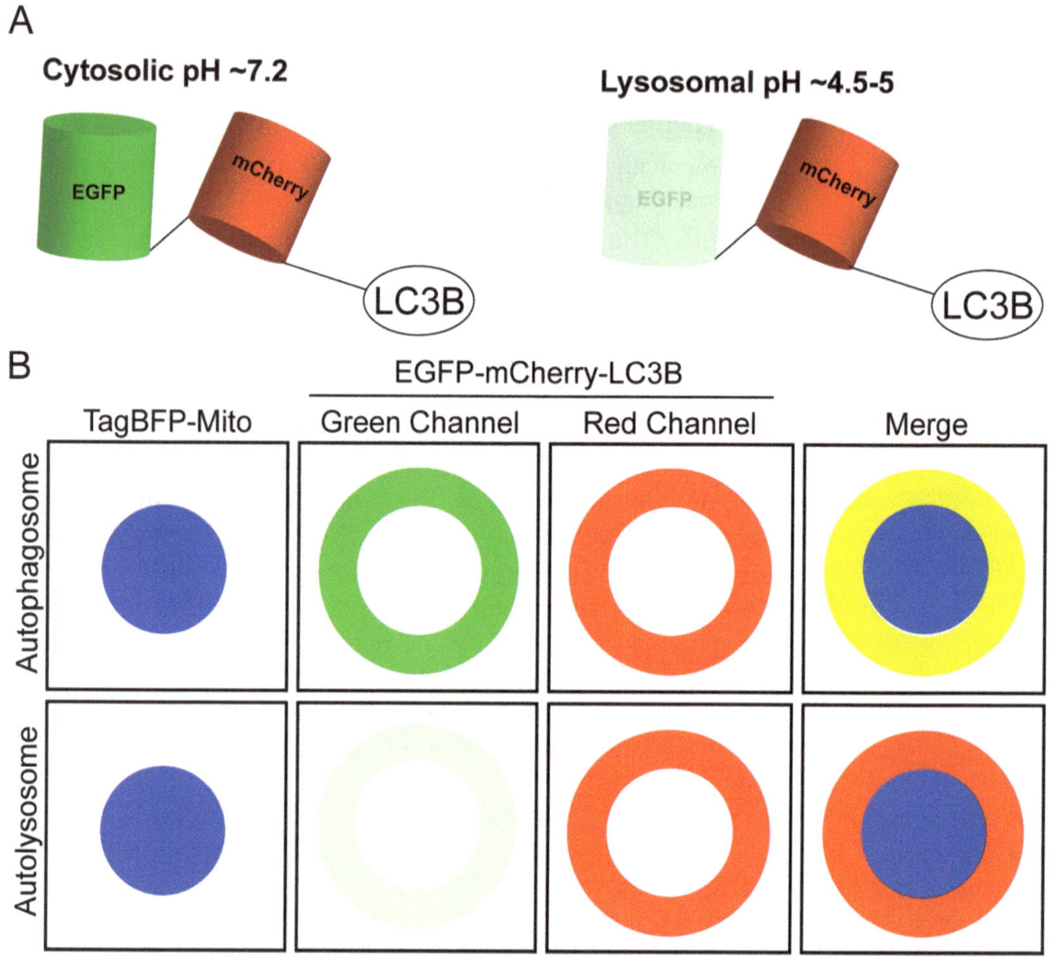

Fig. 2 (**a**) At cytosolic pH, both EGFP and mCherry brightly fluoresce. However, at the lower lysosomal pH, only mCherry fluorescence persists. (**b**) Newly formed autophagosomes display both EGFP and mCherry fluorescence. Upon fusion with lysosomes, EGFP fluorescence is quenched, and the autolysosome can only be visualized in the red channel

3. For each time point, measure the mean intensity for the EGFP, mCherry, and LysoTracker Deep Red channels.
4. Plot the ratio of EGFP to mCherry intensity. Decreases in this ratio indicate quenching of EGFP fluorescence due to lysosomal acidification (Fig. 2).
5. Simultaneously, plot the intensity of LysoTracker for each time point. Increased LysoTracker intensity indicates colocalization of lysosomes with mitochondria-positive autophagosomes.

4 Notes

1. We have used this protocol to visualize mitophagy in HeLa cells and Cos7 cells. We expect this protocol can be adapted to numerous other cell lines.
2. All of these constructs use CMV promoters, which allows for robust, constitutive expression in mammalian cells.
3. HeLa cells express low levels of Parkin; thus, we express exogenous Parkin to facilitate mitophagy. It is also possible to generate stable, Parkin-expressing cell lines using retroviral infection of untagged Parkin (Addgene, 89,299) followed by limiting dilution to isolate monoclonal colonies.
4. HaloTag fusion proteins can be labeled by various cell-permeant fluorescent ligands. For this assay, we used a far-red ligand (JF646 [8]) to avoid spectral overlap with our other fluorescent probes.
5. We generally observe very high levels of co-transfection. If you observe low levels of co-transfection, be sure to thoroughly mix the plasmid solution before addition of FuGENE 6.
6. At these concentrations, TMRE and JF646 can be used without a subsequent washing step. HaloTag-LC3 can also be labeled with a red ligand, but TMRE must be omitted from the assay.
7. HEPES-buffered imaging medium allows for long-term (~4 h) imaging at atmospheric CO_2 levels. After addition of imaging medium, cells should not be returned to a 5% CO_2 incubator.
8. To avoid artifacts associated with overexpression in transient transfection assays, select cells with lower fluorescence intensity. We typically scan the plate to determine the range of expression levels for each channel. Then, we select cells in the lowest quartile of fluorescence intensities, provided the signal to noise ratio of each channel is sufficient to visualize structures of interest. If necessary, expression time after transfection can be reduced to as little as 16 h. Alternatively, monoclonal stable cell line can be generated to provide more consistent expression levels for analysis.
9. TMRE import into mitochondria is dependent on membrane potential. When TMRE is used in non-quenching mode (we find that 30 nM works well), depolarization by CCCP should result in decreased TMRE intensity in mitochondria [9]. Other chemicals, such as a combination of antimycin with oligomycin, can also robustly depolarize mitochondria and trigger mitophagy.

10. In HeLa cells, autophagic engulfment of mitochondria occurs approximately 45 min to 1 h after damage. In order to visualize acidification of autophagosomes, it is necessary to image over a longer time window. To reduce the phototoxic effects of prolonged imaging, we recommend imaging at a slower frame rate.
11. We classify mitochondria as OPTN- or LC3-positive only when we can observe a clear ring of fluorescence around the rounded, fragmented mitochondrion.

References

1. Nguyen T, Padman B, Lazarou M (2016) Deciphering the molecular signals of PINK1/Parkin mitophagy. Trends Cell Biol 26:733
2. Wong Y, Holzbaur E (2014) Optineurin is an autophagy receptor for damaged mitochondria in parkin-mediated mitophagy that is disrupted by an ALS-linked mutation. Proc Natl Acad Sci U S A 111:E4439–E4448
3. Lazarou M et al (2015) The ubiquitin kinase PINK1 recruits autophagy receptors to induce mitophagy. Nature 524:309–314
4. Moore A, Holzbaur E (2016) Dynamic recruitment and activation of ALS-associated TBK1 with its target optineurin are required for efficient mitophagy. Proc Natl Acad Sci U S A 113: E3349–E3358
5. Wang Y, Nartiss Y, Steipe B, McQuibban AG, Kim PK (2012) ROS-induced mitochondrial depolarization initiates PARK2/PARKIN-dependent mitochondrial degradation by autophagy. Autophagy 8:1462–1476 http://www.tandfonline.com/doi/abs/10.4161/auto.21211
6. Pankiv S et al (2007) p62/SQSTM1 binds directly to Atg8/LC3 to facilitate degradation of ubiquitinated protein aggregates by autophagy. J Biol Chem 282:24131
7. Schindelin J, Arganda-Carreras I, Frise E, Kaynig V (2012) Fiji: an open-source platform for biological-image analysis. Nat Methods 9 (7):676–682 https://www.nature.com/articles/nmeth.2019
8. Grimm JB, English BP, Chen J, Slaughter JP, Zhang Z (2015) A general method to improve fluorophores for live-cell and single-molecule microscopy. Nat Methods 12(3):244–250 https://www.nature.com/articles/nmeth.3256
9. Perry S, Norman J, Barbieri J, Brown E, Gelbard H (2011) Mitochondrial membrane potential probes and the proton gradient: a practical usage guide. BioTechniques 50:98–115

Chapter 40

Triggering Mitophagy with Photosensitizers

Cheng-Wei Hsieh and Wei Yuan Yang

Abstract

One can utilize light illumination to stimulate mitochondrial reactive oxygen species production through the use of mitochondria-specific photosensitizers. By proper tuning of the light dosage, the methodology permits probing of a multitude of mitochondrial damage responses, including mitophagy. This light-controllable trick offers unique opportunities for the investigation of mitophagy—one can spatiotemporally define mitochondrial damage, alter the number of impaired mitochondria, as well as modulate the severity of the mitochondrial injury. This light-activated mitophagy can be adapted not only to single-cell imaging techniques but also to cell population-based biochemical assays.

Key words Autophagy, Mitophagy, Photosensitizer, MitoTracker Deep Red

1 Introduction

Mitophagy is a cellular surveillance system that delivers mitochondria into autophagosomes for lysosomal removal [1]. The process allows cells to critically control the quality and quantity of mitochondria. The most studied form of mitophagy is one that is initiated through PINK1—this mitochondrial sensor accumulates on the outer membrane of to-be-turned-over mitochondria [2] and summons the ubiquitin E3 ligase Parkin to facilitate the mitophagy process through mitochondrial ubiquitination.

Approaches commonly utilized to initiate and study PINK1/Parkin-mediated mitophagy include the use of carbonyl cyanide m-chlorophenyl hydrazine (CCCP) [3], a protonophore that dissipates the mitochondrial membrane potential; valinomycin, a K+ ionophore; and oligomycin/antimycin A [4], inhibitors of the OXPHOS complexes III and V. While robust, these small molecules globally target all cellular mitochondria at once and sometimes exhibit effects outside of mitochondria. For example, CCCP can interfere with lysosomal functions independent of its activity in dissipating the mitochondrial membrane potential [5].

Nicholas Ktistakis and Oliver Florey (eds.), *Autophagy: Methods and Protocols*, Methods in Molecular Biology, vol. 1880, https://doi.org/10.1007/978-1-4939-8873-0_40, © Springer Science+Business Media, LLC, part of Springer Nature 2019

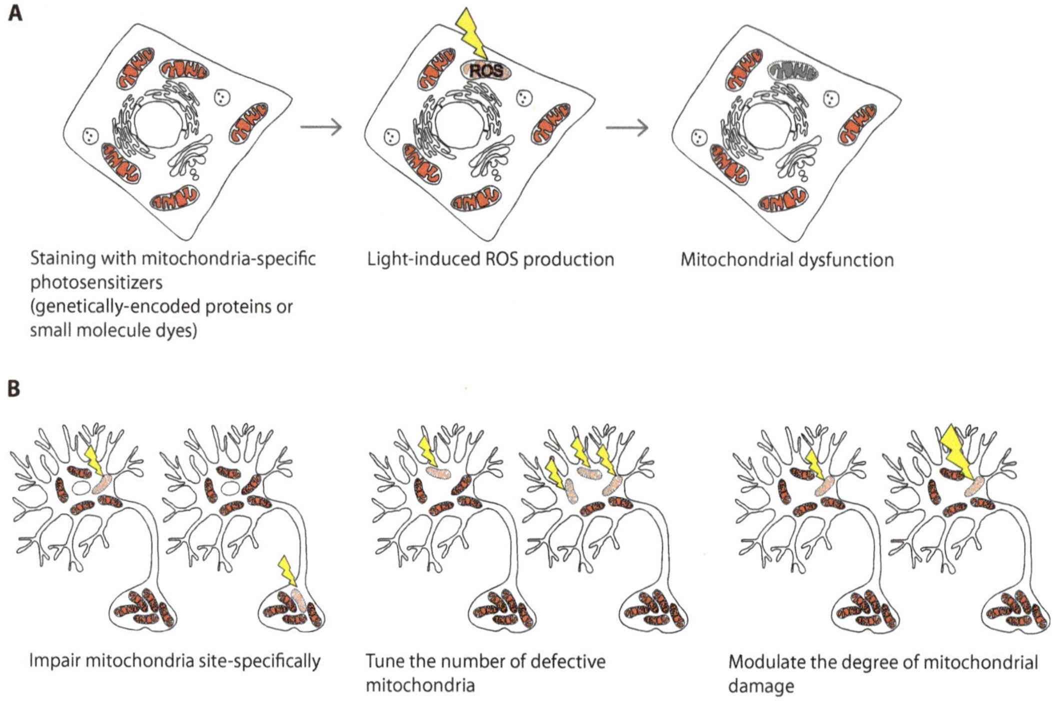

Fig. 1 Impairing mitochondria through light-activated ROS generation. (**a**) Light illumination on cells stained with mitochondria-targeted photosensitizers results in ROS generation within mitochondria, which leads to local oxidation of biomolecules and mitochondrial dysfunction. (**b**) Using this light-activated mitochondrial impairment scheme, one can easily specify the site(s) for mitochondrial injuries to occur (*left*), control the number of defective mitochondria in a cell (*middle*), or modulate the degree of mitochondrial damage (*right*; by altering the light dosage used, e.g., stronger light leads to higher degree of damage)

Alternatively, mitochondria-resident photosensitizers can also be used to trigger mitophagy [6–12]. Through light illumination, photosensitizers mediate reactive oxygen species (ROS) production [13]. These highly reactive ROS are capable of oxidizing biomolecules such as lipids, amino acids, and nucleic acids, causing mitochondrial dysfunction, and elicit an array of mitochondrial damage responses, one of which being mitophagy. In the crowded cellular environment, ROS have very limited diffusion radius and only affect molecules approximated to the illumination area. Therefore, mitochondrial matrix-targeting photosensitizers allow matrix-restricted ROS generation and precise mitochondrial damage without affecting other cellular components (Fig. 1a). A wide range of photosensitizers can all be used for this purpose, including the genetically encoded red fluorescent proteins KillerRed [14] and SuperNova [15], the orange fluorescent protein KillerOrange [16], the green fluorescent proteins miniSOG [17] and SOPP [18], and the near-infrared FAP-TAP [19]. Small-molecule mitochondrial dyes such as MitoView 633, MitoTracker dyes, rhodamine 800, and tetramethylrhodamine have also been demonstrated to be effective in eliciting mitophagy [6, 7]. These photosensitizers make it possible to trigger mitophagy using varying colors of light. In particular, by adjusting the light illumination settings (e.g., the subcellular regions that are going to be exposed to light, as well as the dosage of light used), this methodology allows the possibility to precisely determine when and where mitochondria are to be impaired, the number of mitochondria to be damaged, as well as the severity of the mitochondria injuries (Fig. 1b).

In this chapter we detail the steps needed for triggering mitophagy using MitoTracker Deep Red FM. We provide protocol on how to perform this using a standard laser-scanning confocal microscope, which allows the triggering and monitoring of mitophagy with subcellular precision (Fig. 2a, b). Combined with the use of automatic microscope stages, this can allow for simultaneous, multiplexed recording of mitophagy within a large number of single living cells (Fig. 2c, d). In addition, we also describe how to utilize photosensitizers to globally impair mitochondria within the entire cell culture using collimated light sources (Fig. 3a, b). This makes it possible to use the methodology in high-content screening and the various biochemical assays such as Western blotting, immunoprecipitation, and mass-spectrometry-based molecular identification.

2 Materials

2.1 Cell Culture Reagents

1. Dulbecco's modified Eagle's medium.
2. Dulbecco's modified Eagle's medium without phenol red.
3. Fetal bovine serum.
4. Penicillin–streptomycin (P/S).
5. L-Glutamine.
6. A model system for monitoring mitophagy (e.g., HeLa cells expressing Parkin). Maintain HeLa cells in Dulbecco's modified Eagle's medium supplemented with 10% fetal bovine serum and 1% P/S and maintained at 37 °C, 5% CO_2. Seed HeLa cells onto 35 mm glass-bottom dish (*see* **Note 1**), and then incubate the seeded cells in a 5% CO_2 incubator at 37 °C overnight (*see* **Note 2**). During the second day, transfect cells with. EBFP2-Parkin using Lipofectamine 2000 (*see* **Note 3**), and further incubate the transfected cells in a 5% CO_2 incubator at 37 °C overnight for experiment on the third day.
7. Poly-L-lysine.
8. Hemacytometer.
9. Trypan blue.

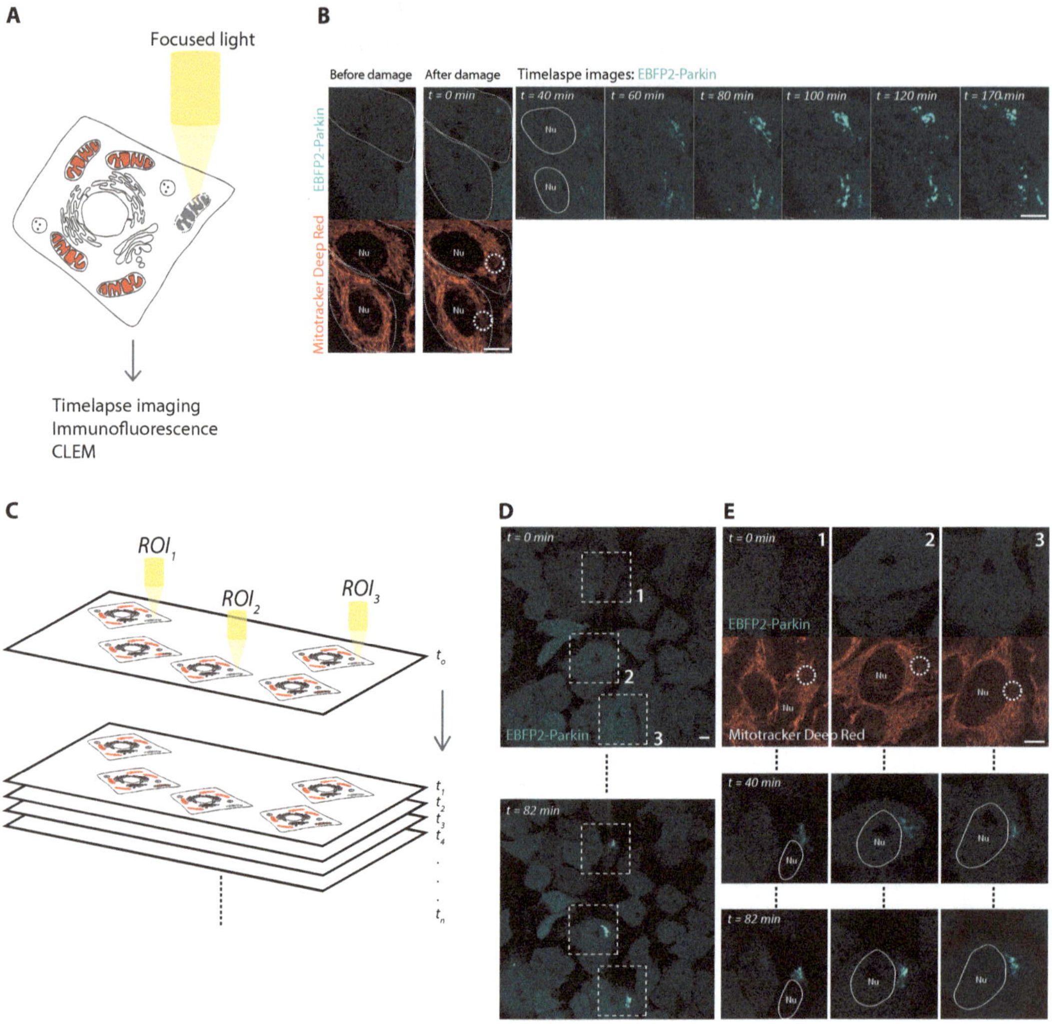

Fig. 2 Subcellular activation of mitophagy. (**a**) Relying on the light-scanning capability of a laser-scanning confocal microscope, one can specifically target a mitochondrion (or a group of) within a cell for impairment, followed by single-cell experiments. CLEM, correlative light and electron microscopy. (**b**) HeLa cells expressing EBFP2-Parkin were stained with MitoTracker Deep Red FM. Before 635 nm illumination, EBFP2-Parkin distribute evenly in the cytoplasm. Mitochondria within the white dotted circles were then selectively damaged through 635 nm illumination (100 μW for 30 s). Non-illuminated mitochondria maintained normal morphology and membrane potential. Time-lapse imaging showed that EBFP2-Parkin selectively labeled onto damaged mitochondria. Nu, nucleus. (**c**) Multiplexed recording of single cells undergoing mitophagy. (**d**) HeLa cells expressing EBFP2-Parkin were stained with MitoTracker Deep Red FM. An automatic stage was used for simultaneous high-resolution time-lapse imaging of multiple cells. Three distant cells were chosen (#1–3) for mitophagy activation and selectively displayed Parkin translocation onto damaged mitochondria ($t = 82$ min). (**e**) Enlarged views of cells 1–3 from (**e**). White dotted circles indicate 635 nm illuminated regions within each cell (100 μW for 30 s). Scale bars, 10 μm

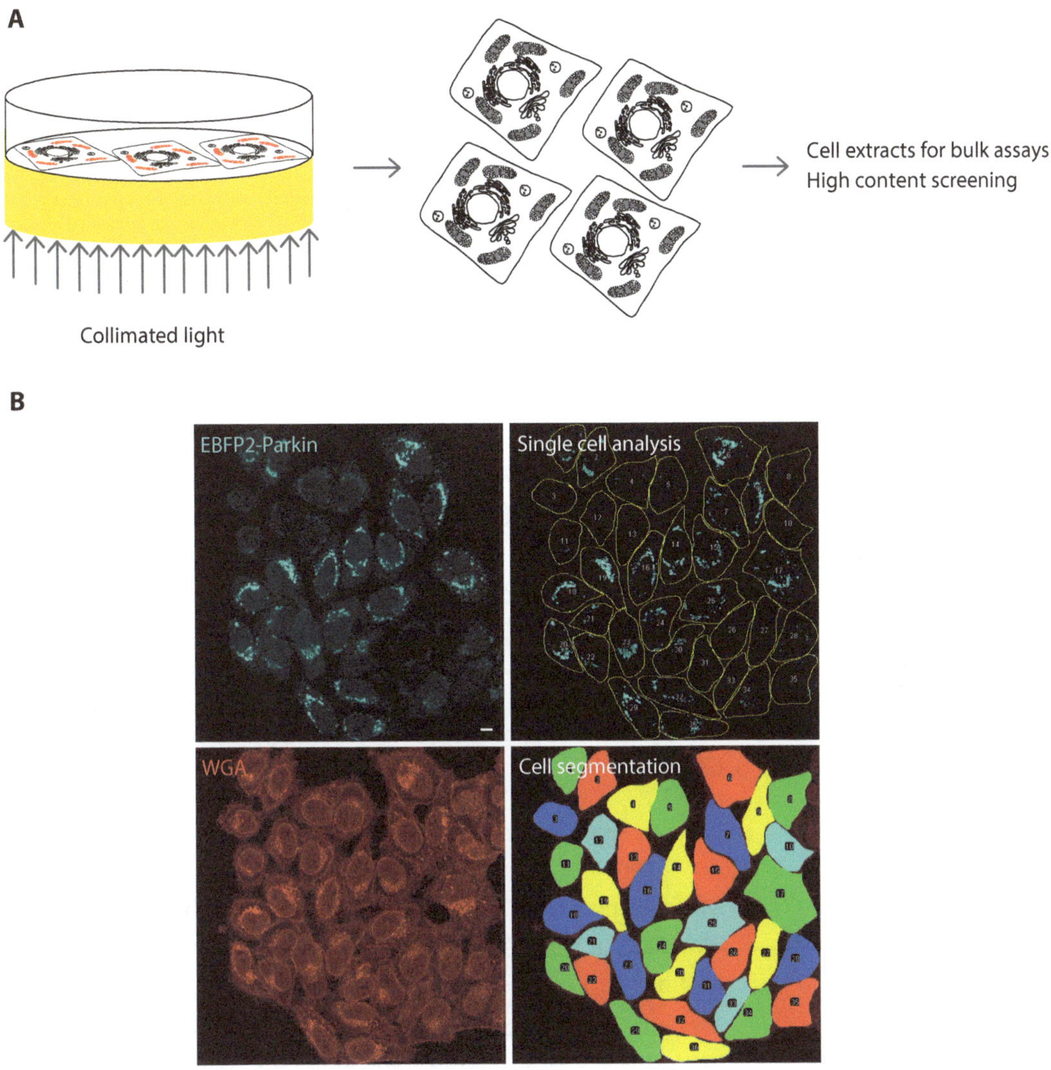

Fig. 3 Global activation of mitophagy. (**a**) Mitochondria in all cells are simultaneously impaired through the use of a collimated light source. This mode of impairment is suited for bulk biochemical assays and high-throughput experiments. (**b**) EBFP2-Parkin expressing HeLa cells stained with MitoTracker Deep Red FM grown on a glass-bottom dish were illuminated with a 660 nm LED (23 mW/cm^2 for 5 min). Cells were fixed and then stained with Alexa Fluor 594 WGA 6 h after illumination for image analysis. Alexa Fluor 594 WGA fluorescence was used to perform cell segmentation (for defining the cell boundary). Scale bar, 10 μm

10. Glass-bottom culture dishes: Cells are cultured on 35 mm glass-bottom dish imaging. Use ones with No. 1.5 glass coverglasses (0.16–0.19 mm) as they are optimal for high-resolution imaging.

2.2 Staining Reagents

1. DMSO.
2. Phosphate-buffered saline (PBS): Use ultrapure water (18 MΩ-cm at 25 °C) to prepare PBS stocks (10×) by dissolving 80 g NaCl, 2 g KCl, 14.4 g Na_2HPO_4, and 2.4 g KH_2PO_4 in 800 mL of double-distilled H_2O. Adjust pH to 7.4 and add ddH_2O to 1 L. Sterilize by autoclaving. Dilute to 1× with double distilled ddH_2O before use.
3. MitoTracker Deep Red FM (Life Technologies, M22426). Prepare MitoTracker Deep Red FM stock solutions first by allowing the dehydrated dye to warm to room temperature. Dissolve the dye into high-quality DMSO to a final concentration of 1 mM, and store this stock solution at −20 °C, protected from light, for further usage.
4. Lipofectamine 2000 (Thermo Fisher Scientific, 11,668).
5. Mitophagy indicator construct (e.g., EBFP2-Parkin).
6. Cell fixation buffer: Dilute 16% paraformaldehyde (PFA) stock solution with pH 7.4 PBS to 4% for cell fixation.
7. Alexa Fluor 594 wheat germ agglutinin (WGA) for cell segmentation purpose in high-content screening. Prepare a 1000× concentrated stock solution (1 mg/mL) in pH 7.4 PBS, and store at −20 °C, protected from light, for further usage.

2.3 Illumination Equipment

1. Laser-scanning confocal microscope equipped with an automatic stage.
2. Microscope stage-top incubator (e.g., LCI Chamlide IC): for maintaining cells at 37 °C and under 5% CO_2 during imaging.
3. 660 nm LED (Thorlabs, M660 L3-C1).
4. Medium for live-cell imaging: Dulbecco's modified Eagle's medium without phenol red supplemented with 10% fetal bovine serum, 2 mM L-glutamine, and 1% P/S.

3 Methods

3.1 Subcellular Activation of Mitophagy

1. Dilute the 1 mM MitoTracker Deep Red FM stock solution to a final concentration of 100 nM in pre-warmed cell culture growth medium.
2. Stain cells (e.g., EBFP2-Parkin expressing HeLa cells from Subheading 2.1, **item 5**) with 100 nM MitoTracker Deep Red FM for 30 min in the 5% CO_2 incubator at 37 °C (*see* **Note 4**).
3. Wash the cells with 2 mL pre-warmed culture medium three times (*see* **Note 5**).

4. Before activating mitophagy using a laser-scanning confocal microscope, adjust the correction collar setting on the high-NA objective lens to match the thickness of glass-bottom dish used.
5. Add a drop of immersion medium on the objective lens and fix the dish onto the stage-top incubator.
6. Allow the dish to thermo-equilibrate on the microscope stage.
7. Image, and select a cell expressing lower levels of EBFP2-Parkin by using 405 nm excitation (*see* **Note 6**).
8. Set the image acquisition parameters for the target cell (*see* **Note** 7).
9. View the selected cell through 635 nm excitation (through imaging MitoTracker Deep Red FM). Choose a circular region of interest (radius = 1.5 μm) over mitochondria to be impaired and scan with 100 μW of 635 nm light for 30 s (an example is shown in Fig. 2b; *see* **Note 8**).
10. Acquire time-lapse images to monitor EBFP2-Parkin labeling on damaged mitochondria (*see* **Note 9**).
11. For multiplexed time-lapse recording of single cells, repeat **steps 6** and 7 on multiple cells while registering their XYZ positions in the imaging software. Then sequentially perform **step 8** on all selected cells. During time-lapse imaging, the automatic stage cycles between all selected positions for multiplexed, high-resolution recording of single cells (an example is shown in Fig. 2d).

3.2 Population-Wise Impairment of Cellular Mitochondria

1. Dilute the 1 mM MitoTracker Deep Red FM stock solution to a final concentration of 500 nM in pre-warmed cell culture growth medium.
2. Stain cells (e.g., EBFP2-Parkin expressing HeLa cells from Subheading 2.1, **item 5**) with 500 nM MitoTracker Deep Red FM for 30 min in the 5% CO_2 incubator at 37 °C (*see* **Note 4**).
3. Wash the cells with 2 mL pre-warmed culture medium three times (*see* **Note 5**).
4. Expose the entire cell culture dish to a collimated 660 nm LED light source (13–23 mW/cm^2 intensity) for 5 min (*see* **Note 10**).
5. Incubate the cells at 5% CO_2, 37 °C for 6 h (or for any desired length of time depending on the observations to be made).
6. Wash the cells three times with PBS.
7. Fix the cells with 4% PFA at room temperature for 10 min.
8. Wash the cells three times with PBS.

9. Stain the cells with 1 μg/mL Alexa Fluor 594 WGA at room temperature for 10 min.
10. Wash the cells three times with PBS.
11. Conduct image acquisition and quantification (an example is shown in Fig. 3b).

4 Notes

1. The glass-bottom dishes should be coated with poly-L-lysine for enhanced cell attachment.
2. Cells should be around 70% confluent following overnight incubation.
3. Do not overexpress too much EBFP2-Parkin. Cells highly overexpressing EBFP2-Parkin show defective Parkin translocation onto damaged mitochondria.
4. Optimize the working concentration of the dye according to the cell type and illumination setting used.
5. Wash the cells with culture medium instead of PBS as Mito-Tracker Deep Red FM is a membrane potential-sensitive dye.
6. Avoid using high intensities of light (either arc lamp or laser) when viewing cells to avoid unnecessary photo-bleaching and ROS generation.
7. To obtain high-quality images for image quantification, higher zoom factor and scan numbers (for image averaging) are recommended. However, one should also take photo-bleaching into account when performing time-lapse experiments.
8. Adjust the 635 nm illumination setting proportionally when choosing a different-sized ROI.
9. EBFP2-Parkin will accumulate on damaged mitochondria within 40–60 min following 635 nm illumination. Using grid dishes, one can mark the cells of interest for downstream immunofluorescence or correlative light and electron microscopy applications.
10. Optimize the working concentration of the dye based on cell density, LED power, and illumination duration.

Acknowledgments

This work was supported by the MOST 104-2628-B-001-001-MY4 research grant from the Ministry of Science and Technology in Taiwan.

References

1. Youle RJ, Narendra DP (2011) Mechanisms of mitophagy. Nat Rev Mol Cell Biol 12(1):9–14
2. Narendra D, Tanaka A, Suen DF et al (2008) Parkin is recruited selectively to impaired mitochondria and promotes their autophagy. J Cell Biol 183(5):795–803
3. Lazarou M, Narendra DP, Jin SM et al (2013) PINK1 drives Parkin self-association and HECT-like E3 activity upstream of mitochondrial binding. J Cell Biol 200(2):163–172
4. Lazarou M, Sliter DA, Kane LA et al (2015) The ubiquitin kinase PINK1 recruits autophagy receptors to induce mitophagy. Nature 524(7565):309–314
5. Padman BS, Bach M, Lucarelli G et al (2013) The protonophore CCCP interferes with lysosomal degradation of autophagic cargo in yeast and mammalian cells. Autophagy 9 (11):1862–1875
6. Hsieh CW, Chu CH, Lee HM et al (2015) Triggering mitophagy with far-red fluorescent photosensitizers. Sci Rep 5:10376
7. Yang JY, Yang WY (2011) Spatiotemporally controlled initiation of Parkin-mediated mitophagy within single cells. Autophagy 7 (10):1230–1238
8. Wong YC, Holzbaur EL (2014) Optineurin is an autophagy receptor for damaged mitochondria in parkin-mediated mitophagy that is disrupted by an ALS-linked mutation. Proc Natl Acad Sci U S A 111(42):E4439–E4448
9. Ashrafi G, Schlehe JS, LaVoie MJ et al (2014) Mitophagy of damaged mitochondria occurs locally in distal neuronal axons and requires PINK1 and Parkin. J Cell Biol 206 (5):655–670
10. Wang Y, Nartiss Y, Steipe B et al (2012) ROS-induced mitochondrial depolarization initiates PARK2/PARKIN-dependent mitochondrial degradation by autophagy. Autophagy 8(10):1462–1476
11. Choubey V, Safiulina D, Vaarmann A et al (2011) Mutant A53T alpha-synuclein induces neuronal death by increasing mitochondrial autophagy. J Biol Chem 286 (12):10814–10824
12. Yang JY, Yang WY (2013) Bit-by-bit autophagic removal of parkin-labelled mitochondria. Nat Commun 4:2428
13. Wojtovich AP, Foster TH (2014) Optogenetic control of ROS production. Redox Biol 2:368–376
14. Bulina ME, Chudakov DM, Britanova OV et al (2006) A genetically encoded photosensitizer. Nat Biotechnol 24(1):95–99
15. Takemoto K, Matsuda T, Sakai N et al (2013) SuperNova, a monomeric photosensitizing fluorescent protein for chromophore-assisted light inactivation. Sci Rep 3:2629
16. Sarkisyan KS, Zlobovskaya OA, Gorbachev DA et al (2015) KillerOrange, a genetically encoded photosensitizer activated by blue and green light. PLoS One 10(12):e0145287
17. Shu X, Lev-Ram V, Deerinck TJ et al (2011) A genetically encoded tag for correlated light and electron microscopy of intact cells, tissues, and organisms. PLoS Biol 9(4):e1001041
18. Westberg M, Holmegaard L, Pimenta FM et al (2015) Rational design of an efficient, genetically encodable, protein-encased singlet oxygen photosensitizer. J Am Chem Soc 137 (4):1632–1642
19. He J, Wang Y, Missinato MA et al (2016) A genetically targetable near-infrared photosensitizer. Nat Methods 13(3):263–268

Chapter 41

Investigating Mitophagy and Mitochondrial Morphology In Vivo Using *mito*-QC: A Comprehensive Guide

Thomas G. McWilliams and Ian G. Ganley

Abstract

Autophagy evolved as a mechanism to sustain cellular homeostasis during instances of nutrient deprivation. Mounting evidence has also clarified that under basal and stress conditions, selective autophagy pathways can target the destruction of specific organelles. Mitochondrial autophagy, or mitophagy, has emerged as a key quality control (QC) mechanism to sustain the integrity of eukaryotic mitochondrial networks. We recently reported the development of *mito*-QC, a novel reporter mouse model that enables the high-resolution study of mammalian mitophagy with precision, in fixed and live preparations. This model holds significant potential to transform our understanding of mammalian mitophagy pathways in vivo, in a variety of physiological contexts. We outline a detailed protocol for use of our recently described *mito*-QC mouse model, including tips and troubleshooting advice for those interested in monitoring mitophagy in vitro and in vivo.

Key words Mitophagy, Autophagy, Mitochondria, Mouse models, Neurodegeneration, Metabolism, Cancer, Immunology, Cardiology, Vascular biology, Nephrology, Developmental biology, Histology, Microscopy

1 Introduction

Mitochondria lie at the heart of eukaryotic metabolism. Mitochondrial damage and dysfunction have been implicated in the pathogenesis of many human diseases, ranging from inherited rare metabolic disorders to widespread conditions such as cancer and neurodegeneration [1]. The seminal discovery that the Parkinson's disease-related proteins PINK1 (*PARK6*) and Parkin (*PARK2*) modulate a distinct type of mitochondrial quality control known as mitophagy, placed this cellular pathway at the center of efforts to understand mitochondrial-related neurodegeneration [2]. We now know that many other types of mitophagy exist, for example, in response to hypoxia, metabolic remodeling, and iron chelation [3]. Although in vitro biochemistry and cell biology approaches have formed the basis for much of these insights, little was known

Nicholas Ktistakis and Oliver Florey (eds.), *Autophagy: Methods and Protocols*, Methods in Molecular Biology, vol. 1880, https://doi.org/10.1007/978-1-4939-8873-0_41,

about the physiological basis of mitophagy until the recent advent of three mitophagy mouse models: *mito*-QC, mt-Keima, and mito-Timer [4–6]. This is because monitoring mitophagy is difficult without a suitable reporter probe in vivo. Antibodies to mitochondrial proteins in vivo are not reliable, and dye-based approaches are incompatible with fixed tissues and electron microscopy. Although electron microscopy remains the gold standard, it is arduous and difficult to execute in a consistent fashion without dedicated on-site expertise. These limitations are summarized in Table 1. We recently described a reporter mouse model engineered to monitor this type of mitochondrial quality control (QC), called "*mito*-QC" [4]. *mito*-QC is based on a tandem-fusion protein (mCherry-GFP) that is targeted to the outer mitochondrial membrane (OMM) *via* the mitochondrial targeting sequence (MTS) of the OMM protein, FIS1. Under steady-state conditions, all mitochondria fluoresce in both green and red, which when merged gives a yellow color. However, when damaged or superfluous mitochondria are targeted to the lysosome for elimination by mitophagy, GFP fluorescence is quenched due to its acid-labile properties—yet mCherry fluorescence remains unaffected. This has the advantage of being able to visualize a mitophagic "end point," as red puncta can be scored or counted to provide an index of mitophagy in cells and tissues. Additionally, researchers can also benefit from being able to study mitochondrial morphology—which is a key determinant of mitochondrial and cellular homeostasis (see Fig. 1). There are several major advantages of *mito*-QC over other existing mitophagy models [1]. It is compatible with fixation and a variety of histochemical approaches. This enables researchers to study mitophagy in labeled subsets of cells in vivo with ease. At first glance, this advantage may seem irrelevant; however this distinguishing feature of *mito*-QC from other available models is crucially important. This is because not all cells are equally susceptible to dysfunction in vivo. This is particularly true in the case of neurodegenerative diseases, where neural subpopulations exhibit multifactorial pathology accompanied by a high degree of selective vulnerability. Common examples of this include A9 dopaminergic neurons that degenerate in Parkinson's disease (PD) [2], GABAergic medium spiny neurons in Huntington's disease (HD), entorhinal and hippocampal CA1 neurons in Alzheimer's disease (AD), ventral motoneurons in amyotrophic lateral sclerosis (ALS), and cerebellar Purkinje neurons that degenerate in the rare lysosomal storage disorder, Niemann-Pick Type C1 (NPC1). In the case of mitochondrial DNA (mtDNA)-associated diseases, patients present with a vast degree of clinical and tissue-specific heterogeneity. This compatibility with fixation also enables a high-throughput approach to assessing mitophagy and mitochondrial biology in experiments where large sample sizes are required, without the inheirent uncertainty, low-throughput nature, and labor associated

Table 1
Comparison of available methods to monitor mitophagy in vitro and in vivo

Method/Technique	Advantages	Disadvantages
Transition electron microscopy (TEM)	• Gold-standard method of detection • Enables direct visualization of autophagosome membranes, autophagic bodies, and other cellular compartments • Immunolabeling can localize proteins of interest to mitophagosomes	• Laborious—many steps where error/inconsistencies can be introduced • Requires high degree of technical expertise to implement and execute • Requires depth of experience to interpret • Not adaptable for high-throughput screening • Difficult to optimize for tissues • Often requires perfusion fixation for tissues • Immunolabeling can be challenging due to loss of epitopes during sample processing
Colocalization of mitotracker and lysotracker	• Rapid, easy to use	• Incompatible to detect mitophagy in vivo • Documented problems with specificity and variability in labelling • Off-target effects in other membrane-bound organelles
Anti-DNA antibodies	• Enables visualization of loss of mtDNA nucleoids following stimulus-induced mitophagy in vitro	• Expensive • Antibody used in field (Progen) is highly inefficient in tissues: manufacturer recommends 1:10 dilution • Yields variable results in vivo
Biochemical analysis of mitophagic flux: • Immunoblotting • Citrate Synthase Assay	• Rapid, easy to perform	• Low sensitivity, i.e., high levels of mitophagy required • Lysosomal inhibitors required to interpret flux (e.g., Bafilomycin A1, hydroxychloroquine) and can have detrimental effects in vivo
Immunofluoresence-based colocalization of mitochondria with LC3-positive autophagosomes	• Rapid, easy to perform	• Difficult to implement for *in vivo* mitophagy • Nonspecific autophagosomes thought to form at mitochondria • Not all mitophagy may be LC3-dependent

(continued)

Table 1
(continued)

Method/Technique	Advantages	Disadvantages
Mito-Keima (mt-Keima)	• Provides end-point readout of mitophagy in vitro and in vivo • Compatible with intravital imaging	• mt-Keima protein is incompatible with fixation and thus directs immunolabeling, only enables a regional analysis of mitophagy in complex tissues • Analysis must be conducted in live tissue slices, not amenable to high-throughput in vivo analysis and may necessitate further verification • mt-Keima signal is lost upon freezing—cryosectioning not possible • Long-term storage of tissue specimens and slices for reference is not possible • No antibodies yet available to Keima protein, thus electron microscopy-based verification not possible • Spectral overlap can complicate interpretation and quantitation Level of reporter expression must be controlled to prevent mislocalization
Mito-Timer (DsRed1-E5)	• Enables a pulse-chase monitoring of mitochondrial half-life and mitophagy in vitro and in vivo • Useful model to understand biogenesis • Compatible with fixation and immunolabeling	• Conditional model requires the use of doxycycline to activate in vitro • Constitutive mito-Timer expression in mice reported to be heterogeneous in tissue (the heart: expression in ventricles, reduced in aorta) • Maturation of mito-Timer protein may be different across cells and tissues in vivo
mito-QC (mCherry-GFP-mtFIS$^{101-152}$)	• Provides end-point readout of mitophagy in vitro and in vivo • OMM labeling enables monitoring/characterization of mitochondrial network and study of dynamics • Constitutive expression facilitates constant monitoring • Compatible with tissue fixation • Immunolabeling enables resolution of mitophagy in a vast	• Fixation must be conducted using formaldehyde at pH 7.0: any deviation in pH of fixative will result in de-quenching of GFP (yellow mitolysosomes instead of red-only) – In this instance, LAMP1 immunostaining can be used to easily verify the lysosomal nature of presumptive mitophagic structures.

(continued)

Table 1 (continued)

Method/Technique	Advantages	Disadvantages
	array of cell subtypes in vivo • Compatible with freezing and samples stored long-term • No specialist microscopy setup required—compatible with simple and sophisticated systems • Antibodies to GFP and mCherry enable immunogold TEM • Compatible with iDISCO for volume imaging (although some de-quenching may occur due to use of solvents) • Ease of measurement: enables multiparametric quantitation—mitolysosome number, size, mitochondrial shape, etc.	• For fluorescent immunolabeling, secondary fluorophores limited to those outside of GFP/mCherry spectra • Incompatible with heat-mediated antigen retrieval – SDS-mediated antigen retrieval possible • Level of reporter expression must be controlled to prevent mislocalization

with probes that require live-imaging in tissues [3]. *mito*-QC affords researchers the ability to visualize end-point turnover in the context of the entire mitochondrial network in vitro and in vivo. Thus in addition to mitophagy, mitochondrial dynamics and organelle crosstalk can also be investigated [4]. As *mito*-QC is a tandem-tag protein, it enables the ratiometric measurement of mCherry-only signal to GFP. This ratio is useful for a variety of analyses, ranging from microscopy to FACS [5]. As excellent antibodies are available to detect GFP, researchers can obtain gold-standard verification in their specimens by immunogold electron microscopy [6]. GFP and mCherry work as an excellent tandem pair, due to their fast maturation times and photostability [7]. Given that the delivery of mitochondria to lysosomes may occur over several hours, maturation times should not be a concern as mitolysosomes are identified as distinct mCherry-only structures. Furthermore, even in the case of suspected dequenching of the reporter signal - the distinct morphology of mitolysosomes enables a bona-fide readout of mitophagy in concert with LAMP1 (or equivalent) immunolabelling of lysosomes. Recently, we used mito-QC to demonstrate that basal mitophagy is unaffected in mammalian tissues lacking a functional PINK1-Parkin signalling pathway [8]. Demonstrating the power of this approach to monitor mitophagy in vivo, mito-QC was also used to demonstrate the evolutionarily-conserved and PINK1/Parkin-independent nature of basal mitophagy [9]. We hope this protocol will serve as a useful reference for both experienced researchers and newcomers to the exciting world of in vivo cell biology.

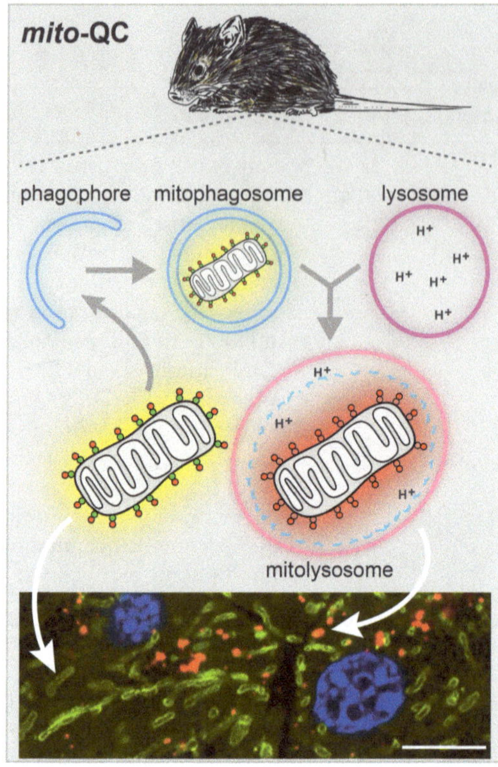

Fig. 1 Basic principle of the *mito*-QC mouse model. The *mito*-QC assay is based on a tandem tag construct of mCherry-GFP. This tag is targeted to mitochondria through the addition of a 51 amino acid mitochondrial targeting sequence derived from the outer mitochondrial membrane protein, FIS1 ($FIS1^{mt101-152}$). The resultant mCherry-GFP-$FIS1^{mt101-152}$ (*mito*-QC) construct is observable as a red and green fluorescent signal, which when merged appears yellow. The engulfment of mitochondria by the phagophore and concomitant fusion of the mitophagosome with the lysosome result in the appearance of mCherry-only puncta, due to the acid-labile properties of GFP. This provides quantitative and facile end-point assessments of mitophagy and mitochondrial morphology both in vitro and in vivo. The bottom panel shows a micrograph of pancreatic acinar cells from *mito*-QC, highlighting actual mitochondria and mitolysosomes. Scale bar, 10 μm

2 Materials

Important: Obtain ethical and institutional approval for all procedures involving the use of animal subjects. In this study, all experiments were subjected to ethical approval from the University of Dundee, in addition to being performed by licensed and trained individuals in accordance with the UK Animals Act (ASPA) 1986, on a Home Office Project License. Please note: a detailed description of surgery and perfusion is beyond the scope of this article and such proceedures should only be performed by trained and experienced professionals.

2.1 Basic Isolation of Tissues

1. *mito*-QC mice at desired age.
2. 1× PBS (Gibco) Life Technologies.
3. Fixative: 3.7% formaldehyde in 200 mM HEPES, at pH 7.0.
   (a) Prepare fresh on day or 24 h before; store at 4°C on ice.
4. 70% EtOH (for sterilization).
5. Autoclaved, sterile surgical instruments (Fine Science Tools).

2.2 Isolation of Adult Tissues Following Trans-cardial Perfusion (or Perfusion Fixation)

1. As above, but with additional appropriate surgical implements for procedure.
2. Automated perfusion: Suitable perfusion system (Perfusion One—Leica or equivalent) or peristaltic pump. For manual delivery of perfusate, use a 20 mL syringe.
3. 1× PBS (Gibco) Life Technologies.
4. For *trans-cardial perfusion with PBS*:
   (a) Perfuse with 1× PBS (Gibco, Life Technologies) until blood runs clear, rapidly excise organs of interest, and place into labeled tubes containing 3.7% formaldehyde in 200 mM HEPES, at pH 7.0—make fresh on day or 24 h before, store at 4 °C.
5. For *trans-cardial perfusion fixation*:
   (a) Perfuse with 1× PBS (Gibco, Life Technologies) and switch to 3.7% formaldehyde in 200 mM HEPES, at pH 7.0—make fresh on day or 24 h before, and store at 4 °C. Rapidly excise organs of interest and postfix in labeled tubes containing 3.7% formaldehyde in 200 mM HEPES, at pH 7.0.
6. Fixation times will vary depending on the age of the animal and type and size of specimen. For adult organs, we typically postfix for 24 h at 4 °C. For embryos, fixation times will vary according to developmental stage.

2.3 Tissue Sectioning

1. Sucrose (D-saccharose).
2. Peel-A-Way histology molds (Ted Pella) or equivalent.

A range of different molds is available. Select the one best for the tissue of interest, depending on size and sectioning setup.

3. Cold room or fridge at 4 °C.
4. OCT—Sakura (cryoprotectant—for cryosectioning).

5. Superglue (for vibrasectioning).
6. For sectioning:
   (a) Thaw-mounted sections—cryostat, e.g., Leica CM3080.
   (b) Free-floating sections—vibratome, e.g., Leica VT2000.
7. Leica Surgipath slides.
8. Ice-cold 1× PBS (if sectioning using vibratome).
9. 12- or 24-well plastic dishes containing 1× PBS, sable paintbrush (to collect free-floating vibrasections).
10. Cryoboxes with desiccant, suitable for slide storage (cryosections) (*see* **Notes 1–4**).

2.4 Nuclear Counterstaining

1. We typically always use DAPI or Hoechst (blue); a far-red stain such as ToPro-3 may also suffice, although we have not tested this extensively.

2.5 Mounting of Tissue Sections

1. No. 1.5 coverslips (Menzel-Glaser).
2. Vectashield H-1000 (*see* **Note 5**).
3. Suitable quick-drying sealing agent: Rimmel 60-second shine (transparent) or TopCoat.
4. Kimwipes/absorbent lint-free tissues.
5. Slide Folder, storage space at 4 °C.

2.6 Imaging of mito-QC Tissues and Cells

1. Zeiss LSM 710 META or LSM880 with Airyscan; laser scanning confocal microscopes or equivalent, with 4×, 40×, 63×, and 100× objectives.
2. Multiphoton laser for deep tissue imaging.
3. For live-cell imaging: Spinning-disk confocal microscopes may enable a faster and milder platform for live-cell imaging experiments.

2.7 Verification of Mitolysosomes by LAMP1 Immunohistochemistry and Immunocytochemistry

1. Antibody to LAMP1: Rat anti-LAMP1, clone 1D4B, Developmental Studies Hybridoma Bank.

2.8 Establishment of Primary MEF Cultures from mito-QC Embryos

1. *mito*-QC embryos at desired/appropriate stage of gestation (E13.5-E17.5).
2. 10 cm tissue culture-treated petri dishes.
3. Autoclaved, sterile surgical instruments (Fine Science Tools).
4. Sterile, disposable scalpels.

5. DMEM containing 10% FBS, 2 mM L-glutamine, 50 U/mL penicillin, 50 μg/mL streptomycin, 1× sodium pyruvate, 1× non-essential amino acids.

2.9 Immuno-histochemical Labeling of Tissue Sections

1. *mito*-QC tissue sections.
2. Slide staining tray/system.
3. 1× PBS.
4. Disposable Pasteur pipettes/squeegee bottle.
5. Primary and secondary antibodies.
6. Detergent: Triton-X100.
7. Optional: 1% SDS for antigen retrieval.

3 Methods

3.1 Preparation of Fixative (3.7% Formaldehyde)

1. Add paraformaldehyde to water and heat to 60 °C with stirring. (All steps must be performed in a designated chemical fume hood as formaldehyde gas will be produced which has been designated as carcinogenic.)
2. When temperature has been reached, add potassium or sodium hydroxide in a dropwise fashion to clarify solution. Remove from heat and allow to cool. (The addition of 1 M HEPES at pH 7.0 gives a final concentration of 200 mM and 3.7% formaldehyde.)
3. Filter sterilise solution and store at 4 °C in dark until use. For experiments involving freshly excised tissues or perfusion, formaldehyde must be prepared fresh and used within 24 h for best results. For cell-based experiments, aliquots from a freshly prepared batch of formaldehyde stored at −20 °C will suffice for single use (*see* **Notes 6–8**).

3.2 Establishment of MEF Cultures to Study Mitophagy from the mito-QC Mouse Model

1. With the aid of a dissection microscope in a laminar flow cabinet, remove all extraembryonic tissues and decapitate embryos using a scalpel in cold sterile PBS or L-15 medium. (A separate scalpel may be required for each individual embryo if different genotypes are a concern.)
2. Eviscerate embryos and remove a sample biopsy for genotyping by diagnostic PCR.
3. In a 10 cm petri dish, homogenize eviscerated tissue with scalpels into fine cubes. Change scalpels between embryos. NOTE: this is essential when working with more than one genotype.
4. Incubate tissue with 4 mL 0.025% trypsin (Gibco) for 5–10 min at 37 °C.

5. Neutralize reaction with 7.5 mL complete MEF medium and transfer to a 15 mL falcon tube.
6. Centrifuge at 500 × *g* for 5 min at room temperature.
7. Carefully aspirate supernatant and resuspend the tissue pellet in 10 mL of media.
8. Triturate briefly and transfer to a 10 cm tissue culture dish. Gently distribute cells by mixing in a figure-of-eight motion, incubate at 37 °C.
9. Monitor health of cultures and change media every 2–3 three days as required.
10. When cells are confluent, they are further propagated for experiments/maintenence or frozen down for future use.
11. Seed cells for analyses in appropriate dishes i.e. plastic dishes containing glass coverslips for microscopy, or suitable tissue culture dishes with microscopy-grade glass/optical-quality plastic (*see* **Notes 9–14**).

3.3 Isolating Tissues from the mito-QC Mouse to Perform Histological Analyses (See Note 15)

For direct isolation of tissues, e.g., visceral tissues

1. Perform euthanasia by an approved/licensed method and confirm death.
2. Sanitize cadaver using 70% EtOH and prepare for laparotomy.
3. Using surgical scissors and forceps, make a longitudinal incision along the abdomen.
4. Identify organ(s) of interest and excise rapidly, taking care not to damage any other tissues.
5. Wash excess blood from isolated tissue(s) using chilled tissue culture grade PBS and perform further subdissection if required.
6. Fix isolated tissue by immersion fixation (*see* **Notes 16–18**).
7. Following fixation: wash tissues in 3× in 1× PBS at 4 °C.
8. Prior to cryosectioning, cryoprotect tissues in 30% w/v sucrose/1× PBS at 4 °C. Cryoprotected tissues will sink to the bottom of the tube over time. At this point, fixed tissues can be stored at 4 °C. Long-term storage may require the addition of a low concentration of sodium azide (NaN_3) to sucrose to prevent microbial growth and putrefaction (*see* **Note 19**).
9. For cryosectioning: remove tissue from sucrose and quickly remove excess sucrose by blotting with a lint-free tissue (e.g., kimwipe). Place tissue in a labeled cryomold with OCT; let equilibrate for 5–20 min at room temperature. Use a permanent marker to label cryomold. Ensure no bubbles are present near the tissue—these can be removed using a hypodermic needle. Orient specimen as desired. The bottom of the cryomold will be the primary sectioning surface (*see* **Notes 20–22**).

10. Freeze the mold containing OCT and tissue into a cryo-ready sectioning block. Partially lower the mold into a suitable crucible containing chilled isopentane. The OCT will turn white, indicating a liquid to solid transition. (This first begins at the periphery and continues around the tissue.) Keep temperature constant at this step. When the entire block turns white (including the surface of OCT), the mold may be fully submerged into chilled isopentane for >1–2 min (*see* **Note 23**).
11. Remove frozen block from isopentane, blot off excess liquid, and place in cryostat chamber, ready for sectioning (*see* **Note 24**).
12. Cryosectioning protocols are beyond the scope of this article.

3.3.1 Isolating Tissues and Trans-cardial Perfusion, e.g., Adult Brain (See Note 25)

1. Commence terminal anesthesia using I.P. administration of pentobarbital (Euthetal) (we have not tested the effect of different agents on *mito*-QC, and so users who wish to utilize different compounds should be mindful of any potential effects on downstream analyses).
2. Proceed with trans-cardial perfusion once animal has reached a surgical plane of anesthesia.
3. Use temperature-equilibrated buffer (1× PBS or equivalent) as the initial perfusate.
   (a) If you are not performing perfusion fixation and are only using PBS, perfuse animal with PBS until blood is removed.
   (b) Note, other physiological solutions may be used, e.g., Krebs, Ringers/Tyrodes etc.; however we do not know the effect of these with the reporter.
4. When perfusate becomes clear, switch to fixative.
5. Observe for standard signs of fixation throughout surgery, e.g., tremor, clearing of the liver.
6. For the brain, decapitate animal and perform craniotomy, taking care not to damage tissue. Tissues should appear white, with vasculature minimally (if at all) apparent. If the perfusion has worked correctly, tissues should appear devoid of blood and characteristic vascular morphology.
7. Remove brain and proceed to processing for sectioning (*see* **Note 26**).

For discussion on sectioning with mito-QC see ***Note 27.***

3.4 Standard Immunostaining Protocol Using mito-QC Sections on Slides

1. Defrost slides with tissue sections in a slide tray at room temperature (30 min–1 h).
2. Rehydrate and wash off residual OCT using 3 × 5 min washes with 1× PBS. At this point, detergent can be included in the

PBS (e.g., 0.3% Triton-X100) to permeablise sections. Some immunolabelling experiments may necessitate alternate strategies i.e. different detergents, concentrations, etc.

3. Incubate using block containing detergent for defined amount of time (*see* **Note 28**).
4. Incubate with primary antibody for defined amount of time, in block solution containing detergent (*see* **Note 29**).
5. Wash primary antibody off with PBS containing detergent at room temperature −3 × 5 min washes.
6. Incubate with secondary antibody in appropriate sera and detergent, following manufacturers concentration of 1–1.5 h max. We find 30–45 min using 1:500 secondary antibody can yield satisfactory results in most cases.
7. Wash secondary antibody off with PBS containing detergent at room temperature −3 × 5 min washes.
8. If possible, perform nuclear counterstaining to aid anatomical identification. We typically use DAPI/Hoechst at 1:5000 for 5 min at RT, followed by brief washing. (We advise against the use of mounting media containing nuclear counterstains.)
9. Blot off excess PBS and mount slides using Vectashield H-1000 and Deckgläser, Menzel-Glaser 1.5 coverslips.
10. Remove excess Vectashield with a kimwipe and seal neatly using transparent TopCoat or nail polish. Drying can be accelerated by placing slides on a protected surface inside a chemical fume cupboard with appropriate airflow.
11. To minimize refraction index artifacts, wait at least 3 h before imaging. We typically mount slides and leave overnight to equilibrate.
12. Place slides in a folder at 4 °C for storage. Slides can also be stored at −20 °C for long term storage.

For discussion on for multiplex immunolabeling using *mito*-QC, *see* **Note 30**.

For discussion mouse-on-mouse immunolabeling using *mito*-QC, *see* **Note 31**.

For discussion on imaging in vivo mitophagy with *mito*-QC, *see* **Note 32**.

For discussion on quantitation of mitophagy using *mito*-QC, *see* **Note 33**.

For discussion on investigating mitochondrial network morphology with *mito*-QC, *see* **Note 34**.

For discussion on modeling selective autophagy in space, volume imaging and 3D rendering of mitolysosomes and mitochondrial network patterns in vivo, *see* **Note 35**.

4 Notes

1. We typically section adult brains on a vibratome; section thickness depends on orientation and desired downstream application. For horizontal (transverse) vibrasections, we use 100–200 micron sections. Serial free-floating cryosections may be obtained using a sledge microtome with a freezing stage attachment.
2. On the cryostat, sections ranging from 5 to 50 microns can be acquired. It can be difficult to obtain consistent sections below 8 microns, although such thin tissue sections may be required for specialist microscopy applications. As a general rule of thumb, we typically acquire good cryosections between 12 and 20 microns. Tissue type is an important factor to consider.
3. For cryosections, optimization of tissue section acquisition/collection may vary between instruments. Consult an experienced histologist for practical guidance, as many factors (tissue type, chamber and specimen core temperature, anti-roll plate, adequate fixation and freezing of specimen, blades and technical expertise) can greatly influence the quality of tissue sections obtained. Air-dry sections and store at −20 to −80 °C or proceed to IHC.
4. For vibrasections, free-floating sections are collected in appropriate 12–24-well dishes at 4 °C. The addition of sodium azide to PBS will prevent putrefaction and microbial growth over time, without interfering with fluorescent signal.
5. For all experiments with *mito*-QC tissues, Vectashield H-1000 provides excellent and consistent results, without any detectable loss of signal for up to 1 year. We have encountered unusual imaging artifacts with Vectashield "hard-set" formulations that contain a hard-setting agent. For this reason, we advise against the use of mounting media containing a hard-setting agent.
6. *mito*-QC exploits the acid-labile properties of GFP and the stability of mCherry to provide an end-point readout of mitophagy. The most critical aspect to the success and interpretation of the assay is the correct preparation of fixative at the appropriate pH. For all experiments, we use 3.7% formaldehyde in 200 mM HEPES buffer, pH 7.0. Experiments using mito-QC in cultured cells have demonstrated de-quenching of GFP in specimens processed with fixative that is not at pH 7.0. Heat-mediated de-quenching of GFP can also occur, and as such,

heat-mediated antigen retrieval is currently not possible with mito-QC. For those wishing to optimize antibody staining, we recommend SDS-mediated antigen retrieval, as described in [10]. Researchers should also be mindful when using reagents that require heating, e.g., when using agarose it is better to use the low-melting point variety and make sure that it has cooled sufficiently. Nonetheless, even in a de-quenched sample, the morphology of mitolysosomes is distinct from that of mitochondria, and users can verify occurrence of mitophagy by simply using IHC to a lysosomal marker such as LAMP1, which provides an excellent and facile way to validate the lysosomal nature of mCherry-only puncta.

7. For best results, we tend to avoid the use of commercial fixatives, which often contain up to 10% (v/v) methanol (CH3OH) as a stabilization agent.

8. *mito*-QC is expressed from the Rosa26 locus and is found ubiquitously in all tissues. Heterozygote animals with one copy of the transgene demonstrate fluorescence in all tissues; however homozygous animals (two copies) exhibit noticeably pronounced fluorescence. We recommend that for analyses of neural tissue (particularly brain), researchers utilize homozygous *mito*-QC mice. It is also worth noting that although *Rosa26*-mediated expression affords ubiquitous inter-tissue expression, we observe a mosaic-type pattern of expression in the liver and a prominent expression of the reporter in blood vessels. This does not affect the ability of the reporter to provide an accurate readout of mitophagy, but users should optimize laser settings in order to obtain consistent results from these particular tissues. Endogenous signal from *mito*-QC usually provides excellent results without the need for amplification. However, as with all transgenic models that use fluorescent proteins such as GFP and mCherry, researchers can avail of commercial reagents (anti-GFP or anti-mCherry/RFP antibodies) to enhance signal should this be required. For immunofluorescence, we favor chicken anti-GFP (ab13909) from Abcam, rabbit anti-GFP (A11122) from Life Technologies/Thermo Fisher Scientific, and chicken anti-GFP (GFP-1020) from Aves. For immunoblotting experiments, mouse anti-GFP (1181460001) from Roche yields excellent results.

9. Despite being one of the most heterogeneous cell populations described in mammals [11], MEFs are routinely and widely used as a tractable source of primary cells to interrogate cell biology and signaling mechanisms in myriad genetically-altered mouse lines. Indeed, their ease of culture and propagation makes them attractive for a variety of assays. We have established MEFs from *mito*-QC embryos at a range of

gestational ages from E12 to E17.5. This is a general protocol that should suffice for the routine generation and maintenance of *mito*-QC MEFs and the study of mitophagy and mitochondrial dynamics in vitro. Positive controls to assess stimulus-induced mitophagy are available. To study PINK1-dependent mitophagy, researchers can overexpress the RBR E3 ubiquitin ligase Parkin and use mitochondrial depolarization agents to trigger mitophagy (data not shown). To study PINK1/Parkin-independent mitophagy, MEFs can be stimulated with the iron chelation agent deferiprone as described in [4, 10].

10. Heterozygote matings will yield wild-type (reporter negative) heterozygote and homozygote (reporter positive) genotypes. Mixed cultures can be established to perform comparative analyses in a variety of cell biology paradigms as described in McWilliams et al. [4]. For general analyses, we tend to favor the use of HET and HOM embryos. Wild-type embryos can be distinguished from *mito*-QC embryos using a microscope equipped with epifluorescence. However, diagnostic end-point PCR is required to distinguish between HET and HOM embryos.
11. The isolation and dispatch of embryos is a regulated procedure, and thus we will not outline this protocol here. Perform in accordance with your institutional and national guidelines governing the ethics and treatment of animal subjects.
12. Embryo staging practices may vary between institutes; however all of our experiments are performed with embryos staged according to the criteria of Theiler (1976).
13. MEFs may be immortalized using T-cell antigen (SV40); however, be mindful that this transition may affect metabolic and mitochondrial status which could impact on the induction of mitophagy.
14. In the event that mouse maintenance is not possible, lentiviral and retroviral *mito*-QC constructs are available for the generation of stable cell lines as described in [12].
15. Tissues may be isolated fresh (i.e., after cervical dislocation or CO_2); however perfusion remains the gold standard to remove blood-containing immunogenic components and provides superior results for downstream analyses. This protocol will detail both methods; however users should be advised that the latter method is recommended to achieve consistent results, especially in the case of immunohistochemistry, and to avoid hematopoietic breakdown products (e.g., lipofuscin) which can confound analyses due to autofluorescence.

16. In keeping with good histological practice, the volume of fixative should be 20× that of the organ to ensure appropriate penetration of fixative. Fixative should be prepared fresh on the day and used within 24 h.
17. Cutting tissues into smaller pieces will aid the fixation process; however this is at the expense of anatomical accuracy and can compromise the precise spatial evaluation of mitophagy in a given region.
18. Depending on your cellular population of interest, fixation times may vary. It is important not to overfix tissues.
19. The w/v percentage of sucrose may be varied depending on type, size, and age of tissue specimen. Many histological protocols stipulate concentrations ranging from 10 to 30% (w/v). For consistency, we have used 30% throughout; however, increasing gradients of sucrose may also be used if preferred.
20. Histological practices and styles often vary between laboratories, and there is no "single correct method." In our experience, equilibration steps at this stage may be shortened or omitted without compromising *mito*-QC signal.
21. We advise researchers to be mindful at the embedding step. As the bottom of the cryomold will be the sectioning start point, the correct orientation of fixed tissue at this point is crucial to obtain satisfactory results downstream. Be mindful of tissue orientation and anatomical plane, e.g., sectioning mito-QC tissues in different planes will yield different mitochondrial network morphology. This will not affect the quantitation of mitophagy, but care is advised to achieve consistent results.
22. Many researchers often remove excess tissue at this point, e.g., visceral fat overlying organs, etc. Generally speaking, we do not do this as comparative analyses are extremely valuable. Embedding multiple organs or organ biopsies within the same block also provides a high-throughput way of sectioning, in addition to valuable comparative analysis.
23. Isopentane may be chilled using dry ice—some researchers favor the use of crushed dry ice for packing and more controlled freezing. Liquid nitrogen may also be used, although freezing may occur in a more rapid and unpredictable pattern. Tissue and OCT have been reported to crack using this method.
24. If sectioning immediately, tissue block will need at least 20 min to equilibrate to cryochamber. If sectioning at a later date, store blocks on dry ice until finished. For long-term storage, wrap blocks individually in tinfoil, and label with tape. Place in sealed bag in a cryobox and store at −20 °C for short-term storage or −80 °C for long-term storage.

25. Trans-cardial perfusion fixation remains the gold standard for immunohistochemical analyses. However, in the absence of a suitable setup (appropriately vented chemical fume cupboard, equipped for surgery), trans-cardial perfusion with PBS combined with immersion fixation provides an equally effective way to assess mitophagy combined with immunohistochemical labeling. This procedure should only be conducted by experienced professionals licensed to perform anesthesia and surgery, and thus we will not detail the specifics of perfusion surgery. Record any adverse side effects or events during anesthesia that may inform results obtained from downstream analyses. Obtain ethical approval and consult your institutional named veterinary surgeon and compliance officer before proceeding.
26. For vibrasectioning, the brain should be postfixed as usual and then washed in PBS. Vibrasectioning will take place in PBS, although vibrasectioning can also be performed on cryoprotected specimens. For the brain or other organs, biopsies can be removed and snapped frozen in liquid N2 from freshly excised organs. This facilitates parallel biochemical/other measurements from a single animal. Organs may be bisected or cut to enable penetration of fixative if desired; however we have found this is not necessary.
27. Typically, cryosections should suffice for the majority of investigations. Excellent resolution can be obtained in the majority of tissues using cryosections from 12 to 20 μm. Very thin cryosections (5–6 μm) can be useful with tissues that can be difficult to immunolabel. However, in the case of brain tissue where thicker sections combined with free-floating immunohistochemistry can facilitate the resolution of entire axon tracts and neural pathways, we have found vibratome sectioning to provide excellent results. We routinely use Leica Surgipath slides for all experiments.
28. Detergent is a key factor in the success of immunostaining. We find that 0.3% Triton-X100 works well in the majority of cases. However, researchers should optimize according to their needs. Different detergents, combinations, and concenrations may yield better results, e.g., NP-40, Saponin, Digitonin, Tween-20, etc. Please experiment and optimise accordingly for your antigen of interest.
29. Practices vary according to both the laboratory and antibody in question. For most antigens, room temperature incubation overnight will yield satisfactory results. More controlled staining can be conducted at 4 °C or for shorter incubation times at RT. Incubation of slides at 37 °C can also be used in the case of certain antibodies. Blocking concentration may be reduced, e.g., 5% to 1% BSA. Consult the relevant literature to

determine the optimal concentration in your field, bearing in mind that a high concentration of primary antibody over a long incubation period may result in suboptimal bleed-through. Take time to titrate and optimize conditions according to your needs. If incubating overnight, use a humified chamber (e.g. StainTray). Ensure slides do not dry out as this will be detrimental to the success of the staining protocol. Use strips of parafilm to prevent drying out. This is one staining method; however staining systems exist that exploit capillary action to use less antibody. Furthermore, PAP pens can be used to reduce amounts of primary antibody and perform multiplex labeling using different primary antibodies on the same slide. Many factors can influence staining success, including section thickness. Detergent concentration may also play a factor in mediating antibody penetration. Consider free-floating sections for troublesome antigens. In addition to detergent, some investigators favor adding a low concentration of inert serum (e.g., BSA) to their wash buffers to help reduce background.

30. As the *mito*-QC tandem tag emits fluorescence in the green and red spectra, users are restricted to using blue and far-red spectra to selectively label other cellular components. We have had success with Alexa Fluor-conjugated antibodies, particularly with Pacific Blue, 633 and 647 nm fluorophores, that we routinely use at manufacturers recommended concentration (1:500). Users are advised to optimize the concentration of their primary antibody to avoid spectral overlap between blue/green and far-red/red channels. Failure to do this can pose a problem during image acquisition and prevent the reliable identification of subcellular structures. Single-labeling pilot experiments in wild-type (non-*mito*-QC) tissues can be useful to determine the dilution required and the pattern of labeling before proceeding with multiplex IHC.
31. In certain circumstances, it may be necessary to use a primary antibody raised in mouse to detect an antigen of interest in tissue. Mouse-on-mouse staining is typically associated with higher background fluorescence, due to cross-reactivity of mouse primary antibodies with endogenous Ig in tissues. Although perfusion aims to deplete tissues of blood and thus reduce immunogenicity, the complete elimination of all erythrocytes from an adult mammal can never be guaranteed. In this instance, we have found that mouse-on-mouse blocking serum (Vector) yields satisfactory results.
32. No special imaging setups are required to obtain visually striking images with *mito*-QC. In general, confocal microscopy provides the gold standard with additional advances such as Airyscan (Zeiss) and Hyvolution (Leica) aiding ultraclear

resolution of mitochondria and mitolysosomes in vivo. When imaging GFP and mCherry using identical laser settings, mitochondrial networks of cultured cells and tissue specimens will appear yellow. Equal intensities will usually result in saturation of the red-only signal in mitolysosomes, which then appear larger than normal. This can compromise and complicate quantitation and can obscure mitochondrial network morphology. For this reason, we recommend users take time to optimize laser and image acquisition settings using a range indicator with their particular microscope. It is always advised to consult experienced microscopists to obtain the best performance from your particular imaging setup.

33. A variety of quantitative approaches are compatible using *mito*-QC, and these range in sophistication, speed, and complexity. As there is no spectral overlap between GFP and mCherry, no corrections have to be applied using our model. The most basic readout using our model is to monitor mitophagy by simply counting mitolysosomes per cell or per field. This can be performed manually or using an object counter application, although this method is laborious and not amenable to high-throughput adaption. We routinely analyze mitophagy using Volocity software (PerkinElmer). Defining a protocol in this software enables consistent, semiautomated batch processing and a multiparametric measurement of mitophagy in tissue sections. Large numbers of tissues and conditions can be assessed in this way. For example, numbers, shape factors, and size of mitolysosomes can be quantitated in tissue. This is particularly valuable when quantifying mitophagy in labeled cells or structures, e.g., in a particular population of neurons or quantitating overlap of mCherry-only puncta with LAMP1-positive lysosomes to obtain numbers of bona fide mitolysosomes in vivo. Tissue expression levels can sometimes vary between animals, and as such, normalizing mCherry-only signal to GFP area provides a way to obtain a consistent measurement of mitophagy between tissues and subjects. We advise users to experiment with the approach best suited and directly available to them. Quantitation is also possible using freeware such as NIH ImageJ/FIJI. Obtaining the size of mitolysosomes may provide valuable information about lysosomal biology in vivo.
34. Mitophagy is not a singular event, and thus it is important to investigate its regulation in the wider context of mitochondrial network homeostasis. Mitochondria are dynamic and functionally pleiotropic organelles, and the modulation of mitochondrial morphology is believed to be important in understanding its function. This is particularly important given our recent discovery of mitophagic heterogeneity in vivo. Abnormal

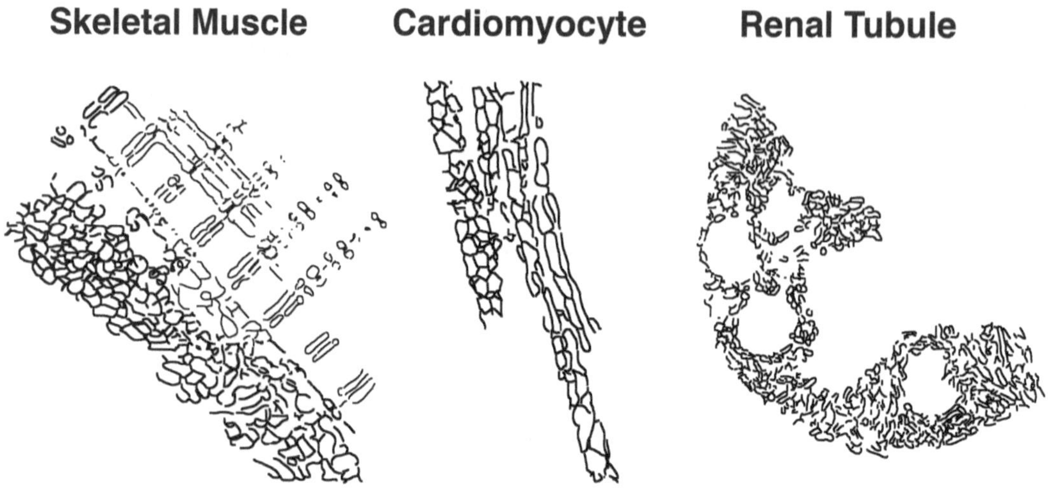

Fig. 2 *mito*-QC enables studies of mitochondrial network morphology in vivo. The outer mitochondrial membrane localization and ubiquitous expression of *mito*-QC enable the study of mitochondria and mitochondrial network morphology in vivo. Shown are camera lucida-style traces of mitochondrial networks from the skeletal muscle, heart, and kidney generated in Adobe photoshop, using the GFP channel as a mask. Pictures are modified from McWilliams et al. [4]

mitochondrial dynamics are associated with defective metabolism and a range of pathophysiological conditions. When laser signals have been optimized accordingly, the OMM signal of *mito*-QC enables the facile quantitation of mitochondrial network architecture in vitro and in vivo. We recently exploited this feature of our model to reveal the mitochondrial reticulum in skeletal muscle using light microscopy, previously only resolved by FIB-SIM. Tracing using appropriate software can highlight the diversity of mitochondrial morphologies within different tissues. Further analysis and quantitation of mitochondrial length and shape factor are also possible using Volocity. A representative figure demonstrating the range of mitochondrial networks is shown in Fig. 2. This aspect of our model is an important one, as such ultrastructural characterizations were previously only possible by EM.

35. We previously employed *mito*-QC with iDISCO to visualize selective autophagy in the adult kidney [4]. iDISCO and their variants (iDISCO$^+$) are tissue clearing techniques that enable volume imaging of optically cleared specimens [13]. Users should be aware that the use of solvents can result in de-quenching in *mito*-QC tissue specimens, although for regions that exhibit high levels of mitophagy this should not be problematic (e.g., proximal tubules of the kidney contain an abundance of morphologically-distinct mitolysosomes). The acquisition of z- stacks from immunolabeled tissue sections enables users to obtain a major amount of information in

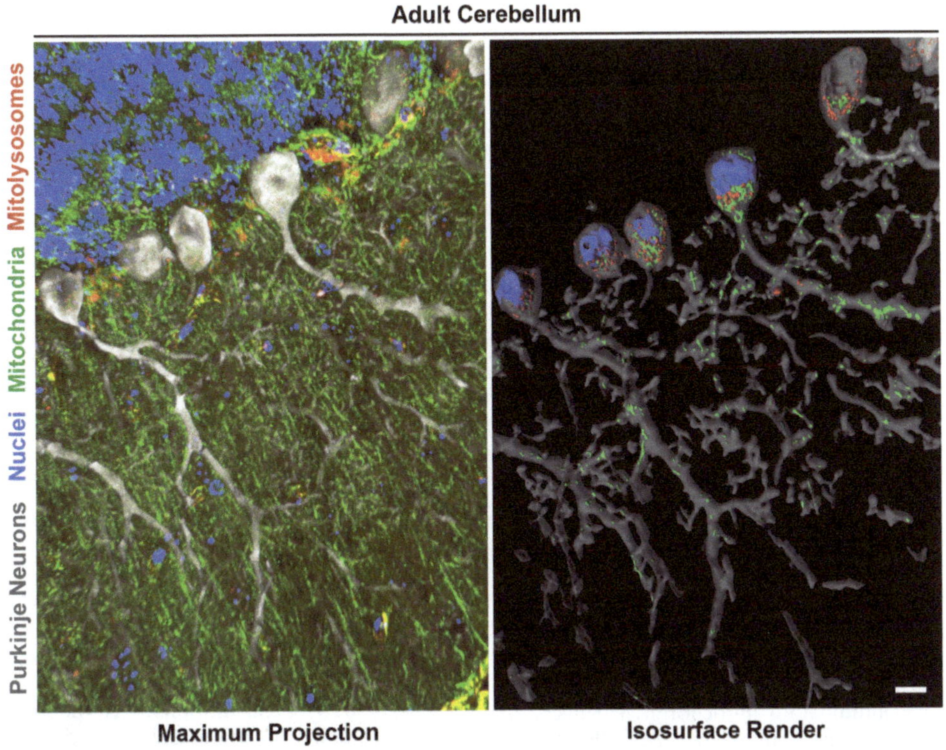

Fig. 3 Simplifying the visualization of in vivo mitophagy in space within complex tissues. In many instances, multiplex immunolabeling in complex tissues such as the brain can prove challenging to interpret for even the most experienced investigators. The compatibility of *mito*-QC with fixation enables immunolabeling of cellular subsets in vivo as shown in a section of mouse cerebellum (left). We used 3D volume rendering of this z-stack to create an isosurface or volume render (right). This enables investigators to create an impactful and informative graphic that can simplify the interpretation of mitophagy within complex structures in vivo. Shown is the concentration of mitolysosomes present in Purkinje cell somata in vivo, as described in McWilliams et al. [4]. With sufficient optimization and z-resolution, this technique can be successfully applied to profile the spatial nature of mitophagy in both cleared and non-cleared preparations. Scale bar, 10 μm

space, which is vital to obtain a comprehensive understanding of how mammalian mitochondrial homeostasis is orchestrated in vivo. 3D Volume Image Analysis Software (Imaris, Bitplane) can be used to reveal cell-specific mitochondrial networks in vivo in ever-increasing detail. Using the isosurface rendering function, it is possible to emphasize the 3D differences between mitochondrial networks of different cell types in vivo in a variety of contexts, e.g., during development, between genotypes, in different treatments, etc. This approach should prove particularly valuable in understanding how mitochondrial homeostasis is regulated during health and disease (Fig. 3).

Acknowledgments

We gratefully acknowledge the outstanding support from our colleagues and collaborators who have enabled us to develop and refine our methods using *mito*-QC. In particular, we wish to thank Dr. Alan R. Prescott for his expertise in microscopy and Dr. Graeme Ball for his expertise in image analysis. We thank Dr. Ngaire Dennison for her guidance on animal experiments. We thank Lambert Montava-Garriga, Dr. Jin-Feng Zhao, and Dr. François Singh for their critical reading of the manuscript. This work was funded by Medical Research Council, UK (IGG; MC_UU_12016/4).

References

1. Nunnari J, Suomalainen A (2012) Mitochondria: in sickness and in health. Cell 148 (6):1145–1159
2. McWilliams TG, Muqit MMK (2017) PINK1 and Parkin: emerging themes in mitochondrial homeostasis. Curr Opin Cell Biol 45:83–91
3. McWilliams TG, Ganley IG (2016) Life in lights: illuminating mitochondrial delivery to the lysosome in vivo. Autophagy 12 (12):2506–2507
4. McWilliams TG, Prescott AR, Allen GFG, Tamjar J, Munson MJ, Muqit MMK, Ganley IG (2016) Mito-QC illuminates mitophagy and mitochondrial architecture in vivo. The Journal of Cell Biology 214(3):333–345
5. Sun N, Yun J, Liu J, Malide D, Liu C, Rovira II, Holmstrom KM, Fergusson MM, Yoo YH, Combs CA, Finkel T (2015) Measuring in vivo mitophagy. Mol Cell 60:685–696
6. Stotland A, Gottlieb RA (2016) α-MHC MitoTimer mouse: in vivo mitochondrial turnover model reveals remarkable mitochondrial heterogeneity in the heart. J Mol Cell Cardiol 90:53–58
7. Nowotschin S et al (2009) Live-imaging fluorescent proteins in mouse embryos: multidimensional, multi-spectral perspectives. Trends Biotechnol 27(5):266–276
8. McWilliams TG, Prescott AR, Montava-Garriga L, Ball G, Singh F, Barini E, Muqit MMK, Brooks SP, Ganley IG (2018) Basal mitophagy occurs independently of PINK1 in mouse tissues of high metabolic demand. Cell Metab 27(2):439–449.e5. https://doi.org/10.1016/j.cmet.2017.12.008. Epub 2018 Jan 11
9. Lee JJ, Sanchez-Martinez A, Zarate AM, Benincá C, Mayor U, Clague MJ, Whitworth AJ (2018) Basal mitophagy is widespread in Drosophila but minimally affected by loss of Pink1 or parkin. J Cell Biol 217 (5):1613–1622. https://doi.org/10.1083/jcb.201801044. Epub 2018 Mar 2
10. Brown D, Lydon J, McLaughlin M, Stuart-Tilley A, Tyszkowski R, Alper S (1996) Antigen retrieval in cryostat tissue sections and cultured cells by treatment with sodium dodecyl sulfate (SDS). Histochem Cell Biol 105 (4):261–267
11. Singhal PK, Sassi S, Lan L, Au P, Halvorsen SC, Fukumura D, Jain R, Seed B (2016) Mouse embryonic fibroblasts exhibit extensive developmental and phenotypic diversity. Proc Natl Acad Sci 113:122–127
12. Allen GFG, Toth R, James J, Ganley IG (2013) Loss of iron triggers PINK1/Parkin-independent mitophagy. EMBO Rep 14 (12):1127–1135
13. Renier N, Wu Z, Simon DJ, Yang J, Ariel P, Tessier-Lavigne M (2014) iDISCO: a simple, rapid method to immunolabel large tissue samples for volume imaging. Cell 159:896–910

Chapter 42

Assays to Monitor Mitophagy in *Drosophila*

Panagiotis Tsapras, Anne-Claire Jacomin, and Ioannis P. Nezis

Abstract

Autophagy is a central pathway utilized by many eukaryotic cells in order to recycle intracellular constituents, particularly under periods of nutrient scarcity or cellular damage. The process is evolutionarily conserved from yeast to mammals and can be highly selective with regard to the contents that are targeted for degradation. The availability of *Drosophila* transgenic lines and fluorophore-labeled autophagic markers allows nowadays for the more effortless visualization of the process within cells. Herein, we provide two protocols to prepare *Drosophila* samples for confocal and transmission electron microscopy for in vivo monitoring of mitophagy, a specific type of autophagy for the clearance of damaged or superfluous mitochondria from cells.

Key words Autophagy, Mitophagy, *Drosophila*, Fat body, Confocal microscopy, Electron microscopy, Complex V

1 Introduction

In the microworld of the cell, recycling is a fundamental process for living healthy. As such, cells may utilize many different processes for effectively salvaging old or defective parts within themselves to ensure their survival, one of which is autophagy (from the Greek for "self-eating"). Autophagy was initially characterized as a nonspecific degradation process of cytosolic constituents through lysosomes. Nevertheless, a surge of studies over the last decades have established the prominent selectivity of autophagy in the removal of intracellular materials. Various substrates including protein aggregates, damaged organelles, invading pathogens, and lipids have been shown to be selectively degraded by autophagy [1, 2]. A well-studied type of selective autophagy is mitophagy, which is utilized for the efficient clearance of damaged or excessive mitochondria from cells [3]. Specific targeting of mitochondria for autophagic degradation is mediated by Atg32 in yeast [4, 5] and NIX in mammals during red blood cell differentiation [6, 7]. In

Nicholas Ktistakis and Oliver Florey (eds.), *Autophagy: Methods and Protocols*, Methods in Molecular Biology, vol. 1880, https://doi.org/10.1007/978-1-4939-8873-0_42,

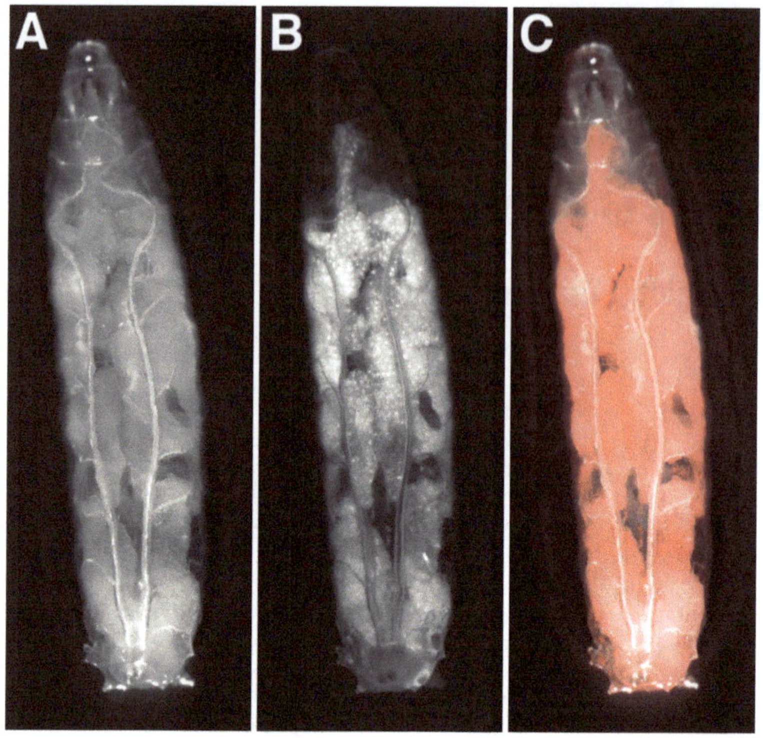

Fig. 1 *Drosophila* larval fat body. Picture of a third instar larvae expressing mCherry-Atg8a in the fat body. (**a**) Transillumination, (**b**) mCherry-Atg8a fluorescence, (**c**) merged channels

addition, mitophagy is regulated by PINK and Parkin in most metazoan cell types [3].

Research in *Drosophila* has contributed to our current understanding of the selective autophagy mechanism in higher eukaryotes. The humble fruit fly is an excellent model to monitor autophagy in vivo as it is easier to handle compared to mammalian models, can be kept in large numbers, and requires little in terms of maintenance [8, 9]. With respect to visualizing autophagy in vivo, a tissue of interest for such undertakings is the *Drosophila* larval fat body (Fig. 1). Its large availability, ease of dissection, and manipulation, coupled with the high autophagic activity observed after starvation, allow for reliable monitoring of the process [10–13].

In this chapter, we describe two protocols for the visualization of mitophagy in the larval fat body using fluorescence confocal microscopy, and transmission electron microscopy techniques.

2 Materials

2.1 *Drosophila* Stocks and Maintenance

Flies are raised on a yeast/cornmeal diet and are kept at 25 °C and 70% humidity, at a 12 h light-dark cycle. The fly stocks used in this chapter are available from the Bloomington Drosophila Stock

Center (Indiana University): Cg-GAL4 (#7011) and UAS-m-Cherry-Atg8a (#37750) [14] (*see* **Note 1**).

1. Fly food: 1 L H_2O, 42 g inactive dry yeast, 60 g yellow cornmeal, 130 g sucrose, 5.5 g agar, and 15 mL Nipagin 10%.
2. Wet yeast paste is prepared fresh, by dissolving 5 g of dry active yeast in 15 mL of water (*see* **Note 2**).
3. FCCP (carbonyl cyanide 4-(trifluoromethoxy) phenylhydrazone) stock solution is prepared at 100 mM in ethanol. The solution appears yellow and can be stored at −20 °C for 2 months protected from light. Working solution is made by diluting the appropriate volume of stock solution in water to 500 μM (*see* **Note 3**).
4. 20% (w/v) sucrose solution in dH_2O.
5. Nutri-Fly Instant Drosophila Medium.

2.2 Equipment for FCCP Feeding and Dissection of Larval Fat Bodies

1. Dissection microscope/stereoscope.
2. Petri dishes.
3. Dark contrast silica pad for better contrast (*see* **Note 4**).
4. Tubes with FCCP fly food.
5. 2× fine forceps (style Dumont #5), spatula, and tweezers.

2.3 Equipment and Reagents for Immunostaining and Mounting Samples on Slides for Confocal Microscopy Investigation

1. 4% methanol-free paraformaldehyde (PFA) prepared in 1.5 mL aliquots in PBS from a 16% solution stock. The aliquots can be kept at −20 °C for a few months.
2. Mouse Anti-Complex V alpha-Subunit Monoclonal Antibody (Invitrogen, #439800).
3. Goat anti-Mouse IgG Secondary Antibody, Alexa Fluor 488 (Invitrogen, #R37120).
4. Hoechst 33342 1 mM (1000×) stock solution in ultrapure water, which can be stored in a dark vial at 4 °C. Working solution is made by diluting in PBS before use.
5. Phosphate-buffered saline (PBS) (1× PBS pH 7.4).
6. 1× block-permeabilization (BP) buffer: 1× PBS, 0.1% Triton-X100, 0.3% bovine serum albumin (BSA). A stock solution of 10x BP buffer can be prepared and kept at 4 °C or −20 °C for longer storage. From this stock solution, a 1:10 dilution can be made in PBS as required before use.
7. 1× PBST: 1× PBS, 0.1% Tween-20.
8. Mounting medium: 70% glycerol, 2% w/v propyl gallate, 1× PBS. This can be stored for up to 6 months in a dark vial at 4 °C.

9. Microscope slides, coverslips 22 × 22 mm, and transparent nail varnish (*see* **Note 5**).
10. Confocal microscope.

2.4 Equipment and Reagents for Ultra-microsectioning and Labeling of Larval Fat Bodies in Preparation for EM Microscopy

Same as in 2.2 plus:

1. Diamond or glass knife.
2. Copper grids 200-mesh.
3. Ultramicrotome.
4. Eyelash glued to a wooden stick (for precise maneuvering of sections).
5. Thin wire with one end glued to a wooden stick and the other made into a loop (for fishing out sections).
6. 2% glutaraldehyde in PBS.
7. 2% osmium tetroxide (OsO4), 1.5% potassium ferricyanide ($K_4[Fe(CN)_6]$), made in dH_2O.
8. 4% and 7% uranyl acetate (UA) made in dH_2O (*see* **Note 6**).
9. Graded series of ethanol concentrations: 30%, 50%, 70%, 85%, 95%, 100%, and 100% ethanol absolute dehydrated.
10. Pure propylene oxide (PO) (*see* **Note 7**).
11. Resin mix A: 25gr Epon-812, 20gr Araldite, 60gr dodecenyl-succinic anhydride (DDSA) (*see* **Note 8**).
12. Resin mix B: 8 drops of epoxy accelerator 2,4,6-Tris dimethy-laminomethyl phenol (DMP-30) are added in 10gr of Resin mix A (*see* **Note 9**).
13. 0.4% lead citrate made in dH_2O (*see* **Note 10**).
14. Rotator.
15. Flat embedding molds.
16. 60 °C oven
17. Electron microscope.

3 Methods

3.1 Visualization of Mitophagy Using Fluorescence Confocal Microscopy

3.1.1 Preparing Fly Crosses, Feeding of Larvae with FCCP, and Starvation

1. Cross ~7 virgins Cg-GAL4 with 4 to 7 males UAS-mCherry-Atg8a (*see* **Note 11**).
2. Transfer the parents into new tubes every 12 h to stage the embryos (*see* **Note 12**).
3. When the larvae reach 4 days after egg laying (4 dAEL), prepare the FCCP-supplemented food by soaking a small amount of Nutri-Fly Instant *Drosophila* Medium in 500 μM FCCP solution. A control tube is made with the same volume of vehicle only (ethanol).

4. To collect larvae, scoop out a small portion of the fly food with a spatula and rinse it off with water on a petri dish. Most larvae that are present in the food will be dispersed by the water, and they can then be picked up from the petri dish.
5. FCCP feeding: transfer the larvae into the FCCP- and vehicle-supplemented fly food. Place back the tubes at 25 °C for 10 h.
6. Starvation: instead of FCCP feeding, larvae can be starved for 4 h in 20% sucrose to induce autophagy.

3.1.2 Dissecting Fat Bodies from Larvae

1. Scoop out the larvae in a petri dish containing water; transfer them twice in petri dishes containing PBS to clean them from food residues.
2. Put each larva into a drop of PBS on the black dissection pad.
3. Rip off the tip of the posterior end of the larvae, taking extra care not to grab the rest of the body to avoid damaging the fat body (*see* **Note 13**).
4. Hold the larva by the head and gently squeeze in a rolling manner the fat body along with other tissues out of the posterior opening (*see* **Note 14**).
5. Carefully separate the fat body from the rest of the tissues (mostly the gut, Malpighian tubules, and trachea).
6. Place the isolated fat bodies in the baskets, halfway immersed in PBS in a small petri dish (*see* **Notes 15** and **16**).

3.1.3 Fixation and Immunostaining

1. Pick each basket and sink into 300 μL of 4% PFA in a 48-deep well plate.
2. Let the tissue fix for 20–30 min at room temperature (RT), protected from light (*see* **Note 17**).
3. Use an average of 300 μL for each subsequent step.
4. Rinse three times in PBS by transferring the basket to new wells. Each washing step is 5–10 min at room temperature.
5. Block for 1 h at RT in 1× BP buffer.
6. Incubate overnight (O/N) at 4 °C with anti-complex V antibody, diluted 1:1000 in BP buffer.
7. Rinse three times in PBST for 10–20 min each at RT.
8. Incubate with goat anti-mouse A468 diluted 1:500 in BP buffer O/N at 4 °C (*see* **Note 18**).
9. Rinse three times in PBST for 10–20 min each at RT.
10. Incubate for 15 min in Hoechst solution (diluted 1:1000 in PBS).
11. Rinse once in PBS and prepare samples for mounting on slides.

3.1.4 Sample Mounting on Slides for Confocal Microscopy Observation

1. Transfer baskets to a small PBS-containing petri dish, so that they are halfway immersed in the solution. This transfer is performed to make picking of the fat bodies easier for placing in the mounting medium.
2. Put a drop (~15 μL) of mounting medium on the center of a glass slide or two drops in reasonable distance of each other if one wants to mount two samples per glass slide.
3. Carefully transfer fat bodies with fine tweezers or forceps to the mounting medium (*see* **Note 19**).
4. Ensure fat bodies are as spread as possible.
5. Apply coverslip and seal periphery with transparent nail varnish as soon as the mounting medium is evenly spread under the coverslip. Ensure no bubbles are formed when placing the coverslip on the mounting medium drop.
6. Let slides to dry by placing them in a dark container (to preserve fluorophores). The slides can be kept at 4 °C.
7. The samples are ready for confocal microscopy imaging (Fig. 2).

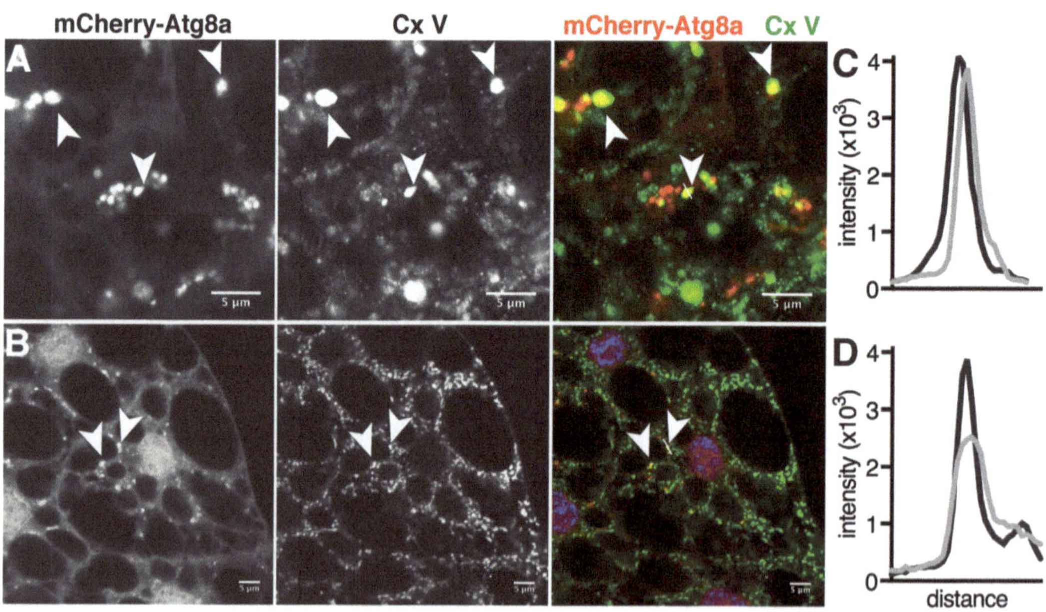

Fig. 2 Visualization of mitophagy using fluorescence microscopy. (**a**, **b**) Confocal section of a fat body cell from larvae expressing mCherry-Atg8a (red) and stained for mitochondrial complex V (Cx V, green). Arrowheads show some of the mitochondria colocalizing with autophagosomes. (**c**, **d**) Intensity plots for the red and green channel where colocalization between an autophagosome and an mitochondrion is observed. Larvae were fed with FCCP for 10 h (**a**, **c**) or starved for 4 h in 20% sucrose (**b**, **d**)

3.2 Observation of Mitochondria by Electron Microscopy (See Note 20)

3.2.1 Sample Fixation and Embedding in Preparation for TEM Observation

1. Dissect larval fat bodies as in Subheading 3.1.2.
2. Fix the tissue in 2% glutaraldehyde for 1.5 h at RT.
3. Wash three times in 1 × PBS for 5 min per wash. Ensure fixation solution has been thoroughly removed before advancing to the washing steps.
4. Wash three times in dH_2O for 5 min per wash, at RT.
5. Incubate for 1 h in 2% OsO4, 1.5% $K_4[Fe(CN)_6]$ made in dH_2O, at 4 °C (*see* **Note 21**).
6. Wash three times in dH_2O for 5 min per wash, at RT.
7. Stain samples "en bloc" (i.e., before embedding with sucrose infusion) with 4% UA in the dark, for 30 min at RT.
8. Wash three times in dH_2O for 5 min per wash, at RT.
9. Incubate samples in graded series of 30%, 50%, 70%, 85%, 95%, 100%, and 100% EtOH absolute dehydrated concentrations for 10 min each, at 4 °C (*see* **Note 22**).
10. Use glass vials (if not already). Prepare the samples for resin infiltration by incubating twice with PO for 15 min at RT.
11. Infiltrate samples with resin by incubating in a 3:1 solution of PO/Resin mix A for 1 h at RT. Ensure homogenous infiltration by rotating the samples.
12. Repeat **step 11** by changing the PO/Resin mix A ratio to 1:1.
13. Repeat **step 11** by changing the PO/Resin mix A ratio to 1:3.
14. Continue infiltration by incubating the samples in pure Resin mix A solution, overnight at RT and constantly rotating.
15. The next day, continue with infiltration by incubating the samples twice in previously prepared pure Resin mix B, for 2 h each at RT. Ensure samples are continuously rotated.
16. Coat flat embedding molds with Resin mix B, and mount the samples in an alignment that will allow their ultramicrosectioning during the following step.
17. Polymerize the mounted samples by incubating for 24–48 h at 60 °C.

3.2.2 Ultra-microsectioning of Samples and Staining for TEM Investigation

1. Using an ultramicrotome equipped with either a diamond or glass knife, mount the sample-containing mold in the holder, and start sectioning. Aim on obtaining continuous ribbons of sections. Slices should be 60–80 nm thick (*see* **Note 23**).
2. Place the sections on uncoated 200-mesh copper grids.
3. Stain with 7% UA for 5–10 min at RT.
4. Wash grids 2–3 times with dH_2O for 5–10 min each, to remove excess UA.
5. Stain with 0.4% lead citrate for 1–2 min at RT.

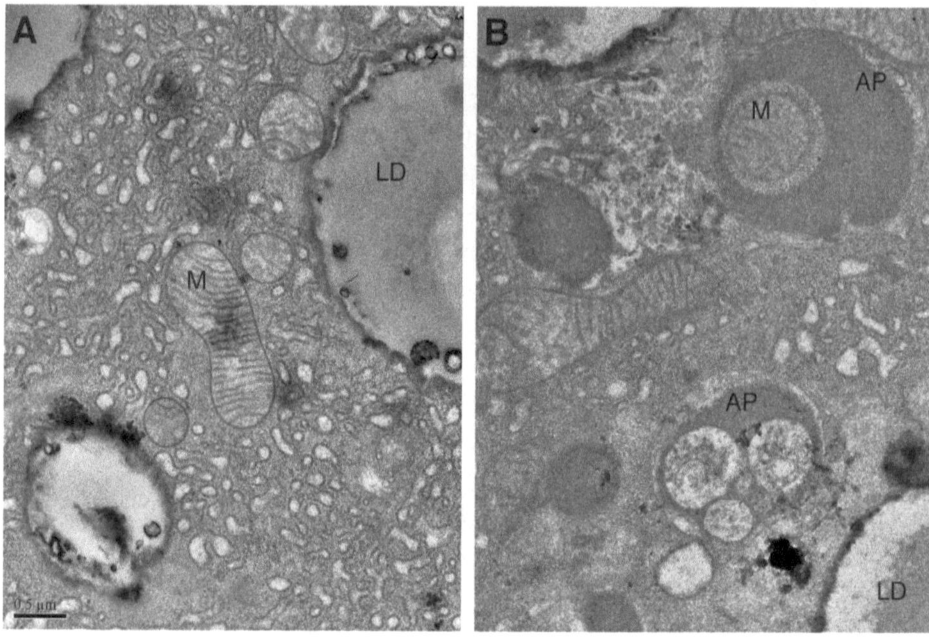

Fig. 3 Visualization of mitophagy using TEM. Representative region of fat body from fed (**a**) and starved (**b**) third instar larvae. (**a**) No evidence of mitophagy is visible in fat body cells of fed larvae. (**b**) An autophagosome, which has engulfed a mitochondrion, can be seen after amino acid deprivation of larvae by incubation in 20% sucrose solution for 4 h at 25 °C. *LD* lipid droplet, *M* mitochondrion, *AP* autophagosome

6. Wash grids thoroughly 3–4 times with dH_2O for 5–10 min each, to remove excess lead citrate.
7. Dry sections and observe under a TEM (Fig. 3).

4 Notes

1. mCherry-Atg8a is used as a marker for autophagosomes and autolysosomes. *Drosophila* Atg8a is homolog to mammalian LC3 and GABARAP proteins.
2. Put only a small drop of wet yeast paste on the solidified fly food; too much yeast affects the larval development. Let the wet yeast paste dry for a couple of hours before transferring the flies to avoid getting them stuck.
3. FCCP is toxic and must be handled wearing personal protective equipment. FCCP may cause long-lasting harmful effects to aquatic life and must not be drained.
4. Alternatively, one may prefer using translucent pad or deep well glass.
5. Transparent nail varnish is important as it will not interfere with the imaging during fluorescence microscopy acquisition.

6. UA is highly sensitive to light and precipitates upon exposure, so keep it in dark as much as possible. It also precipitates during aging, so the supernatant of the solution must be used. UA is additionally highly toxic and mildly radioactive, so take precautions to minimize direct contact with skin or accidental ingestion.
7. PO is corrosive to plasticware, so use glassware for storage and after dehydrating samples in graded series of ethanol. Also, it is very volatile, so one must be quick between the transfers to prevent the solution from drying. Finally, it is a probable carcinogen, so take precautions to avoid skin contact or accidental ingestion.
8. Ensure Resin mix A components are mixed thoroughly. For this, one may use an electric mixer and a 100 mL plastic beaker.
9. After adding the epoxy accelerator, remove gas from the container under a vacuum and store Resin mix B at −20 °C. This will stall the polymerization process of the resin, which is otherwise rapidly rendered very viscous if left at RT after preparation. To use, let Resin mix B to reach RT after thawing, as this will prevent dew condensation.
10. Exercise great caution when weighing the lead citrate powder as the compound is highly toxic. Also, it precipitates rapidly in the presence of carbon dioxide either in the air or the water, resulting in the formation of a white water-insoluble toxic precipitate in the solution and the appearance of white dots in the sample under an EM. Store and handle with extreme care and try to minimize exposure to CO_2 as much as possible. To make 0.4% lead citrate, dissolve 0.4 gr lead citrate in 100 mL freshly distilled CO_2-free water. After the compound is dissolved, store in an airtight container at RT, and aim to make a new solution whenever needed for an experiment. During contrast enhancement with lead citrate, add a few pellets of NaOH in the petri dish and keep the lid tightly closed to remove CO_2 from the dish.
11. At least one male per three females should be used when setting crosses to ensure optimal mating.
12. Let the flies mate for 1–2 full days before transferring twice a day. The very first tube is not used for further processing as the embryos are not staged.
13. It is essential to open the larva as close as possible to the posterior end to avoid damaging the fat body too much. Alternatively, it is possible to rip off the head of the larva.
14. It is possible to proceed by inverting the larvae by pushing the head inside out (roughly like a sock).

15. Dissected fat bodies can be gently picked up to be transferred. Make sure not to pinch the tissue as that may damage and break it.
16. Dissected tissues must not remain in PBS for more than 20–25 min before fixation.
17. To avoid fading of the mCherry-Atg8a fluorescence, all the subsequent steps must be performed protected from light as much as possible. The plate can be placed in a dark box or wrapped in foil.
18. The secondary antibody can also be incubated for 2 h at RT.
19. Six to ten fat body lobes are a reasonable amount of tissue to mount per coverslip. If more fat bodies need to be observed, duplicate the coverslip or slide.
20. Generally, during sample preparation for TEM, glass vials with plastic snap caps are recommended, as they minimize potential risks of contamination and evaporation of solution, which would otherwise leave samples to dry up. 1–2 mL final volume is usually sufficient for each solution.
21. Store osmium tetroxide at 4 °C in the dark and protect from direct exposure to light during handling. Also, exercise extreme caution as osmium tetroxide is highly toxic and volatile.
22. Grading concentrations of ethanol are essential to ensure thorough dehydration of samples. By starting with low ethanol and progressively increasing the concentration, samples are infiltrated much more efficiently than by starting with a high ethanol concentration.
23. The thickness of each section can be estimated by consulting a color reference chart for ultramicrotome sections in EM. Generally, sections of 60–80 nm thick should appear silver to gold on the knife edge when cut with an ultramicrotome.

Acknowledgments

This work was supported by BBSRC grants BB/L006324/1 and BB/P007856/1 awarded to Dr. Ioannis Nezis and the BBSRC MIBTP research training grant allocated to Panagiotis Tsapras as an MIBTP Ph.D. student.

References

1. Kimura T, Mandell M, Deretic V (2016) Precision autophagy directed by receptor regulators – emerging examples within the TRIM family. J Cell Sci 129(5):881–891
2. Rogov V, Dotsch V, Johansen T, Kirkin V (2014) Interactions between autophagy receptors and ubiquitin-like proteins form the

molecular basis for selective autophagy. Mol Cell 53(2):167–178
3. Youle RJ, Narendra DP (2011) Mechanisms of mitophagy. Nat Rev Mol Cell Biol 12(1):9–14
4. Okamoto K, Kondo-Okamoto N, Ohsumi Y (2009) Mitochondria-anchored receptor Atg32 mediates degradation of mitochondria via selective autophagy. Dev Cell 17(1):87–97
5. Kanki T, Wang K, Cao Y, Baba M, Klionsky DJ (2009) Atg32 is a mitochondrial protein that confers selectivity during mitophagy. Dev Cell 17(1):98–109
6. Novak I, Kirkin V, McEwan DG, Zhang J, Wild P, Rozenknop A, Rogov V, Lohr F, Popovic D, Occhipinti A, Reichert AS, Terzic J, Dotsch V, Ney PA, Dikic I (2010) Nix is a selective autophagy receptor for mitochondrial clearance. EMBO Rep 11(1):45–51
7. Sandoval H, Thiagarajan P, Dasgupta SK, Schumacher A, Prchal JT, Chen M, Wang J (2008) Essential role for Nix in autophagic maturation of erythroid cells. Nature 454 (7201):232–235
8. Zirin J, Perrimon N (2010) Drosophila as a model system to study autophagy. Semin Immunopathol 32(4):363–372
9. Mulakkal NC, Nagy P, Takats S, Tusco R, Juhasz G, Nezis IP (2014) Autophagy in Drosophila: from historical studies to current knowledge. Biomed Res Int 2014:273473
10. McPhee CK, Baehrecke EH (2009) Autophagy in Drosophila melanogaster. Biochim Biophys Acta 1793(9):1452–1460
11. Scott RC, Schuldiner O, Neufeld TP (2004) Role and regulation of starvation-induced autophagy in the Drosophila fat body. Dev Cell 7(2):167–178
12. Mauvezin C, Ayala C, Braden CR, Kim J, Neufeld TP (2014) Assays to monitor autophagy in Drosophila. Methods 68(1):134–139
13. Nagy P, Varga A, Kovacs AL, Takats S, Juhasz G (2015) How and why to study autophagy in Drosophila: it's more than just a garbage chute. Methods 75:151–161
14. Nezis IP, Lamark T, Velentzas AD, Rusten TE, Bjorkoy G, Johansen T, Papassideri IS, Stravopodis DJ, Margaritis LH, Stenmark H, Brech A (2009) Cell death during Drosophila melanogaster early oogenesis is mediated through autophagy. Autophagy 5(3):298–302

Chapter 43

Mitophagy Dynamics in *Caenorhabditis elegans*

Konstantinos Palikaras, Eirini Lionaki, and Nektarios Tavernarakis

Abstract

Mitochondrial selective autophagy (mitophagy) is a critical cellular process for mitochondrial homeostasis and survival both under basal and stress conditions. Distinct cell types display different requirements for mitochondrial turnover depending on their metabolic status, differentiation state, and environmental cues. This points to the necessity of developing novel tools for real-time, tissue-specific assessment of mitophagy. *Caenorhabditis elegans* is an invaluable model organism for this kind of analysis providing a platform for simultaneous monitoring of mitophagy in vivo in different tissues and cell types, during development, stress conditions, and/or throughout life span. In this chapter we describe three versatile, noninvasive methods, developed for monitoring in vivo early and late mitophagic events in body wall muscles and neuronal cells of *C. elegans*. These procedures can be readily used and/or provide insights into the generation of novel imaging methods to investigate further the role of mitophagy at the organismal level under normal and pathological conditions.

Key words Aging, Autophagosome, Autophagy, *Caenorhabditis elegans*, DsRed, Green fluorescent protein (GFP), Lysosomes, Fluorescent microscopy, Mitochondria, Mitophagy, mtRosella

1 Introduction

Mitochondria represent central hubs of cellular metabolism and play pivotal roles in a variety of cellular processes, including ATP generation, iron metabolism, production of several metabolites and cofactors, Ca^{+2} buffering and signaling, among others. Their importance for cellular physiology predicts the adverse effects of mitochondrial perturbations for cellular and organismal homeostasis. Damaged mitochondria can negatively impact cellular survival through the generation of toxic levels of reactive oxygen species and pro-apoptotic factors. Mitochondrial dysfunction has been associated with multiple pathologies in humans including cancer, type II diabetes, cardiovascular disorders, myopathies, and neurodegenerative diseases [1, 2].

Konstantinos Palikaras and Eirini Lionaki contributed equally to this work.

Nicholas Ktistakis and Oliver Florey (eds.), *Autophagy: Methods and Protocols*, Methods in Molecular Biology, vol. 1880, https://doi.org/10.1007/978-1-4939-8873-0_43,

Given their important role for cellular homeostasis, mitochondrial welfare is monitored and controlled at several levels. Mitochondria fuse to dilute toxic agents and refresh their components. An arsenal of local chaperones and proteases, whose coordinated expression consist the mitochondrial unfolded protein response, tackles mild proteotoxic stress. When mitochondrial damage exceeds the threshold of protein quality control mechanisms, dysfunctional organelles are removed by mitophagy, a selective type of autophagy [3]. Mitochondrial fission primes mitochondria for autophagic degradation, while the fission/fusion machinery is closely implicated in mitophagy initiation [4, 5]. As distinct cell types experience different metabolic needs, the mitochondrial network varies significantly between cell types or among different developmental states of a cell lineage [6]. Therefore, mitochondrial quality control mechanisms may have different impacts depending on the variable cellular context.

Mitophagy has been at the spotlight of research for more than a decade. Mitochondrial selective autophagy is also induced under nonpathogenic conditions, when the cell needs to adapt its metabolic status to environmental changes, such as hypoxia [7], or to developmental changes, like in the maturing reticulocytes, the precursors of red blood cells [8, 9]. These findings highlight the importance of assessing mitophagy in specific cell types, developmental stages, and/or under defined stress conditions.

Methods to monitor mitophagy include applications of biochemical and imaging techniques on distinct model organisms, in vitro and in vivo. Biochemically, mitophagy can be assessed by the lysosomal-dependent loss of mitochondrial proteins or mitochondrial-associated enzymatic functions (citrate synthase activity). These approaches cannot distinguish mitophagic activity between different cell types within tissues and usually cannot assess subtle differences in the rate of mitophagy. Moreover, electron microscopy has yielded images of mitochondria surrounded by autophagosomal/lysosomal membranes. Its limitations with regard to sample preparation and quantification of the results have restricted the broad use of this method. Fluorescence microscopy has gained increasing attention when it comes to monitoring of dynamic processes. Mitophagy is monitored by colocalization of mitochondria-targeted probes with autophagosomal/lysosomal markers or by mitochondria-targeted ratiometric and dual pH-sensitive probes [10–12]. The latter have revolutionized our understanding of mitophagy process in vivo.

C. elegans is a genetically modifiable small nematode with a transparent body, which comprises different tissues and organs, with high functional similarities to the mammalian counterparts. These nematodes can be monitored alive under the fluorescent microscope for real-time in vivo assessment of dynamic cellular processes, without dissection or other intrusive methods prior to microscopy. Therefore, although amenable to all the aforementioned approaches, *C. elegans* was used for the establishment of tools for monitoring in vivo real-time mitophagy, both systemically and in a tissue-specific manner. In this chapter, we describe the development of complementary systems for monitoring mitophagy in vivo, which are used to uncover several conditions that either promotes or inhibits mitophagy in *C. elegans* [11, 13]. In the first approach, we generated transgenic animals expressing the pH-sensitive *Rosella* biosensor in mitochondria of either body wall muscles or neurons. mtRosella is a mitochondria-targeted dual fluorescent probe, which comprises a pH-sensitive GFP and a pH-insensitive DsRed moiety. Thus, mitophagy is signified by a reduction in the ratio of GFP/DsRed fluorescent signal. We examined transgenic animals expressing mtRosella either in body wall muscle cells or in neurons under normal and mitophagy-inducing conditions, such as mitochondrial and oxidative stress. Mitochondrial stress, induced by RNAi against *isp-1* and *frh-1* genes, results in decreased GFP/DsRed ratio of mtRosella, highlighting mitophagy stimulation (Fig. 1a). Furthermore, oxidative stress triggers elimination of impaired mitochondria in both soma and axons of neurons (Fig. 2). Neuronal mitophagy is of particular importance as neurons rely heavily on their mitochondrial network, and even mild changes may affect their functionality and survival [14]. In addition to mtRosella, we developed a method for monitoring earlier mitophagic events by generating transgenic animals that express the mitophagy receptor protein DCT-1, the homologue of the mammalian BNIP3/NIX, fused with GFP together with the autophagosomal marker LGG-1, the homologue of the mammalian cytosolic microtubule-associated protein 1 light chain 3 (MAP 1LC3/LC3), fused to DsRed in body wall muscle cells [11, 13, 15, 16]. We examined animals carrying both the mitophagy receptor and autophagosomal markers, under normal and mitophagy-inducing conditions. Mitophagy stimulation induces the formation of autophagosomes that extensively colocalize with DCT-1 (Fig. 2).

The investigation of the molecular mechanisms and physiological role of mitochondrial elimination in cellular and organismal homeostasis demands the development of reliable, non-invasive, quantitative methods for mitophagy assessment in vivo. In the following sections, we describe detailed protocols for in vivo imaging of mitochondrial selective autophagy in *C. elegans*, using three versatile imaging tools.

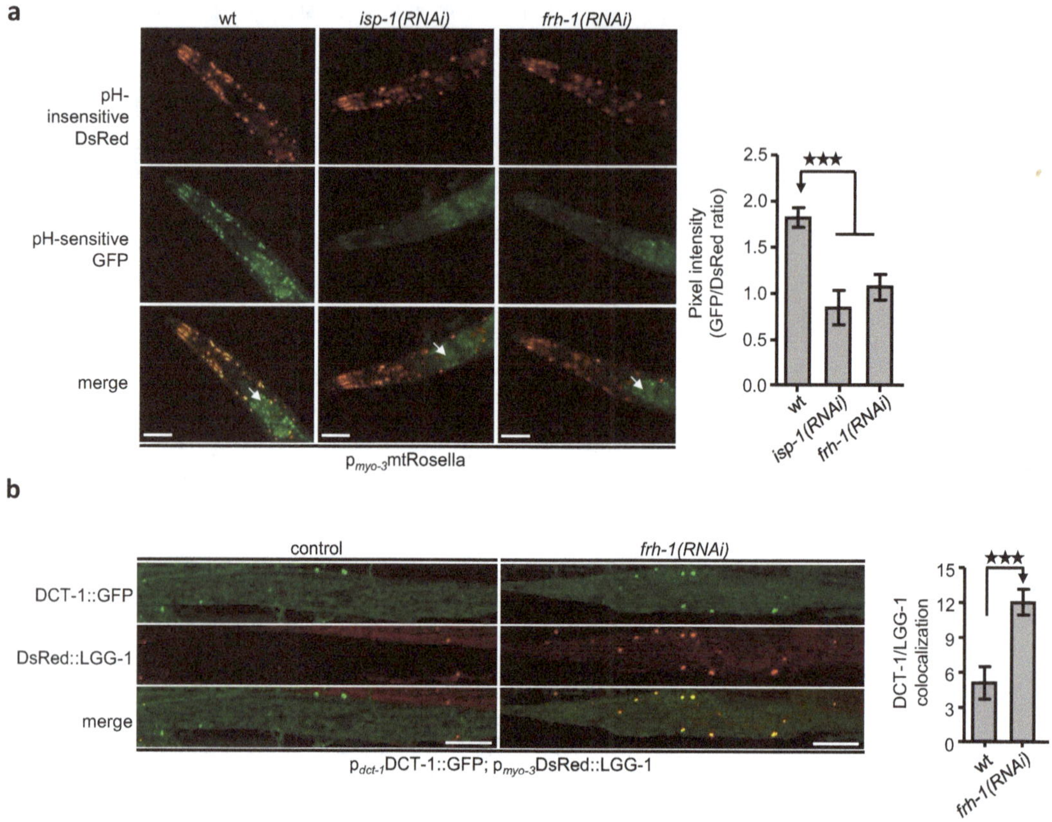

Fig. 1 Mitophagy activation in response to mitochondrial dysfunction. (**a**) Transgenic animals expressing mtRosella in body wall muscles were subjected to *isp-1(RNAi)* and *frh-1(RNAi)*. The decreased ratio between pH-sensitive GFP to pH-insensitive DsRed signifies mitophagy stimulation ($n = 40$, $^{***}P < 0.001$; one-way ANOVA). Arrows point out intestinal autofluorescence. Acquisition information: exposure time, 200 ms; contrast, medium. Images were acquired using 10× objective lens. (**b**) Transgenic nematodes expressing the mitophagy receptor DCT-1 fused with GFP together with the autophagosomal marker LGG-1 fused with DsRed were subjected to RNAi against *frh-1* gene. Mitophagy stimulation is indicated by the increased colocalization events between DCT-1::GFP and DsRed::LGG-1 ($n = 50$, $^{***}P < 0.001$; unpaired *t*-test). Acquisition information: resolution, 1024 × 1024; master gain, Track1, 562, and Track2, 804; emission filters, Track1 Channel1, 575–703, and Track2 Channel2, 493–545; laser intensity, Track1 (543 nm), 12.9%, and Track2 (488 nm), 25%. Images were acquired using 63× objective lens. Scale bars, 20 μm. Error bars, SEM values

2 Materials

2.1 Reagents

1. Wormpick[a] or eyebrow/eyelash hair[b] (*see* **Note 1**).
2. Cholesterol stock solution, 5 mg/mL: dissolve cholesterol in absolute ethanol by stirring. Store at 4 °C. Do not flame or autoclave.
3. Nystatin stock solution, 10 mg/mL: dissolve nystatin in 70% (V/V) ethanol. Store at 4 °C and shake prior to use as this is a suspension. Do not autoclave.

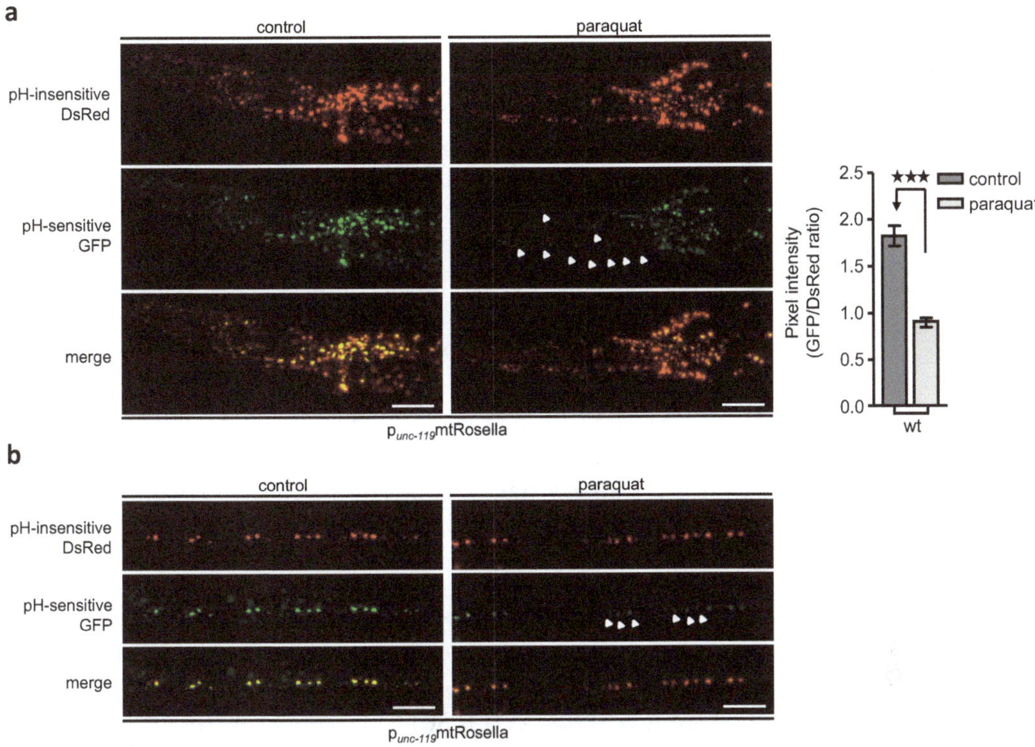

Fig. 2 Assessment of mitophagy in *C. elegans* neurons. Transgenic nematodes expressing mtRosella in neuronal cells were exposed to paraquat. (**a**) Induction of neuronal mitophagy is underlined by the reduced ratio between pH-sensitive GFP to pH-insensitive DsRed ($n = 30$, $^{***}P < 0.001$; unpaired *t*-test). Error bars, SEM values. (**b**) Local elimination of impaired mitochondria in neuronal axons. Acquisition information: resolution, 1024 × 1024; master gain, Track1, 658, and Track2, 714; emission filters, Track1 Channel1, 548–703, and Track2 Channel2, 493–550; laser intensity, Track1 (543 nm), 7.9%, and Track2 (488 nm), 6.2%. Images were acquired using 40× objective lens. Scale bars, 20 μm

4. 1 M phosphate buffer, pH 6: 102.2 g KH_2PO_4, 57.06 g K_2HPO_4 in 1 L distilled water. Autoclave and keep at room temperature.
5. 1 M $MgSO_4$.
6. 1 M $CaCl_2$.
7. Levamisole.
8. Petri dishes (60 mm × 15 mm).
9. LB medium.
10. LB plates: 100 μg/mL ampicillin, 10 μg/mL tetracycline.
11. Nematode growth medium (NGM) agar plates: 3 g NaCl, 2.5 g Bacto Peptone, 0.2 g streptomycin, 17 g agar, and add 900 mL distilled water. Autoclave. Let cool to 55–60 °C. Add 1 mL cholesterol stock solution, 1 mL 1 M $CaCl_2$, 1 mL 1 M $MgSO_4$, 1 mL nystatin stock solution, 25 mL 1 M phosphate

buffer, pH 6.0, and distilled sterile water up to 1 L. Pour about 8 mL medium per Petri dish and allow for solidification. Keep the plates at 4 °C until used.

12. RNAi agar plates: 3 g NaCl, 2.5 g Bacto Peptone, 17 g agar, and add 900 mL distilled water. Autoclave. Let cool to 55–60 °C. Add 1 mL cholesterol stock solution, 1 mL 1 M $CaCl_2$, 1 mL 1 M $MgSO_4$, 1 mL nystatin stock solution, 25 mL 1 M phosphate buffer, pH 6.0, 100 μg/mL ampicillin, 4 mL 1 M isopropyl β-D-1-thiogalactopyranoside (IPTG), and distilled sterile water up to 1 L. Pour about 8 mL medium per Petri dish, and leave to solidify. Keep the plates at 4 °C until used (*see* **Note 2**).
13. NGM plates seeded with *Escherichia coli* (OP50 strain): use a single colony of *E. coli* (OP50), and inoculate a 25 mL culture using Luria-Bertani (LB) liquid medium (10 g Bacto Tryptone, 5 g Bacto Yeast Extract, 5 g NaCl, and distilled water up to 1 L, and sterilize by autoclaving). Allow inoculated culture to grow for 8–10 h at 37 °C with shaking. Seed NGM plates with 150 μL *E. coli* (OP50) solution, and incubate the plates at room temperature overnight to allow the growth of the bacterial lawn. Store the *E. coli* (OP50) solution at 4 °C. Prepare freshly new *E. coli* (OP50) culture every 4 days.
14. Ampicillin.
15. Tetracycline.
16. M9 buffer: 3 g KH_2PO_4, 6 g Na_2HPO_4, 5 g NaCl, in 1 L distilled water. Autoclave and add 1 mL 1 M $MgSO_4$. Store at 4 °C.
17. 0.5 M paraquat: 1 g paraquat in 8 mL distilled water. Prepare aliquots of 400 μL to avoid contamination, and store them at 4 °C (*see* **Note 3**).
18. 0.5 M levamisole: 1.2 g levamisole in 10 mL distilled water. Store levamisole solution at 4 °C (*see* **Note 4**).
19. 2% agarose pads: 0.5 g agarose into a glass beaker. Add 25 mL M9 buffer. Heat the mixture in a microwave until the agarose is completely dissolved. Stir the mixture periodically, and keep it warm on a heating plate. Add 2–3 drops of 2% agarose on a glass slide, and cover it quickly with a second glass slide, so as to form a thin agarose film (pad). Wait for 1 min until the agarose pad solidifies, and remove the top slide (*see* **Note 5**).
20. Nematode strain expressing mtRosella biosensor in body wall muscle cells and neurons; IR1631: N2;*Ex003* [p_{myo-3}TOMM-20::Rosella; *rol-6(su1006)*] and IR1864: N2;*Ex001* [$p_{unc-119}$TOMM-20::Rosella; *rol-6(su1006)*] (*see* **Notes 6–8**).
21. Nematode strain co-expressing the mitophagy receptor DCT-1 fused with GFP together with the autophagosomal marker

LGG-1 fused with DsRed; R1511: N2; *Ex*[p_{dct-1}DCT-1::GFP; *rol-6(su1006)*]; *Ex011*[p_{myo-3}DsRed::LGG-1; p_{myo-2}GFP] (*see* **Notes 6, 7, 9**).

2.2 Equipment

1. Incubators for stable temperature (20 and 37 °C).
2. Dissecting stereomicroscope.
3. UV-dissecting stereomicroscope.
4. UV cross-linker.
5. Microwave.
6. Zeiss AxioImager Z2 epifluorescence microscope (Zeiss, model: Zeiss AxioImager Z2).
7. Zeiss AxioObserver Z1 confocal microscope (Zeiss, model: Zeiss AxioObserver Z1).

2.3 Software

1. ImageJ image processing software: freely available at http://rsb.info.nih.gov/ij/-[17].
2. Zeiss ZEN 2012 software.
3. Microsoft Office 2011 Excel (Microsoft Corporation, Redmond, USA).
4. GraphPad Prism software package (GraphPad Software Inc., San Diego, USA).

3 Methods

3.1 Preparation of RNAi Agar Plates Seeded with dsRNA-Expressing Bacteria

1. Streak *E. coli* (HT115) bacteria expressing dsRNA *isp-1* and *frh-1* genes or bearing an empty vector (EV), onto LB agar plates containing 100 μg/mL ampicillin, 10 μg/mL tetracycline.
2. Incubate the plates at 37 °C overnight.
3. Use a sterilized toothpick to isolate few bacterial colonies of each condition (EV, *isp-1* and *frh-1*) and place them in separate bacteriological culture tubes containing 5 mL LB medium, 5 μL of 100 mg/mL ampicillin, and 5 μL of 10 mg/mL tetracycline.
4. Incubate cultures at 37 °C overnight.
5. Prepare three different tubes, and add 5 mL LB medium and 5 μL of 100 mg/mL ampicillin.
6. Add 300 μL (50 μL per 1 mL LB/amp) of each overnight culture (EV, *frh-1* and *isp-1*) into separate bacteriological culture tubes containing 5 mL LB medium, 5 μL of 100 mg/mL ampicillin, and 5 μL of 10 mg/mL tetracycline.

7. Incubate cultures at 37 °C until OD_{600} is between 0.5 and 0.8 (approximately 4 h).
8. Place 200 μL of each culture (EV, *frh-1* and *isp-1*) described in Subheading 3.1, **step 7**, on RNAi agar plates.
9. Let the plates dry at room temperature overnight.

3.2 Maintenance, Expansion, and Synchronization of Transgenic Nematodes Population

Use the nematode strains expressing mtRosella, in body wall muscles (IR1631) and neurons (IR1864), and DCT-1::GFP together with autophagosomal marker LGG-1 fused with DsRed (IR1511) to monitor mitophagy.

1. Use a regular or a UV-dissecting stereomicroscope to select transgenic L4 larvae based on the selection marker (*see* **Notes 8** and **9**).
2. Use a wormpick to select and transfer 10 L4 larvae of transgenic nematodes on a freshly *E. coli* (OP50)-seeded NGM agar plate. Use at least five plates for each strain to expand transgenic population (*see* **Note 1**).
3. Incubate the nematodes at 20 °C.
4. Four days later, the plates (described in Subheading 3.2, **step 2**) contain mixed larval populations.
5. Synchronize worm population by selecting transgenic L4 larvae from the NGM agar plates to start the experiments.

3.3 Mitophagy-Inducing Conditions

3.3.1 Mitophagy Stimulation by RNAi Against isp-1 and frh-1 Genes

1. Pick and transfer 15–20 L4 larvae of transgenic animals expressing either mtRosella in body wall muscles (IR1631) or DCT-1::GFP together with DsRed::LGG-1 (IR1511) onto separate RNAi agar plates seeded with dsRNA-expressing bacteria. For each experimental condition, use at least three plates containing transgenic nematodes.
2. Keep and grow the animals at 20 °C.
3. After 2 days, either use 2-day-old adult worms directly or transfer them to freshly seeded RNAi plates to avoid progeny and starvation (*see* **Note 10**). After 2 days, the latter would represent 4-day-old adult worms.
4. Nematodes are ready for microscopic examination. Censored or dead animals are eliminated from the imaging process (*see* **Notes 11** and **12**).

3.3.2 Mitophagy Stimulation in Response to Oxidative Stress

1. Place six NGM *E. coli* (OP50) bacteria-seeded plates in a UV irradiation chamber.
2. Irradiate with UV light (254 nm) the NGM bacteria-seeded plates for 15 min (*see* **Note 13**).
3. Use paraquat as a chemical inducer of oxidative stress. Add 100 μL 0.5 M paraquat on top of OP50-seeded NGM plates

(8 mM final concentration/total agar volume). Add equivalent volume of water, which is the solvent of paraquat, to the three remaining plates that will serve as control.

4. Spread the drug over the entire surface by swirling the plates.
5. Cover the plates with foil or place them in a dark space and let them dry at room temperature.
6. Select and transfer 10–15 2- or 4-day-old adult transgenic animals expressing mtRosella in neuronal cells (IR1864) on plates containing paraquat or control plates using a wormpick (*see* **Note 14**).
7. Incubate the animals at 20 °C for 2 days.
8. Nematodes are ready for microscopic examination. Censored or dead animals are eliminated from imaging process (*see* **Notes 11** and **12**).

3.4 Prepare and Mount the Samples for Imaging

1. Prepare 2% agarose pads by adding 25 mL M9 buffer to 0.5 g agarose into a glass beaker. Heat the mixture in a microwave until the agarose is completely dissolved. Stir the mixture periodically, and keep it warm on a heating plate. Add 2–3 drops of 2% agarose on a glass slide, and cover it quickly with a second glass slide, so as to form a thin agarose film (pad). Wait for 1 min until the agarose pad solidifies and remove the top slide. Only use freshly prepared 2% agarose pads (*see* **Note 5**).
2. Add 10 μL of 20 mM M9-levamisole buffer on the agarose film (*see* **Note 15**).
3. Use an eyebrow/eyelash hair to transfer the transgenic nematodes in M9-levamisole drop (*see* **Note 1**). Place 15–30 worms per drop.
4. Gently place a coverslip on top of nematodes.
5. Seal the agarose pads with nail polish to maintain humidity during imaging.
6. Proceed to microscopic examination of the samples.

3.5 Acquisition Process

3.5.1 Acquisition Process Using Nematode Strains (IR1631 and IR1864) Expressing mtRosella in Body Wall Muscle and Neuronal Cells

1. Detect single transgenic animals expressing mtRosella either in body wall muscles or neurons using an epifluorescence microscope.
2. Use 10x objective lens and capture images of entire transgenic nematodes by using a microscope-attached camera (*see* **Note 16**).
3. Use the same imaging settings (lens and magnifier used, filters exposure time, resolution, laser intensity, gain, etc.) throughout the imaging process.
4. Save and collect the acquired images.

3.5.2 Imaging and Data Analysis Using Nematode Strain (IR1631) Expressing mtRosella in Body Wall Muscle and Neuronal Cells

1. Download and install ImageJ software (https://imagej.nih.gov/ij/).
2. Open images obtained in Subheading 3.5.1, **step 4** with ImageJ program.
3. Convert images to grayscale images with a pixel depth of 8 bit by selecting the "split channel" command via the "image" and "color" drop-down menu.
4. Use the "freehand selection" too to manually set the fluorescent area of interest.
5. Measure emission intensity by selecting the "measurement" command via the "analyze" drop-down menu to perform pixel intensity analysis.
6. Copy the displayed data from the separate "results" window.
7. Paste and import the data by using any software package, such as the Microsoft Office 2011 Excel software package (Microsoft Corporation, Redmond, USA).
8. Normalize pixel intensity values to the selected area.
9. Measure GFP to DsRed ratio. The levels of GFP/DsRed ratio underline mitophagy upregulation since GFP fluorescent signal is quenched upon the fusion with the acidic environment of lysosomes, whereas DsRed fluorescent signal remains stable.

3.5.3 Acquisition Process Using Nematode Strain (IR1511) Co-expressing Mitophagy Receptor and Autophagosomal Marker in Body Wall Muscle Cells

1. Detect single body wall muscle cells of transgenic animals co-expressing the mitophagy receptor DCT-1::GFP together with autophagosomal marker DsRed::LGG-1 using a confocal microscope.
2. Use 63× objective lens.
3. Image an entire single body wall muscle cell by performing z-stack scanning method (*see* **Note 16**). Keep the same imaging and acquisition settings (lens and magnifier used, filters exposure time, resolution, laser intensity, gain, etc.) during imaging process.
4. Save and collect the acquired images.

3.5.4 Data Analysis Using Nematode Strain (IR1511) Co-expressing Mitophagy Receptor and Autophagosomal Marker in Body Wall Muscle Cells

1. Open and process images acquired in Subheading 3.5.3, **step 4** with any confocal software.
2. Analyze mitophagy levels by manually counting the colocalization events between mitophagy receptor (DCT-1::GFP) and autophagosomal marker (DsRed::LGG-1) in each stack of body wall muscle cell (*see* **Note 17**).
3. Document the obtained data by using the Microsoft Office 2011 Excel (Microsoft Corporation, Redmond, USA).

3.5.5 Statistical Analysis

Report the significance of each experiment by using any statistical analysis software.

1. Open the desired statistical analysis software.
2. Create a new "table and graph."
3. Select a specific type of graph to display the data (e.g., scatter-plot, column graph bar, etc.) and import data.
4. Decrease the variability of mitophagy levels between animals and cells by increasing the sample size. For each experimental procedure, examine at least 30 transgenic animals or 40 body wall muscle cells for each strain and treatment. Each assay should be repeated at least three (3) times.
5. Suggested statistical analysis tests: (a) Student's t-test with a significance cutoff level of $p < 0.05$ for comparisons between two groups and (b) one-factor (ANOVA) variance analysis corrected by the post hoc Bonferroni test for multiple comparisons. Examples of such experiments are given in Figs. 1 and 2.

4 Notes

1. (a) Cut 2–3 cm of platinum wire (90% platinum, 10% iridium wire, 0.010 in. diameter; e.g., Tritech Research, Los Angeles, CA). Break off the thin part of a glass Pasteur pipette, and melt the glass at the site of breakage on a Bunsen burner. Attach the end of the platinum wire. Flatten the wire tip using pincers or a light hammer. Before using the wormpick, always sterilize the tip over flame. (b) Take a toothpick, and glue an eyebrow/eyelash hair to the tip of it. Let it dry at room temperature. Before using the eyebrow/eyelash hair, always sterilize it in 70% of EtOH.
2. Prepare fresh IPTG-containing RNAi plates every 2 weeks. IPTG efficiency diminishes over time.
3. Paraquat is a photosensitive chemical. Protect stock solutions, aliquots, and plates from light by enwrapping them with foil or placing them in a dark space.
4. Prepare fresh levamisole stock solution every 6 months.
5. Agarose pads have to be freshly made every time.
6. For basic *C. elegans* culture, maintenance, and manipulation techniques, see WormBook, http://www.wormbook.org/). Follow standard procedures for *C. elegans* strain maintenance. Nematode rearing temperature was kept at 20 °C, unless noted otherwise.
7. The nematode strains are available upon request by Professor Tavernarakis N. (tavernarakis@imbb.forth.gr).

8. Maintain the following transgenic nematode strains by selecting rollers under a regular dissecting stereomicroscope, IR1631: N2;*Ex003* [$p_{myo\text{-}3}$TOMM-20::Rosella; *rol-6 (su1006)*] and IR1864: N2;*Ex001* [$p_{unc\text{-}119}$TOMM-20:: Rosella; *rol-6(su1006)*].
9. Maintain the following transgenic nematode strain by selecting rollers with GFP-positive pharynx under a UV-dissecting stereomicroscope, R1511: N2; *Ex*[$p_{dct\text{-}1}$DCT-1::GFP; *rol-6 (su1006)*]; *Ex011*[$p_{myo\text{-}3}$DsRed::LGG-1; $p_{myo\text{-}2}$GFP].
10. Caloric restriction and starvation promotes autophagy and mitophagy elevation [11]. Thus, well-fed and non-starved nematodes should be assessed for mitophagy induction under desired conditions.
11. Animals are characterized as censored when they display defects that interfere with normal physiology or have been compromised by experimental mishandling. Censored and dead animals are excluded from analysis. To avoid increased censoring and lethality due to excessive internal egg hatching (bag-of-worms phenotype or worm bagging), under mitophagy-inducing conditions:
    (a) Incubate specimens for shorter period in the presence of each drug.
    (b) Decrease the concentration of paraquat.
    (c) Use NGM plates containing fluorodeoxyuridine (FUdR), an inhibitor of DNA synthesis that blocks egg hatching.
    (d) Use older adult hermaphrodites (e.g., 4-day-old worms) that display reduced egg production.
12. Contaminations may appear during the experimental procedure. Contamination of NGM plates with bacteria not indented for feeding or with fungi may have a detrimental impact on animal survival and mitophagy stimulation. Contaminated plates and animals should be removed from the study.
13. Bacteria may metabolize chemical compounds diminishing their efficacy. Therefore bacteria should be killed by UV irradiation prior to drug application. *Caution*: UV-killed dsRNA-expressing bacteria will display also decreased gene silencing efficiency.
14. The appropriate developmental stage, age, drug concentration, and duration of oxidative stress should be experimentally determined each time, when animals of different genetic backgrounds, that might be sensitive to stress, are used. L1–L4 larvae are hypersensitive to paraquat leading to severe lethality.

15. M9-levamisole buffer (final levamisole concentration 20 mM). Use M9 buffer instead of water to ensure a favorable osmotic environment. Protect the animals from drying out during the imaging process.
16. In *C. elegans*, intestinal autofluorescence increases with age. Therefore, body wall muscles or neurons close to the intestine should be avoided during the imaging process. Focus on body wall muscles and neurons in the pharyngeal area to avoid intestine-derived autofluorescence.
17. Mitophagy events are defined by the colocalization of GFP and DsRed signals, which correspond to the mitophagy receptor (DCT-1::GFP) and autophagosomes (DsRed::LGG-1), respectively.

Acknowledgments

We thank A. Pasparaki for expert technical support. We thank R. Devenish for providing the pAS1NB-CS-Rosella plasmid. Some nematode strains used in this work were provided by the *Caenorhabditis* Genetics Center, which is funded by the National Center for Research Resources (NCRR) of the National Institutes of Health (NIH). We thank A. Fire for plasmid vectors. This work was funded by grants from the European Research Council (ERC) and the European Commission 7th Framework Programme. K.P. is supported by an AXA Research Fund long-term fellowship. E.L. is supported by a Scholarship for Strengthening Post-Doctoral Research from The Greek State Scholarships Foundation (IKY) within the framework of the Operational Programme "Human Resources Development Program, Education and Life-Long Learning".

References

1. Lopez-Otin C, Blasco MA, Partridge L, Serrano M, Kroemer G (2013) The hallmarks of aging. Cell 153(6):1194–1217. https://doi.org/10.1016/j.cell.2013.05.039
2. Springer MZ, Macleod KF (2016) In Brief: Mitophagy: mechanisms and role in human disease. J Pathol 240(3):253–255. https://doi.org/10.1002/path.4774
3. Youle RJ, Narendra DP (2011) Mechanisms of mitophagy. Nat Rev Mol Cell Biol 12(1):9–14. https://doi.org/10.1038/nrm3028
4. Hamacher-Brady A, Brady NR (2016) Mitophagy programs: mechanisms and physiological implications of mitochondrial targeting by autophagy. Cell Mol Life Sci 73(4):775–795. https://doi.org/10.1007/s00018-015-2087-8
5. Twig G, Elorza A, Molina AJ, Mohamed H, Wikstrom JD, Walzer G, Stiles L, Haigh SE, Katz S, Las G, Alroy J, Wu M, Py BF, Yuan J, Deeney JT, Corkey BE, Shirihai OS (2008) Fission and selective fusion govern mitochondrial segregation and elimination by autophagy. EMBO J 27(2):433–446. https://doi.org/10.1038/sj.emboj.7601963
6. Collins TJ, Berridge MJ, Lipp P, Bootman MD (2002) Mitochondria are morphologically and functionally heterogeneous within cells. EMBO J 21(7):1616–1627. https://doi.org/10.1093/emboj/21.7.1616

7. Liu L, Feng D, Chen G, Chen M, Zheng Q, Song P, Ma Q, Zhu C, Wang R, Qi W, Huang L, Xue P, Li B, Wang X, Jin H, Wang J, Yang F, Liu P, Zhu Y, Sui S, Chen Q (2012) Mitochondrial outer-membrane protein FUNDC1 mediates hypoxia-induced mitophagy in mammalian cells. Nat Cell Biol 14(2):177–185. https://doi.org/10.1038/ncb2422
8. Kundu M, Lindsten T, Yang CY, Wu J, Zhao F, Zhang J, Selak MA, Ney PA, Thompson CB (2008) Ulk1 plays a critical role in the autophagic clearance of mitochondria and ribosomes during reticulocyte maturation. Blood 112 (4):1493–1502. https://doi.org/10.1182/blood-2008-02-137398
9. Schweers RL, Zhang J, Randall MS, Loyd MR, Li W, Dorsey FC, Kundu M, Opferman JT, Cleveland JL, Miller JL, Ney PA (2007) NIX is required for programmed mitochondrial clearance during reticulocyte maturation. Proc Natl Acad Sci U S A 104(49):19500–19505. https://doi.org/10.1073/pnas.0708818104
10. McWilliams TG, Prescott AR, Allen GF, Tamjar J, Munson MJ, Thomson C, Muqit MM, Ganley IG (2016) mito-QC illuminates mitophagy and mitochondrial architecture in vivo. J Cell Biol 214(3):333–345. https://doi.org/10.1083/jcb.201603039
11. Palikaras K, Lionaki E, Tavernarakis N (2015) Coordination of mitophagy and mitochondrial biogenesis during ageing in C. elegans. Nature 521(7553):525–528. https://doi.org/10.1038/nature14300
12. Sun N, Yun J, Liu J, Malide D, Liu C, Rovira II, Holmstrom KM, Fergusson MM, Yoo YH, Combs CA, Finkel T (2015) Measuring in vivo mitophagy. Mol Cell 60(4):685–696. https://doi.org/10.1016/j.molcel.2015.10.009
13. Schiavi A, Maglioni S, Palikaras K, Shaik A, Strappazzon F, Brinkmann V, Torgovnick A, Castelein N, De Henau S, Braeckman BP, Cecconi F, Tavernarakis N, Ventura N (2015) Iron-starvation-induced mitophagy mediates lifespan extension upon mitochondrial stress in C. elegans. Curr Biol 25(14):1810–1822. https://doi.org/10.1016/j.cub.2015.05.059
14. Martinez-Vicente M (2017) Neuronal mitophagy in neurodegenerative diseases. Front Mol Neurosci 10:64. https://doi.org/10.3389/fnmol.2017.00064
15. Fang EF, Kassahun H, Croteau DL, Scheibye-Knudsen M, Marosi K, Lu H, Shamanna RA, Kalyanasundaram S, Bollineni RC, Wilson MA, Iser WB, Wollman BN, Morevati M, Li J, Kerr JS, Lu Q, Waltz TB, Tian J, Sinclair DA, Mattson MP, Nilsen H, Bohr VA (2016) NAD+ replenishment improves lifespan and healthspan in ataxia telangiectasia models via mitophagy and DNA repair. Cell Metab 24 (4):566–581. https://doi.org/10.1016/j.cmet.2016.09.004
16. Fang EF, Waltz TB, Kassahun H, Lu Q, Kerr JS, Morevati M, Fivenson EM, Wollman BN, Marosi K, Wilson MA, Iser WB, Eckley DM, Zhang Y, Lehrmann E, Goldberg IG, Scheibye-Knudsen M, Mattson MP, Nilsen H, Bohr VA, Becker KG (2017) Tomatidine enhances lifespan and healthspan in C. elegans through mitophagy induction via the SKN-1/Nrf2 pathway. Sci Rep 7:46208. https://doi.org/10.1038/srep46208
17. Schneider CA, Rasband WS, Eliceiri KW (2012) NIH image to ImageJ: 25 years of image analysis. Nat Methods 9(7):671–675

Chapter 44

Methods for Studying Mitophagy in Yeast

Panagiota Kolitsida and Hagai Abeliovich

Abstract

Under some experimental conditions, eukaryotic cells, from yeast to man, will digest a portion of their mitochondrial cohort through an autophagic process termed mitophagy. In humans, defects in mitophagy have been proposed to play a causative role in a number of late-onset degenerative diseases such as Parkinson's disease and type II diabetes. As a consequence the study of mitophagy, as a quality control process in eukaryotic cells, has become an increasingly important focus in contemporary cell biology. When faced with the task of assaying mitophagy in yeast, the experimentalist has at his or her disposal a variety of induction conditions and assay systems to choose from. Here, we survey several well-established protocols for inducing and monitoring mitophagy in the yeast *Saccharomyces cerevisiae* and discuss their relative merits, limitations, and potential pitfalls.

Key words Mitophagy, Autophagy, Membrane trafficking, Quality control, Western blotting, Assay, Enzyme kinetics, Fluorescence microscopy

1 Introduction

Autophagy is a general name for trafficking pathways in eukaryotic cells which take material from the cytosol into the lumen of the lytic organelle, variously known as the lysosome or the vacuole, depending on cell type. Macroautophagy is a type of autophagy in which the cytosol-derived components are first packaged, while in the cytosol, into a double-membrane intermediate called the autophagosome. Fusion of the outer membrane of the autophagosome with the limiting membrane of the vacuole, or with elements of the endosomal/lysosomal system, leads to destruction of the inner autophagosomal membrane, which in turn exposes the biopolymers imported from the cytosol to the various hydrolases that reside in the lysosome. One of the first indications that something was afoot was the observation in the early 1960s of classically cytoplasmic elements such as mitochondria, inside what was believed to be lysosomes [1]. This observation, which threatened to undermine De Duve's concept of "lysosomes" as distinct from the cytosol, was ultimately explained by postulating and later

Nicholas Ktistakis and Oliver Florey (eds.), *Autophagy: Methods and Protocols*, Methods in Molecular Biology, vol. 1880, https://doi.org/10.1007/978-1-4939-8873-0_44, © Springer Science+Business Media, LLC, part of Springer Nature 2019

demonstrating macroautophagic delivery of cytosolic material into the lumen of the lysosome [2].

The cytosolic content that was identified morphologically inside lysosomes in these studies was mitochondria. However, it was several decades before mitophagy was conceptually defined as a special, quality control version of autophagy which culls defective mitochondria [3]. Studies in *S. cerevisiae*, in the mid-2000s, revealed several conditions under which yeast cells can be coaxed into degrading a portion of their own mitochondria through mitophagy, and the elucidation of the mechanism by which defective components are identified and degraded in this experimental system is currently ongoing.

Autophagy is a type of trafficking mechanism, in which cellular material flows from one compartment (in general, the cytosol) into another (the vacuole or lysosome). Classically, to quantitatively or semiquantitatively measure such a transport step, one needs to have [1] an irreversible covalent modification that takes place in the destination compartment, which can be assayed to quantify the percentage of molecules that have been transported, relative to the total pool and [2] a situation under which all the detectable molecules are in the origin compartment, which provides a baseline where we can get a quantitative assessment of the degree of molecular trafficking which took place over a given time interval (for a prototypical example of such an assay, see [4]). In principle, two covalent modifications of reporter molecules have been used to analyze mitophagy in yeast: vacuolar clipping of mitochondrially targeted chimeric GFP fusion proteins [5–7] and proteolytic vacuolar activation of mitochondrially mislocalized alkaline phosphatase [8]. In the former, Western blotting with anti-GFP antibody can identify arrival of the reporter in the vacuole, which is observed as the release of free GFP from the chimera. In the latter, an enzymatic reaction is used to assay active alkaline phosphatase.

In this survey, we will cover the mainstream methods which are currently used to assay for mitophagy in *S. cerevisiae*, and we will explain their respective advantages and disadvantages.

1.1 Overall Considerations

General macroautophagy in yeast is usually assayed on cells that are logarithmically growing in a glucose-based medium. To induce classical macroautophagy, these cells are washed in water or nitrogen starvation medium and then resuspended in nitrogen starvation medium and incubated for 2–4 h before harvesting and assaying various indicators of autophagy. Remarkably, when assaying specifically for mitophagy under these conditions, no signal is observed in our assay systems, at least not in any robust and meaningful way that lends itself to analysis. In fact, one can continue the incubation in nitrogen starvation medium for a full week with no induction of mitophagy. In the presence of glucose, *S. cerevisiae* shuts down oxidative phosphorylation, a phenomenon known as the Crabtree effect [9, 10], and it is conceivable that this is the reason for the lack of mitophagy under these conditions. However, one can also show that starving log phase cells growing in respiratory conditions (i.e., using glycerol or lactate as carbon source) is also not an efficient way to induce mitophagy in yeast. In the following, we will describe conditions under which mitophagy is induced in robust fashion in yeast cells and assays which allow quantitative and semiquantitative measurement of mitophagic flux under these conditions. We will discuss potential pitfalls and the limitations inherent in each method.

1.1.1 Stationary Phase Mitophagy

When yeast cells growing in a nonfermentable medium (using a nonfermentable carbon source) are incubated for very long periods of time—three days or longer—one can observe the occurrence, starting with day 3, of mitophagy in these cells [11]. If a fermentable carbon source—glucose—is used, then one still will see mitophagy, but the results are potentially confounded by the occurrence of a diauxic shift [12]. The stationary-phase protocol works robustly in the SEY6210 genetic background but is inefficient, or does not work, in the BY4742, BY4741, or S288C backgrounds, which are known to harbor mutations that appear to affect mitochondrial biology (see https://www.yeastgenome.org/). We have not tested other popular genetic backgrounds, such as W303.

1.1.2 Starvation-Induced Mitophagy

When yeast cells growing in rich nonfermentable medium (yeast extract and peptone-based) are washed and transferred to SD-N medium, mitophagy has been reported to occur at shorter time scales, ranging from 6 h to overnight. This protocol is inefficient in SEY6210 cells but occurs in other genetic backgrounds, such as BY4742/1 and related strains [5].

2 Materials

2.1 Materials for Inducing Mitophagy

Reagents:

(a) SEY6210 or BY yeast strain expressing an appropriate reporter system (see below).

(b) SD medium.

(c) SL medium.

(d) SD-N medium.

(e) Sterile DDW.

2.2 Materials for Assaying Mitophagy Using Fluorescence Microscopy

(a) Yeast strains expressing a mitochondrially targeted GFP protein and a vacuole localized RFP (the mitochondrial marker emission color must be distinguishable from the vacuole marker). Alternatively, the vacuole can be stained by FM4–64 [13].

(b) Fluorescence microscope with DIC optics and filter cubes for GFP and RFP.

(c) Microscope slides.

2.3 Materials for Assaying Mitophagy by Western Blotting of GFP-Tagged Mitochondrial Proteins with Anti-GFP Antibody

(a) Yeast strain expressing a mitochondrial GFP fusion protein.

(b) 0.5 M tris pH 6.8.

(c) 10% SDS.

(d) Urea.

(e) 0.5 M EDTA, pH 8.

(f) Acid-washed glass beads, 400–600 μM diameter.

(g) Acetone.

(h) Trichloroacetic acid (TCA), 6.1 N.

(i) β-mercaptoethanol.

2.4 Materials for Assaying Mitophagy by Alkaline Phosphatase Activity Assay

(a) 0.5 M PIPES, pH 7.

(b) 2 M KCl.

(c) 5 M K-acetate.

(d) 1 M Mg_2SO_4.

(e) 100 mM $ZnSO_4$.

(f) PMSF.

(g) Glycerol.

(h) Acid-washed glass beads, 400–600 μM diameter.

(i) Glycine.

(j) Para-nitrophenyl phosphate (Sigma cat N9389).

3 Methods

3.1 Method for Induction of Mitophagy in Stationary Phase

(a) Streak cells from frozen stock onto SD plates.

(b) Transfer cells to 10 mL starter cultures in selective (or nonselective, depending on the reporter system) minimal medium, shaking at 220 RPM, 26 °C, for at least overnight.

Calculate the amount of cells required per time point (typically, 10 OD_{600} units per assay point) in your assays and, from that, the culture volume you need to set up for the experiment. Collect 0.8 OD_{600} units of starter culture per 10 mL final experimental culture, and wash them 2× with distilled water. Resuspend the cells in the final volume of SL medium with or without selection. Then incubate the cells at 26 °C, shaking at 220 RPM. Take samples at days 1, 2, 3, 4, etc. (depends on the reporter system; see below).

3.2 Method for Inducing Mitophagy by Nitrogen Starvation

1. Streak cells from a frozen stock onto SD or YPD. Transfer to a liquid SD medium, and grow overnight. Calculate the amount of cells per time point (typically, 10 OD_{600} units per assay point) in your assays and, from that, the culture volume you need to set up for the experiment.
2. Collect 0.8 OD_{600} units of starter culture per 10 mL final experimental culture, and wash them 2× with distilled water.
3. Finally, resuspend the cells in the desired volume of YPL medium, as dictated by the endpoint assay used. Incubate overnight at 30 °C with 220 RPM shaking. Next day, when the OD_{600} of the culture reaches 0.4–0.8, collect the appropriate amount of cells, wash them 2× in sterile DDW, and resuspend them in SD-N. Incubate the cells at 30 °C with 220 RPM in SD-N, and take samples (10 OD_{600} units) at time 0, 6 h, and 18 h of incubation in SD-N.

4 Assay Systems for Mitophagy in Yeast Cells

4.1 Immunoblotting for Mitochondrially Targeted, Chimeric GFP Fusion Proteins

Starting point: 10 OD_{600} units of cells collected from a culture (*see* Subheading 3, above).

1. Pellet the cells by centrifugation at 4200 × *g* for 3 min at 26 °C.
2. Resuspend the pellet in 1 mL 10% freshly diluted, ice-cold TCA, and transfer to an eppendorf tube.
3. Spin the cells at 3500 × *g* for 1 min.
4. Wash the pellet 3× with cold acetone (−20 °C).
5. Dry the pellet at room temperature or in a SpeedVac.
6. Add 100 μL glass beads (400–600 μM).
7. Add 100 μL cracking buffer (50 mM tris pH 6.8, 1% SDS, 6 M urea, 1 mM EDTA) at room temperature.
8. Vortex 15–30 min, room temperature.
9. Add 100 μL 2× SDS-PAGE loading buffer (100 mM tris pH 6.8, 4% (w/v) SDS, 20% (v/v) glycerol, 5% (v/v) β-mercaptoethanol).
10. Spin 5 min 17,000 × *g*.
11. Load 0.5 OD_{600} unit equivalents per time point on an SDS-PAGE gel, and immunoblot with anti-GFP antibody.

4.2 Fluorescence Microscopy of Mitochondrial GFP Fusion Proteins

Since the chimeric protein contains GFP, which is relatively resistant to degradation in the yeast vacuole, appearance of free GFP on blots should correlate with appearance of GFP fluorescence in the vacuole. Cells should be viewed at time 0 and at subsequent time points which are then also assayed in parallel by Western blotting. The vacuole can be identified by using DIC optics [14], by co-expression of a vacuolar RFP chimera such as Vph1-RFP [11], or by FM4-64 staining [15]. Exclusive reliance on fluorescence imaging should be avoided, since the properties of fluorophores vary with local environmental parameters such as pH, ionic strength, redox potential, etc., leading to potential false-positive and false-negative observations.

4.2.1 Rosella FP

A variant of the GFP fusion microscopy method, which was pioneered by Rod Devenish, consists of a tandem GFP-RFP couple fused C-terminally to a mitochondrially targeted protein [16]. While RFP fluorescence is stable under vacuolar pH, GFP is much more sensitive and loses quantum efficiency under these conditions. Thus, transport of the chimeric protein to the vacuole can be identified as loss of green fluorescence and appearance of red-only fluorescence which overlaps with the vacuole. The same controls and caveats that were introduced for the GFP microscopy protocol are relevant for rosella as well. In addition to those caveats, the use of *Rosella* precludes the use of red dyes (e.g., FM4-64) and red fluorescent proteins for the independent verification of the vacuole. One should also be aware that any experimentally induced or physiological shifts in intracellular or vacuolar pH could generate a false signal or obscure a bona fide one. Concomitant Western blotting to verify clipping of the chimeric protein is therefore highly encouraged in the case of *Rosella* but also in the standard GFP clipping assay.

4.3 Mitochondrially Targeted Pho8Δ60

The use of Pho8Δ60 to assay autophagic trafficking dates back to an important publication by Noda and Ohsumi [17]. The principle of the assay relies on the fact that vacuolar alkaline phosphatase (Pho8) is synthesized as a type II membrane protein with a transmembrane domain (TMD) that is proximal to the N-terminus. The N-terminus and the TMD encode the localization signals that direct the nascent protein to the secretory pathway [18]. Upon truncation of the TMD and the N-terminus, encoded in the first 60 codons of the reading frame, the resultant protein (Pho8Δ60) is cytosolic. In addition, it is catalytically inactive, as further processing by proteolytic cleavage of a pro-sequence in the C-terminus is required to generate an active enzyme. Under autophagy-inducing conditions, a small proportion of this pro-alkaline phosphatase zymogen is delivered to the vacuole (approximately 10% of the population over a 24 h incubation) and is activated. This can be detected by an alkaline phosphatase assay (see below). If one adds a mitochondrial targeting sequence to Pho8Δ60, the system becomes a mitophagy reporter [14].

5 Notes

5.1 Notes for the Free GFP Release Assays

Since the assay purports to measure the transport of a mitochondrial protein to the lumen of the vacuole, an important point is to establish that the free GFP signal does not arise from chimeric protein molecules that could be targeted by nonselective autophagy from the cytoplasm. The simplest way to ensure this is to demonstrate that no apparent free GFP is observed in *atg32Δ* cells. Atg32 is the yeast mitochondrial autophagy receptor and is essential for all currently known mitophagy events in yeast. At the same time, Atg32 is not known to be necessary for any other cellular function. [If the free GFP signal, observed either via fluorescence microscopy or Western blotting, is Atg32-independent, then it either arises from mistargeted chimeric protein in the cytoplasm or does not reflect mitophagy, or you may have discovered a novel mitophagic pathway. In the latter case, you would need to identify a novel receptor that could replace Atg32 in your assay and provide the same kind of assurance.] A second potential problem, which we have run into more frequently, is the appearance of a weak, cross-reactive non-GFP band that is co-induced under mitophagy-inducing conditions and runs in the same MW range of 25–28 kDa. This artifact can be identified by employing a control comprised of cells harboring empty vector (i.e., not expressing GFP) or by using non-tagged cells in the case of an integrated reporter. The detection system should then be calibrated (this is usually carried out by increasing the blocking stringency) so that the untagged control exhibits no "free GFP" signal in blots. It is also highly recommended not to resort to separately displaying the chimeric protein and the free GFP band with separate exposures, as this obscures the actual signal to noise ratio and increases the likelihood of this type of artifact (see Fig. 1). The choice of tagged protein is also important; for example, mitochondrial outer-membrane proteins are subject to mitophagy, as well as OMMAD, an ERAD-like degradation pathway which involves the proteasome, and therefore does not spare the GFP moiety of the chimera [12]. Thus, GFP chimeras derived from mitochondrial outer-membrane proteins may not be good quantitative reporters of mitophagy, since the full-length chimera could be underrepresented due to OMMAD-dependent degradation, which does not release stable free GFP.

5.1.1 Effects of the Identity of the "Host" Protein on the Mitophagic Efficiency of Individual GFP-Tagged Reporters

Chromosomal tagging of different mitochondrial matrix proteins with GFP leads to observed differential efficiencies of mitophagy [19]. For example, Idh2-GFP, although highly expressed, is rarely observed in the vacuole, both by fluorescence microscopy and by Western blotting. In contrast, Idp1-GFP is efficiently converted to free GFP in an Atg32-dependent manner, with approximately

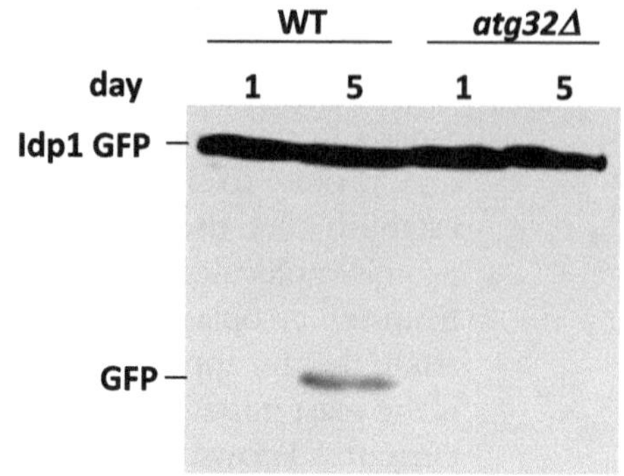

Fig. 1 Demonstration of Atg32-dependent generation of free GFP from Idp1-GFP as a measure of mitophagic trafficking, using Western blotting. Wild type (WT) and *atg32Δ* cells (SEY6210 genetic background) expressing chromosomally integrated Idp1-GFP from the endogenous *IDP1* promoter were incubated for 1 or 5 days in synthetic lactate medium as described in the text. Cells were harvested, and protein extracts were probed with anti-GFP antibody

50–60% of the signal accumulating as free GFP following a 7-day incubation in lactate-based minimal medium. While the mechanism of this selectivity is currently being investigated, the practical implications are that any assay involving a chimeric GFP-tagged reporter that is expressed from its endogenous promoter must take into account that the result does not provide an overall, quantitative assessment of mitophagic clearance but a specific assessment for that particular reporter. Thus, while using an endogenously expressed chimera can provide a binary yes/no answer to the question of whether mitophagy took place or not, the fact that different reporters will relay different results somewhat undermines the concept of a global quantification of mitophagy as a general process. One possibility around this issue is to utilize an overexpressed chimeric protein, as this largely seems to wash out the selectivity effects [19]. An additional possibility is to use a chimeric protein which does not include a yeast mitochondrial moiety, such as mtDHFR-GFP [6].

5.2 Notes for Alkaline Phosphatase-Based Methods

The use of an enzymatic assay to measure trafficking poses several challenges, relative to the free GFP release assay. In principle, since k_{cat} and K_m for Pho8 are known, one can theoretically calculate the number of Pho8 molecules which have been activated, per sample equivalent (e.g., per μg protein). However in order to do this, rigorous enzyme kinetics needs to be carried out. This includes:

(a) The protein concentration in the different extracts tested needs to be measured and normalized.

(b) A time course series must be carried out on each sample (as opposed to just some arbitrary endpoint such as 10 min).

(c) A second series, of varying extract amounts, must be carried out for each sample.

(d) Both series must be linear and overlap. This is called the Selwyn plot [20]. In brief, the Selwyn plot calibrates the assay to ensure that the reaction is linear with both time and enzyme concentration, thus validating the steady-state approximation.

(e) The initial velocity (the linear component of the time-dependent plot of NPP production) needs to be calculated for each sample, at a common protein concentration, for a saturating concentration of pNPP ($>>K_m$).

(f) Using published values of K_m and k_{cat}, one can then calculate an estimate of the total enzyme amount in each sample from the Michaelis-Menten equation.

If one only wishes to obtain a relative comparison of the samples and not an absolute quantification (as one obtains with the free GFP release assay), one can compare equal protein amounts at (linear) initial velocity conditions (points 1, 2, and 5) to obtain a rigorous comparison without calculating the precise amounts of vacuolar Pho8. Simply measuring arbitrary reaction endpoints is not sufficient; *if given enough time, a small amount of active enzyme and a large amount of active enzyme will generate the same amount of product.*

In addition to the above, the same controls that are required for the free GFP release assay also need to be used for the Pho8Δ60 assay: one must demonstrate a dependence on Atg32, to verify that activation of Pho8 does not occur post-lysis and to generate a baseline signal. A second important control is a no-reporter control.

Another important consideration that one has to take into account is that the yeast genome encodes a second, cytoplasmic alkaline phosphatase, Pho13 [21]. Thus in addition to expressing mtPho8Δ60 from a plasmid or from a genomic locus, the cells used in this assay should be *pho8Δ pho13Δ*, in order to eliminate background.

References

1. Ashford TP, Porter KR (1962) Cytoplasmic components in hepatic cell lysosomes. J Cell Biol 12:198–202
2. Deter RL, Baudhuin P, De Duve C (1967) Participation of lysosomes in cellular autophagy induced in rat liver by glucagon. J Cell Biol 35(2):C11–C16
3. Lemasters JJ (2005) Selective mitochondrial autophagy, or mitophagy, as a targeted defense against oxidative stress, mitochondrial dysfunction, and aging. Rejuvenation Res 8(1):3–5
4. Braell WA, Balch WE, Dobbertin DC, Rothman JE (1984) The glycoprotein that is transported between successive compartments of the Golgi in a cell-free system resides in stacks of cisternae. Cell 39(3 Pt 2):511–524
5. Kanki T, Klionsky DJ (2008) Mitophagy in yeast occurs through a selective mechanism. J Biol Chem 283(47):32386–32393

6. Okamoto K, Kondo-Okamoto N, Ohsumi Y (2009) Mitochondria-anchored receptor Atg32 mediates degradation of mitochondria via selective autophagy. Dev Cell 17(1):87–97
7. Journo D, Mor A, Abeliovich H (2009) Aup1-mediated regulation of Rtg3 during mitophagy. J Biol Chem 284(51):35885–35895
8. Kanki T et al (2009) A genomic screen for yeast mutants defective in selective mitochondria autophagy. Mol Biol Cell 20(22):4730–4738
9. De Deken RH (1966) The Crabtree effect: a regulatory system in yeast. J Gen Microbiol 44 (2):149–156
10. Hagman A, Säll T, Piškur J (2014) Analysis of the yeast short-term Crabtree effect and its origin. FEBS J 281(21):4805–4814
11. Tal R, Winter G, Ecker N, Klionsky DJ, Abeliovich H (2007) Aup1p, a yeast mitochondrial protein phosphatase homolog, is required for efficient stationary phase mitophagy and cell survival. J Biol Chem 282(8):5617–5624
12. Heo JM et al (2010) A stress-responsive system for mitochondrial protein degradation. Mol Cell 40(3):465–480
13. Vida TA, Emr SD (1995) A new vital stain for visualizing vacuolar membrane dynamics and endocytosis in yeast. J Cell Biol 128 (5):779–792
14. Kanki T, Kang D, Klionsky DJ (2009) Monitoring mitophagy in yeast: the Om45-GFP processing assay. Autophagy 5(8):1186–1189
15. Kissova I, Deffieu M, Manon S, Camougrand N (2004) Uth1p is involved in the autophagic degradation of mitochondria. J Biol Chem 279 (37):39068–39074
16. Rosado CJ, Mijaljica D, Hatzinisiriou I, Prescott M, Devenish RJ (2008) Rosella: a fluorescent pH-biosensor for reporting vacuolar turnover of cytosol and organelles in yeast. Autophagy 4(2):205–213
17. Noda T, Matsuura A, Wada Y, Ohsumi Y (1995) Novel system for monitoring autophagy in the yeast Saccharomyces cerevisiae. Biochem Biophys Res Commun 210 (1):126–132
18. Klionsky DJ, Emr SD (1990) A new class of lysosomal/vacuolar protein sorting signals. J Biol Chem 265(10):5349–5352
19. Abeliovich H, Zarei M, Rigbolt KT, Youle RJ, Dengjel J (2013) Involvement of mitochondrial dynamics in the segregation of mitochondrial matrix proteins during stationary phase mitophagy. Nat Commun 4:2789
20. Marangoni AG (2003) Enzyme kinetics: a modern approach. John Wiley & Sons, Hoboken, NJ
21. Kaneko Y, Toh-e A, Banno I, Oshima Y (1989) Molecular characterization of a specific p-nitrophenylphosphatase gene, PHO13, and its mapping by chromosome fragmentation in Saccharomyces cerevisiae. Mol Gen Genet 220 (1):133–139

Chapter 45

Measuring Antibacterial Autophagy

Keith B. Boyle and Felix Randow

Abstract

Bacteria that escape from membrane-enclosed vacuoles to the cytosol of cells are targeted by autophagy, which recognizes and captures bacteria into autophagosomes wherein their proliferation is restricted. Here we discuss two means by which antibacterial autophagy is assessed: (1) the visualization and enumeration of autophagy protein recruitment to the vicinity of cytosolic bacteria by means of immunofluorescence microscopy and (2) the measurement of autophagy-dependent restriction of bacterial proliferation by means of colony-forming unit assay.

Key words *Salmonella*, Galectin, Ubiquitin, Colony-forming unit, Immunofluorescence

1 Introduction

Selective autophagy is essential for cell-autonomous defense of the cytosol against invading bacteria [1, 2]. *Salmonella enterica* serovar Typhimurium (*S.* Typhimurium) is a Gram-negative bacterium with a facultative intracellular lifestyle that causes severe enterocolitis in humans. Upon invasion of cells, *S.* Typhimurium establishes its intracellular niche in a specialized vesicular compartment, the *Salmonella*-containing vacuole (SCV). However, the limiting SCV membrane frequently becomes damaged, and access to the cytosol promotes *S.* Typhimurium proliferation, particularly if autophagy is impaired [3]. Because of this marked phenotype, *S.* Typhimurium has become the pathogen of choice for autophagy research. The methods we describe here pertain to the study of infection of mammalian cells with *S.* Typhimurium but can be modified for other bacterial species that access the host cytosol, such as *Mycobacterium tuberculosis* and *Streptococcus pyogenes.* It is notable that certain bacterial species that colonize the cytosol in an obligatory manner, such as *Shigella flexneri*, do so in part by counteracting the host autophagy response [4–6].

The escape of *S.* Typhimurium into the cell cytosol triggers the recruitment of specific host proteins to the vicinity of the bacterium

Nicholas Ktistakis and Oliver Florey (eds.), *Autophagy: Methods and Protocols*, Methods in Molecular Biology, vol. 1880, https://doi.org/10.1007/978-1-4939-8873-0_45, © Springer Science+Business Media, LLC, part of Springer Nature 2019

that coordinate a cellular response to the bacterium. Whether a protein of interest is recruited to the bacterium can be assessed by immunofluorescence, either via a specific antibody or by ectopic expression of the fluorescently tagged protein in the cell. The relatively large size of the bacterium (approximately 1 μm) and its amenability to genetic manipulation, thereby permitting the creation of strains expressing fluorescent proteins, enable the ready identification of infected cells and the visualization of individual bacteria. Bacteria-mediated damage to the SCV membrane, often occurring shortly after infection (approximately 15–60 min post-infection in the case of *S.* Typhimurium), exposes otherwise hidden glycans (glycosylated proteins and lipids) to the cytosol of the cell. These glycans are specifically recognized by members of the Galectin family of proteins [7, 8]. We routinely monitor the localization of Galectin-8, using an anti-Galectin-8 antibody, to identify bacteria that have gained access to the cytosol [8]. Both the damaged membrane and the bacterium itself are recognized by the host ubiquitylation machinery [9–11]. While the identification of the substrates that become ubiquitylated is still progressing, they comprise both bacterial and host proteins. Ubiquitylated proteins can be detected by anti-ubiquitin antibodies that detect most, if not all, ubiquitin linkage types, such as FK2, or that selectively react with only selected linkage types. Ubiquitylated proteins accumulate around the bacterium and early during infection act as a secondary marker for a ruptured SCV. The cytosolic bacteria are either immediately captured via Galectin-8/NDP52 (aka CALCOCO2)-dependent autophagy or they evade capture and proliferate, outgrowing their Galectin-8-positive membranes but retaining their ubiquitin coat [8, 12]. Thus at later time points of infection (after 2 h post-infection), galectins should no longer be used as definitive markers of cytosolic entry. It should also be noted that ubiquitin recruitment cannot be taken as a bona fide identifier of cytosolic bacteria since ubiquitin has been reported to accumulate on intact SCVs around 4–6 h post-infection [13]. Although electron microscopy can be used to definitively determine whether a given intracellular bacterium is surrounded by host membranes, it is much more difficult to ascertain that a given SCV has not suffered damage to its membrane.

Macroautophagy of cytosolic *S.* Typhimurium encloses the bacterium in a double-membrane structure called the autophagosome, a process that can be visualized by the recruitment of autophagy proteins. Members of the ATG8 family of proteins, comprising three LC3 and three GABARAP genes in humans, provide particularly good markers due to their prolonged association with autophagosomes, although other autophagy proteins can be used for the study of dynamic aspects of antibacterial autophagy [14, 15]. Due to the shortage of well-defined antibodies against individual LC3 and GABARAP proteins, we favor the generation of

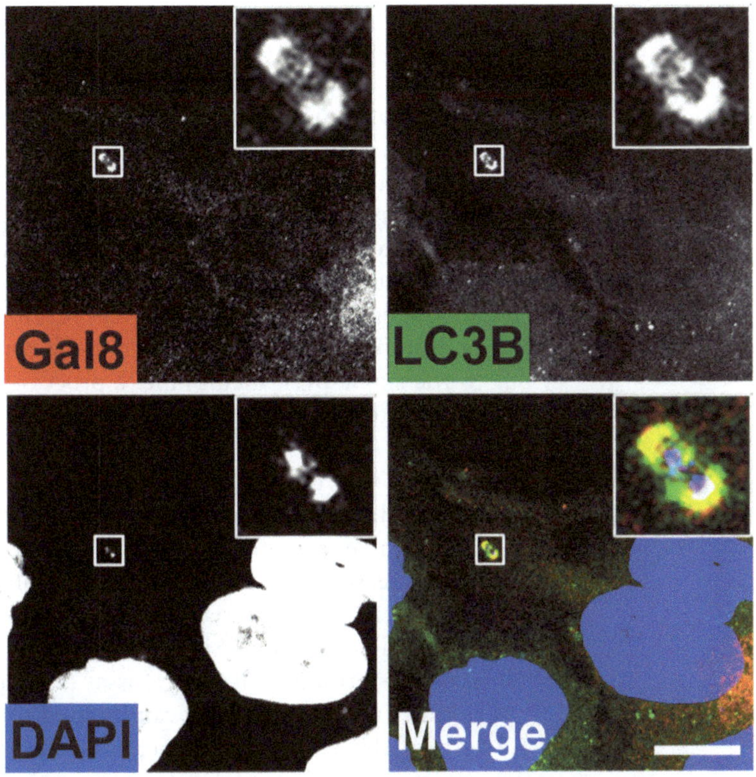

Fig. 1 Galectin-8-positive *S.* Typhimurium are attacked by autophagosomes. HeLa cells were transduced with a Moloney murine leukemia virus-based retrovirus harboring a GFP-LC3B fusion gene together with a blasticidin resistance gene and a stable population of cells selected with blasticidin. Cells were infected with *S.* Typhimurium, fixed with paraformaldehyde 1 h post-infection and stained with anti-Galectin-8 antibody. The sample was mounted in 4′,6-diamidino-2-phenylindole dihydrochloride (DAPI)-containing mounting medium to label the DNA of both the HeLa cell and bacteria. Micrograph was acquired on a laser-scanning confocal microscope. The inset depicts two *S.* Typhimurium to which both Galectin-8 and GFP-LC3B have been recruited, indicative that these bacteria were exposed to the cytosol and are attacked by macroautophagy. Scale bar, 10 μm

stable cell lines specifically expressing fluorescently tagged individual LC3 isoforms. GFP-LC3B is recruited to the same cytosolic bacteria to which either or both Galectin-8 and ubiquitin are recruited (Fig. 1). While GFP-LC3B is commonly used in the antibacterial autophagy literature, we have shown that LC3B recruitment to *Salmonella* in human cells is dependent on LC3C [16] so we suggest that investigators consider using fusion proteins of both LC3B and LC3C. However, LC3C and its binding partner, the autophagy cargo receptor NDP52, are nonfunctional in both mouse and rat cells. We prefer stable transgenic cell lines rather than transfection-based approaches as the former yield populations of cells with homogenous expression of the marker proteins and avoid the need to perform a transfection for each experiment. We recommend the use of a HIV or Moloney murine leukemia virus-based expression construct, harboring both a GFP-LC3 fusion and a drug selection marker to transduce and select the cells of interest [17]. During macroautophagy, LC3 is conjugated to double-membrane phagophores, a process that also occurs during antibacterial autophagy. However, macroautophagy is not the only process recruiting LC3 to intracellular bacteria. Certain cell types, such as murine embryonic fibroblasts, and certain signals, including activation of Toll-like receptors, permit the conjugation of LC3 directly to the single SCV membrane, a process related to LC3-associated phagocytosis (LAP) [18–20]. As such, the frequency of recruitment of LC3 to *S.* Typhimurium does not always correlate with the restriction of bacterial proliferation [18]. Thus caution should be exercised when interpreting the recruitment of LC3 to bacteria.

The majority of the proliferation of *S.* Typhimurium in cultured cells, at least during the first 10 h post-infection, is accounted for by cytosol-exposed bacteria [21]. Thus the determination of the number of viable bacteria inside cells at specified time points by means of a colony-forming unit (CFU) assay (essentially, their ability to form colonies on an agar plate) can be used to assess whether perturbation of the cells, by genetic, chemical, or other means, directly affects the antibacterial autophagy pathway. (The assay is alternatively known as the gentamicin protection assay, after the antibiotic that selectively kills only extracellular bacteria.)

To investigate whether a gene of interest affects antibacterial immunity, cells can be treated with specific siRNAs, or knockouts can be created using CRISPR technology before assessing the ability of *S.* Typhimurium to proliferate in those cells [22]. Whether an observed phenotype is autophagy-dependent can be determined by epistatic analysis in autophagy-proficient and autophagy-deficient cells, such as ATG5−/− cells (Fig. 2). These experiments should be complemented with an assessment of the recruitment of the autophagy marker proteins discussed above. When used together, the combination of approaches described here will allow investigators to determine whether a gene of interest contributes to antibacterial autophagy and, if so, at which stage of the response its function lies.

2 Materials

Prepare all solutions using ultrapure water (prepared by purifying deionized water, to attain a sensitivity of 18.2 MΩ-cm at 25 °C).

S. Typhimurium National Tissue Culture Collection strain 12023 is used for both assays. Bacteria are transformed with a low-copy plasmid encoding a fluorescent protein such as mCherry

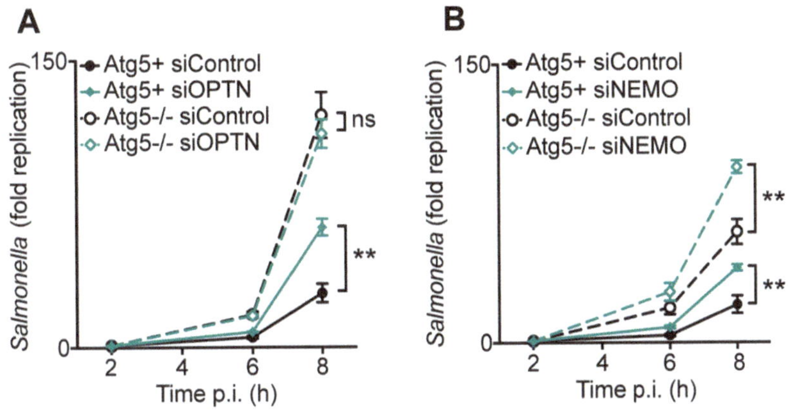

Fig. 2 The antibacterial role of optineurin, but not NEMO, is autophagy-dependent. Mouse embryonic fibroblasts (MEFs) harboring a genetic deletion in the ATG5 gene (ATG5$^{-/-}$; autophagy deficient) or complemented with an ATG5-expressing retrovirus (ATG5^{+}; autophagy competent) were transfected with siRNA against the autophagy cargo receptor optineurin (OPTN), the NF-kappa-B essential modulator (NEMO), or control siRNA. At 72 h post-transfection, cells were infected with *S.* Typhimurium, and a colony-forming unit assay was performed. Data show the kinetics of intracellular bacterial proliferation for each sample relative to 1 h. Note that siRNA-mediated depletion of OPTN increased bacterial proliferation in ATG5+ but not ATG5−/− cells (**a**), while depletion of NEMO enhanced proliferation in either setting (**b**), indicating that Optineurin but not NEMO restricts bacterial proliferation via autophagy. Figure reproduced from [5]

and an ampicillin resistance cassette. This permits both the identification of mCherry-positive bacteria by fluorescent light microscopy and selection of ampicillin-resistant bacterial colonies on LB agar plates during CFU assays.

2.1 Immunofluorescence Microscopy

1. HeLa (American Tissue Culture Collection) are kept as frozen stocks in 90% fetal bovine serum (FBS)/10% dimethylsulfoxide in liquid nitrogen. A fresh vial of cells is thawed every 3–4 months, i.e., after approximately 35 passages (*see* **Note 1**).
2. HeLa growth medium: 500 mL Iscove's Modified Dulbecco's Medium, 10% heat-inactivated FBS, 300 μL of the antibiotic gentamycin (available at 50 mg/mL). Inactivate FBS by submersing the thawed bottle in a preheated water bath at 56 °C for 30 min.
3. Cell culture dishes and plates: 10 cm diameter tissue culture-treated dishes and 24-well tissue culture treated plates.
4. Coverslips: 13-mm 1.5 grade glass coverslips.
5. siRNA oligonucleotides (*see* **Note 2**).
6. Lipofectamine RNAiMAX transfection reagent and OptiMEM buffer for preparing siRNA transfection mix.
7. 4% Paraformaldehyde (PFA): Heat 1.8 L of PBS to between 50 and 60 °C, while stirring, ensuring that the temperature does not exceed 60 °C. Add 80 g crystalline paraformaldehyde,

and leave for 10 min. Add 4 M NaOH dropwise until the solution clears. Cool to 25 °C and adjust pH to 7.4. Filter-sterilize and store in 13 mL aliquots at −20 °C (*see* **Note 3**).

8. Quenching buffer: 100 mM glycine in 500 mL PBS; add 0.5 mL of a 10% solution of sodium azide prepared in water (*see* **Note 4**).
9. Permeabilization solution: 0.2% Triton X-100 in PBS. Make a 20% Triton X-100 stock solution by diluting 10 mL Triton X-100 in 40 mL water. Mix thoroughly and filter sterilize. For the working solution, add 100 μL of 20% Triton X-100 to 10 mL PBS freshly each time.
10. Blocking buffer: 2% BSA in PBS. Dissolve 1 g bovine serum albumin (BSA) in 50 mL PBS. Add 50 μL of 10% sodium azide. Filter sterilize and store at 4 °C.
11. Mounting medium: VECTASHIELD HardSet with 4′,6-diamidino-2-phenylindole dihydrochloride (DAPI).
12. Antibodies: goat anti-Galectin-8 (BD Biosciences; AF-1305) use at 1:50 dilution or mouse anti-ubiquitin FK2 (Enzo Bioscience; BML-PW8810-0500) use at 1:400 dilution.

2.2 CFU Assay

1. 5 cm sterile plates containing 15 mL LB/agar and the appropriate antibiotic.
2. 3 mm sterile glass beads.
3. Lysis buffer: 0.1% Triton X-100 in PBS. Add 1.25 mL of 20% Triton X-100 stock solution to 250 mL PBS.
4. Sterile 96-well deep-well plates.

3 Methods

These procedures will require Category 2 containment facilities, depending on local regulations.

3.1 Cell Culture and RNA Interference

1. Culture HeLa cells in cell culture medium in a humidified incubator at 37 °C and 5% CO_2. Cells should be split every 2–3 days (*see* **Note 5**).
2. Trypsinize cells and seed at a density of 15,000 cells per well of 24-well plate in 500 μL cell culture medium, ensuring cells are seeded onto 13 mm, 1.5 thickness coverslips for immunofluorescence analysis (*see* **Note 6**). Prepare each sample in triplicate for each time point. Allow the cells to attach for 18–24 h. Transfect cells with relevant siRNAs according to the manufacturer's protocol. Aspirate medium after 48 h, and replace with 500 μL fresh cell culture medium.

3.2 Culturing of S. Typhimurium and Infection of Cells

1. Pick a single colony of *S.* Typhimurium into 1 mL LB broth, and incubate at 37 °C with shaking at 210 rpm until saturation of the culture, at least 8 h (*see* **Note 7**).
2. Subculture 150 μL of saturated culture of *S.* Typhimurium into 5 mL LB broth, and incubate at 37 °C with shaking at 210 rpm for 3.5 h (*see* **Note 8**).
3. Aspirate medium from cells and replace with 1 mL pre-warmed medium, not containing gentamycin antibiotic, 30 min prior to infection.
4. Add 20 μL of *S.* Typhimurium culture per well, shake the plate back and forth several times, and incubate in a humidified incubator at 37 °C and 5% CO_2 for 15 min (*see* **Note 9**).
5. Wash the cells twice with 500 μL pre-warmed PBS, and add 500 μL cell culture medium containing 100 μg/mL gentamycin. Place cells back into the incubator. At 1 h post-infection, aspirate the medium, and add 500 μL cell culture medium containing 20 μg/mL gentamycin (*see* **Note 10**).

3.3 Staining of Cells for Immunofluorescence

1. At the desired time point post-infection, wash the cells in 500 μL PBS, and add 250 μL of 4% PFA per well. Incubate at room temperature for 20 min, rinse three times in 500 μL quenching buffer, and leave in 500 μL quenching buffer (*see* **Note 11**).
2. Aspirate buffer from the coverslips, add 250 μL permeabilization buffer, and incubate for 10 min. Rinse the wells twice in PBS, and add 250 μL blocking buffer for 30 min (*see* **Note 12**).
3. Prepare appropriate dilution of the primary antibody in blocking buffer, place 20 μL per sample on a piece of flat Parafilm, and carefully place the coverslip facedown on top. Incubate in a humidified atmosphere for 1 h (*see* **Note 13**).
4. Wash coverslip by dipping it 10 times in a beaker with at least 100 mL PBS, dab off excess PBS on a paper towel, and incubate with relevant fluorescently conjugated secondary antibody for 30 min. Wash coverslip as above, finally dipping once in water to remove traces of PBS.
5. Mount coverslip by placing facedown on a drop of mounting medium on a microscope slide, and allow to dry in a dark place overnight. Store the slide in the fridge (*see* **Note 14**).

3.4 Fluorescent Cell Microscopy

To enumerate the fraction of *S.* Typhimurium around which a protein of interest localizes, we use an epifluorescent microscope equipped with a 100x oil immersion objective lens, and filter sets for at least GFP, mCherry, and DAPI.

1. Place the microscope slide on the microscope, and visualize the cell nuclei via DAPI fluorescence (*see* **Note 15**).

2. The bacteria will be visible as DAPI-positive structures in the cell. If the bacteria carry a fluorescence protein, these DAPI-positive structures will colocalize with that fluorescent signal (*see* **Note 16**).
3. Count the number of bacteria visible in the field of view.
4. Move to the relevant fluorescent channel, and score the number of those bacteria around which there lies a fluorescent signal above that of the cellular background (*see* **Note 17**).
5. Enumerate at least 100 bacteria per coverslip, with three coverslips per condition (*see* **Note 18**).
6. To generate a publication-quality micrograph of a cell containing a bacterium surrounded by a fluorescence signal, acquire a confocal image on an appropriate laser-scanning confocal microscope.

3.5 Colony-Forming Unit Assay

The CFU assay measures the number of viable intracellular bacteria inside a population of cells at a given time point. It relies on the ability of individual bacteria extracted from mammalian cells to grow into a colony on an agar plate. To ensure the assay truly measures only intracellular bacteria and not those that have remained extracellular, we use the fast-acting antibiotic gentamycin to kill extracellular bacteria. The mammalian cells must also be lysed to free the intracellular bacteria, with the greater sensitivity of mammalian cells to detergents such as Triton X-100 compared to bacteria permitting the use of such reagents. By performing the assay at distinct time points post-infection, a kinetic analysis of growth of the bacterial population can be determined. The first time point is chosen as the "input" for the assay with subsequent time points normalized to that value, yielding a fold proliferation of bacteria over time. The "input" values across different samples should broadly be quite similar. This first time point should minimize the chance for bacteria to proliferate but permits the gentamycin to kill those on the outside of the cells—we find that 1 h post-infection is optimal. Robust proliferation of the population of bacteria is usually evident at 6 h, approximately 5–15-fold proliferation relative to 1 h, and increases further at 8 h post-infection.

If the bacteria under study harbor an antibiotic resistance marker, we suggest to perform CFU assays on plates containing antibiotics to avoid the potential of cross-contamination with environmental bacteria. We use sterile deep-well 96-well plates to carry out serial dilution of the lysates in convenient 1.0 mL volumes and a multichannel pipette, capable of dispensing 0.1–1.0 mL volumes.

1. Prepare and label sufficient 5 cm agar plates containing 15 mL LB/agar and the appropriate antibiotic. Place 10–20 sterile glass beads on each plate (*see* **Note 19**).

2. Add 0.9 mL PBS to the wells of the 96-well plate to be used for the serial dilutions (*see* **Note 20**).
3. Aspirate the medium from the cells, and add 0.5 mL of PBS to wash the cells.
4. Aspirate the PBS, add 1.0 mL lysis buffer, and incubate for 5 min (*see* **Note 21**).
5. Pipette the lysate in each well up and down 10 times to fully detach the bacteria, and transfer 0.85 mL to a well of a deep-well 96-well plate (*see* **Note 22**).
6. Once all lysates have been transferred to the 96-well plate, pipette each lysate up and down 10 further times, and transfer 0.1 mL to the first diluent well containing 0.9 mL PBS (*see* **Note 23**).
7. Complete all dilutions.
8. Immediately before plating out the relevant dilution series, mix the diluted lysate again 10 times.
9. Transfer 0.1 mL of the diluted bacteria to two agar plates. Shake the plates back and forth, and discard the glass beads (*see* **Note 24**).
10. Allow the plates to dry before transferring them to a 37 °C incubator overnight.
11. Remove the plates from the incubator, and enumerate the number of colonies per plate (*see* **Note 25**).
12. Using appropriate computer software, calculate the fold change of proliferation of each sample compared to the 1 h post-infection input sample.

4 Notes

1. HeLa cell cultures are usually split every 2 (at a 1:4 dilution) or 3 (at a 1:8 dilution) days, and cells should be tested regularly for the absence of mycoplasma contamination. We ensure all our cell lines do not carry mycoplasma by routine monitoring using the MycoAlert assay kit.
2. We prefer Silencer Select siRNAs, used at 6 pmol per well of a 24-well plate.
3. PFA is a toxic substance, and the solution should be prepared using a heated stirrer in a fume hood. Not all PFA will initially dissolve, thus leaving a cloudy solution. 4 M NaOH should be added dropwise, with approximately 5 s between drops until the majority of the solution clears. A small number of insoluble particles may remain. We store 4% PFA at −20 °C, permitting at most two freeze-thaw cycles.

4. Sodium azide prevents growth of microorganisms in the quenching buffer.
5. Other cell types can also be used. Their normal growth medium should be used and culture conditions determined empirically.
6. Cells should be approximately 20% confluent at the time of transfection.
7. If available, ampicillin-resistant mCherry-positive *S.* Typhimurium can be used.
8. This equates to an optical density at 600 nm of approximately 1.0. We use a programmable shaking incubator with variable temperature control for the convenience of beginning the experiment at a desired time in the morning, i.e., the subculture can be kept at 4 °C before automatic switching to 37 °C at the desired time.
9. After infection under these conditions, approximately 50–75% of cells are infected with between one and five bacteria each. The optimal number of bacteria and infection time varies between cell lines. For example, we find that mouse embryonic fibroblasts are more easily infected than HeLa cells so an infection time of 7 min is sufficient.
10. For convenience and swiftness, we use a positive displacement repeater pipette with sterile, disposable Combitips.
11. Coverslips can be processed immediately, or the plate can be stored in the fridge for up to 2 weeks. Ensure that the buffer does not evaporate from the wells during this time.
12. Cells expressing GFP-LC3 do not require staining with antibody to visualize GFP-LC3.
13. We find that a convenient way to do this is to place the Parafilm on top of a larger piece of wet paper towel to ensure a humid atmosphere. We then place an opaque upturned box on top.
14. We use VECTASHIELD HardSet mounting medium with DAPI. This permits the identification of non-fluorescent *S.* Typhimurium by means of DAPI labelling of the bacterial DNA. We fit up to four coverslips on a standard microscopic slide.
15. We advise that another member of the laboratory relabels slides so that the samples can be counted in a blinded manner.
16. If cell lines harbor contamination with mycoplasma, they will be visible in a non-infected specimen as small DAPI-positive structures.
17. The fluorescent signal may either completely or only partially surround the bacterium. We use tally counters to keep track of the counting.

18. We enumerate a greater number of bacteria if the frequency of marker-positive bacteria is low, say less than 10%.
19. We usually plate out the following dilutions for each time point: 1 h, 1:10 and 1:100; 6 h, 1:100 and 1:1000; 8 h, 1:100 and 1:1000. Each sample is prepared in triplicate wells, with the relevant dilution of each well plated on two agar plates, giving a total of six agar plates per experimental condition per time point. We find that using "dry" LB/agar plates works best so we prepare the agar plates 2–3 days in advance of the experiment.
20. We use sequential rows of the 96-well plate for each dilution.
21. 0.1% Triton X-100 is sufficient to lyse the mammalian but not the bacterial cells.
22. We pipette up and down 0.85 mL as a larger volume can enter and clog the filter of the pipette tips.
23. This further mixing should be done so as to prevent any settling of the bacteria that may occur over time.
24. We recommend transferring PBS containing the bacteria to up to 24 agar plates before shaking them in batches. Waiting longer than this can result in the PBS drying on the agar and compromise even spreading of the bacteria.
25. We use an automated colony counter for this purpose. Ensure that the agar plates selected for each time point have a number of colonies within the linear range of the assay. In our hands, this is between 10 and 200 colonies per 5 cm agar plate.

Acknowledgments

This work was supported by the MRC (U105170648) and the Wellcome Trust (WT104752MA).

References

1. Randow F, MacMicking JD, James LC (2013) Cellular self-defense: how cell-autonomous immunity protects against pathogens. Science 340:701–706. https://doi.org/10.1038/nm.3108
2. Deretic V, Saitoh T, Akira S (2013) Autophagy in infection, inflammation and immunity. Nat Rev Immunol 13:722–737. https://doi.org/10.1038/nri3532
3. Boyle KB, Randow F (2013) The role of "eat-me" signals and autophagy cargo receptors in innate immunity. Curr Opin Microbiol 16:339–348. https://doi.org/10.1016/j.mib.2013.03.010
4. Huang J, Brumell JH (2014) Bacteria-autophagy interplay: a battle for survival. Nat Rev Microbiol 12:101–114. https://doi.org/10.1038/nrmicro3160
5. Noad J, von der Malsburg A, Pathe C et al (2017) LUBAC-synthesized linear ubiquitin chains restrict cytosol-invading bacteria by activating autophagy and NF-κB. Nat Microbiol 2:17063. https://doi.org/10.1038/nmicrobiol.2017.63
6. Ogawa M, Yoshimori T, Suzuki T et al (2005) Escape of intracellular Shigella from autophagy. Science 307:727–731. https://doi.org/10.1126/science.1106036

7. Dupont N, Lacas-Gervais S, Bertout J et al (2009) Shigella phagocytic vacuolar membrane remnants participate in the cellular response to pathogen invasion and are regulated by autophagy. Cell Host Microbe 6:137–149
8. Thurston TLM, Wandel MP, von Muhlinen N et al (2012) Galectin 8 targets damaged vesicles for autophagy to defend cells against bacterial invasion. Nature 482:414–418. https://doi.org/10.1038/nature10744
9. Perrin AJ, Jiang X, Birmingham CL et al (2004) Recognition of bacteria in the cytosol of mammalian cells by the ubiquitin system. Curr Biol 14:806–811. https://doi.org/10.1016/j.cub.2004.04.033
10. Fiskin E, Bionda T, Dikic I, Behrends C (2016) Global analysis of host and bacterial ubiquitinome in response to Salmonella Typhimurium infection. Mol Cell 62:967–981. https://doi.org/10.1016/j.molcel.2016.04.015
11. Fujita N, Morita E, Itoh T et al (2013) Recruitment of the autophagic machinery to endosomes during infection is mediated by ubiquitin. J Cell Biol 203:115–128. https://doi.org/10.1083/jcb.201304188
12. Thurston TLM, Ryzhakov G, Bloor S et al (2009) The TBK1 adaptor and autophagy receptor NDP52 restricts the proliferation of ubiquitin-coated bacteria. Nat Immunol 10:1215–1221. https://doi.org/10.1038/ni.1800
13. Patel JC, Hueffer K, Lam TT, Galan JE (2009) Diversification of a Salmonella virulence protein function by ubiquitin-dependent differential localization. Cell 137:283–294. https://doi.org/10.1016/j.cell.2009.01.056
14. Mizushima N, Yoshimori T, Ohsumi Y (2011) The role of Atg proteins in autophagosome formation. Annu Rev Cell Dev Biol 27:107–132. https://doi.org/10.1146/annurev-cellbio-092910-154005
15. Shibutani ST, Yoshimori T (2014) Autophagosome formation in response to intracellular bacterial invasion. Cell Microbiol 16:1619–1626. https://doi.org/10.1111/cmi.12357
16. von Muhlinen N, Akutsu M, Ravenhill BJ et al (2012) LC3C, bound selectively by a noncanonical LIR motif in NDP52, is required for antibacterial autophagy. Mol Cell 48:329–342. https://doi.org/10.1016/j.molcel.2012.08.024
17. Randow F, Sale JE (2006) Retroviral transduction of DT40. In: Buerstedde J-M, Takeda S (eds) Reviews and protocols in DT40 research: subcellular biochemistry. Springer, Dordrecht, pp 383–386
18. Kageyama S, Omori H, Saitoh T et al (2011) The LC3 recruitment mechanism is separate from Atg9L1-dependent membrane formation in the autophagic response against Salmonella. Mol Biol Cell 22:2290–2300. https://doi.org/10.1091/mbc.E10-11-0893
19. Sanjuan MA, Dillon CP, Tait SWG et al (2007) Toll-like receptor signalling in macrophages links the autophagy pathway to phagocytosis. Nature 450:1253–1257. https://doi.org/10.1038/nature06421
20. Mehta P, Henault J, Kolbeck R, Sanjuan MA (2014) Noncanonical autophagy: one small step for LC3, one giant leap for immunity. Curr Opin Immunol 26:69–75. https://doi.org/10.1016/j.coi.2013.10.012
21. Malik-Kale P, Winfree S, Steele-Mortimer O (2012) The bimodal lifestyle of intracellular Salmonella in epithelial cells: replication in the cytosol obscures defects in vacuolar replication. PLoS One 7:e38732. https://doi.org/10.1371/journal.pone.0038732
22. Boettcher M, McManus MT (2015) Choosing the right tool for the job: RNAi, TALEN, or CRISPR. Mol Cell 58:575–585. https://doi.org/10.1016/j.molcel.2015.04.028

Chapter 46

Quantitative Phosphoproteomics of Selective Autophagy Receptors

Thomas Juretschke, Petra Beli, and Ivan Dikic

Abstract

Selective autophagy enables degradation of specific cargo such as protein aggregates or organelles and thus plays an essential role in the regulation of cellular homeostasis. Cargo specificity is achieved on the level of autophagy receptors that concurrently bind the cargo and the autophagosomal membrane. Recent studies have demonstrated that selective autophagy is tightly regulated by posttranslational modifications of autophagy receptors, in particular protein phosphorylation. Phosphorylation of autophagy receptors by different kinases, including Tank-binding kinase (TBK1), can increase their affinity toward the cargo or autophagosomes and thereby regulate the specificity and activity of selective autophagy depending on the cellular condition.

Here, we report an approach for quantitative analysis of phosphorylation sites on autophagy receptors using mass spectrometry-based proteomics. In this protocol, GFP-tagged autophagy receptors are purified based on the high-affinity binding between GFP and GFP-Trap agarose. Interaction partners and background binders are subsequently removed by washes under denaturing conditions to obtain a pure fraction of the bait protein, thereby reducing the complexity of the analyzed sample. The bait protein is then digested on-bead, and peptides are analyzed by liquid chromatography-tandem mass spectrometry (LC-MS/MS). The described approach permits systematic identification and quantification of phosphorylation sites on autophagy receptors and other autophagic components. In addition to phosphorylation, this protocol is suitable for investigating other posttranslational modifications, including protein ubiquitylation.

Key words Autophagy receptors, Selective autophagy, Phosphorylation, Mass spectrometry-based proteomics, SILAC

1 Introduction

Autophagy is an evolutionary conserved degradation pathway that plays essential roles in cellular homeostasis [1, 2]. Although previously considered as an unselective process that sequesters portions of cytoplasm during periods of starvation, it is now clear that autophagy can also degrade specific cargo (e.g., protein aggregates or organelles) with the help of autophagy receptors [3, 4]. A dozen of autophagy receptors were identified in human cells, which can recognize degradation signals in the form of ubiquitylation on the

Nicholas Ktistakis and Oliver Florey (eds.), *Autophagy: Methods and Protocols*, Methods in Molecular Biology, vol. 1880, https://doi.org/10.1007/978-1-4939-8873-0_46, © Springer Science+Business Media, LLC, part of Springer Nature 2019

cargo and concurrently bind to autophagosomes [5–7]. Specific autophagy receptors that recognize protein aggregates, mitochondria, endoplasmic reticulum, and intracellular pathogens have been identified so far [7]. Autophagy receptors share structural features including the presence of an ubiquitin-binding domain (UBD) that binds ubiquitin on the cargo and an LC3-interacting region (LIR) that binds LC3/GABARAP proteins on autophagosomal membranes [8]. An increasing number of studies link genetic alterations in autophagy receptors to neurodegenerative disorders highlighting their important role in cellular homeostasis [7]. Due to their essential function in regulating selective autophagy and thereby cellular homeostasis, it is not surprising that the function of autophagy receptors is tightly regulated. The activity of autophagy receptors is controlled on multiple levels, such as expression, cellular localization, and posttranslational modifications (PTMs), in particular protein phosphorylation [4]. Phosphorylation plays a central role in the regulation of the autophagy receptors OPTN, p62 (SQSTM1), BNIP3, NBR1, and Nix. Phosphorylation of these autophagy receptors occurs in autophagy-relevant domains, such as the UBD and LIR, and can increase their affinity toward the cargo or the autophagosomal membrane [9–15]. The protein kinases Unc-51 like autophagy activating kinase 1 (ULK1), Tank-binding kinase 1 (TBK1), and casein kinase 2 (CK2) were shown to phosphorylate and regulate autophagy receptors [9–15]. Mass spectrometry-based experiments showed that TBK1 phosphorylates the autophagy receptors OPTN, p62, NDP51, and TAX1BP1 [10, 12]. OPTN is one of the most studied autophagy receptors that functions in the degradation of protein aggregates, mitochondria, and intracellular pathogens such as *Salmonella* [4]. Phosphorylation of OPTN on serine 177 in the LIR increases the affinity of OPTN to LC3, which has been shown to be essential for the clearance of cytosolic *Salmonella* [9]. TBK1 also regulates OPTN during mitophagy by phosphorylation on serine 473 in the UBAN domain, which increases its affinity toward ubiquitin chains on mitochondria. In combination with phosphorylation of serine 177 and 513, this posttranslational modification promotes the recruitment and retention of OPTN to damaged mitochondria [10, 12]. In addition to phosphorylation, OPTN can also be ubiquitylated on lysine 193 by the HECT domain and Ankyrin repeat-containing ubiquitin ligase HACE1 [16], pointing to an intricate regulation of autophagy receptors by different PTMs.

Protein phosphorylation can be detected using phosphorylation-specific antibodies. However, this approach is limited by the available antibodies and does not permit unbiased identification of posttranslational regulatory events. Mass spectrometry-based proteomics has delivered key insights into the regulation of autophagy receptors by phosphorylation [9, 10, 12]. Advances in sample preparation techniques and mass spectrometry instrumentation enabled analysis of protein phosphorylation on individual proteins and proteome-wide [17].

Here, we describe a protocol for quantitative analysis of phosphorylation on autophagy receptors. GFP-tagged autophagy receptors are transiently or stably expressed in human cells. Proteins are extracted from cells and subjected to affinity purification using GFP-Trap agarose. The pulldowns are subsequently washed under denaturing conditions to remove interaction partners and to obtain a pure fraction of the bait protein. The bait protein is then digested on-bead using trypsin, and the resulting peptides are analyzed by high-pressure liquid chromatography coupled to tandem mass spectrometry (LC-MS/MS). Quantitative proteomics approaches, such as stable isotope labeling with amino acids in cell culture (SILAC), chemical labeling (TMT, iTRAQ), or absolute quantification (AQUA), can be employed to quantify the relative or absolute amount of the phosphopeptides in different conditions, respectively [18, 19]. For instance, quantitative phosphoproteomics can be used to monitor the relative phosphorylation site abundance after activation of specific autophagy pathways (e.g., induction of mitophagy) or to determine protein kinase-substrate relations (e.g., overexpression of wild-type and kinase-dead TBK1). The reported protocol is not limited to autophagy receptors and can also be used to analyze other posttranslational modifications, including protein ubiquitylation.

2 Materials

All solutions should be prepared with Milli-Q water of 18.2 MΩ·cm resistivity at 25 °C.

2.1 Stable Isotope Labeling with Amino Acids in Cell Culture (SILAC)

1. Dulbecco's phosphate-buffered saline (DPBS).
2. Cell culture media without arginine and lysine (SILAC media).
3. Amino acids: L-arginine (Arg0), L-lysine (Lys0), L-arginine-U-^{13}C6 99% (Arg6), L-lysine-4,4,5,5,-D4 96–98% (Lys4), L-arginine-U-^{13}C6-^{15}N4 99% (Arg10), L-lysine-U-^{13}C6-^{15}N2 99% (Lys8).
4. Fetal bovine serum (FBS) dialyzed by ultrafiltration with a 10,000 Da molecular weight cutoff.
5. Antibiotics for cell culture (penicillin, streptomycin).
6. L-glutamine.
7. Sodium pyruvate.

2.2 Transfection of Cells

1. Transfection medium: SILAC media without FBS.
2. Polyethylenimine (PEI) "Max," (Mw 4,000*)—high potency linear PEI (Polysciences, product number 9002-98-6) (*see* **Note 1**).

2.3 Cell Lysis

1. Modified RIPA buffer: 50 mM Tris-HCl, pH 7.5, 150 mM NaCl, 1 mM EDTA, 1% IGEPAL CA-630, 0.1% sodium deoxycholate (*see* **Note 2**).
2. Phosphatase inhibitor stock solutions (100×): 500 mM β-glycerophosphate, 500 mM sodium fluoride, 100 mM sodium orthovanadate (*see* **Note 3**).
3. Protease inhibitor cocktail.
4. Quick Start Bradford 1× dye reagent.

2.4 Pulldown with GFP-Trap Agarose

1. Denaturation buffer: 8 M urea in 1× PBS.
2. GFP-Trap_A beads (ChromoTek, product number gta-200).

2.5 On-Bead Trypsin Digestion

1. Sequencing grade trypsin: 0.5 μg/μL in 50 mM acetic acid.
2. Digestion buffer: 2 M urea in 25 mM ABC.
3. Acetic acid.
4. Ammonium bicarbonate (ABC).
5. 1 mL syringe with 27-gauge needle.
6. Chloroacetamide (CAA).
7. Thermomixer.
8. Dithiothreitol (DTT).

2.6 Desalting and Concentration of Peptides

1. C_{18} 47 mm extraction disk (3M, Empore).
2. Hamilton syringe with 16-gauge needle.
3. Buffer A: 0.1% formic acid.
4. Buffer B: 80% ACN, 0.1% formic acid.
5. C_{18} elution buffer: 50% ACN, 0.1% formic acid.
6. Buffer A*: 5% ACN, 0.1% TFA.
7. 96-well plate.
8. Vacuum concentrator.
9. Acetonitrile (ACN).
10. Formic acid.
11. Methanol.
12. Trifluoroacetic acid (TFA).

2.7 Analysis of Peptides by LC-MS/MS

1. Reprosil-Pur Basic C18, 1.9 μm (Dr. Maisch HPLC GmbH).
2. 30 cm fused silica emitter, 360 μm outer diameter, 75 μm inner diameter, 8 μm laser-pulled tip (PicoTip) (New Objective, Inc.).
3. Pressure injection cell.
4. EASY-nLC 1000 liquid chromatograph (Thermo Scientific).
5. Hybrid Quadrupole-Orbitrap mass spectrometer: Q Exactive Plus (Thermo Scientific).

3 Methods

3.1 Stable Isotope Labeling with Amino Acids in Cell Culture (SILAC)

1. Collect cells growing in culture and dispense equal amounts of cells to three tubes, pellet by centrifugation, and wash twice with DPBS to remove residual media.
2. Resuspend cells in light, medium, or heavy SILAC media (*see* **Note 4**).
3. Grow cells in SILAC media for 5–10 doublings (*see* **Note 5**).

3.2 Transfection of Cells

1. Adherent cells should be 60–70% confluent on the day of transfection. Seed cells 1 day before transfection accordingly.
2. Prepare the transfection mix, containing transfection medium, plasmid DNA, and PEI, in a 2 mL tube. Use DNA-to-PEI ratio of 1:10 for HEK293T cells. For transfection of 150 mm plates, use 1500 μL transfection medium, 20 μg DNA, and 200 μg PEI (*see* **Note 6**).
3. Mix gently and incubate the transfection mix for 15 min at room temperature (RT).
4. Replace the medium and reduce the volume to 50% (10 mL for 150 mm plate) to increase transfection efficiency.
5. Add the transfection mix dropwise to the cells.
6. Replace medium 24 h after transfection.

3.3 Cell Lysis

1. 48 h post-transfection, treat cells according to the experimental design. Mock-treated cells should be used as control. A possible experimental design for identification and quantification of autophagy-relevant phosphorylation sites is outlined in Fig. 1a (*see* **Note 7**).
2. Wash adherent cells twice with DPBS. In case of suspension cells, collect cells by centrifugation, and wash twice with DPBS. The following steps should be performed on ice.
3. Add ice-cold modified RIPA buffer freshly supplemented with phosphatase and protease inhibitors to the cells (~500 μL lysis buffer per 1×10^7 cells).

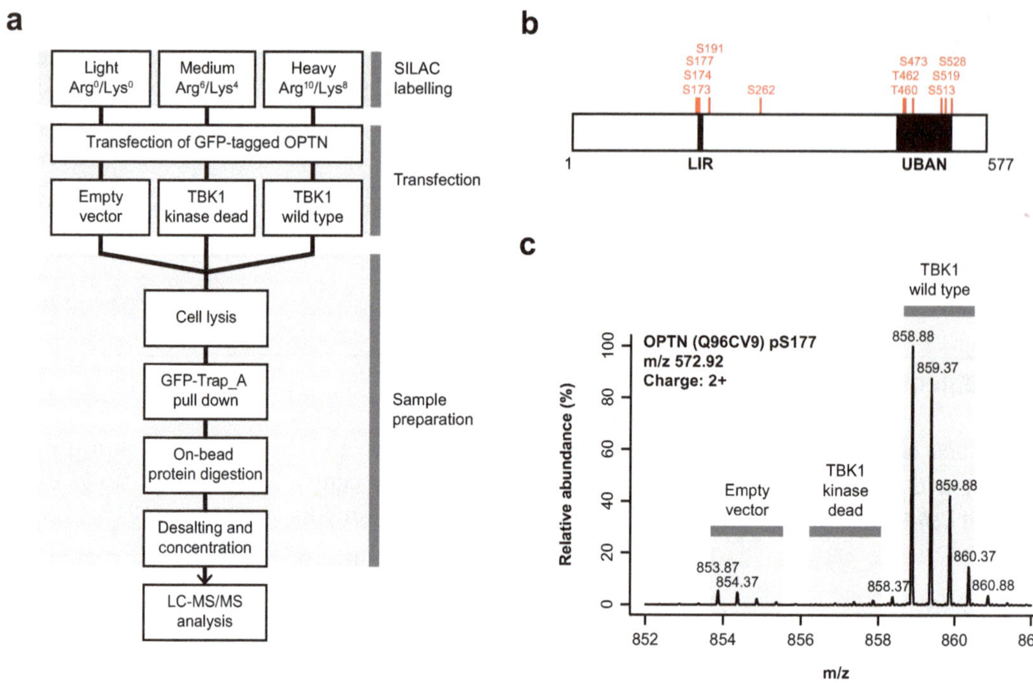

Fig. 1 Mapping and quantification of TBK1-dependent sites on the autophagy receptor OPTN. (**a**) Stable isotope labeling with amino acids in cell culture (SILAC) is employed to determine the relative abundance of the phosphopeptides in different conditions. GFP-tagged OPTN is transiently or stably expressed in all SILAC conditions. Light-labeled cells are co-transfected with an empty vector, medium-labeled cells with kinase-dead TBK1, and heavy-labeled cells with wild-type TBK1. Proteins are extracted from cells and subjected to GFP-Trap agarose pulldowns. After washing of the pulldowns under denaturing conditions, the bait protein (GFP-OPTN) is digested on-bead with trypsin, and peptides are analyzed by LC-MS/MS. (**b**) Schematic representation of the OPTN domain organization and identified phosphorylation sites. TBK1-dependent phosphorylation sites are clustered in the LC3-interacting region (LIR) and in the ubiquitin-binding domain in ABIN proteins and NEMO (UBAN). (**c**) The full scan shows the relative abundance of the phosphorylated peptide corresponding to OPTN phosphorylated on serine 177 in different conditions. Expression of the wild-type TBK1 in heavy-labeled cells results in increased abundance of the phosphorylated peptide

4. Collect cells using a cell scraper in a 15 mL tube (adherent cells).
5. Incubate cell lysates on ice for 10 min.
6. Pellet cell debris by high-speed centrifugation (16,000 × *g*) at 4 °C for 15 min (*see* **Note 8**).
7. Measure protein concentration, and combine equal amounts of protein from each SILAC condition in a 15 mL tube (*see* **Note 9**).

3.4 Pulldown with GFP-Trap Agarose

1. Wash 30 μL of GFP-Trap_A beads three times with 1 mL modified RIPA buffer in a 1.5 mL tube (centrifuge at 2500 × *g* for 45 s in between washes) (*see* **Note 10**).

2. Add the combined protein lysate to the GFP-Trap_A beads, and incubate on a rotation wheel for 1 h at 4 °C in a 15 mL tube.
3. Pellet the beads (1000 × *g*) in a swinging bucket rotor for 1 min, and remove the supernatant.
4. Resuspend the beads in the remaining supernatant, and transfer into a 1.5 mL tube (*see* **Note 11**).
5. Wash the beads once with 1 mL modified RIPA buffer (2500 × *g* for 1 min).
6. Wash the beads three times with denaturation buffer (2500 × *g* for 1 min).

3.5 On-Bead Trypsin Digestion

1. Remove excess liquid from beads using a syringe with 27-gauge needle, and add 100 μL of digestion buffer (*see* **Note 12**).
2. Add DTT to a final concentration of 1 mM, and incubate the samples for 45 min at RT.
3. Add CAA to a final concentration of 5.5 mM. Incubate at RT for 45 min in the dark with gentle shaking (*see* **Note 13**).
4. Digest proteins overnight using 300 ng of trypsin at room temperature with gentle shaking (*see* **Note 14**).
5. Pellet the beads (1000 × *g*, 1 min) and collect carefully the supernatant.
6. Acidify the sample to pH 2 with 0.1% TFA.

3.6 Desalting and Concentration of Peptides

7. Use a 16-gauge Hamilton syringe to cut out two disks from a C_{18} 47 mm extraction disk, and place into a 200 μL pipette tip [20].
8. Wash C_{18} tips once with 25 μL methanol by centrifugation at 400–800 × *g*.
9. Wash C_{18} tips once with 25 μL Buffer B.
10. Wash C_{18} tips twice with 25 μL Buffer A.
11. Load acidified samples onto the C_{18} tips.
12. Wash C_{18} tips once with 50 μL Buffer A (*see* **Note 15**).
13. Elute the peptides directly into a 96-well plate by passing 50 μL C_{18} elution buffer slowly through the C_{18} tips using a syringe.
14. Vacuum concentrate peptides at 45 °C for 25 min to reduce the sample volume to 4.5–5 μL.
15. Add 0.5–1 μL Buffer A* to the sample.

3.7 Analysis of Peptides by LC-MS/MS

1. Pack a nanospray column (15 cm length, 75 μm inner diameter) with C_{18} reversed-phase chromatography material (1.9 μm bead size) using a pressure injection cell [21].

2. Use a nano-flow UHPLC system to load the peptide sample onto the C_{18} column.
3. Elute the peptides with a linear gradient from 5 to 50% ACN in 2 h.
4. The mass spectrometer is operated in data-dependent mode automatically switching between MS and MS^2 acquisition [22, 23].
5. MS spectra (m/z 300–1650) are acquired in the Orbitrap mass analyzer with a resolution of 70,000 at $m/z = 200$ after accumulation of ions to a target value of 3e6 estimated based on predictive automatic gain control from the previous full scan.
6. The 10 most intense ions are isolated using the quadrupole mass filter (maximum injection time 120 ms, isolation window 2.6 m/z, AGC target 1e5) and subsequently fragmented in the higher-energy C-trap dissociation (HCD) cell [24].
7. MS^2 spectra are acquired in the Orbitrap mass analyzer with a resolution of 35,000 at $m/z = 200$.

3.8 Data Analysis

1. Analyze raw MS data using the MaxQuant software [25].
2. Search MS^2 spectra against a database containing protein sequences obtained from the UniProtKB using the Andromeda search engine [26]. Spectra are searched with a parent ion mass tolerance of 6 ppm, fragment ion mass tolerance of 20 ppm, strict trypsin specificity and allowing up to three miscleavages. Cysteine carbamidomethylation is searched as fixed modification, whereas protein N-terminal acetylation, methionine oxidation, and phosphorylation of serine, threonine, and tyrosine are searched as variable modifications.
3. Filter reverse hits and potential contaminants from the MaxQuant output table containing all identified phosphorylation sites.

4 Notes

1. Prepare 1 mg/mL PEI stock solution in water, adjust pH to 7.5, aliquot, and store at −20 °C.
2. Other lysis buffers that are more suitable for a studied cell line or protein can also be used. Consult the GFP-Trap_A datasheet regarding buffer compatibility.
3. Commercial phosphatase inhibitor cocktails can be used as well.
4. For preparation of SILAC media, supplement lysine- and arginine-free media with dialyzed FBS and antibiotics.

L-glutamine and sodium pyruvate can be added if required. Add amino acids to a final concentration of 28 and 48.7 mg/L of arginine and lysine, respectively. This concentration corresponds to 1/3 of the concentration of arginine and lysine in DMEM media and is well tolerated by commonly used cell lines such as HEK293T, HeLa, and U-2 OS. For other cell lines, the optimal conditions have to be determined experimentally [27]. Chemical labeling strategies based on isobaric tags such as TMT and iTRAQ can also be used for the relative quantification of phosphopeptides [12]. Absolute quantification approaches (e.g., AQUA) that use isotopically labeled reference peptides can be employed to determine the absolute amounts of phosphopeptides in different conditions [12, 19].

5. The time required for complete labeling of cellular proteins with isotope-containing amino acids depends on the cell line. Testing the incorporation of isotope-containing amino acids by LC-MS/MS is recommended before proceeding with the experiment [28].
6. DNA-to-PEI ratio should be optimized for each cell line. For HEK293T and HCT116 cells a ratio of 1:10 and for U-2 OS cells a ratio of 1:3 yield good transfection rates. PEI transfection is cost-effective for large-scale transfections. However, other transfection reagents and methods might be more suitable depending on the used cell line. Cells stably expressing a GFP-tagged protein of interest can be used as well.
7. Biological replicate experiments should be performed to be able to determine reproducibly regulated phosphorylation sites.
8. To test the expression and correct molecular weight of the GFP-tagged protein, take a small aliquot of protein lysate for Western blotting.
9. The recommended starting amount is 7–9 mg of protein (~3 mg of protein per SILAC condition in case of a triple SILAC experiment). The amount of the starting material depends on the expression level of the bait protein.
10. Resuspend the slurry properly using a cut pipette tip.
11. To minimize the loss of beads during the transfer, rinse the 15 mL tube with modified RIPA buffer.
12. If other proteases than trypsin are used, the concentration of urea in digestion buffer should be adjusted.
13. Alkylation of cysteines with CAA is required to prevent unspecific side reactions.
14. Before the start of the experiment, use in silico digestion tools (e.g., UniProt PeptideCutter) to test that tryptic digestion of the protein of interest yields peptides that can be analyzed

using electrospray ionization mass spectrometry (peptide length ~5–40 amino acids). In addition to trypsin, other proteases such as GluC or LysC can be used to cover protein regions that are not adequately digested by trypsin.

15. Peptide-loaded C_{18} tips can be stored for several months at 4 °C.

Acknowledgments

This work is supported by the German Research Foundation (Emmy Noether Program, BE 5342/1-1 and SFB 1177 on Selective Autophagy).

References

1. Levine B, Kroemer G (2008) Autophagy in the pathogenesis of disease. Cell 132:27–42. https://doi.org/10.1016/j.cell.2007.12.018
2. Mizushima N, Yoshimori T, Ohsumi Y (2011) The role of Atg proteins in autophagosome formation. Annu Rev Cell Dev Biol 27:107–132. https://doi.org/10.1146/annurev-cellbio-092910-154005
3. Yang Z, Klionsky DJ (2010) Eaten alive: a history of macroautophagy. Nat Cell Biol 12:814–822. https://doi.org/10.1038/ncb0910-814
4. Stolz A, Ernst A, Dikic I (2014) Cargo recognition and trafficking in selective autophagy. Nat Cell Biol 16:495–501. https://doi.org/10.1038/ncb2979
5. Rogov V, Dötsch V, Johansen T et al (2014) Interactions between autophagy receptors and ubiquitin-like proteins form the molecular basis for selective autophagy. Mol Cell 53:167–178. https://doi.org/10.1016/j.molcel.2013.12.014
6. Behrends C, Sowa ME, Gygi SP, Harper JW (2010) Network organization of the human autophagy system. Nature 466:68–76. https://doi.org/10.1038/nature09204
7. Deng Z, Purtell K, Lachance V et al (2017) Autophagy receptors and neurodegenerative diseases. Trends Cell Biol 27:491. https://doi.org/10.1016/j.tcb.2017.01.001
8. Wild P, McEwan DG, Dikic I (2014) The LC3 interactome at a glance. J Cell Sci 127:3–9. https://doi.org/10.1242/jcs.140426
9. Wild P, Farhan H, McEwan DG et al (2011) Phosphorylation of the autophagy receptor optineurin restricts Salmonella growth. Science 333:228–233. https://doi.org/10.1126/science.1205405
10. Richter B, Sliter DA, Herhaus L et al (2016) Phosphorylation of OPTN by TBK1 enhances its binding to Ub chains and promotes selective autophagy of damaged mitochondria. Proc Natl Acad Sci 113:4039–4044. https://doi.org/10.1073/pnas.1523926113
11. Rogov VV, Suzuki H, Marinković M et al (2017) Phosphorylation of the mitochondrial autophagy receptor nix enhances its interaction with LC3 proteins. Sci Rep 7:1131. https://doi.org/10.1038/s41598-017-01258-6
12. Heo J-M, Ordureau A, Paulo JA et al (2015) The PINK1-PARKIN mitochondrial Ubiquitylation pathway drives a program of OPTN/NDP52 recruitment and TBK1 activation to promote mitophagy. Mol Cell 60:7–20. https://doi.org/10.1016/j.molcel.2015.08.016
13. Matsumoto G, Wada K, Okuno M et al (2011) Serine 403 phosphorylation of p62/SQSTM1 regulates selective autophagic clearance of ubiquitinated proteins. Mol Cell 44:279–289. https://doi.org/10.1016/j.molcel.2011.07.039
14. Pilli M, Arko-Mensah J, Ponpuak M et al (2012) TBK-1 promotes autophagy-mediated antimicrobial defense by controlling autophagosome maturation. Immunity 37:223–234. https://doi.org/10.1016/j.immuni.2012.04.015
15. Lim J, Lachenmayer ML, Wu S et al (2015) Proteotoxic stress induces phosphorylation of p62/SQSTM1 by ULK1 to regulate selective autophagic clearance of protein aggregates.

PLoS Genet 11:e1004987. https://doi.org/10.1371/journal.pgen.1004987

16. Liu Z, Chen P, Gao H et al (2014) Ubiquitylation of autophagy receptor optineurin by HACE1 activates selective autophagy for tumor suppression. Cancer Cell 26:106–120. https://doi.org/10.1016/j.ccr.2014.05.015
17. Cox J, Mann M (2011) Quantitative, high-resolution proteomics for data-driven systems biology. Annu Rev Biochem 80:273–299. https://doi.org/10.1146/annurev-biochem-061308-093216
18. Bantscheff M, Lemeer S, Savitski MM, Kuster B (2012) Quantitative mass spectrometry in proteomics: critical review update from 2007 to the present. Anal Bioanal Chem 404:939–965. https://doi.org/10.1007/s00216-012-6203-4
19. Ordureau A, Münch C, Harper JW (2015) Quantifying ubiquitin signaling. Mol Cell 58 (4):660–676. https://doi.org/10.1016/j.molcel.2015.02.020
20. Rappsilber J, Mann M, Ishihama Y (2007) Protocol for micro-purification, enrichment, pre-fractionation and storage of peptides for proteomics using StageTips. Nat Protoc 2:1896–1906. https://doi.org/10.1038/nprot.2007.261
21. Ishihama Y, Rappsilber J, Andersen JS, Mann M (2002) Microcolumns with self-assembled particle frits for proteomics. J Chromatogr A 979:233–239
22. Michalski A, Damoc E, Hauschild JP et al (2011) Mass spectrometry-based proteomics using Q Exactive, a high-performance benchtop quadrupole Orbitrap mass spectrometer. Mol Cell Proteomics 10:M111.011015. https://doi.org/10.1074/mcp.M111.011015
23. Kelstrup CD, Young C, Lavallee R et al (2012) Optimized fast and sensitive acquisition methods for shotgun proteomics on a quadrupole orbitrap mass spectrometer. J Proteome Res 11:3487–3497. https://doi.org/10.1021/pr3000249
24. Olsen JV, Macek B, Lange O et al (2007) Higher-energy C-trap dissociation for peptide modification analysis. Nat Methods 4:709–712. https://doi.org/10.1038/nmeth1060
25. Cox J, Mann M (2008) MaxQuant enables high peptide identification rates, individualized p.p.b.-range mass accuracies and proteome-wide protein quantification. Nat Biotechnol 26:1367–1372. https://doi.org/10.1038/nbt.1511
26. Cox J, Neuhauser N, Michalski A et al (2011) Andromeda: a peptide search engine integrated into the MaxQuant environment. J Proteome Res 10:1794–1805. https://doi.org/10.1021/pr101065j
27. Bendall SC, Hughes C, Stewart MH et al (2008) Prevention of amino acid conversion in SILAC experiments with embryonic stem cells. Mol Cell Proteomics 7:1587–1597. https://doi.org/10.1074/mcp.M800113-MCP200
28. Cox J, Matic I, Hilger M et al (2009) A practical guide to the MaxQuant computational platform for SILAC-based quantitative proteomics. Nat Protoc 4:698–705. https://doi.org/10.1038/nprot.2009.36

Chapter 47

Analysis of Chaperone-Mediated Autophagy

Y. R. Juste and A. M. Cuervo

Abstract

Chaperone-mediated autophagy (CMA) is a selective type of autophagy whereby a specific subset of intracellular proteins is targeted to the lysosome for degradation. These proteins are identified by a chaperone that targets them to lysosomes. There, they are translocated into the organelle lumen through a lysosomal membrane receptor/translocation complex. CMA plays an important role in maintaining cellular proteostasis by eliminating damaged and altered proteins. CMA also participates in the control of the cellular energetic balance through recycling of amino acids resulting from lysosomal proteolysis of the substrate proteins. Lastly, due to the intrinsic protein selectivity of CMA, this type of autophagy exerts regulatory functions by mediating timely degradation of key cellular proteins that participate in processes such as lipid and glucose metabolism, cell cycle, DNA repair, and cellular reprogramming, among others. Dysfunctional CMA occurs with age and has now been described in a growing list of human pathologies such as metabolic disorders, neurodegeneration, cancer, immunodeficiency, and diabetes. In this chapter, we describe current methodologies to quantitatively analyze CMA activity in different experimental models.

Key words Chaperones, Lysosomes, Proteolysis, Subcellular fractionation

1 Introduction

Chaperone-mediated autophagy (CMA) is a multistep process that results in selective degradation of intracellular soluble proteins [1]. Selectivity is driven by the presence of a CMA-targeting motif—a pentapeptide sequence sharing biochemical similarity to KFERQ—in the substrate proteins [2]. This motif is recognized by a constitutively expressed intracellular chaperone, the heat shock-cognate chaperone of 70 kDa (hsc70) (Fig. 1) [3]. Once hsc70 binds the substrate proteins, they are targeted to the surface of a subset of lysosomes active for this autophagic pathway. At the lysosomal membrane, the substrate/chaperone complex docks at the cytosolic tail of a monomeric single-span protein termed lysosome-associated membrane protein type-2A (LAMP2A) [4]. Binding of substrates to LAMP2A initiates its multimerization at the lysosomal membrane, which comprises the basis of the CMA translocation complex [5]. The unfolding of the substrate protein is

Nicholas Ktistakis and Oliver Florey (eds.), *Autophagy: Methods and Protocols*, Methods in Molecular Biology, vol. 1880, https://doi.org/10.1007/978-1-4939-8873-0_47,

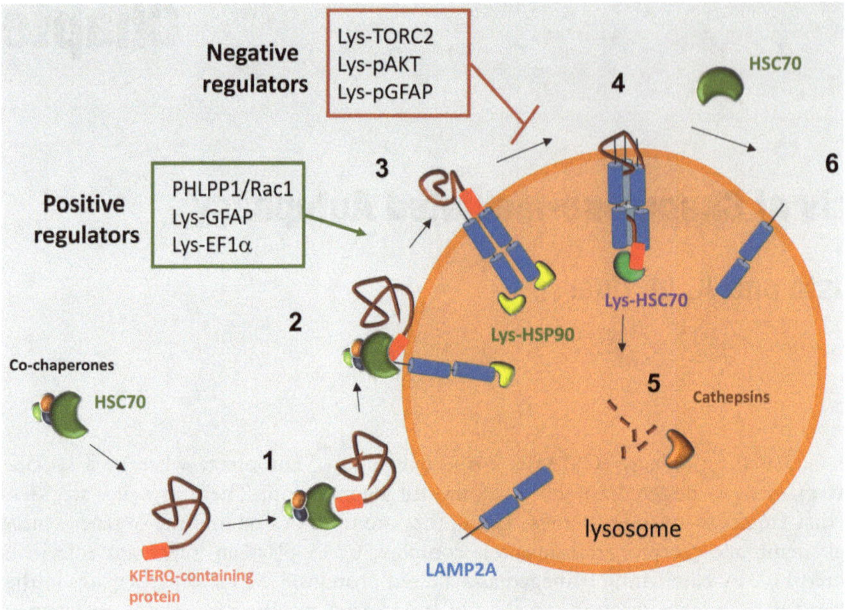

Fig. 1 Chaperone-mediated autophagy: schematic model of the steps in chaperone-mediated autophagy. 1. Substrate binding by HSC70 and cochaperones and targeting to lysosomes. 2. Binding of the substrate to LAMP2A at the lysosomal membrane. 3. HSP90 binds to LAMP2A to stabilize it while it organizes into higher molecular weight complexes. 4. Substrate crosses the lysosomal membrane through a LAMP2A-enriched translocation complex, and translocation is complete by the action of luminal HSC70. 5. The substrate is rapidly degraded by luminal proteases (cathepsins). 6. Once substrate translocation is complete, LAMP2A dissociates into monomers in a process dependent on cytosolic HSC70. Red box: negative regulators of CMA at the lysosomal membrane. Green box: positive regulators of CMA at the lysosomal membrane

not required for its binding to the chaperone or to the lysosomal surface, but it is a prerequisite for substrate translocation across the lysosomal membrane. Substrate internalization is mediated by a lysosome-resident hsc70 (lys-hsc70) [6], and it is rapidly followed by complete degradation of the substrate protein into its constitutive amino acids by lysosomal luminal proteases, also known as cathepsins.

In addition to these proteins that interact directly with the substrate protein—hsc70 on both side of the lysosomal membrane and LAMP2A—the lysosomal membrane also hosts regulators that modulate CMA activity directly at this compartment. A lysosomal resident form of hsp90 is involved in maintaining the stability of LAMP2A during its multimerization (Fig. 1) [5]. A pair of regulators GFAP/EF1α controls the stability of the LAMP2A translocation complex in a GTP-dependent manner [7]. GFAP exists in the membrane in two forms: unmodified GFAP that associates with the LAMP2A multimer to stabilize the translocation complex and phosphorylated GFAP (pGFAP) that is masked by EF1α [7]. In the presence of GTP, EF1α is released from the membrane allowing pGFAP to become accessible to unmodified GFAP [7]. Affinity of

GFAP to form dimers with pGFAP is higher than its binding to LAMP2A; therefore, it dissociates from the translocation complex and binds to pGFAP, resulting in the disassembly of the translocation complex. A second mechanism that controls the dynamics of LAMP2A assembly into the translocation complex involves mTOR complex 2 or TORC2, the kinase AKT, and the phosphatase PHLPP1 [8]. TORC2 provides a continuous inhibitory regulation on CMA through phosphorylation of AKT, that is, the kinase of GFAP at the lysosomal membrane [8]. Continuous phosphorylation of GFAP makes the speed of assembly/disassembly of LAMP2A from the translocation complex constitutively slow (Fig. 1). Whenever maximal activation of CMA is required, this inhibitory effect of TORC2 needs to be released and that is attained through the recruitment of PHLPP1, in a Rac-1-dependent manner, to the lysosomal membrane [8]. PHLPP1 dephosphorylates AKT and that accelerates LAMP2A assembly/disassembly and consequently the rate(s) of substrate(s) uptake (Fig. 1) [8].

In contrast with other types of autophagy, such as macroautophagy or microautophagy that requires formation of vesicles for the uptake of the substrates to be degraded, in CMA, substrate proteins are recognized individually, and they are translocated into the lysosomal lumen independent of vesicular trafficking [1]. Selective targeting of individual proteins for degradation has been reported in a type of microautophagy, known as endosomal microautophagy (e-MI) [9]. In this process, the chaperone hsc70 also recognizes the same KFERQ-like motif in the substrate proteins but delivers them to the surface of late endosomes. There, cargo is internalized in multivesicular bodies (MVB) that form on the surface of this organelle using components of the ESCRT complex [9]. An important difference with CMA is that, despite the abundance of LAMP2A in the late endosomal membrane, e-MI does not use this membrane protein for substrate internalization. Additionally, e-MI substrates do not need to undergo unfolding prior to internalization into MVB luminal vesicles [9], while CMA substrates must be unfolded for internalization into the lysosomal lumen [10].

The selectivity of CMA seems beneficial under conditions in which discrimination between different types of proteins for degradation is required. For example, an increase in CMA activity is observed during prolonged starvation [11]. Degradation of proteins through CMA will provide cells with free amino acids required to sustain protein synthesis under these conditions [11, 12]. Likewise, activation of CMA during mild oxidative stress or after exposure to compounds that decrease proteostasis allows the selective removal of the proteins damaged or altered under these conditions [13]. In addition, selective removal of proteins through CMA has been shown to exert important regulatory functions in metabolic pathways, DNA repair pathways, and cell cycle, among others [14–16].

Malfunctioning of CMA plays a key role in an increasing amount of severe human disorders [17–20]. In many instances, the mechanisms underlying CMA failure in these pathologies involve perturbations in the functioning of the CMA translocation complex in lysosomes. The interplay of CMA with several neurodegenerative diseases is bidirectional, whereby CMA contributes to the elimination of pathogenic proteins, but it eventually becomes a casualty of the toxic effect of these aberrant proteins [18, 20]. Given this dual role of CMA in neurodegeneration, and the growing number of diseases associated to CMA failure, it has become important to analyze its status in disease conditions to understand if enhancing CMA activity could be a worthwhile endeavor in treating these diseases.

Analysis of the levels of CMA effectors and modulators in lysosomes can yield useful insights into CMA activity in tissues and cells in culture. However, to get an accurate picture of CMA activity, it is necessary to track the targeting and translocation of CMA substrates into lysosomes. To that effect, both in vitro systems with isolated lysosomes and a photoswitchable CMA reporter that works in intact cells are the gold standard methods in the field. In this chapter, we describe the battery of markers that can be used to obtain information on the steady-state status of CMA as well as CMA functional assays and their applicability to different experimental models.

2 Materials

2.1 Isolation of Rat Liver Lysosomes

1. Wistar rats (200–250 g): Male or female can be selected depending on the purpose of the experiment. To enrich in CMA-active lysosomes and reduce liver glycogen content (as it can interfere with lysosomal isolation), rats can be starved for 24–48 h before liver dissection but should be maintained with water ad libitum.
2. Tools: Dissection instruments (forceps, scissors, clamps), double cloth gauze, funnel, Teflon/glass homogenizer (for motorized homogenizer).
3. Centrifugation supplies: Polycarbonate tubes (30 mL), ultraclear tubes for SW41 rotor (Beckman, Fullerton, CA), and SW41 rotor (Beckman).
4. Homogenization solution: 0.25 M sucrose (American Bioanalytical, Natick, MA) in double-distilled water (ddH_2O). Prepare fresh or the day before, and store at 4 °C.

5. Centrifugation media: Metrizamide (Fitzgerald Industries International, Acton, MA). Prepare as 85.6% (w/v) stock in ddH_2O, and adjust to pH 7.2 with NaOH. Store in aliquots at −20 °C protected from light (*see* **Note 1**).
6. Lysosome resuspension buffer: 10 mM 3-(N-Morpholino) propanesulfonic acid (MOPS), 0.25 M sucrose, pH 7.2 (adjusted with NaOH). Store at 4 °C for 1 week maximum.

2.2 Lysosome Purity and Integrity

1. Millipore multiscreen assay system (Millipore, Bedford, MA), 0.22 μm Durapore filter 96-well plates, vacuum manifold and polystyrene flat-bottom 96-well plates.
2. Acetate buffer: 0.4 M sodium acetate, pH 4.4.
3. β-hexosaminidase substrate solution: 4 mM 4-methylumbelliferyl-N-acetyl-B-D-glucopyranoside in ddH_2O. Sonicate to dissolve and store a −20 °C protected from light. Before use, the solution can be sonicated again and kept at 37 °C until use.
4. 10% Triton X-100 (Bio-Rad, Hercules, CA) in ddH_2O and store at room temperature.
5. Reaction mixture: 10 mL acetate buffer, 10 mL β-hexosaminidase substrate solution, 0.5 mL 10% Triton X-100, 19.5 mL ddH_2O. Store at −20 °C and thaw by placing in a 37 °C before use.
6. Stop solution: 0.5 M glycine, 0.5 M Na_2CO_3, in ddH_2O.
7. Blocking solution: 20 mg/mL bovine serum albumin (BSA) in ddH_2O.
8. 0.25 M sucrose (American Bioanalytical, Natick, MA): Prepare fresh in ddH_2O.

2.3 Lysosomal Binding/Uptake Assay

1. Incubation buffer: 10 mM MOPS, 0.25 M sucrose in ddH_2O adjusted to pH 7.3. Prepare fresh and store at 4 °C for 1 week maximum.
2. Chymostatin (Sigma): prepare as 10 mM stock, store at −20 °C, and dissolve in incubation buffer before use.
3. Proteinase K (Sigma): prepare as 5 mg/mL stock in 10 mM Tris–HCl pH 7.5, 1 mM $CaCl_2$. Store at −20 °C, and dissolve in incubation buffer before use.
4. 4-(2-Aminoethyl)benzenesulfonyl fluoride hydrochloride (AEBSF, American Analytical): dissolve in ddH_2O as 1 mM stock and store at −20 °C.
5. Reagents for standard SDS-PAGE and immunoblot.

2.4 Protein Degradation Assay

1. Millipore multiscreen assay system: 0.22 μm Durapore filter 96-well plates, vacuum manifold, and polystyrene flat-bottom 96-well plates.

2. Proteolysis buffer: 10 mM MOPS pH 7.3, 1 mM DTT, 5.4 μm cysteine, 0.25 M sucrose in ddH_2O adjusted to pH 7.3. Prepare fresh and store at 4 °C for 1 week maximum. Cysteine and DTT should be added right before use.
3. Trichloroacetic acid (America Bioanalytical): 20% in ddH_2O and store at room temperature.
4. BSA dissolved at 20 mg/mL in in ddH_2O and stored at 4 °C.

2.5 Dynamics of CMA Translocation Complex

1. Electrophoresis apparatus: Invitrogen XCell Sure Lock Running Apparatus.
2. NativePAGE 3–12% Bis-Tris gel (Invitrogen).
3. Anode buffer: prepare 1 L by adding 50 mL of 20× NativePAGE running buffer (Invitrogen) to 950 mL of ddH_2O.
4. Cathode buffer: prepare 200 mL by adding 10 mL of running buffer, 10 mL of NativePAGE cathode additive, and 180 mL of ddH_2O. For a lighter shade of blue, 1 mL of additive can be used instead.
5. NativePAGE Sample Prep Kit (Invitrogen).
6. High molecular weight native marker, Amersham (GE Healthcare Life Science).
7. Solubilizing solution: 20 mM MOPS, 150 mM NaCl, and 1% octylglucoside powder, pH 7.4 in ddH_2O. Store at 4 °C.
8. Coomassie fixing solution: 40% (v/v) methanol, 10% (v/v) acetic acid; fill to 100 mL with ddH_2O. Store at room temperature.
9. Coomassie and BLOT destaining solution: 8% acetic acid in ddH_2O. Store at room temperature.

2.6 Cell Immunofluorescence

1. Cell culture medium: Dulbecco's modified Eagle's medium (DMEM) (Sigma) supplemented with 10% newborn calf serum (NCS). The medium should be adapted depending on the cell type requirements.
2. Microscope cover slips (22 × 22 mm).
3. Phosphate-buffered saline (PBS): 1.37 M NaCl, 0.03 M KCl, 0.07 M Na_2HPO_4, 0.11 M K_2HPO_4 pH 7.4. Store at room temperature.
4. Methanol fixing solution: 100% methanol placed at −20 °C at least 4 hours prior to fixing cells.
5. Blocking solution: 0.2 (w/v) powdered nonfat milk, 2% NCS, 0.1 M glycine, 1% BSA, and 0.01% Triton X-100 in PBS. Prepare fresh and maintain at 4 °C until use.
6. Primary antibodies: *See* Table 1 for source and dilutions. Please note that IgG rabbit anti-LAMP2A was originally developed in

Table 1
Antibodies or CMA-related proteins and recommended dilutions

Antigen	Type	Source/cat#	Dilution (IB)	Dilution (IF)
AKT	Rabbit pAb IgG	Cell signaling [9272]	1:1000	1: 200
pAKT (ser473)	Rabbit pAb IgG	Cell signaling [9271]	1:1000	1: 25
Cath A (A-19)	Goat pAb IgG	Santa Cruz [sc-26049]	1:500	N/A
Cath D	Goat pAb IgG	Santa Cruz [sc-6486]	1:500	N/A
EF1α	Mouse mAb IgG	Millipore [05-235]	1:1000	N/A
GAPDH	Rabbit mAb IgG	Cell signaling [2118]	1:1000	1:100
GFAP	Mouse mAb IgG	Millipore [MAB360]	1:1000	1:200
pGFAP	Rabbit pAb IgG	ABGENT [AP3562a]	1:1000	N/A
HSC70 (13D3)	Mouse mAb IgM	Novus biological [NB120-2788]	1:5000	1:500
HSP90	Rat mAb IgG	ENZO	1:10,000	N/A
LAMP-1 [H4A3]	Mouse mAb IgG	Abcam [ab25630]	1:3000	1:100
LAMP2A	Rabbit pAb IgG	ThermoFisher [51-2200]	1:1000	1:200
mTOR	Rabbit pAb IgG	Cell signaling [2972]	1:1000	N/A
Rictor	Rabbit mAb IgG	Cell signaling [2114]	1:1000	N/A
Rac1	Mouse mAb IgG	Millipore	1:1000	N/A
Ribonuclease A	Rabbit pAb IgG	Rockland immunochemicals	1:10,000	N/A

pAb polyclonal antibody, *mAb* monoclonal antibody

our laboratory [4] and is now available through Invitrogen (cat# 51-2200) (note that most commercial antibodies are developed against the luminal part of LAMP2 and recognize the three isoforms (LAMP2A, B, and C)). For hsc70, we recommend using IgM mouse monoclonal anti-hsc70 antibody clone 13D3 (available through several vendors) because most commercial antibodies recognize both hsp70 and hsc70, but clone 13D3 has been well characterized as specific for hsc70.

7. Secondary antibodies: Fluorophores are selected depending on the combination of primary antibodies used, but common ones used in these procedures are Alexa Fluor488 goat-conjugated anti-mouse IgM antibody (ThermoFisher Scientific) (for anti-hsc70) and Alexa Fluro555 goat anti-rabbit IgG (ThermoFisher Scientific) (for anti-LAMP2A).
8. Mounting media: SlowFade Diamond Antifade Kit with DAPI (ThermoFisher scientific).

2.7 Photoconvertible CMA Reporter

1. Transfection/transduction reagents: For transient transfection with the plasmid containing the KFERQ-reporter use Lipofectamine 2000 (ThermoFisher scientific) and follow manufacturer's instructions. For lentiviral-mediated stable expression, transduce cells using polybrene/transfection reagent (Sigma) (10 mg/mL stock solution), store at −20 °C, and dilute 1:1000 in culture media before use.
2. DMEM supplemented with 10% NCS.
3. Light-emitting diode (LED) at 405 nm wavelength.
4. Microscope cover slips (22 × 22 mm).
5. Phosphate-buffered saline (PBS) 1.37 M NaCl, 0.03 M KCl, 0.07 M Na_2HPO_4, 0.11 M K_2HPO_4 pH 7.4. Store at room temperature.
6. Paraformaldehyde fixing solution (PFA): Prepare as 4% PFA in PBS.
7. Mounting media: SlowFade Diamond Antifade Kit with DAPI (ThermoFisher Scientific).

2.8 Modulation of CMA in Cultured Cells

1. Serum deprivation: Dulbecco's modified Eagle's medium (DMEM) (Sigma) without additions. Washing solution (PBS).
2. Oxidative stress: H_2O_2 prepared fresh to a final concentration of 100 μM (dilute with culture media); paraquat prepared fresh to a final concentration of 40 μM (dilute with culture media). Note: final concentration varies depending on the cell type. Concentrations indicated here effectively induce CMA in mouse fibroblasts.
3. CMA chemical activator: Atypical retinoid 7 (AR7) (originally developed by our laboratory [21] and now commercially available (Sigma). Prepare as 10 mM stock in DMSO, and store at −20 °C until use. Dilute in serum-free DMEM to working solution for a final concentration of 5–20 μM (depending on the cell type).
4. Inhibitors of lysosomal proteolysis: 2 M NH_4Cl prepared fresh in ddH_2O for a final concentration of 10–20 mM; 10 mM

leupeptin stock solution in ddH_2O, store at −20 °C until use, and dilute in culture media to a final concentration of 100–200 μM.

2.9 Measurement of CMA In Vivo

1. Rats (Wistar or any other strain of interest) or mice (C57BL/6 mice or any other strain of interest).
2. 1 mL TB syringe, slip tip with BD PrecisionGlide Needle (Sigma).
3. 25–30 gauge needles.
4. Leupeptin prepared in sterile saline (9 g/L NaCl) for a final concentration of 2 mg per 100 g body weight. To avoid injecting large volumes, prepare at a concentration that requires injection of 200–300 μL of solution. Prepare fresh.

3 Methods

The two most common reasons that motivate the study of CMA are (1) the analysis of changes in the activity of this autophagic pathway in different conditions or in response to different interventions and (2) the interest in determining if a specific protein undergoes degradation through this autophagic pathway. In this chapter, we first detail methods to directly assess CMA activity (independently of the substrate degraded), and in the last section, we briefly summarize the array of procedures to test if a protein is a CMA substrate.

3.1 Measuring CMA Activity In Vitro

3.1.1 Isolation of Rat Liver Lysosomes

1. Rinse the liver from a 24-h-starved rat extensively with 4 °C cold 0.25 M sucrose to remove any residual blood (*see* **Note 2**). Weigh the liver, and mince in a 50 mL plastic conical tube, making sure you act quickly and keep the tube on ice to prevent unwanted proteolysis. Add 3 volumes of cold 0.25 M sucrose/g of liver.
2. Transfer the minced liver into the glass homogenizer, and homogenize using a Teflon pestle and a motorized homogenizer with 8–10 strokes at maximum speed (this is done, if possible, in a cold room or alternatively keep the glass homogenizer inside an ice bucket).
3. Filter the homogenate through a double cotton gauze, and add 4 volumes/g liver of cold 0.25 M sucrose. Save a small aliquot (100–200 μL) to use as reference of total liver homogenate.
4. Centrifuge the homogenate at 6800 × *g* for 5 min at 4 °C, and collect the resulting supernatant into a clean tube (be careful to not collect the white layer above the pellet, as these are mainly heavy mitochondria). Discard the post nuclear pellet that contains unbroken cells, plasma membrane, nuclei, and heavy

mitochondria. The best way to collect the supernatant is to directly decant it to the new tube.

5. Centrifuge the supernatant at 17,000 × *g* for 10 min at 4 °C, and resuspend the pellet with a "cold finger" (a glass tube with ice inside and dry outside to avoid including water in the sample). Add 3.5 volume/g starting liver of 0.25 M sucrose solution, and centrifuge again at 17,000 × *g* for 10 min at 4 °C to wash the resuspended pellet (mitochondria/lysosomal fraction), and ensure that any additional cytosolic components incorporated in the initial pellet are released into the supernatant.
6. Discard the supernatant from the previous centrifugation. Resuspend the pellet using the "cold finger" and 0.25 M sucrose until the final volume (including the volume of the resuspended pellet) is 1.1 mL. Add 2 volumes of 85.6% (w/v) metrizamide (2.2 mL) to reach a final concentration of 57% metrizamide, and mix everything carefully with a 5 mL plastic pipette. Load the sample at the bottom of a 14 × 95 mm, thin-wall, ultraclear ultracentrifugation tube (make sure to not touch the walls of the tube) (Fig. 2).
7. Generate a discontinuous metrizamide gradient on top of the 57% metrizamide containing the light mitochondrial/lysosome fraction by overlaying: 2 mL of 32.8% metrizamide, 3.3 mL of 26.3% of metrizamide, and 3.5 mL of 19.8% metrizamide (all diluted in ddH_2O). Fill the tube with 0.25 M sucrose up to 2 mm of the top edge of the tube. Generate the gradient while the bottom part of the tube (the one containing the lysosomes) is inside an ice bucket.
8. Centrifuge in SW 41 swinging bucket rotor at 141,000 × *g* at 4 °C for 1:09 h (setting acceleration to 4 and deceleration to 9). The time of centrifugation at maximal speed is 1 h, and the 9 min are required for acceleration and deceleration.
9. After centrifugation, white to light brown material is clearly visible in each of the interphases. Mitochondria are mostly retained at the lowest interphase of the gradient; the following interphase contains a mixture of mitochondria and lysosomes. A mix population of lysosomes is present in the third interphase from the bottom, and lysosomes with high CMA activity migrate to the top interphase. Collect each of the interphases separately (approx. 3–4 mL) with a Pasteur pipet in 30 mL polycarbonate tubes.
10. Add 5–10 volumes of 0.25 M sucrose solution, and wash by centrifugation at 37,000 × *g* for 15 min at 4 °C.
11. Resuspend the pellet coming from the third interphase using a glass rod to avoid lysosomal damage (a Pasteur pipet with the tip blunted with a flame and left to cool down can be used).

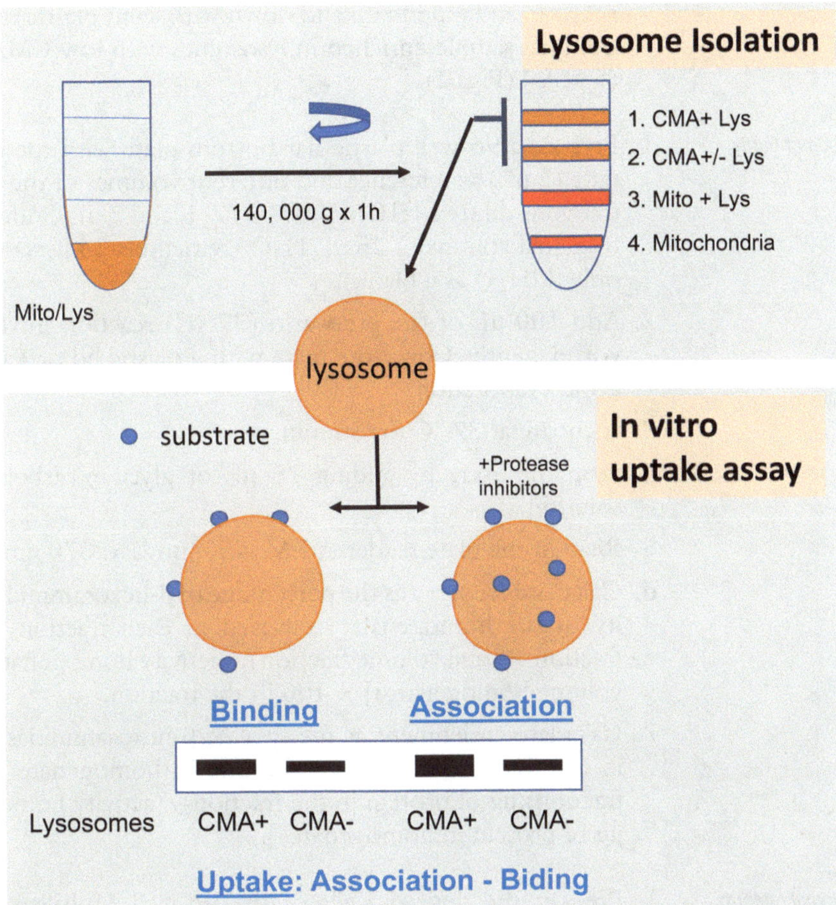

Fig. 2 Measurement of CMA in vitro. Top: lysosomes active for CMA can be isolated by flotation in a discontinuous gradient of metrizamide. Bottom: incubation of CMA substrates with intact lysosomes pre-treated or not with inhibitors of lysosomal proteases allows to quantify substrate binding and translocation (uptake) inside the lysosomal lumen

Once the pellet is disaggregated, add 300 μL of 0.25 M sucrose, pipette up and down with a plastic tip with the end cut to avoid friction during pipetting, and place it into a microfuge tube to centrifuge at 10,000 × *g* for 5 min at 4 °C.

12. Use the supernatant obtained from the centrifugation of the third interphase (containing CMA+ lysosomes) to resuspend the pellet coming from the first interphase applying the same procedure (disaggregate the pellet quickly with a glass rod, and then add the supernatant from the third interphase centrifugation and pipette up and down with a cut plastic tip). This will be the final sample enriched in CMA-active lysosomes (CMA+).
13. Resuspend the pellet from the centrifugation of the third interphase (with a glass rod), and then add 300 μL of 0.25 M

sucrose, and pipette up and down with a cut plastic tip. This is the final sample enriched in lysosomes with low CMA activity (CMA-) (Fig. 2).

3.1.2 Lysosomal Purity

1. Prepare a 96-well plastic flat-bottom plate with the homogenate (2 μL) as reference and different volumes of the collected fractions diluted 1:10 in ddH_2O (2, 5 and 7 μL). Add ddH_2O to a final volume of 25 μL H_2O (remember to use a well with only ddH_2O as a blank).
2. Add 100 μL of the pre-warm (37 °C) reaction mixture, and vortex gently. Cover the plate with a plastic lid or Parafilm to avoid evaporation.
3. Incubate at 37 °C for 30 min.
4. Stop the assay by adding 75 μL of glycine/carbonate stop solution.
5. Read in the plate reader at EM: 450 nm/Ex: 370 nm.
6. Calculate recovery as the percentage of β-hexosaminidase activity of the homogenate recovered in each fraction [(activity fraction × total volume fraction)/(activity homogenate × total volume homogenate)] × 100 in the fraction.
7. Calculate enrichment as the specific β-hexosaminidase activity in the fractions relative to the one in homogenate [(activity fraction/μg of protein in the fraction)/(activity homogenate/μg of protein in homogenate)].

3.1.3 Lysosomal Integrity

1. Pre-wet the desired wells of the 96-well MultiScreen-Mesh filter plate with a Durapore (PVDF) filter (0.22 μm) by filling with ddH_2O (5 min, RT).
2. Shake the water out, and block with 20 mg/mL BSA in ddH_2O (approx. 100 μL) for 30 min at RT.
3. Filter the blocking solution with the vacuum manifold.
4. Add 30 μL of 0.25 M sucrose per well.
5. Add 15 μL of lysosomes (directly from the gradient), and include a blank with only ddH_2O.
6. Apply vacuum for 30 s to 1 min until all the solution is filtered and collected in a 96-well plastic plate.
7. Take the flow through, and assay for β-hexosaminidase as in the previous section (samples can be stored at −20 °C and assayed for activity later).

3.2 In Vitro Reconstitution of CMA

3.2.1 Lysosomal Binding/Uptake

1. Incubate freshly isolated rat liver lysosomes with 100 μM chymostatin for 10 min on ice. Depending on the substrate, a protease inhibitor cocktail can be used instead (recommended combination: 100 μM leupeptin, 100 μM AEBSF, 10 μM pepstatin and 1 mM EDTA).
2. Carry out transport assay in 500 μL microfuge tubes by adding freshly isolated rat liver lysosomes treated or not with protease inhibitor (50–100 μg protein in a volume of 10 μL) and GAPDH or another protein of interest (50 μg) in a final volume of 30 μL of MOPS buffer (*see* **Note 3**). Incubate samples for 20 min at 37 °C (Fig. 2).
3. At the end of the incubation (*see* **Note 4**), centrifuge samples at 25,000 × *g* for 5 min at 4 °C.
4. Aspirate the supernatants, and wash the pellets with 100 μL of incubation buffer carefully to not disturb the pellets.
5. Resuspend the pellet in 20 μL of Laemmli sample buffer with protease inhibitors, boil for 5 min at 95 °C, and analyze by SDS-PAGE and immunoblot for GAPDH. A lane with 1/10 of the amount of GAPDH should be included in the immunoblot for quantification purposes.
8. Perform densitometric analysis on the immunoblots, and use a generated standard curve for each antibody using increasing concentrations of antigen to determine the linear range. These values allow the calculation of (1) binding as the percentage of total added GAPDH associated to lysosomes untreated with protease inhibitors, (2) association as the percentage of GAPDH recovered in lysosomes treated with protease inhibitors, and (3) uptake, calculated as the difference between association and binding.

3.2.2 Intact Lysosomes Protein Degradation Assay

1. This assay is carried out in 96-well MultiScreen-Mesh filter plate with a Durapore (PVDF) filter (0.22 μm) pre-wet with ddH_2O for 10 min at room temperature. After aspirating the water, add to the wells the 20 μL MOPS/DTT proteolysis buffer (*see* **Notes 3** and **5**).
2. Add freshly isolated rat liver lysosomes (25 μg protein) (10 μL of a 1:4 dilution in proteolysis buffer) per well and 10 μL of the radiolabeled protein cocktail (2000 dpm/μL), and adjust the final volume to 60 μL with proteolysis buffer. Incubate for 30 min at 37 °C. One well should contain the same reagents, except for the lysosomes, to determine the amount of spontaneous proteolysis (autolysis) that will be used as a blank.
3. At the end of the incubation, add 90 μL of 20% TCA and 30 μL of 20 mg/mL BSA to each well to stop the reaction. Incubate at 4 °C for at least 30 min to allow for protein precipitation.

4. Collect the acid-soluble radioactivity in a polystyrene 96-well plate using the multiscreen vacuum system. Collect the flow through (acid-soluble) from each sample and the filter retained material (acid-precipitable) in separate scintillation vials for dpm counting.
5. Proteolysis is calculated as the percentage of total protein (radioactivity precipitated by acid at time 0) transformed into amino acids and small dipeptides (radioactivity that remains soluble after adding the acid stop solution) at the end of the incubation. The amount of spontaneous hydrolysis (calculated from the blank well) needs to be discounted following the formula: [((dpm flow through sample − dpm flow through blank)/dpm pellet time 0) × 100] (*see* **Note 6**).

3.2.3 Dynamics of the CMA Translocation Complex

1. Resuspend freshly isolated lysosomes (pellet from the centrifugation at 37,000 × *g* for 15 min) in 1% octyl glycoside in MOPS pH 7.3 supplemented with protease inhibitors, and incubate on ice for 15 min (*see* **Note 7**).
2. Spin at 16,000 × *g* for 15 min at 4 °C to collect the solubilized lysosomal membrane complexes in the supernatant.
3. Discard pellet and add to the supernatant the following components from the NativePAGE Sample Prep Kit: 2.5 μL of NativePAGE 4× sample buffer, 1 μL of G-250 sample additive, and 10 μL of ddH_2O per sample.
4. Load gel in a cold room (4 °C) and after adding the anode buffer (clear) in the outer chamber and the cathode buffer (blue) in the inner chamber, run the gel at 150 Volts (constant) for ~2.5–3.5 h or until dye collects at the bottom.
5. Cut out the molecular weight marker lane, and stain with Coomassie Blue (*see* **Note 8**).
6. Remove gel, put it into transfer buffer for at least 2 min, and proceed with the wet transfer into a methanol pre-wet PVDF membrane. Transfer overnight at a constant current of 0.25 Amps.
7. After removing from the transfer, shake the membrane in 8% acetic acid for 15 min, and wash 3 times for 5 min each time with TBST.
8. Block with 5% milk in TBST for 1 h. Incubate overnight with the primary antibody against LAMP2A (*see* Table 1) diluted in 3% BSA in TBST. Incubate with secondary antibody (1:10,000) in 5% milk in TBST for 1 h, wash 3 times (15 min each time) with TBST, and develop using standard chemiluminescence detection (*see* **Note 9**).
9. Monomeric LAMP2A is detected as a wide band in the 90–110 KD range (depending on the tissue), and the

multimeric complex of LAMP2A shown to participate in substrate uptake into lysosomes via CMA is observed as a well-resolved band of 700 KD. Changes in the percentage of total lysosomal LAMP2A detected in this multimeric complex provide information on CMA activity in lysosomes.

3.3 Measurement of CMA Activity in Intact Cells

3.3.1 Modulation of CMA in Cultured Cells

1. CMA activation: CMA can be upregulated in cells in culture by two methods:

   (a) Inducing stresses known to activate CMA [1]: the stressors used are not selective for CMA and activate other arms of the cellular response to stress but are useful when quantifying the ability of cells to activate CMA. The best characterized are:

   Serum removal: Starvation is a well-established inducer of CMA [11] and can be mimicked in cells by removing the serum from the culture media [22] as follows:

   - Plate cells to 80% confluency in media containing 10% newborn calf serum.
   - After 12 h wash the cells extensively with PBS or Hanks solution (×3 washes with half of the volume of the plate or well).
   - Add complete medium but without addition of serum for 16–24 h. CMA activation reaches exponential kinetics about 10–16 h (depending on the cell type) and plateaus by 24 h after serum removal.

   Oxidative stress: Conditions that induce mild oxidative stress can be used to upregulate CMA [13].

   - Plate cells to 80% confluency in media containing 10% newborn calf serum.
   - After 12 h change media to fresh media containing 10% newborn calf serum and 100 μM of H_2O_2 or 40 μM of paraquat (*see* **Note 10**).
   - After 4 h with the oxidizing agent, aspirate the culture media, and replace by fresh media containing 10% newborn calf serum.
   - Monitor CMA activity at 12–24 h after the oxidative stress.

   Genotoxic stress: CMA is activated during the recovery from stressors inducing double-strand DNA breaks [15].

   - Plate cells to 80% confluency in media containing 10% newborn calf serum.
   - After 12 h change media to fresh media containing 10% newborn calf serum and any of the following agents: 10 μM etoposide, 225 μM MMS, or 20 μM cisplatin.

- After 12 h aspirate the culture media, and replace by fresh media containing 10% newborn calf serum.
- Monitor CMA activity at 12 h after removing the DNA damaging agent.

(b) Chemical activation of CMA: To date, there is only one type of compound tested and validated to selectively activate CMA [21] (*see* **Note 11**).

- Plate cells to 80% confluency in media containing 10% newborn calf serum.
- After 12 h change media to fresh media containing 10% newborn calf serum and 5–10 μM AR7 (CAS number 80306–38-3; #SML0921, Sigma) (*see* **Note 11**).
- Monitor CMA activity at 8–16 h after adding AR7.

2. CMA inhibition:

(a) Genetic downregulation of CMA: The most selective way to block CMA is through genetic knockdown of LAMP2A, [23, 24] since other components involved in CMA are also shared by other cellular processes. To knock down LAMP2A using lentivirus-mediated shRNA in cultured cells:

- Plate approximately 500,000 cells/well in a 6-well plate a day before transduction.
- Replace the media with 0.5 mL of culture media containing 200 μg/mL of polybrene, and add 0.5 mL of lentiviral particles containing shRNA against LAMP2A (the original construct is in a backbone vector that also expresses GFP) [24].
- After 24 h add an additional 1 mL of serum-supplemented media.
- After 48–72 h, check for GFP expression with a fluorescence microscope to determine transduction efficiency.
- Check for LAMP2A knockdown efficiency by immunoblotting after 7–10 days (half-life of LAMP2A protein in most cell types under basal conditions is about 3.5 days).

(b) Chemical downregulation: To date, there is no chemical compound capable to selectively inhibit CMA without affecting other lysosomal degradation pathways. Compounds that block degradation in the lysosomal lumen (either by increasing the lysosomal pH or by inhibiting the catalytic activity of lysosomal proteases) will also block protein degradation dependent on CMA.

3.3.2 Measuring Protein Degradation by CMA

Degradation of proteins in lysosomes via CMA in cultured cells can be determined using metabolic labeling and tracking degradation of intracellular proteins as a pool as follows (*see* **Note 12**):

1. Plate cells in 12-well plate to 60% of confluency, and supplement the media with 10% serum and ^{3}H-leucine to 2 μCi/ml.
2. After 48 h wash cells extensively, and add 300 μL of fresh media supplemented with an unlabeled excess of leucine (for cells growing in standard DMEM, add 2.8 mM leucine).
3. After 1 h aspirate the media, and replace with fresh media with an excess of unlabeled leucine as above (this allows eliminating short half-life proteins usually degraded by the proteasome). Leave triplicate wells without additions, and supplement 3 wells with 20 mM NH_4Cl and 100 μM leupeptin (to block all lysosomal degradation) and 3 wells with 10 mM 3-methyladenine (to block macroautophagy).
4. Take aliquots of 35 μL of the media every 6 h for 24 h, and precipitate them with TCA and BSA as described in the previous sections.
5. At the last time point, aspirate the media and add 500 μL of solubilizing buffer. After 24 h count the radioactivity in 20 μL of the solubilized wells.
6. Proteolysis is calculated as the percentage of total protein labeled at time 0 converted into free radiolabeled amino acids at each time point. Lysosomal proteolysis is calculated by discounting the proteolysis remaining in the wells treated with lysosomal proteolysis inhibitors to the total proteolysis. CMA-dependent degradation is calculated as the percentage of lysosomal proteolysis insensitive to 3-methyladenine (*see* **Note 13**).

3.3.3 Cell Immunofluorescence

Quantification of changes in the number of lysosomes active for CMA can be used as a complementary measurement of CMA (Fig. 3). Lysosomes with the capability to perform CMA contain HSC70 in their lumen and can be identified as HSC70- and LAMP2A-positive vesicles as follows:

1. Plate cells on sterile cover slips at the bottom of a 24-well plate until they reach semi-confluence in media with 10% NCS.
2. Warm serum-free media and PBS at 37 °C.
3. Remove culture medium from cells, and wash once with the media and a second time with PBS.
4. Remove PBS and add −20 °C chilled 100% methanol for 4 min to extract the fraction of soluble HSC70 and preserve the one associated to membranes.
5. Wash 3 times with PBS to remove methanol.

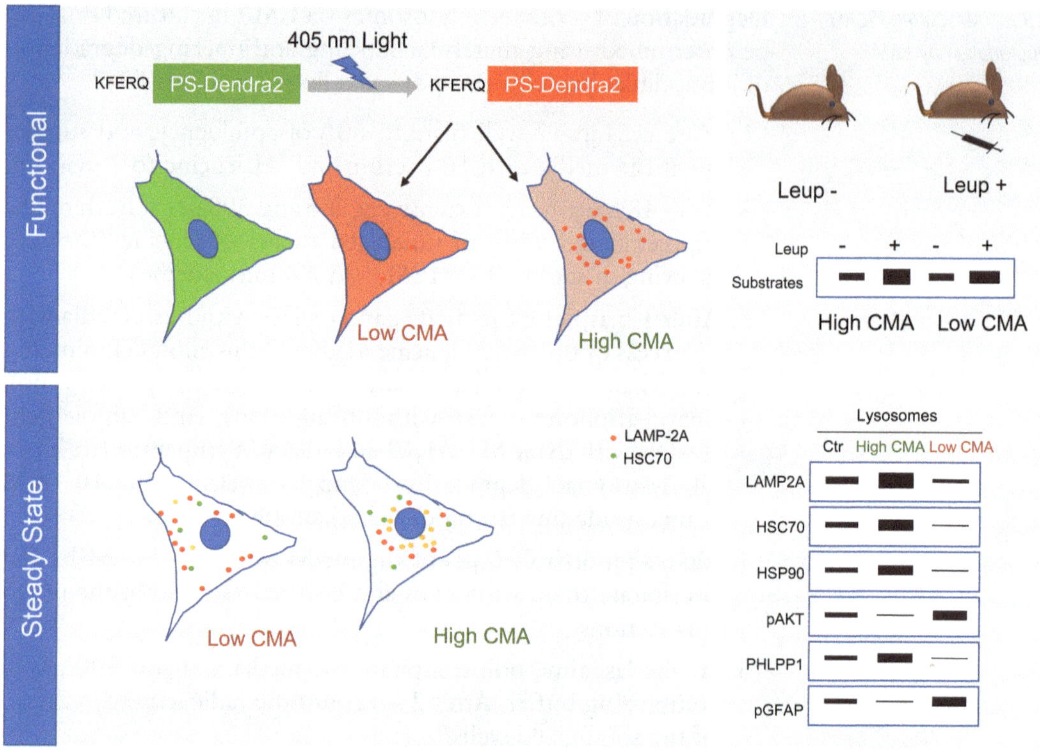

Fig. 3 Measurement of CMA in cultured cells and in vivo. Top: functional assays, expression of photoswitchable reporters (left) allows quantification of CMA in cultured cells as the number of fluorescent puncta (lysosomes) per cell. Injection of leupeptin in vivo (to block lysosomal proteolysis) and immunoblot for CMA substrate of lysosomes from injected and not injected mice allows quantification of substrate flux by CMA (right). Bottom: steady-state assays, quantification of number of lysosomes positive for LAMP2A and HSC70 allows for detecting changes in lysosomes capable of performing CMA (left). Immunoblot of isolated lysosomes for well-characterized CMA activator and inhibitor proteins. Profile of differences in specific protein levels in lysosomes with upregulated and downregulated CMA activity are shown

6. Block/permeabilize for 30 min at room temperature.
7. Wash 3 times with PBS, and add the two primary antibodies for 1 h.
8. Wash 3 times with PBS, and add both secondary antibodies.
9. Wash off secondary antibodies with PBS and mount for imaging as usual to visualize in a fluorescent microscope and determine the number of lysosomes capable to perform CMA as the percentage of LAMP2A-positive vesicles that also label for HSC70.

3.3.4 Photoconvertible CMA Reporter

CMA activity can be measured in cells in culture using photoswitchable (PS) artificial substrates. Three different versions have been generated by our group and tested and validated by other groups: KFERQ-PS-CFP [25], KFERQ-photoactivable (PA)-mCherry [25], and KFERQ-PS-Dendra (unpublished). These differ in the color they fluoresce but follow the same principle and can be used indistinguishably. When exposed to a 405 nm LED light, KFERQ-PS-CFP protein changes from blue to green fluorescence, KFERQ-PA-mCherry that normally does not fluoresce changes to red fluorescence, and KFERQ-PS-Dendra changes from green to red fluorescence. Change in color allows performing pulse/chase type experiments where the switched fluorescence can be followed as it is delivered to lysosomes (Fig. 3) [25].

1. Transduce cells with lentivirus carrying the reporter following the same steps as for shRNA delivery (*see* previous section).
2. Twenty-four hour after transduction, the cells can be photo-activated by exposure to a 3.5 mA (current constant) and 90 V light-emitting diode (LED: Norlux, 405 nm) for 4 min.
3. Split photoswitched cells, and plate on cover slip at the bottom of 24-well plates.
4. Once attached (2–3 h) add the desired treatments (physiological stimuli, drugs, etc.).
5. Fix cells at different times or at a single end time point (16 h after treatment) with 4% paraformaldehyde in PBS for 15 min.
6. Capture images using a fluorescent microscope. CMA can be quantified as the number of fluorescent puncta per cell (*see* **Note 14**).

3.4 Measurement of CMA Activity In Vivo

3.4.1 Changes in CMA Components

Using CMA-active lysosomes isolated from mice, changes in CMA mediators and modulators can be measured via immunoblotting (Fig. 3). When possible, two to three mice livers can be pooled to get enough material for multiple immunoblots. Increase in LAMP2A, HSC70, PHLPP1, and GFAP in lysosomes is supportive of CMA activation; increase in phospho-AKT and phospho-GFAP is supportive of CMA inactivation. Increase in levels of known CMA substrates such as GAPDH, ribonuclease A, alpha-synuclein in the isolated lysosomes is also a good indication of CMA activation.

3.4.2 Substrate Degradation

Degradation of substrate proteins in lysosomes (flux) can be measured upon blocking their proteolysis inside the lysosomal lumen as follows (Fig. 3):

1. Inject leupeptin prepared in saline intraperitoneally to mice or rats (2 mg/100 g of body weight) or only saline to the control animals.
2. After 3 h isolate lysosomes active for CMA following the procedures described in the previous section.
3. Immunoblot lysosomes for well-known CMA substrates such as GAPDH, ribonuclease A, alpha-synuclein (for a more

complete list *see* [1]). Calculate CMA flux as the increase in levels of the substrates in lysosomes isolated from leupeptin-treated animals relative to levels in those from untreated animals.

4 Determining if a Protein Is a CMA Substrate

Investigators interested in testing whether a protein is a bona fide CMA substrate should perform as many of the following tests as possible in their system of interest:

1. Sequence analysis: every CMA substrate protein contains in its amino acid sequence a KFERQ-like pentapeptide motif that is both necessary and sufficient for its targeting to the lysosome [2]. The initial step in determining if a protein is a CMA substrate is to identify the presence of this motif. This consensus motif has a very specific amino acid composition, and any variation from the original set of criteria needs to be experimentally confirmed as a bona fide targeting motif. Briefly, the motif is based on the physical properties of the amino acid residues which have been identified as a combination of one or two of the positively charged residues K, R; one of the negatively charged residues D, E; and one or two of the hydrophobic residues F, L, V, I and a glutamine (Q) that either initiates or terminates the sequence [2] (*see* **Note 15**).
2. Association with lysosomes, preferentially with those that have higher CMA activity (positive for lys-HSC70) [26]: this can be determined using co-localization by immunofluorescence (*see* Subheading 3.3) or by immunoblot in isolated lysosomes (*see* Subheading 3.1).
3. Degradation in lysosomes: detected as an increase in the cellular levels of the protein of interest upon treatment of cells with inhibitors of lysosomal proteolysis (20 mM NH_4Cl/100 μM leupeptin) for 12 h. Note that this property is also shared with proteins degraded by other autophagic pathways (macro- and microautophagy) and by endocytosis.
4. Interaction with CMA components: detectable by co-immunoprecipitation of the protein of interest with HSC70 in cytosolic fractions [3] or with the cytosolic tail of LAMP2A in isolated lysosomes [4]. Note that because protein interaction with HSC70 also occurs in many other cellular pathways, confidence of a relation with CMA degradation will be higher if the HSC70/protein interaction can be completed by addition of KFERQ-containing proteins (i.e., ribonuclease A (RNase A)). Because KFERQ-dependent binding to HSC70 also mediates degradation of proteins by e-MI, additional

criteria need to be used to differentiate between both autophagic pathways.

5. Changes in intracellular levels upon CMA modulation: increase in the intracellular levels of the candidate protein in cells that lack LAMP2A (although activation of other autophagy pathways to compensate for reduced CMA prevents substrate accumulation in many instances).
6. Capability to directly translocate into isolated lysosomes: this is the most definitive evidence of a protein being a CMA substrate; as in this type of in vitro assay, there is no contribution of any other proteolytic system to the observed lysosomal translocation and degradation (*see* Subheading 3.2, for details).
7. Mutagenesis of the targeting motif: given that the KFERQ-like targeting motif is both a necessary and sufficient sequence that is required for HSC70 to target substrate proteins to the lysosomal membrane, alterations in this motif can be utilized to support the involvement of this chaperone in their degradation. Mutation of the Q residue in the motif to alanine (A) disrupts the interaction of the protein with HSC70 and its subsequent targeting to lysosomes. Elimination or alteration of the targeting motif will result in a decreased association of the substrate with lysosomes in the case of CMA or with late endosomes/MVB in the case of e-MI. Furthermore, in some instances, interaction with HSC70 persists even in the absence of the CMA-targeting motif. HSC70 binding in these cases often has switched to another region in the protein, due to the ability of HSC70 to bind hydrophobic protein patches. However, that type of interaction does not target the protein for CMA degradation and can be easily identified by demonstrating lack of competition for HSC70 binding with other proteins bearing the CMA-targeting motif such as RNase A.

5 Notes

1. Metrizamide is light sensitive and needs to be dissolved in the dark (beaker wrapped with aluminum foil) and slowly to avoid solid clump formation. Best to start with half of the final volume of water and slowly add metrizamide powder while stirring. To adjust the pH of the solution use 0.01 M NaOH once the metrizamide is dissolved.
2. Liver perfusion is not strictly necessary. Extensive washing of the livers with sucrose removes most of the blood remaining in the tissue.
3. In the transport and degradation assays with intact lysosomes, supplementation with HSC70 and ATP may be necessary or

not depending on the origin of the lysosomes. In our experience, lysosomes from rat liver, when prepared following this protocol, contain enough HSC70 associated to the cytosolic side of their membrane, and supplementation is not required. However, lysosomes from cultured cells or from some tissues need to be supplemented with additional recombinant HSC70 (10 μg/mL final concentration) and ATP (5 mM) to maximize uptake. Although we recommend an incubation of 30 min at 37 °C to measure proteolysis in intact lysosomes, in those instances in which lysosomal fragility is suspected, the incubation can be performed instead for 45–60 min at room temperature (25 °C) to reduce lysosomal breakage. In this case, it is advisable to use an ATP regenerating system in the incubation media [ATP (10 mM), $MgCl_2$ (10 mM), phosphocreatine (2 mM) and creatine phosphokinase (50 pg/mL)].

4. In some instances, uptake of substrates by lysosomes can also be determined using proteinase K treatment to remove the protein associated with the cytosolic side of the lysosomal membrane. In that case, after incubation with the substrate, all samples are cooled down on ice (1 min), and the samples treated with chymostatin are exposed to proteinase K (5 μL of 1 mg/mL solution). Samples are incubated on ice for 10 min, and the reaction stopped by adding 5 μL AEBSF (100 μM). All samples are collected by centrifugation at 25,000 × *g* for 5 min at 4 °C. Uptake is determined as the percentage of GAPDH added that is associated with the proteinase K-treated samples.
5. Presence of DTT and cysteine in the proteolysis buffer is absolutely necessary to obtain maximal degradation. DTT needs to be added fresh to the solution.
6. Although the use of multiple technical replicates is always encouraged, it is especially important in this assay since due to the small volume of lysosomes added to the reaction and the fact that lysosomes are in suspension, there are higher chances of variability in the total amount of lysosomes added per well. The short half-life of lysosomes once purified (membrane breakage and loss of luminal pH start to increase exponentially 1 h after isolation) makes it not possible to determine the concentration of lysosomal protein added per well. When working with lysosomes of different origins, it is necessary to save a small amount frozen to determine both protein concentration and β-hexosaminidase activity later and correct proteolysis rates.
7. If the sample subjected to solubilization is in solution (instead of as a pellet), adjust the final concentration of detergent in the solution using a concentrated stock.

8. For fast Coomassie staining of the molecular weight marker, place the strip of gel in Coomassie fixing solution, and microwave for 45 s. Shake 15 min in this same solution. Remove solution, and place gel in destaining solution. Microwave for 45 s, and shake in destaining overnight at room temperature. The next day, remove destaining, and store in water.
9. We have found that using concentrations of LAMP2A antibody higher than recommended in Table 1 has a negative impact on the detection of the 700 KD multimeric complex of LAMP2A and therefore strongly recommend using a 1:5000 dilution (instead of the standard 1:1000 used for regular immunoblot as indicated in Table 1). In some instances, in which only a small fraction of LAMP2A is organized into the multimeric complex, the high abundance of monomers in lysosomes could make detection of the multimers challenging when developing the membrane. This problem can be resolved by doing a short exposure of the membrane (to identify the LAMP2A monomer) and then proceed to a long exposure covering the bottom of the membrane containing the LAMP2A monomer with foil paper to eliminate this signal that may otherwise cover the considerably weaker signal of the multimer.
10. Concentrations of H_2O_2 and paraquat need to be optimized for each cell type. These concentrations work well in mouse and human fibroblasts in culture and were determined after doing viability assays. In these cells, we have found that concentrations of paraquat up to 100 μM result in less than 10% cell death. Concentrations inducing more than 15% cell death in the cell type of interest should be avoided. Paraquat should be handled carefully with appropriate safety personal protective equipment (mask, gloves) as it is highly toxic when in powder form.
11. There has been some level of confusion regarding methods to chemically activate CMA due to the misinterpretation of a study from the late James Dice group [27]. In this study, it was described how compounds that modify other biological processes such as anisomycin or cycloheximide (that inhibit protein synthesis) or 6-aminonicotinamide (that inhibits pentose phosphate metabolism) lead to activation of CMA (among many other pathways, including macroautophagy). Consequently, the use of these drugs for selective activation of CMA is highly discouraged, and they should be used more under the category of cellular stressors that elicit a CMA response. Activation of other cellular stress pathways, including other proteolytic systems, by these drugs, makes interpretation of findings complex. Use of AR7 or related derivatives [21] is recommended as for these compounds; lack of effect on

proteasomal or lysosomal degradation pathways (other than CMA) has been confirmed. Concentrations of AR7 resulting in maximal activation of CMA should be determined experimentally for each cell type. The range indicated here has been proven effective in mouse fibroblasts, astrocytes, macrophages, T cells, and kidney epithelial cells.

12. It is also possible to determine CMA activity in cells in culture by tracking the degradation of previously well-characterized substrates by immunoblot after addition of the inhibitors in the same conditions. However, because degradation of specific proteins by CMA may depend on the cellular conditions, following a pool of proteins provides a better representation of CMA activity than a single protein.
13. Other inhibitors of macroautophagy can be used, but times and conditions in this protocol have been optimized for 3-methyladenine. Acute chemical blockage of macroautophagy is preferred to genetic blockage for these experiments as that will prevent activation of compensatory mechanisms.
14. Using PS reporters allows measuring degradation of the switched proteins without having to inhibit protein synthesis. When using the fluorescent reporters, it is important to use all the controls as described in [25] to confirm that the fluorescent puncta are indeed lysosomes and not protein aggregates.
15. Note that the KFERQ-like motif is used for interaction with HSC70, and consequently the presence of the motif is not sufficient to differentiate CMA substrates from e-MI substrates [9].

References

1. Kaushik S, Cuervo AM (2012) Chaperone-mediated autophagy: a unique way to enter the lysosome world. Trends Cell Biol 22:407–417
2. Dice JF (1990) Peptide sequences that target cytosolic proteins for lysosomal proteolysis. Trends Biochem Sci 15:305–309
3. Chiang H et al (1989) A role for a 70-kilodalton heat shock protein in lysosomal degradation of intracellular proteins. Science 246:382–385
4. Cuervo AM, Dice JF (1996) A receptor for the selective uptake and degradation of proteins by lysosomes. Science 273:501–503
5. Bandyopadhyay U et al (2008) The chaperone-mediated autophagy receptor organizes in dynamic protein complexes at the lysosomal membrane. Mol Cell Biol 28:5747–5763
6. Agarraberes F, Terlecky S, Dice J (1997) An intralysosomal hsp70 is required for a selective pathway of lysosomal protein degradation. J Cell Biol 137:825–834
7. Bandyopadhyay U et al (2010) Identification of regulators of chaperone-mediated autophagy. Mol Cell 39:535–547
8. Arias E et al (2015) Lysosomal mTORC2/PHLPP1/Akt Regulate Chaperone-Mediated Autophagy. Mol Cell 59:270–284
9. Sahu R et al (2011) Microautophagy of cytosolic proteins by late endosomes. Develop Cell 20:131–139
10. Salvador N et al (2000) Import of a cytosolic protein into lysosomes by chaperone-mediated autophagy depends on its folding state. J Biol Chem 275:27447–27456

11. Cuervo AM et al (1995) Activation of a selective pathway of lysosomal proteolysis in rat liver by prolonged starvation. Am J Phys 269: C1200–C1208
12. Wing SS et al (1991) Proteins containing peptide sequences related to KFERQ are selectively depleted in liver and heart, but not skeletal muscle, of fasted rats. Biochem J 275:165–169
13. Kiffin R et al (2004) Activation of chaperone-mediated autophagy during oxidative stress. Mol Biol Cell 15:4829–4840
14. Schneider JL, Suh Y, Cuervo AM (2014) Deficient chaperone-mediated autophagy in liver leads to metabolic dysregulation. Cell Metab 20:417–432
15. Park C, Suh Y, Cuervo AM (2015) Regulated degradation of Chk1 by chaperone-mediated autophagy in response to DNA damage. Nat Commun 6:6823
16. Valdor R et al (2014) Chaperone-mediated autophagy regulates T cell responses through targeted degradation of negative regulators of T cell activation. Nat Immunol 15:1046–1054
17. Cuervo AM, Dice JF (2000) Age-related decline in chaperone-mediated autophagy. J Biol Chem 275:31505–31513
18. Cuervo AM et al (2004) Impaired degradation of mutant alpha-synuclein by chaperone-mediated autophagy. Science 305:1292–1295
19. Kiffin R et al (2007) Altered dynamics of the lysosomal receptor for chaperone-mediated autophagy with age. J Cell Sci 120:782–791
20. Orenstein SJ et al (2013) Interplay of LRRK2 with chaperone-mediated autophagy. Nat Neurosci 16:394–406
21. Anguiano J et al (2013) Chemical modulation of chaperone-mediated autophagy by retinoic acid derivatives. Nat Chem Biol 9:374–382
22. Berger JJ, Dice JF (1985) Effect of serum deprivation and replacement on proteolysis in cultured human fibroblasts. Prog Clin Biol Res 180:479–481
23. Massey AC et al (2006) Consequences of the selective blockage of chaperone-mediated autophagy. Proc Natl Acad Sci U S A 103:5905–5910
24. Massey AC et al (2008) Early cellular changes after blockage of chaperone-mediated autophagy. Autophagy 4:442–456
25. Koga H et al (2011) A photoconvertible fluorescent reporter to track chaperone-mediated autophagy. Nat Commun 2:386
26. Cuervo AM, Dice JF, Knecht E (1997) A population of rat liver lysosomes responsible for the selective uptake and degradation of cytosolic proteins. J Biol Chem 272:5606–5615
27. Finn P et al (2005) Effects of small molecules on chaperone-mediated autophagy. Autophagy 1:141–145

Chapter 48

Interactive Autophagy: Monitoring a Novel Form of Selective Autophagy by Macroscopic Observations

Jana Petri and Roland L. Knorr

Abstract

Traditional lectures and cookbook laboratory exercises are today's standard tools in scientific teaching and learning. However, these conventional methods are suboptimal. Combining active learning techniques with physical experiences can improve educational success significantly. Still, hands-on material which supports active and physical teaching concepts is rare. Here, we introduce an interactive, performance-based method.

As an example, we studied autophagosome formation. We observed assembly of the phagophore by membrane fusion, cargo isolation by bending the phagophore and membrane scission. We extracted characteristic time scales of autophagosome formation. Moreover, we observed capturing the autophagic cargo within a single membrane for the first time. In this chapter, we provide an easy tool to engage participants in the process of scientific perception. We are convinced that "hands-on" experiments and interactive analyses will encourage students to participate more actively in classes and thus, will improve learning. Moreover, we anticipate that the approach enhances translation of scientific concepts between different fields by providing scientists with a fresh view on, e.g., membrane-bound processes and can improve communication of science to the public.

Key words Autophagy, Degradation, Selective, Membrane, Bending, Remodeling, Fusion, Scission, Active learning, Yoshinori Ohsumi, Dance, Performance

1 Introduction

Traditional teaching methods provide plenty of numbers and facts. Such methods are suboptimal for participants as they often fail to engage their intellect [1] and may even contribute, for example, to students leaving science [2]. One key to improve teaching outcome is to engage participants in the learning process [3]. Such instructional methods are referred to as active learning. Typically, they include a variety of resources and tools such as clicker use, regular tests, interactive apps, and cooperative learning [3]. Especially in combination with physical experiences, active involvement can improve learning significantly [2, 4–10]. Faculty may accept the need for change towards engaging their students into the learning

Nicholas Ktistakis and Oliver Florey (eds.), *Autophagy: Methods and Protocols*, Methods in Molecular Biology, vol. 1880, https://doi.org/10.1007/978-1-4939-8873-0_48,

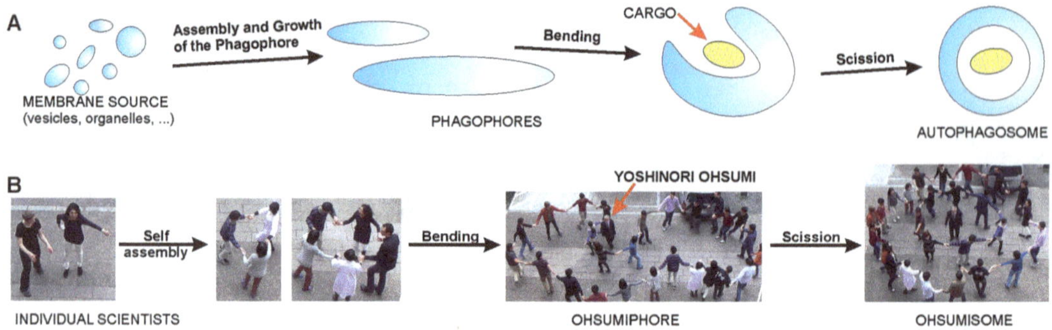

Fig. 1 Formation of autophagosomes and Ohsumisomes. (**a**) Membrane assembly, remodeling, and dynamics during autophagy. (**b**) Our novel macroscopic method reproduced major membrane remodeling processes during the formation of autophagosomes. The specific cargo captured was Yoshinori Ohsumi

process and may favor teaching skills but is overwhelmed by the challenge to incorporate new teaching methods [11]. Moreover, hands-on concepts which support interactive teaching of complex biological pathways are rarely published [12].

Autophagy is a very complex intracellular pathway, leading to the membrane-bound degradation of cytoplasmic material. The mechanistic bases of autophagosome formation, such as the concerted activities of the autophagy-related (*Atg)* proteins, have been the focus of sustained research for over two decades [13, 14]. However, many intricacies of this membrane-bound process [15], Fig. 1, are still unknown. Because of the small thickness of the membranes, they can be considered as fluid, quasi-two-dimensional (2D) materials. Macroscopically, quasi-two-dimensional materials are not known which are characterized by material properties similar to membranes (e.g., considering area and volume conservation). During autophagy, such fluid membranes form complex and dynamic structures by a combination of morphological and topological changes [15, 16]. In publications, typically autophagic membranes are represented by selected, static, 2D images, either in the form of 2D-sections (e.g., obtained by electron microscopy, sketched in Fig. 1a) or as 2D-projections, such as fluorescent snapshots. Inherently, static snapshots cannot represent the fluid nature of the membrane and resulting dynamics appropriately. Thus, one of the challenges in the field is to understand the connection between static 2D-representations, the actual shapes of autophagosomes in cells and their dynamic 3D-remodeling. For example, the closure of the phagophore by membrane scission was interpreted as a membrane fusion process in a significant number of publications [16]. Topologically, this is not correct [15]. Recently, it was pointed out that bending of the phagophore inherently is coupled to the removal of its strongly bend rim, Fig. 1a. The latter process contributes a major amount of energy required for phagophore bending [17].

In this chapter we present an experiment which was implemented within a seminar covering membrane remodeling during autophagy. Together with an international team of biochemists, we physically performed the dynamics of membrane shape transformations as observed during autophagy. We reproduced the initial stages of autophagosome formation, whereby a membrane-bound organelle, the autophagosome, was generated de novo. Analyzing the data revealed that our pre-autophagosomal membranes were characterized by dynamic instability, with a stability time of about 9 s. Within 21.8 ± 1.9 s, a continuous, phagophore-like membrane structure formed. Typically, autophagic membrane scission occurred after additional 4.7 ± 2.5 s. Unexpectedly, we also observed enclosure of the autophagic cargo within a single membrane. In all studied cases, Yoshinori Ohsumi was successfully captured as the specific cargo under the experimental regime employed. Consequently, we suggest naming this mode of autophagy "Ohsumiphagy" and the corresponding organelle the "Ohsumisome".

We argue that the interactive components of the experiment presented can be expanded to multiple levels of scientific education including skills, such as the design of the experiment, risk management, data analysis, and presentation. Doubtless, applying similar tutorials in lectures or seminars will engage participants intellectually and physically. We hypothesize that learning success will improve significantly compared to traditional teaching or the use of popular tools such as clickers. In summary, we provide a tool to interactively access dynamics of membrane remodeling and also a low-threshold exercise for science education and interdisciplinary awareness in general.

2 Materials

- Camera.
- Stand for mounting the camera in elevated position (e.g., larger building).
- Participants.
- Marker to indicate recording area, e.g., adhesive tape or chalk.

3 Methods

In this section, we list the methods used for the particular experiment in detail.

3.1 Camera and Setup

To be able to identify individual participants, we installed the video recording system well above the recording area, *see* Fig. 2. A standard commercial camcorder was used (HC-V180 from Panasonic,

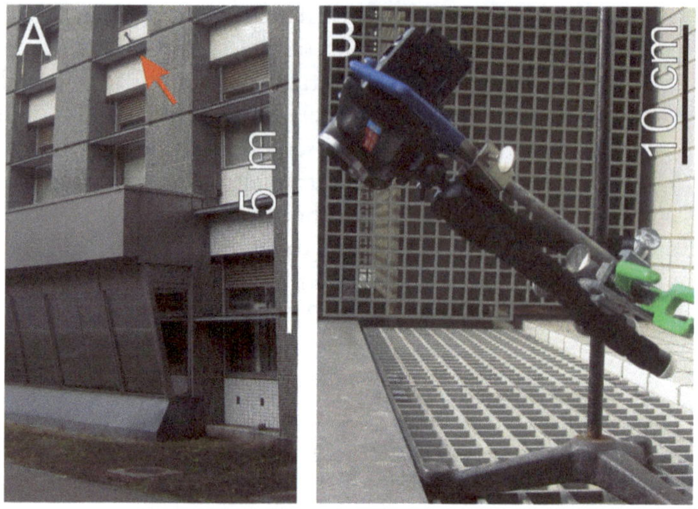

Fig. 2 Installation of the camera. (**a**) Macro positioning of the camcorder on the third floor of the Tokyo Institute of Technology. (**b**) Micro positioning and mounting of the camcorder

Fig. 3 Assembly of autophagosomal membranes. To highlight the dynamics of the membrane, representative time points were chosen, and the membranes were tracked and color coded. (**a**) Initially, the individual scientists ($n = 26$, marked with yellow circles) were randomly scattered. The position of the cargo (red) was arbitrary. The mobility within the system was high as seen by the movement of the cargo over time. (**b**) Within about 10 s, most single scientists assembled into larger membrane-like structures, here four. After about 20 s only one continuous membrane remained, indicating a strong preference for spontaneous membrane consolidation. (**c**) At 26 ± 2 s, a double membrane structure had formed by scission in the majority of trials. The separation of the outer (yellow) and the inner (pink) membrane was clearly visible. Colored lines were added to assist visualization of membrane dynamics. Yellow, growing autophagosomal membrane and outer vesicle; red, cargo; purple, inner vesicle

Osaka/Japan). It was equipped with a 28 mm objective, 10 megapixels, full high-definition video capabilities and was mounted on a standard laboratory stand, Fig. 2.

A vertical alignment of the video system with respect to the recording area seemed ideal to us to decrease distortion of the images and thus, to improve analysis of the data. In practice, a vertical alignment was rather difficult to achieve. We installed the camcorder on the third floor of the Tokyo Institute of Technology, Suzukakedai Campus, building S2, Fig. 2. An angle of approximately 50° was sufficient to identify individual participants throughout the experiment, Fig. 3.

3.2 Preparation of the Recoding Area

To make participants aware of the area where the experiment was recorded, we marked the field of view of the camcorder on the ground by adhesive tape. Other materials such as chalk might be used instead.

3.3 Interactive Seminar

The experiments were conducted as part of an interactive seminar in the lab of Prof. Yoshinori Ohsumi at Tokyo Institute of Technology on "Membrane remodeling during autophagy." Before listening to a talk in the traditional format, all participants ($n = 26$) were introduced to the concept of performing autophagosome formation physically as a team, whereby each participant represented a small vesicle initially. We presented the goals of our experiments: (1) to show that a macroscopic, performance-based method can be used to resemble the dynamics of membrane remodeling. As a cellular process, we chose autophagy with Yoshinori Ohsumi as the specific cargo. (2) To reveal reproducibility, we planned to repeat our experiment three times. (3) We aimed to extract a characteristic time scale of membrane remodeling by video analysis. (4) We hypothesized that cooperative learning influences the time required to form autophagosomes: we expected that autophagosomes form much faster in the last experiment.

While the group was guided outside of the building and scattered randomly within the experimental area, Fig. 3, the camera was switched to live recording mode. Single experiments started by an acoustic signal (for the participants) and an optical signal, which was used later to exactly define $t = 0$ during image analysis. The participants self-assembled into larger, membranous structures. Initially, these structures were mainly circular, but they obtained noncircular shapes including membrane ruffles as the number of assemblies decreased by processes resembling membrane fusion (Figs. 3). In all experiments, the cargo moved steadily. The growing membrane structures engulfed the cargo. Cargo isolation was followed by a membrane remodeling event reminiscent of autophagic membrane scission [16]: the continuous, phagophore-like membrane formed a double-membrane vesicle. The inner and outer autophagic vesicle were separated. Our observations suggests that different steps of autophagosome formation are mimicked with our performance-based approach. The cargo, Yoshinori Ohsumi, was captured in all experiments successfully. Therefore, we suggest naming this particular mode of autophagy "Ohsumiphagy" and the corresponding organelle the "Ohsumisome."

Surprisingly, we found that at one occasion, a single-membrane vesicle isolated the cargo (Fig. 5 and *see* **Note 1**). This unconventional membrane morphology led to discussions within the participants about the extension of the study by one additional test trial. Whereas some participants were interested in studying this unconventional phenomenon in more detail, others suggested ignoring

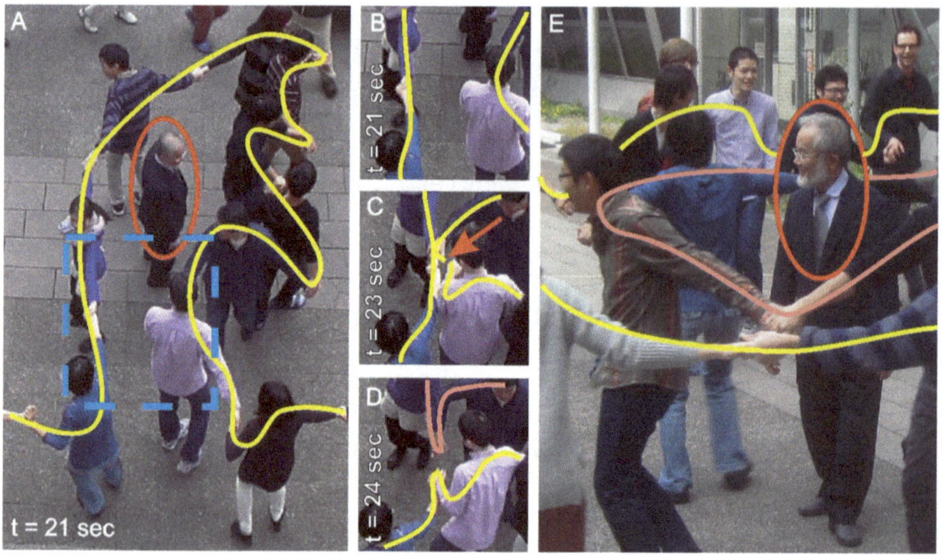

Fig. 4 Membrane scission. (**a**) At $t = 21$ s, a clear membrane invagination had formed around the cargo. (**b**) The marked area (dashed blue line) is followed over time. (**c**) At $t = 23$ s. a short-lived membrane pore was formed (arrow) which resealed within less than 1 s. (**d**) Sealing of the pore resulted in completion of the scission process, with distinctive separation of both membranes. (**e**) Lateral snapshot where the cargo, Yoshinori Ohsumi, can be clearly resolved as isolated within the double bilayer structure. The time is given relative to the start of the experiment, color code as in Fig. 3

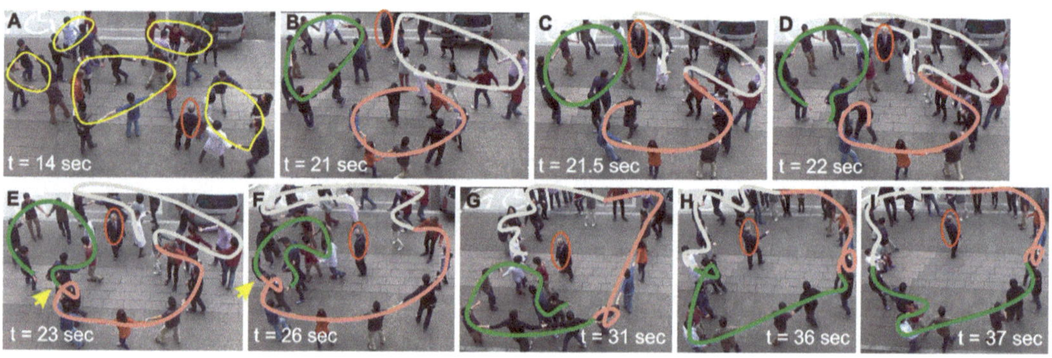

Fig. 5 Formation of unconventional Ohsumisomes. (**a**) Until $t = 20$ s. Assembly of membranes proceeded similar as shown in Figs. 3 and 4. (**b–i**) The three assemblies are labeled in individual colors and within 1 s, all assemblies formed pores, but did not reseal as expected. (**e**) At $t = 23$ s, the pink assembly had resealed with one end of the two other assemblies (gray and green), leaving their open edges several meters apart from each other. Thus, resealing of the edges became sterically hindered and the edges were stable for up to 15 s. (**e–f**) At the exact position where the membrane resealed initially, (arrow) it ruptured again and resealed with the other edge of the green strand (cis-trans conversion). (**f–i**) Membrane reorganization and final resealing, resulted in capturing of the cargo. In contrast to conventional Ohsumisomes, only a single membrane was observed in this experiment

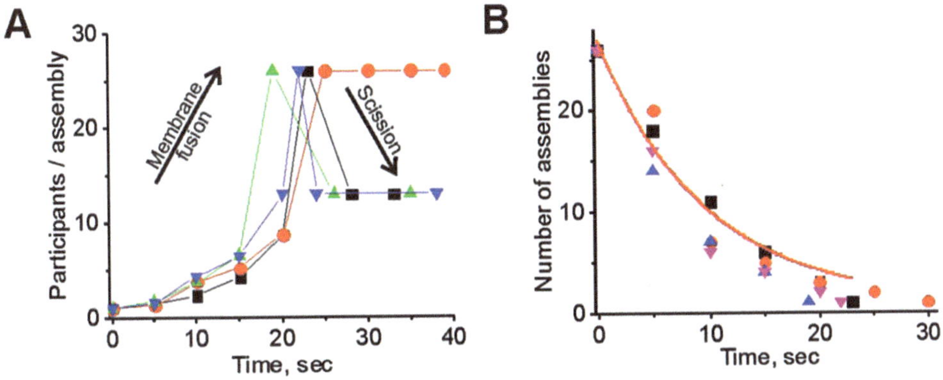

Fig. 6 Time scales of Ohsumisome formation. (**a**) As the number of individual participants decreased, their average number per assembly increased over time by membrane fusion. After 21.8 ± 1.9 s, only one continuous membrane remained. Conventionally, membrane scission occurred within 4.7 ± 2.5 s (*see* also Fig. 4). In a single case, we observed a yet uncharacterized single membrane structure (*see* Fig. 5). (**b**) The formation of Ohsumiphores were characterized by a low stability of small assemblies. We fitted the number of assemblies over time with an exponential decay (red line). The stability time was found to be 9.5 ± 1.1 s

the outlier (*see* **Note 1**). Finally, the majority of the participants decided to extend our study by another experiment.

After the actual talk, the videos were screened and discussed among the participants. The close inspection of the movies revealed that cargo isolation by a single membrane was possible by propagation of membrane pores via a sequence of incorrect membrane-resealing processes, **Note 1**. This resulted in a rather random, uncoordinated enclosure of the cargo within the dynamic membrane structure, Fig. 5.

3.4 Analysis of the Experiments

We quantified the assembly of autophagosomes by counting the number of assemblies and their sizes by analyzing the individual video sequences at fixed time points of 5 s, Fig. 6. A visual signal (waiving hands) was used to define $t = 0$ for each experiment. We found that over time, the average size of the assembly increased, whereas the number of individual membrane segments decreased, indicating membrane fusion. Eventually, membrane scission occurred and the single, continuous phagophore membrane split into two autophagic vesicles. At the same time, the cargo was isolated within the two membranes, Fig. 4. Clearly, membrane scission did not occur during one experiment, Fig. 6.

The stability of pre-autophagosomal membranes followed similar characteristics in all four experiments with a characteristic mean stability time of 9.5 ± 1.1 s as obtained from fitting the data in Fig. 6b with an exponential decay. Surprisingly, the time scales of our first and forth experiment did not change significantly. This suggests that cooperative learning might be less influential in our particular group of highly specialized participants than we anticipated initially, **Note 2**. More research with different cohorts could help to understand whether the strong autophagic background of the participants influenced this result, **Note 3**.

4 Notes

1. Here, we present a tool where theoretical knowledge and practice from laboratory experiments are combined by physically approaching the topic of remodeling biological membranes. Such complex biological processes can, therefore, be understood in an unconventional and straightforward way.

   Conventional laboratory exercises for students follow a strict protocol, and experiments are supposed to work. Thus, they do not reflect the rather complex, nonlinear scientific process in which an experiment fails or shows an unexpected outcome. Learning processes might profit from experiments that failed, Table 1.
2. The key to improve learning will be to actively involve participants in the processes of developing the experimental plan, running the experiments and interpreting the results. Multiple directions and questions can be included in future experiments, for example: What is the aim of the study? Which is the biological process of interest? How to achieve a mutual goal

Table 1
Research questions and possible experiments to compare learning success in various performance-based settings/experiments

Research questions	Experimental design
Do students perform better by (a) conventional active learning techniques, (b) integrated, performance-based learning?	Split participants into two groups following approach (a) or (b). Quantifying the learning success.
Do participants learn differently when a performance proceeds as planned or is interrupted/fails (learning from defeat, **Note 1**)?	Interfere with some of multiple groups performing (e.g., crossing the area with a group of people). Quantify the learning success.
Can guessing the underlying biological process by watching another group performing improve learning?	Split participants into two or multiple groups which perform a biological process of their choice. Test the students.
How does a varying number of participants, multiple or dynamic cargoes, influence the experiment?	For example, mark participants as cargoes with red shirts and apply different experimental settings.
Do dynamics differ between groups with different scientific backgrounds?	Split participant according to their background at e.g. conferences, summer schools, or open science days.

and how to allocate single tasks to participants? How many participants take part and which materials are required and/or are available? Which are the appropriate controls of the experiment? Which location will be chosen for the experiment? Is it necessary to include a contingency plan? How to analyze, interpret, and present the results?

3. Whether active teaching combined with physical methods improves learning can be tested in future experiments. Here, we list putative questions and experiments addressing the educational improvement by the tool introduced in this chapter.

Acknowledgments

We thank Alexander May (Institute of Innovative Research/Tokyo Institute of Technology) for translations, cultural advice, and technical assistance. We further thank Yoshinori Ohsumi, Hitoshi Nakatogawa, and Nobuo Noda (Institute of Innovative Research/ Tokyo Institute of Technology and Microbial Chemistry Research Foundation/Institute of Microbial Chemistry) and their lab members for strong support and participation in the experiment. We specifically thank Yoko Hara for motivation of the participants. We also thank Reinhard Lipowsky (MPI-CI) for institutional and financial support, and Ben Wiggins (University of Washington) and Constance Scharff (FU Berlin) for critical reading and discussion of the manuscript.

References

1. Fischer CN (2011) Changing the science education paradigm: from teaching facts to engaging the intellect: Science Education Colloquia Series, Spring 2011. Yale J Biol Med 84:247–251
2. Handelsman J et al (2004) Scientific teaching. Science 304:521–522
3. Knight JK, Wood WB (2005) Teaching more by lecturing less. Cell Biol Educ 4:298–310
4. Bowen CW (2000) A quantitative literature review of cooperative learning effects on high school and college chemistry achievement. J Chem Educ Easton 77:116–119
5. Bradforth SE et al (2015) University learning: improve undergraduate science education. Nat News 523:282
6. Freeman S et al (2014) Active learning increases student performance in science, engineering, and mathematics. Proc Natl Acad Sci 111:8410–8415
7. Graham MJ, Frederick J, Byars-Winston A, Hunter A-B, Handelsman J (2013) Increasing persistence of college students in STEM. Science 341:1455–1456
8. Haak DC, HilleRisLambers J, Pitre E, Freeman S (2011) Increased structure and active learning reduce the achievement gap in introductory biology. Science 332:1213–1216
9. Kontra C, Lyons DJ, Fischer SM, Beilock SL (2015) Physical experience enhances science learning. Psychol Sci 26:737–749
10. Lorenzo M, Crouch CH, Mazur E (2006) Reducing the gender gap in the physics classroom. Am J Phys 74:118–122
11. Coil D, Wenderoth MP, Cunningham M, Dirks C (2010) Teaching the process of science: faculty perceptions and an effective methodology. CBE-Life Sci Educ 9:524–535
12. Rezende-Filho FM, da Fonseca LJS, Nunes-Souza V, Guedes da SG, Rabelo LA (2014) A

student-centered approach for developing active learning: the construction of physical models as a teaching tool in medical physiology. BMC Med Educ 14:189
13. Mizushima N, Yoshimori T, Ohsumi Y (2011) The role of Atg proteins in autophagosome formation. Annu Rev Cell Dev Biol 27:107–132
14. Noda NN, Inagaki F (2015) Mechanisms of autophagy. Annu Rev Biophys 44:101–122
15. Knorr RL, Mizushima N, Dimova R (2017) Fusion and scission of membranes: ubiquitous topological transformations in cells. Traffic 18:758–761
16. Knorr RL, Lipowsky R, Dimova R (2015) Autophagosome closure requires membrane scission. Autophagy 11:2134–2137
17. Knorr RL, Dimova R, Lipowsky R (2012) Curvature of double-membrane organelles generated by changes in membrane size and composition. PLoS One 7:e32753

INDEX

Nicholas Ktistakis and Oliver Florey (eds.), *Autophagy: Methods and Protocols*, Methods in Molecular Biology, vol. 1880, https://doi.org/10.1007/978-1-4939-8873-0, © Springer Science+Business Media, LLC, part of Springer Nature 2019

G

H

I

K

L

M

N

O

P

FSC
www.fsc.org
MIX
Papier | Fördert gute Waldnutzung
FSC® C083411

Kolibri 360 GmbH
Ferdinand-Jühlke-Straße 7
99095 Erfurt, Deutschland
produktsicherheit@kolibri360.de